Fish & Fisheries Series

Volume 38

Series Editor

David L. G. Noakes, Fisheries & Wildlife, Oregon State University, Corvallis, USA

The volumes in the Fish & Fisheries series will cover topics ranging from the biology of individual species or groups of fishes, to broader concepts in fisheries science, conservation and management. The series is directed to professionals and researchers in fish biology, senior undergraduate and postgraduate students, and those concerned with commercial production or harvest of fishes.

Prospective authors and/or editors should consult the **Series Editor or Publishing Editor** for more details. The series editor also welcomes any comments or suggestions for future volumes:

David L.G. Noakes
Fisheries & Wildlife Department
Oregon State University
Corvallis, USA
FIFI@oregonstate.edu
Alexandrine Cheronet
Publishing Editor
alexandrine.cheronet@springer.com

More information about this series at http://www.springer.com/series/5973

Margaret F. Docker

Editor

Lampreys: Biology, Conservation and Control

Volume 2

 Springer

Editor
Margaret F. Docker
Department of Biological Sciences
University of Manitoba
Winnipeg, MB, Canada

Fish & Fisheries Series
ISBN 978-94-024-1682-4   ISBN 978-94-024-1684-8   (eBook)
https://doi.org/10.1007/978-94-024-1684-8

Library of Congress Control Number: 2019935983

This Springer imprint is published by the registered company Springer Nature B.V.
The registered company address is: Van Godewijckstraat 30, 3311 GX Dordrecht, The Netherlands

Pacific lamprey *Entosphenus tridentatus* (Photo David Herasimtschuk © Freshwaters Illustrated)

In memory of our friend and colleague,
Phil Cochran

Foreword

This volume completes a monumental task. Lampreys are a remarkable group of species; however, all but one are little noted and not well known. It is ironic that we should learn so much from a combination of those attempting to eradicate one species and others working in relative isolation on what are generally regarded as peculiar evolutionary remnants.

The one well-known species has given all the lampreys a very bad name. Lampreys are mostly defined by the combination of negative terms for all the features they lack: no jaws, no bony skeleton, no paired fins, no scales, no true teeth and only a single nostril; and by their habits as blood-sucking vampires. Of course, it is quite inappropriate to categorize, classify or recognize any organism on the basis of features that it does not possess. It would indeed be odd if we were to construct a dichotomous identification key based upon the lack of features in each taxon.

That one species is a textbook example of the negative effects of an invasive, non-native species on the native fauna. The negative effects associated with invasive sea lamprey *Petromyzon marinus* in the upper Laurentian Great Lakes are the classic example in almost every textbook of animal ecology or fisheries management. But studies of chemical communication, especially pheromones, are remarkably well known for lampreys as a result of attempts for integrated control of sea lamprey in the Laurentian Great Lakes. Furthermore, we have gained remarkable insights into the life history, growth, sexual development and behavior of lampreys as a consequence of studies directed to control sea lamprey.

This volume provides a remarkable compilation and combination of conservation and control. The obvious advantages of lampreys as model species are quite clear. Life history, sex determination and perhaps even sex reversal for lampreys are clearly elaborated in this volume. There are no other species where the contrast in life histories can rival that of parasitic and non-parasitic lampreys. Lampreys are ideal model species to study the combined effects of genetic and environmental factors on early development and life history. Whether one accepts the operational definition of (some) lampreys as true parasites remains an intriguing question for those interested in community ecology. The question of the evolution of the parasitic life history of lampreys is profoundly intriguing, and it would be difficult to postulate the origin and evolution of a more dramatic life history pattern.

Of course, any significant focus on lampreys must include the extensive body of information on the control of sea lamprey in North American lakes. This volume provides what must be close to a definitive compilation of that complex situation. However, this moves beyond the usual historical summary to a critical evaluation of control programs, and, most importantly, to an assessment of emerging control techniques. The contrast with the consideration of attempts to conserve and restore native lampreys in western North America is at the same time ironic and informative.

The efforts to manage the recovery of threatened native lampreys in the Pacific Northwest have particular significance for indigenous peoples in the region. Typically, lampreys were a first food and they are still recognized for their cultural significance. That is the basis for some of the most dedicated efforts to propagate lampreys as part of conservation and restoration programs. The future prospects for lampreys are laid out in this volume for interests as diverse as taxonomy, conservation, control and restoration.

Corvallis, OR, USA   David L. G. Noakes
                                              Editor, Springer Fish and Fisheries Series

                                                  Professor of Fisheries & Wildlife
                                          Director, Oregon Hatchery Research Center
                                                          Oregon State University

Acknowledgements

I would like to express my deep gratitude to series editor David L. G. Noakes, and Alexandrine Cheronet and Judith Terpos at Springer, for allowing me the opportunity to expand this book project into two volumes. I must also thank the many authors who have contributed to this volume. Without their considerable expertise, hard work and patience, this book would not have been possible. I would also like to acknowledge the peer reviewers who likewise contributed their time and expertise: Chris T. Amemiya, Tyler J. Buchinger, John B. Hume, Mary L. Moser, Stewart B. Reid, Todd B. (Mike) Steeves, C. Michael Wagner and Murray D. Wiegand. The additional insights they provided are greatly appreciated. Elizabeth Docker provided invaluable assistance in the final editing and proofreading of the manuscript, and Kristian Sattelberger helped with references.

On a personal note, I would like to express my heartfelt gratitude to my parents (Sandy and Hilda) who always showed their support and enthusiasm for my chosen career; my siblings (Elizabeth, Susan and Ian) and extended family (Jason, Michael, Benjamin, Nicholas, Jonathan, Stuart, Henry and Josephine); and most of all, my daughter, Sylvia, for her patience, love and support.

Contents

Chapter 1
The Lamprey Gonad

Margaret F. Docker, F. William H. Beamish, Tamanna Yasmin, Mara B. Bryan and Arfa Khan

Abstract Understanding gonadal development in lampreys is complicated by their complex life cycle, the long period during which their gonads remain histologically undifferentiated, and their lack of any close living relatives. This chapter synthesizes the available information related to lamprey sex determination, sex differentiation, sexual maturation, and sex steroids, and it identifies key research needs. A detailed review of lamprey sex ratios shows that: (1) adult lampreys (i.e., during the upstream migration or at spawning) exhibit a small but consistent excess of males in virtually all species studied (with significantly female-biased sex ratios noted only in sea lamprey in the three upper Great Lakes following initiation of sea lamprey control); (2) larval sex ratios are generally at parity or with an excess of females; (3) transformers collected above barriers or following lampricide treatment tend to be male biased in the earliest age classes to metamorphose; and (4) there is spatial and temporal variation in sex ratio during the parasitic feeding phase, but overall sex ratio is less male biased than during the adult phase, suggesting that females suffer higher mortality just prior to or during sexual maturation. The shift in sex ratio observed in the upper Great Lakes following initiation of control led to suggestions of environmental sex determination (ESD), specifically density-dependent sex determination, but evidence for ESD in lampreys is equivocal. Sex ratios did not become female biased

---

M. F. Docker (✉) · T. Yasmin · A. Khan
Department of Biological Sciences, University of Manitoba, 50 Sifton Road, Winnipeg, MB R3T 2N2, Canada
e-mail: Margaret.Docker@umanitoba.ca

T. Yasmin
e-mail: yasmint@myumanitoba.ca

A. Khan
e-mail: khana349@myumanitoba.ca

F. W. H. Beamish
Environmental Science Program, Burapha University, Saen Suk, Bangsaen 20131, Chonburi, Thailand
e-mail: billbeamish@hotmail.com

M. B. Bryan
Caribou Biosciences, 2929 7th Street, Suite 105, Berkeley, CA 94720, USA
e-mail: mbryan@cariboubio.com

© Springer Nature B.V. 2019
M. F. Docker (ed.), *Lampreys: Biology, Conservation and Control*,
Fish & Fisheries Series 38, https://doi.org/10.1007/978-94-024-1684-8_1

in the lower Great Lakes, and all five lakes now show a slight excess of males even though abundance has been low and relatively stable for the past several decades. Furthermore, although a significant relationship between larval density and sex ratio has been observed in two non-parasitic species in the southeastern United States, significant relationships between larval density and sex ratio are not evident among contemporaneous sea lamprey populations (i.e., before or after the initiation of control). ESD, usually temperature-dependent sex determination, has been reported in a number of fish species, but no fish species with exclusively ESD have been identified to date. Skewed sex ratios may result from environmental influences on genetic sex determination rather than strict ESD, and the nature of the genotype × environment interactions can differ among populations and over time. However, apart from ruling out "the usual suspects" (i.e., genes implicated in sex determination in other vertebrates), nothing is known regarding the possible genetic basis of sex determination in lampreys. In contrast, many of the genes involved in the sex differentiation process (i.e., development of the undifferentiated gonad into an ovary or testis) tend to be conserved among vertebrates, and initial studies suggest that at least some of the same genes are involved in gonadal development in lampreys. Understanding the factors influencing lamprey sex determination and differentiation has been complicated by lack of knowledge regarding the critical sex differentiation period. Lampreys are sometimes said to pass through an initial female stage or female intersexual stage, because mitosis and meiosis appear to occur in most larvae regardless of future sex. However, meiosis and oocyte growth are more synchronized and extensive in female larvae, and the extent to which oocytes develop and regress in presumptive males either varies among individuals and species or reflects differences in the degree to which these transient processes are detected. Ovarian differentiation is generally thought to be complete at ~1 and 2–3 years of age in non-parasitic and most parasitic lamprey species, respectively, and at 4–5 years in the anadromous sea lamprey. Later and more prolonged mitosis in parasitic species permits elaboration of a larger stock of oocytes, and persistence of a limited number of undifferentiated germ cells in some parasitic species may allow further oocyte recruitment in large larvae. In all species, testicular differentiation occurs at or around the onset of metamorphosis, at which time resumption of mitosis in the remaining undifferentiated germ cells produces spermatogonia. In vivo biopsy studies showed that sea lamprey gonads can remain labile as long as undifferentiated germ cells remain in the gonads (i.e., after the apparent completion of ovarian differentiation, but up until differentiation of the remaining germ cells at the end of the larval stage). The presence of "atypical" gonads (which often developed into typical males in biopsied larvae) in sea lamprey from both the Great Lakes and Atlantic drainages is consistent with delayed gonadal differentiation but requires further study. Despite the apparent lability of the lamprey gonad, hormonal sex control has not been successful. Non-parasitic lampreys begin maturing during the latter stages of metamorphosis; in contrast, parasitic species remain sexually immature until they approach the end of the juvenile feeding phase, and sexual maturation proceeds during the non-trophic spawning migration. Although the rate of maturation varies among species, depending on the duration of migration, all species and life history types appear to converge again during final maturation ~1–2 months before spawning. Oocytes begin to approach their size at maturity (~1 mm in virtually all species), and, at ovulation, the oocytes (now typically called eggs) are synchronously released into the body cavity. The mature ovary constitutes ~25–35% of a female's total body weight, regardless of species, but the total number of eggs (fecundity) increases approximately with the cubic power of body length so that fecundity in the largest anadromous parasitic species (e.g., mean 172,000 in sea lamprey) is almost two orders of magnitude higher than that of the much smaller non-parasitic species. The mature testis constitutes ~2–10% of a male's body weight, with gonadosomatic index (GSI) appearing to be higher in males of non-parasitic species compared to parasitic species, although absolute testis size is still much higher in parasitic species. The study of lamprey steroidogenesis and steroid receptors is contributing to our understanding of the evolution of steroid hormones as transcriptions factors in vertebrates, but much still needs to be learned regarding the role of sex steroids in lamprey sex differentiation and sexual maturation.

Keywords Atresia · Egg size · Environmental sex determination · ESD · Fecundity · Genetic sex determination · Gonadogenesis · Gonadosomatic index · GSI · Hormonal sex control · Intersex · Life history type · Ovarian differentiation · Sex differentiation · Sex ratio · Sex reversal · Sex steroids · Sexual maturation · Spawning · Spermatogenesis · Steroidogenesis · Testicular differentiation · Upstream migration · Vitellogenesis

1.1 Introduction

As one of only two surviving groups of ancient jawless vertebrates, lampreys are of enduring evolutionary interest. Study of lamprey biology, for example, continues to provide invaluable insight into the events that occurred at the dawn of vertebrate evolution and is helping us understand the degree to which various traits and processes are conserved across vertebrate lineages (see Docker et al. 2015). Research related to lamprey biology is also helping to inform efforts directed at controlling sea lamprey *Petromyzon marinus* in the Laurentian Great and Lake Champlain (see Chap. 5) and initiatives to manage or conserve native lampreys (Maitland et al. 2015). In particular, understanding reproduction is important for effective control and conservation. Sea lamprey control is primarily achieved through use of the selective lampricide 3-trifluoromethyl-4-nitrophenol (TFM) and barriers which largely prevent upstream-migrating adults from reaching their spawning grounds (Chap. 5), but alternative methods that reduce sea lamprey numbers by interfering with sex determination (e.g., leading to highly skewed sex ratios) or other aspects of gonadal development and reproduction could further enhance control. Strategies that disrupt sea lamprey reproduction are already being developed (e.g., the sterile-male-release technique and use of pheromones to disrupt upstream migration or mating; see Twohey et al. 2003; Li et al. 2007; Johnson et al. 2015a; Bravener and Twohey 2016), and there are others to explore (Sower 2003; Docker et al. 2003; Bergstedt and Twohey 2007). For species of conservation concern, efforts are being expended to improve our knowledge of the reproductive physiology of these lampreys (e.g., Mesa et al. 2010; Farrokhnejad et al. 2014), including studies to optimize artificial fertilization and propagation methods (see Chap. 2) and to better understand the reproductive ecology of lampreys in their natural environments (e.g., Jang and Lucas 2005; Johnson et al. 2015b; Whitlock et al. 2017).

However, despite great interest in the reproduction of lampreys, study of the many aspects of their reproductive biology is complicated by a number of factors, one of which is their complex life cycle and long generation times. There are at least 41 recognized species of extant lampreys (Potter et al. 2015); all are semelparous, dying after a single spawning season (although not necessarily after a single spawning event), and all spawn in fresh water (Johnson et al. 2015b). All lampreys pass through a freshwater filter-feeding larval stage that lasts approximately 3–8 years (Fig. 1.1), although the duration is variable among species and populations (see Dawson et al. 2015) and may also differ between the sexes (Docker 2009; Manzon et al. 2015). Consistent with a prolonged larval stage, there is also a prolonged period of sexual indeterminacy, and the gonadal differentiation process is asynchronous in males and females: ovarian differentiation occurs during the larval stage (at 1–3+ years of age), but testicular differentiation does not occur until metamorphosis, several years later (Fig. 1.1). Following metamorphosis (see Manzon et al. 2015), 18 species are parasitic, feeding on the blood or tissue of other fishes in marine or freshwater systems (see Chaps. 3 and 4). Some of the anadromous species (e.g., sea lamprey and Pacific lamprey *Entosphenus tridentatus*) can reach total lengths (TL) in excess of 600–800 mm (see Chap. 3) and can migrate several hundreds of kilometers to headwater streams where they spawn (Moser et al. 2015). In contrast, the freshwater-resident parasitic lampreys (e.g., sea lamprey in the Great Lakes, silver lamprey *Ichthyomyzon unicuspis*) are smaller at maturity. The remaining 23 species are non-parasitic "brook" lampreys that bypass the post-metamorphic feeding phase and thus reach maturity at even smaller sizes (~100–150 mm TL; see Docker 2009). Parasitic lampreys remain sexually immature during the feeding phase (and, hence, are technically considered juveniles rather than adults at this point; see Docker et al. 2015). In contrast, sexual maturation in the non-parasitic brook lampreys begins during the latter stages of metamorphosis (Fig. 1.1). Brook lampreys remain within their natal streams and spawn and die the following spring, that is, within 6–10 months of metamorphosis. In contrast, sexual maturation in parasitic species is delayed for 1–4 years following metamorphosis (Docker 2009; Chap. 4).

Studies of lamprey reproduction have also been hindered by their divergence from other vertebrates ~500 million years ago (Docker et al. 2015), with no other extant vertebrates to "bridge the gap." Lampreys share the general organization of the hypothalamic-pituitary-gonadal (HPG) axis with all other vertebrates, and much has been learned about the evolution of the vertebrate HPG axis by studying the lamprey HPG axis (Sower 2015, 2018). However, the hormones that coordinate the axis and regulate reproductive physiology are often different among vertebrate groups. Similarly, although the study of lamprey steroidogenesis and steroid receptors has contributed greatly to our understanding of the evolution of steroid hormones as

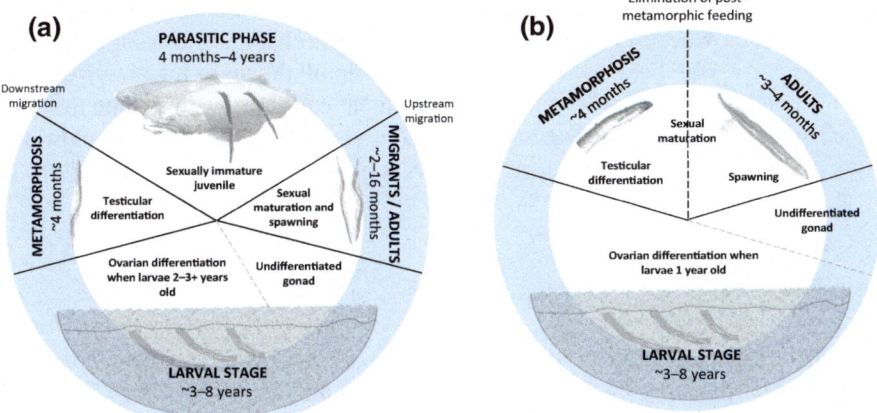

Fig. 1.1 The lamprey life cycle, showing the timing of key events in the development of the gonad relative to the stages in the life cycle in **a** parasitic and **b** non-parasitic (brook) lampreys; a detailed timeline of gonadal processes is shown in Fig. 1.6

transcriptions factors in vertebrates (Thornton 2001; Baker 2004; Baker et al. 2015), the idiosyncrasies of lamprey steroids have often been perplexing. Lampreys produce gonadal steroids that differ from those of other vertebrates by possessing an additional hydryoxyl group at the C15 position (Kime and Rafter 1981; Kime and Callard 1982; Bryan et al. 2003, 2004; Lowartz et al. 2003), and these novel steroids have presented many technical challenges. Much of our initial knowledge regarding steroid synthesis in lampreys was inferred from studies that incubated radiolabeled precursors with gonadal or other tissue extracts, but results were often inconclusive because several of the products could not be identified through comparison to known steroid standards (e.g., Weisbart and Youson 1975, 1977; Weisbart et al. 1978). Likewise, the lack of commercially available 15α-hydroxylated radiolabeled steroids has hindered binding experiments to detect receptors for lamprey-specific 15α-hydroxylated steroids. Elucidating the genetic basis of sex determination and sex differentiation in lampreys will provide insights into the degree to which the genes involved in these processes are conserved among vertebrates. However, the prolonged period of sexual indeterminacy in lampreys and their long divergence from other vertebrates make it difficult to make extrapolations based on what is known in other vertebrates, and identification of homologs of key genes in other vertebrates can be challenging (Spice et al. 2014; Khan 2017). Nevertheless, the publication of the sea lamprey genome (Smith et al. 2013, 2018) is leading to a wealth of new knowledge of the genes and gene networks that control many aspects of lamprey biology (McCauley et al. 2015; see Chap. 6) and is expected to contribute substantially to our understanding of lamprey reproduction as well.

In this chapter, we provide a wide-reaching review of topics related to the lamprey gonad, ranging from an in-depth discussion of lamprey sex ratios to a synthesis of the literature related to lamprey sex determination, sex differentiation, sexual maturation, and sex steroids. Because one of the main goals of this chapter is to inspire lamprey biologists and researchers in other disciplines to help fill the many remaining gaps in our understanding of lamprey gonadal development and function, we hope that this chapter will provide both the necessary background and the appropriate stimulus for future research into the many intriguing aspects of lamprey reproductive biology.

1.2 Sex Ratios

A small but variable excess of males has long been observed among upstream migrants and spawning adult lampreys (e.g., Dean and Sumner 1898; Young and Cole 1900; Wigley 1959; Zanandrea 1961). Sex ratio data are less readily available for other stages, when individuals are more disperse and harder to catch and when males and females can only be distinguished following internal examination (see Sect. 1.4.1.5), but these other stages generally do not show the same bias toward males. However, the complex lamprey life cycle makes interpretation of stage-specific sex ratios difficult, and it is not yet known whether these apparent differences might be the result of: (1) sex-specific differences in mortality (e.g., higher mortality of females during the post-larval stages); (2) sampling bias (e.g., due to sex-specific differences in age at metamorphosis or differences in the temporal or spatial distribution of the sexes during feeding and migration); or (3) an environmental influence on the sex differentiation process. In this section, we review the available stage-specific sex ratio data for lampreys and attempt to determine the extent to which each of these three factors may be operating. A full discussion of a possible extra-genetic influence on sex differentiation (i.e., environmental or density-dependent sex determination) is provided in Sect. 1.3.2.2.

1.2.1 Sex Ratio of Upstream Migrants and Adults

Sex ratio data are available for 10 of the 18 parasitic lamprey species during their upstream migrations or at spawning. Most species show roughly even sex ratios or a small excess of males (Table 1.1): 45–52% male in adult Caspian lamprey *Caspiomyzon wagneri*, 68% male in Vancouver lamprey *Entosphenus macrostomus*, 48–57% male in Pacific lamprey, 61–65% male in pouched lamprey *Geotria australis*, 42% male in chestnut lamprey *Ichthyomyzon castaneus*, 49–59% male in silver lamprey, 48–61% male in anadromous European river lamprey *Lampetra fluviatilis*, 50–63% male in all but the praecox anadromous form of Arctic lamprey *Lethenteron camtschaticum*, 49–57% male in the short-headed lamprey *Mordacia mordax*, 44–65% male in anadromous sea lamprey, and 50–68% male in freshwater-resident sea lamprey prior to the initiation of sea lamprey control.

Table 1.1 Sex ratio of adult lampreys (i.e., during the upstream migration or at spawning); significantly male- and female-biased sex ratios (calculated using the Chi test function in Excel) are identified with a superscript m and f, respectively. An asterisk indicates that individuals that could not be identified as either male or female were omitted from the sex ratio; values in bold signify sex ratios observed following initiation of lampricide (TFM) treatment in the respective basin; rkm = river km

Species and location	Year	n	% Male	Reference	Comments
PARASITIC SPECIES					
Caspiomyzon wagneri Caspian lamprey (anadromous)					
Shirud R, Iran	2006	211	51.7	Nazari and Abdoli (2010)	Spring migrants; 59, 42, 49, 55, 53% male in weeks 1–5, respectively
Shirud R, Iran	2008	104	51.0	Ahmadi et al. (2011)	Fall migrants 47% male; spring 52% male
Shirud R, Iran	2009	53	45.3	Shirood Mirzaie et al. (2017)	Fall migrants 20% male; spring 55% male
Shirud R, Iran	2012	401	46.7	Abdoli et al. (2017)	Spring migrants
Entosphenus macrostomus Vancouver (or Cowichan Lake) lamprey (freshwater)					
Bear and Cowichan L tributaries, BC	2017	28	67.9	Wade et al. (2018)	Late June
Entosphenus tridentatus Pacific lamprey (anadromous)					
Oregon streams		108	57.4	Kan (1975)	Pre-spawners
		113	54.9		Spawners
Chemainus R, BC	1978	124	48.4	R. J. Beamish (1980)	Early migrants; males and females roughly equal throughout sampling period
North Umpqua R, OR @ Winchester Dam (rkm 11.2)	2009	24	33.3*	Lampman (2011)	% male + % unknown = 65% (2009) and 59% (2010)
	2010	45	38.2*		
Willamette R, OR @ Willamette Falls (rkm 205)	2007	206	49.0	Clemens et al. (2016), Clemens (pers. comm.)	50, 56, 61, 44, 50% male in Apr, May, June, July, Sept, respectively
	2008	143	56.6		33, 39, 56, 53, 71% male in Apr, May, June, July, Aug
Willamette R, OR @ Willamette Falls (rkm 205)	2016	269	50.6	Porter et al. (2017)	51, 51, 50% male in June, July, Aug
Snake R basin, WA @ rkm 589 and 635	2006	50	39.9*	McIlraith et al. (2015)	July–Oct; sex unidentifiable in 4, 20, and 14% lamprey in 2006, 2007, 2008
	2007	46	45.9*		
	2008	50	46.0*		
Geotria australis pouched lamprey (anadromous)					
Warren R @ rkm 32 and 61, Australia	1976–1980	379	65.4[m]	Potter et al. (1983)	First 4 months of spawning run (mid-July to mid-Nov)
Donnelly R estuary, Australia	1981	71	60.6	Potter et al. (1983)	Estuary at onset of spawning migration (July, Aug)
	1982	125	63.2[m]		
Ichthyomyzon castaneus chestnut lamprey (freshwater)					
Muskegon R, MI (L Michigan)	1981	38	42.1	Schuldt et al. (1987)	Upstream migration (May)

(continued)

Table 1.1 (continued)

Species and location	Year	n	% Male	Reference	Comments
Ichthyomyzon unicuspis silver lamprey (freshwater)					
Peshtigo R, WI	1978	–	59.0	Schuldt et al. (1987)	Upstream migration (May)
Menominee R, MI/WI	1980	51	58.8		
Oconto R, WI (L Michigan)	1980	47	48.9		
Lampetra fluviatilis European river lamprey (anadromous)					
Severn Estuary, UK	1972–1976	621	48.6	Abou-Seedo and Potter (1979)	52, 47, 71% male in early, mid, late run
Firth of Forth, UK	1980	206	60.7^m	Maitland et al. (1984)	Upstream migrants Sept–Oct
	1981	430	56.7^m		Upstream migrants Aug–Nov
R Teith @ Deanston, UK	1984	53	56.6	Morris (1989)	Spawning adults (Apr, May)
R Derwent @ Stamford Bridge Weir, UK	2003	76	72.4^m	Jang and Lucas (2005)	Upstream migrants Mar–Apr
		1,284	48.4		On spawning grounds; 23% male during nest building, 73% during spawning, 21% post-spawning
Lampetra fluviatilis European river lamprey (anadromous praecox form)					
Severn Estuary, UK	1972–1976	621	48.6	Abou-Seedo and Potter (1979)	42, 52, 45% male in early, mid, late run
Lampetra fluviatilis European river lamprey (dwarf freshwater form)					
Endrick Water (L Lomond), UK	1983	38	81.6^m	Morris (1989)	Upstream migrants (Nov)
	1984	16	31.3		Spawning adults (Apr, May)
Lethenteron camtschaticum Arctic lamprey (anadromous)					
Utkholok R, Russia	2005	142	63^m	Kucheryavyi et al. (2007)	Upstream migrants and mature adults
Lethenteron camtschaticum Arctic lamprey (anadromous praecox form)					
Utkholok R, Russia	2005	38	92^m	Kucheryavyi et al. (2007)	Upstream migrants and mature adults
Lethenteron camtschaticum Arctic lamprey (freshwater non-parasitic form)					
Utkholok R, Russia	2005	632	50	Kucheryavyi et al. (2007)	Mature adults
Lethenteron camtschaticum Arctic lamprey (freshwater form)					
Slave R @ Fort Smith and Hay R, NWT	1967	37	59.5	Nursall and Buchwald (1972)	Upstream migrants (June–Aug)
Mordacia mordax short-headed lamprey (anadromous)					
Dandenong Cr, near Melbourne, Australia	1963	57	49.1	Potter et al. (1968)	Upstream migrants (Nov 1963 and 1964, Sept 1965)
	1964	63	47.6		
	1965	81	53.1		
Derwent R, Tasmania, Australia	1967	60	56.7	Potter et al. (1968)	Upstream migrants (Jan)

(continued)

Table 1.1 (continued)

Species and location	Year	n	% Male	Reference	Comments
Petromyzon marinus sea lamprey (anadromous)					
Sheepscot R, ME	1949	52	44.1	Applegate (1950)	Upstream migrants
Barrows Stream, ME	1960–1964	64	65[m]	Davis (1967)	Upstream migrants
St John R, NB @ Mactaquac Dam (rkm 120)	1974	285	54.1	Beamish and Potter (1975)	Upstream migrants (June–July)
St John R, NB @ Mactaquac Dam (rkm 120)	1974–1977	393	57.5[m]	Beamish et al. (1979)	Upstream migrants; 73% male in early May, 55% in late June
Keswick R, NB	1974–1977	63	65.1[m]	Beamish et al. (1979)	Spawning and spent adults (late June–early July)
Connecticut R @ Holyoke Dam (rkm 135), MA	1981	484	56[m]	Stier and Kynard (1986)	Upstream migrants (May–June); 55–59% male early in run, 59–67% late in run
	1982	404	62[m]		
Connecticut R @ Holyoke Dam (rkm 135), MA	2013	97	45.4	Castro-Santos et al. (2017)	Upstream migrants (May–June)
Dordogne R, France	2003	101	47.5	Beaulaton et al. (2008)	Upstream migrants; 56% male in Jan–Feb, 43% in Mar–May
	2004	124	49.2		
Garonne R, France	2003	149	49.0	Beaulaton et al. (2008)	Upstream migrants
	2004	49	45.0		
Ulla R Estuary, Spain	2010	133	60.9[m]	Silva et al. (2016)	Upstream migrants; 83, 60, 41% male in Jan, Feb, Mar
Petromyzon marinus sea lamprey (freshwater)					
Cayuga Inlet, NY	1886	745	64.4[m]	Meek (1889)	Upstream migrants (May, June)
Cayuga Inlet, NY	1898	1,140	51.7	Surface (1899)	Upstream migrants
Cayuga Inlet, NY	1950	372	61.0[m]	Wigley (1959)	Upstream migrants; % male reasonably consistent over run
	1951	1,820	60.8[m]		
	1952	1,306	53.8[m]		
Ocqueoc R, MI (L Huron)	1947	679	53.6	Applegate (1950)	Upstream migrants
	1949	24,643	68.2[m]		
Carp R, MI (L Superior)	1947	1,600	62.3[m]	Applegate (1950)	Upstream migrants; 75% male at end of run (early July) in 1947
	1948	2,931	62.9[m]		
	1949	2,763	67.5[m]		
L Superior tributaries	1954–1978	1,911–50,975	28[f]–71[m]	Heinrich et al. (1980)	Upstream migrants; % males peaked 1961–1964, declined thereafter (see Fig. 1.2)
L Michigan tributaries	1954–1978	774–18,043	21[f]–68[m]	Heinrich et al. (1980)	Upstream migrants; % males peaked 1963 (see Fig. 1.2)
L Huron tributaries	1947–1978	197–24,643	31[f]–71[m]	Smith (1971), Heinrich et al. (1980)	Upstream migrants; % males peaked 1950–1955 (see Fig. 1.2)

(continued)

Table 1.1 (continued)

Species and location	Year	n	% Male	Reference	Comments
Humber R (L Ontario)	1968–1978	1,191–6,848	**42**[f]**–58**[m]	Heinrich et al. (1980)	Upstream migrants (see Fig. 1.2)
Humber R (L Ontario)	1968–1972	1,223–4,387	50–**56**[m]	Potter et al. (1974)	Upstream migrants
L Superior	1995–2016	45–1,880	**33**[f]**–66**[m]	GLFC (1996–2017)[a]	Upstream migrants (see Fig. 1.2)
L Michigan	1995–2016	228–2,225	**38**[f]**–55**[m]	GLFC (1996–2017)[a]	Upstream migrants (see Fig. 1.2)
L Huron	1995–2016	136–12,231	**49–67**[m]	GLFC (1996–2017)[a]	Upstream migrants (see Fig. 1.2)
L Ontario	1995–2016	397–5,154	**48–62**[m]	GLFC (1996–2017)[a]	Upstream migrants (see Fig. 1.2)
L Erie	1995–2016	20–1,982	**51–73**[m]	GLFC (1996–2017)[a]	Upstream migrants (see Fig. 1.2)

BROOK LAMPREYS

Ichthyomyzon fossor northern brook lamprey

Brule R, WI (L Superior)	1945	17	58.8	Churchill (1945)	Prior to spawning (June)
Sturgeon R, MI (L Superior)	1960	24	75[m]	Purvis (1970)	Spawners (June)
Little Cedar R, MI	1980	24	54.2	Schuldt et al. (1987)	Pre-spawners (April, May)
Walla Walla Cr, WI	1980	16	56.3		
Little Wolf R, WI (L Michigan)	1980	31	64.5		

Ichthyomyzon gagei southern brook lamprey

10 river systems in AL, FL, GA, LA, OK, TX	1930–1951	98	59.1	Dendy and Scott (1953)	Adults; pooled sex ratio from 18 collections
Little and Choclafaula Cr, AL	1980–1981	110	60.9[m]	F. W. H. Beamish (1982)	Pre-spawners (early March, 1–2 km downstream from spawning site); earliest migrants 80% male
Hodnett and Choclafaula Cr, AL	1980–1982	567	45.1[f]	Beamish and Thomas (1984)	Transformers and adults; no differential migration
19 streams in AL, AR, LA, MS, TX	1988–1992	5–87	25–65[m]	Beamish et al. (1994)	Transformers and adults (see Table 1.2)

Lampetra aepyptera least brook lamprey

7 streams in MD, DE, KY, TN, AL	1980, 1988	20–38	46.2–79.2[m]	Docker and Beamish (1994)	Transformers and adults (Oct–Feb; see Fig. 1.3)

Lampetra lanceolata Turkish brook lamprey

Iyidere Stream, Turkey	2005–2006	54	53.7	Gözler et al. (2011)	Pre-spawning and mature adults

Lampetra planeri European brook lamprey

R Yeo, UK	1947–1960	57–240	54.4–77.0[m]	Hardisty (1961a)	Spawning season; 78, 68, 66, and 62% male in weeks 1–4, respectively

(continued)

Table 1.1 (continued)

Species and location	Year	n	% Male	Reference	Comments
R Stensån, Sweden	1976	120	51.5	Malmqvist (1978)	Spawning season
Rörum South R, Sweden	1976	52	63.3	Malmqvist (1978)	Spawning season
Stampen Stream, Sweden	1976	44	61.3	Malmqvist (1978)	Spawning season
Länsmansbäcken, Sweden	1977	163	60.1^m	Malmqvist (1980)	Spawning season (Mar–June); % males highest (65–76%) in early May
	1978	192	60.9^m		
Endrick Water (L Lomond), UK	1984	39	56.4	Morris (1989)	Adults (Apr, May)
River Teith, UK	1984	28	64.3	Morris (1989)	Adults (Apr, May)
Lampetra richardsoni western brook lamprey					
Morrison Cr, BC	1987	22	50.0	Beamish et al. (2016)	Mature or spent adults (May–July); 100% male after mid-June
Lampetra richardsoni western brook lamprey (var. *marifuga*)					
Morrison Cr, BC	1984	24	79.2^m	Beamish (1985)	
Morrison Cr, BC	1987	42	88.1^m	Beamish et al. (2016)	Mature or spent adults (May–July)
Lampetra zanandreai Po brook lamprey					
Italy	<1951	1,314	59.1^m	Zanandrea (1961)	~50% male during maturation; >50% during spawning
Lethenteron appendix American brook lamprey					
Huron R tributaries (L Erie)	1899	259	78.4^m	Young and Cole (1900)	Spawners (Apr)
Wednesday Br, NH	1959	13	69.2	Sawyer (1960)	Near end of spawning (May)
Big Cr, ON (L Erie)	1970	85	67.1^m	Kott (1971)	Pre-spawners (Apr–May); no differential migration
Buffalo Cr, TN	1973	126	54.8	Seagle and Nagel (1982)	Transformers and adults
Fox R, WI	1980	30	66.7	Schuldt et al. (1987)	Pre-spawners (April and May)
Betsie R, MI (L Michigan)	1980	139	79.1^m		
Tetrapleurodon geminis Mexican brook lamprey and/or *T. spadiceus* Mexican lamprey					
Michoacan, Mexico	1961–1962	76	53.9	Álvarez del Villar (1966)	Spawning adults

[a] Annual reports to the GLFC for previous calendar year; authors as follows (for publication year): Schleen LP, Young RJ, Klar GT (1996, 1998); Klar GT, Schleen LP, Young RJ (1997); Klar GT, Schleen LP (1999, 2001, 2003); Schleen LP, Klar GT (2000, 2002); Young RJ, Klar GT (2004, 2006); Klar GT, Young RJ (2005); Adair RA, Young RJ (2007, 2009); Young RJ, Adair R (2008); Sullivan P, Adair R (2010, 2012, 2014); Adair R, Sullivan P (2011, 2013, 2015); Sullivan P, Adair R, Woldt A (2016); Mullett K, Sullivan P (2017); see http://www.glfc.org/annual-reports.php

Extreme male-biased sex ratios (>80% male) were observed only in the freshwater form of European river lamprey (Morris 1989) and the praecox form of Arctic lamprey (i.e., an anadromous form with a reduced marine feeding phase; Kucheryavyi et al. 2007) (see Chap. 4). A similarly male-biased sex ratio has been observed in the rare parasitic form of the western brook lamprey *Lampetra richardsoni* (i.e., the "marifuga" variety; Beamish 1985; Beamish et al. 2016). Docker (2009) suggested that alternative life history types such as these might be dominated by males because transitions related to feeding and migratory type may occur more readily in males. Because the trajectories associated with ovarian development diverge in different life history types years before the paths associated with testicular development diverge, female life history type may be less flexible (see Sect. 1.4.1).

Significantly female-biased adult sex ratios have been reported only in the Great Lakes sea lamprey following initiation of sea lamprey control and only in the three upper Great Lakes (Superior, Michigan, and Huron; Fig. 1.2). The observed shift to female-biased sex ratios after the onset of sea lamprey control suggested that lamprey sex ratio was correlated with abundance (see Sect. 1.2.6). However, in the two lower Great Lakes, a significant excess of females was observed in only a single year (1978) in Lake Ontario, despite similar (albeit somewhat later) declines in abundance, and Lake Erie exhibited only even or male-biased sex ratios. Furthermore, sex ratio in all three upper Great Lakes returned to parity or an excess of males by the mid- to late 1990s.

No significantly female-biased sex ratios have been observed among any of the nine non-parasitic lamprey species for which adult sex ratio data are available. Again, most species show an excess of males (Table 1.1): northern brook lamprey *Ichthyomyzon fossor* 54–75% male, least brook lamprey *Lampetra aepyptera* 46–79% male, Turkish brook lamprey *L. lanceolata* 54% male, European brook lamprey *L. planeri* 54–73% male, the typical parasitic form of western brook lamprey 50% male, Po brook lamprey *L. zanandreai* 59% male, American brook lamprey *Lethenteron appendix* 55–79% male, and Mexican brook lamprey *Tetrapleurodon geminis* (or Mexican lamprey *T. spadiceus*) 54% male. Sex ratio of adult and metamorphosing southern brook lamprey *I. gagei* in 19 populations ranged from 25 to 65% male, but none were significantly female biased (Table 1.2).

It is unknown if sex-specific differences in capture efficiency, particularly with different gear types, produce biased sex ratio data. Preliminary data from mark-recapture studies show no evidence of sex-specific differences in trapping efficiency in upstream-migrating sea lamprey in the Great Lakes (Sean Lewandoski, U.S. Fish and Wildlife Service, Marquette, MI, personal communication, 2018). In contrast, Beaulaton et al. (2008) suggested that anadromous sea lamprey trapped in pots showed a slight excess of females relative to those collected in nets. However, it is well known that sex ratios measured only during a restricted part of the spawning run can be biased, presumably as the result of behavioral differences between male and female lampreys (see Sect. 1.2.5). Furthermore, external sex determination may be vulnerable to observational error in early season trap captures when sexually dimorphic characteristics are not as readily identifiable (Johnson et al. 2015b). Nevertheless, many of the studies performed to date have monitored sex ratio over

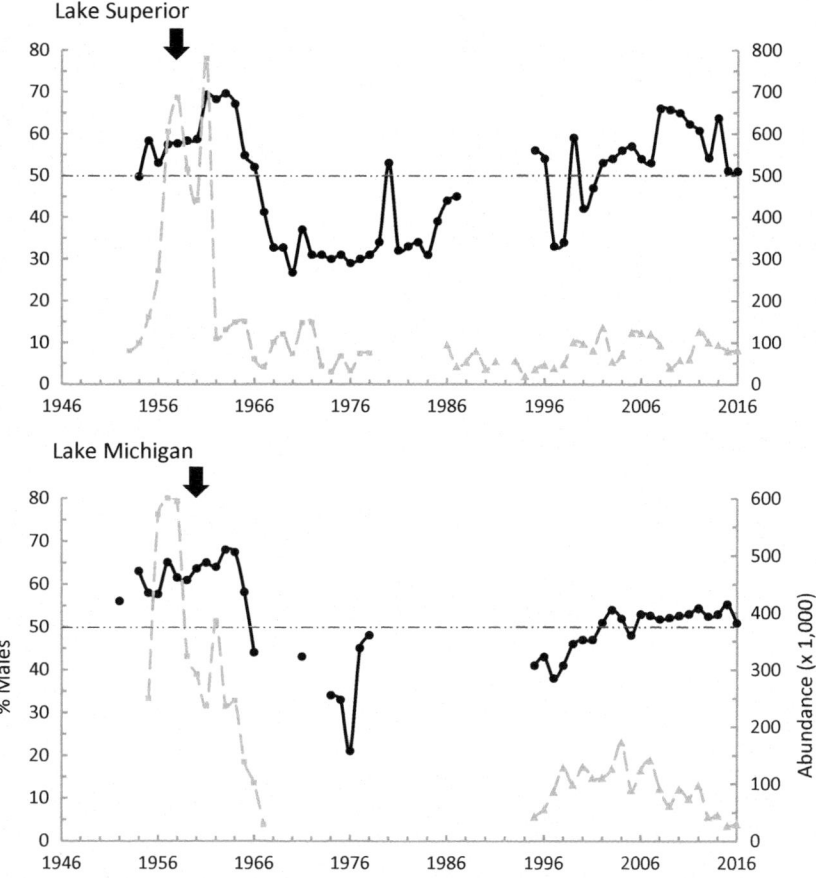

Fig. 1.2 Sex ratio (percent males; *solid line*) of sea lamprey *Petromyzon marinus* in the three upper Great Lakes (Superior, Michigan, and Huron) and lower Great Lakes (Ontario and Erie) prior to and following initiation of sea lamprey control compared to adult (spawner) abundance (*gray broken line*). Parity (50:50 sex ratio) is indicated by a *horizontal dashed line*, and *vertical arrows* show year of first lampricide (TFM) treatment in each basin. Sex ratio data for 1947–1987 were collected from Smith (1971), Heinrich et al. (1980), and Houston and Kelso (1991); sex ratio data for 1995–2016 were collected from annual reports to the Great Lakes Fishery Commission (GLFC 1996–2017; see Table 1.1 for list of report authors); in Lake Erie, data were excluded if n < 20. Recent adult abundance estimates (*gray triangles*) are based on standardized sea lamprey index values that have been scaled to the lake-wide level (Jess Barber, U.S. Fish and Wildlife Service, Marquette, MI, personal communication, 2018); Lake Erie values represent 3-year averages (see Chap. 5). Historical lake-wide abundance estimates (*gray squares*) were obtained from Sullivan et al. (2003; Erie) or were approximated by scaling abundance in index streams (Smith 1971; Heinrich et al. 1980) to peak historical values estimated in GLFC (2015): 780,000 (Superior), 600,000 (Michigan), 700,000 (Huron), and 450,000 (Ontario)

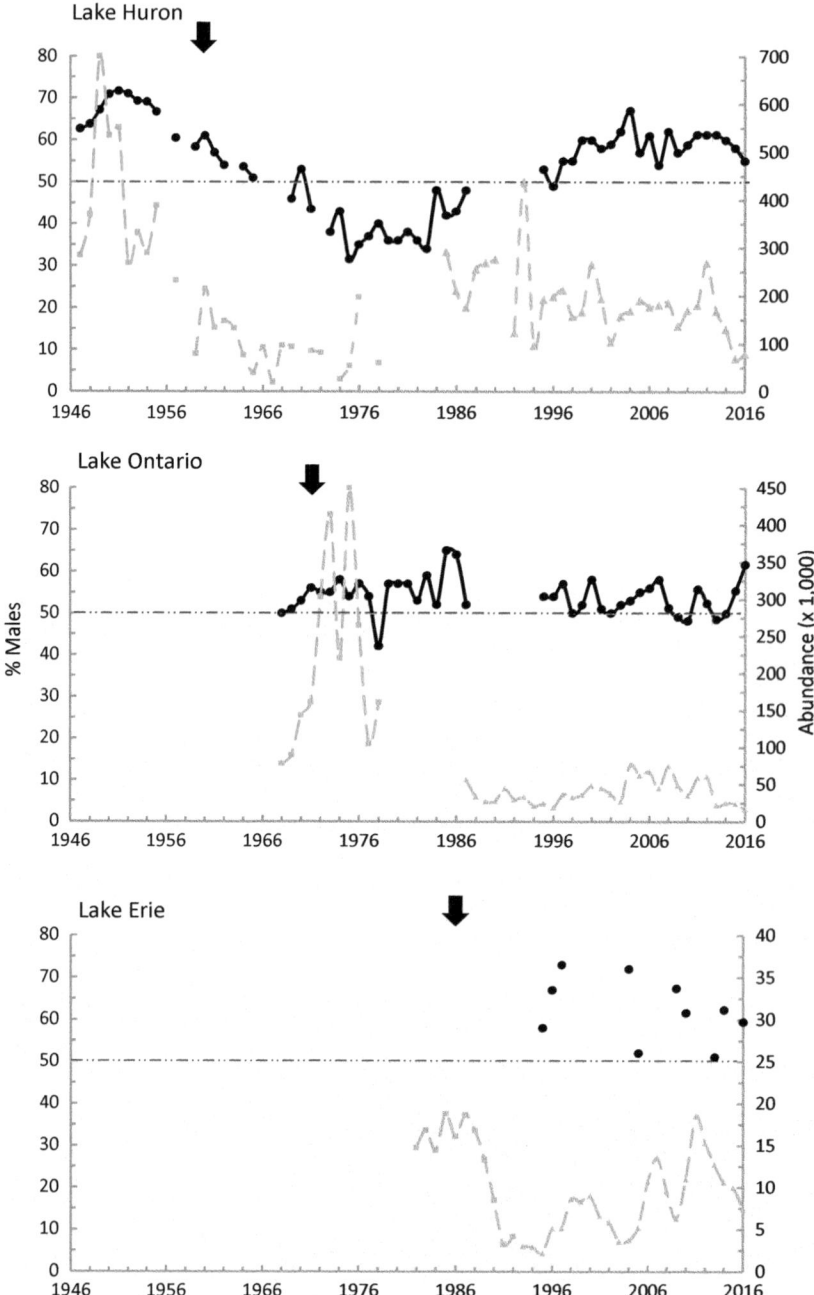

Fig. 1.2 (continued)

Table 1.2   Sex ratio of larval and post-larval (metamorphosing and adult) southern brook lamprey *Ichthyomyzon gagei* in 20 streams in the southeastern U.S., and the excess of males in the post-larval stages relative to the larval stage; significantly male- and female-biased sex ratios are identified with a superscript m and f, respectively. Larval data are from Beamish (1993); post-larval data are from Beamish et al. (1994)

Stream	Larval				Post-larval			Post-larval % male − larval % male
	Years sampled (total)	n	% Male	Density (larvae per m^2)	Years sampled	n	% Male	
Beaver Cr, LA	1989	60	20^f	0.26	1991	40	38	18
Big Cr, AL	1989	119	30^f	1.52	1989, 1992	33	64	34
Binion Cr, AL	1987–1990 (4)	120	37^f	0.17	1988	14	57	20
Choclafaula Cr, AL	1980–1991 (11)	1707	41^f	0.68	1980, 1981, 1991	5	40	− 1
Clear Cr, LA	1989	185	39^f	1.2	1989, 1992	49	65^m	26
Dry Prong Cr, LA	1989	80	36^f	0.99	1989	4	25	− 11
Dyson Cr, LA	1989	95	9^f	0.5	1992	24	42	33
Eden Cr, AL	1989–1990 (2)	121	19^f	0.29	1991	20	60	41
Hell Hole Cr, AL	1987–1991 (4)	219	49	1.75	1988, 1991	80	46	− 3
Keisler Cr, AR	1989	119	40^f	0.04	1992	21	62	22
Legg Cr, TX	1989	117	29^f	0.24	1989	9	33	4
Little Cypress Cr, TX	1989	157	41^f	0.23	1989	25	48	7
South Fork Saline R, AR	1989	51	29^f	1.9				
Spring Cr, LA	1989	91	40		1989, 1991	22	45	5
Teel Cr, AL	1989–1990 (2)	170	27^f	0.13	1991	17	41	14
Ten Mile Cr, AR	1989	148	26^f	1.55	1992	17	47	21
Terry's Cr, LA	1989	137	28^f	0.5	1989, 1991, 1992	87	45	17
Thomas Cr, AR	1989	127	46	0.18	1989, 1992	45	58	12
Uspohoa Cr, MS	1989	40	38		1991, 1992	63	63^m	25
Water Prong Cr, MS	1987–1990 (4)	219	45	1.13	1988	48	56	11

most, or all, of the spawning run, confirming that there are very few "exceptions to the rule" that adult lamprey populations are male biased. That there appears to be a fairly consistent excess of males in the adult stage relative to the larval stage among both parasitic and non-parasitic lampreys (see Sect. 1.2.2) suggests that females may experience higher mortality in the post-larval stages. Limited data from feeding-phase sea lamprey also show a higher proportion of females during this stage relative to upstream migrants (see Sect. 1.2.4) which would suggest that females are dispro-portionately lost from the population during or after the feeding phase. However, nothing is known regarding sex-specific differences in mortality, and this requires further study.

1.2.2   Larval Sex Ratios

Sex ratio data for larval lampreys are more limited than during the adult stage, because sex identification in larvae generally requires lethal sampling and histological prepa-ration prior to examination under a light microscope or, for larger larvae, internal examination under a dissecting microscope (see Sect. 1.4.1.5). Nevertheless, sex ratio data are available for at least four parasitic and four brook lamprey species (Table 1.3). Larval sex ratios appear to be more variable among populations and species than adult sex ratios but, in general, they are at parity or with an excess of females. For example, among brook lampreys, significantly female-biased larval sex ratios were observed in European brook lamprey from the River Yeo (Hardisty 1960a), in 15 of the 20 southern brook lamprey populations examined by Beamish (1993), and in three of the 12 least brook lamprey populations examined by Docker and Beamish (1994). Only one significantly male-biased larval brook lamprey population has been observed (in the least brook lamprey; Fig. 1.3), despite even or male-biased sex ratios among adults in these three species (Hardisty 1961a, b; Beamish et al. 1994; Docker and Beamish 1994; Table 1.2; Fig. 1.3).

Likewise, among parasitic species, few streams have been found with a significant excess of males during the larval stage, even in Great Lakes sea lamprey when adult sex ratios were significantly male biased. In sea lamprey larvae collected prior to or during initial lampricide treatments in 28 tributaries to Lakes Huron, Superior, and Ontario, a significant excess of males was observed in only four rivers: two on the north shore of Lake Huron (Echo and Garden) and two on the east shore of Lake Superior (Batchawana and Michipicoten; Fig. 1.4). Larval sex ratios were significantly female biased in 15 of the 28 rivers and at parity in nine. Most notably, significantly female-biased sex ratios (9–30% male) were observed among sea lam-prey collected during initial treatments on all five tributaries surveyed on the north shore of Lake Superior, in four of the five tributaries examined in the Georgian Bay region of Lake Huron (8–41% male), and in five of the nine Lake Ontario tributaries surveyed (13–39% male; Torblaa and Westman 1980; Fig. 1.4). There was no evi-dence of a sex-specific bias in the larvae collected during lampricide treatment, based on comparison with the sex ratio of samples collected by other survey means (Purvis

Table 1.3 Sex ratio of larval lampreys; significantly male- and female-biased sex ratios are identified with a superscript m and f, respectively. An asterisk indicates that individuals that could not be identified as either male or female were omitted from the sex ratio; values in bold signify sex ratios observed following initiation of lampricide (TFM) treatment in that stream; italics signify populations isolated above barriers

Species and location	Year	n	% Male	Reference	Comments
PARASITIC SPECIES					
Entosphenus tridentatus Pacific lamprey (anadromous)					
Oregon drainages		24–101	28.6[f]–41.7	Kan (1975)	Larvae >90 mm
Lethenteron camtschaticum Arctic lamprey (all forms)					
Utkholak R, Russia	2005	63	52	Kucheryavyi et al. (2007)	
Mordacia mordax short-headed lamprey (anadromous)					
Bunyip R, Diamond Cr, and Plenty R, Australia	1986–1987	303	40–45	Hardisty et al. (1992)	Larvae >90 mm
Petromyzon marinus sea lamprey (anadromous)					
Petitcodiac R, NB	1998	55	52.7*	Barker and Beamish (2000)	Larvae >120 mm; 49% atypical (see Sect. 1.4.1.4)
Petromyzon marinus sea lamprey (freshwater)					
Big Garlic R, MI (L Superior)	1959	141	19[f]	Manion and Smith (1978)	Collected during first TFM treatment
Big Garlic R, MI (L Superior)	1966	289	27[f]	Manion and Smith (1978)	1960 year class above barrier dam
	1967	407	19[f]		
	1969	672	15[f]		
	1970	924	18[f]		
	1971	298	22[f]		
	1972	357	22[f]		
Little Garlic R, MI (L Superior)	1965	644	**23[f]**	Purvis (1979)	TFM treatment 1960
Potato R, MI (L Superior)	1966–1969	363	**21.5[f]**	Purvis (1979)	TFM treatment 1965
Sturgeon R, MI (L Superior)	1970	394	**29[f]**	Purvis (1979)	TFM treatment 1967
Ocqueoc R, MI (L Huron)	<1965	267	47.6	Hardisty (1965b)	
Ocqueoc R, MI (L Huron):					
Below falls	1968	525	45.3[f]	Purvis (1979)	TFM treatment 1968
	1973	162	**22[f]**		

(continued)

18 M. F. Docker et al.

Table 1.3 (continued)

Species and location	Year	n	% Male	Reference	Comments
Above falls	1968	120	24.2^f		
Silver Cr (L Huron), above dam	1969	–	*17*	Purvis (1979)	Larvae ≥9 years old
L Superior (9 streams):					
First TFM treatment	1958–1964	996	9^f–72^m	Torblaa and Westman (1980)	See Fig. 1.4
Subsequent treatments	1962–1978	1,811	**11^f–57**		
L Huron (10 streams):					
First TFM treatment	1960–1967	3,207	8^f–70^m	Torblaa and Westman (1980)	See Fig. 1.4
Subsequent treatments	1966–1975	4,425	**7^f–44**		
L Ontario (9 streams):					
First TFM treatment	1971–1972	2,474	13^f–54	Torblaa and Westman (1980)	See Fig. 1.4
Subsequent treatments	1973–1978	1,551	**19^f–56**		
Brown's Cr, ON (L Huron)	1998	49	**55.1***	Barker and Beamish (2000)	56% atypical
12 streams: L Superior (2), L Michigan (1), L Huron (4), L Ontario (5)	1995, 1996	1,149	**9.0–81.7***	Wicks et al. (1998a)	8–100% atypical (see Sect. 1.4.1.4)
BROOK LAMPREYS					
Ichthyomyzon fossor northern brook lamprey					
Sturgeon R, MI	1960	261	48.7	Purvis (1970)	TFM treatments 1960, 1963, 1966
	1966	366	**49.2**		
Ichthyomyzon gagei southern brook lamprey					
Little and Choclafaula Cr, AL	1980–1981	486	49.5	F. W. H. Beamish (1982)	
20 streams in AL, AR, LA, MS, TX	1980–1991	40–1,707	9^f–49	Beamish (1993)	
Lampetra aepyptera least brook lamprey					
12 streams in MD, DE, KY, TN, AL	1987, 1988	66–297	28.7^f–70.9^m	Docker and Beamish (1994)	See Fig. 1.3
Lampetra planeri European brook lamprey and/or *L. fluviatilis* European river lamprey					
R Yeo, UK		281	42.3^f	Hardisty (1960a)	
R Usk, UK		61	44.3	Hardisty (1960a)	European brook and river lampreys
R Teifi, UK		49	49.0		
R Stensån, Sweden	1976	978	45.0^f	Malmqvist (1978)	
Rörum South R, Sweden	1976	912	44.2^f	Malmqvist (1978)	
Stampen Stream, Sweden	1976	564	46.7	Malmqvist (1978)	

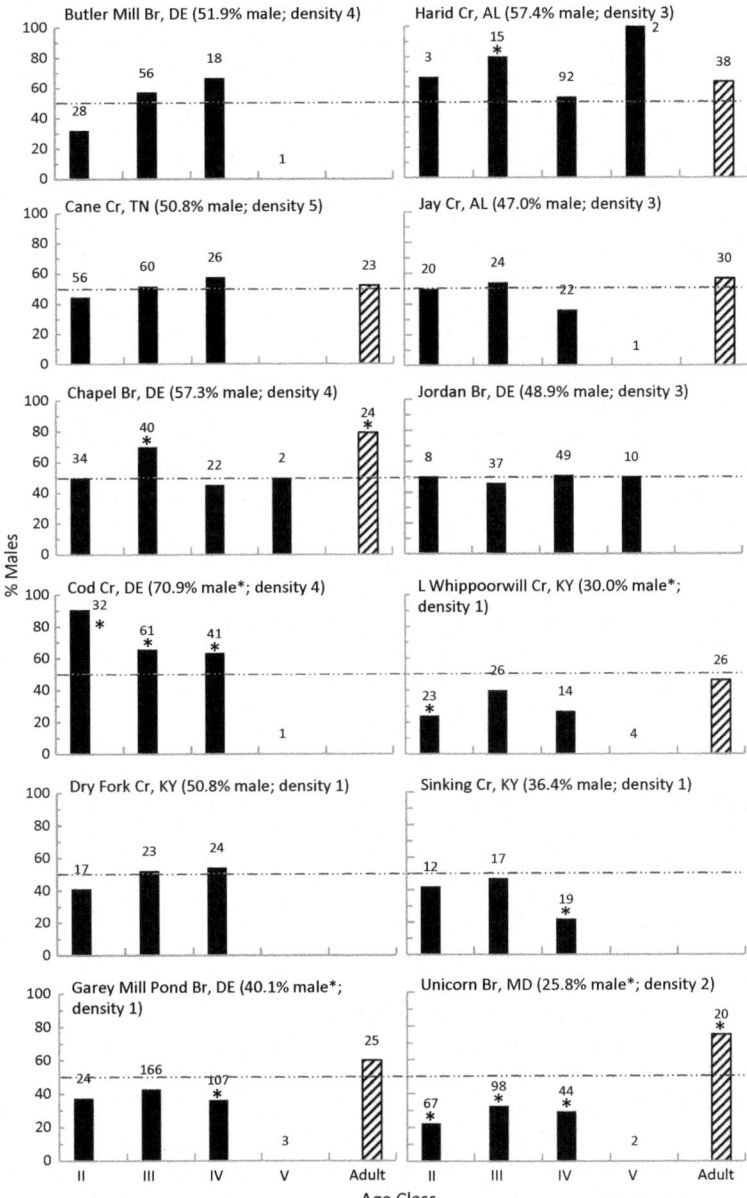

Fig. 1.3 Sex ratio (percent males) of least brook lamprey *Lampetra aepyptera* in length-frequency derived larval age classes II–V (*solid bars*) and post-larval individuals (transformers and adults; *hatched bars*); parity is indicated by a *horizontal dashed line*. Sex ratios that are significantly different from parity are identified with an *asterisk*; sample size for each age class is given above the bar. Overall sex ratios and estimates of relative larval density (1 lowest to 5 highest) are given for each of the 12 streams. Data are from Docker and Beamish (1994) and Margaret F. Docker (unpublished data)

1979; Torblaa and Westman 1980). After initiation of lampricide treatments, no significantly male-biased sex ratios were observed (Purvis 1979; Torblaa and Westman 1980), although a wide range of values was still observed among tributaries (7–57% male; Table 1.3). A general trend toward increased femaleness following initiation of lampricide treatments was observed when averaged across all streams (Fig. 1.5), but the relationship between abundance and larval sex ratio is far from clear (see Sect. 1.2.6).

Interpretation of larval sex ratios is complicated by potential sex-specific differences in age of metamorphosis and technical difficulties associated with determining sex in smaller larvae. There is evidence that males, at least in some species or populations, metamorphose at younger ages (and smaller sizes) than females. In European brook lamprey from three Swedish streams, for example, Malmqvist (1978) estimated 7–9 year classes in females versus 5–7 year classes in males, and he suggested that female-biased larval sex ratios are the result of more year classes of females being present. In least brook lamprey, Docker and Beamish (1994) found that, with only two exceptions (Cod Creek and Butler Mill Branch), sex ratio was consistent among age classes II–IV, but the relatively small number of age class V individuals were disproportionately female (Fig. 1.3). These authors concluded that males were likely under-represented in larval age class V due to their earlier recruitment to the adult population. Earlier metamorphosis in male northern brook lamprey was demonstrated by Purvis (1970) when he found that the earliest-metamorphosing age class of a cohort re-established following lampricide treatment was almost exclusively male (see Sect. 1.2.3). Delayed metamorphosis in females is a relatively well-understood phenomenon in non-parasitic species. In these species, because individuals cease feeding at the onset of metamorphosis, increased body size and thus fecundity (see Sect. 1.6.3) can only be achieved by larger size at metamorphosis (see Docker 2009; Manzon et al. 2015). However, this phenomenon is less well understood in parasitic species which continue to feed and grow following metamorphosis. Nevertheless, there is evidence that Great Lakes sea lamprey also may show sex-specific differences in age at metamorphosis. For example, larval sea lamprey populations isolated for years above barriers (i.e., after most individuals are thought to have transformed and left the population) typically show female-biased sex ratios. Purvis (1979) reported that a population of larval sea lamprey isolated for 9 years above a lamprey-proof dam in Silver Creek, a tributary to Lake Huron, was only 17% male. Similarly, larval sex ratio in the Ocqueoc River was 24% male above falls which limited annual recruitment but 45% male below the falls. That these isolated populations represent older residual females is borne out by the observation that the earliest transformers were disproportionately male (see Sect. 1.2.3). Malmqvist (1978) predicted that postponed metamorphosis in non-parasitic species would provide a selective advantage for females if the mortality rate during the final larval years is low. Clearly, however, a scenario of low larval mortality rates would not apply in Great Lakes tributaries subject to regular lampricide treatments. There is evidence for selection of younger age at metamorphosis in Great Lakes sea lamprey following initiation of sea lamprey control (e.g., Morkert et al. 1998), but whether there has been selection against later metamorphosis in females relative to males is unknown.

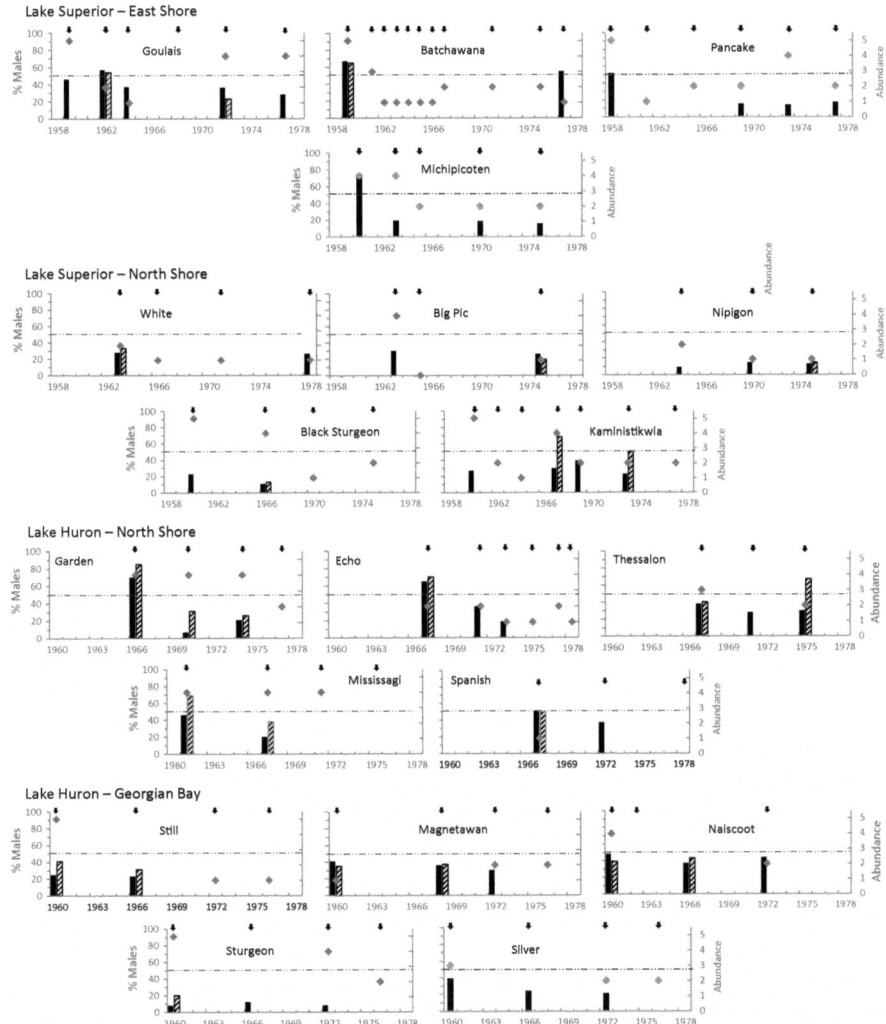

Fig. 1.4 Sex ratio (percent males) of larval (*solid bars*) and transformed (*hatched bars*) sea lamprey *Petromyzon marinus* collected during initial and subsequent lampricide treatments in 28 tributaries to Lakes Superior, Huron, and Ontario. *Vertical arrows* show timing of TFM treatments within each tributary; *gray diamonds* represent qualitative estimates of larval abundance (1 lowest to 5 highest) prior to each treatment. Sample sizes per collection averaged 188 and 220 (ranges 16–906 and 10–1,173) for larvae and transformers, respectively. Sex ratio data are from Torblaa and Westman (1980); relative abundance estimates and treatment dates were collected from Great Lakes Fishery Commission annual reports (GLFC 1960–1978)

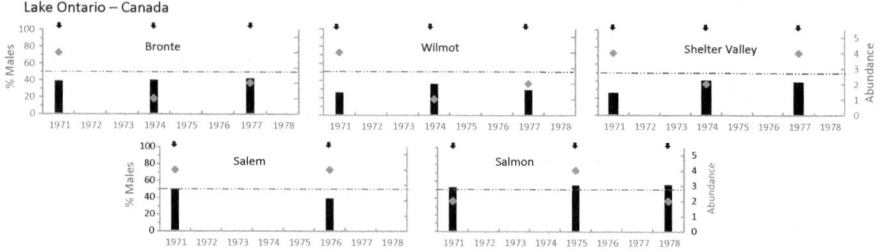

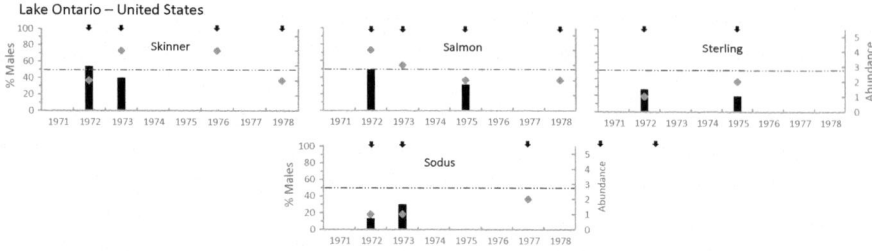

Fig. 1.4 (continued)

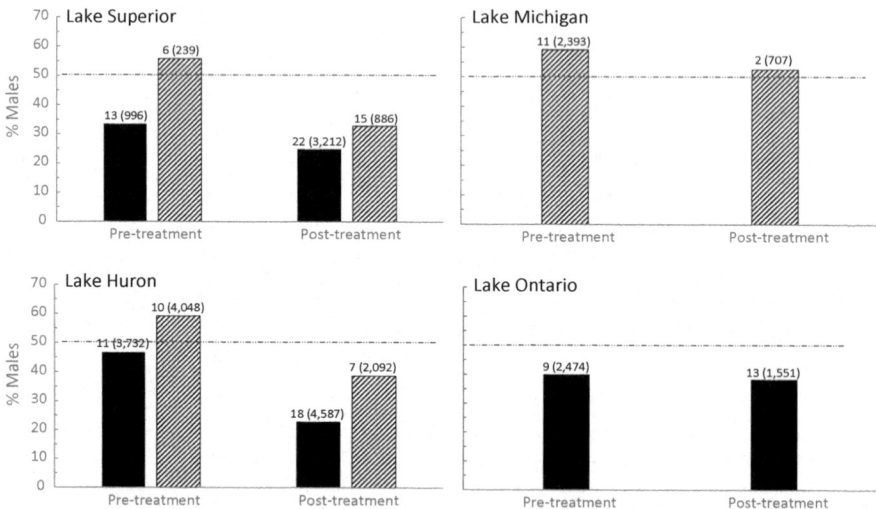

Fig. 1.5 Mean sex ratio (percent males) of larval (*solid bars*) and transformed (*hatched bars*) sea lamprey *Petromyzon marinus* prior to and after initiation of lampricide treatments in tributaries to Lakes Superior, Michigan, Huron, and Ontario; pre-treatment includes sea lamprey collected prior to and during initial lampricide treatments; post-treatment includes sea lamprey collected in all subsequent treatments. Number of collections (and total sample size) in each mean is given above the bar. Parity (50:50 sex ratio) is indicated by a *horizontal dashed line*. Data are from Purvis (1979) and Torblaa and Westman (1980)

Therefore, a disproportionate number of females among older and larger larvae (and therefore female-biased larval sex ratios overall as the result of more year classes of females being present) may result from features of lamprey biology. However, an excess of females in the older and larger age classes, in turn, may produce observational biases in studies when, for ease of sex identification, only large larvae are included in the analyses. For example, sex in Great Lakes sea lamprey can generally be distinguished histologically by ~90 mm TL (see Sect. 1.4.1.5), but studies evaluating sea lamprey sex ratios before and after initiation of sea lamprey control typically sampled only the very largest larvae (e.g., >119 mm; Purvis 1979). Docker and Beamish (1994) found that a small female-biased age class V did not greatly influence overall sex ratio in least brook lamprey, given the much larger number of individuals sexed from age classes II–IV, but the degree to which bias will be introduced will increase when only the largest age classes are sampled. Further study is also required to determine the extent to which lampricide treatment schedules bias larval sex ratios. If male lampreys are, on average, metamorphosing and leaving the streams earlier than females and prior to the next round of lampricide treatment, the large larvae that are killed and recovered during treatment will be disproportionately female relative to the sex ratio of the younger (largely unsampled) age classes and the larval population as a whole. However, if lampricide is applied at intervals short enough to kill both males and females prior to metamorphosis or long enough to allow both sexes to recruit to the parasite population with equal frequency, less bias is expected.

Nevertheless, it should be noted that sexing individuals when they are too small or too young may also be problematic. In lampreys, ovarian differentiation occurs during the larval stage, but testicular differentiation is delayed until the onset of metamorphosis (see Sect. 1.4.1). Male larvae are thus identified as those individuals that are not yet female by the stage at which females are expected to be clearly identifiable. As a result, "slow" differentiating females might be misdiagnosed as presumptive males. Moreover, at least a few small oocytes are present in most presumptive testes, particularly in smaller larvae at the onset of sex differentiation, resulting in some future males being misdiagnosed as females. Although the size at which sex appears to be confidently identifiable has been established for many species, and reversal past this point is understood to be rare, Lowartz and Beamish (2000) used a gonadal biopsy technique to monitor the gonad of individual larvae over time and demonstrated that sex reversal did occasionally occur following primary differentiation (see Sect. 1.4.1.4). Other exceptions to normal sex differentiation include the occurrence of "intersex" or otherwise atypical gonads, and this also may prevent accurate evaluation of larval sex ratios. Among 12 Great Lakes streams, Wicks et al. (1998a) classified 8–100% of larvae >90 mm TL as intersexes. Omitting these individuals, sex ratios were 9–82% male, but it is impossible to know whether some of these individuals would subsequently develop as normal males or females. Similarly, Barker and Beamish (2000) reported that 56 and 49% of larvae in one sea lamprey population from the Great Lakes and one anadromous population, respectively, had atypical gonads, although equal sex ratios were detected among those identifiable as definitive males or females (Table 1.3). Greater study is required to understand whether these atypical individuals would eventually develop as functional males or females and how larval sex ratio in general is related to adult sex ratio.

1.2.3 Sex Ratio of Transformers and Downstream Migrants

Sex ratio data for metamorphosing and downstream-migrating lampreys are available for five parasitic species and four non-parasitic species (Table 1.4), and sex ratio during this stage appears to be highly variable. Significantly skewed sex ratios were observed in three of the non-parasitic species studied: a female-biased sex ratio was reported in the southern brook lamprey during the early stages of metamorphosis (F. W. H. Beamish 1982), and male-biased sex ratios were reported in the northern brook lamprey and mountain brook lamprey *Ichthyomyzon greeleyi* (Purvis 1970 and Beamish and Medland 1988, respectively). However, the highly male-biased sex ratio observed in northern brook lamprey transformers appears to have been a consequence of sea lamprey control practices (see below).

In parasitic species, significantly skewed transformer sex ratios have been reported in only the Great Lakes sea lamprey, where both male- and female-biased sex ratios have been observed. Among sea lamprey populations in several tributaries to the upper Great Lakes surveyed prior to or during initial lampricide treatments, Purvis (1979) found that the sex ratio of transformers was generally at parity or male-biased. However, exceptions were observed. Most notably, males comprised only 4% of all transformers collected in Rock River, a tributary on the south shore of Lake Superior. Applegate and Thomas (1965) reported a slight but significant excess of females in transformers collected during initial lampricide treatment in the Ogontz River in 1960 and in downstream migrants in the Pere Marquette and Ocqueoc rivers in 1962–1963 (Table 1.4). The only significantly male-biased sex ratio observed by Applegate and Thomas (1965) was in metamorphosed sea lamprey outmigrating from the Carp Lake River in 1960–1961. However, the male-biased sex ratio of this transformer cohort might be an artifact of the barrier constructed in 1955 that prevented subsequent recruitment (see below). Purvis (1979) found that sex ratios of transformers collected following initiation of sea lamprey control were generally female-biased or equal, but there was still considerable variation (12–54% male). Torblaa and Westman (1980) also reported highly variable sex ratios (20–85% male) in transformed sea lamprey collected during initial treatments (Fig. 1.4). After initiation of control, there was an overall shift toward fewer males (Fig. 1.5), but there was still considerable among-stream variation (13–68%). In Lewis Creek, a tributary to Lake Champlain, where lampricide treatments were not initiated until 1990 (see Chap. 5), the sex ratio of transformers collected during the initial treatment was not significantly different from parity; a female-biased sex ratio (35% male) was observed in the first collection following initial treatment, but sex ratio returned to parity thereafter (Zerrenner and Marsden 2005). Sex ratio may vary temporally during downstream migration (see Sect. 1.2.5), biasing collections made during a restricted part of the run. However, many of the studies above either collected transformers within the stream prior to

Table 1.4 Sex ratio of metamorphosing and downstream-migrating lampreys; significantly male- and female-biased sex ratios are identified with a superscript m and f, respectively. Values in bold signify sex ratios observed following initiation of lampricide (TFM) treatment in that stream; italics signify populations isolated above barriers

Species and location	Year	n	% Male	Reference	Comments
PARASITIC SPECIES					
Geotria australis pouched lamprey (anadromous)					
Tributaries of the Donnelly and Warren R, Australia	1976–1979	218	54.1	Potter et al. (1980)	Transformers and downstream migrants
Entosphenus tridentatus Pacific lamprey (anadromous)					
Oregon drainages		12–46	57.6–58.7	Kan (1975)	
Lampetra fluviatilis European river lamprey (anadromous)					
Teme R, UK		142	54.9	Bird and Potter (1979)	Stage 6 metamorphosis
Firth of Forth, UK	1980	93	47.3	Maitland et al. (1984)	Downstream migrants
	1981	42	57.1		
Mordacia mordax short-headed lamprey					
Bunyip R, Diamond Cr, and Plenty R, Australia	1986–1987	89	50.6	Hardisty et al. (1992)	Transformers
Petromyzon marinus sea lamprey (freshwater)					
Ocqueoc R, MI (L Huron)	<1965	76	52.6	Hardisty (1965b)	Transformers
Ocqueoc R, MI:					
Below falls	1968	995	52	Purvis (1979)	Collected during first TFM treatment
Above falls	1968	84	18[f]		
Ogontz R, MI (L Michigan)	1960	527	44.2[f]	Applegate and Thomas (1965)	Collected during first TFM treatment
Sturgeon R, MI (L Michigan)	1961	266	61[m]	Purvis (1979)	Mean TL 137–140 mm (males) and 140–142 mm (females); first TFM treatment 1963
	1962	382	66[m]		
	1963	30	70[m]		
Ford R, MI (L Michigan)	1961	476	76[m]	Purvis (1979)	Mean TL 133–135 mm (male) and 141–149 mm (female); first TFM treatment 1964
	1962	56	75[m]		
	1963	13	69		
Cedar R, MI (L Michigan)	1961	343	52	Purvis (1979)	Mean TL 138–145 mm (male) and 150–155 mm (female); first TFM treatment 1964
	1962	86	55		
Whitefish R, MI (L Michigan)	1961	56	50	Purvis (1979)	TFM treatments 1962, 1965
	1962	174	67[m]		
	1963	685	**54[m]**		
	1966	22	**9f**		
Bark R, MI (L Michigan)	1961	511	39[f]	Purvis (1979)	

(continued)

Table 1.4 (continued)

Species and location	Year	n	% Male	Reference	Comments
Huron R, MI (L Superior)	1958	25	64	Purvis (1979)	First TFM treatment 1958
	1961	42	**48**		
	1965	38	**16^f**		
Middle R, MI (L Superior)	1958	56	73^m	Purvis (1979)	First TFM treatment 1958
	1962	22	**18^f**		
Chocolay R, MI (L Superior)	1961	34	**29^f**	Purvis (1979)	First TFM treatment 1958
Rock R, MI (L Superior)	1958	24	4^f	Purvis (1979)	First TFM treatment 1958
	1961	52	**12^f**		
Ontonagon R, MI (L Superior)	1960	32	69^m	Purvis (1979)	Collected during first TFM treatment
Sturgeon R, MI (L Superior)	1970	32	**53**	Purvis (1979)	TFM treatment 1966
Potato R, MI (L Superior)	1969	190	**28.9^f**	Purvis (1979)	TFM treatment 1965
Little Garlic R, MI (L Superior)	1965	209	**44**	Purvis (1979)	TFM treatment 1960
Big Garlic R, MI (L Superior)	1966	46	54	Manion and Smith (1978)	Isolated 1960 year class; downstream migrants 24–33% male Sept–Dec, 14% male Jan–May
	1967	172	31^f		
	1969	314	23^f		
	1970	541	21^f		
	1971	313	21^f		
	1972	298	15^f		
L Superior (9 streams):					
First TFM treatment	1959–1963	20–45	33^f–65	Torblaa and Westman (1980)	See Figs. 1.4, 1.5
Subsequent treatments	1962–1975	10–64	**13^f–68^m**		
L Huron (10 streams):				Torblaa and Westman (1980)	
First TFM treatment	1960–1967	43–1,173	20^f–85^m		See Figs. 1.4, 1.5
Subsequent treatments	1966–1975	26–1,060	**26^f–67^m**		
Lewis Cr, VT (L Champlain):					
Below falls	1990	127	47	Zerrenner and Marsden (2005)	First TFM treatment 1990
	1994	71	**35^f**		
	1999	18	**56**		
	2000	23	**48**		
Above falls	1990	150	25^f		
Carp Lake R, MI (L Michigan)	1956–1957	370	46.2	Applegate and Thomas (1965)	Downstream migrants; 50% male Nov–Dec, 40% male Mar–Apr

(continued)

Table 1.4   (continued)

Species and location	Year	n	% Male	Reference	Comments
Carp Lake R, MI (L Michigan)	1960–1961	8,780	76.4[m]	Applegate and Thomas (1965)	Isolated since 1955; downstream migrants 82% male Oct, 64% male Apr
Pere Marquette R, MI (L Michigan)	1962–1963	1,862	41.2[f]	Applegate and Thomas (1965)	Downstream migrants; 40% male Nov, 45% male Feb–Apr
Ocqueoc R, MI (L Huron)	1962–1963	408	42.6[f]	Applegate and Thomas (1965)	Downstream migrants; 45% male Dec, 42% male Mar–Apr
BROOK LAMPREYS					
Ichthyomyzon fossor northern brook lamprey					
Brule R, WI (L Superior)	1944	68	54.4	Churchill (1945)	Transformers; mean TL 110 mm (males) and 125 mm (females)
Sturgeon R, MI (L Superior)	1966	31	**97[m]**	Purvis (1970)	TFM treatment 1963
Ichthyomyzon gagei southern brook lamprey					
Little and Choclafaula Cr, AL	1980–1981	45	31.1[f]	F. W. H. Beamish (1982)	Early stages of metamorphosis
Ichthyomyzon greeleyi mountain brook lamprey					
Bent Cr, NC	1980–1986	86	64.0[m]	Beamish and Medland (1988)	Stages 1–7 metamorphosis
Lampetra aepyptera least brook lamprey					
7 streams in MD, DE, KY, TN, AL	1980, 1988	20–38	46.2–79.2[m]	Docker and Beamish (1994)	Transformers and adults (Oct–Feb; see Fig. 1.3)
Lampetra planeri European brook lamprey					
R Honddu, UK		55	52.7	Bird and Potter (1979)	Stage 6 metamorphosis

outmigration (Purvis 1979; Torblaa and Westman 1980) or collected virtually all individuals throughout the period of downstream migration (e.g., Applegate and Thomas 1965).

However, the observed differences in transformer sex ratios might be, in part, an artifact of sea lamprey control (i.e., dependent on time since isolation above barrier dams and on lampricide treatment frequency). Nevertheless, these "manipulations" (and populations isolated above natural barriers) also have provided much of the evidence for sex-specific differences in age at metamorphosis. Applegate and Thomas (1965), observing a heavily male-biased (76% male) sex ratio among 8,870 sea lamprey outmigrating from Carp Lake River in 1960–1961, concluded that females metamorphosed at a younger age than males (Table 1.4). Because virtually no recruitment had occurred in this river since construction of a barrier in 1955 (i.e., no individuals younger than age V–VI were present), these authors concluded that the majority of females had already metamorphosed and outmigrated prior to monitoring in 1960–1961. However, other studies suggest the opposite (i.e., that females metamorphose at an older age). Manion and Smith (1978) monitored (over the course of multiple years, 1966–1972) the 1960 sea lamprey year class in the Big Garlic River after it was isolated above a barrier dam. In this study, the sex ratio of downstream migrants was 54% male in 1966 (age class V), but progressively decreased to 15% male by 1972. This is strong evidence that female sea lamprey metamorphose at an older age than males and that they can remain in the larval stage for 12 years or more. Metamorphosing sea lamprey collected above natural barriers also appear to be female-biased if sampled after they have been isolated without recruitment for some time (Zerrenner and Marsden 2005). Later metamorphosis in females is consistent with the observation by Purvis (1979) that metamorphosing female sea lamprey are, on average, larger than males (although the size difference varied among streams; Table 1.4), and Applegate and Thomas (1965) likewise found that female downstream migrants in the Carp Lake River were slightly larger than the males (148 vs. 145 mm TL). Clearly, duration since isolation (i.e., whether the leading or trailing edge of metamorphosis is being surveyed) will influence whether, and in which direction, the sex ratio will be skewed. Similarly, sex ratio of metamorphosing lampreys may also be influenced by the frequency of lampricide treatment. Purvis (1970) monitored the 1963 year class of northern brook lamprey in the Sturgeon River following re-establishment after lampricide treatment. He found that only 6% of the individuals captured in 1966 (age class III) had metamorphosed, but 31 out of 32 of these transformers (i.e., representing the leading edge of metamorphosis) were male. Churchill (1945) found that male northern brook lamprey transformers were smaller than female transformers, likewise suggesting that males metamorphose at a younger age and smaller size. However, in the absence of lampricide treatments (i.e., where survival to metamorphosis was not prevented in all but the earliest-transforming individuals), the sex ratio was not significantly different from parity (Churchill 1945).

1.2.4 Sex Ratio During the Parasitic Feeding Phase

Given the difficulty sampling lampreys during the parasitic feeding phase at sea or in large lakes, relatively little information is available regarding sex ratios during this stage. Nevertheless, sex ratio data during this stage are available for four species (Table 1.5). Sex ratios were not significantly different from parity in the limited collections available for Pacific lamprey, western river lamprey *Lampetra ayresii*, and freshwater-resident Arctic lamprey; in anadromous sea lamprey, sex ratios were equal or male-biased. In contrast, in the Great Lakes, everything from an equal sex ratio to highly female- or highly male-biased sex ratios has been reported. In Lakes Superior, Michigan, and Huron, when collections were pooled over multiple years in the 1970s, the percentage of males in catches of parasitic-phase sea lamprey (23–35%) was consistently 8% lower than in adults (31–43; Johnson and Anderson 1980). Recent data from Lake Huron (2000–2015) show an identical 8%

under-representation of males during the parasitic feeding stage relative to upstream migrants (Gale Bravener, Fisheries and Oceans Canada, Sea Lamprey Control Centre, Sault Ste. Marie, ON, personal communication, 2018). Across all years, sex ratio during the parasitic feeding phase averaged 53% male but upstream migrant sex ratios averaged 61%, and males were under-represented during the feeding phase in 13 of the 16 years monitored. Nevertheless, despite these consistent differences between stages, the same general return to "normal" sex ratios was observed by the mid-1990s. Sex ratio of parasitic-phase sea lamprey collected by commercial fishers in Lake Huron (n = 18,352) averaged 41% male in 1983–1995 but 54% in 1996–2014 (n = 21,357; Gale Bravener, personal communication, 2018).

However, available data suggest considerable potential for sampling bias introduced by spatial and temporal differences in the distribution of the sexes. In general, males appear over-represented in collections made earlier in the season and under-represented later in the season and in deeper offshore waters versus surface or inshore areas (see Sect. 1.2.5). Also, in Great Lakes sea lamprey, it appears that females are over-represented in larger size classes; this was most noticeable in Lake Superior where males made up 31–35% of individuals <300 mm TL but only 14% of those ≥500 mm TL (Johnson and Anderson 1980). A consistent over-representation of females during the feeding stage (and the larval stage) relative to the adult stage would suggest increased mortality of females after the feeding phase, but spatial and temporal variations in sex ratio could be confounding observations.

1.2.5 Spatial and Temporal Variation in Sex Ratio

As mentioned above, there appear to be sex-specific behavioral (e.g., timing of upstream and downstream migration) and life history (e.g., older average age at metamorphosis in females) differences in lampreys that result in spatial and temporal variation in sex ratio. These differences can result in biased sex ratio estimates in lampreys, but biased sex ratios can, in turn, help inform our understanding of lamprey biology.

With respect to upstream migration, a number of studies report sex ratio over the course of the spawning run (Table 1.1), and several studies suggest that males initiate upstream migration earlier than females, at least in terms of time during the season. There is no evidence to suggest that females feed for one or more seasons longer than males (with the exception of the praecox form of Arctic lamprey; Kucheryavyi et al. 2007; see Sect. 1.2.1). In anadromous sea lamprey, most researchers report a greater proportion of males earlier in the run (Beamish et al. 1979; Beaulaton et al. 2008; Silva et al. 2016), although Stier and Kynard (1986) report a slight increase in the proportion of males later in the run. In sea lamprey in the Great Lakes and Cayuga Lake, Applegate (1950) and Wigley (1959) found that the proportion of males was reasonably consistent over the duration of the run. Studies in other species also fail to show a consistent over-representation of males earlier in the run. In the Caspian lamprey, sex ratio was similar in fall versus spring migrants (Ahmadi et al. 2011)

Table 1.5 Sex ratio of lampreys during the parasitic feeding phase; significantly male- and female-biased sex ratios are identified with a superscript m and f, respectively

Species and location	Year	n	% Male	Reference	Comments
Entosphenus tridentatus Pacific lamprey (anadromous)					
Pacific Ocean off Oregon		23	43.5	Kan (1975)	
Strait of Georgia, BC	1975–1979	39	58.3	R. J. Beamish (1980)	
Lampetra ayresii western river lamprey (anadromous)					
Strait of Georgia, BC	1975, 1976	50	60.0	R. J. Beamish (1980)	46–50% male in surface waters Jul–Aug, 24–36% male Aug–Sept
Lethenteron camtschaticum Arctic lamprey (anadromous)					
Great Slave Lake, NWT	1966	72	45.8	Nursall and Buchwald (1972)	Late Aug–Sept
Petromyzon marinus sea lamprey (anadromous)					
Washademoak L, NB	1975	141	72.4^m	Potter and Beamish (1977)	May (115–155 mm TL)
Washademoak L, NB	1974–1977	134	61.2^m	Beamish et al. (1979)	May (mean TL 133 mm males; 143 mm females)
St John R, NB @ Mactaquac Dam (rkm 120)		88	48.9		June–July (mean TL 267 mm males; 272 mm females)
Petromyzon marinus sea lamprey (freshwater)					
Canadian waters of Great Lakes, esp. L Huron	1967 1968	2,530 3,022	$\sim0^f$–80^m $\sim0^f$–50^m	Johnson (1969)	Males rare in offshore areas by Aug–Dec (see Sect. 1.2.5)
L Superior: from US commercial fishermen	1970–1978	2,800	22.7^f	Johnson and Anderson (1980)	31–35% male <300 mm, 14% male ≥500 mm TL
L Superior: from Canadian fishermen	1967–1976	9–73	~3–22^f	Johnson and Anderson (1980)	30% male in Jun–Jul, 5–13% male Aug–Dec
L Michigan: from US fishermen	1971–1978	7,082	34.5^f	Johnson and Anderson (1980)	34–37% male <300 mm, 30–32% male ≥400 mm TL
L Huron: from US fishermen	1971–1978	1,351	30.9^f	Johnson and Anderson (1980)	29–38% male <300 mm, 25% male ≥500 mm TL
L Huron: from Canadian fishermen	1967–1976	123–1,900	~5–40^f	Johnson and Anderson (1980)	25–30% male in Mar–June, 4–6% male Aug–Dec; 10% male offshore, 20% male inshore
L Ontario: from Canadian fishermen	1967–1976	13–1,815	$\sim4^f$ –55	Johnson and Anderson (1980)	25–49% in Jan–Apr, 8–18% May–Dec; 9% male offshore, 24% male inshore
L Erie: from Canadian fishermen	1969–1972	27–160	~10–33^f	Johnson and Anderson (1980)	32–43% male in Apr–Jun, 7–28% male July–Dec

and relatively stable over the course of the spring run (Nazari and Abdoli 2010; Table 1.1). In the Pacific lamprey, the proportion of males was either similar over the course of upstream migration (R. J. Beamish 1980; Porter et al. 2017) or increased or decreased, depending on the year (Clemens et al. 2016; Benjamin Clemens, Oregon Department of Fish and Wildlife, Corvallis, OR, personal communication, 2017). In the European river lamprey, the proportion of males appeared to increase late in the run in the typical anadromous form but was reasonably consistent throughout the run in the praecox form (Abou-Seedo and Potter 1979). Brook lampreys exhibit much more limited spawning migrations (i.e., 1–2 km; Malmqvist 1980; F. W. H. Beamish 1982), and no consistent differences between the sexes are evident. Studies report either no evidence of sex-specific differences in migration timing (Kott 1971; Beamish and Thomas 1984), a higher proportion of males in the early part of the spawning "run" (Hardisty 1961a; F. W. H. Beamish 1982) or nesting period (Young and Cole 1900), or a higher proportion of males somewhere in the middle (Malmqvist 1980). Some of the observed differences among species and studies may be related to the duration of the run that was monitored. Some studies did not include initial inshore movement in parasitic species, and some may not have monitored the full spawning period (e.g., nesting, spawning, and post-spawning periods; see Jang and Lucas 2005). In parasitic species, males may predominate in the earliest part of the run if they cease feeding and move inshore first (see below), and males in both parasitic and non-parasitic species may potentially increase again at the end of the spawning period if males spend longer on the spawning grounds (Malmqvist 1978) or survive longer after spawning (Beamish et al. 2016; see Sect. 1.5.2). There also may be species- or situation-specific differences that have yet to be clarified. Nevertheless, it is important to be aware that sex ratios obtained over a restricted portion of the spawning run have the potential to be biased.

With respect to downstream migration, studies that have monitored the entire cohort of outmigrating Great Lakes sea lamprey have found that the proportion of male outmigrants was generally higher in the fall than in winter and early spring (Applegate and Thomas 1965; Manion and Smith 1978; Table 1.4).

During the parasitic feeding phase, both temporal and spatial differences have been observed in the distribution of the sexes. In anadromous sea lamprey, males were over-represented (61%) among small sea lamprey captured shortly after initiation of feeding in May relative to larger individuals captured in June–July (49%), suggesting that males start feeding earlier than females (Potter and Beamish 1977; Beamish et al. 1979). In the Great Lakes sea lamprey, data suggest that males may also cease feeding and move to inshore areas in preparation for upstream migration earlier than females. Johnson (1969) found that both sexes were captured by commercial fishermen between April and July, but males virtually disappeared from offshore areas by August and males encountered in the fall generally came from inshore catches. In a more detailed follow-up study, Johnson and Anderson (1980) reported that sex ratio varied temporally and spatially in each of Lakes Superior, Huron, Ontario, and Erie. Precise timing varied among lakes (e.g., the proportion of males peaked between January and April in Lake Ontario and between June and July in Lake Superior; Table 1.5). Nevertheless, in all cases, males were noticeably more prevalent in the spring and early summer (25–49%) than in the late summer and fall (4–18% in Lakes Superior, Huron, and Ontario; 7–28% in Lake Erie). In Lakes Huron and Ontario, females were more prevalent in gill nets in deeper offshore waters relative to inshore trap nets or seines (Johnson and Anderson 1980). A similar pattern was observed in the western river lamprey feeding in the Strait of Georgia: the percentage of males (46–50%) was higher in surface waters in mid-summer when the species was most abundant near the surface, but sex ratio decreased to 24–36% male by late summer (R. J. Beamish 1980). Thus, considering these observations alongside the temporal variation observed to date in the sex ratio of downstream- and upstream-migrating lampreys, one could infer that male lampreys generally outmigrate, start feeding, and cease feeding earlier than females. However, further study is required before broad generalizations can be made.

With respect to larval sex ratios, temporal and spatial variation that may be related to an environmental effect (e.g., following initiation of sea lamprey control or variation among locations within a species) is discussed in Sect. 1.2.6. In addition, there is limited evidence for some longitudinal segregation of the sexes within a stream. There is no evidence to suggest that larval male and female lampreys actively select for specific habitat types. However, the largely passive downstream drift that occurs during the prolonged larval stage is expected to result in an accumulation of older larvae as distance from the spawning grounds increases, although the degree to which this happens may be related to specific features of the stream (e.g., gradient or frequency and magnitude of flooding; Dawson et al. 2015). If females tend to metamorphose at older ages than males (see Sect. 1.2.3), older larvae in downstream reaches should be female biased relative to more upstream reaches. However, few studies have examined spatial differences in larval sex ratios and the results are conflicting. In the Little Garlic River, a tributary on the south shore of Lake Superior, Purvis (1979) found that the percentage of males collected near the mouth (17%) was about half that collected 1.6 km upstream (30%), as predicted. In contrast, in Shelter Valley Creek, a tributary of Lake Ontario, Lowe (1972) found a much higher proportion of males near the mouth (53%) than in the upper reaches (11%).

1.2.6   Environmental Influences on Sex Ratio

It has often been observed that lamprey sex ratio is correlated with relative abundance. Meek (1889) and Wigley (1959) reported that the proportion of male sea lamprey in Cayuga Lake was positively correlated with population size, and Hardisty (1961b) likewise reported a direct relationship between the proportion of male adult European brook lamprey and relative spawner abundance. Most notably, in the sea lamprey in the three upper Great Lakes, significantly male-biased adult sex ratios coincided with the peak of abundance that preceded the initiation of sea lamprey control; sex ratios then shifted to an excess of females as population abundance dramatically declined following implementation of control (Smith 1971; Purvis 1979; Torblaa and Westman 1980; Fig. 1.2). However, the relationship between sea lamprey abundance and sex ratio is not as clear as generally suggested. First, a similar decline in the proportion of males following the initiation of sea lamprey control was not observed in either of the lower Great Lakes, despite a similar rapid decline in abundance following initiation of lampricide treatments (see below). With one exception, sex ratios of adult sea lamprey in Lakes Ontario and Erie have not deviated from the variable, but consistent, excess of males typical of adult lampreys. Secondly, variable, but comparatively low, abundances have been maintained in the upper Great Lakes since these initial declines (see Chap. 5); yet the sex ratio in all three lakes returned to parity or an excess of males by the mid- to late 1990s.

The methods by which sea lamprey abundance is estimated in the Great Lakes have changed over time, but, given the wealth of data available from the sea lamprey control program, the trends are clear. A standardized method to assess spawner abundance was implemented in 1977 in Lakes Michigan and Huron, in 1978 in Lake Ontario, and in 1980 in Lakes Superior and Erie, and subsequent refinements have been adopted (Mullett et al. 2003; see Chap. 5). In brief, these more recent abundance estimates are derived from mark-recapture studies conducted on a subset of streams with traps, and abundance in non-sampled streams is modeled using stream drainage area and other factors that allow sea lamprey index values to be scaled up to lake-wide abundance estimates using a lake-specific correction factor (Mullett et al. 2003; Jessica Barber, U.S. Fish and Wildlife Service, Marquette, MI, personal communication, 2018; Fig. 1.2). Abundance estimates prior to, and immediately following, initiation of sea lamprey control were based on counts of upstream migrants at selected barriers (Smith 1971; Heinrich et al. 1980) or nest counts (Sullivan et al. 2003); these estimates permit comparison among years but do not represent lake-wide estimates. Nevertheless, historical lake-wide abundances have been estimated; peak values were ~780,000 in Lake Superior; 600,000 in Lake Michigan; 700,000 in Lake Huron; 450,000 in Lake Ontario; and 40,000 in Lake Erie (GLFC 2015). Thus, scaling pre-control abundance in the index streams to these peak lake-wide values allows comparison with recent abundance data. For example, Smith (1971) reported a peak abundance of 69,584 upstream migrants in 24 index streams to Lake Superior in 1961. Therefore, we assumed that lake-wide abundance peaked at 780,000 in 1961, and index values from other years were scaled accordingly (e.g., a count of 9,614 adults in 1962, or 14% of the peak value from 1961, was assumed to represent a lake-wide abundance of 109,758).

Although not precise (particularly where a small number of index streams was monitored), this approach allows us to compare historical and recent data. It shows that the decline in the proportion of males strongly paralleled the approximated declines in spawner abundance in the three upper Great Lakes (Fig. 1.2). However, it also shows that female-biased sex ratios have failed to persist in the upper Great Lakes despite lake-wide abundance levels over the past two decades that have remained at a fraction of their historic peaks (11, 16, and 28% in 1997–2016 in Lakes Superior, Michigan, and Huron, respectively). Even in Lake Ontario, where sex ratio has varied little over the past 50 years, the significant shift to female excess coincided with the most abrupt decline in abundance (1978). Nevertheless, sex ratio remained at parity or with a slight excess of males despite abundance having been reduced to ~10% of peak values (Fig. 1.2). Historic abundance data are sparse for Lake Erie, but they suggest less dramatic changes in sea lamprey abundance. Sea lamprey were present in the lake for more than 50 years before becoming sufficiently abundant to cause noticeable effects on the fish community (Sullivan et al. 2003). Recent estimates indicate that spawner abundance fell dramatically immediately after initiation of sea lamprey control in 1986 (e.g., falling from 19,372 in 1988 to 2,633 in 1989), but abundance started increasing again by 1996 and has remained relatively high since (Sullivan et al. 2003; Chap. 5). Unfortunately, sex ratio data are not readily available for Lake Erie prior to and during the initial rapid decline in abundance in the mid-1980s.

Therefore, it has been challenging to understand the possible mechanisms responsible for these changes (or lack thereof) in adult sex ratios. If abundance affects sex ratio through a direct effect on the sex differentiation process (i.e., if crowding during the larval stage favors male differentiation; see Sect. 1.3.2.2), it is not clear why a relationship between adult sex ratio and abundance was not observed in Lake Ontario and why sex ratios in the upper Great Lakes returned to parity by the mid-1990s despite continued low abundances. Furthermore, evidence of a relationship between abundance and larval sex ratio is even more equivocal. A shift to slightly more female-biased larval sex ratios was reported following initiation of sea lamprey control, but only when averaged across streams (Torblaa and Westman 1980; Fig. 1.5). This relationship was far less evident within individual streams. Significantly female-biased sex ratios (8–41% male) were observed in 15 of 28 tributaries to Lakes Huron, Superior, and Ontario even before these streams were treated with lampricide and when lake-wide spawner abundance was still high. Furthermore, there was no apparent correlation between qualitative assessment of larval abundance (GLFC 1960–1978) and sex ratio (Fig. 1.4). Sex ratio within streams having larval abundance ranked 1, 2, 4, and 5 (where 1 is lowest and 5 is highest) averaged 33, 41, 46, and 35% male (n = 4, 7, 10, 7), respectively. In subsequent treatments, sex ratios averaged 30, 31, and 26% male in streams where larval abundance was ranked at 1, 2, and 4, respectively. A decline in the proportion of males was also reported in metamorphosing sea lamprey following initiation of sea lamprey control, but, as with larvae, even initial sex ratios were highly variable (20–85% male) so that a pattern was evident only when averaged across streams (Torblaa and Westman 1980; Fig. 1.5). Transformer sex ratio was similarly not correlated with qualitative assessments of larval abundance: during initial treatments, sex ratios averaged 43, 48, 64, and 42% male in streams where larval abundance was ranked at 1, 2, 4, and 5, respectively, and averaged 17, 57, and 33% male during subsequent treatments at relative abundances of 1, 2, and 4.

Interestingly, however, geographic patterns in larval sex ratio were evident (Torblaa and Westman 1980). Most notably, female-biased sex ratios were observed in all five tributaries surveyed on the north shore of Lake Superior and in four of five tributaries to Georgian Bay (Fig. 1.4). Given the variable sex ratios present in streams with divergent physical and chemical characteristics, Torblaa and Westman (1980) suggested that environmental factors play a role in lamprey sex differentiation. Specific environmental characteristics that might be shared by the streams with female-biased sex ratios were not identified by these authors, but the fact that these streams include most of those on the north shore of Lake Superior and in Georgian Bay (but not those on the east shore of Lake Superior or the north shore of Lake Huron) is interesting and deserving of further study.

As was observed with adults, a relationship between sea lamprey larval abundance and sex ratio has not been observed in recent decades. In 1994 and 1995, Wicks et al. (1998a) compared larval sex ratios in a total of 12 streams tributary to Lakes Superior, Michigan, Huron, and Ontario where the highest density was ~35× higher than the lowest density. Interpretation of sex ratios was complicated by the observation that 8–100% of the larvae >90 mm TL were classified as intersexes. Nevertheless, including only individuals that were recognizable as male or female, sex ratios were significantly female biased (9–37% male) in eight streams where densities ranged from 0.1 to 4.2 larvae/m^2 and significantly male biased (82% male) in only one stream (Gordon's Creek, tributary to Lake Huron) where larval density was estimated at 0.3 larvae/m^2. Including the intersex individuals, females outnumbered males in eight streams (mean density 1.1 larvae/m^2), and males outnumbered females in three streams (mean density 0.2 larvae/m^2). The predominance of females in the majority of streams is consistent with a shift toward female-biased sex ratios following initiation of sea lamprey control, but adult data suggest that spawner sex ratios were returning to parity by this time (e.g., 56, 41, 53, and 54% male in Lakes Superior, Michigan, Huron, and Ontario, respectively, in 1995; Fig. 1.2). Female-biased sex ratios in larval populations and equal or slightly male-biased sex ratios in adult populations is the general pattern observed in most lamprey species, regardless of abundance (see Sects. 1.2.1 and 1.2.2). Furthermore, the relationship between sex ratio and larval density was not significant, although Wicks et al. (1998a) observed a trend toward an increase in the proportion of females with increasing density (i.e., the opposite of what has been predicted). These authors suggested that sea lamprey control measures have lowered larval densities in streams to a point that density no longer has a significant effect on sex differentiation (i.e., that contemporary densities, even over the range studied, are all relatively low as the result of frequent lampricide treatment). Wicks et al. (1998a) also measured or calculated stream pH, alkalinity, hardness, temperature (as degree days), and larval growth rate, and they did not detect relationships between any of these characteristics and sex ratio. However, the high proportion of intersexes in these populations makes it very difficult to interpret the relationship between sex ratio and environmental factors. Wicks et al. (1998a) observed that the proportion of intersex larvae in a population increased with larval growth rate, but the cause and the impact on the population sex ratio is unknown (see Sect. 1.4.1.4). More recently, Johnson et al. (2017) suggested that larval growth rate, rather than density per se, influenced sex determination in sea lamprey, but sex ratio was evaluated only at upstream migration and results may have been confounded by sex-specific differences in rates of metamorphosis (see Sect. 1.3.2.2). More work is needed to understand contemporary larval sex ratios in Great Lakes sea lamprey.

Hardisty (1960a), upon finding that larval sex ratios did not correlate with adult sex ratio or abundance, concluded that the environment did not have a direct influence on the sex differentiation process, and he concluded instead that sex-specific differences in rates of metamorphosis or mortality produced the skewed adult sex ratios. The only evidence that larval lamprey sex ratios are correlated with in-stream abundance comes from two brook lamprey species in the southeastern United States. Sex ratio variations were observed among least brook lamprey populations in Maryland, Delaware, Kentucky, Tennessee, and Alabama; the proportion of males ranged from 29 to 71% and was found to increase significantly with relative larval density (Docker and Beamish 1994). There was no evidence for sex-specific differences in mortality (i.e., sex ratio differences were established in the earliest age classes in which sex could be identified and remained relatively consistent thereafter). Furthermore, although a disproportionate representation of females in the oldest age class suggested that females recruit to the adult population at older ages than males (Fig. 1.3), this age class was small and had little influence on the overall sex ratio. In this study, larval sex ratio was not significantly related to water hardness, pH, annual thermal units, or latitude. In the southern brook lamprey, Beamish (1993) observed a positive relationship between larval density and the proportion of males when conditions for larval growth were favorable, but he found that under poor growth conditions, higher densities were associated with fewer males.

We also considered whether the female-biased sex ratios in these two brook lamprey species could be a response to exploitation (i.e., given that female-biased sex ratios in sea lamprey in the upper Great Lakes followed the population "crash" that occurred after initiation of control measures). F. W. H. Beamish (1980) suggested that a slight excess of males is typical of established lamprey populations, and an increase in the proportion of females has been suggested as a compensatory response to low abundances or rapid decreases in abundance (Jones et al. 2003). Some of the southern brook lamprey populations examined by Beamish (1993) have been sampled repeatedly over several years. For example, Choclafaula Creek in Alabama was sampled in 11 successive years by Beamish and coworkers (Beamish 1993; Table 1.2) and by other researchers in the 1940s and 1950s (see Dendy and Scott 1953). However, there was no correlation between the frequency of known sampling events and population sex ratio, and the frequently sampled populations still exhibited high larval densities. Furthermore, overexploitation has been reported in other lamprey species (see Maitland et al. 2015), but there are no reports of similar compensatory shifts in sex ratio. Granted, abundance and especially sex ratio data for these species are far more limited than with the well-studied sea lamprey in the Great Lakes, but there is no evidence for female biases in any of the more heavily exploited species. For example, abundance of adult Pacific lamprey in much of the Pacific Northwest has decreased exponentially following peak returns in the 1950s and 1960s. Counts of upstream migrants at Winchester Dam in the coastal Umpqua River decreased from a high of ~46,800 in 1966 to only 34 in 2000; at Ice Harbor Dam in the Snake River, a tributary to the Columbia River, counts decreased from a peak of 49,450 in 1963 to 203 in 2001 (Close et al. 2002). Likewise, at Bonneville Dam in the mainstem Columbia River, counts averaged 103,700 during 1939–1969 but only 38,700 in 1997–2010 (Murauskas et al. 2013). Sex ratio data are patchy, but equal or slightly male-biased sex ratios have been reported during periods of both high and low abundance. Upstream migrants were 55–57% male in two Oregon streams pre-1975 (Kan

1975) and 49–57% male in the Willamette River in 2007–2016 (Clemens et al. 2016; Porter et al. 2017). Potential female-biased sex ratios were reported by Lampman (2011) and McIlraith et al. (2015), but the proportion of males was likely underestimated, especially in the early stages of maturity, because many of the unidentifiable lamprey were likely males (Table 1.1). However, greater study is required, particularly if shifts in sex ratio are transitory following perturbation.

Finally, to try to understand the ultimate cause of sex ratio variations in lampreys, it is important to also consider other changes that accompanied the initial increase in the percentage of males observed following invasion of the upper Great Lakes and the subsequent shift to female-biased sex ratios in the ensuing population "crash." Decreases in the availability of prey (e.g., lake trout *Salvelinus namaycush*) and decreases in sea lamprey size at maturity were observed as sea lamprey abundance increased in each lake, and these trends were reversed as sea lamprey numbers declined again following the initiation of sea lamprey control (see Chap. 5). Houston and Kelso (1991) found that sea lamprey sex ratio was more closely related to prey availability in Lake Superior (commercial catch and stocking rates, 1954–1987) and Lake Huron (stocking rates, 1949–1987) than it was to spawner abundance. The proportion of males increased in close association with declines in prey availability and then waned again as prey abundance increased once more. In Lake Superior, for example, severe sea lamprey predation beginning in the late 1940s, combined with intensive commercial fishing in the early 1950s, resulted in lake trout stocks being at an all-time low by 1960 (Heinrich et al. 1980). The proportion of males in Lake Superior peaked in 1961–1964 (69%). Lake trout numbers began to rebound in Lake Superior by about 1962, resulting from a combination of sea lamprey control measures, intensive stocking, and commercial fishing restrictions (Heinrich et al. 1980; Smith and Tibbles 1980), and sea lamprey sex ratios decreased to an average of 53% male by 1965–1966 (Fig. 1.2). In Lake Huron, commercial catch of lake trout in U.S. waters declined from 177 t in 1947 to less than 0.5 t in 1959, and the proportion of sea lamprey males peaked in 1950–1955 (70%). Stocking efforts began in 1963 (Heinrich et al. 1980; Smith and Tibbles 1980) and sex ratio was 54% male by 1964. Although not included in the Houston and Kelso (1991) study, the same pattern was seen in Lake Michigan: lake trout had been almost extirpated by 1950 and the proportion of males first exceeded 60% in 1954 and remained high (averaging 63%) until 1964. The subsequent decline in the proportion of males (58 and 44% in 1965 and 1966, respectively) corresponded closely with the initiation of lake trout and Pacific salmon (coho and Chinook salmon, *Oncorhynchus kisutch* and *O. tshawytscha*, respectively) stocking efforts in 1965–1966 (Heinrich et al. 1980; Smith and Tibbles 1980).

Therefore, as an alternative explanation to density-dependent sex determination acting during the larval stage, one could speculate that the changes in adult sea lamprey sex ratio observed during this time period were a response to dramatic changes in prey availability and were mediated largely during the feeding phase. There was certainly little lag time between the sharp declines and recoveries noted in prey abundance and the peaks and valleys observed in the proportion of male sea lamprey. A short lag time would be more consistent with differential mortality acting during the feeding and adult stages rather than density-dependent sex determination in the larval stage. During the initial population explosion, as prey levels declined dramatically, it is possible that female mortality rates during the late parasitic and adult stages were disproportionately higher than in males given the much higher energetic demands of ovarian maturation (F. W. H. Beamish 1980; see Sects. 1.5.3 and 1.5.4). Resource limitations were certainly apparent as sea lamprey abundance increased and prey availability decreased, and these changes paralleled the initial shifts in sex ratio. In Lake Superior, size at maturity declined throughout the 1950s, from a reported high in 1953–1954 (455 mm, 227 g) to a low (412 mm, 140 g) in the early 1960s, but it began to increase modestly throughout the 1960s–1980s. By the late 1980s, it had almost returned to the maximum size recorded pre-control (Houston and Kelso 1991), and it appears to have more or less stabilized since then (averaging 434 mm and 202 g since 1955; GLFC 1996–2017). In Lakes Huron and Michigan, sea lamprey remained small until the 1960s (414 mm, 133 g in 1947–1960 in Lake Huron; 433 mm, 164 g in 1954–1962 in Lake Michigan) but showed pronounced increases in the 1960s–1980s, and TL and weight have averaged 475 mm and 235 g (Huron) and 488 mm and 259 g (Michigan) in the last two decades (Smith 1971; Heinrich et al. 1980; Houston and Kelso 1991; GLFC 1996–2017). The shift to an excess of female adult sea lamprey, as prey availability rebounded and sea lamprey size increased again, would then suggest that survival of females subsequently became higher than that of males during the parasitic and adult phases as the result of improved feeding conditions. There are no data available to test this conjecture, but it is important to remain aware of the many population-level changes that occurred during and following colonization and control of sea lamprey in the Great Lakes.

Furthermore, Lake Ontario showed a similar recovery of the prey base and increase in sea lamprey size following initiation of sea lamprey control, but, as mentioned, it did not show the pronounced shift to female-biased sea lamprey sex ratios that were evident in the three upper Great Lakes. Commercial salmonid catches and salmonid stocking rates were increasing by the mid-1970s, and sea lamprey size, which averaged 412 mm and 154 g in 1968–1970 (prior to initiation of sea lamprey control), increased thereafter; TL and weight reached ~480 mm and 260 g by the late 1980s (Houston and Kelso 1991) and has remained at this level during the last two decades (484 mm, 257 g; GLFC 1996–2017). Yet, despite evidence that sea lamprey in Lake Ontario were no longer resource-limited, a significant excess of females was detected in 1978 only, and the highest proportion of males (64–65%) was observed in 1985–1986. One possible difference is the pattern of colonization in Lake Ontario compared to the upper Great Lakes. Whether sea lamprey invaded Lake Ontario via manmade canals in historical times (Eshenroder 2014) or whether they colonized post-glacially but remained rare until ecological changes in the mid-1800s served as a "release" (Christie and Kolenosky 1980; Waldman et al. 2009) has long been debated (see Chap. 4). There is also some debate regarding the first credible report of sea lamprey in Lake Ontario (i.e., as early as 1835 or as late as 1888; see Christie and Kolenosky 1980; Eshenroder 2014), but it is nevertheless evident that they have been present in Lake Ontario for at least 50–100 years longer than in Lakes Michigan, Huron, and Superior (where they were first observed in 1936,

1937, and 1938, respectively), and they have been slower to reach pest proportions in Lake Ontario (Christie and Kolenosky 1980; Larson et al. 2003; see Chap. 5). These differences require further exploration to help understand whether the response in the upper Great Lakes may have been related to rapid population expansion and collapse (compared, perhaps, to a less dramatic perturbation from equilibrium conditions in Lake Ontario) or other factors (see Sect. 1.3.2). Our understanding of the mechanisms responsible for the sex ratio shifts in the upper Great Lakes is still far from complete.

1.3   Sex Determination

Sex determination is the event that predisposes a bipotential gonad to develop as an ovary or a testis (Sandra and Norma 2010), and these predisposing events can be genetic or environmental (Sarre et al. 2004; Siegfried 2010; Parma and Radi 2012). All birds and mammals exhibit genetic sex determination (GSD) where the master sex-determining genes are conserved within each taxon (Ellegren 2010; Cutting et al. 2013). In other vertebrates, mechanisms of sex determination vary, and both GSD (with a variety of different master sex-determining genes even among closely related species) and environmental sex determination (ESD) are known (Bulmer 1987; Takada et al. 2005; Heule et al. 2014). The factors that influence sex determination in lampreys continue to elude biologists, although new genomic technologies are now being used to try to resolve this previously intractable problem and recent discoveries related to mechanisms of sex determination in other vertebrates are guiding the way. Therefore, we begin here with an overview of sex determination in other vertebrates, particularly the variable nature of sex-determining mechanisms in the so-called "lower vertebrates" and the complicated interplay between genetic and environmental factors in these taxa. This broader taxonomic overview provides the background information necessary for understanding the possible sex-determining mechanisms at play in lampreys. For example, earlier studies suggesting strict ESD in lampreys (e.g., Docker and Beamish 1994) may have been premature, and future researchers should be aware of the complexity of vertebrate sex-determining mechanisms.

1.3.1   Sex Determination in Other Vertebrates

1.3.1.1   Genetic Sex Determination

In many vertebrates, the "master switch" that activates the sex-specific developmental cascade directing the undifferentiated gonad to develop into an ovary or testis is genetic. The master sex-determining genes are highly conserved within birds and mammals (*DMRT1* and *SRY*, respectively), and, in these well-studied vertebrates, sex chromosomes subsequently evolved from a pair of autosomes after acquisition of the master sex-determining gene. This triggered a cascade of neutral and adaptive processes that caused the once identical chromosomes to diverge from each other in size, gene content, and structure (Charlesworth et al. 2005; Graves and Peichel 2010; Wright et al. 2016). Mammals are said to have an XY/XX sex determination system, wherein the males are the heterogametic sex, inheriting different sex chromosomes (X and Y) at fertilization, and the females are homogametic. Birds have a ZZ/ZW sex determination system, wherein the females (inheriting Z and W sex chromosomes) are the heterogametic sex (Wallis et al. 2008; Cutting et al. 2013). In birds, *DMRT1* is located on the Z chromosome and is required in a dosage-dependent manner for male development (Smith et al. 1999; Shetty et al. 2002; Yano et al. 2012; Cutting et al. 2013). In mammals, *SRY* (the Sex-determining Region Y) initiates development of the male phenotype; its absence leads to female development (Cutting et al. 2013). Thus, although the sex-determining genes are conserved within each taxon, different sex-determining genes have evolved independently in birds and mammals. In fact, although many of the key genes involved in the subsequent differentiation of the gonads are conserved among vertebrates (see Sect. 1.3.2), their relative positions in the ovarian and testicular cascades—including which genes represent the master switch at the top of the cascade—often differ (Cutting et al. 2013). Turnover of sex chromosomes can lead to the evolution of novel sex determination mechanisms, and genes involved in sex determination in the "lower" vertebrates (i.e., reptiles and non-amniotes) are far less conserved (see Cutting et al. 2013; Graves 2013; Wright et al. 2016; Capel 2017).

Sex determination mechanisms are poorly known in most fishes and can be highly variable even among closely related species (Siegfried 2010). GSD, or sex determination with a significant genetic component, has been inferred in many teleost fish species (Devlin and Nagahama 2002; Ospina-Álvarez and Piferrer 2008; see Sect. 1.3.2.1). However, many of these species do not have morphologically distinct sex chromosomes, and the sex-determining genes have been identified in very few species. As of 2001, of the more than 1,700 fish species that had been cytogenetically characterized, just over 10% were found to have heteromorphic sex chromosomes (Devlin and Nagahama 2002). In species with recognizable sex chromosomes (e.g., Chen and Reisman 1970; Peichel et al. 2004; Chen et al. 2008), males are the heteromorphic sex in some cases (e.g., rainbow trout *Oncorhynchus mykiss* and medaka *Oryzias latipes*; Thorgaard 1977; Matsuda et al. 2002), but females are heteromorphic in others (e.g., blue tilapia *Oreochromis aureus*; Mair et al. 1991a).

The master sex-determining genes have been identified in only a limited number of fish species, and it is clear that these genes are highly variable even among closely related species. Most dramatically, different species in the genus *Oryzias* appear to use a number of different sex-determining genes. *DMY*, a homolog of the bird *DMRT1* gene, acts as the testis-determining gene in the medaka and Malabar ricefish *Oryzias curvinotus* (Matsuda et al. 2002; Nanda et al. 2002). In contrast, the Indian ricefish *O. dancena* uses *SOX3* as the male-determining factor, and the Luzon ricefish *O. luzonensis* uses the $Gsdf^Y$ gene (gonadal somatic derived factor on the Y chromosome; Myosho et al. 2012). In the latter species, $Gsdf^Y$ is present in both males and females, but males have 12 silent nucleotide mutations relative to females that alter upstream promoter regulation to direct male development (Myosho et al. 2012). Conversely, sex-determining genes may be shared among taxa that are not closely related. In the half-smooth tongue sole *Cynoglossus semilaevis*, *DMRT1* appears to be required for male development in the same dosage-dependent manner as is seen in birds (Chen et al. 2014). In some salmonid fishes, *sdY* (sexually dimorphic on the Y chromosome), a homolog of the mammalian *SRY*, determines maleness (Yano et al. 2012, 2013). *sdY* appears to be male-specific in at least 10 species in the genera *Oncorhynchus*, *Salmo*, *Salvelinus*, *Thymallus*, *Hucho*, and *Parahucho*. However, *sdY* was found in both sexes of the European whitefish *Coregonus lavaretus* and the lake whitefish *C. clupeaformis* (Yano et al. 2012, 2013), suggesting that it is male specific in subfamilies Salmoninae and Thymallinae, but not in subfamily Coregoninae. In the tiger pufferfish or fugu *Takifugu rubripes*, sex is determined by sex-specific nucleotide differences in the *Amhr2* gene (Kamiya et al. 2012). To date, eight different genes have been implicated in sex determination in teleost fishes (Table 1.6).

Furthermore, in addition to species that have monogenic sex determination (i.e., with a single sex-determining gene), other species are known that have polygenic sex determination, where sex is determined by the combined effects of multiple loci (Ohno 1974; Devlin and Nagahama 2002; Vandeputte et al. 2007; Sandra and Norma 2010; Heule et al. 2014; Liew and Orban 2014).

1.3.1.2 Environmental Sex Determination

In addition to more variable mechanisms related to GSD, sex determination in "lower" vertebrates can also include ESD. According to evolutionary theory, ESD should be favored when an environmental factor is more advantageous to one sex or the other but offspring disperse randomly and are unable to choose their environment. Under these conditions, for example, ESD could ensure that an individual of a relatively large size will become the sex in which the rewards for being large are greater (Charnov and Bull 1977; Conover 1984). Thus, the environmental variables to which sex determination is sensitive may act as cues to indicate conditions of favorable growth.

ESD, specifically temperature-dependent sex determination (TSD), has been particularly well studied in reptiles (see Charnov and Bull 1977; Janzen and Phillips 2006; Warner 2011). All crocodiles and alligators and most turtles appear to use ESD exclusively (Janzen and Krenz 2004; Warner 2011). In the American alligator *Alligator mississippiensis*, sex determination is highly sensitive to temperature changes; offspring are all female when eggs are incubated at ≤ 30 °C and all male when incubated at ≥ 34 °C (Ferguson and Joanen 1982). In some species of turtles, equal sex ratios are produced at intermediate temperatures, with a higher prevalence of males and females being produced at low and high temperatures, respectively (Woolgar et al. 2013; Mork et al. 2014). Interestingly, however, although the master sex-determining switch in these reptiles is temperature, the resulting developmental cascade appears to use some of the same genes that are at the top of the cascade in

Table 1.6 Genes implicated in vertebrate species with known genetic mode of sex determination; the sex-determining genes (*DMRT1* and *SRY*, respectively) are conserved in birds and mammals, but at least eight different genes have been implicated in sex determination in fishes

Gene symbol*	Gene name	Species	Sex specificity	References
Amhr2	Anti-Müllerian hormone receptor type II	Pufferfish *Takifugu rubripes*	Gene present in both male and females; single nucleotide polymorphism determines sex	Kamiya et al. (2012)
amhy	Y-linked anti-Müllerian hormone gene	Patagonian pejerrey *Odontesthes hatcheri*	Found on Y chromosome in XY males	Hattori et al. (2012)
DMRT1	Doublesex and mab-3-related transcription factor 1	Birds, half-smooth tongue sole *Cynoglossus semilaevis*	Male specific, two copies needed for ZZ/ZW system	Smith et al. (1999), Shetty et al. (2002), Chen et al. (2014)
DMY	*DM*-domain gene on the Y chromosome	Medaka *Oryzias latipes*, Malabar ricefish *Oryzias curvinotus*	Present on Y chromosome of XY males	Matsuda et al. (2002), Nanda et al. (2002)
Gsdf^Y	Gonadal soma derived growth factor on the Y chromosome	Luzon ricefish *Oryzias luzonenesis*	Male-specific factor on Y-chromosome	Myosho et al. (2012)
gdf6Y	TGF-b family growth factor	Turquoise killifish *Nothobranchius furzeri*	Male-specific region on the Y-chromosome	Reichwald et al. (2015)
SOX3	SRY-box 3	Indian ricefish *Oryzias dancena*	Male determining factor on Y-chromosome	Takehana et al. (2014)
sdY	Sexually dimorphic on the Y-chromosome	10 salmonid species (genera *Oncorhynchus*, *Salmo*, *Salvelinus*, *Thymallus*, *Hucho*, and *Parahucho*)	Male-specific gene found on Y-chromosome	Yano et al. (2012, 2014)
SRY	Sex determination region on the Y chromosome	Therian mammals	Male-specific gene found on Y-chromosome	Wallis et al. (2008)

*Formatting conventions for gene names depend on the type of organism; in general, symbols for genes are italicized, but the proteins they encode are not italicized; gene names written out in full are generally not italicized (although they are in fishes); capitalization of gene symbols varies considerably among organisms (e.g., all in upper case in primates, chickens, and domestic species; all in lower case in fishes; with only the first letter in upper case in mice and rats). In this chapter, to avoid confusion, we use the formatting employed by the authors cited above (or used most commonly by the authors cited in the text) for each gene in question (i.e., we use consistent formatting for each gene throughout the chapter, but not consistent formatting among genes

vertebrates with GSD. For example, temperature-dependent expression of *DMRT1* is seen in the developing genital ridge in red-eared slider turtle *Trachemys scripta elegans* males (Kettlewell et al. 2000). GSD predominates in most (but not all) lizards and snakes (Charnov and Bull 1977; Bulmer 1987; Schwanz et al. 2016), and ESD and GSD are not necessarily mutually exclusive. In the eastern three-lined skink *Bassiana duperreyi*, for example, sex is genotypically determined by the inheritance of heteromorphic sex chromosomes, but extreme incubation temperatures are able to override the genetic sex (Radder et al. 2009). An interaction of GSD and ESD is also seen in the central bearded dragon *Pogona vitticeps*. In this species, GSD operates under most conditions, but, under extremely high temperatures, the populations can consist of 100% phenotypic females despite their genotype (Quinn et al. 2007). Some authors have suggested that ESD is ancestral in amniotes (e.g., Uller et al. 2007; Pokorná and Kratochvíl 2016), while others argue that GSD is ancestral (e.g., Alam et al. 2018). However, in general, recent studies suggest that sex determination systems exist across a continuum of genetic and environmental influences and that the classical dichotomy between GSD and ESD—at least in the lower vertebrates—is "blurrier" than once thought (e.g., Holleley et al. 2015, 2016).

ESD, usually TSD, has also been reported in a number of fish species. In the Atlantic silverside *Menidia menidia*, arguably the best-studied fish species with respect to TSD, low fluctuating temperatures characteristic of the early breeding season produce a high proportion of females, while a predominance of males are produced as a result of higher temperatures experienced later in the season. Females, having a longer growing season, are consequently the larger sex (Conover and Kynard 1981; Baumann and Conover 2011). Sex determination in the California grunion *Leuresthes tenuis* (which, like the silverside, belongs to the order Atheriniformes) appears to be influenced by both temperature and photoperiod; a higher prevalence of females is produced at cooler temperatures and longer day lengths (Brown et al. 2014).

However, no fish species with exclusively ESD have been identified to date. In contrast to the steep 100% change in sex ratio observed in many reptiles over a narrow temperature range, genotype × environment interactions in teleost fishes produce a gradual sex ratio shift with temperature (Conover 2004; Duffy et al. 2015). The Atlantic silverside employs both TSD and GSD; the degree to which TSD or GSD predominates varies among populations, but no populations display pure TSD or GSD (Lagomarsino and Conover 1993; Duffy et al. 2015; see Sect. 1.3.2.2). In a third atheriniform fish species, the pejerrey *Odontesthes bonariensis*, sex ratios of 100% female or 100% male can be achieved when embryos are reared at low and high temperatures, respectively (Yamamoto et al. 2014). However, both sexes are produced at intermediate temperatures, and there is a high, although not complete, correlation between phenotypic sex and *amhy* genotype, the master sex-determining gene in the closely related Patagonian pejerrey *Odontesthes hatcheri* (Hattori et al. 2012). Similarly, in the European sea bass *Dicentrarchus labrax*, sex determination depends both on genetic factors and temperature, and there is no known temperature regime that produces 100% males or 100% females (Palaiokostas et al. 2015). Ospina-Álvarez and Piferrer (2008) concluded that many cases of skewed sex ratios observed at extreme temperatures might, in fact, be the consequence of thermal effects on GSD (e.g., growth-dependent sex differentiation; Piferrer et al. 2005) rather than the result of strict TSD.

1.3.2 Sex Determination in Lampreys

1.3.2.1 Genetic Sex Determination

Sex determination mechanisms in basal vertebrates are poorly understood, and the factors that influence sex determination in lampreys continue to elude biologists (Docker 1992; McCauley et al. 2015). At what point in the life cycle sex is determined is still unknown, and whether there is a genetic component has yet to be resolved. Karyological studies in lampreys have failed to identify heteromorphic sex chromosomes (e.g., Ishijima et al. 2017), although this is not surprising given the difficulty associated with counting the large number of very small "dot-shaped microchromosomes" found in this group (e.g., 84 pairs in sea lamprey; Potter and Rothwell 1970; McCauley et al. 2015). Furthermore, identification of sex chromosomes (e.g., by karyotyping or banding patterns) will depend on the sensitivity of the method used to search for them (Ospina-Álvarez and Piferrer 2008). A recent study using reduced-representation genotyping (i.e., Restriction site Associated DNA Sequencing or RAD-Seq, which sequences ~0.1–10% of the genome) failed to find genomic differences between male and female European brook lamprey (Mateus et al. 2013). Although RAD-Seq lacks the power to identify subtle genetic differences between the sexes, these results suggest that physically extensive genomic differentiation (i.e., X- or W-linked loci on sex chromosomes) does not exist between male and female lampreys. Thus, as is typical of many other fishes, if sex determination in lampreys is genetically based, the underlying system evolved without major chromosome divergence (Mateus et al. 2013).

Sex-determining systems can be inferred from the sex ratios of large numbers of progeny from single-pair matings, hormonally sex-reversed individuals, gynogens, and triploids (e.g., Mair et al. 1991a, b), but these approaches are either not possible or not practical in lampreys. Hormonal sex control has not been successfully achieved in lampreys (Docker 1992; see Sect. 1.4.1.6), and, although meiotic gynogens have been generated in sea lamprey (Rinchard et al. 2006), the length of time until progeny sex ratios could be determined makes this approach impractical without improvements to larval rearing protocols (see Chap. 2).

More recently, Khan (2017) used a candidate gene approach to test whether 19 genes implicated in sex determination or sex differentiation in other vertebrates were present in lampreys in a sex-specific manner. Candidate genes tested included putative master sex-determining genes identified in birds (*DMRT1*), mammals (*SRY*), and teleost fish species (*Amhr2, amh, gsdf, sdY*; Table 1.6), as well as key genes involved elsewhere in the sex differentiation cascade (*SOX2, SOX8, SOX9, SOX10, SOX17, SF1, TRA-1, RSPO1, WT1, WNT3, WNT5, FOXL2,* and *FEM1*; see Sect. 1.4.2.1).

Khan (2017) searched for homologs of these genes in the sea lamprey genome (Smith et al. 2013) and tried to amplify (using polymerase chain reaction, PCR) fragments of these genes in male and female sea lamprey and Pacific lamprey. Two lamprey species from divergent genera of Northern Hemisphere lampreys (Potter et al. 2015) were included because sex-determining genes in fishes can be highly variable even among closely related species. Homologs of *SRY, amh, Amhr2, GSDF, sdY,* and *TRA-1* were not found in the sea lamprey reference genome; *SOX1* and *SOX17* had homologs in the sea lamprey reference genome, but primers designed for them repeatedly failed to amplify them from genomic DNA. The remaining 11 genes (*SOX2, SOX8, SOX9, WT1, FEM-1, SF1, DMRT1, FOXL2, WNT3, WNT5,* and *RSPO1*) were successfully amplified from sea lamprey genomic DNA, and seven of these also amplified in Pacific lamprey (*SOX8, SOX9, FEM-1, DMRT1, FOXL2, WNT3* and *WNT5*). However, sex-specific differences (i.e., in terms of presence or absence of the gene or sex-specific sequence differences) were not apparent in any of them. Given the wide and unpredictable variation in sex-determining mechanisms seen in other fishes, it is not surprising that a putative sex-determining gene in lampreys was not found using this approach. Sex-associated loci in lampreys, if they exist, may be unique to lampreys. A genome-wide association study (GWAS) using whole-genome resequencing is currently being used to test for the genetic basis of sex determination in sea lamprey (Margaret F. Docker, unpublished data).

It is also possible that some of the candidate genes not found in the sea lamprey somatic genome are among the ~20% of the genome that is "jettisoned" during the programmed genome rearrangement that occurs in lampreys during the very early stages of development (Smith et al. 2009, 2013; Bryant et al. 2016; see Chap. 6). Some of these genes might be found in the newly available germline genome (Smith et al. 2018). Interestingly, programmed DNA elimination has also been reported in the zebra finch *Taeniopygia guttata* (Pigozzi and Solari 1998). In this species, the germline-restricted chromosome (GRC) is eliminated from mature sperm and transmitted only through the oocyte (Pigozzi and Solari 2005), and a recently identified GRC-limited gene is more highly expressed in the ovary than in the testis (Biederman et al. 2018). In sea lamprey, the germline-specific genes are eliminated from the somatic genome within the first few days of embryonic development (Bryant et al. 2016), that is, at least 2 years before histological signs of sex differentiation (see Sect. 1.4.1). Therefore, it is not clear how somatically eliminated genes could be involved in lamprey sex differentiation, but it is fascinating to contemplate.

1.3.2.2   Environmental Sex Determination

Environmental sex determination (in particular, density-dependent sex determination) has been proposed in lampreys to explain the observation that sea lamprey adult sex ratios in the three upper Great Lakes became highly male biased when their abundance peaked in each of these lakes and then dramatically shifted to a significant excess of females as abundance declined following implementation of control measures (Smith 1971; Purvis 1979; Torblaa and Westman 1980; see Sects. 1.2.1 and

1.2.6). However, evidence for ESD in lampreys—or at least an exclusive or dominant environmental component to sex determination—is equivocal. Similar shifts in adult sex ratio were not observed in the lower Great Lakes, and sex ratio returned to the typical slight excess of males in all the lakes by the mid-1990s (Fig. 1.2). Furthermore, there is little evidence that larval sex ratio and abundance are correlated in these populations (Hardisty 1960a; Torblaa and Westman 1980; Wicks et al. 1998a; see Sect. 1.2.6).

There is no support for the suggestion that exposure to the lampricide TFM produced a direct effect on sea lamprey sex ratios. First, there is no evidence that TFM produced female-biased sex ratios by preferentially killing male sea lamprey larvae (Purvis 1979; Torblaa and Westman 1980). Secondly, it appears unlikely that TFM caused direct feminization of larval sea lamprey. Interestingly, TFM exposure in other fishes resulted in changes in the level of plasma sex steroids during laboratory trials (Munkittrick et al. 1994), and TFM (or impurities associated with its field formulations) was found to act as an estradiol agonist in rainbow trout hepatocytes (i.e., binding to the estrogen receptor and inducing vitellogenin production in vitro; Hewitt et al. 1998a). However, the effect was less dramatic in live caged rainbow trout, white sucker *Catostomus commersonii*, and longnose dace *Rhinichthys cataractae* monitored following exposure to TFM during a normal field treatment (Hewitt et al. 1998b). Elevated mixed function oxidase (MFO) activity was detected in livers, particularly in fish held closest to the lampricide application points, but MFO had declined to low levels within 18 days of treatment, and there was no induction of vitellogenin in live fish (see Sect. 1.5.5.2). Therefore, the authors concluded that the weak estrogenic activity of TFM and the transient exposure produced only slight in vivo effects. TFM treatment periods are short (~12 h; Hubert 2003), as is its persistence in the water column (e.g., the estimated half-life of TFM is 16–32 h; McConville et al. 2016). Thus, it is unlikely that feminization of the sea lamprey gonad occurred during very short exposure to TFM relative to the long period of sexual lability (see Sect. 1.4.1) and that females so produced would survive treatment. Because Wicks et al. (1998a) observed a high proportion of atypical larvae in anadromous as well as Great Lakes sea lamprey larvae, their occurrence is not related to TFM treatment (see Sect. 1.4.1.4).

The most compelling evidence for ESD in lampreys comes from studies on two non-parasitic species in the southeastern United States, the southern brook lamprey (Beamish 1993) and least brook lamprey (Docker and Beamish 1994). These species occur outside of the Great Lakes basin, and thus they have never been exposed to TFM. In southern brook lamprey, comparison among 20 populations showed a positive relationship between the proportion of males and larval density when larval growth was good, but higher densities were associated with fewer males under poor growth conditions (Beamish 1993). Among 12 least brook lamprey populations, the proportion of male larvae was significantly correlated with larval density (Docker and Beamish 1994; see Sect. 1.2.6). Because ESD has been proposed as a mechanism by which large individuals become the sex benefiting most from large size (or those developing under growth-limiting conditions become the sex that is penalized least by small size; see Sect. 1.3.1.2), density-dependent sex determination in lampreys would indeed be expected to produce more females at low densities (i.e., under growth-enhancing conditions). Large females are more fecund than small females (see Sect. 1.6.3), and larval growth rates are generally higher under conditions of low density (Murdoch et al. 1992; see Dawson et al. 2015). However, contrary to this expectation, female least brook lamprey within each age class were not larger than males from the same cohort, and, in particular, there were no consistent sex-specific differences in size at the approximate time of sex differentiation. Females were consistently the larger sex only in the oldest larval age class and in the adult population, but due to delayed metamorphosis relative to males (see Sect. 1.2.2) rather than an obvious tendency for large larvae to develop as female. It should be noted that the Docker and Beamish (1994) study determined larval age using both statoliths and length-frequency aging methods (see Dawson et al. 2015). Using only length to infer age classes would prevent unbiased testing for size differences between the sexes at the approximate time of sex differentiation. It would be interesting to further test this hypothesis using a single larval age class (e.g., in artificially propagated larvae or those obtained in the wild from a known parental cohort; Dawson et al. 2015; Hess et al. 2015; see Chap. 7).

Experimental evidence for density-dependent sex determination is inconclusive at best. Docker (1992) reared wild-caught sea lamprey larvae at four experimental densities for >3 years and found no significant relationship between density and sea lamprey sex ratio, but such long-term laboratory studies are fraught with experimental difficulties. Sea lamprey larvae were collected from Lewis Creek, Vermont, prior to any histological signs of sex differentiation (i.e., <60 mm TL), and they were reared in outside experimental tanks for 39 months (i.e., until sex determination was complete at TL ~ 90–100 mm). Larvae were fed brewer's yeast and exposed to natural photoperiods and water temperatures that approximated natural stream temperatures. Nominal densities (in duplicate) were 10, 20, 50, and 100 larvae per 0.3-m^2 tank, although, in anticipation of mortality during the study, more larvae (18, 24, 58, and 137 larvae) were initially stocked into each tank (giving initial densities of 60–457 larvae/m^2 and 22.5–206 g/m^2). At the end of the 39 months, there were 11, 7, 41, and 57 larvae per 0.3-m^2 tank (23–190 larvae/m^2 and 38–207 g/m^2), excluding one tank at each of the three lowest densities where all larvae died before the experiment's conclusion. Sex ratios at the four densities were 33, 35, 27, and 46% male (i.e., hinting at the expected increase in the proportion of males at higher densities), but the relationship between sex ratio and density was only significant at the 10% level. Given the relatively low survival rates, it was not possible to exclude differential mortality between the sexes (i.e., higher female mortality at high densities), although Murdoch et al. (1992) found no evidence of sex-specific mortality in sea lamprey larvae reared for 9 months in the laboratory. Growth rates were also low, averaging only 8.9 mm per year. Unfortunately, all of the experimental densities used in this study were high (60–470 larvae/m^2 at the study's outset) compared to natural densities (see Dawson et al. 2015). Slade et al. (2003) found an average of 0.01–10.4 larvae/m^2 in patches of preferred habitat in tributaries to Lakes Superior and Michigan, and sea lamprey densities >5 larvae/m^2 are now considered moderate to high in the Great Lakes (Steeves et al. 2003). Clearly, considerably larger tanks would be required to rear a sufficient number of larvae at realistic densities to achieve sufficient statistical power, and replication of this study using large-scale propagation methods being developed for other lamprey species would be valuable (see Chap. 2).

A subsequent laboratory study showed indications of sex reversal (i.e., even after initial sex differentiation) in sea lamprey held at medium and high larval densities, but not consistently towards males as would be expected (Beamish and Griffiths 2001; F. W. H. Beamish, unpublished data). Using the gonadal biopsy technique developed by Lowartz et al. (1999) (see Sect. 1.4.1.4), larvae >110 mm TL were categorized as presumptive males, females, or atypical. No changes were seen in larvae reared at low density (10 larvae/m^2) for 22 weeks. In contrast, two of 13 female larvae held at medium density (30 larvae/m^2) showed complete oocyte atresia (i.e., suggesting complete sex reversal to male), although one of the two intersex lamprey showed an increase (rather than the expected decrease) in oocyte density, and the gonads of the other larvae remained unchanged. In the high density tanks (70 larvae/m^2), two of the 16 females showed a complete loss of oocytes (suggesting a reversal to males), and one masculine female changed to a male; however, one male reversed sex to a female. Thus, a number of lamprey apparently changed sex during the study, but not in directions that supported the hypothesis that low larval density shifts development toward femaleness and high density to maleness. Furthermore, the extent to which sex is typically reversed following initial differentiation is still unknown (see Sect. 1.4.1).

In an exciting recent study, Johnson et al. (2017) suggested that sex determination in sea lamprey is directly influenced by larval growth rate rather than density per se. Tagged sea lamprey larvae stocked into unproductive lentic environments grew more slowly than those in productive stream environments, and, when recaptured as upstream migrants 2–7 years later, the sex ratio of sea lamprey from the lentic environments showed a higher proportion of males (79% overall) relative to those from the more productive stream environments (66%). However, larval sex ratios were not determined, so Johnson et al. (2017) were unable to exclude the possibility that the differences in adult sex ratio were established after sex differentiation (e.g., as the result of differential mortality between the sexes or differential rates of metamorphosis). In fact, the changes observed in sex ratios over time—particularly the observation that the sex ratio from stream environments became progressively less male biased between Years 2 and 6 (82 and 50% male, respectively)—suggest that sex ratio differences are the result of differential rates of metamorphosis (Table 1.7). Capture of upstream migrants derived from larvae stocked into streams tapered off by Year 5, but larvae stocked into lentic areas continued to be captured as upstream migrants in Year 7, and there may have been continued recovery of individuals from lentic areas in subsequent years (i.e., after the study's conclusion). Thus, it is conceivable that delayed metamorphosis in females, particularly under growth-limiting lentic conditions, produced an initial excess of adult males, but that collections in subsequent years might have revealed more females (see Manion and Smith 1978; Sect. 1.2.3). It is also possible that females experience higher mortality under growth-limiting conditions (e.g., because of higher energetic demands during gonadal development; see Sect. 1.2.6). Nevertheless, this is the only study to date that has monitored the sex ratio of individuals stocked into different natural environments. Similar studies

Table 1.7 Sex ratio of sea lamprey *Petromyzon marinus* that were tagged and stocked as larvae into stream or lentic environments and recovered 2–7 years later as upstream migrants (data from Johnson et al. 2017)

Year recovered	Stream environments (n = 5)		Lentic areas (n = 3)	
	Number recovered	% Male	Number recovered	% Male
Year 2	11	81.8	4	50.0
Year 3	63	69.8	7	42.9
Year 4	101	67.3	65	86.1
Year 5	29	51.7	60	85.0
Year 6	4	50.0	27	63.0
Year 7	0	–	8	75.0
Average		66.3		78.9

in more confined areas (i.e., where larvae could be recovered and sexed prior to metamorphosis) would be very interesting. Sea lamprey residing in lentic habitats near river mouths have been observed in the Great Lakes basin since at least the 1960s (Hansen and Hayne 1962; Wagner and Stauffer 1962), and recent advances in sampling methods have also revealed deepwater riverine larval populations (e.g., Fodale et al. 2003; Schleen et al. 2003; Arntzen and Mueller 2017; see Chap. 7). It is not clear to what extent these habitats contribute to recruitment, but a better understanding of the demographics of these larval populations (e.g., sex ratio, growth rate, age at metamorphosis) is important.

However, it should be kept in mind that a signal of ESD in lampreys may be hard to detect and interpret. The growing body of knowledge regarding ESD in other fishes suggests that lamprey sex determination could involve both genetic and environmental components (see Sect. 1.3.1.2). Many purported cases of ESD in other fishes appear to be the result of environmental influences on GSD rather than strict ESD (Ospina-Álvarez and Piferrer 2008), and the nature of the genotype × environment interactions can differ among populations and even over time within populations. For example, in some northern populations of Atlantic silverside, sex determination is controlled by major genetic factors that are largely temperature-insensitive (Lagomarsino and Conover 1993; Duffy et al. 2015). In more southerly populations, TSD prevails and sex is determined by the interaction of temperature-sensitive and polygenic factors, although even in these populations, 20–30% of individuals appear less sensitive or completely insensitive to temperature (Lagomarsino and Conover 1993; Duffy et al. 2015). Geographically intermediate populations exhibit sex determination that is a more balanced mixture of TSD and GSD (Lagomarsino and Conover 1993). Moreover, of potential relevance to lamprey sex determination, experiments in the Atlantic silverside demonstrate that sex-determining mechanisms are capable of rapid evolution. Conover et al. (1992) used thermal manipulations during the sex-determining period to create highly skewed sex ratios over 8–10 generations in the laboratory, and they found that two of these populations evolved

GSD from TSD after only 2–3 generations. Under extreme temperatures, selection for temperature-insensitive sex-determining genes and GSD presumably prevented production of highly male- or female-biased sex ratios. Thus, the persistence of temperature-insensitive genotypes would be beneficial in that it permits an adaptive response to extreme swings in sex ratio that would generate frequency-dependent selection for the minority sex (Duffy et al. 2015). In the sea lamprey in the upper Great Lakes, sex ratios of upstream migrants remained within the range 25–75% male, and more extreme sex ratios were not observed even at the peak of abundance or following rapid declines in abundance (Fig. 1.2).

Therefore, we hypothesize that the rapid colonization and population explosion of sea lamprey in the upper Great Lakes initially produced highly skewed sex ratios as the result of a relatively strong environmental influence on sex determination. Subsequently, by the mid-1990s (after ~4–6 generations), population levels may have returned to equilibrium conditions followed by a stabilization of the sex-determining mechanism (e.g., with selection for individuals less sensitive to density). There are reports of other invasive fish species showing male-biased sex ratios following initial invasion. For example, Gutowsky and Fox (2011) observed that round goby *Neogobius melanostomus* populations in recently invaded areas were male biased, and male biases were even more evident in the freshly colonized upstream segments of the river (69% male) compared to the area of first introduction (58% male). Similar male-dominated sex ratios were observed in the invaded Gulf of Gdansk in the Baltic Sea (75% male), in the western basin of Lake Erie and the Detroit River (86% male; Corkum et al. 2004), and in Hamilton Harbor, Ontario (Young et al. 2010). In contrast, sex ratios close to parity have been reported in this species' native range (Kovtun 1979). Although the mechanism of sex determination is likewise unknown in this species, like sea lamprey in the upper Great Lakes, round goby invasion was rapid, and perturbation of the system was dramatic. In the lower Great Lakes, where sea lamprey reached pest proportions more gradually, a similar perturbation of the sex-determining system may not have occurred (see Sect. 1.2.6).

Clearly, our understanding of a possible environmental influence on sex determination in lampreys is still incomplete. Elucidation of the effect of density or other environmental factors on larval sex ratios (i.e., as established at the time of sex differentiation) has been complicated by the very long period during which the gonad is histologically undifferentiated and presumably still labile to influence of the environment. In most other fishes, the sexually labile period lasts for only a few weeks to a few months (e.g., Conover and Fleisher 1986), making them much more amenable to study. More refined larval rearing methods will help (see Chap. 2). Furthermore, should sex-specific loci be identified in lamprey populations (at least loci that strongly, if not completely, correlate with phenotypic sex at intermediate sex ratios; see Sect. 1.3.2.1), environmental effects on sex differentiation could be recognized more readily by identifying conditions that produced significant mismatches between phenotypic and genotypic sex.

1.4 Gonadogenesis and Sex Differentiation

Gonadogenesis is the generation and development of the gonad, and sex differentiation is the process by which the undifferentiated gonad develops into a recognizable ovary or testis (Piferrer and Guiguen 2008; Parma and Radi 2012). Lampreys possess a single elongated gonad, which remains histologically undifferentiated for up to several years (Hardisty 1971). Ovarian differentiation occurs during the larval stage (at ~1–3 years of age in most species), but testicular differentiation (at least the first production of spermatogonia) does not occur until metamorphosis (Figs. 1.1 and 1.6). Later sexual differentiation in males is characteristic of many fish species (e.g., Yoshikawa and Oguri 1978; Nakamura et al. 1998; Saito et al. 2007), although, with few exceptions (e.g., anguillid eels; Beullens et al. 1997a, b), the delay is far less pronounced than in lampreys and the entire process occurs over the course of weeks or months rather than years. Presumably, this delay in gonadogenesis is the result of the evolution of metamorphosis and prolongation of the larval phase in lampreys (Evans et al. 2018; see Chap. 4). Eye development in larval lampreys likewise appears to "pause" after reaching a very immature stage and is only resumed at later larval stages and metamorphosis (Suzuki and Grillner 2018).

The histological process of gonadogenesis and sex differentiation in different lamprey species has been described in detail by previous authors (e.g., Okkelberg 1921; Hardisty 1965a, b, 1971; Hughes and Potter 1969; Fukayama and Takahashi 1982, 1983; Hardisty et al. 1986, 1992; see Table 1.8). Here, we present an overview of the process, with an emphasis on: (1) aspects of the process that differ between future males and females, particularly those in the early stages of differentiation that might presage subsequent differentiation; (2) features that differ among species and life history types; and (3) facets that continue to elude researchers (e.g., the extent to which the gonad remains labile during the larval stage). Trying to interpret the dynamic process of gonadal differentiation from a number of static observations has long been a challenge in such studies, although a gonadal biopsy technique has been developed to monitor gonadal development in individuals over time (Lowartz and Beamish 2000; Beamish and Barker 2002). We also review what is known to date regarding the genes involved in sex differentiation in lampreys which, compared to sex determination (see Sect. 1.3.1.1), are relatively conserved among vertebrates (Siegfried 2010; Cutting et al. 2013). Elucidating the genetic basis of sex differentiation in lampreys will show how deeply conserved these genes are across all vertebrates and could provide early molecular markers predictive of a lamprey's future sex.

1.4.1 Sex Differentiation

Lampreys have been described as possessing a long period of sexual indeterminacy or sexual lability (e.g., Hardisty 1965a, b; Fukayama and Takahashi 1982; Docker

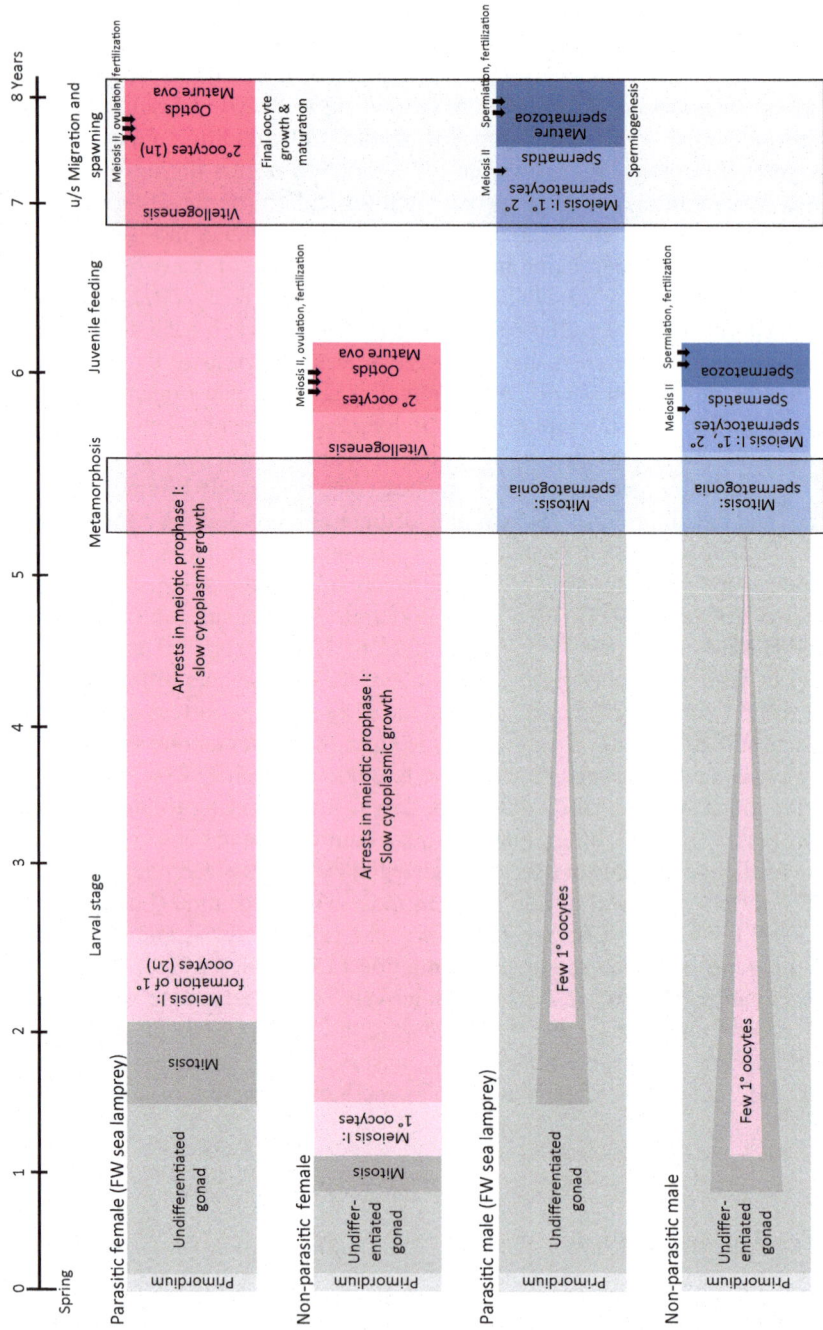

◄**Fig. 1.6** Timing of key events during lamprey gonadal development in a representative parasitic lamprey (Great Lakes sea lamprey *Petromyzon marinus*) and in non-parasitic lampreys. Later (i.e., at a larger size) and more prolonged mitosis in parasitic species produces more oocytes during the larval stage relative to non-parasitic lampreys. Ovarian differentiation (*pink*) begins when the germ cells synchronously enter meiosis (e.g., Fig. 1.7b, c), after which the resulting oocytes gradually increase in diameter (see Fig. 1.9). Non-parasitic females, with the elimination of the juvenile feeding phase, initiate sexual maturation during metamorphosis; parasitic lampreys remain sexually immature during the post-metamorphic feeding phase. In future males, limited oogenesis may occur during the larval stage (e.g., Fig. 1.7a), but these small oocytes are generally eliminated by atresia, leaving only a few residual undifferentiated germ cells until shortly before or during metamorphosis (e.g., Fig. 1.7e). Testicular differentiation (*blue*) is characterized by renewed mitotic divisions, which produce nests of spermatogonia (e.g., Fig. 1.7f)

1992). However, what is generally meant is that the gonad remains histologically undifferentiated for a prolonged period. The terms "sexual indeterminacy" and "sexual lability" imply that individuals, prior to histological differentiation, are not yet committed to develop as males or females, but we do not know yet if this is the case (see Sect. 1.3.2). Undifferentiated germ cells appear to be bipotential (Hardisty 1965b), thus retaining the ability to become oocytes or spermatocytes, but when the fate of these cells is determined and when the gonad itself is irreversibly committed to develop as an ovary or testis is unresolved. It has generally been thought that the fate of the gonad is set once the majority of undifferentiated germ cells have become oocytes (Hardisty 1965a; Docker 1992), but there is evidence that sex may remain labile—in at least some individuals and some species—as long as undifferentiated germ cells (i.e., reserve "stem" cells) remain (Lowartz and Beamish 2000; Beamish and Barker 2002).

Different sex differentiation strategies have been described for gonochoristic teleost species (i.e., where individuals develop only as males or females and remain the same sex throughout life; Yamamoto 1969; Devlin and Nagahama 2002). In those species referred to as differentiated gonochoristic species, ovaries and testes develop directly from the undifferentiated gonad; examples include coho salmon, muskellunge *Esox masquinongy*, common carp *Cyprinus carpio*, and European sea bass (see Devlin and Nagahama 2002). In contrast, in undifferentiated gonochoristic species (e.g., zebrafish *Danio rerio*, tiger barb *Puntigrus tetrazona*), all gonads initially develop as ovaries, but, in approximately half of the population, the ovarian tissue degenerates and the gonad is invaded by somatic cells, producing an intersexual gonad that ultimately resolves into a testis (Devlin and Nagahama 2002). In other species, all gonads appear to be intersexual prior to differentiation into either ovaries or testes (Devlin and Nagahama 2002). In the Nassau grouper *Epinephelus striatus*, all or most males appear to develop from an intersexual or bisexual gonad (i.e., possessing both oocytes and spermatocytes; Sadovy and Colin 1995). The juvenile gonad of the European eel *Anguilla anguilla* is sometimes referred to as an intersexual "Syrski organ," but it is not clear that all individuals go through a transitory intersexual stage (e.g., Colombo and Grandi 1996; Beullens et al. 1997a, b). Recent studies suggest that female European eel develop directly from the undifferentiated

Table 1.8 Studies examining sex differentiation in 13 lamprey species, showing age and size (total length, TL) at which ovaries are recognizable in each species and oocyte size and number by the end of the larval stage. Larvae were considered to be age class 0 from the time of hatch until 31 December of that year, age class I in the following year, and so forth. An asterisk indicates the dominant year class in which ovarian differentiation typically occurs. In all species, testicular differentiation occurs at metamorphosis; max TL indicates the size at which ovarian differentiation is thought to be complete (i.e., after which point undifferentiated gonads generally represent presumptive males)

Species	Recognizable ovaries first evident in:		Oocyte diameter (μm)	Number oocytes per cross-section		References	Comments
	Age class	Min TL (mm)	Max TL (mm)		Mean	Range		
PARASITIC SPECIES								
Geotria australis pouched lamprey (anadromous)	II		50–59	31.5		60–80	Hardisty et al. (1986)	Very few males with growing oocytes; no evidence of indirect development
Ichthyomyzon castaneus chestnut lamprey (freshwater)			80–88	134		~55	Beamish and Thomas (1983)	
	I–II*				63.1	25–114	Spice and Docker (2014)	Only 6.2% larvae recognizable as differentiated females by age I
Lampetra fluviatilis European river lamprey (anadromous)	II	50	65–70	76–110[a]	74.9	35–110	Hardisty (1970)	

(continued)

Table 1.8 (continued)

Species	Recognizable ovaries first evident in:			Oocyte diameter (μm)	Number oocytes per cross-section		References	Comments
	Age class	Min TL (mm)	Max TL (mm)		Mean	Range		
Lethenteron camtschaticum Arctic lamprey (anadromous)	II–III*	70–90	100	100–120		~100–110	Fukayama and Takahashi (1982)	Male differentiation mostly direct, but 10–20% indirect (i.e., following extensive degeneration of growing oocytes)
Mordacia mordax short-headed lamprey (anadromous)		60–79	90	67.8	23.5	16–30	Hughes and Potter (1969), Hardisty et al. (1992)	Four examples of apparent sex reversal from comparatively well-developed ovary (indirect development)
Petromyzon marinus sea lamprey (anadromous)	IV–V		120–130	44–56	200–396	110–340	Hardisty (1969, 1971), Barker and Beamish (2000)	Barker and Beamish (2000) found 49.1% of larvae >120 mm TL to have "atypical" gonads, suggesting even longer period of sexual indeterminacy (see Sect. 1.4.1.4)
Petromyzon marinus sea lamprey (freshwater)	III	90	90–100	61–79	170–250	94–280	Hardisty (1965b), Lewis and McMillan (1965), Hardisty (1969), Docker (1992), Barker and Beamish (2000)	Barker and Beamish (2000) found 55.9% of larvae >100 mm TL to have "atypical" gonads

(continued)

Table 1.8 (continued)

Species	Recognizable ovaries first evident in:			Oocyte diameter (μm)	Number oocytes per cross-section		References	Comments
	Age class	Min TL (mm)	Max TL (mm)		Mean	Range		
BROOK LAMPREYS								
Ichthyomyzon fossor northern brook lamprey	I*–II				31.2	20–43	Spice and Docker (2014)	25.7% all larvae recognizable as differentiated females by age I
Ichthyomyzon gagei southern brook lamprey	I		50–90	100–120		~10	Beamish and Thomas (1983)	Ovarian differentiation complete by 12–17 months
Lampetra aepyptera least brook lamprey	I		55			16	Docker and Beamish (1994)	
Lampetra planeri European brook lamprey	I	41–60	60–70	100–120	34.2	25–50	Hardisty (1960b, 1961c, 1965a)	

(continued)

Table 1.8 (continued)

Species	Recognizable ovaries first evident in:			Oocyte diameter (μm)	Number oocytes per cross-section		References	Comments
	Age class	Min TL (mm)	Max TL (mm)		Mean	Range		
Lethenteron appendix American brook lamprey		50	70				Okkelberg (1921)	Male differentiation thought to be mostly indirect
Lethenteron reissneri Far Eastern brook lamprey	I–II	30–50	60–70	110–140		~12–20	Fukayama and Takahashi (1983)	Male differentiation mostly direct, but 10–20% indirect (derived from "sex reversal" of typical ovaries)
Mordacia praecox Australian brook lamprey				90.1[b]	21[b]	16–25[b]	Hughes and Potter (1969)	

[a]76 μm in 3- and 4-year old larvae (mean TL 100 mm), 110.4 μm at metamorphosis
[b]During early metamorphosis; not available for large larvae

gonad, but there appear to be two routes to produce males: direct male differentiation from the undifferentiated gonad or delayed indirect male development via an intersexual stage (Geffroy et al. 2013, 2016).

Similar differences in the pattern of sex differentiation may exist among and even within lamprey species. However, because the early stages of oogenesis are seasonal and relatively transitory (Hardisty 1965a, 1969), some apparent differences may be the result of observational biases. Furthermore, inconsistent terminology used by different authors to describe the early stages of sex differentiation sometimes confounds comparisons. Some authors refer to the early stage of gonadogenesis in lampreys as "intersexual" or "bisexual" or "hermaphroditic" because oocytes appear to develop in some or all presumptive male larvae during the early stages of differentiation (e.g., Okkelberg 1921; Lewis and McMillan 1965; Hardisty et al. 1986). However, in other fishes, the term intersexual generally refers to the simultaneous presence of male and female gonadal tissue in gonochoristic species (Bahamonde et al. 2013), and the undifferentiated cell nests observed in larval lamprey gonads (which are not homologous with the cysts of germ cells in the mature testis) should not be viewed as male elements in an intersexual or hermaphroditic larval gonad (Hardisty 1965a). Other studies refer to lampreys passing through an initial female stage or female intersexual stage (Hardisty 1965b, 1971). In most cases, this initial stage is short-lived in future males, requiring atresia of only a modest number of early stage oocytes (Hardisty 1965b; Hardisty et al. 1992). However, in some cases, initial female differentiation proceeds further, and future males are thought to result from atresia of the entire stock of oocytes beyond the typical age of ovarian differentiation (Hardisty 1965a; Fukayama and Takahashi 1982, 1983). To avoid confusion, we use the term "intersexual" sparingly, using it only to refer to gonads that are more obviously intermediate between male and female (see Sect. 1.4.1.4). In the initial stages of gonadogenesis, cells undergoing meiosis are considered oocytes only following the onset of cytoplasmic growth (Hardisty 1971), and we consider the "mixed" larval gonads that possess both cell nests and oocytes a transitional stage of differentiation and not as an expression of intersexuality (Hardisty 1965b). Individuals or gonads that appear to be developing as males are referred to as "presumptive" or "putative" males or testes prior to differentiation of male germ cells at or following the onset of metamorphosis. We use the term "indirect" male differentiation to refer to presumptive males that appear to develop following large-scale oocyte atresia prior to testicular differentiation (Fukayama and Takahashi 1982, 1983) and "direct" male differentiation where there is little evidence that oocytes develop en masse or to relatively large sizes in future males (Hardisty et al. 1986). However, we recognize that there are likely different degrees to which oocytes develop and regress in presumptive males (or different degrees to which they are detected) rather than two discrete strategies.

Hardisty et al. (1992) theorized that indirect development is more prominent in species with low fecundity (i.e., non-parasitic lampreys) than in those with greater fecundity. However, fecundity appears to be determined largely by the phasing of mitosis and meiosis. The onset of meiotic prophase usually marks the end of the proliferative (mitotic) phase, after which point additional oocytes generally are not created. Thus, earlier (i.e., at a smaller body size) and less extensive mitosis prior to meiosis limits the number of oocytes produced in non-parasitic species (Hardisty 1965b). In contrast, delayed onset and more prolonged mitotic division of germ cells in parasitic species permit elaboration of a large stock of oocytes (Hardisty 1960b, 1964, 1965b, 1970; see Sect. 1.6.2). To some extent, however, persistence of a limited number of undifferentiated germ cells in some species may allow further oocytes to be added following completion of ovarian differentiation (Hardisty et al. 1986).

1.4.1.1 Initial Stages of Gonadogenesis

In lampreys, primordial germ cells first appear during embryonic development and migrate to the genital ridge to form the gonadal primordium (Lewis and McMillan 1965; Hardisty 1965a, 1971). In European brook lamprey prolarvae measuring ~7 mm TL, Hardisty (1965a) counted a total of 10–94 primordial germ cells measuring 18–30 μm in diameter. In freshwater-resident ("landlocked") sea lamprey larvae measuring 18–35 mm TL, Lewis and McMillan (1965) and Hardisty (1965b) found 5–15 and 0–3 germ cells per cross-section, respectively, and each germ cell was surrounded by its own envelope of follicle cells. Pouched lamprey larvae measuring 15–20 mm TL did not yet possess a distinct genital ridge (i.e., they possessed only isolated primordial germ cells along the mid-dorsal surface of the body cavity), but a distinct gonad was visible by 20–39 mm TL, generally with <10 germ cells per cross-section (Hardisty et al. 1986). At this stage of development, the germ cells are typically referred to as protogonia (primary gonia) if they occur singly. The protogonia are smaller (~10–16 μm in diameter) and more numerous than the primordial germ cells, suggesting that some mitosis has already occurred (Hardisty 1965a, b). Subsequent division of the protogonia produces groups of 2–4 deuterogonia (secondary gonia) measuring ~10–11 μm in diameter (Hardisty 1965b, 1971). Although gonia (i.e., germ cells during the mitotic phase) that give rise to primary oocytes and primary spermatocytes are typically called oogonia and spermatogonia, respectively, the terms gonia, protogonia, and deuterogonia are often used instead in the lamprey literature before the future sex of the gonad is clear (e.g., Hardisty 1965a, b, 1971). Besides the occasional mitosis of protogonia to produce deuterogonia, there is little subsequent development of the gonad for several months to several years, depending on the species and life history type.

In non-parasitic lampreys, mitosis is initiated within the first 6 months of larval life. For example, in the European brook lamprey, germ cell proliferation begins in the late summer or autumn when larvae measure ~15–25 mm TL, and it reaches its peak in the early spring (February–March) when larvae are almost 1 year old and measure ~25–40 mm TL (Hardisty 1965a). These mitotic proliferations start to produce cell nests or cysts, although isolated protogonia and small groups of deuterogonia may still persist. However, the mitotic stage is relatively short-lived in this species, with meiosis being initiated almost simultaneously in both the cysts and isolated gonia. In five European brook lamprey larvae 31–40 mm TL that were just over 1 year old, a maximum of 19 germ cells were evident per cross-section, and all were already in meiotic prophase and arranged in small groups of up to 5 germ cells per cyst (Hardisty 1965a). Similarly, in the Far Eastern brook lamprey *Lethenteron reissneri*, mitosis was first apparent in larvae ≥30 mm TL and was relatively limited prior to the onset of meiosis (Fukayama and Takahashi 1983). In this species, cysts were evident in only 18% of individuals 30–50 mm TL.

In contrast, in most parasitic species, mitosis is delayed until larvae are larger and older, and it is more extensive prior to the onset of meiosis. In the gonads of landlocked sea lamprey, mitotic proliferation was not evident until larvae were 51–65 mm TL (during September of presumably their second year of larval life; Hardisty 1965b). The dividing germ cells remained together to form cysts (e.g., Fig. 1.7a), and ~20% of all individuals 51–60 mm TL possessed cystic gonads, although some isolated gonia and small groups of 2–4 cells were still evident. Mitotic activity peaked at 61–70 mm TL (between their second and third years), producing 4–13 cysts per cross-section, with an average of 20–50 germ cells per cyst. By 71–80 mm TL, ~70% of the gonads were in the cystic stage, and the number of germ cells per cross-section averaged 322 (Hardisty 1965b, 1969). By 81–90 mm TL, the proportion of individuals with cystic gonads had decreased to ~20%, and cysts were seen only rarely in landlocked sea lamprey ≥90 mm TL, indicating that these nests of undifferentiated germ cells are a transitory developmental feature of the early stages of differentiation (Hardisty 1969). The observation that well over 50% of sea lamprey larvae developed cystic gonads (in a population where sex ratio was close to parity; Table 1.3) indicates that germinal proliferation at this stage does not just occur in future females (i.e., it appears to be more indicative of the stage of differentiation than the sex orientation of the gonad; Hardisty 1965b), but it is worth noting that it appears not to occur in all future males either (see Sect. 1.4.1.3). There is no obvious bimodality at this stage pointing to future female and future male development, although subtle sex-specific differences in the onset or extent of germ cell proliferation cannot be ruled out (Hardisty 1969). In the Arctic lamprey, mitosis produced cystic gonads by the time larvae reached ~55 mm TL (Fukayama and Takahashi 1982). In larvae 60–90 mm TL, 75% of the gonads were in the cystic stage, again suggesting that mitotic proliferation occurs during the early larval stage in many—but not all—presumptive males. Only a few solitary gonia were present, and gonads often exhibited at least 8–10 cysts per cross-section. It was estimated that each cyst contained 8–512 germ cells, suggesting that mitosis had occurred at least nine times (Fukayama and Takahashi 1982). Germ cell proliferation in the anadromous sea lamprey, the largest and most fecund lamprey (see Sect. 1.6.3), appears to be even more delayed and more prolonged. In sea lamprey larvae collected in the U.K., mitosis was rarely observed in individuals <70 mm TL, and peak mitotic activity was observed at 81–90 mm TL (Hardisty 1969). In this size class, >50% of all gonads were in the cystic stage and cystic gonads continued to be found in larvae measuring 120–130 mm TL (Hardisty 1969).

The pouched and short-headed lampreys from the Southern Hemisphere (which belong to separate families distinct from each other and from the Northern Hemisphere lampreys; Potter et al. 2015) show somewhat different patterns of mitosis during the initial stages of gonadogenesis. In the pouched lamprey, mitotic divisions are initiated at the size expected in a relatively fecund parasitic lamprey (i.e., com-

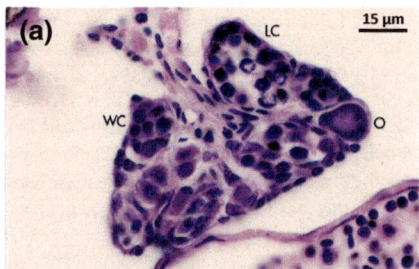

Undifferentiated gonad

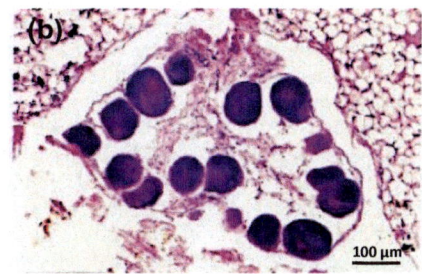

Ovary, larval least brook lamprey (non-parasitic)

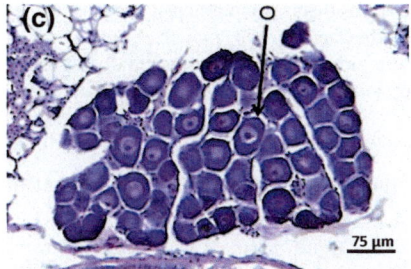

Ovary, larval chestnut lamprey (parasitic)

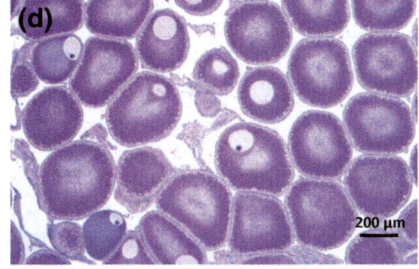

Ovary in mid-vitellogenesis, upstream migrant

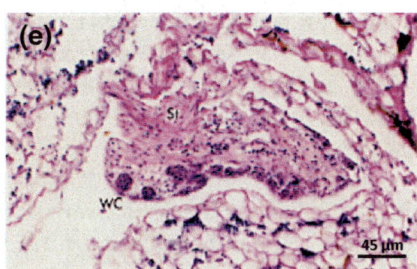

Presumptive testis, larval lamprey

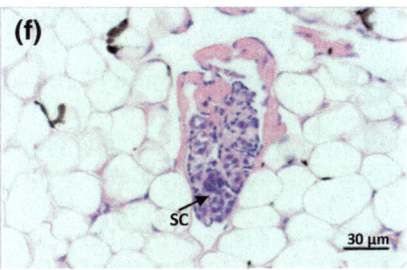

Differentiating testis, early metamorphosis

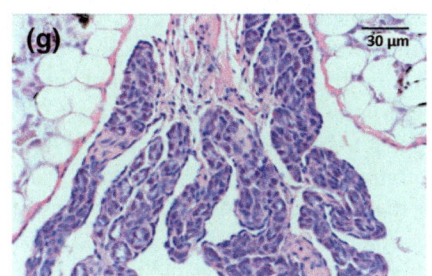

Differentiated testis, late metamorphosis

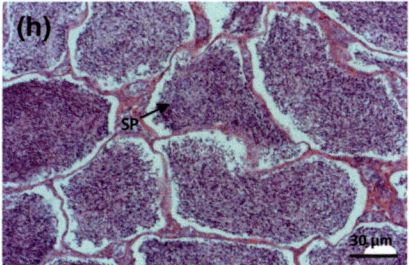

Testis with mature spermatozoa

◀**Fig. 1.7** Representative stages of sex differentiation and development in lampreys: **a** undifferentiated gonad (sea lamprey *Petromyzon marinus*, 70 mm TL) containing both well-defined cysts (*WC*) and loose clusters (*LC*) of undifferentiated germ cells as well as the occasional small oocyte (*O*); differentiated ovary in **b** least brook lamprey *Lampetra aepyptera* larva (118 mm TL) and **c** chestnut lamprey *Ichthyomyzon castaneus* larva (58 mm TL), where the only germ cells are oocytes (*O*) undergoing gradual cytoplasmic growth; **d** oocytes of upstream-migrating Pacific lamprey *Entosphenus tridentatus* (~595 mm TL) during mid-vitellogenesis; **e** gonad of future male least brook lamprey larva (104 mm TL) showing persistence of only undifferentiated germ cells (*WC*) amid stromal tissue (*St*); **f** differentiating testis of early metamorphosing sea lamprey (123 mm TL) showing appearance of spermatogonial cysts (*SC*); **g** testis of sea lamprey in late metamorphosis (127 mm TL) showing rapid increase in size of testis and development of lobular structure; and **h** mature spermatozoa (*SP*) in pre-spermiating sea lamprey (454 mm TL). The photomicrographs **a, b** and **e** were originally published in Docker (1992); **c** was originally published in Spice (2013); and **f, g** and **h** were originally published in Khan (2017); they are reproduced with permission of M. F. Docker, E. K. Spice and A. Khan, respectively. The photomicrograph **d** was originally published in Clemens et al. (2013) and is reproduced with permission of the Canadian Journal of Zoology

parable to landlocked sea lamprey and Arctic lamprey), but germinal proliferation during the larval stage only appears to occur in presumptive females. Hardisty et al. (1986) found that mitosis in pouched lamprey produced nests of germs cells in just over 30% of all gonads by 50–59 mm TL. The proportion of cystic gonads increased to 42% by 60–69 mm and nearly 50% in larvae >80 mm TL, but it never exceeded 50%. Moreover, compared to sea and Arctic lampreys, mitosis appeared more limited, and mean germ cell count per cross-section was only ~15–30 when larvae measured 50–70 mm TL. However, much like the anadromous sea lamprey, these mitotic cysts appeared to persist throughout the remainder of the larval stage. Mean germ cell count in both mitotic and meiotic phase gonads continued to increase throughout larval life, reaching ~100 and 300 germ cells, respectively, by 100 mm TL. Hardisty et al. (1986) suggested that oogenesis in this species may occur in seasonal waves of mitosis and meiosis, thereby adding oocytes to the ovary throughout the larval stage (see Sect. 1.4.1.2). The short-headed lamprey also shows onset of mitosis at ~50 mm TL, as expected, but mitosis appears much more limited throughout the larval stage (Hardisty et al. 1992). Only small cell nests were evident, and germ cells averaged only 7.1 cells per section by 60–69 mm TL and 10–11.4 cells in larvae measuring ≥75 mm TL. The limited mitotic proliferation in this species may account for its relatively low fecundity compared to other anadromous lampreys (see Sect. 1.6.3).

Meiosis and germ cell degeneration also appear to be initiated in most larval lampreys regardless of future sex. Thus, to avoid confusion, the term auxocyte (i.e., any cell undergoing meiosis) is often used rather than oocyte for the earliest stages of meiosis (e.g., Hardisty 1965a, b, 1971; Fukayama and Takahashi 1982, 1983), and auxocytes are considered oocytes only at the first phase of cytoplasmic growth (e.g., Hardisty 1971) or at meiotic prophase (Lewis and McMillan 1965). We will use the definition employed by Hardisty (1971) because it is the most obvious to discern histologically (see Sect. 1.4.1.2). As mentioned above, meiosis is initiated almost simultaneously with mitosis in the non-parasitic species that have been studied, occurring in the spring as larvae approach the end of their first full year of larval life (Hardisty

1965a; Fukayama and Takahashi 1983). In European brook lamprey, auxocytes were observed in >80% of the gonads in larvae 50–60 mm TL, at which point, meiotic cysts numbered ~3–8, with no evidence of bimodality (Hardisty 1965a). The first sign of divergence between future male and female European brook lamprey is the more synchronized oogenesis and oocyte growth that occurs in presumptive females later that summer (Sect. 1.4.1.2) and the appearance of morphological changes that are thought to "betray" the future male character of the gonad (Hardisty 1965a; Sect. 1.4.1.3).

In parasitic species, meiosis is generally initiated after a longer period of mitotic proliferation. In the landlocked sea lamprey, meiosis was first evident in the 51–60 mm size class and was at its peak in May (Hardisty 1969). By 71–80 mm TL, auxocytes were observed in 85% of all larvae and even growing oocytes were apparent in most of these (e.g., Fig. 1.7a). More synchronized meiosis and oocyte growth peaked in June and August in presumptive females, and the proportion of mixed gonads (i.e., with auxocytes and cysts) began to decrease and mixed gonads were rarely found in larvae >100 mm TL (Hardisty 1969). Cysts began to break up following invasion of follicle cells, and germ cell degeneration was evident by 61–70 mm TL; at this point, degenerating germ cells averaged 50–280 per section and represented 50–90% of the total germ cell count. By September, the majority of cysts contained degenerating germ cells, presumably representing regression of auxocytes or oocytes in the gonads of future males and degeneration of undifferentiated germ cells in future females. By 71–80 mm TL, Hardisty (1965b) found two distinct groups beginning to emerge based on the cross-sectional area of the gonad, but there was substantial overlap between presumptive males and females until 91–100 mm TL so that reliable identification of future males and females based on the presence of small and large gonads, respectively, would not be possible (see Sects. 1.4.1.2, 1.4.1.3 and 1.4.1.5). In anadromous sea lamprey, the smallest larva with auxocytes was 70 mm TL; meiosis was evident in 10% of larvae 71–80 mm TL and in >50% of larvae by 91–100 mm TL, and mixed gonads persisted into the 121–130 mm size class (Hardisty 1969). Meiosis was evident in Arctic lamprey at 70–90 mm TL, and gonads could be divided into three groups based on the degree to which the process appeared synchronous: (1) in 50% of the 14 larvae examined, germ cells in almost all the cysts had synchronously entered into meiotic prophase; (2) in 21% of the larvae, germ cells in a given cyst were seen to enter into meiotic prophase simultaneously, but, overall, cysts of mitotic germ cells were more numerous than those of meiotic germ cells; and (3) in 29% of the larvae, onset of meiosis was not synchronous even in the same cysts (Fukayama and Takahashi 1982). Whether these different patterns are an early indication of the future sex of the larvae is unknown. In the short-headed lamprey, onset of meiosis was first seen in ~10% of larvae <50 mm TL, and meiosis was evident in ~25 and 60% of the larvae by 50–59 and 60–69 mm TL, respectively (Hardisty et al. 1992). By 70–79 mm TL, these authors concluded that 96% of short-headed lamprey exhibited at least some germs cells that appeared to have differentiated in a "female direction." In contrast, in the pouched lamprey, the proportion of cystic gonads showing meiosis never exceeded 23% (at 70–79 mm TL) and declined thereafter to only 5% at 90–99 mm TL (Hardisty et al. 1986). Even in future females, the early meiotic changes in cystic gonads appeared to be transient (proceeding rapidly to the cytoplasmic growth phase in oocytes) and meiotic gonia were seen in only 17% of future males.

1.4.1.2   Ovarian Differentiation

Oogenesis (i.e., the process by which eggs are produced) starts with transformation of oogonia into primary oocytes with the onset of meiosis I (Fig. 1.6). Because the onset of meiosis also occurs in at least some future males (see Sect. 1.4.1.1), lamprey gonads with auxocytes or oocytes are usually not immediately characterized as ovaries. Rather, the differentiation of the ovary is typically recognized by the synchronous transition of germ cells into meiotic prophase (Fukayama and Takahashi 1982), and ovarian differentiation is generally considered complete after the rapid growth of oocytes ceases (Hardisty 1971). Ovarian differentiation is generally complete at 1 and 2–3 years of age in non-parasitic and most parasitic lamprey species, respectively, and at 4–5 years in the anadromous sea lamprey (Hardisty 1969, 1971; Table 1.8). However, intraspecific variation likely exists in many species. First, although age is often inferred from TL, it is not yet clear whether the onset of ovarian differentiation is triggered by size or age or a combination of the two. For example, in the landlocked sea lamprey, ovarian differentiation is generally complete at 90–100 mm TL and 3 years of age, but Docker (1992) observed that a 73-mm female estimated to be ~5.5 years old had a fully differentiated ovary; this individual had been maintained for 39 months at high density in the laboratory and grew only 6 mm during this time. Conversely, in fast-growing sea lamprey populations, completion of ovarian differentiation may be delayed well past 90–100 mm TL (Wicks et al. 1998a, b; Barker and Beamish 2000; see Sect. 1.4.1.4). Furthermore, because mitosis and meiosis are highly seasonal processes, their onset will presumably depend on larvae reaching the appropriate ("threshold") size or age by key times of the year. Intraspecific variation in the presumed age at ovarian differentiation has been reported in chestnut and northern brook lampreys also (Spice and Docker 2014; Table 1.8).

In presumptive ovaries, synchronous and extensive meiosis leads to the rapid replacement of cell nests by growing oocytes (Hardisty 1965a, b, 1971; Fig. 1.8a). The cysts are invaded and broken up by follicle cells that stretch to enclose each germ cell (Lewis and McMillan 1965), and cytoplasmic growth (i.e., during the first oocyte growth stage) is quite rapid while the cysts are breaking up (Hardisty 1965a). It is estimated that 25–30% of germ cells survive the cystic and early meiotic stages to progress to the cytoplasmic growth phase (Hardisty 1971; Hardisty et al. 1986). Early stage oocytes can be distinguished from other cells by their greater amounts of basophilic cytoplasm and larger size surrounded by follicle cells and a large space or nuclear vesicle within the oocyte (Hardisty 1965a, b; Lewis and McMillan 1965; e.g., Fig. 1.7b). The oocytes remain arrested in meiotic prophase throughout the rest of the larval stage, and, during the second stage of oocyte growth, the cytoplasm becomes basophilic and continues growing at a more gradual rate (Hardisty 1971).

The third phase of oocyte growth does not occur until the onset of vitellogenesis and sexual maturation (see Sect. 1.5.1).

The process of ovarian differentiation is similar in all species, although the timing of the process and the resulting number and size of oocytes varies dramatically (Table 1.8). The earlier onset and "curtailment" of mitosis prior to oogenesis in non-parasitic species results in fewer oocytes compared to parasitic species (Fig. 1.7b, c). The size of oocytes at the end of the larval stage (with means ranging from 32 μm in the pouched lamprey to 100–120 μm in most non-parasitic species) is directly related to the time between initiation of oogenesis and the onset of metamorphosis and inversely related to the duration and extent of somatic growth during the post-larval phase (Fig. 1.9). Egg size at maturity is remarkably consistent among all lamprey species (see Sect. 1.6.1).

In the well-studied European brook lamprey, although some growing oocytes were evident in the majority of gonads, more and larger oocytes developed in presumptive females, and the nests of undifferentiated germ cells were generally eliminated by the end of the summer (Hardisty 1965a). In larvae measuring 50–100 mm TL, the number of oocytes per cross-section was bimodally distributed, with 0–4 and 15–60 oocytes per section in presumptive males and females, respectively (Hardisty 1965a). In the rapid phase of cytoplasmic growth, oocyte diameter increased from a mean of 13.3 μm to 61 μm the following June when TL averaged 65 mm (Hardisty 1970). The mean number of oocytes per cross-section at the completion of ovarian differentiation was estimated to be 31–34, and they measured ~100 μm in diameter (Hardisty 1961c, 1964, 1965a; Fig. 1.9b). Other brook lamprey species show very similar patterns.

In parasitic species, ovarian differentiation has been studied most extensively in the landlocked sea lamprey (Hardisty 1965b, 1969; Lewis and McMillan 1965; Docker 1992; Wicks et al. 1998b; Barker and Beamish 2000). As detailed above, meiosis is generally initiated when larvae reach ~51–60 mm TL, and auxocytes and even growing oocytes are observed in most larvae during the initial stages of sex differentiation (e.g., Fig. 1.7a). In presumptive females, cysts begin to break up and germ cell degeneration begins at ~61–70 mm TL, and growing oocytes first appear in appreciable numbers at 71–80 mm (Hardisty 1965b). In an 80-mm presumptive female, Hardisty (1965b) counted 60 oocytes with cytoplasmic growth, 450 cells in early meiosis, 50 degenerating cells of various kinds, and 120 undifferentiated germ cells. The first stage of cytoplasmic growth occurred at 80–100 mm, at which point, oocytes increased from ~12 to 40 μm in diameter (Hardisty 1965b, 1971; Barker and Beamish 2000; Fig. 1.10a). Definitive ovaries could be recognized in 20 and 45% of all larvae by 81–90 and 91–100 mm TL, respectively. Thereafter, the number of undifferentiated germ cells continued to decrease with TL and the number of oocytes increased, so that ovaries contained only oocytes and virtually no remaining undiffer-entiated germ cells by 100 mm TL (Hardisty 1969; Docker 1992; Fig. 1.8a). There was a clear bimodality in oocyte numbers in larvae 71–100 mm TL, with modes of 21–40 oocytes in presumptive males and 121–140 (with a maximum of 200) in females (Hardisty 1965b). By 100–119 and 120–139 mm TL, mean oocyte count had increased further (to 153 and 250, respectively), suggesting that further germ cell pro-liferation produced some additional oocytes; this would imply that a few remaining

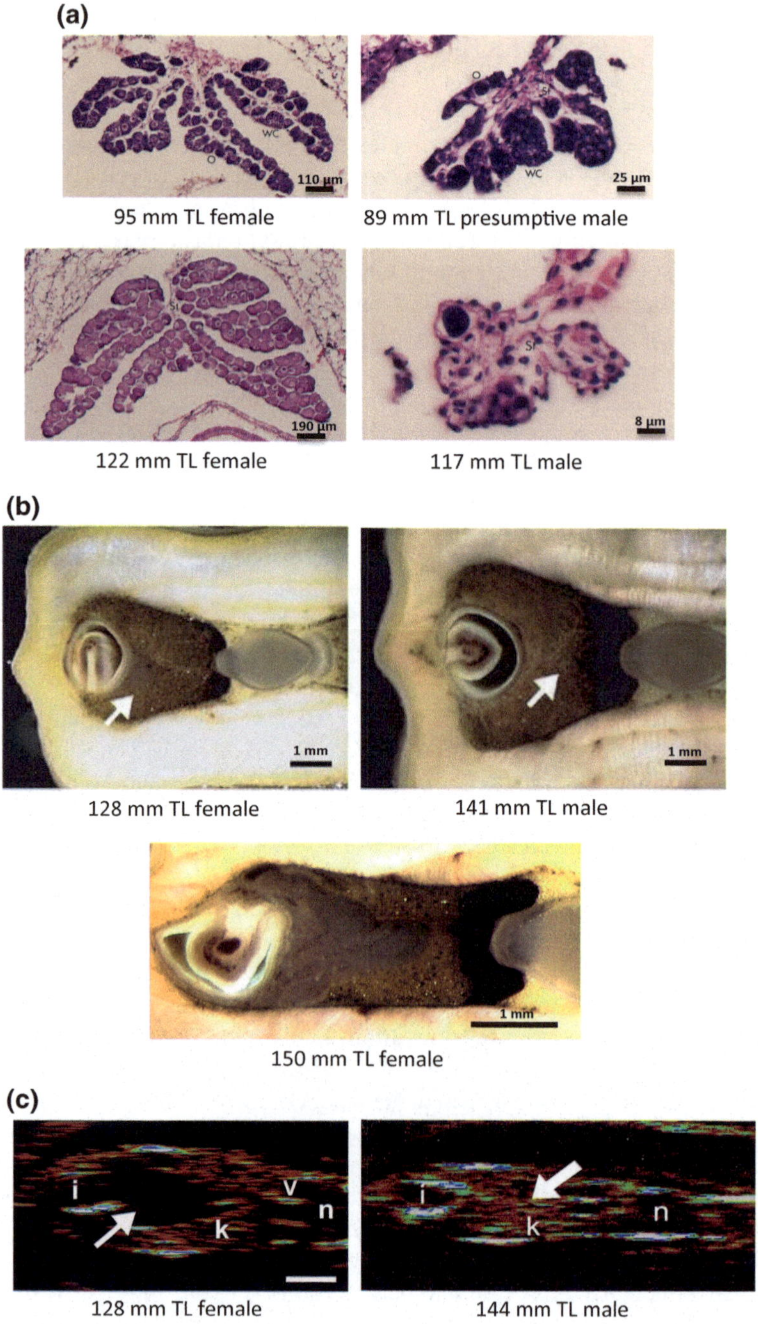

(a)

95 mm TL female

89 mm TL presumptive male

122 mm TL female

117 mm TL male

(b)

128 mm TL female

141 mm TL male

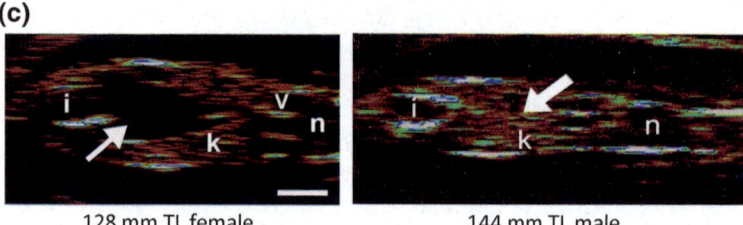

150 mm TL female

(c)

128 mm TL female

144 mm TL male

◄**Fig. 1.8** Identification of female and male larval Great Lakes sea lamprey *Petromyzon marinus*: **a** following histological preparation (stained with hematoxylin and eosin) and viewed under a compound microscope; **b** under a dissecting microscope; and **c** using acoustic microscopy (see Maeva et al. 2004). In **a** and **b**, larvae were cross-sectioned at the midpoint of their TL; in **c**, live larvae were anaesthetized and placed on their side. Sex can be reliably identified histologically in most Great Lakes sea lamprey by 90–100 mm TL: ovaries are recognized by their finger-like lobes and a large number of developing oocytes (*O*) despite the possible persistence of well-defined cysts of undifferentiated germ cells (*WC*); in contrast, at this size, presumptive testes are small with at most a few small oocytes. The ovary is sufficiently large by ~120 mm TL (with few or no remaining undifferentiated germ cells) that it is recognizable even under a dissecting microscope, but the future testis is still small, with a few undifferentiated germ cells (often no longer organized into distinct cysts) amid reticulate stromal tissue (*St*). In **c**, females are recognizable by the prominent ovary (*arrow*) which, because it is considerably less reflective to the acoustic signal than the surrounding kidney tissue (*k*), appears dark; males are recognizable by the absence of this large dark region. The location of the intestine, posterior cardinal veins, and notochord are indicated by *i*, *v*, and *n*, respectively. The photomicrographs in **a** were originally published in Docker (1992) and are reproduced with permission of M. F. Docker; the photographs in **b** and **c** were originally published in Maeva et al. (2004) and are reproduced with permission of The Fisheries Society of the British Isles

undifferentiated germ cells persisted beyond 100 mm TL (Hardisty 1965b). Barker and Beamish (2000) likewise reported increases in the number of oocytes with TL well past the size at which ovarian differentiation is thought to be complete (i.e., with means of 141 and 170 oocytes per section at 120–139 and ≥140 mm TL, respectively). During the second stage of oocyte growth, the mean oocyte diameter increases with female TL to ~53–65 μm by 120–139 mm and 61–75 μm by ≥140 mm (Hardisty 1965b, 1971; Barker and Beamish 2000; Figs. 1.8a, 1.9b and 1.10a). As a result of the increase in the number and size of oocytes, ovary size began to increase rapidly by the time larvae measured 90–100 mm TL, so that the subtle bimodality observed by Hardisty (1965b) in gonadal cross-sectional area at 71–80 mm TL became increasingly more pronounced, and there was little or no overlap in larvae >100 mm TL (Hardisty 1965b). In large larvae, the cross-sectional area of the ovary can be in excess of 1.0 mm^2, and the ovaries are characterized by finger-like lobes containing double rows of oocytes separated by a central vascular core (Hardisty 1965b; Docker 1992; Barker et al. 1998; Barker and Beamish 2000; Fig. 1.8a, b).

In anadromous sea lamprey, initiation of ovarian differentiation is even more delayed than in the landlocked form, but an even greater number of oocytes are produced. In sea lamprey larvae from the U.K., the earliest stages of oogenesis were not evident until presumptive females reached 81–90 mm TL (Hardisty 1969). In this size class, 56% of the gonads were observed to be in the initial stages of oogenesis in June–August, but oogenesis progressed rapidly and in synchrony, and it was rarely observed before or after the summer. Breakdown of the cysts and replacement by oocytes took place more gradually than in the landlocked form, and ovarian differentiation was generally not complete until 120–130 mm TL at ~5 years of age (Hardisty 1969). The number of oocytes per section averaged 322 and 396 at 120–139 and ≥140 mm TL, respectively, in sea lamprey larvae from the U.K. (Hardisty 1969)

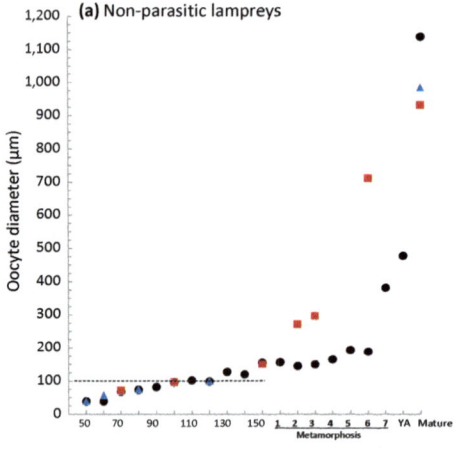

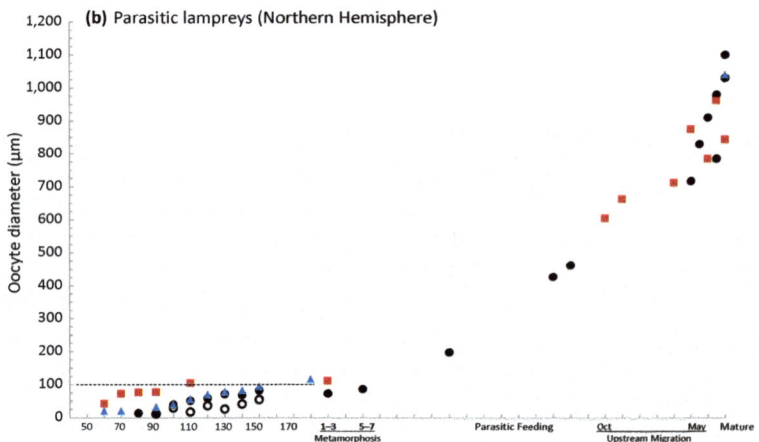

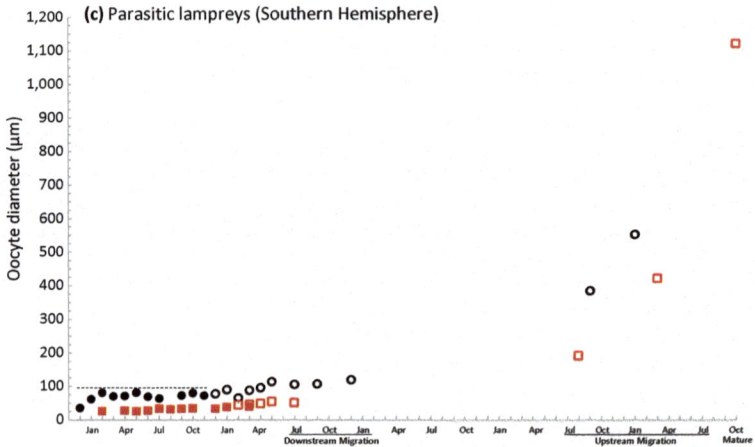

◄**Fig. 1.9**   Oocyte diameter (μm) during development in: **a** non-parasitic lampreys; **b** parasitic lampreys from the Northern Hemisphere; and **c** parasitic lampreys from the Southern Hemisphere. In **a**, *solid circles* represent mean oocyte diameter in Far Eastern brook lamprey *Lethenteron reissneri* for 10-mm larval size classes ranging from 50–60 to 150–160 mm TL, metamorphosing stages 1–7, and young adults (data from Fukayama and Takahashi 1983) and mature adults from *Lethenteron* sp. N (Yamazaki et al. 2001); *solid squares* represent southern brook lamprey *Ichthyomyzon gagei* from larval age groups I, II, and III, metamorphosing stages 2, 3 and 6, and adults (Beamish and Thomas 1983); and *solid triangles* represent mean oocyte diameter in larvae and adults of European brook lamprey *Lampetra planeri* (Hardisty 1961c and 1964, respectively). In **b**, *solid circles* and *open circles* represent Great Lakes and anadromous sea lamprey *Petromyzon marinus*, respectively (Applegate 1949; Lewis and McMillan 1965; Hardisty 1969; Barker and Beamish 2000); *solid squares* represent European river lamprey *Lampetra fluviatilis* (Hardisty 1961c, 1970; Witkowski and Jęsior 2000; Dziewulska and Domagała 2009); and *solid triangles* represent Arctic lamprey *Lethenteron camtschaticum* (Fukayama and Takahashi 1982; Yamazaki et al. 2001). In **c**, *solid circles* and *open circles* represent larval (>95 mm TL) and post-larval, respectively, short-headed lamprey *Mordacia mordax* (Hughes and Potter 1969; Hardisty et al. 1992); and *solid squares* and *open squares* represent larval (75–99 mm TL) and post-larval pouched lamprey *Geotria australis* (Potter et al. 1983; Hardisty et al. 1986). For ease of comparison of larval oocyte diameter among panels, a *dotted line* is drawn at 100 μm

and 159, 189, and 200 oocytes per section at 100–119, 120–139, and ≥140 mm TL, respectively, in sea lamprey larvae from New Brunswick, Canada (Barker and Beamish 2000). Reasons for the apparently higher number of oocytes per section in the European population relative to the North American population have not been explored, but the increase in oocyte number with TL in both populations suggests again that recruitment of additional oocytes can continue after ovarian differentiation is considered complete. As expected, given the later onset of oogenesis, oocytes in the anadromous form are smaller at a given female larval size relative to lampreys which initiate oogenesis earlier; mean oocyte diameter was measured to be 24 and 44 μm at 120–139 mm and ≥140 mm TL, respectively, in the U.K. population (Hardisty 1969) and 51 and 56 μm at the same sizes in anadromous sea lamprey females from North America (Barker and Beamish 2000; Fig. 1.9a).

In other parasitic species from the Northern Hemisphere, ovarian differentiation generally occurs at 2–3 years of age (at ~70–90 mm) and results in a moderate number of oocytes (~50–100 per cross-section) that are moderately sized by the onset of metamorphosis (~60–75 μm; Figs. 1.7c and 1.9b). However, ovarian differentiation in the short-headed and pouched lampreys shows some significant differences from these other parasitic species. In the short-headed lamprey, the timing of oogenesis is very similar to that of other moderately sized lampreys, but far fewer oocytes are produced. As expected, ovarian differentiation was first observed (in about ~10% of all larvae) by 60–69 mm TL and appeared to be complete by 80–89 mm TL when definitive ovaries made up 50% of larvae (Hardisty et al. 1992). Cysts had mostly degenerated by 90–109 mm TL, at which point residual undifferentiated germ cells were evident in only ~20% of females. In contrast to other parasitic species, however, mean number of oocytes ranged from only 15 per cross-section at 70 mm TL to 20 per cross-section in larvae >110 mm (Table 1.8). This number is even less than that seen in most non-parasitic lampreys and is presumably the result of more restricted mitotic

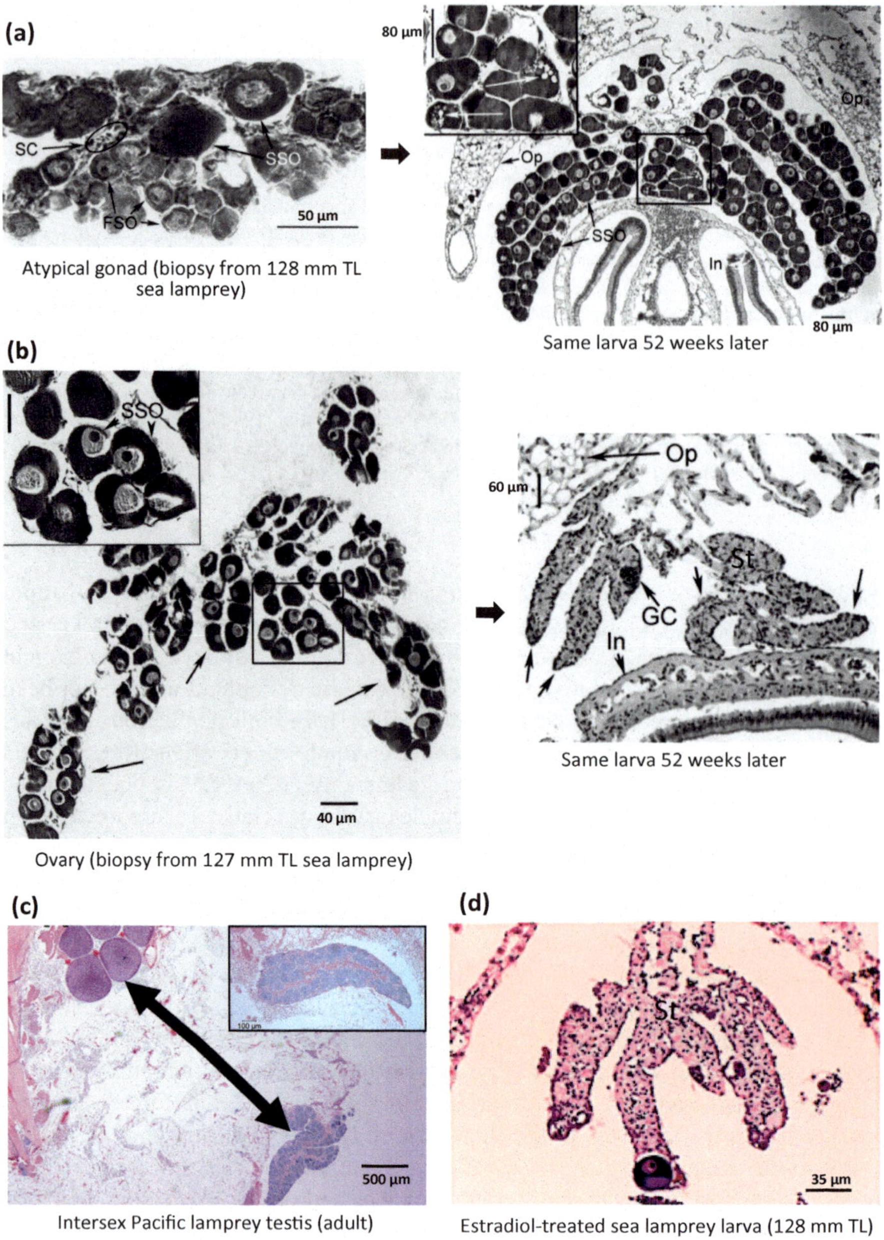

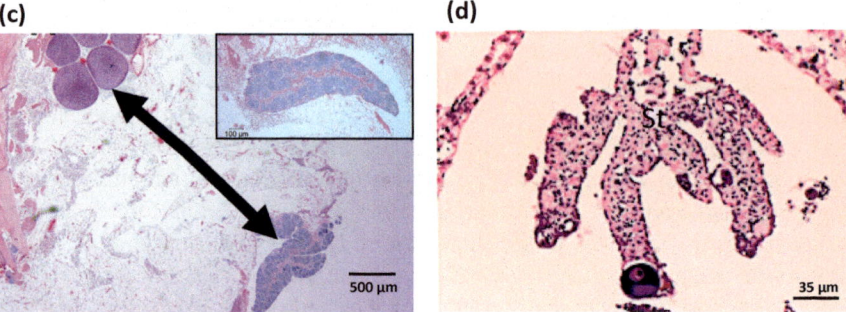

(a)
SC
SSO
FSO
50 µm
Atypical gonad (biopsy from 128 mm TL sea lamprey)
80 µm
Op
SSO
In
80 µm
Same larva 52 weeks later
(b)
SSO
40 µm
Ovary (biopsy from 127 mm TL sea lamprey)
Op
60 µm
St
GC
In
Same larva 52 weeks later
(c)
500 µm
Intersex Pacific lamprey testis (adult)
(d)
St
35 µm
Estradiol-treated sea lamprey larva (128 mm TL)

◄**Fig. 1.10** Atypical, intersex, and sex-reversed lamprey gonads: **a** development of a type I atypical sea lamprey *Petromyzon marinus* gonad into an ovary 52 weeks later, showing oocytes in
the first (*FSO*) and second (*SSO*) stages of oocyte growth and somatic cells (*SC*) in the atypical gonad, and showing only large SSO and a few atretic oocytes (*inset*) with vacuoles in their
cytoplasm (*white arrows*) in the ovary; the intestine (*In*) and opisthonephros (*Op*) are also shown;
b sex reversal of a typical sea lamprey ovary into a presumptive testis, showing a single cyst of
undifferentiated germ cells (*GC*), retention of the ovary's finger-like lobes (*black arrows*) despite
a complete loss of oocytes, and somatic cells in the stroma (*St*); **c** gonadal biopsy from a male
Pacific lamprey *Entosphenus tridentatus* during upstream migration (TL 520 mm, Sept) showing both mid-vitellogenic oocytes (*top arrowhead*) and spermatogonia and spermatocytes (*bottom
arrowhead* and *inset*); and **d** severe inhibition of oocytes in a female sea lamprey larva following treatment with 0.01 mg/L estradiol (E_2) for 21 weeks and subsequent rearing for 39 months.
The photomicrographs **a** and **b** were originally published in Lowartz and Beamish (2000) and are
reproduced with permission of The Fisheries Society of the British Isles; **c** was originally published
in Clemens et al. (2012) and is reproduced with permission of the National Research Council of
Canada; **d** was originally published in Docker (1992) and is reproduced with permission of M. F.
Docker

proliferation prior to the onset of meiosis (see Sect. 1.4.1.1), although the increase
in oocyte number with TL suggests some limited additional oocyte recruitment even
after ovarian differentiation is complete (Hardisty et al. 1992). Oocyte size was
comparable to that of most other parasitic lampreys during the larval stage (e.g.,
25–50 μm in larvae measuring >95 mm TL) and had increased to 63–90 μm by
stage 1 of metamorphosis (Hardisty et al. 1992; Fig. 1.9c). In contrast, the pouched
lamprey is characterized by its "exceptional state of immaturity" relative to other
larval lampreys, but the number of oocytes is what would be expected based on
body size at maturity (Hardisty et al. 1986). Growing oocytes were first evident in a
small proportion (6%) of larvae at 50–59 mm, and those showing a putative female
orientation (i.e., premeiotic cysts and auxocytes or growing oocytes) increased slowly
over the larval period up to a maximum of 39%. This suggests that at least some
future females possessed only cystic gonads even as large larvae. Moreover, ovarian
differentiation (i.e., with only growing oocytes in the gonad) was rarely complete,
and oocyte diameter in even the largest larvae (mean TL 88 mm) was only 32 μm
(Hardisty et al. 1986). In metamorphosing females, gonads possessing only cysts
were no longer apparent (and mean oocyte diameter had increased to 43 and 52 μm
by early and late metamorphosis, respectively), but 8% of females still exhibited
premeiotic cysts and auxocytes or growing oocytes rather than fully differentiated
ovaries. Hardisty et al. (1986) estimated that ~75% of germ cells underwent atresia
between the cystic stage and metamorphosis (i.e., based on a mean of 285 germ cells
per section in gonads with cysts or cysts and auxocytes and 60–80 oocytes in the
largest larvae and metamorphosing pouched lamprey), but he suggested that retention
of some undifferentiated germ cells may be an important mechanism for additional
recruitment of oocytes even after the onset of meiosis and oogenesis. This suggestion
was supported by the observation that mean oocyte count was substantially higher
(>160 per section) in larvae and metamorphosing individuals that were >70 mm TL
compared to the number of oocytes in larvae <70 mm TL (≤80 oocytes per section).

Additional recruitment of oocytes after the onset of oogenesis and perhaps even after the onset of metamorphosis (coupled with the exceptionally small size of the oocytes in larvae) could be particularly important in pouched lamprey, because, despite its large size at maturity, this species' small size at metamorphosis might otherwise limit the number of oocytes it could elaborate (see Sect. 1.6.3). Therefore, based on the degree of variation in the stage of ovarian maturity at metamorphosis observed among lamprey species, Hardisty et al. (1986) concluded that metamorphosis must not be dependent on larvae attaining a specific stage of gonadal development.

It has long been thought that sex in lampreys is irreversible following completion of ovarian differentiation (e.g., Hardisty 1965a), although a few cases of reversal of fully differentiated ovaries to presumptive testes have been shown using gonadal biopsy (see Sect. 1.4.1.4). Such cases of sex reversal presumably also depend on retention of residual undifferentiated germ cells.

1.4.1.3   Testicular Differentiation

Although mitosis and the early stages of meiosis appear to be initiated in the gonads of most lampreys regardless of future sex (see Sect. 1.4.1.1), these processes are halted in presumptive males, and the majority of cysts and auxocytes or occasional oocytes degenerate (Hardisty 1965a, b). As a result of this atresia, there is generally a reduction in the size of presumptive testes at this point, and only small numbers of undifferentiated germ cells that have not entered meiotic prophase remain (Fig. 1.7e). These remaining stem cells proliferate to produce spermatogonia only at the end of the larval stage or the onset of metamorphosis (Hardisty 1965a, b, 1971; Fig. 1.7f, g). Because the undifferentiated germ cells appear to remain bipotential throughout the larval stage, testicular differentiation is generally not considered complete until production of spermatogonia. The primary gonial cysts formed during the early stages of differentiation are not homologous to these secondary cysts produced on resumption of mitosis at metamorphosis. When mitosis resumes, the outline of the testis starts to become lobed as groups of germ cells are pinched apart by follicle cells; with further increases in mitotic activity, the testis starts to gain finger-like extensions that contain maturing cysts of spermatogonia (Hardisty 1971; Fig. 1.7g). The rate at which the subsequent stages of spermatogenesis occurs differs among life history types (i.e., it is accelerated in non-parasitic lampreys toward the end of metamorphosis, but parasitic species remain sexually immature until the end of the parasitic feeding phase; see Sect. 1.5.2). However, during the larval stage (given the relative inactivity of the presumptive testis at this point), there is less interspecific variation related to testicular development than there is with respect to ovarian development. Variation among species tends to be related mostly to the extent to which mitosis and the early stages of meiosis are initiated in future males (i.e., whether male development is direct or indirect) and the degree to which morphological differentiation of the testis might precede cytological differentiation.

Several lamprey species appear to show both direct and indirect male development. With the exception of the pouched lamprey, auxocytosis is generally observed in the gonads of 70–80% of all larvae (i.e., in all females and some, but not all, future males) during the initial phase of gonadal differentiation (see Sect. 1.4.1.1). This suggests that, even within a species, male differentiation may occur either directly (i.e., without passing through an initial but abortive "female" stage) or indirectly following extensive oocyte atresia. However, a closer examination suggests that the "pathway to male development" may be more of a continuum than two discrete categories. For example, in European brook lamprey, Hardisty (1965a) suggested that male differentiation proceeds via three routes: (1) testes that develop from gonads which, at an early stage, possess some of the somatic characteristics associated with the definitive male gonads (presumably the ~20% of larvae observed without auxocytes; i.e., those showing direct male development); (2) those that are composed almost or entirely of premeiotic cysts and auxocytes beyond the second summer of larval life, the subsequent degeneration of which leaves a few potentially male germ cells amid fibrous connective tissue (i.e., individuals that appear to show a "somewhat indirect" path of male development); and (3) those that differentiate, generally beyond the typical size of ovarian differentiation at a later stage, from predominantly ovarian-type structures, following atresia of all oocytes (i.e., individuals showing even more indirect male development appearing almost as sex reversals). Hardisty (1965a) observed evidence of this third route of male development in 17% of all European brook lamprey larvae 51–60 mm TL up to maximum of 21% at 61–70 mm TL. Initial meiotic activity and oocyte growth were more extensive and synchronous than in the second category, and oocyte numbers (4–46) were similar to those of the definitive ovaries. Indirect male differentiation appeared to involve infiltration of somatic cells and fibrous tissue into a primarily ovarian-type structure, and isolated germ cells sometimes resembling protogonia were seen in the cortical somatic region immediately below the peritoneal epithelium. Hardisty (1965a) concluded that the male germ cell line would subsequently be derived from these residual germ cells. He indicated that the higher oocyte numbers observed in these putative males likely represented those that would degenerate early in development and the lower numbers those that would degenerate later. Therefore, there were rarely numerous large oocytes past the usual point of ovarian differentiation, and the number and proportion of degenerating oocytes decreased with TL. Further, he noted that even the apparent reversals were not "sharply marked off" from the two other types, and he considered these cases extreme examples of delayed differentiation rather than sex reversal. However, in one severe case, the gonad of an 81-mm larva contained 60–70 oocytes, many of them in an advanced state of degeneration, but the shape of the gonad lobes and width of the mesogonial area were characteristic of a testis (see below). Busson-Mabillot (1967) also suggested two pathways for male differentiation in this species, observing that ~20% of all larvae (~40% of males, assuming a 50:50 sex ratio) developed directly into presumptive testes. She concluded that the remaining 80% of larvae developed an ovary-like structure and only secondarily produced presumptive males following oocyte atresia.

Direct and indirect male differentiation appears to occur, to different degrees, in other lamprey species as well. In Arctic and Far Eastern brook lampreys, Fukayama and Takahashi (1982, 1983) reported that development of future testes appeared to occur through degeneration of cysts and auxocytes in most larvae, particularly in 10–20% of the individuals that exhibited signs of widespread oocyte atresia during or after the cytoplasmic growth phase. However, in the Arctic lamprey, oocytes in the presumptive testes were often smaller than those in presumptive ovaries, suggesting that male development in this species is somewhat less indirect than in European and Far Eastern brook lampreys. In the short-headed lamprey, Hardisty et al. (1992) reported that >90% of larval gonads developed some meiotic cells and growing oocytes in the initial stages of sex differentiation, but extensive atresia of oocytes in an ovarian-type gonad was evident in only four of 303 larvae examined. Hardisty (1965b) considered differentiation in sea lamprey to be more direct than in the European brook lamprey, although growing oocytes were observed in 63% of landlocked sea lamprey larvae measuring 71–80 mm TL. This suggests that oocytes developed in at least some future males, although Hardisty (1965b) did not observe second stage oocytes undergoing degeneration. He indicated that the apparent absence of atretic oocytes was almost certainly due to a lack of histological observations throughout the entire year, because degeneration in the early meiotic prophase is extensive only in the spring and late autumn and usually affects all the elements of the cyst. Atresia is also thought to occur rapidly, so that static observations are less likely to capture this transitory process. For example, in the fetal and neonatal rat ovary, most degenerating germ cells were eliminated within 24 h of the onset of degeneration (Beaumont and Mandl 1962). Extensive atresia has been demonstrated in at least some future male sea lamprey using a gonadal biopsy technique that showed that presumptive testes can develop belatedly (i.e., TL > 118 mm) following atresia of both first and second stage oocytes (Lowartz and Beamish 2000; Beamish and Barker 2002; see Sect. 1.4.1.4). The pouched lamprey is the only species known to date that may show only direct male differentiation (Hardisty et al. 1986). Follow-up work is required to determine if there are distinctly different routes of male differentiation in lampreys (among and within species), or whether individual variation or observational biases are at play. Interestingly, almost 100 years ago, Okkelberg (1921) viewed sex differentiation in lampreys as consisting of a continuum ranging from pure females to pure males and including various intersexual forms.

Despite the possible differences seen among species during the early stages of male differentiation, further development of the presumptive testis appears to be more consistent. There is a general decrease in the number of cell nests and the number of germ cells within each nest, although considerable individual variation is often observed. In the European brook lamprey, for example, by 70–90 mm TL, a high proportion of the testes contain only single isolated germ cells or small groups of germ cells per section (Hardisty 1965a). At this point, there remains little evidence of previous meiotic stages or growing oocytes, although they are occasionally found even in large larvae (i.e., at 130 mm TL). Resumption of mitosis can be observed in some larvae >90 mm (i.e., prior to the onset of metamorphosis), although this "pro-spermatogonial" proliferation occurs only slowly in the later periods of larval life. Nevertheless, Hardisty (1965a) noted distinct differences in the size and cytological characteristics in the germ cells of the pre-metamorphic testis compared to the undifferentiated deuterogonia of earlier stages, especially the presence of a single nucleolus in the pro-spermatogonia compared with the double nucleolus in the deuterogonia. He also noted that cytological changes occurring in the presumptive testis are accompanied by morphological differentiation. A wave of increased activity in the peritoneal epithelium covering the surface of the testis forms indentations where the epithelial cells insinuate themselves between the gonia (Hardisty 1965a). As a result, the larger cell nests are continually broken up into smaller groups or separate cells, each invested by its own follicle cells.

Interestingly, Hardisty (1965a) observed that these morphological differences in European brook lamprey, unlike in most other species, often preceded the more dramatic cytological differences, and he thought that the male character of the gonad is "betrayed" at an early stage by these features. He suggested that "morphologically differentiated testes" could often be distinguished from early stage undifferentiated gonads and differentiating ovaries, even if "female development" was occurring in the germ cells, by four characteristics: (1) the shape and character of the gonad, where vertical clefts in the peritoneal epithelium often separate the presumptive testis into a number of relatively low lobes that have a flattened rectangular appearance rather than the rounded outline of ovarian lobes; (2) a broader area of attachment between the presumed testis and the dorsal wall of the body cavity (i.e., a wider mesogonial stalk) compared to the slender mesogonium of the ovary; (3) more crowded nuclei in the peritoneal epithelium covering the testis compared to the relatively sparser epithelial cells on the surface of the ovaries (presumably because the rapid growth of the oocytes outpaces proliferation of the ovarian epithelium); and (4) more developed fibrous connective tissue in the testis, particularly in the hilar region where the blood vessels and nerves enter the gonad (e.g., Fig. 1.7e), although this latter character tended to be more variable. Hardisty (1965a) suggested that the somatic elements of the gonad might, in fact, induce male development (e.g., by inhibiting further meiosis).

In sea lamprey, cysts and auxocytes similarly degenerate in presumptive males, so that the future testes are often smaller than the undifferentiated gonad. Following initiation of ovarian differentiation in future females, Hardisty (1965b) considered those gonads that still possessed a high number of germ cells (316 and 248 at 81–90 mm and 91–100 mm TL, respectively) to be undifferentiated and those that possessed much lower numbers (mean of 68 at 91–100 mm) to be presumptive testes. Docker (1992) counted 3–31 cysts per section and 1–103 cells per cyst in presumptive male sea lamprey ≥90 mm TL; cyst number was unrelated to TL, but number of cells per cyst decreased with TL. In the largest male larvae, isolated germ cells were common amid extensive connective stromal tissue (Fig. 1.8a). Occasional small basophilic oocytes (~12–14 μm) were evident in presumptive males, but the number per section decreased with TL and diameter did not increase with TL (Docker 1992). Hardisty (1965b) found that the proportion of presumptive males with oocytes decreased from 88% (with an average of 21 oocytes per section) at 71–90 mm TL to 50% (15 per section) at 91–100 mm, 28% (7 per section) at 111–130 mm, and 27% (6 per section) in larvae >130 mm TL. Docker (1992) reported very similar proportions of presumptive males with oocytes (48, 27, and 27% at 99–109, 110–129, and ≥130 mm TL, respectively), but she rarely observed >6 oocytes per section.

Nevertheless, germ cell number and organization in presumptive male sea lamprey were highly variable even after ovarian differentiation appeared complete in females, and presumptive testes were often indistinguishable from early undifferentiated gonads. The somatic characters that distinguished the early presumptive testes in European brook lamprey (Hardisty 1965a) appear to be less developed in sea lamprey (Hardisty 1965b), and they appear to emerge after (not before) the extensive germ cell regression and reduction in gonad size. Hardisty (1965b) reported a hint of bimodality in gonad cross-sectional area by the time sea lamprey larvae reached 71–90 mm TL, but differences between the sexes appeared diagnostic only once larvae reached ~100 mm TL (see Sect. 1.4.1.2). In large presumptive male larvae (110–130 mm TL), gonad cross-sectional area was ~0.003–0.08 mm^2, once again approaching the small size of the early undifferentiated gonad.

Slow germ cell proliferation may resume in the latter part of the larval period, and both the size of the gonad and germ cell count tend to increase slightly in premetamorphic males. For example, Hardisty (1965a) observed a few large male European brook lamprey larvae (141–150 mm TL) showing a noticeable increase in the cross-sectional area of the gonad (~0.16–0.45 mm^2), which is almost certainly indicative of a resumption of mitosis prior to metamorphosis. In most lamprey species, a marked increase in the rate of cell division occurs at the onset of metamorphosis. Cysts of spermatogonia become evident (Fig. 1.7f), and, as mitotic proliferation of spermatogonia continues, the entire gonad becomes occupied by closely packed nests of germs cells. It is at this point that the clefts and lobes described in European brook lamprey testes become well developed in male sea lamprey (Fig. 1.7g).

Unlike other lampreys, future male pouched lamprey undergo little mitosis in the initial stages of differentiation (see Sect. 1.4.1.1). As a result, the presumptive testes retain the low germ cell numbers and morphological appearance of smaller larvae throughout the larval stage (Hardisty et al. 1986). Unlike the Northern Hemisphere species, there also is no evidence that mitotic activity in future testes accelerates in the period preceding or even during metamorphosis. Undifferentiated gonads with only isolated germ cells were found to persist in a small proportion of large larvae (13 and 3% in the 80–89 and 90–99 mm size classes, respectively). In cystic gonads, a linear relationship between the number of germ cells and TL indicated a constant rate of proliferation, and germ cell numbers per cross-section were similar in metamorphosing and larval males at the same TL. In fact, mean cell counts actually decreased during metamorphosis, from 132 and 106 germ cells per section in stages 1–2 and 3–4 of metamorphosis, respectively, to 72 per section in stages 5 and above (Hardisty et al. 1986; see Manzon et al. 2015 re: stages of metamorphosis). In downstream-migrating pouched lamprey, 40–60 gonial cells were found per section, and there was still no evidence of mitosis (Potter and Robinson 1991). It appears that, in this species, spermatogonial proliferation is not initiated until the marine feeding phase (see Sect. 1.5.2).

In the short-headed lamprey, the future testis is even less well developed at downstream migration (Hughes and Potter 1969; Hardisty et al. 1992). During the larval stage, presumptive testes remain small and difficult or impossible to distinguish from earlier undifferentiated stages (Hardisty et al. 1992). These authors found that germ cell proliferation resumed at metamorphosis, but, even then, the future testes averaged only 38–73 μm in diameter during stages 1–3 of metamorphosis (area ~0.004–0.005 mm^2). There were generally only a few sporadic gonia located in the cortical zone of the rounded, oval, or fusiform gonad (Hardisty et al. 1992). Even in late metamorphosing stages and young adults, gonad diameter in presumptive males measured only 38–83 μm. A few individuals had gonads measuring up to 181 μm in diameter (~0.03 mm^2), but there was no correlation between gonad size and germ cell number; the two largest gonads had only 8 and 9 germ cells per section and were almost entirely composed of connective tissue.

1.4.1.4 Atypical or Intersex Gonads and Sex Reversal

The gonad in larval lampreys is frequently referred to as being hermaphroditic (e.g., Okkelberg 1921; Lewis and McMillan 1965) or intersexual (Hardisty 1971; Hardisty et al. 1992), because, in most species, some or all future males appear to pass through an initial but brief "female" stage as part of normal male development. Even more dramatically, a small proportion of presumptive male testes may differentiate from an ovarian-like structure following atresia of the entire stock of oocytes (see Sect. 1.4.1.3). In general, however, it has typically been thought that sex is definitive in most individuals once ovarian differentiation is complete (e.g., Hardisty 1965a). In this section, we review more recent reports suggesting that intersexual or highly atypical gonads can persist far beyond the length at which ovarian differentiation is normally complete (e.g., Barker and Beamish 2000) and, even more surprisingly, that complete sex reversal is possible after primary sex differentiation (e.g., Lowartz and Beamish 2000). However, because testicular differentiation (i.e., development of spermatogonia) does not occur until the onset of metamorphosis, it is important to note that intersex larvae do not possess the sex cells of both males and females. Intersex (or "atypical") gonads in larval lampreys refer to those where the morphological characteristics of the gonad (e.g., area, shape) are intermediate between females and presumptive males or where the morphological characters do not match the cytological characters (e.g., type, number, or size of the germ cells). True intersex gonads (i.e., in post-metamorphic lampreys) have been reported (Beard 1893; Okkelberg 1921; Hardisty 1965a; Clemens et al. 2012), but they are much rarer.

Atypical or intersexual gonads have been reported in a number of sea lamprey larvae from several rivers tributary to the Great Lakes (Wicks et al. 1998a, b; Barker and Beamish 2000) and in anadromous sea lamprey larvae collected from the Petit-codiac River in New Brunswick (Barker and Beamish 2000). In sea lamprey from 12 streams in the Great Lakes basin, Wicks et al. (1998a) found that sex could be identified in at least some individuals measuring 90–100 mm TL, but 8–100% of larvae measuring 90–160 mm TL were categorized as intersexes. Growth rates varied among streams, but, overall, 2- and 3-year-old larvae were estimated to be 54–93 and 72–128 mm TL, respectively, when aged using statoliths, or 58–109 and 86–163 mm TL, respectively, when larval age was estimated using length-frequency histograms. Wicks et al. (1998a) observed that the proportion of intersex larvae in a population increased with larval growth rate, and they suggested that, as a response to TFM treatments, sea lamprey may allocate a disproportionate amount of energy to somatic growth (at the expense of gonadal development) in order to shorten the larval period. In these situations, gonadal development would presumably resume following metamorphosis. Wicks et al. (1998b) conducted more detailed histological analysis on larvae from three Great Lakes streams. Using larvae from one stream, they first established statistical tolerance limits for various morphological criteria (gonad perimeter, which indicates the degree to which the margin is either smooth or crenulated, and cross-sectional area) and cytological criteria (germ cell number, oocyte number, and oocyte diameter) for typical male and female larvae from four size classes (90–105, 106–120, 121–135, and >136 mm TL). Atypical gonads were then identified as those where one or more characteristic fell outside these tolerance limits or where some gonadal characteristics fell within male tolerance limits and some fell within the typical female range. Atypical larvae comprised 52 and 33% of the larvae in Cobourg Brook and Farewell Creek (Lake Ontario), respectively, in collections from May, June, and September, and they made up 80% of the larvae collected from Little Gravel River (Lake Superior) in July. Atypical larvae were atypical in different ways, but such gonads usually included a high number of undifferentiated germ cells (4–1,372 per section versus 4–598 in typical males and 0–159 in typical females), 0–167 oocytes per section measuring 16–79 μm in diameter (vs. 0–8 oocytes measuring 13–22 μm in males and 75–90 oocytes measuring 46–77 μm in females), and up to 84 atretic oocytes per section (when none were evident in typical males or females).

This same pattern of atypical gonads was found in other Great Lakes tributaries by Barker et al. (1998): 9–82% of the gonads of sea lamprey larvae >90 mm TL were deemed atypical, and these authors suggested that there might be annual variation in the proportion of atypical larvae. In Gordon's Creek (Lake Huron), 82% of the gonads appeared atypical in June 1995, but only 14% of the gonads were atypical in October 1996. However, these differences could also represent seasonal differences: 43% of the larvae collected in June 1995 from Cobourg Brook (Lake Ontario) had atypical ovaries, but only 9 and 19% of those collected in September 1995 from Cannon Creek (Lake Huron) and Lynde Creek (Lake Ontario), respectively, had atypical gonads. It could be that more gonads appear atypical in the spring as the result of intense mitotic or meiotic activity, and that these processes then "settle down" or are followed by rapid atresia in late summer and early fall. However, virtually all other studies examining lamprey gonadal histology include larvae sampled in the spring and summer without reporting a large proportion of atypical larvae. Hardisty (1965b, 1971), for example, did not observe first-stage oocytes in female sea lamprey larvae after the larvae reached 90–100 mm TL, regardless of season (see Sect. 1.4.1.2). As was observed by Wicks et al. (1998b), however, atypical larvae were atypical in different ways, and Barker et al. (1998) classified them into four categories based on morphological and cytological characteristics. Typical ovaries (in larvae 115–165 mm TL) were horseshoe-shaped, with prominent lobes, no atresia, and second-stage oocytes (56–88 μm in diameter) arranged in pairs. In contrast, category 1 atypical gonads (120–129 mm) were small, angular in shape and without lobes, and they possessed only first-stage oocytes (15–18 μm) around the gonad's perimeter with no evidence of atresia. Category 2 atypical gonads (116–137 mm TL) were horseshoe-shaped and without lobes, and they had second-stage oocytes (33–46 μm) scattered individually throughout the gonad, many germ cell clusters, and 2,000–10,000 atretic oocytes. Category 3 gonads (122–146 mm TL) were also horseshoe-shaped but with lobes, no atresia, excess stromal tissue, and only a few oocytes (52–58 μm diameter) in some lobes. Category 4 gonads (119–141 mm TL) were also horseshoe-shaped with lobes, but with many germ cell clusters, first- and second-stage oocytes (39–56 μm), and 5,000–40,000 atretic oocytes. Oocyte diameter and ovarian cross-sectional area in typical female larvae increased with TL (see Sect. 1.4.1.2), but neither feature was correlated with TL in atypical larvae. The estimated potential fecundity of atypical gonad categories 2 and 4 was well above those for typical gonads, but this was predominantly the result of large numbers of undifferentiated germ cells in these gonads (up to 5.3 and 4.1 million in category 2 and 4, respectively). In contrast, the total number of oocytes and the total number of undifferentiated germ cells in typical gonads was 19,000–65,000 and 500–80,000, respectively (see Sect. 1.6.2). Potential fecundity for atypical gonad types 1 and 3 was consistent with those for typical gonads, but these values likewise included undifferentiated germ cells and small oocytes. Wicks et al. (1998a) suggested that atypical gonads observed in Great Lakes sea lamprey >90 mm TL may result from a slowing of gonadogenesis as a result of selection for rapid somatic growth. Histological observations made in the early years of sea lamprey control may not have detected this phenomenon. Wicks et al. (1998b) and Barker et al. (1998) further suggested that the unusually high number of undifferentiated germ cells per section in these Great Lakes sea lamprey may extend the period of sex differentiation, during which time the gonad may remain labile and be susceptible to influence from abiotic or biotic factors (see Sect. 1.3.2.2). Alternatively, the atypical gonads may represent a transition in sex, perhaps induced by cyclic changes in larval density or growth rate resulting from periodic TFM treatments (Wicks et al. 1998b).

However, atypical gonads were also common in larvae from the anadromous sea lamprey population examined by Barker and Beamish (2000), indicating that their occurrence is not related to chemical treatment of streams or a population response to TFM treatment (see Sect. 1.2.6). Atypical gonads were reported in sea lamprey larvae collected from the Petitcodiac River in New Brunswick at the end of June (120–140 mm TL), and they were histologically similar to those collected from Brown's Creek (Lake Huron) in May (100–140 mm TL). Although many larvae from both the anadromous and Great Lakes populations could be easily distinguished as male or female (see Sect. 1.4.1.5), gonads were atypical in 49 and 56% of the larvae, respectively. As observed previously (see above), the atypical gonads were characterized by intermediate or inconsistent morphological characters, an unusually high number of undifferentiated germ cells (0–703 and 0–988 per section in anadromous and landlocked sea lamprey, respectively), a variable number of oocytes (0–222 and 0–197 per section), oocytes of variable sizes (0–49 μm and 11–92 μm in diameter), and often the occurrence of atretic oocytes (0–40 and 0–35, respectively) where none were found in typical males or females. The reason why such a high proportion of atypical sea lamprey larval gonads were observed in these studies but not previously—from both Great Lakes and anadromous populations—remains elusive.

An in vivo biopsy technique developed by Lowartz et al. (1999) and Lowartz and Beamish (2000) permitted a non-lethal means of examining lamprey gonadal histology in a single individual over time, allowing these researchers to learn the fate (at least during the larval stage) of atypical gonads. Results indicated that atypical gonads often developed into typical males (albeit at a larger size than is usually associated with sexual differentiation), but a few cases of full sex reversal (i.e., from a typical female to a typical male) were also observed. Gonads from sea lamprey larvae (TL >118 mm) collected from a tributary to Lake Huron in May and June were biopsied, and larvae were then reared for another 1, 2, 4, 8, 16, and 52 weeks, at which point they were sacrificed for histological analysis. Because the biopsied tissues were inconsistent in size and orientation (i.e., full gonadal cross sections could not be taken), the criteria previously developed for typical males and females and atypical gonads (Wicks et al. 1998b) were modified. Tolerance limits for typical males and females in two size categories (121–135 mm and >136 mm TL) were based on the number of oocytes and undifferentiated germ cells per unit area and oocyte diameter.

At the time of biopsy, 17% of the 87 examined larvae possessed atypical gonads that were divided into two basic types (Lowartz and Beamish 2000): 33% were type I atypical gonads showing asynchronous oocyte development (i.e., with both first- and second-stage oocytes averaging 20.4 and 40.8 μm, respectively; Fig. 1.10a), and 67% were type II atypical gonads that, despite the large size of the larvae, resembled an indifferent gonad with both undifferentiated germ cells and predominantly first-stage oocytes, although second-stage oocytes were occasionally present (overall mean diameter 16.4 μm). The exciting aspect of this study was the ability to follow the fate of these atypical gonads over the next 1–52 weeks. Gonadal composition remained relatively stable over the first 4 weeks, but significant changes were observed by week 8. Of the 15 initially atypical gonads, only one remained atypical or indifferent after week 8. One of the type I atypical gonads (from a larva 128 mm TL at biopsy) developed into a presumptive testis by 8 weeks, but the remaining four became ovaries with atretic oocytes by week 52 (Fig. 1.10a). Of the 10 type II atypical gonads, two remained indifferent (at 1 and 52 weeks), one developed into an ovary with atretic oocytes, and seven became presumptive testes after 8 or 52 weeks. This observation is consistent with delayed sex differentiation (through delayed atresia of first-stage oocytes) in these individuals (e.g., Wicks et al. 1998b; see above).

Nevertheless, it appeared that gonadal differentiation was not delayed or atypical in all individuals because, at the time of the biopsy, 63% of individuals had typical ovarian tissue and 20% had gonads resembling typical presumptive testes (Lowartz and Beamish 2000). However, over time, the proportion of typical ovaries declined, and the proportion of presumptive testes increased due to oocyte atresia in 16 previously typical ovaries, development of presumptive testes from atypical gonads in eight individuals (see above), and complete sex reversal in three typical ovaries (Fig. 1.10b). These observations thus provide experimental support for previous suggestions by Hardisty (1971) and Fukayama and Takahashi (1982, 1983). Based on interpretation of the histological appearance of the gonads at one point in time, these authors suggested that presumptive testes can develop through atresia of an ovary's entire oocyte stock even after ovarian differentiation appears to be complete. The biopsy results also provide insights into how quickly such transitions can occur. The only change noted in typical ovaries during the first 4 weeks was very modest atresia (of $\leq 7\%$ of the total oocyte stock) in three individuals. By week 8, four of six typical ovaries showed oocytes undergoing more significant atresia: on average, 41% of oocytes were atretic, and the number of oocytes per unit area had decreased. By week 16, remarkable changes to gross gonadal morphology were observed in three of the five previously typical ovaries: 18% of the oocytes of one female were atretic, and complete sex reversal was seen in the other two females. These sex-reversed individuals showed 100% oocyte atresia and occurrence of undifferentiated germ cells, but the finger-like lobes characteristic of ovaries were retained. By week 52, complete sex reversal from a typical ovary had been demonstrated in a third individual (Fig. 1.10b), and nine other ovaries (of the initial 22) displayed oocyte atresia. Nevertheless, Lowartz and Beamish (2000) did not observe any sex reversals from presumptive testes to ovaries, suggesting that ovarian development is precluded once oocytes fail to develop or are entirely lost to atresia. In the female-to-male sex reversals, germ cells would occasionally appear in gonads which previously exhibited only oocytes. In these cases, it is likely that a few germ cells were initially present but not included in the biopsied tissue, because there is no evidence that oocytes would "revert" to undifferentiated germ cells. These transitions were considered female-to-male sex reversals (as opposed to transition of the ovary to a sterile gonad), because it was assumed that the remaining few undifferentiated germ cells would undergo mitotic proliferation at the onset of metamorphosis (see Sect. 1.4.1.3). By the end of the study, typical ovaries and presumptive testes made up 46 and 23% of all individuals, respectively, with ovaries with atretic oocytes and atypical gonads making up the remaining 29 and 3%, respectively (Lowartz and Beamish 2000). If individuals in the latter two categories became presumptive males, the sex ratio would be 54% male, which is very much in line with the adult sex ratio observed in Lake Huron and the other Great Lakes since the mid-1990s (Fig. 1.2). However, the ultimate fate of the atretic ovaries and atypical gonads is unknown.

The Lowartz and Beamish (2000) study was groundbreaking in demonstrating that sex differentiation in a substantial proportion of sea lamprey larvae is labile for most or all of the larval stage and that primary sex differentiation is not definitive in all lampreys. However, Lowartz and Beamish (2000) expressed some concern that manipulation of the gonad during surgery could have been responsible for the observed oocyte atresia and sex reversal, because mechanical manipulation of the ovary of Siamese fighting fish *Betta splendens* resulted in the generation of testicular tissue (Becker et al. 1975). In the Becker et al. (1975) study, however, the ovary was removed, squashed, and replaced into the abdominal cavity. In comparison, the method employed by Lowartz and Beamish (2000) seemed far less invasive, and development proceeded normally in many of the ovaries (e.g., oocytes continued to increase in diameter at rates seen in wild populations; see Sect. 1.4.1.2). Nevertheless, the concern was addressed in a follow-up study by Beamish and Barker (2002).

This latter study included a sham control (i.e., where the larvae underwent the same operation as the biopsy group but without the removal of gonad tissue). Furthermore, in the biopsy group, only ~3% of the total gonad length was removed and it was gently dissected from the dorsal wall of the coelom. At the time of sacrifice, cross-sections for histological examination were taken from three regions along the length of the ovary to ensure that any observed changes were not confined just to the biopsy area. Another important advance in the study by Beamish and Barker (2002) was the inclusion of metamorphosing sea lamprey (in addition to larvae 92–156 mm TL). At the time of sacrifice 32–49 weeks later, gonad cytology did not differ between the three regions examined and sex ratios did not differ significantly among the three groups (e.g., atypical gonads were found in 3, 11, and 4% of individuals subjected to the biopsy, sham, and control treatments, respectively), suggesting that biopsy and surgery did not affect subsequent development. Nevertheless, significant changes in gonadal morphology and composition were still observed in 27% of the 30 biopsied individuals: two atypical ones became presumptive males; one female experienced extensive oocyte atresia; four females reversed to males; and one male reversed to female. This is the first (and, to date, only) report of male-to-female reversal. In contrast to the larvae, all metamorphosing and juvenile lamprey examined at both the beginning and end of the study were classified as typical males or females, and none underwent sex reversal. Significant increases in cross-sectional area of the testes and ovaries during and after metamorphosis suggested normal testicular and ovarian growth despite surgery. Therefore, lability of the sea lamprey gonad may extend in at least some individuals until the end of the larval stage, but sex differentiation appears to be complete and fixed by the time metamorphosis has begun. Retention of even small numbers of undifferentiated germ cells may permit sex reversal, but sex is no longer labile once undifferentiated germ cells begin spermatogonial proliferation (see Sect. 1.4.1.3).

There are a few reports of intersexuality in post-metamorphic lampreys (e.g., Beard 1893; Okkelberg 1921). In addition, Holčík and Delić (2000) mention two Ukrainian brook lamprey *Eudontomyzon mariae* that appear to be hermaphrodites, but it seems that this conclusion was based on the presence of intermediate secondary sex characteristics rather than internal examination. More recently, Clemens et al. (2012) described the simultaneous presence of both oocytes and spermatogonia or spermatocytes in adult Pacific lamprey. During their 2007 and 2008 sampling seasons, Clemens et al. (2012) classified two of the 427 Pacific lamprey that were sampled during their upstream migration in the Willamette River in Oregon as male intersexes. Their gonads resembled normal testes macroscopically, but histological examination of a biopsy sample showed the presence of a small number of distinct oocytes. One individual collected from Willamette Falls in August possessed only pre-vitellogenic oocytes (~20–30 μm diameter) interspersed throughout the spermatogonia-filled testis (see Sect. 1.5.2). In the second individual collected in September, at least six mid-vitellogenic oocytes (~600 μm diameter) were evident, and they were separate from the testicular tissue which contained both spermatogonia and early stage spermatocytes (Fig. 1.10c). Clemens et al. (2012) concluded that these two males would be unlikely to self-fertilize or spawn viable eggs, but it is unknown if they would be able to produce viable sperm at maturity (~10 months hence). Intersexuality in post-metamorphic lampreys is thought to be rare. However, because detection of the two intersexes by Clemens et al. (2012) required histological examination, it is possible that a "touch of intersexuality" in adult lampreys is more common than currently thought.

1.4.1.5   Sex Identification in Larval Lampreys

As detailed above, the age and size at which larval lampreys can be "sexed" (i.e., when females can be reliably identified) will depend on the species, and, to some extent, on the population or individual (Table 1.8). Because male lampreys remain undifferentiated throughout the larval stage, they are generally identified as presumptive males when they are not yet female at the point when female differentiation should be complete (i.e., males are inferred by default). However, it should be noted that sex reversal has been suggested (Fukayama and Takahashi 1983) or reported (Lowartz and Beamish 2000) in some individuals even after the point at which ovarian differentiation has occurred (see Sect. 1.4.1.4).

In brief, non-parasitic species can generally be sexed histologically following the summer of their first full year of larval life (i.e., in age class I or at ~14–16 months of age) or at ~50–70 mm TL (Table 1.8). After this point, females should be clearly identifiable when distinct oocytes (diameter >40 μm) make up most or all of the germ cells (numbering ~15–35 in cross-section; Fig. 1.7b). Presumptive males are identified by the absence of these features; the future testis remains small and still retains undifferentiated germ cells (Fig. 1.7e).

Parasitic species generally cannot be sexed until they are at least 2 years old, but the age and size at which individual species can be sexed is more variable than in brook lampreys. In Northern Hemisphere species, ovarian differentiation is typically complete at smaller sizes and younger ages in lampreys with smaller adults (e.g., chestnut and European river lampreys; Fig. 1.7c) and at progressively larger sizes and older ages in large-bodied species (Table 1.8). Landlocked sea lamprey can generally be sexed histologically by 90–100 mm TL (Fig. 1.8a), but anadromous sea lamprey usually cannot be sexed until they are 120–130 mm TL. At these sizes, females should be clearly identifiable by their large ovary, consisting of finger-like lobes containing a large number (~25–200+) of large (diameter >40 μm) oocytes. In contrast, males are identified by the absence of these features, even if a few small oocytes persist (Fig. 1.8a). For example, the gonad of presumptive male sea lamprey is much smaller in cross-sectional area than the developing ovary. It generally has a smooth or shallowly cleft, angular shape and is comprised of stromal tissue and undifferentiated germ cells occurring either singly or clustered in cell nests. If oocytes are present, they are generally few (≤6 per section) and small (<20 μm; Docker 1992; Wicks et al. 1998a).

However, given apparent variability even within species, the appropriate "cut-off" point (i.e., the size at which female differentiation is deemed complete and presumably irreversible) should be verified for each population. If individuals are sexed prior to completion of ovarian differentiation, they may be erroneously called male when they are not yet obviously female, or, conversely, they may be prematurely diagnosed as female if the oocytes have yet to undergo atresia. In European brook lamprey, for example, Hardisty (1965a) reported an excess of females (35% male) at 41–60 mm TL, but sex ratios approached parity when individuals were sexed at 61–80 mm (45% male) and 81–100 mm (48% male). Alternatively, individuals with delayed differentiation may appear atypical or intersexual. In fast-growing sea lamprey populations, for example, ovarian differentiation may not be complete until individuals are well past 90–100 mm (Wicks et al. 1998a). In contrast, ovarian differentiation might be complete at 70–80 mm TL in slow-growing individuals (Docker 1992; see Sect. 1.4.1.2).

In addition to using histological examination, large larval lampreys can often be sexed under a dissecting microscope, because, near the end of the larval stage, there is a considerable difference in the size, shape, and texture of the ovary compared to the testis (Fig. 1.8b). Differences in the size and composition of the ovary and testis also allow for live larvae to be sexed using acoustic microscopy. Conventional low-frequency ultrasound (3.5–15 MHz) has long been used to non-lethally determine sex and stage of maturity in adult fishes (e.g., Martin et al. 1983; Colombo et al. 2004), but the high-resolution ultrasound technique developed by Maeva et al. (2004) was sufficiently sensitive to determine sex in live larval lampreys >110 mm TL. By using a focusing lens to concentrate high-frequency ultrasound (15–100 MHz), female sea lamprey larvae could be identified in ~30 s per animal by the presence of a relatively large (1–1.5 mm diameter) ovary which was considerably less reflective to the acoustic signals than the surrounding kidney tissue (Fig. 1.8c). Males could sometimes be recognized by the appearance of a small (0.2–0.3 mm) testis with slightly stronger reflective properties than the kidney, and they could always be identified by the absence of an ovary. The only other non-lethal method currently known for identifying sex in larval lampreys is the gonadal biopsy method developed by Lowartz and Beamish (2000). Non-lethal sexing techniques are important for studies that need to monitor the gonad over time (e.g., for evidence of sex reversal; Lowartz and Beamish 2000; see Sect. 1.4.1.4) or that require live larvae of known sex for subsequent studies (e.g., to examine sex-specific differences in mortality or sex-specific differences in endocrine profiles or gene expression patterns; see Sect. 1.4.2.2).

1.4.1.6   Effect of Hormone Treatments on Sex Differentiation

Hormonal sex control is the manipulation of an individual's gonadal sex by the administration of hormones (e.g., androgens or estrogens) before or during sex differentiation. In this manner, sex differentiation has been partially or completely redirected (i.e., where the inherent sex differentiation process is overridden so that the gonads develop as testes or ovaries regardless of genetic sex) in a number of teleost fishes (e.g., Donaldson and Hunter 1982; Yamazaki 1983; Piferrer 2001). Both 100% males and 100% females have been produced, and several studies have shown hormonal sex reversal to be both permanent and functional (Hunter et al. 1982). The relative ease with which gonadal steroids control sex in previously undifferentiated embryos led Yamamoto (1969) to conclude that androgens and estrogens were the respective male and female sex inducers in fishes. However, there is debate whether steroidogenesis precedes (e.g., Feist et al. 1990) or follows (e.g., van den Hurk et al. 1982; Rothbard et al. 1987) gonadal differentiation, and whether the high doses sometimes used are within the physiological capabilities of the animal. Nevertheless, the ease with which hormonal sex control can be achieved in fishes is thought to indicate labile sex determination (i.e., that sex differentiation can be influenced by environmental factors even in species with a genetic component to sex determination; see Sect. 1.3.1).

Despite the apparent lability of the larval lamprey gonad (see Sect. 1.4.1.4), hormonal sex control in lampreys has been unsuccessful to date. Knowles (1939) found that the gonads of larval European river lamprey were not noticeably affected by injections of the androgen testosterone (T) propionate or the estrogen estrone (see Sect. 1.7.2). Likewise, sex reversal was not achieved in previously undifferentiated European brook lamprey larvae immersed in T propionate or estradiol (E_2) benzoate for 6 months (Hardisty and Taylor 1965). However, in the latter study, more larvae immersed in T propionate contained cysts of undifferentiated germ cells relative to the controls, and fewer larvae possessed oocytes. Nevertheless, because T propionate impaired growth, treated larvae were also smaller than the control larvae, and it is possible that presumptive females simply had not yet completed ovarian differentiation (Hardisty and Taylor 1965). Immersion in E_2 benzoate caused an apparent degeneration of oocytes, rather than the expected feminization. In this case, however, the treated larvae were larger than the controls. It is possible that degeneration of oocytes occurred as part of the normal progression toward testicular differentiation in future males (Hardisty and Taylor 1965; see Sect. 1.4.1.3) or represented a paradoxical or pharmacological effect (see below).

Similar results have been observed in sea lamprey: gonadal steroids were shown to be generally ineffective in altering larval sex ratios, but they often produced gonadal abnormalities (Docker 1992). However, the precise results differed depending on the initial size of the larvae. Larvae were divided into three size classes that reflected their presumed stage of gonadal development at the onset of treatment: 1) undifferentiated (i.e., initial TL <60 mm); 2) in the process of ovarian differentiation (60–89 mm TL); and 3) following completion of ovarian differentiation ($\geq$90 mm TL). Larvae were immersed twice weekly in T, E_2, and 17α-methyltestosterone (MT) at concentrations of 0.01, 0.1, or 1.0 mg/L for 21 weeks, and they were then maintained without further treatment for another 25 months until most were large enough for identification of sex. Gonads of the initially undifferentiated larvae (<60 mm TL) were the least affected, which is counter to the assumption that they would be the most susceptible to hormonal influence. Sex ratios were not significantly different from the controls, and few histological differences were noted. Intersex gonads (see Sect. 1.4.1.4) were observed in 13 and 27% of the larvae treated with the lowest doses of MT and T, respectively, and in 11 and 17% of the larvae treated with the two higher doses of E_2, but as many as 12% of the control larvae were also intersexual.

However, growth was significantly impaired in the T-treated larvae, and virtually all small larvae died at the medium and high MT doses. Hormone treatments also failed to alter sex ratios in larvae that were in the process of ovarian differentiation (60–89 mm TL), but these treatments often resulted in histological abnormalities that were suggestive of incomplete sex reversal. For example, E_2 treatment appeared to cause slight "feminization" of males; relative to the controls, presumptive testes showed increased cross-sectional area and finger-like lobes (i.e., showing superficial morphological resemblance to ovaries), but with an increase in the amount of stromal tissue rather than in the number or size of oocytes (i.e., without corresponding cytological changes). Following treatment with T, females had more and larger well-defined cysts than control females, and intersexes were observed with larger gonads and larger oocytes than comparable control larvae. Docker (1992) suggested that these individuals might represent incompletely masculinized females. A significant effect on sex ratio was observed only in larvae that were ≥ 90 mm TL at the initiation of treatment, but the gonads were often abnormal in appearance. The medium dose of E_2 and the lower two doses of MT produced more females than were evident in the control tanks, but individuals mostly showed evidence of inhibition of germ cell growth rather than masculinization or feminization per se, and the survival rate of MT-treated larvae was low. In females treated with T after completion of ovarian differentiation, there was a decrease in the size and abundance of the remaining cysts of undifferentiated germ cells and an increase in oocyte diameter and ovarian cross-sectional area. Paradoxical feminization following treatment with androgens has been reported in other fishes (e.g., Hackmann and Reinboth 1974; Goudie et al. 1983; Davis et al. 1990) and may be the result of aromatization of T and MT to compounds with estrogenic properties (Davis et al. 1990; see Sect. 1.7.1). Most notable was the drastic reduction in oocyte number and size in large females treated with E_2, often producing near-sterile gonads (Fig. 1.10d). Oocyte inhibition in already-differentiated females suggests a pharmacological effect caused by direct toxic action on the gonad (Tsuneki 1976) or by inhibition of pituitary gonadotropin secretion (Gorbman 1983), although it should be noted that the inhibitory effects were least pronounced at the highest E_2 dose (Docker 1992).

The lack of success to date in producing normal sex-reversed lampreys does not necessarily mean that hormonal sex control is not possible in lampreys. Successful hormonal sex control in different teleost fish species is the result of considerable experimentation to refine treatment protocols (e.g., Donaldson and Hunter 1982; Hunter et al. 1982; Yamazaki 1983; Piferrer 2001). Developing the right protocols for lampreys is complicated by our current lack of understanding of the extent to which lamprey gonads can be "atypical" even without treatment with exogenous hormones (see Sect. 1.4.1.3), uncertainties regarding the physiologically relevant sex steroids in lampreys (see Sect. 1.7.2), the extraordinarily long period during which the gonad remains indifferent, and a clear understanding of the "window of lability." So far, the effect of hormone treatments has been evaluated only by comparing the sex ratio and gonadal histology of treated and control lamprey larvae. This allows us to only infer the changes that were produced in the treated individuals. Therefore, one improvement would be to use the gonadal biopsy method developed by Lowartz and

Beamish (2000) so that "before and after" comparisons of each individual could be made, and individuals could potentially be followed over time. This method would allow researchers to better determine if atypical gonads (e.g., with the cytological features of one sex and the morphological features of the other sex) had been typical prior to treatment, suggesting that treatment caused partial sex reversal, and to detect complete sex reversal in a small number of individuals. Identification of the "true" sex steroids is not necessary for successful sex control, and synthetic hormones are often effective (e.g., Piferrer 2001). Nevertheless, some steroids are not as effective as others, and some may produce toxicological rather than physiological effects (Hunter et al. 1983; Piferrer 2001). Considerable trial and error is also required to determine appropriate doses (e.g., Yamazaki 1983; Piferrer 2001).

With respect to determining the window of lability, fishes are generally suscepti- ble to exogenous hormones prior to phenotypic sex differentiation (Yamazaki 1983). In most teleost fishes, this generally occurs within a few weeks to months of hatch- ing, and testicular and ovarian differentiation occur at the same time or very close together (Patiño and Takashima 1995; Wang et al. 2007; Sandra and Norma 2010). In lampreys, however, the differentiation process is delayed, prolonged, and asyn- chronous in males and females (see Sects. 1.4.1.2 and 1.4.1.3), and we do not know if the lamprey gonad is open to exogenous influence as long as undifferentiated germ cells persist (see Sect. 1.4.1.4), or whether sex-specific differences not yet visible by light microscopy are established even prior to initiation of ovarian differentiation. An apparent lack of histological differentiation does not necessarily indicate that the ger- minal and somatic elements are not differentiated at a molecular level (Hardisty et al. 1992). In the Docker (1992) study where observed histological changes suggested incomplete sex reversal, hormone treatments may have been initiated too late or ter- minated too soon. In the larvae that were presumed to be initially undifferentiated, average TL was still only 66 mm by the time hormone treatments ceased; ovarian dif- ferentiation would not have been complete yet. Hormone treatment during only the early stages of sex differentiation might have resulted in transitory changes that were completely or partially reversed by the time of histological examination. Although successful sex control has been achieved in coho salmon by a single 2-h treatment (Piferrer and Donaldson 1989), the timing is critical, and treatments of insufficient duration either have little or no effect on sex differentiation (e.g., Hackmann and Reinboth 1974; Takahashi 1975) or produce intersexual or sterile fish (e.g., Boney et al. 1984; Komen et al. 1989). However, longer treatments are not necessarily more effective, because they can also result in intersexuality and sterility, impaired growth, or high mortality (e.g., Hunter et al. 1983; Sower et al. 1984).

In lampreys, hormonal treatment throughout the entire undifferentiated stage does not guarantee success. In an unpublished study by L. H. Hanson at the Hammond Bay Biological Station in Michigan (cited in Docker 1992), several hundred sea lamprey larvae were immersed twice-weekly in estrone, E_2, diethylstilbestrol, progesterone, or methyltestosterone for 3–5 years following hatch. Gonadal differentiation was complete prior to cessation of treatment, and the lampreys were sexed as large larvae (≥ 120 mm TL) or during metamorphosis. Mortality was high throughout the study, and all the treated larvae exhibited very thin gonads classified either as aberrant testes or sterile gonads. However, 80% of the surviving control larvae were also male or sterile, and the cause of such abnormalities in untreated larvae is unknown (see Sect. 1.4.1.4).

Hormonal sex control in other fishes has had a profound impact on aquaculture. For example, producing monosex stocks (e.g., favoring the sex that shows the greatest growth) can have significant economic advantages (Solar et al. 1991; Piferrer 2001). It can also have important applications for the control of invasive fish species (e.g., Gutierrez and Teem 2006; Thresher et al. 2014). For example, sex-ratio distortion systems that induce an extreme male bias can be particularly effective for the control of pest species (Senior et al. 2015). Extreme male bias can be achieved using "Trojan sex chromosomes" in species with predominantly genetic, but hormonally reversible, sex determination (Gutierrez and Teem 2006; Thresher et al. 2014). In fish species with an XY sex-determining system, viable females carrying two Y chromosomes can be created over two generations using estrogen treatments during early development, and then they can be released into the wild population. Mating of these YY females with normal XY males produces only males (XY and YY), and the male bias increases in subsequent generations. Therefore, fewer individuals need to be released compared to the sterile-male-release technique (see Chap. 6), because the effects of this method extend beyond the life of the released individuals (Cotton and Wedekind 2007; Schill et al. 2016). A similar sex-ratio distortion approach could be an effective and highly species-specific alternative to lampricides (Thresher et al. 2019), but considerably more work would be required to develop such a system in lampreys (see Sect. 1.3.2.1). Nevertheless, the Trojan Y approach could represent a "friendlier" alternative to sex-ratio distortion gene drives, because it does not require the release of genetically modified organisms into the environment (Senior et al. 2015). The consequences of YY female additions are non-permanent (as long as XX females still exist), so undesirable effects can be reversed by cessation of YY input (Cotton and Wedekind 2007).

1.4.2   Genes Involved in Sex Differentiation

Although sex-determining genes are highly variable in reptiles and non-amniotes (see Sect. 1.3.1), many of the genes involved in the sex differentiation process tend to be conserved among vertebrates (Piferrer and Guiguen 2008; Sandra and Norma 2010; Siegfried 2010; Piferrer et al. 2012; Cutting et al. 2013; Forconi et al. 2013). Many studies have examined whether genes known to be involved in mammalian sex differentiation are expressed during gonadal differentiation in model and commercially valuable fish species (see Piferrer et al. 2012). Although such a candidate gene approach has been less useful for identifying the sex-determining genes in fishes, often doing little more than ruling out "the usual suspects" (see Sects. 1.3.1.1 and 1.3.2.1), this approach has generally worked well to identify at least some of the genes involved in sex differentiation in different taxa.

Genes involved in gonadal differentiation tend to be present in both sexes during the early stages of development, but the expression becomes sex-biased during the critical period of gonadal differentiation. Bimodal expression patterns in the developing gonads can therefore be indicators of a gene's role in sex differentiation. Even better, in species where the genetic basis of sex determination has been identified or where monosex populations can be produced, gene expression can be studied in individuals of known sex even before the earliest signs of histological differentiation (e.g., Baron et al. 2005; Tong et al. 2010; Tao et al. 2013). Once sex-specific gene expression patterns are identified—with or without a known genetic basis of sex determination—they can be used as early molecular markers for identification of future sex (e.g., Geffroy et al. 2016; Ribas et al. 2016). Study of the genes involved in sex differentiation in lampreys is still in its infancy. It is complicated by their evolutionary divergence from other vertebrates (e.g., making it more difficult to recognize homologs of the genes of interest), their anatomical differences (e.g., the lack of Müllerian ducts and Sertoli cells; see Sects. 1.5.1 and 1.5.2), and uncertainty regarding when—during the long period during which their gonads remain histologically undifferentiated—the critical gonadal differentiation period is (see Sects. 1.4.1.5 and 1.4.1.6). Nevertheless, initial studies implicate at least some of the same genes as other vertebrates in ovarian and testicular development in lampreys.

1.4.2.1 Genes Involved in Sex Differentiation in Other Vertebrates

We briefly review a few of the key sex differentiation genes that have been well studied in other vertebrates, because they provide the list of candidates for study in lampreys. Genes involved in the sex differentiation process include those which encode steroidogenic enzymes, hormone receptors and their ligands, and transcription factors (or sequence-specific DNA-binding factors) that control the rate of transcription of key genes to ensure that they are expressed in the right amount at the right time.

One of the key steroidogenic enzyme genes involved in gonadal differentiation appears to be aromatase *CYP19a1* (see Table 1.6 for guidelines regarding the formatting of gene names). CYP19A1 is the enzyme responsible for the conversion of androgens to estrogens (see Sect. 1.7.1), and it appears to be essential for ovarian differentiation in virtually all vertebrate species examined (Piferrer and Guiguen 2008). In rainbow trout, for example, although sex is not histologically identifiable until ~67 days post-fertilization (dpf) at 10 °C, *CYP19a1a* expression was 10× higher in developing ovaries relative to developing testes by 35 dpf, and expression levels were 60–100× higher at 45 dpf (Vizziano et al. 2007). Early expression of *CYP19a1* before ovarian morphological differentiation has also been demonstrated in the Nile tilapia *Oreochromis niloticus* (Nakamura et al. 1998; D'Cotta et al. 2001; Tao et al. 2013) and turbot *Scophthalmus maximus* (Ribas et al. 2016), although some studies have paradoxically shown higher *CYP19a1a* expression during testicular differentiation (e.g., in Siberian sturgeon *Acipenser baerii*; Berbejillo et al. 2012). Other steroidogenic enzyme genes showing sex differences in expression during gonadal differentiation in fishes are 3β-hydroxysteroid dehydrogenase (*HSD3b1*), which was found to be overexpressed in female rainbow trout at ~40 dpf; and 11-hydroxylase (*CYP11b2.1*), which showed up to 60× higher expression in males compared to females at 45 dpf (Vizziano et al. 2007). Later expression of *CYP11b2.1* is consistent with the observation in many fishes that testicular differentiation is delayed relative to ovarian differentiation (see Sect. 1.4.1). In Nile tilapia, Tao et al. (2013) measured gene expression in XX (female) and XY (male) gonads at 5, 30, 90, and 180 days after hatching (dah). They found several steroidogenic enzyme genes, including *CYP19a1a*, to be upregulated in XX gonads at 5 dah, the critical time for sex determination and differentiation. In contrast, in XY gonads, the steroidogenic enzyme genes (including *CYP11b2*, which encodes the aldosterone synthase enzyme) were not significantly upregulated until 90 dah. These results suggest that, at the time critical to sex determination, the XX tilapia produced estrogen, but the XY fish did not produce androgens. Consistent with this finding, genes encoding both estrogen and androgen receptors were expressed in XX gonads at 5 dah, but only estrogen receptors were expressed in XY gonads. Expression of steroidogenic enzyme genes was most pronounced at 30 and 90 dah for XX and XY gonads, respectively, which corresponded to the initiation of meiosis and oogenesis in females and meiosis or spermatogenesis in males (Tao et al. 2013). In some species, male development involves inhibition of aromatase production (Devlin and Nagahama 2002). In the European sea bass, males have twice the amount of methylation in the aromatase promoter region as females (i.e., repressing gene expression) to decrease the production of estrogen and promote testis rather than ovary development (Navarro-Martin et al. 2011). In tilapia, significant reduction of estrogens as a result of a decrease in aromatase can lead to oocyte atresia and eventual sex reversal (Li et al. 2013).

Two well-studied genes encoding hormones and their receptors are the Anti-Müllerian hormone and the Anti-Müllerian hormone receptor 2 genes (*amh* and *Amhr2*, respectively). *Amh* exerts its male-specific action by causing the regression of the Müllerian ducts that would otherwise develop into the female reproductive organs and tract (Josso et al. 2001). Teleost fishes (but not cartilaginous and other bony fishes) lack Müllerian ducts (Adolfi et al. 2019), but, interestingly, in Nile tilapia, *amh* expression was localized to the testes and it was detected sooner than other male-specific genes (Ijiri et al. 2008). *Amhr2* plays a role in sex determination in the tiger pufferfish (see Sect. 1.3.1.1). It also appears to play a role in sex differentiation; in XY (male) tilapia, mutations within *Amhr2* can lead to drastic sex reversals (Morinaga et al. 2007).

Transcription factor genes known to be important in sex differentiation include *FOXL2* (forkhead box L2), which is involved in ovarian development, and *DMRT1, SOX9*, and *SF1* (doublesex and mab-3 related transcription factor 1, sex-determining region Y-related high mobility group containing box 9, and steroidogenic factor 1, respectively), which are involved in testicular development (Bulun et al. 2003; Wilhelm et al. 2007; Sandra and Norma 2010). *FOXL2* is the activator of aromatase, and it is an antagonist of *DMRT1* in mice and various fish species (Nakamoto et al. 2006; Ijiri et al. 2008; Barrionuevo et al. 2016). In rainbow trout, *FOXL2a* and *CYP19a1* show the same temporal expression patterns in presumptive females (Baron et al. 2004; Vizziano et al. 2007), and, in the medaka, *FOXL2* expression is localized in all the somatic cells also expressing *CYP19a1* (Nakamoto et al. 2006). Deficiencies in *FOXL2* have been associated with a decrease in aromatase activity and ovary-to-testis sex reversal (Li et al. 2013; Barrionuevo et al 2016). *DMRT1*, a transcription factor belonging to the *DMRT* family of genes, is largely known as the master sex-determining gene in birds (Smith et al. 1999; Kikuchi and Hamaguchi 2013; see Sect. 1.3.1). During testicular differentiation, it also plays the important role of inhibiting genes essential for female gonadal development (Graves 2013). In several fish species, *DMRT* is upregulated during testicular differentiation, and its expression is localized to somatic cells surrounding the testis (Matsuda et al. 2002; Kikuchi and Hamaguchi 2013; Adolfi et al. 2015). Even in turtle species that show TSD, *DMRT1* is detected in the developing genital ridge at low (i.e., male-producing) temperatures (Kettlewell et al. 2000; Woolgar et al. 2013; Mork et al. 2014). In mammals, *SOX9* is one of the earliest genes to be upregulated in pre-Sertoli cells following the expression of *SRY*, and *SOX9* is known to then activate downstream genes such as *amh* (Bowles and Koopman 2001; Brennan and Capel 2004). In birds and some fish species, *SOX9* is largely associated with initial development in both sexes but becomes exclusive to males during testis development (Takada et al. 2005; Vizziano et al. 2007). In mammals, birds, and fishes, *SOX9* also appears to be required for subsequent testicular maintenance and spermatogenesis (Morais da Silva et al. 1996; Barrionuevo et al. 2016). Mutations in *SOX9* can lead to ovary-to-testis sex reversals (Wagner et al. 1994; Vidal et al. 2001; Takada et al. 2005). *SF1* is a transcription factor found in Leydig and Sertoli cells that is required for the activation and upregulation of *amh* in the developing male by promoting the regression of the Müllerian ducts (Josso et al. 2001; Kato et al. 2012). Its role in fishes is not well known. However, in mammals, *SF1* works synergistically with *SRY* and *amh* to activate *SOX9*, and it is essential for spermatogenesis (Schepers et al. 2003; Takada et al. 2005; Sekido and Lovell-Badge 2008).

Sex differentiation involves multiple genes, acting in concert or in sequence, and this is becoming particularly evident with studies that use a transcriptomics approach to sequence and quantify the complete set of genes that are expressed during gonadal differentiation (see Sandra and Norma 2010; Siegfried 2010; Piferrer et al. 2012; Cutting et al. 2013; Ribas et al. 2016). Using this approach to identify all the genes that are differentially expressed in the gonads of male and female turbot prior to, during, and after histological differentiation, Ribas et al. (2016) were able to measure the simultaneous expression patterns of 18 candidate genes implicated in sex differentiation in other vertebrates, and they also identified 56 other genes that had not been previously related to sex differentiation in fish but that were found to have sex-specific expression patterns at 3 months of age (i.e., ~1.5 months prior to histological identification). Of these 56 genes, 44 were associated with ovarian differentiation and 12 were associated with testicular differentiation. Despite this complexity, Ribas et al. (2016) found that expression levels of *CYP19a1a* alone at 3 months of age allowed early accurate identification of sex.

CYP19a1 expression is likewise an effective early molecular marker for ovarian differentiation in the Nile tilapia (Nakamura et al. 1998; D'Cotta et al. 2001; Tao et al. 2013), and upregulation of *DMY* (*DM*-domain gene on the Y chromosome) has been found to be an early indicator of testicular differentiation in medaka (Kobayashi et al. 2004; see Sect. 1.3.1). In some species, a small set of genes can be used together to predict whether an individual's gonads are in the early stages of ovarian or testicular differentiation. In the European eel, four genes—*DMRT1*, *amh*, *Gsdf* (gonadal soma derived factor), and *pre-miR202* (pre-microRNA 202)—showed a testis-specific expression pattern, and three genes—*zar1* (zygotic arrest 1), *zp3* (zona pellucida 3), and *foxn5* (forkhead box N5)—were specific to ovarian differentiation. Interestingly, gene expression in the gonad of intersexual eels was similar to that of males, supporting previous suggestions that the intersexual gonad represents a transitional stage in the indirect development of males (Geffroy et al. 2016; see Sect. 1.4.1).

1.4.2.2   Genes Involved in Sex Differentiation in Lampreys

Little is known regarding genes involved in sex differentiation in lampreys, although a few recent studies have used a candidate gene approach to test whether genes implicated in gonadal differentiation in other vertebrates show sex-specific patterns of expression in lampreys as well (Spice et al. 2014; Khan 2017; Mawaribuchi et al. 2017). Efforts are also being made to use a transcriptomics approach to examine the expression of these and other genes during sex differentiation and sexual maturation in lampreys (Ajmani 2017). Identification of early molecular markers for ovarian or testicular differentiation in lampreys would prove very useful. In other fishes, *CYP19a1* expression is one of the most common early molecular markers for ovarian differentiation (see Sect. 1.4.2.1). CYP19 activity has been demonstrated in lampreys (Callard et al. 1980; see Sect. 1.7.1), but expression of *CYP19* has not yet been studied in lampreys.

Spice et al. (2014) examined the expression of eight other candidate genes during ovarian differentiation in the chestnut and northern brook lampreys: 17β-hydroxysteroid dehydrogenase (*HSD17b*); dehydrocholesterol reductase 7 (*dhcr7*); estrogen receptor β (*erβ*); Wilm's tumor suppressor protein 1 (*WT1*); germ cell-less (*gcl*); deleted in azoospermia associated protein 1 (*dazap1*); insulin-like growth factor 1 receptor (*igfr1*); and cytochrome c oxidase subunit III (*coIII*). The target genes were identified and primers for quantitative reverse-transcriptase PCR (qRT-PCR) were designed using sequence data from the sea lamprey genome (Smith et al. 2013) or transcriptome data from chestnut and northern brook lamprey ovaries (Spice 2013). These eight genes were chosen because they were known to be involved in sex differentiation and related processes in other vertebrates (Hsu et al. 2008; Labrie et al. 1997; Maekawa et al. 2004; Li et al. 2006; Hale et al. 2011; see Sect. 1.4.2.1) or because they were found to be differentially expressed during ovarian development in a small sample of chestnut and northern brook lampreys using transcriptome sequencing (*dhcr7*, *coIII*; Spice 2013). Primers for other target genes were designed from the sea lamprey genome (e.g., *DMRT1*, *DMRTa2*, *SF1*, and gonadotropin releasing hormone receptor 1, *GnRH1*) or northern brook lamprey transcriptome (e.g., *FOXL2*, *GnRH2*, progestin receptor 1, *SOX9*, and *HSD3b*; Spice 2013), but these genes amplified poorly. Spice et al. (2014) tested for differential expression of the eight candidate genes among four stages of histological gonadal development: (1) undifferentiated or presumptive male stage with few undifferentiated germ cells; (2) cystic stage during or following initial mitotic proliferation; (3) first stage of oocyte growth; and (4) differentiated females in the second stage of oocyte growth (see Sect. 1.4.1). They also tested for differences in gene expression between the two species, because the chestnut lamprey has delayed ovarian differentiation and higher potential fecundity relative to the northern brook lamprey, and for differences related to intraspecific variation in fecundity (i.e., number of oocytes per cross-section).

Spice et al. (2014) found that *HSD17β* expression was higher in differentiated ovaries in the second stage of oocyte growth than in undifferentiated gonads, and expression of this gene was directly correlated with fecundity. Along with CYP19, 17β-hydroxysteroid dehydrogenase helps regulate the levels of active androgens and estrogens (Labrie et al. 1997; see Sect. 1.7.1). *Igf1r* expression was almost 100× higher in chestnut lamprey relative to northern brook lamprey during the first phase of oocyte growth, and expression level was directly related to number of oocytes per section (Spice et al. 2014). Insulin-like growth factor 1 is associated with increased growth and fecundity (but reduced lifespan), and it also stimulates the production of sex steroids (Dantzer and Swanson 2012). Therefore, increased expression of *igf1r* in chestnut lamprey may be related to their greater size and fecundity as adults. *CoIII* expression was 54–70× higher in northern brook lamprey compared to chestnut lamprey during all stages of development, inversely related to oocyte number within species, and highest during the cystic stage of gonadal development in both species (Spice et al. 2014). Therefore, because there is evidence to suggest that this gene plays a role in regulating apoptosis in other vertebrates (Wu et al. 2009), it is tempting to speculate that *coIII* upregulation is correlated with germ cell degeneration during ovarian differentiation and reduced fecundity in non-parasitic lampreys. However, far more research is required.

With respect to genes implicated in testicular differentiation, Mawaribuchi et al. (2017) examined *DMRT1* expression patterns in larval and post-metamorphic Far Eastern brook lamprey, and they found that *DMRT1* expression was significantly greater in post-metamorphic testes than in ovaries. Further investigation using in situ hybridization with *DMRT1* showed a significant level of detection in spermatogonial cysts of post-metamorphic males, but no detection in females (Mawaribuchi et al. 2017). Khan (2017) compared expression of seven candidate genes (*DMRTA2, SF1, SOX8, SOX9, WT1, dazap1,* and *gcl*) in ovaries and presumptive testes from larval, metamorphosing, and adult sea lamprey (i.e., between males and females and among stages of gonadal development in males), and she found that upregulation of *DMRTA2, SOX9, WT1,* and *dazap1* corresponded with an increase in germ cells in the testes during spermatogenesis and spermiogenesis. The increase in *SOX9* expression in males preceded the increase in *DMRTA2* expression. *DMRTA2* and *SOX9* expression was consistent with expression patterns in many other vertebrate species (see Sect. 1.4.2.1), and *WT1* is likewise upregulated in male rainbow trout embryos shortly prior to sex differentiation (Hale et al. 2011). Similarly, upregulation of *dazap1* expression is consistent with observations that the deleted in azoospermia

(DAZ) family of genes is involved in male fertility in humans and that these genes are required for germ cell formation, differentiation, and maturation in other species (Yen 2004). However, their role in fish development is uncertain, and they tend to be expressed in the gonads of both sexes in other fish species (Xu et al. 2007; Peng et al. 2009; Li et al. 2011). In sea lamprey, *SF1* was expressed at all stages in both males and females (Khan 2017), although it is not certain what role it plays because lampreys have neither Sertoli cells nor Müllerian ducts (see Sect. 1.4.2.1). Homologs of *amh* and *amhr2* have not been found in the sea lamprey and Arctic lamprey genomes (Khan 2017; Adolfi et al. 2019). In teleost fishes, *amh* and *amhr2* appear to have been retained as genes associated with male differentiation, but teleosts have secondarily lost their Müllerian ducts; in contrast, agnathans never had Müllerian ducts (Adolfi et al. 2019). Therefore, with some exceptions, many of the genes involved in testicular differentiation appear to be conserved across vertebrates. Nevertheless, until wider transcriptomic analysis is conducted, the involvement of additional lamprey-specific genes cannot be ruled out.

1.5 Sexual Maturation

Reproduction in lampreys is a seasonal and highly synchronized process (see Johnson et al. 2015b). Because all lampreys are semelparous, sexual maturation represents the culmination of their life cycle, and resources are put into maximizing reproductive effort without regard for future survival. At maturity, the single elongate gonad constitutes ~25–35% of a female's total body weight and ~2–10% of a male's body weight. Sexual maturation in parasitic lampreys is generally initiated near the end of the juvenile feeding phase, and it is completed during the non-trophic spawning migration (Fig. 1.6). Because parasitic lampreys differ in the duration of the spawning migration (ranging from a few months to >1 year; Moser et al. 2015), the stage of maturity observed at the start of migration and the rate of maturation during migration vary among and within species. In non-parasitic lampreys, these same processes are greatly accelerated: sexual maturation is initiated immediately after metamorphosis and occurs over a period of ~3–4 months (Docker 2009). In non-parasitic lampreys, the non-trophic period of metamorphosis coalesces with the non-trophic period of sexual maturation, so that both processes are entirely "financed" using energy reserves accumulated during the larval stage (Hardisty 2006; Docker 2009).

In this section, we present an overview of the maturational changes observed in the gonads of lampreys during sexual maturation and a brief discussion of the most conspicuous extra-gonadal change observed during this process, that is, the body shrinkage required to fuel maturation and migration. Histological details of gonadal maturation are reviewed elsewhere (e.g., Lewis and McMillan 1965; Afzelius et al. 1968; Nicander et al. 1968; Larsen 1970; Hardisty 1971; Hughes and Potter 1969; Dziewulska and Dogmała 2009). The spawning migration and reproductive behavior of lampreys are reviewed by Moser et al. (2015) and Johnson et al. (2015b), respectively; the role of the hypothalamic-pituitary axis and gonadal steroids in lamprey reproduction are reviewed by Sower (2015) and in Sect. 1.7, respectively.

1.5.1 Vitellogenesis and Oocyte Maturation

In all lamprey species, oocytes that arrested in meiotic prophase I during ovarian differentiation undergo slow cytoplasmic growth for the duration of the larval stage (Hardisty 1971; Fig. 1.6). At metamorphosis, there are no undifferentiated germ cell nests remaining in the ovary, and all oocytes are deeply basophilic and show the same degree of development (Lewis and McMillan 1965; see Sect. 1.4.1.2). However, among species and life history types, oocyte size at metamorphosis varies. In non-parasitic species, given the long period between their earlier oogenesis and their delayed metamorphosis relative to most parasitic species, oocyte diameter measures up to 150 μm at the onset of metamorphosis (e.g., Beamish and Thomas 1983; Fukayama and Takahashi 1983; Fig. 1.9a). At the other extreme, in the pouched lamprey, oocytes undergo only limited cytoplasmic growth between the relatively late onset of oogenesis and early metamorphosis. Pouched lamprey oocytes average 32 μm in the largest larvae and 43 μm in downstream migrants (Potter et al. 1983; Fig. 1.9c). In the landlocked sea lamprey and most other moderately sized parasitic species, oocyte diameter measures ~80–100 μm at metamorphosis (e.g., Hardisty 1961c, 1969; Lewis and McMillan 1965; Fukayama and Takahashi 1982; Fig. 1.9b).

Cytoplasmic (pre-vitellogenic) growth of the primary oocyte continues after metamorphosis, gradually in parasitic species but more rapidly in non-parasitic species. At metamorphosis, parasitic and non-parasitic lampreys experience a dramatic "parting of the ways" in terms of the phasing of oocyte growth and development (Hardisty 1971, 2006). In the southern brook lamprey, for example, oocytes measure almost 300 μm in diameter by stage 3 of metamorphosis (i.e., ~1 month after its onset; Beamish and Thomas 1983; Fig. 1.9a), but sea lamprey (290–400 mm TL) captured in Lake Huron between mid-May and mid-January (i.e., ~1.5 years after metamorphosis) still had oocytes measuring only 150–250 μm in diameter. Cytoplasmic growth results from accumulation of substrates secreted by the follicular cells and by the incorporation of nurse cells into the oocytes (Lewis and McMillan 1965). As growth of the oocyte continues (whether rapidly or slowly), basophilic granules continue to fill the amorphous cytoplasm, but, because these granules are now widely dispersed, the overall degree to which the cytoplasm appears basophilic decreases (Lewis and McMillan 1965). Some of the follicular cells also acquire basophilic granules in their cytoplasm, and they increase in size until the cell membrane between the oocyte and the nurse cell appears to break down and the cells merge (Lewis and McMillan 1965).

The vitellogenic stage of primary oocyte growth is also accelerated in non-parasitic species, both in terms of its onset relative to metamorphosis and the rate at which it proceeds (Hardisty 1971). Vitellogenins (precursors of the major egg yolk proteins) are synthesized in the liver, delivered via the bloodstream to the growing oocyte, taken up into the oocyte by receptor-mediated endocytosis, and processed into derivative yolk proteins which are then stored as granules, globules, or platelets (Wiegand 1982; Reading et al. 2017). Hepatic synthesis of vitellogenins is induced in females by estrogen (Mewes et al. 2002; Reading et al. 2017; see Sect. 1.5.5.2). In Far Eastern brook lamprey females, vitellogenesis was evident by stage 4 of metamorphosis: extensive structural alterations of hepatocytes indicative of vitellogenin synthesis were observed, and a female-specific protein (presumably vitellogenin) was detected in the blood serum (Fukayama and Takahashi 1985; Fukayama et al. 1986). This evidence of vitellogenin synthesis and transport preceded the first dramatic increase in oocyte diameter which was observed at the end of metamorphosis (~220–620 μm by stage 7; Fukayama and Takahashi 1983). In the southern brook lamprey, oocyte diameter increased dramatically between stage 3 (~295 μm, in late September) and stage 6 (~710 μm, in early December; Beamish and Thomas 1983), and intense vitellogenesis and oocyte growth in both species has been shown to continue until shortly before spawning (Fig. 1.9a).

In most parasitic species, vitellogenesis is initiated near the end of the parasitic feeding phase and continues during upstream migration, proceeding much more slowly than in non-parasitic lampreys (Hardisty 1971). In landlocked sea lamprey, for example, Lewis and McMillan (1965) reported that yolk granules begin to appear in the peripheral regions of the oocyte in feeding-phase individuals measuring ~200–400 mm TL. At this point, oocyte diameter is 390–480 μm (Fig. 1.9b). As the eosinophilic yolk granules increase in number, the basophilic granules observed during pre-vitellogenic growth begin to withdraw toward the nucleus and eventually dissolve, after which a thin eosinophilic, hyaline vitelline membrane appears around the periphery of the oocyte (Lewis and McMillan 1965). Radial striations corresponding to the zona radiata may be seen just inside the thickened vitelline membrane, and immediately inside it is a thin, non-granular layer adjacent to the yolk. These two layers constitute the cortical zone, although the non-granular layer will disappear as the yolk granules grow larger and the oocytes increase in size (Lewis and McMillan 1965). In the Arctic lamprey, Fukayama et al. (1986) found that two females (mean 477 mm TL) captured at sea in July showed histological evidence of vitellogenesis and appreciable levels of the female-specific serum protein presumed to be vitellogenin. Vitellogenesis in European river lamprey also begins during the marine feeding phase (Zanandrea 1959; Larsen 1970), and mean oocyte diameter is already ~600 μm at freshwater entry (~6 months year prior to spawning; Fig. 1.9b).

Vitellogenesis also begins during the marine feeding phase in Pacific lamprey, although this species typically enters fresh water ~1 year prior to spawning, and its oocytes are less well developed at the onset of the spawning migration than in species with more condensed migrations. Pacific lamprey captured at the mouth of the Klamath River in California (i.e., at freshwater entry) possessed oocytes that were mostly in early and mid-vitellogenesis and measured only ~400 μm diameter (Fig. 1.7d; Clemens et al. 2013). In the Willamette River in Oregon, by the time upstream migrants reached Willamette Falls 205 km from the ocean, oocytes in late vitellogenic and early maturational stages were also observed; by April and May, all oocytes were in the late stages of vitellogenesis or in the early stages of maturation and measured ~700 μm in diameter (Clemens et al. 2013). Interestingly, in addition to this river-maturing type, Clemens et al. (2013) also noted a few "ocean-maturing" Pacific lamprey. At freshwater entry, these individuals were already in late vitellogenesis and possessed larger gonads, with gonadosomatic index (GSI) averaging 5.5% rather than 1.2–2.8% in the river-maturing form (see Sect. 1.5.3). These ocean-maturing individuals (which are relatively rare and known to date only from the Klamath River) generally enter fresh water in late winter and likely spawn only weeks or months after freshwater entry (Clemens et al. 2013; Parker 2018). Parker (2018) found that there was a genetic basis for maturation timing in this population (see Chap. 4).

The pouched lamprey and short-headed lamprey from the Southern Hemisphere appear to represent extreme examples of "river-maturing" lampreys. Although the initial stages of testicular maturation appear to occur at sea in these two species (see Sect. 1.5.2), vitellogenesis does not occur until after freshwater entry. The pouched lamprey enters fresh water 15–16 months before spawning, at which point the oocytes are still very small (mean diameter 190 μm; Fig. 1.9c) and without evidence of yolk platelets (Potter et al. 1983). Likewise, the short-headed lamprey has small oocytes without conspicuous yolk granules at the onset of its long spawning run (Hughes and Potter 1969).

Despite differences in the timing of vitellogenesis relative to metamorphosis or upstream migration and the rate at which it proceeds, all species and life history types appear to converge again during final ovarian maturation. Vitellogenesis has produced oocytes that are approaching their size at maturity (~1,000 μm; Fig. 1.9; see Sect. 1.6.1), and final maturation happens rapidly, usually within a few weeks before spawning (Wigley 1959; Lewis and McMillan 1965; Hardisty 1971; Larsen 1970; Farrokhnejad et al. 2014). The oocytes, which have remained in the diplotene stage of meiotic prophase I for years, resume meiosis, as is evidenced by the migration of the nuclear envelope or germinal vesicle (GV) and its subsequent breakdown (Yaron and Sivan 2005). In Caspian lamprey from the Shirud River in Iran, Ahmadi et al. (2011) found evidence that meiosis I had resumed in >85% of female spring migrants captured between late March and mid-May. Spawning in this river usually occurs in May and June (Nazari et al. 2010; Ahmadi et al. 2011). Interestingly, 75% of the autumn migrants (captured between late September and early November) also exhibited oocytes with migrating and peripheral GVs. Relatively high GSI values (compared to spring migrants; Table 1.9) are also consistent with unusually early maturation in these autumn migrants, although presumably they still overwinter before spawning in the spring (Ahmadi et al. 2011; see Sect. 1.5.3). In European river lamprey, Larsen (1973) reported that the GV was in the peripheral position a few months before spawning, around the time that secondary sex characteristics started to develop (see Sect. 1.5.5.3). In all species, following completion of meiosis I, the resulting haploid secondary oocytes immediately initiate meiosis II (Fig. 1.6). The secondary oocytes arrest in meiosis II (at the metaphase stage) until fertilization. An ootid (i.e., an immature ovum) is formed shortly after fertilization, and it rapidly (within minutes) matures into the mature ovum (Gilbert 2000). This short-lived ootid stage is the female counterpart of the male spermatid (see Sect. 1.5.2). Hardisty (1971) used the term "egg" to refer to oocytes following ovulation, although this distinction is not universally made; in most cases, the oocytes, ootids, and ova are all referred to as eggs or egg cells regardless of their stage of maturity. Nevertheless, for convenience, we adopt Hardisty's terminology and use egg to refer to oocytes in the final stages of maturation and fertilization.

Other changes involved in final ovarian maturation also appear similar among lamprey species. We will briefly describe them here in a few well-studied representative species. In European river lamprey, the follicle cells surrounding the oocyte begin to form a thin granular layer covering its vegetal pole (i.e., the hemisphere with large yolky cells) by about January, and a slight elevation of the thecal interna is observed at the animal (non-yolky) pole (Larsen 1970). The follicle cells grow in height and become separated from each other, reaching a maximum height and separation by about March. The elevation of the theca also becomes more pronounced with time, and, in the mature oocyte, it forms a conical projection (Hardisty 1971). Ovulation, which is the release of the oocyte from the theca when the follicular layer ruptures, occurs rapidly. Larsen (1970) concluded that all the oocytes are released into the body cavity synchronously (within a few hours), because she never observed partially ovulated ovaries, and she observed that the process appeared to start in the posterior of the ovary (like spermiation; see Sect. 1.5.2). The follicle cells appear to contain neutral and acid mucopolysaccharides, and Larsen (1970) speculated that the enzymatic breakdown of the acid mucopolysaccharides might help rupture the follicle at ovulation (i.e., by increasing the colloid osmotic pressure in the follicle, thus causing an uptake of water and increase in hydrostatic pressure). Remnants of follicle cells that remain on the ovulated egg appear to be identical to the adhesive layer which, after spawning, attaches the egg to the gravel in the nest (Larsen 1970; see Johnson et al. 2015b; Chap. 2). Little is known about when spermiation and ovulation occur relative to the time active spawning begins (Johnson et al. 2015b), but Larsen (1970) found that palpation of the abdomen of European river lamprey in February and March (when the secondary sex characteristics had developed) revealed softness in the ovary corresponding to a gradual loosening of the connective tissue, and ovulated eggs could be pressed out in March or April. Ovulated eggs have a small tuft of fibrous jelly at the animal pole (Larsen 1970; Hardisty 1971). A large amount of fluid accumulates in the body cavity; the eggs are suspended in it, and it likely facilitates their expulsion during spawning (Lewis and McMillan 1965; Larsen 1970).

During mating, the eggs are forced to the exterior through a pair of genital pores, first into the urogenital sinus and then out through the pore on the urogenital papilla. No Müllerian ducts (nor vasa efferentia in males) are present (Applegate 1949; Adolfi et al. 2019). Kille (1960) and Kobayashi and Yamamoto (1994) described the fertilization process in European river and brook lampreys and in Arctic lampreys, respectively. In brief, lamprey eggs do not possess a micropile; the spermatozoa penetrate the chorion of the egg in the region of the tuft of fibrous jelly at the animal pole. The function of the tuft seems to be to orient the spermatozoan head so that it strikes the chorion at about 90°. Spermatozoa that strike the chorion at a smaller angle are unsuccessful at penetrating the chorion, and eggs are more difficult to fertilize if the tuft has been destroyed. Lamprey eggs retain their capacity for fertilization for only a few minutes after contact with fresh water, although they may remain fertilizable

Table 1.9 Gonadosomatic index (GSI) (mean ± one standard deviation, SD, and range) in nine parasitic and nine non-parasitic (brook) lamprey species at maturity or near maturity; range or maximum is for individual values at or near maturity (i.e., not the range during maturation); rkm = river km

Species	Females (%) Mean ± SD	Range or maximum	Males (%) Mean ± SD	Range or maximum	Reference	Comments
PARASITIC SPECIES						
Caspiomyzon wagneri Caspian lamprey (anadromous)						
Shirud R, Iran @ rkm 0.2 (spring migrants)	20.5 ± 6.8	Max 31.4			Nazari and Abdoli (2010)	Mid-Apr; 8.9% in late March shortly after river entry
Shirud R, Iran (autumn migrants)	15.3 ± 2.9		8.6 ± 4.5		Ahmadi et al. (2011)	Late Sept–early Nov
Shirud R, Iran (spring)	12.1 ± 2.8		6.5 ± 2.1		Ahmadi et al. (2011)	Late Mar–mid-May
Shirud R, Iran (spring)	30.0 ± 9.0	25.3–35.1	9.5 ± 8.0	8.5–13.2	Farrokhnejad et al. (2014)	April
Shirud R, Iran (autumn)	12.6 ± 2.8	7.1–17.9	5.0 ± 1.8	2.9–6.4	Shirood Mirzaie et al. (2017)	Oct–Nov
Shirud R, Iran (spring)	10.1 ± 1.1	7.1–11.7	6.3 ± 1.5	2.4–9.4	Shirood Mirzaie et al. (2017)	Mar–Apr
Entosphenus tridentatus Pacific lamprey (anadromous)						
Columbia R, WA @ rkm 235	33.7 ± 18.6		2.3 ± 0.2		Robinson et al. (2009)	June (see Fig. 1.11c)
Willamette R, OR @ rkm 205	~25	Max 43			Clemens et al. (2013)	Apr–May: GSI calculated using ovary-free body weight
Geotria australis pouched lamprey (anadromous)						
Donnelly R, Australia	27.1		0.5		Potter et al. (1983), Potter and Robinson (1991)	Oct (~1 month prior to spawning); 0.4% (females) and 0.05% (males) on river entry

(continued)

Table 1.9 (continued)

Species	Females (%) Mean ± SD	Range or maximum	Males (%) Mean ± SD	Range or maximum	Reference	Comments
Ichthyomyzon castaneus chestnut lamprey (freshwater)						
Muskegon R, MI (L Michigan)	18.3 ± 4.5				Schuldt et al. (1987)	Apr–May
Ichthyomyzon unicuspis silver lamprey (freshwater)						
8 tributaries of St. Lawrence R, QC	19.3 ± 13.1	9.6–34.1			Vladykov (1951)	Apr–May
Menominee R, MI/WI (L Michigan)	14.3 ± 4.7				Schuldt et al. (1987)	Apr–May
Peshtigo R, WI (L Michigan)	18.8 ± 3.4				Schuldt et al. (1987)	Apr–May
Oconto R, WI (L Michigan)	15.9 ± 2.8				Schuldt et al. (1987)	Apr–May
Lampetra fluviatilis European river lamprey (anadromous)						
Severn Estuary, UK	~7.5		~6		Abou-Seedo and Potter (1979)	March
Ijssel Sea, The Netherlands	22.7			5.1–5.3	Mewes et al. (2002)	April
Rega R @ rkm 20, Poland	22.8				Dziewulska and Domagala (2009)	May
Lampetra fluviatilis European River lamprey (praecox)						
Severn Estuary, UK	8.4 ± 1.0*		4.6 ± 0.6*		Abou-Seedo and Potter (1979)	March
Lampetra fluviatilis European River lamprey (freshwater)						
R Endrick (L Lomond), UK	18.2		12.6		Maitland et al. (1994)	March (see Fig. 1.11b)
Lethenteron camtschaticum Arctic lamprey (anadromous)						
Ishikari R, Japan	~32		3		Fukayama and Takahashi (1985)	May

(continued)

Table 1.9 (continued)

Species	Females (%) Mean ± SD	Range or maximum	Males (%) Mean ± SD	Range or maximum	Reference	Comments
Lethenteron camtschaticum Arctic lamprey (freshwater)						
Slave R, NWT	~22	~16–26	~4	~2.5–5.5	Nursall and Buchwald (1972)	Late June
Petromyzon marinus sea lamprey (anadromous)						
St Lawrence R, QC	12.9 ± 0.9	11.6–14.7			Vladykov (1951)	Various dates; egg diameter (900–1,050 µm) suggests near maturity
St John R @ rkm 120, NB	14.6 ± 2.1		1.6 ± 0.6		Beamish and Potter (1975)	June–early July
St John R @ rkm 120, NB	16.8 ± 1.5*		1.3 ± 0.4*		Beamish et al. (1979)	Late June
Minho R @ rkm 65, Spain	14.8		1.2		Araújo et al. (2013)	Early May: 7.7% (females) and 1.7% (males) at river mouth (Jan–Apr)
Petromyzon marinus sea lamprey (freshwater)						
Ocqueoc R and Carp Cr, MI (L Huron)	21.6 ± 4.6	13.6–29.5			Applegate (1949)	June–early July; 10.7% at river mouth (mid-Apr)
Hibbards Cr, WI (L Michigan)	24.9 ± 4.4	17.7–32.9			Vladykov (1951)	June–July
Little Thessalon R, ON (L Huron)	20.2 ± 2.2	15.5–23.2			Vladykov (1951)	Early June
Cayuga L, NY	17.6		2.9		Wigley (1959)	Mid-May: 12.9% at river mouth (late Apr)
Chocolay R, MI (L Superior)	19.5 ± 4.9	11.2–31.0			Manion (1972)	Mid-June: 10.1% late May

(continued)

Table 1.9 (continued)

Species	Females (%) Mean ± SD	Range or maximum	Males (%) Mean ± SD	Range or maximum	Reference	Comments
Peshtigo and Menominee R, MI/WI (Green Bay, L Michigan)	18.3				Johnson (1982)	May–June
Duffins Cr and Humber R, ON (L Ontario)	19.7				O'Connor (2001)	May–June
Duffins Cr and Humber R, ON (L Ontario)	19.3				Gambicki and Steinhart (2017)	May and June
Tahquamenon and Betsy R, ON (L Superior)	19.1				Gambicki and Steinhart (2017)	May and June
Tetrapleurodon spadiceus Mexican lamprey (freshwater)						
Celio R	21.8	16.2–25.5			Álvarez del Villar (1966)	Nov (on spawning grounds)
NON-PARASITIC SPECIES						
Eudontomyzon hellenicus Macedonian brook lamprey						
Kefalárion Br, Greece	18.7				Lapierre and Renaud (2015)	Oct, May
Ichthyomyzon fossor northern brook lamprey						
Yamaska R, QC	33.7 ± 5.4	27.8–46.7			Vladykov (1951)	On spawning grounds
Little Cedar R, MI (L Michigan)	19.5 ± 1.7				Schuldt et al. (1987)	Apr–May
Walla Walla Cr, WI (L Michigan)	16.9 ± 1.6				Schuldt et al. (1987)	Apr–May
Little Wolf R, WI (L Michigan)	19.0 ± 3.1				Schuldt et al. (1987)	Apr–May

(continued)

Table 1.9 (continued)

Species	Females (%) Mean ± SD	Range or maximum	Males (%) Mean ± SD	Range or maximum	Reference	Comments
Ichthyomyzon gagei southern brook lamprey						
Choclafaula Cr, AL	26.3 ± 2.2		~8		F. W. H. Beamish (1982)	Late March (see Fig. 1.11a)
Ichthyomyzon greeleyi mountain brook lamprey						
4 tributaries of French Broad R, NC	21.5 ± 1.6*				Beamish and Medland (1988)	Adults (early May); 0.8% stage 1–3 metamorphosis, 12.6% stage 7
Lampetra aepyptera least brook lamprey						
8 streams in MD, DE, KY, TN, AL	~18.5	25.5			Docker and Beamish (1991)	Adults; ~2% stage 2 metamorphosis, 6–18% stages 5–7
Lampetra hubbsi Kern brook lamprey						
Merced R, CA	31.1				Lapierre and Renaud (2015)	Feb–March
Lampetra planeri European brook lamprey						
R Endrick (L Lomond), UK	21.5		11.9		Maitland et al. (1994)	March (see Fig. 11c)
Lethenteron appendix American brook lamprey						
Veuillette R & Noire R, QC	20.6 ± 4.8	13.2–30.0			Vladykov (1951)	Late May
Fox R, WI (L Michigan)	14.5 ± 3.4				Schuldt et al. (1987)	Apr–May
Betsie R, MI (L Michigan)	19.7 ± 2.5				Schuldt et al. (1987)	Apr–May
Tetrapleurodon geminis Mexican brook lamprey						
Celio R, Mexico	24.8				Álvarez del Villar (1966)	Nov (on spawning grounds)

* 95% confidence limit

for up to several days in lamprey Ringer solution or lamprey peritoneal fluid at temperatures of about 4 °C (Hardisty 1971; see Chap. 2). Female lampreys generally die within a week of spawning (Applegate 1950; Hagelin and Steffner 1958; Pletcher 1963), although Baker et al. (2017) documented both male and female pouched lamprey in New Zealand surviving for more than 3.5 months after spawning.

Studies to date suggest that female lampreys generally release most or all of their mature eggs. Applegate (1949), Manion and McLain (1971), and Manion and Hanson (1980) concluded that the percentage of unspawned eggs in sea lamprey from the upper Great Lakes is generally <5%. Similarly, in sea lamprey from Cayuga Lake in New York, Wigley (1959) estimated that only 1% of the ovary (by weight) remains after spawning, and Maitland et al. (1994) reported that GSI in female European brook lamprey was only ~1% following spawning (Fig. 1.11c). However, GSI in spent European river lamprey was still ~14% (Fig. 1.11b). Applegate (1949) found that some sea lamprey females, largely those collected near the end of the spawning season, retained up to 37% of their eggs, and O'Connor (2001) reported egg retention rates up to 70% (see Sect. 1.6.3).

However, there is still a great deal of uncertainty regarding the extent to which oocytes may be "lost" prior to maturation. Hughes and Potter (1969) and Hardisty (1971) have suggested that atresia occurring immediately before or at the onset of vitellogenesis may be significant in non-parasitic lamprey species (see Sect. 1.6.2). Intensely basophilic cytoplasm (at a time when normal oocytes tend to become less basophilic), followed by initial hypertrophy of the cell and degeneration of the granulosa cells so that the follicle becomes a mass of hyaline globules surrounded by a contracted basal membrane, was taken by these authors as histological evidence of impending atresia. In addition, atretic oocytes often have irregular margins and non-ovoid shapes and exhibit hypertrophy of the nucleus and nucleolus (Hardisty 1971). Lewis and McMillan (1965) likewise observed atresia following initiation of vitellogenesis in the landlocked sea lamprey. The first signs of atresia during this stage included concentration of yolk particles that resulted in more intense staining and phagocytes derived from follicular cells that began to congregate at the periphery of the oocyte and ingest the yolk. The result was an irregular mass of inward-moving phagocytes surrounding a diminishing ball of yolk, eventually leaving only a ball of cuboidal and ovoid follicular and stromal cells. Applegate (1949) reported the occurrence of small undeveloped or partially developed ova in some sea lamprey (particularly smaller-bodied individuals) that appeared consistent with atresia during vitellogenesis or a cessation of vitellogenesis in some oocytes, but he saw no indication that oocyte atresia occurred during the final stages of maturation. In European river lamprey, Dziewulska and Domagała (2009) likewise found that atretic oocytes occupied, on average, <1% of the ovary during the upstream migration and final maturation (October–May).

Fig. 1.11 Gonadosomatic index (GSI) as percentage of total body weight in males (*solid circles*) and females (*open circles*) in: **a** southern brook lamprey *Ichthyomyzon gagei* during the late larval (*Lv*) stage, metamorphosis (stages *1–7*), sexual maturation, and after spawning (*vertical arrow*) (data from F. W. H. Beamish 1982); **b** European river lamprey *Lampetra fluviatilis* during its upstream migration (Maitland et al. 1994); and **c** Pacific lamprey *Entosphenus tridentatus* during its prolonged upstream migration (Robinson et al. 2009)

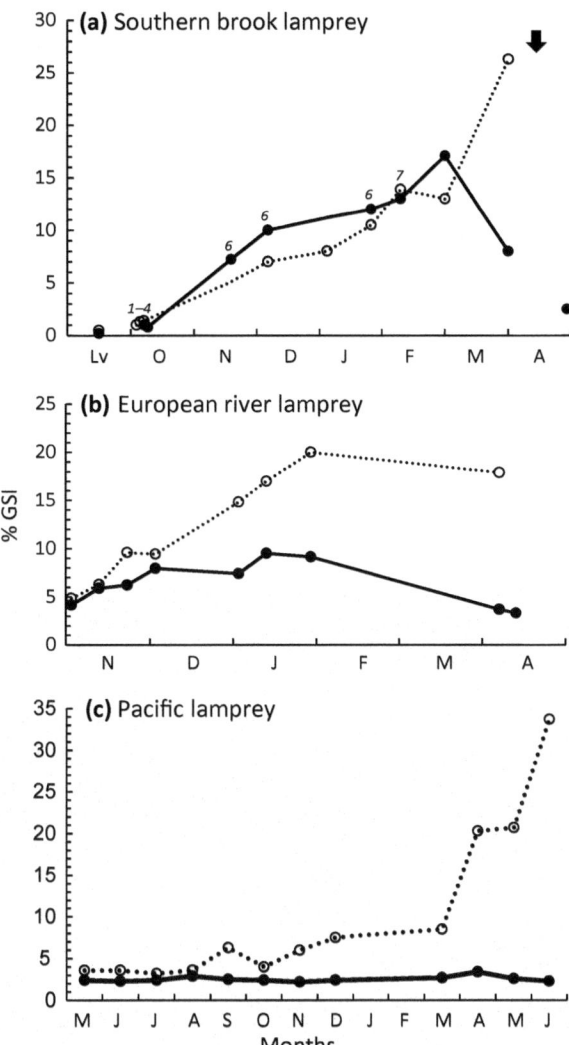

1.5.2 Spermatogenesis

Spermatogenesis in lampreys (as in other non-amniote vertebrates) progresses in cysts, in contrast to the acystic form of spermatogenesis that occurs in the seminiferous tubules of amniote testes (Yoshida 2016). Lampreys also lack Sertoli cells, the somatic cells in the testes of jawed vertebrates that are essential for the progression of germ cells to spermatozoa and for nourishment of the developing sperm (Schulz et al. 2010; Yoshida 2016), although similar functions appear to be performed in lampreys by lobule cells (Hardisty 1971). Despite these structural differences, the general pattern of spermatogenesis is conserved among vertebrates and can be divided into three general steps: (1) mitotic division of the undifferentiated germ cells to produce spermatogonia; (2) meiosis, which first yields primary (diploid) and then secondary (haploid) spermatocytes, and then spermatids; and (3) spermiogenesis, during which spermatids undergo rapid morphological transformation to produce motile, flagellated spermatozoa (de Kretser et al. 1998; Papaioannou and Nef 2010). Mitosis produces both type A and type B spermatogonia; type A spermatogonia (recognizable by their lack of heterochromatin) are responsible for the proliferation and renewal of spermatogonia, and type B spermatogonia (whose nuclei contain heterochromatin) ultimately undergo meiosis to produce spermatocytes (de Rooij and Russell 2000). The number of mitotic divisions may vary among lamprey species. Spermiation, equivalent to ovulation in females, is the process by which mature spermatids are released. In lampreys, this does not occur via ducts or tubules as it does in jawed vertebrates. Rather, when the cysts have completed formation of sperm, they simultaneously rupture and release sperm into the body cavity (Hughes and Potter 1969; Hardisty 1971).

In lampreys, production of spermatogonia is initiated at the onset of (or, in some species, just before or after) metamorphosis (see Sect. 1.4.1.3). Mitotic divisions increase the size of the testis; there is an increase in both the diameter of the cysts (as each begins to contain large numbers of spermatogonia) and the number of cysts (as perilobular connective tissue invades and divides existing cysts) (Hardisty et al. 1970; Hardisty 1971). As with ovarian development and maturation, the onset and rate of progression of the remaining stages depend on the species and life history type (Fig. 1.6). In non-parasitic species, spermatogenesis progresses rapidly following the initiation of metamorphosis. Hardisty et al. (1970) reported that the testes of European river and brook lampreys could not be distinguished from each other in the early stages of metamorphosis on the basis of size alone, but that, even in these early stages, the lobular structure in the testis of the European brook lamprey was better developed and contained a larger number of spermatogonia than those found in the European river lamprey. The "parting of the ways" between the life history types becomes very evident during the latter stages of metamorphosis. By late autumn or early winter, the European brook lamprey testis is a large lobular structure occupying most of the body cavity. Mitosis is complete and spermatogonia enter meiosis I, yielding primary (diploid) and then secondary (haploid) spermatocytes (Hardisty et al. 1970). Meiosis is generally complete by late February or March, at which time spermatids and spermatozoa first appear (Hardisty 1971). Up until the primary spermatocyte stage, development is usually synchronous throughout the entire testis. In contrast, in the later stages of spermatogenesis, spermatocytes, spermatids, and sometimes spermatozoa are all commonly found within the same testis, although development in individual lobules is usually synchronized (Hardisty 1971).

Similarly, in the Australian brook lamprey *Mordacia praecox*, the testis contains only a few presumptive spermatogonia at the onset of metamorphosis in late October to November, but mitotic activity is evident by December, and the cysts contain large numbers of spermatogonia by February and March (Hughes and Potter 1969). A few meiotic divisions have started to occur by this point, and increased meiotic activity during the next month produces all remaining stages of spermatogenesis with the exception of fully mature sperm. Hughes and Potter (1969) reported that cysts located centrally in the testis mature before lateral cysts and those in the middle of the testis mature before cysts in the anterior and posterior regions. By May, meiosis had progressed to the point that there was a greater incidence of primary spermatocytes and spermatids. By June and July, only a few remaining tracts of premeiotic spermatogonia were found in the dorsolateral part of the testis, and the majority of cysts had become elongated and contained spermatids and immature sperm. Spermiogenesis, characterized by an elongation and condensation of nuclear chromatin and the shedding of cytoplasmic material from the head region of the sperm, was first evident in June and July. At this point, mitochondria and extensive vacuoles are located in the cytoplasm of the sperm, a small acrosome cap is found under the plasma membrane at the proximal extremity of the sperm head, and a flagellum arises from a centriole situated deep within the nuclear chromatin (Hughes and Potter 1969). Spawning is thought to occur around July.

In parasitic species, mitotic proliferation and renewal of spermatogonia continues throughout the juvenile phase, and the onset of meiosis generally does not occur until the end of the parasitic feeding phase or the start of the upstream migration. Therefore, the onset of meiosis in males occurs at approximately the same time as the onset of vitellogenesis in females, although the two processes are not precisely aligned within species, and spermatogenesis appears to be less synchronous among and within individuals than is vitellogenesis (see Sect. 1.5.1). For example, in anadromous Arctic lamprey in Japan, vitellogenesis is initiated at the end of the parasitic feeding phase, but production of spermatocytes from meiosis occurs only after the onset of the spawning migration. Fukayama and Takahashi (1985) found only spermatogonia undergoing mitotic proliferation in male Arctic lamprey captured at sea and those captured in October on their upstream migration; spermatocytes were not observed until February. In contrast and somewhat surprisingly, spermatocytes are evident in the testes of pouched lamprey at the start of their upstream migration, even though they enter fresh water 15–16 months in advance of spawning and vitellogenesis has not yet been initiated in females (Potter and Robinson 1991). This relatively early onset of meiosis in male pouched lamprey is also surprising because the onset of mitosis seems relatively late; no spermatogonia were evident by the end of metamorphosis and downstream migration in this species (see Sect. 1.4.1.3). During the marine feeding phase, the pouched lamprey testis increased in size ~20-fold, and, by the end of this phase, spermatogonia were organized into distinct cysts. By freshwater entry, at least some spermatocytes were evident in all males examined (Potter and Robinson 1991). Some spermatids and early spermatozoa were evident shortly thereafter, showing completion of meiosis in at least some cysts. An average of 2–12% of the testicular area was made up of these stages by 0–3 months following freshwater entry, and GSI was still very low (see Sect. 1.5.3). However, by 9 and 15–16 months following freshwater entry, post-meiotic cysts made up an average of 31 and 38% of the testicular area, respectively. These cysts were often more distended than previously, but GSI was still relatively low (0.5%) and mature sperm were still not evident, which is unusual relative to other lamprey species approaching spawning. Spermiogenesis was also not observed in pouched lamprey held in the laboratory for the last 6 months of the spawning run. The proportion of spermatocytes and spermatids increased during this time, but mature sperm were not observed, even when held for 2–3 months past their normal spawning time (Potter and Robinson 1991). In the short-headed lamprey, spermatids have also been reported shortly after the start of the upstream migration (Hughes and Potter 1969), indicating that meiosis in males of this other Southern Hemisphere species is likewise initiated prior to vitellogenesis in females and long before spawning.

In other parasitic species, stage of maturity at the start of the upstream migration is as expected based on time until spawning (i.e., it is more advanced in those that spawn shortly after river entry). Anadromous sea lamprey that enter fresh water only 1–2 months before spawning in late June–early July already possess testes with primary and secondary spermatocytes and spermatids at the start of the upstream migration in mid-May (Fahien and Sower 1990). Spermiogenesis is in progress by late May and early June, producing spermatids, immature sperm, and even some mature sperm. By early July, the majority of testes contain only mature sperm. In European river lamprey that enter fresh water in early autumn (~7–8 months before spawning), males are either still at the spermatogonial (mitotic) stage or just starting meiosis. Zanandrea (1959) observed only spermatogonia in males captured during the marine feeding phase in the Gulf of Gaeta, and Evennett and Dodd (1963) likewise found only spermatogonia in males captured in the River Severn at the start of their upstream migration in late September. Meiosis was evident in most testes by October, when the majority of migrants had primary spermatocytes, and the remaining stages progressed rapidly. Timing of the final stages of sexual maturation converged with that of the European brook lamprey. By late winter or early spring, secondary spermatocytes, spermatids, and spermatozoa were evident, and, as with all the species discussed above, development was asynchronous among and within individuals (Evennett and Dodd 1963; Hardisty 1971). Abou-Seedo and Potter (1979) reported a slightly earlier onset of meiosis in typical European river lamprey males captured in the Severn Estuary, with spermatocytes already evident in most (but not all) males captured just prior to freshwater entry in September. The praecox form of this species was even more mature on freshwater entry: spermatids were already evident in one male examined, although primary spermatocytes were also evident (Abou-Seedo and Potter 1979). The praecox European river lamprey in this region appears to reduce its post-metamorphic period by 1 year relative to the typical anadromous form, but it reduces the duration of the parasitic feeding phase by only 6 months by delaying its upstream migration until the winter or spring prior to spawning (see Chap. 4). Meiosis also appears to start before freshwater entry in the Caspian lamprey, which generally enters fresh water 2–8 months prior to spawning. In this species, the testes of early migrants are characterized by the presence of spermatocytes, which are replaced with spermatozoa closer to the spawning period (Ahmadi et al. 2011). In the Shirud River in Iran, autumn migrants showed an advanced stage of maturity relative to the spring migrants. The testes of all autumn migrants were full of spermatozoa by late September to early November, compared to only one-third of spring migrants even by late March to mid-May; GSI was likewise higher in the fall (see Sect. 1.5.3). Although it has been suggested that the autumn migrants may spawn in the fall, they likely represent an initial period of migratory activity that is halted by low winter temperatures (Ahmadi et al. 2011). If so, however, both male and female autumn migrants appear to start maturing unusually early relative to other lampreys, although they likely undergo final gonadal growth and maturation in the spring.

At spermiation, sperm are shed directly into the body cavity when the cysts simultaneously rupture and release sperm into the body cavity (Hardisty 1971). Mature spermatozoa have a cylindrical head ~14 μm in length and 0.5–1 μm in diameter, with a tail that may extend up to 140 μm (Hardisty 1971). Kille (1960) reported that the swimming life of sperm on a glass slide may be <1 min after activation in fresh water. However, Kobayashi (1993) and Ciereszko et al. (2002) found that Arctic and sea lamprey sperm were still motile for up to 4–5 min following activation, particularly in the presence of female coelomic fluid. This is quite long compared to other freshwater fishes (average 2.5 min; Browne et al. 2015) and may permit relatively high fertilization rates in lampreys. In the laboratory, prior to contact with fresh water, lamprey sperm can be stored for up to 1 day (Ciereszko et al. 2000; see Chap. 2). The fertilization process is described in Sect. 1.5.1. Although female lampreys generally die within 1 week of spawning, males have been observed to live for 1–2 months (Pletcher 1963; see Sect. 1.5.1).

1.5.3 Gonadosomatic Index

The gonadosomatic index (GSI), which is the gonad mass as a proportion of the total body mass, is useful as a tool for estimating stage of sexual maturity and for comparing reproductive output among individuals, populations, or species (deVlaming et al. 1982; Lowerre-Barbieri et al. 2011; Zeyl et al. 2014). In lampreys, GSI in females is generally higher than GSI in males, particularly as oocyte size and consequently ovary weight increase rapidly during the final stages of maturation, but interesting differences appear to exist among life history types (Table 1.9; Fig. 1.11).

1.5.3.1 Temporal Changes in GSI During Sexual Maturation

Because the rate at which the ovary and testis mature varies among species (see Sects. 1.5.1 and 1.5.2), change in GSI is a useful indicator of the onset and progression of these processes. This is especially true in females. For example, the initiation of vitellogenesis corresponds with a dramatic increase in both oocyte diameter and GSI. In the southern brook lamprey, F. W. H. Beamish (1982) observed a 10-fold increase in GSI (from 1.3 to 14%) between stage 2 of metamorphosis in October and stage 7 in mid-February (Fig. 1.11a); during this time, oocyte diameter increased from 270 to >710 μm (Fig. 1.9a). More gradual increases in both oocyte diameter (Fig. 1.9b, c) and GSI (Table 1.9; Fig. 1.11b, c) are observed following the onset of vitellogenesis in parasitic species. In anadromous Arctic lamprey, GSI averaged only 2% in early vitellogenic females captured at sea in July, increased to 6–8% after river entry in October–November, and reached 14% near the end of vitellogenesis the following April (Fukayama and Takahashi 1985). The increase in GSI is even more protracted in Pacific and pouched lampreys (Potter et al. 1983; Robinson et al. 2009; Fig. 1.11c). However, all life history types appear to converge again in the final stages of maturation, with rapid increases in oocyte size and GSI occurring during the final few months and even weeks before spawning. GSI in female southern brook lamprey doubled in the final 6 weeks prior to spawning (F. W. H. Beamish 1982). GSI in female European river lamprey likewise rose sharply from ~10 to 20% between early December and late January (Maitland et al. 1994; Fig. 1.11b) and from ~8 to 23% between December and April (Mewes et al. 2002). In sea lamprey from Lake Superior, Manion (1972) found that female GSI increased from only 10% in late May to 20% just 3 weeks later. A sharp increase in GSI just prior to the spawning period was also reported in sea lamprey from Cayuga Lake (Wigley 1959) and in anadromous sea lamprey females (Beamish et al. 1979).

Thus, GSI values for female lampreys "nearing" maturity will generally be lower than values at maturity, so it should be noted that the GSI values presented in Table 1.9 will likely be underestimates in most cases. This appears to be particularly true in populations or individuals that undergo rapid maturation following river entry. European river lamprey entering the Severn River consist of two forms: the typical form with its more protracted upstream migration (~September–March, with a peak in November) and the praecox form, which delays upstream migration until January–March (Abou-Seedo and Potter 1979). However, GSI of both typical and praecox females just entering the estuary in March (~1 month before spawning) was still only ~7 and 8%, respectively. This suggests that oocyte growth in these later-entering migrants will be very rapid in the final month before spawning and that temporal proximity to the spawning period is not always a good indicator of stage of maturity.

Conversely, an increase in GSI well in advance of expected spawning could indicate an earlier or more protracted spawning period than previously thought. GSI in female Macedonia brook lamprey *Eudontomyzon hellenicus* in October (28%) and May (12%) is consistent with either two discrete spawning periods (one at the end of January, one at the end of May) as proposed by Renaud (1982, 1986) or a single protracted spawning period. Caspian lamprey autumn migrants in the Shirud River in Iran showed GSI values slightly higher or comparable to that of spring migrants (Ahmadi et al. 2011; Shirood Mirzaie et al. 2017), although they were still considerably lower than spring values (30%) recorded from females in spawning condition (Farrokhnejad et al. 2014; Table 1.9). Further study is required to confirm when the autumn migrants spawn, but timing of maximum GSI can help deduce the spawning period. Likewise, the exact spawning period has not been reported for the Kern brook lamprey *Lampetra hubbsi*, but, judging from the condition of females collected in February–March (GSI >30%), spawning was likely imminent (Lapierre and Renaud 2015). Because metamorphosing and recently metamorphosed Kern brook lamprey have been collected in mid-February (Vladykov and Kott 1976), this suggests sexual maturation proceeds extremely rapidly in this species (Lapierre and Renaud 2015).

Male lampreys also show an increase in GSI with maturity, but with far more variation among species and without the dramatic rise seen in females during the final stages of sexual maturation (Fig. 1.11; see Sect. 1.5.3.2). Male GSI largely increases as the result of spermatogonial proliferation during and after metamorphosis, and it may actually decrease in the final stages of maturation when the spermatids undergo morphological transformation to produce spermatozoa but without the cell growth observed in females. In anadromous Arctic lamprey, for example, GSI in males rose from ~1 to 4% between July and October when the number and size of spermatogonial cysts were increasing, and it peaked in February (at 7%) when spermatocytes first appeared in the testis (Fukayama and Takahashi 1985). GSI in male southern brook lamprey likewise increased rapidly (to ~10–15%) during spermatogonial proliferation and meiosis during metamorphosis (F. W. H. Beamish 1982; Fig. 1.11a), and maximum GSI in male pouched lamprey (although only 1.2%) was also observed during the peak of spermatogonial proliferation and meiosis (Potter and Robinson 1991). GSI decreased in all three species when the testes were undergoing active spermiogenesis.

Not surprisingly, there is a dramatic decrease in GSI following spawning in both females and males. In tagged sea lamprey from Cayuga Lake, Wigley (1959) found that only 1% of the ovary remained after spawning (i.e., females were completely spent with few unspawned eggs; see Sects. 1.5.1 and 1.6.3), although 26% of the testis remained after spawning. GSI in spent female and male European brook lamprey was ~1 and 3%, respectively (Maitland et al. 1994), and GSI in spent male southern brook lamprey was 3% (F. W. H. Beamish 1982).

1.5.3.2   Interspecific Differences in GSI at Sexual Maturation

In female lampreys, GSI at maturity appears to be reasonably consistent among species. Using the data in Table 1.9, the overall mean GSI for females at or approaching maturity (excluding European river lamprey from the Severn Estuary; see Sect. 1.5.3.1) was 20%. Means in each study ranged from 10% in spring-migrating Caspian lamprey to 34% in Pacific lamprey; GSI values >30% were recorded in seven species and values >40% were recorded in two species. Given that egg and ovary weight increase markedly until maturation, lower means likely indicate that the females sampled were not yet fully mature. Wide ranges observed within samples suggest that there is some asynchrony in the timing of maturation (i.e., with some individuals not fully mature) and/or variable relative reproductive effort among individuals. In the Caspian lamprey, GSI (in both females and males) was highest in the first individuals to mature and decreased gradually thereafter (i.e., 35 and 25% in the first and last female, respectively, to mature; Farrokhnejad et al. 2014). Therefore, lower mean GSI values after mid-April appeared to result from inclusion of slower-maturing females (Nazari and Abdoli 2010).

Thus, the overall mean GSI of 20% is almost certainly an underestimate for most individual females at maturity. Based on available information, a GSI of ~25–35% for female lampreys at maturity seems more reasonable. It is well accepted that finite resources or body space constraints limit egg production and ovary size in female lampreys to this level, but how much variation exists among species, populations, and individuals is not known. With respect to fecundity, Hardisty (1964) suggested that marked differences among species in the relationship between egg number and body length would not be expected, because all lampreys exhibit similar body forms, and there is relatively little variation in egg size among species (see Sect. 1.6.1). However, it is worth noting that mean GSI values were different among females of the three life history types: 13% for anadromous parasitic species, 19% for freshwater parasitic species, and 22% for non-parasitic species. This suggests that brook lamprey females devote a greater proportion of their energy reserves to ovarian development than do parasitic species, especially anadromous species with more arduous migrations. A similar trend was "hinted at" with respect to fecundity. Although absolute fecundity increased approximately with the cubic power of TL regardless of species, anadromous lampreys have slightly lower fecundities than would be predicted based on the general power relationship with TL (see Sect. 1.6.3). Nevertheless, the seven species noted with maximum GSI >30% represented all life history types: three anadromous species (Caspian, Pacific, and Arctic lampreys), two freshwater parasitic species (silver and sea lampreys), and two non-parasitic species (northern brook and American brook lampreys). Therefore, additional study is needed to determine whether there are subtle but consistent differences among species, populations, or individuals in the relative size of the ovary at maturity and whether such differences are related to allocation of energy resources among competing demands or other factors.

GSI in male lampreys at or near maturity was lower than in females and much more variable (Table 1.9). Mean values were as low as 0.5% in pouched lamprey (~1 month before spawning) and as high as 12 and 13% in male European brook and river lampreys, respectively (Maitland et al. 1994; Fig. 1.11b, c). A peak of 17% was recorded in southern brook lamprey males ~1 month before spawning, but GSI was only 8% at maturity (F. W. H. Beamish 1982). Males of parasitic species had an overall lower GSI (mean 4.2 and 3.5% in anadromous and freshwater species, respectively) compared to non-parasitic species (10%), but, given the pronounced differences in body size, absolute testis size is still much greater in the larger-bodied parasitic species. Absolute testis mass averaged 14.0 g in anadromous sea lamprey (Beamish and Potter 1975), 2.6–6.2 g in anadromous Caspian lamprey (Ahmadi et al. 2011; Shirood Mirzaie et al. 2017), 1.3 g in freshwater-resident European river lamprey (Maitland et al. 1994), and only 0.2 and 0.5 g in southern and European brook lampreys, respectively (F. W. H. Beamish 1982; Maitland et al. 1994). Presumably, the large absolute size of the testis in larger-bodied parasitic species produces more sperm to fertilize the much larger number of eggs from their large-bodied conspecifics.

The effect of testis size on fertility has been (and continues to be) studied in a range of animal species, particularly those showing sperm competition, and testis size is often used as a proxy for reproductive investment (Pintus et al. 2015). Larger testes are considered "the quintessential adaptation to sperm competition," although focusing predominantly on testis size ignores other potentially adaptive features such as sperm density and sperm quality (Ramm and Schärer 2014). The GSI of Atlantic salmon *Salmo salar* males that mature as very small-bodied parr is about twice that of anadromous males (Fleming 1998), although absolute gonad size is a limiting factor in mature male parr, and they are capable of only 1–2 successful spawnings (Thomaz et al. 1997). However, mature male parr have higher sperm concentrations and motility and a longer spermatozoa life span (Daye and Glebe 1984; Gage et al. 1995). Whether male lampreys are sperm limited and whether fertility depends merely on testis size or other traits deserves further study, particularly with respect to life history evolution (see Chap. 4).

1.5.4 Body Shrinkage During Sexual Maturation

Because reproduction in lampreys is followed by senescence and death, lampreys devote considerable resources during sexual maturation to maximizing their single reproductive effort (Larsen 1980). Furthermore, lampreys do not feed during maturation and upstream migration. In parasitic species, the intestine of the juvenile parasite atrophies during the upstream migration (e.g., Vladykov and Mukerji 1961; Battle and Hayashida 1965; Dockray and Pickering 1972; Potter et al. 1983; see Sect. 1.5.5.4). Non-parasitic species cease to feed at metamorphosis (like all lampreys), and the poorly developed juvenile intestine atrophies before ever becoming patent (Battle and Hayashida 1965). Thus, in addition to being semelparous, lampreys are "capital breeders," financing their considerable reproductive efforts using stored (rather than incoming) resources (Tammaru and Haukioja 1996; Hardisty 2006). In parasitic species, this terminal period of natural starvation lasts from a few months to >1 year. The upstream migration (ranging from a few kilometers to >1,000 km; Moser et al. 2015) and sexual maturation are fueled largely with lipid and protein reserves accumulated during the parasitic feeding phase (Kott 1971; Beamish et al. 1979; Larsen 1980; Hardisty 2006) and mobilized primarily from the body wall (Larsen 1980; Beamish et al. 1979; O'Connor 2001; Araújo et al. 2013). Non-parasitic species, which initiate sexual maturation "on an empty stomach" immediately after metamorphosis, are non-trophic for 6–10 months prior to spawning, and they must fuel the sexual maturation process and their shorter upstream migration with resources accumulated during the filter-feeding larval stage (Docker 2009; Chap. 4). Therefore, all lampreys experience body shrinkage during sexual maturation, although the degree to which this happens varies among species and between females and males.

Female and male anadromous pouched lamprey decreased 20% and 11%, respectively, in mean TL between freshwater entry in July–August and maturity the following October, and mean weight decreased 13% and 23% in females and males during this period (Potter et al. 1983; Fig. 1.12a). In Pacific lamprey, which similarly enter fresh water ~1 year before spawning, females and males shrunk in TL by ~23 and 15%, respectively (R. J. Beamish 1980). In anadromous sea lamprey from the St. John River in New Brunswick, TL of females and males decreased by an average of 16 and 12%, respectively, between entry into fresh water in May and completion of spawning in late June (Potter and Beamish 1977). F. W. H. Beamish

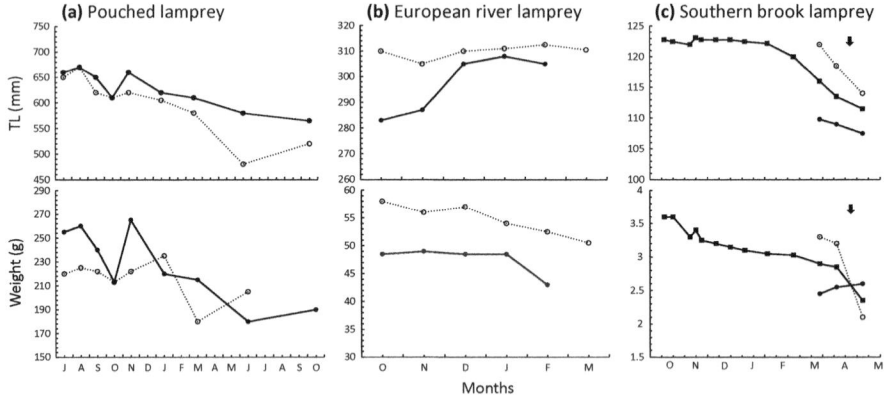

Fig. 1.12 Total length (TL) and weight in males (*solid circles*), females (*open circles*), and both sexes combined (*solid squares*) over the course of the spawning migration in two representative parasitic lamprey species: **a** pouched lamprey *Geotria australis* (data from Potter et al. 1983) and **b** European river lamprey *Lampetra fluviatilis* (Abou-Seedo and Potter 1979); and **c** one representative non-parasitic species, southern brook lamprey *Ichthyomyzon gagei* (F. W. H. Beamish 1982). *Vertical arrows* indicate the time of spawning

(1980) subsequently suggested that values based on TL at first capture in fresh water are underestimates and calculated total shrinkage to be 24% in females and 19% in males. In anadromous sea lamprey from the Minho River in Spain, there was little or no change in TL between river entry in January–April and capture near the spawning grounds (65 km upstream) in early May, but weight decreased by an average of 23 and 20% in females and males, respectively (Araújo et al. 2013). In anadromous European river lamprey from the River Severn, TL either did not change or was found to increase between October and March, but weight decreased by 13 and 11% in females and males, respectively (Abou-Seedo and Potter 1979; Fig. 1.12b). Arctic and Caspian lampreys shrunk in TL by ~25% and 22%, respectively (Holčík 1986a, b), with females showing greater shrinkage than males (Holčík 1986a).

Parasitic species that remain in fresh water typically have shorter spawning migrations, in terms of both duration (2–3 months) and distance (<100 km; Moser et al. 2015). Therefore, we would expect that shrinkage would be less than in anadromous species with long migrations, but whether this is the case is not clear. Comparing TL and weight of 27 males and 37 females tagged at the start of their upstream migration and recaptured after spawning, Wigley (1959) reported a decline in TL of 18 and 11% for females and males, respectively, and, after accounting for the proportion of weight loss due to shed gametes, he estimated that weight loss "from other causes" (predominantly shrinkage) was 17 and 6% for females and males, respectively. However, using data from Applegate (1949), we estimated that female sea lamprey in Carp Creek, a tributary to Lake Huron, decreased in TL and weight by only 6 and 9%, respectively, between mid-April to mid-May and mid- to late June. In contrast, female sea lamprey captured in mid-June during their upstream migration in the Chocolay River, a tributary to Lake Superior, were 16% shorter and weighed 41% less than females captured a mere 3 weeks earlier (Manion 1972). The Vancouver lamprey, which spawns in nearshore lake habitat or in the lower portions of tributaries, is estimated to shrink in TL by only ~6% (R. J. Beamish 1982).

In the non-parasitic northern brook lamprey, Leach (1940) monitored TL and weight of a single female and single male between early September (i.e., in the early stages of metamorphosis) and late March or April. He found that TL decreased by 8.1 and 6.8% in females and males, respectively, in the ~7.5 months between measurements. Weight loss, measured over 6.5 months, was 16 and 12% in females and males, respectively. Because the last measurements were taken a few months prior to spawning (which occurred in late May or early June), total shrinkage during metamorphosis and maturation would be even greater. In a larger sample size, Leach (1940) estimated the total reduction in TL to be ~10%. One particularly small female, which was 92 mm at sexual maturity, died in mid-May as "little more than a swollen bag of eggs" (Leach 1940). Southern brook lamprey held in the laboratory at seasonally adjusted temperatures decreased in TL and weight (sexes combined) by an average of ~8 and 25%, respectively, in the 7 months between early metamorphosis in mid-October and attainment of sexual maturity in mid-April (F. W. H. Beamish 1982; Fig. 1.12c). Rate of shrinkage was greatest in the final 1–2 months prior to spawning, especially in females. Females decreased in TL and weight by 2.9 and 3.1%, respectively, between just February and mid-April. Not surprisingly, given the higher relative weight of the mature ovary compared to the mature testis (see Sect. 1.5.3), females "lost weight" much more precipitously than males following spawning, and spent females weighed only 63% of that of ripe females (F. W. H. Beamish 1982; Fig. 1.12c).

1.5.5   Effect of Hormone Treatments on Sexual Maturation

Recent advances in the study of neuroendocrine hormones (Sower 2015, 2018) and sex steroids (see Sect. 1.7) in lampreys have allowed many inferences to be made regarding the role of these hormones in lamprey reproduction (e.g., by measuring hormone levels in males and females during different stages of maturation). Here, we review some of the early experiments that used hypophysectomy (i.e., surgical removal of the pituitary), gonadectomy (i.e., castration), and hormone replacement therapy to experimentally examine the role of pituitary and gonadal hormones in the sexual maturation in lampreys.

1.5.5.1   Spermatogenesis and Oogenesis

Evennett and Dodd (1963) performed hypophysectomy on upstream-migrating male European river lamprey between early October (when only spermatogonia were present or the germ cells were just entering meiotic prophase) and early March (when the testis contained an increasing number of spermatozoa), and they examined the testes by biopsy in April and June. The results indicated that spermatogenesis in lampreys (unlike in other vertebrates) is catalyzed by, but not entirely dependent on, the pituitary and that the degree to which spermatogenesis is disrupted depends on the timing of hypophysectomy. When the pituitary was removed in early October, maturation was delayed by ~2 months; hypophysectomy in November and December resulted in a shorter delay in sperm maturation, and little, if any, disturbance was observed when hypophysectomy was delayed until January or later. Replacement therapy (using lamprey pituitary extracts, pregnant mares' serum, and mammalian chorionic gonadotropin) appeared to restore normal spermatogenesis (Evennett and Dodd 1963). Delay in sperm formation following hypophysectomy is presumably due to a loss of pituitary stimulation of gonadal steroid production (Gorbman 1983; Sower 2015), and T implants appear to slightly accelerate spermatogenesis in hypophysectomized males (Evennett and Dodd 1963; Dodd and Weibe 1968). Knowles (1939) reported an acceleration of spermatogenesis in intact males injected with T propionate. Larsen (1974) found that spermiation was normal in intact males exposed to E_2 in the water, but implantation of males with E_2 pellets inhibited or delayed spermiation.

In contrast to spermatogenesis, oogenesis in lampreys is severely affected by hypophysectomy. In female European river lamprey, Evennett and Dodd (1963) found that hypophysectomy performed at any point during the spawning migration inhibited the normal increase in size and dry weight of the eggs and prevented ovulation. Loss of pituitary stimulation of gonadal estrogen production presumably impaired vitellogenesis (Sower and Larsen 1991; Mewes et al. 2002; Reading et al. 2017; see Sect. 1.5.5.2). Ovulation was restored in hypophysectomized females that were subsequently treated with lamprey pituitary extracts, pregnant mares' serum, or mammalian chorionic gonadotropin (Evennett and Dodd 1963). Interestingly, there were no signs of oocyte atresia in the hypophysectomized females, even though atresia of yolk-containing eggs "invariably" follows hypophysectomy in other female vertebrates (Evennett and Dodd 1963). Replacement therapy using sex steroids has not been performed on hypophysectomized females. However, in intact European river lamprey females, ovulation was incomplete and sometimes delayed following implantation with or immersion in E_2, and it was incomplete or completely inhibited in females implanted with T (Larsen 1974). It is not yet known what roles pituitary hormones and sex steroids play in the earlier stages of spermatogenesis and oogenesis (see Sect. 1.4.1).

1.5.5.2   Vitellogenesis

During maturation in oviparous vertebrates, estrogen induces the liver of females to synthesize vitellogenins (calcium-rich yolk precursors), which are subsequently released into the blood and deposited in the ovary (see Sect. 1.5.1). As in other vertebrates, the liver of female lampreys becomes significantly hypertrophied at the onset of vitellogenesis, and there is a marked elevation in serum calcium that is correlated with an increase in oocyte dry weight (Pickering and Dockray 1972; Pickering 1976a). In lampreys, however, plasma E_2 levels are generally as high or higher in maturing males compared to females (e.g., Mewes et al. 2002), and E_2 appears to be the major reproductive hormone in both sexes (see Sect. 1.7.2). Thus, studies examining the effect of E_2 administration on intact male lampreys and ovariectomized females have been critical in demonstrating that, despite the apparent lack of sex-specificity in serum E_2 levels, vitellogenesis in lampreys is nevertheless stimulated by the direct effect of estrogen on the liver. In the upstream-migrating European river lamprey, Pickering (1976a) showed that E_2 implantation in intact males resulted in hypertrophy of the liver and marked elevations in serum calcium, protein, and organic phosphorus levels. In contrast, in intact females, E_2 stimulated a small but insignificant rise in serum calcium and a barely significant elevation of organic phosphorus. However, pronounced increases in ovarian size and oocyte dry weight in these E_2-implanted females indicated that vitellogenin levels were not elevated in the plasma because these proteins were being deposited in the ovary (Pickering 1976a). This was further supported by results from ovariectomized females: when vitellogenin deposition in the ovary was prevented, serum calcium levels were elevated, and implantation with E_2 further increased levels of protein, calcium, and phosphorus in the blood. Implantation of intact and gonadectomized males and females with T produced no effect on any of the parameters measured (Pickering 1976a).

Mewes et al. (2002) isolated lamprey vitellogenin from the blood of maturing female European river lamprey, and they showed vitellogenin to be stimulated in males following injection of high doses of E_2 into the coelom. Thus, although the regulation of hepatic vitellogenesis by E_2 appears to be a universal phenomenon in oviparous vertebrates, with the livers of both sexes being capable of synthesizing vitellogenin when stimulated with E_2 (Reading et al. 2017), the female-specificity of vitellogenin synthesis in lampreys (in spite of the presence of E_2 in the blood of both sexes) is somewhat paradoxical. Mewes et al. (2002) have suggested that complex regulatory mechanisms prevent the hepatocytes of male lampreys from synthesizing vitellogenin at physiological E_2 levels. In the African clawed frog *Xenopus laevis*, vitellogenin synthesis appears to be regulated through sex-specific interaction between E_2, the E_2 receptor (ER), and the vitellogenin and ER genes. Low E_2 doses are sufficient to induce transcription of the ER, but a 1,000-fold higher dose is required for activation of vitellogenin gene transcription (Barton and Shapiro 1988). Unless male clawed frogs are treated with high doses of exogenous E_2, the amount of functional ER is too low to induce transcription of the vitellogenin gene.

1.5.5.3 Secondary Sex Characteristics

The secondary sex characteristics of lampreys do not appear until the onset of sexual maturation (Johnson et al. 2015a, b). The external characteristics include swollen cloacal lips in both sexes, enlarged dorsal fins which differ in shape between the sexes, and the appearance of a urogenital papilla which is much more developed in males than in females. Some authors refer to the presence of an anal fin (e.g., Larsen 1974) or post-anal fin (e.g., Evennett and Dodd 1963) in sexually mature female lampreys. However, this is not a true fin because it lacks supporting fin rays; thus, it is more appropriately referred to as an anal fin-like fold (Vladykov 1973; Lapierre and Renaud 2015) or a post-cloacal finfold (Renaud 2011). Internal changes at maturity include the appearance of a pore at the base of each urinary duct for the release of gametes (Dodd et al. 1960; see Sects. 1.5.1 and 1.5.2).

In mammals and birds, the role of the sex hormones in the development of secondary sex characteristics has been established through classic experiments involving castration and hormone replacement (Goldstein and Wilson 1975; Owens and Short 1995). Such a role has been similarly demonstrated in lampreys. Hypophysectomy, whether performed early or late in the spawning migration, completely inhibited secondary sex characteristic development in both male and female European river lamprey (Evennett and Dodd 1963). Replacement therapy, using lamprey pituitary extracts, pregnant mares' serum, and mammalian chorionic gonadotropin, restored development of secondary sex characteristics in both sexes. Secondary sex characteristics were also restored following sex steroid treatment: intraperitoneal implantation with T pellets induced male characteristics in hypophysectomized males (Evennett and Dodd 1963; Larsen 1987), and T implantation in intact and hypophysectomized females produced an enlarged urogenital papilla similar to that of mature males (Evennett and Dodd 1963). Gonadectomy also prevented the appearance of the secondary sex characteristics, and their development was restored following sex steroid treatment (Evennett and Dodd 1963; Larsen 1974). T implants induced male characteristics in gonadectomized males and intact females, and E_2 induced female characteristics in gonadectomized females and intact males (Larsen 1974). In no instance did T stimulate female characteristics or E_2 induce male characteristics. However, different sex characteristics appeared to vary in their sensitivity to these hormones. Changes in the dorsal fins were difficult to induce, but the urogenital papilla and anal fin-like fold characteristic of males and females, respectively, were readily affected by exogenous hormones. In fact, untreated males kept in the same tank as E_2-implanted males received enough E_2 through the water to inhibit growth of the urogenital papilla and stimulate enlargement of the anal fin-like fold (Larsen 1974).

Nevertheless, all studies to date show that exogenous sex hormones are unable to prematurely induce secondary sexual characteristics. Knowles (1939) demonstrated that T propionate and estrone (E_1) were effective in inducing the cloacal swelling and pore development characteristic of maturation in both sexes as European river lamprey approached sexual maturity, but the same treatment elicited only a weak response at best in larvae. Evennett and Dodd (1963) similarly showed that pituitary extract and other injections given in November did not induce secondary sex characteristics in hypophysectomized upstream migrants until February. Sex steroid implants performed in November–February likewise did not accelerate the appearance of the secondary sex characteristics, except for a slightly precocious growth of the urogenital papilla in males (Evennett and Dodd 1963; Larsen 1974). Increased sensitivity of the relevant tissues to hormonal stimulation at the time of maturation is presumably important for normal development of these characteristics (Evennett and Dodd 1963), although the mechanism of this action is, as yet, unknown.

1.5.5.4 Intestinal Atrophy

Sexual maturation in lampreys is accompanied by a marked degeneration of the intestine (Larsen 1969a; Dockray and Pickering 1972; Pickering 1976b), but removal of the gonad reduces the rate at which atrophy occurs (Larsen 1969b; Dockray and Pickering 1972; Pickering 1976b). Vasil'eva (1961) equated the histological changes in the intestine of migrating lampreys to those that occur in the hibernating frog, and she attributed these changes to the effects of starvation. That gonadectomy can prevent normal intestinal degeneration does not necessarily exclude this possibility, because gonadal development during the spawning migration is accomplished at the expense of other body tissues. However, a direct or indirect effect of sex steroids has also been suggested by hormone replacement studies: both E_2 and T promoted intestinal degeneration in gonadectomized (as well as intact) European river lamprey during the early stages of their spawning migration (Pickering 1976a). Hypophysectomy can also reduce the atrophy of the intestine (Larsen 1972), presumably by preventing a pituitary influence on the secretory activity of the gonads. Furthermore, when gonadectomy was delayed until after the gut had already atrophied, castration resulted in re-differentiation and growth of the intestine (Larsen 1972, 1974; Pickering 1976b).

Therefore, it is tempting to conclude that normal intestinal degeneration in parasitic lampreys is triggered by sex steroids produced by the maturing gonads (Larsen 1980). Nevertheless, this hypothesis is probably too simple. For example, the effect of these steroids appears to be dependent upon the time of administration. Sex steroids administered to early upstream migrating European river lamprey (in September) promoted intestinal atrophy, but they were ineffective in January in counteracting the intestinal hypertrophy observed after gonadectomy (Larsen 1974; Pickering 1976b). Furthermore, initiation of sexual maturation and onset of intestinal degeneration do not appear to be well-coordinated in all species. In the pouched lamprey, Potter et al. (1983) found that the intestine underwent rapid atrophy immediately after freshwater entry, even though the gonads were still very immature. In contrast, Youson and Beamish (1991) reported that the parasitic form of western brook lamprey found in Morrison Creek on Vancouver Island retained a functional intestine even while possessing well-developed gonads. The factors that trigger cessation of the parasitic feeding phase and the onset of sexual maturation are as yet unknown in lampreys, but they are of considerable interest in terms of life history evolution (see Chap. 4).

1.6 Egg Size and Fecundity

Potential fecundity is the finite number of oocytes produced during larval development, and actual or absolute fecundity is the number of oocytes that survive to maturity (Beamish and Thomas 1983). All (or almost all) the eggs that survive to maturity are released during the semelparous lamprey's single spawning season, making lifetime reproductive output in female lampreys easily quantifiable. The trade-off between number of offspring and resource allocation per offspring (e.g., egg size) is one of the central tenets of life-history theory (Lack 1954; Smith and Fretwell 1974; Stearns 1989). In lampreys, however, a relationship between egg size and egg number or between egg size and female body size has not emerged.

In contrast, fecundity and its relationship with female body size has been well studied in lampreys. This is not surprising, given its importance in understanding the selective pressures involved in the evolution of different life history traits (see Docker 2009; Chap. 4) and in predicting the reproductive potential of species of conservation and management concern.

1.6.1 Egg Size

Following Hardisty (1971), we use the term "egg" to refer to an oocyte following ovulation. Prior to ovulation, different lamprey species show different rates of oocyte growth (Fig. 1.9; see Sect. 1.5.1), but final egg size appears to be reasonably consistent among species. The average egg diameter across species from all the means in Table 1.11 was 955 μm. Assuming this is a slight underestimate because it includes eggs "nearing maturity" (see below), a good rule of thumb appears to be that egg diameter at maturity is ~1,000 μm (1.0 mm).

Malmqvist (1986) suggested that lamprey egg size increases with adult body length so that eggs are largest in the largest parasitic species, but this does not appear to be a consistent pattern. Although large eggs have been reported in the large-bodied pouched lamprey (mean diameter 1,120 and 1,180 μm; Potter et al. 1983; Baker et al. 2017), mean egg diameter in other large anadromous species appears to be more modest: 940 μm in anadromous sea lamprey, 700–800 μm in Pacific lamprey, and 770–1,040 μm in Arctic lamprey (Vladykov 1951; Kan 1975; Clemens et al. 2013; Yamazaki et al. 2001; Kucheryavyi et al. 2007). Conversely, mean egg diameters in excess of 1,000 μm have been reported in at least six brook lamprey species (Table 1.11). Within species, Witkowski and Jęsior (2000) found a positive relationship between egg diameter and TL in both spring and fall European river lamprey migrants, but Manion (1972) and Kopp (2017) found no relationship between egg size and female size in Great Lakes sea lamprey.

However, methodological differences likely confound comparisons among species and studies. At maturity, the lamprey egg is slightly ovoid, with the long axis being ~1.1–1.2× the length of the short axis (Witkowski and Jęsior 2000; Yamazaki et al. 2001). Some studies measure and report both the long and short axes (e.g., F. W. H. Beamish 1982; Beamish and Thomas 1983; Witkowski and Jęsior 2000; Yamazaki et al. 2001; Dzeiwulska and Domagała 2009); some report only the maximum diameter (i.e., only the long axis; Lapierre and Renaud 2015); some report the mean of the long and short axes (Kopp 2017); and some do not explicitly state which axis was measured but presumably measure and report the maximum diameter (e.g., Applegate 1949; Clemens et al. 2013; Baker et al. 2017). In Table 1.11, we tried to include only the maximum diameter. Furthermore, some studies measure egg size using fresh material (e.g., Manion 1972), but most others use eggs preserved in 10% formalin (e.g., Yamazaki et al. 2001; Nazari and Abdoli 2010; Clemens et al. 2013), 4% formalin (Witkowski and Jęsior 2000), or a solution of formalin, acetic acid, and alcohol (Applegate 1949). An average of 17% shrinkage in diameter has been observed in formalin-preserved fish eggs relative to fresh eggs (Frimpong and Henebry (2012). At least egg diameter appears to be the same among various parts of the ovary (Applegate 1949).

Malmqvist (1986) cautioned that differences in degree of maturity will also make egg size difficult to compare among studies. Table 1.11 excluded samples where oocytes were clearly described as being immature, but rapid maturation in the final weeks before maturity suggests that even small differences in timing can produce substantial differences in egg size (see Sect. 1.5.1). For example, Manion (1972) found a 10% increase in egg diameter in sea lamprey sampled on 16 June compared to just 3 weeks earlier. In some cases, diameter has been reported for eggs free in the coelom just prior to being released (mean 1,000–1,100 μm; Applegate 1949; Piavis 1971), but most studies measured egg diameter in intact ovaries prior to ovulation. Studies measuring fertilized eggs (although not included here) would further confound comparisons, because the egg swells and the volume increases by 20–25% after fertilization (Hardisty 1986).

There are some suggestions that egg weight (mass) in lampreys is more variable among individuals or populations than egg diameter, although fewer studies report egg weight. Furthermore, as with egg diameter, even minor differences in stage of maturity and differences in preservation methods will complicate comparisons. Manion (1972) found a 17% increase in egg weight in sea lamprey between 26 May and 16 June, and Gambicki and Steinhart (2017) found that egg weight decreased by 6–12% after freezing and thawing relative to fresh weight. Overall, mean egg weight at maturity appears to be ~400–580 μg. Average egg weight was 424 μg in European river lamprey in late February (i.e., still 1–2 months prior to spawning) (Witkowski and Jęsior 2000) and 470 and 610 μg in sexually mature Kern brook lamprey and Macedonia brook lamprey, respectively (Lapierre and Renaud 2015). Egg weight in sea lamprey in the Chocolay River in Michigan (a tributary to Lake Superior) was 225–672 μg (mean 390 μg) in mid-June, a few weeks prior to spawning (Manion 1972). There was a positive relationship between egg diameter and egg weight (R^2 = 0.624), but no relationship between egg weight and female TL (Manion 1972). In other studies, average egg weight in sea lamprey from Lakes Superior, Huron, Michigan, and Ontario ranged from 470 to 580 μg, with no relationship among populations between mean egg weight and female TL (Johnson 1982; O'Connor

2001; Gambicki and Steinhart 2017). Gambicki and Steinhart (2017), comparing egg weight from Lake Superior in 2011 to the 1960 values from Manion (1972), concluded that sea lamprey egg weight increased 43% since 1960, corresponding with increases of ~13 and 45% in mean TL and body weight, respectively (see Sect. 1.6.5). These authors caution that the observed differences in egg weight might be the result of methodological differences, but the relationship between female size or body condition and egg weight deserves further study. For example, it would be interesting to examine if the decreases in sea lamprey growth rates that accompanied their high abundance relative to prey abundance in the early 1960s (see Sect. 1.2.6) meant that female sea lamprey in 1960 were less able to develop well-provisioned eggs.

Even fewer studies measure egg dry weight (e.g., dried at 60 °C to a constant weight), although this is likely a more meaningful indicator of nutrient provisioning in the egg. Docker and Beamish (1991) used egg dry weight to investigate intraspecific differences in egg size in least brook lamprey. Within three of the eight populations examined, these authors found that egg dry weight was positively related to female body length. However, among populations, mean egg dry weight was inversely related to size at maturity: in populations where females matured at small sizes (mean ~100–110 mm TL), females produced comparatively fewer but heavier eggs than larger-bodied populations (size at maturity ~130–150 mm TL). Docker and Beamish (1991) concluded that the heavier eggs, which presumably contained more yolk, provided for higher embryonic and larval survival in an unproductive environment. Marsh (1984, 1986) similarly suggested that large eggs of the orangethroat darter *Etheostoma spectabile* are advantageous where food is scarce. It is not known if this pattern is observed in other lamprey species. In widely dispersed parasitic lamprey species that do not home (see Moser et al. 2015), variation in egg size to "match" environmental conditions in rearing streams would not be expected. However, feeding conditions during the parasitic feeding phase might affect egg dry weight, and investigation of inter- and intraspecific variation in the yolk caloric value and biochemical composition of lamprey eggs is also needed. Bird et al. (1993) found marked differences in the fatty acid composition of the ovary of European brook and river lampreys, reflecting differences in their diets during the preceding microphagous and parasitic feeding phases, respectively. Further study is needed to determine if such pronounced differences between non-parasitic and parasitic species (or more subtle differences among parasitic species) affect egg quality and embryo survival.

The effect of maternal attributes on egg properties has been extensively studied in teleost fishes, and a positive relationship between egg size and fish size appears to be nearly universal (see Quinn et al. 1995; Chambers 1997; Kamler 2005). In lampreys, however, the only clear relationship between attributes of females and the properties of their eggs is that egg number increases with body size (Sect. 1.6.3). Given the potentially high embryo mortality rates from sources apparently unrelated to egg size or quality (Cochran 2009; Smith and Marsden 2009; see Dawson et al. 2015), it may be that the lamprey reproductive strategy relies more on devoting resources to increasing fecundity while keeping egg size relatively constant at a "generalized" optimal size. However, better assessments of egg quality are required.

1.6.2   Potential Fecundity

Lampreys produce a finite number of oocytes during the larval stage (see Sect. 1.4.1.2), and these represent their total reproductive potential. Among species, the relative number of oocytes elaborated during the larval stage is often compared by using the number of oocytes per cross-section (Table 1.8), but few estimates have been made of the total number of oocytes in the larvae. This is likely related to both the effort required and concern over the reliability of such estimates. Hardisty (1961c) was among the first, if not the first, to estimate potential fecundity of lamprey larvae. He counted oocyte numbers and measured their diameters from transverse histological sections taken at 1- to 2-mm intervals to extrapolate the total number of oocytes along the length of the ovary. Using this method, he estimated that total oocyte number in 11 European brook lamprey larvae ranged from 4,900 to 10,600 and averaged 7,100 (Hardisty 1961c). Comparing these values to adult fecundity estimates, he concluded that a high proportion of the larval oocytes in this species (up to 90%) must fail to reach maturity (Hardisty 1964; Table 1.10). Hardisty (1971) suggested that atresia in brook lampreys may help provide nourishment to the remaining oocytes during their non-trophic post-metamorphic period. Kuznetsov et al. (2016) counted a subsample of the oocytes under a binocular microscope, and they also suggested that potential fecundity in two Russian populations of this species (mean 6,955) far exceeded absolute fecundity (mean 1,877). In contrast, Hardisty (1961c, 1964) estimated potential fecundity in European river lamprey in the U.K. to be 14,000–26,000 oocytes (mean 19,000; Hardisty 1961c, 1964), suggesting that only ~16% of larval oocytes undergo atresia. Kuznetsov et al. (2016) estimated potential fecundity in larvae of anadromous and freshwater-resident European river lamprey in Russia to be 20,155 and 10,174, respectively, which is unexpectedly lower than mean absolute fecundity from these same populations (21,080 and 12,103, respectively). Either the potential fecundity estimates made by Kuznetsov et al. (2016) are underestimates or these populations gain rather than lose oocytes during the final larval stage (see below). Hardisty estimated potential fecundity in Great Lakes and anadromous sea lamprey at 110,000–165,000 and 182,000–328,000, respectively (Hardisty 1964, 1969, 1971), suggesting that ~45 and 21% of oocytes, respectively, are lost through atresia (Table 1.10).

Beamish and colleagues, using a method designed to count each oocyte in serial transverse sections once, concluded that potential fecundity (and hence atresia) was lower than reported by Hardisty (1961c, 1964) and less variable among life history types. Beamish and Thomas (1983) used measures of maximum oocyte diameter to categorize oocytes in each serial section as representing 25, 50, 75 or 100% of the maximum oocyte diameter. The number of oocytes was then counted in every tenth slide, and cumulative oocyte volume for each category was calculated. Potential fecundity in chestnut and southern brook lampreys was estimated at 8,289–20,641 and 1,035–2,800 oocytes, respectively, suggesting that ~3 and 12% of oocytes, respectively, were lost between the larval and adult stages. Using a similar method in Great Lakes sea lamprey, Barker et al. (1998) estimated potential fecundity to be

Table 1.10 Potential fecundity (mean ± one standard deviation, SD, and range) estimated in six lamprey species; different populations are listed separately. Percentage of oocytes lost to atresia for each species is estimated by comparing mean potential and absolute fecundities from the same geographic regions; absolute fecundity is given in Table 1.11

Species	Potential fecundity				References	Estimated atresia (%)
	n	Larval TL (mm)	Oocyte number			
		Range	Mean ± SD	Range		
PARASITIC SPECIES						
Eudontomyzon danfordi Carpathian lamprey (freshwater)	3	105–125	10,533	7,200–16,000	Renaud and Holčík (1986)	15.5
Ichthyomyzon castaneus chestnut lamprey (freshwater)	10	~94–150	14,542 ± 2,732*	8,289–20,641	Beamish and Thomas (1983)	3.2
Lampetra fluviatilis European river lamprey (anadromous) U.K.	3	61–88	17,567 ± 7,488	11,700[a]–26,000	Hardisty (1961c)	15.8
	4	79–99	19,225 ± 3,690	14,600–21,500	Hardisty (1961c)	
	6	89–110	14,000	8,000–20,000[b]	Hardisty et al. (1970)	
Lampetra fluviatilis European river lamprey (anadromous) Gulf of Finland, Russia	12	63–130	19,480 ± 2,196	12,138–31,164	Kuznetsov et al. (2016)	− 4.6
	23	85–105	20,830 ± 2,125	14,781±25,962	Kuznetsov et al. (2016)	
Lampetra fluviatilis European river lamprey (freshwater) Lake Ladoga, Russia	50	62–167	10,036 ± 2,998	5,434–20,577	Kuznetsov et al. (2016)	− 19.0
	29	63–104	9,283 ± 2,816	3,595–20,602	Kuznetsov et al. (2016)	
	25	72–117	11,203 ± 3,530	3,015–21,449	Kuznetsov et al. (2016)	
Petromyzon marinus sea lamprey (anadromous)	5	117–143	289,000 ± 29,992	255,000–328,000	Hardisty (1969)	21.3
	4	132–152	196,000	182,000–213,000	Hardisty (1969)	
Petromyzon marinus sea lamprey (freshwater) Great Lakes	6	103–114	134,000	114,000–165,000	Hardisty (1964, 1971)	44.9
	8	115–165	~81,000	33,000–129,000[c]	Barker et al. (1998)	8.9

(continued)

Table 1.10 (continued)

Species	Potential fecundity				References	Estimated atresia (%)
	n	Larval TL (mm)	Oocyte number			
		Range	Mean ± SD	Range		
BROOK LAMPREYS						
Lampetra planeri European brook lamprey U.K.	7	53–125	7,243 ± 1,754	5,500–10,600	Hardisty (1961c)	89.2
	4	48–54	5,999 ± 2,530	4,900–9,300	Hardisty (1961c)	
Lampetra planeri European brook lamprey Russia	14	94–154	6,574 ± 1,994	2,962–10,549	Kuznetsov et al. (2016)	73.0
	25	115–165	7,335 ± 2,110	3,468–14,580	Kuznetsov et al. (2016)	
Ichthyomyzon gagei southern brook lamprey	21	45–150	~1,870	1,035–2,800	Beamish and Thomas (1983)	11.8

*95% confidence limit

[a] Oocyte numbers as low as 6,351 were observed in this population, but Hardisty assumed that counts lower than 14,000 oocytes were European brook lamprey

[b] Transformers and early macrophthalmia

[c] Including undifferentiated germ cells; oocytes alone numbered 19,000–65,000, which is lower than mean absolute fecundity (~70,000 eggs)

33,000–129,000, again suggesting that a smaller proportion of oocytes are lost to atresia (~9%) than previously suggested (Table 1.10). Barker et al. (1998) considered Hardisty's potential fecundities to be overestimates, because Hardisty's method did not consider that an oocyte might be sectioned and counted more than once. It should be further noted that Barker et al. (1998) included undifferentiated germ cells in their estimates of potential fecundity. Counting only oocytes, Barker et al. (1998) estimated potential fecundity to be 19,000–65,000, which is less than mean absolute fecundity in Great Lakes sea lamprey (~70,000). As discussed in Sect. 1.4.1.2, there is evidence in some parasitic species that, during the late larval stage, small numbers of residual undifferentiated germ cells may develop into oocytes following completion of ovarian differentiation, potentially compensating for, or even exceeding, oocyte atresia.

Histological evidence for oocyte atresia is likewise inconsistent, and there is some debate regarding the timing of atresia. Barker et al. (1998) saw no atretic oocytes in typical larvae (see Sect. 1.4.1.4). Hardisty (1964) indicated that degenerating oocytes are often seen in European brook lamprey ovaries at all stages of the larval period, but he acknowledged that the proportion undergoing atresia is far too small to account for the inferred reduction in oocyte numbers between larval and adult stages. Some authors suggest that, where atresia does occur, it is complete or largely complete by the onset of metamorphosis (Hardisty 1971; Beamish and Thomas 1983; Docker and Beamish 1991). In contrast, Weissenberg (1927) reported extensive degeneration of oocytes in European brook lamprey ovaries during metamorphosis, and Hughes and Potter (1969) and Hardisty (1971) suggested that atresia occurring immediately before or at the onset of vitellogenesis may be significant in non-parasitic species.

Granted, it is possible that there is simply considerable variation in the extent and timing of atresia among species. For example, it has been suggested that a high degree of atresia should be evident in brook lampreys that have recently diverged from their parasitic ancestor (see Docker 2009; Spice and Docker 2014). That is, recently derived brook lampreys may still elaborate a large number of oocytes during the larval stage and then adjust (through atresia) the final number of oocytes to correspond with their reduced adult body size relative to the parasitic ancestor. Beamish and Thomas (1983) hypothesized that the lack of atresia in southern brook lamprey suggests that it diverged from the parasitic chestnut lamprey some time ago, thus giving natural selection sufficient time in which to reduce the number of oocytes elaborated in the larvae to better match body size at maturity. However, based on genetic evidence, it is not likely that southern brook lamprey is long separated from the chestnut lamprey; both southern and European brook lampreys appear to be recently diverged from their parasitic ancestors (Docker 2009; Chap. 4). Furthermore, when comparing the similar number of oocytes observed per cross-section in southern and European brook lampreys (Table 1.8), it seems unlikely that potential fecundity is so different in the two species. Hardisty (1971) likewise suggested that a high degree of atresia in the Great Lakes sea lamprey is indicative of its recent derivation from an anadromous ancestor. He argued that the ~40% decline in oocyte numbers by maturity is the result of post-metamorphic "adjustment," because the smaller-bodied adult landlocked sea lamprey lacks sufficient energy resources to support all the developing oocytes. It was Hardisty's view that the greater imbalance in oocyte numbers between larval and adult landlocked sea lamprey relative to the anadromous sea lamprey represents an incomplete transition from an anadromous to a fully landlocked form. However, Barker et al. (1998)'s estimate of potential fecundity in the Great Lakes sea lamprey suggests no such imbalance. Uncertainty regarding the reliability of different potential fecundity estimates makes it difficult to correlate time of divergence between life history types and extent of atresia in the derived form.

Nevertheless, it should be noted that potential fecundity is still smaller in land-locked versus anadromous sea lamprey and in non-parasitic versus parasitic species. This means that, even if there is post-larval "tinkering" of oocyte numbers to correspond with adult size, there has still been some reduction in the number of oocytes elaborated during the larval stage (i.e., in anticipation of the smaller adult size) in the derived forms. Lower potential fecundity in these smaller-bodied lampreys appears to be mediated by the earlier onset of oogenesis (see Sect. 1.4.1.2). Thus, as suggested by Hardisty (1964) and Beamish and Thomas (1983), fecundity differences among species are very likely genetically based and largely determined at or before sex differentiation (although there is some evidence of intraspecific variation; see Spice and Docker 2014; Sect. 1.4.1.2). Within each species, individual variations in absolute fecundity may then be related to individual growth or body condition. By this view, final oocyte numbers would be dependent on the resources available to support the developing oocytes and mediated through differences in the extent of atresia. This would provide a mechanism by which the maximum number of oocytes that theoretically could be brought to maturity is produced in the larval stage, and then the final number can be adjusted downward (but not upward) prior to maturity. For lampreys producing more oocytes than they generally mature, the energetic inefficiencies associated with this strategy may be offset by the potential to enhance fitness under conditions that favor growth and the accumulation of energy reserves. However, the extent to which atresia is influenced by energy expenditures related to other demands (e.g., long versus short spawning migrations) and the availability and quality of food remains unknown. Further examination of atresia in multiple large populations, using standardized methods, is required.

1.6.3 Absolute Fecundity

Absolute or actual fecundity of lampreys varies considerably among and within species. It increases with body size, varying at least two orders of magnitude between the large anadromous parasitic species and the much smaller brook lampreys (Table 1.11; Fig. 1.13). For example, fecundity of anadromous sea lamprey (mean TL 743 mm, range 666–841 mm) in the St. Lawrence River and its tributaries ranged from 123,873 to 258,874, with a mean of almost 172,000 (Vladykov 1951). Fecundity of upstream migrants in the St. John River in New Brunswick (mean 729 mm TL) ranged from 151,836 to 304,832 and averaged 210,228 eggs (Beamish and Potter 1975). Pacific lamprey, another large anadromous species, is also highly fecund: mean fecundities of 127,178 and 140,312 eggs have been reported (Clemens et al. 2013 and Kan 1975, respectively), and maximum reported fecundity is close to 240,000 eggs in females measuring ~400–500 mm TL (Kan 1975). In comparison, females of the pouched lamprey (mean TL >500 mm TL) have relatively low fecundity (mean 57,942; Potter et al. 1983). Hardisty et al. (1986) suggested that the number of eggs in this species is limited by its slender trunk. Pouched lamprey is also known to undergo a very prolonged upstream migration (i.e., spending ~16 months in fresh water; Potter et al. 1983), which may limit the energy that can be allocated for egg maturation. In contrast, duration of the spawning migration is typically shorter in anadromous sea lamprey (~3–4 months; see Moser et al. 2015). However, the highly fecund Pacific lamprey also spends >1 year in fresh water prior to spawning (Clemens et al. 2009).

Other anadromous lamprey species are smaller at maturity than the three species above, and they have correspondingly lower fecundities. However, there is considerable variation within and among species. For example, average fecundity of anadromous European river lamprey ranges from 15,900 in small-bodied populations (mean TL 285 mm: Hardisty 1964) to >36,000 in larger-bodied river lamprey from Poland (mean TL 405–432 mm; Witkowski and Jęsior 2000). Fecundity in anadromous Arc-

Table 1.11 Absolute and relative fecundity (mean ± one standard deviation, SD) in 13 parasitic and 16 non-parasitic (brook) lamprey species, and egg diameter at maturity in nine and 11 parasitic and brook lamprey species, respectively; relative fecundity is number of eggs per g female body weight (W). Published fecundity values were included only if total length (TL) of the specimens was provided or could be approximated; for egg diameter, if both long and short axes were measured, only the long axis is given

Species	n	Female size TL (mm) Mean	Range	W (g) Mean	Fecundity Absolute fecundity (total number of eggs) Mean ± SD	Range	Relative fecundity n/g ± SD	Egg diameter (μm) Mean ± SD	Range	References
PARASITIC SPECIES										
Caspiomyzon wagneri Caspian lamprey (anadromous)			370–410			20,000–32,000				Berg (1948)
	59	387	310–485	106.7	41,924 ± 5,382	31,800–51,200	398 ± 93	920 ± 81	780–1,150	Nazari and Abdoli (2010)
Entosphenus tridentatus Pacific lamprey (anadromous)	25		~403–496		140,312 ± 12,925*	98,300–238,400	486 ± 37*	800	600–1,000	Kan (1975)
	30	~562		314.6	127,178 ± 33,405		397 ± 71	700[a]		Clemens et al. (2013)
Entosphenus minimus Miller Lake lamprey (freshwater)	10		~70–95		604	503–727	525			Kan and Bond (1981)
Eudontomyzon danfordi Carpathian lamprey (freshwater)	2	213	210–215		8,900	7,500–10,300				Renaud and Holčík (1986)
Geotria australis pouched lamprey (anadromous)	8	~526[b]		211	57,943	48,004–68,212	295	1,120		Potter et al. (1983)
	1	370[c]		162	56,100		346	1,180		Baker et al. (2017)
Ichthyomyzon castaneus chestnut lamprey (freshwater)								1,070		Beamish and Thomas (1983)
	22	246		38.7	14,078 ± 3,506		377 ± 87	900 ± 50		Schuldt et al. (1987)

(continued)

Table 1.11 (continued)

Species	n	Female size TL (mm) Mean	Range	W (g) Mean	Fecundity (total number of eggs) Absolute fecundity Mean ± SD	Range	Relative fecundity n/g ± SD	Egg diameter (μm) Mean ± SD	Range	References
Ichthyomyzon unicuspis silver lamprey (freshwater)	10	253	201–312	45.1	19,012 ± 5,917	12,006–29,412	498 ± 272	775 ± 61[a]	700–850	Vladykov (1951)
	21	290		63.0	13,403 ± 7,106		210 ± 89	990 ± 70		Schuldt et al. (1987)
	12	287		56.8	21,259 ± 7,583		375 ± 99	910 ± 70		Schuldt et al. (1987)
	24	313		80.8	22,820 ± 4,971		287 ± 60	940 ± 40		Schuldt et al. (1987)
Lampetra ayresii western river lamprey (anadromous)	2	203	175–230		24,343	11,398–37,288	1,160	700[c]		Vladykov and Follett (1958)
Lampetra fluviatilis European river lamprey (anadromous)	54		260–400		21,000		467		700–900	Berg (1948)
		285	185–350		15,900	7,500–28,200				Hardisty (1964)
			~341–385			26,000–41,000			880–1,030	Hardisty (1986)
	10[d]	432	385–460	129	36,868 ± 8,640	26,530–52,072	265 ± 47			Witkowski and Jesior (2000)
	16[e]	405	337–462	136	37,177 ± 6,721	23,460–49,479	283 ± 65	961 ± 38	878–1,016	Witkowski and Jesior (2000)
	13	297	297–350		21,080 ± 3,281	15,878–28,172				Kuznetsov et al. (2016)
Lampetra fluviatilis European river lamprey (anadromous praecox)			~180–250			650–10,000			650–720	Hardisty (1986)
			~212–214			10,000–16,000				Hardisty (1986)

(continued)

Table 1.11 (continued)

Species	n	Female size			Fecundity			Egg diameter (μm)		References
		TL (mm)		W (g)	Absolute fecundity (total number of eggs)		Relative fecundity			
		Mean	Range	Mean	Mean ± SD	Range	n/g ± SD	Mean ± SD	Range	
Lamperta fluviatilis European river lamprey (freshwater)	15	226	200–247		10,135		661			Tsimbalov et al. (2015)
	20		264–348		12,103 ± 2,773					Kuznetsov et al. (2016)
Lethenteron camtschaticum Arctic lamprey (anadromous)			≤434		84,200	50,000–117,000				Berg (1948)
			≤625			80,000–107,000				Berg (1948)
	2		300–350	54.6	51,500		943			Morozova (1956)
	12		351–400	86.2	81,478		945			Morozova (1956)
	29		401–450	113.7	94,685		833			Morozova (1956)
	3		451–500	167.0	101,933		609			Morozova (1956)
									520–600	Holčík (1986a)
	15		353–442			62,936–119,180		1,040 ± 100	850–1,230	Yamazaki et al. (2001)
	10	279	170–330		13,669	12,272–34,586	289	770	max 1,250	Kucheryavyi et al. (2007)
Lethenteron camtschaticum Arctic lamprey (freshwater)	18		~170–300		21,415 ± 5,261	9,790–29,780				Nursall and Buchwald (1972)
Lethenteron camtschaticum Arctic lamprey (freshwater non-parasitic form)			100–165		1,478	468–3,341	346	600		Kucheryavyi et al. (2007)
Mordacia mordax short-headed lamprey (anadromous)	20	366			9,794 ± 89	5,462–13,372				Hughes and Potter (1969)
	20	326			5,992 ± 65	3,789–8,053				Hughes and Potter (1969)

(continued)

Table 1.11 (continued)

Species	n	Female size TL (mm) Mean	Female size TL (mm) Range	W (g) Mean	Fecundity Absolute fecundity (total number of eggs) Mean ± SD	Fecundity Absolute fecundity (total number of eggs) Range	Relative fecundity n/g ± SD	Egg diameter (µm) Mean ± SD	Egg diameter (µm) Range	References
Petromyzon marinus sea lamprey (anadromous)	10	743	666–841	842	171,590 ± 46,986	123,873– 258,874	205 ± 36	940 ± 50		Vladykov (1951)
	40	729	~640–840		210,228	151,836–304,832	236		770–1,070	Beamish and Potter (1975)
Petromyzon marinus sea lamprey (freshwater)										
L Huron	70	434	320–536	181.3	62,014 ±18,428	21,000–107,138	353 ± 68	1,070[a]	1,030–1,110	Applegate (1949)
L Michigan	10	359	291–439	127.2	62,870 ± 18,219	38,678–85,712	549 ± 182	860 ± 47	800–950	Vladykov (1951)
L Huron (North Channel)	10	384	330–435	135.7	56,913 ± 13,833	28,891–74,023	420 ± 73	850 ± 86	750–1,000	Vladykov (1951)
Cayuga L	29	395	297–510	147.0	45,602 ± 20,134	13,974–85,162	316 ± 76			Wigley (1959)
L Superior	29	405	340–511	158.3	68,599 ± 12,971	43,997–101,931	463 ± 121	860 ± 80[a]	680–1,030	Manion (1972)
L Michigan (Green Bay)	14	485	410–569	267.4	97,016 ± 29,398	48,974–146,132	363	980		Johnson (1982)
L Superior	20	454		199.5	74,515		374			Morse (unpublished data)[f]
L Michigan	30	452		264.0	81,748		310			Morse (unpublished data)[f]
L Huron	20	436		232.0	77,184		333			Morse (unpublished data)[f]
L Ontario	30	468		286.0	107,429		376			Morse (unpublished data)[f]
L Erie	20	481		277.0	94,344		341			Morse (unpublished data)[f]

(continued)

Table 1.11 (continued)

Species	n	Female size TL (mm)		W (g)	Fecundity Absolute fecundity (total number of eggs)		Relative fecundity	Egg diameter (μm)		References
		Mean	Range	Mean	Mean ± SD	Range	n/g ± SD	Mean ± SD	Range	
L Ontario	33	468		274	92,971 ± 482		336			O'Connor (2001)
L Champlain	29	456	364–550	173.8	67,660 ± 6,870*		389			Smith and Marsden (2007)
L Superior	30	459		228.8	80,228		351			Gambicki and Steinhart (2017)
L Ontario	49	503		268.1	95,212		355			Gambicki and Steinhart (2017)
Tetrapleurodon spadiceus Mexican lamprey (freshwater)	5	210	187–225	24.9	7,884 ± 983	6,617–9,095	317 ± 29			Álvarez del Villar (1966)
BROOK LAMPREYS										
Eudontomyzon mariae Ukrainian brook lamprey	1	147			2,429					Berg (1948)
Eudontomyzon hellenicus Macedonia brook lamprey	4	103	99–105		994 ± 106	897–1,138	645	1,080 ± 140	900–1,230	Lapierre and Renaud (2015)
Ichthyomyzon fossor northern brook lamprey	2	117	112–122		1,095	1,050–1,140		1,200		Leach (1940)
	9	138	128–150	4.3	1,524 ± 311	1,115–1,979	355 ± 36	1,062 ± 91	950–1,200	Vladykov (1951)
	11	133		4.4	1,634 ± 229		378 ± 35	990 ± 20		Schuldt et al. (1987)
	7	129		4.4	1,668 ± 302		386 ± 22	920 ± 40		Schuldt et al. (1987)
	11	124		3.8	1,475 ± 282		389 ± 54	940 ± 50		Schuldt et al. (1987)

(continued)

Table 1.11 (continued)

Species	n	Female size TL (mm) Mean	Range	W (g) Mean	Fecundity Absolute fecundity (total number of eggs) Mean ± SD	Range	Relative fecundity n/g ± SD	Egg diameter (μm) Mean ± SD	Range	References
Ichthyomyzon gagei southern brook lamprey	10		101–153		1,787 ± 754	1,000–3,264		907 ± 40	840–960	Dendy and Scott (1953)
	14	128	116–143		1,681 ± 269	1,102–2,114		950 ± 78	830–1,120	F. W. H. Beamish (1982)
	14		~115–147			1,000–3,264		930		Beamish and Thomas (1983)
	310	118	98–142[l]		1,480	~475–2,922				Beamish et al. (1994)
Lampetra aepyptera least brook lamprey	1	~104			1,164					Seversmith (1953)
	75[h]	134	101–152		2,390	681–3,721[e]	546			Docker and Beamish (1994)
Lampetra hubbsi Kern brook lamprey	11	99	89–108		1,865 ± 413	1,433–2,762	358	1,020 ± 50	960–1,100	Lapierre and Renaud (2015)
Lampetra lanceolata Turkish brook lamprey	12		~150–170		1,899	1,567–2,290				Gözler et al. (2011)
Lampetra planeri European brook lamprey	14	133	115–156		1,443 ± 382	950–2,087	440	984 ± 33	930–1,040	Hardisty (1964)
	27	~110			873 ± 221	330–1,276	333			Malmqvist (1978)
	12	~125			1,043 ± 255	700–1,573	316			
	16	~150			1,356 ± 261	983–1,833	309			
	6		132–144		1,740 ± 529	993–2,259				Kuznetsov et al. (2016)
	18		134–189		2,014 ± 471	1,288–3,355				Kuznetsov et al. (2016)
Lampetra zanandreai Po brook lamprey			~139–148		1,850	1,790–1,904				Zanandrea (1951)
Lethenteron alaskense Alaskan brook lamprey	4		150–175		2,846 ± 542	2,188–3,477		900[a]		Vladykov and Kott (1978)
Lethenteron appendix American brook lamprey	10	143	116–158	5.1	2,339 ± 756	1,085–3,648	453 ± 73	895 ± 45	830–950	Vladykov (1951)
	16	180	159–194		3,787	2,698–5,185	357			Kott (1971)
		124			1,691	1,327–2,070				Rohde et al. (1976)
	27	~186			2,833 ± 212*					Seagle and Nagel (1982)
	10	134		4.2	1,617 ± 360		388 ± 59	900 ± 30		Schuldt et al. (1987)
	29	130		4.2	2,020 ± 464		484 ± 60	890 ± 50		Schuldt et al. (1987)

(continued)

Table 1.11 (continued)

Species	n	Female size TL (mm)		W (g)	Fecundity — Absolute fecundity (total number of eggs)		Relative fecundity n/g ± SD	Egg diameter (μm)		References
		Mean	Range	Mean	Mean ± SD	Range		Mean ± SD	Range	
Lethenteron kessleri Siberian brook lamprey	8		166–230			1,820–5,800			730–1,120	Holčík (1986c)
	1	114	116–133		2,125	1,387–1,644		1,150 ± 40	1,120–1,220	Yamazaki et al. (2001)
	10	159	152–166		1,895 ± 329	1,301–2,348[i]		1,120		Yamazaki et al. (2001); Yamazaki and Koizumi (2017)
Lethenteron reissneri Far Eastern brook lamprey	15					1,720–3,360			680–840	Renaud and Naseka (2015)
Lethenteron sp. N	12	144	97–129		2,498	697–990		1,140 ± 100	1,070–1,340	Yamazaki et al. (2001)
	1							980		Yamazaki et al. (2001)
Lethenteron sp. S	11		105–127			723–1,371		1,150 ± 40	1,100–1,210	Yamazaki et al. (2001)
	19	135	105–147		2,942	495–1,545		1,240 ± 80	1,080–1,370	Yamazaki et al. (2001)
	1	130						1,010		Yamazaki et al. (2001)
	1				1,431			1,170		Yamazaki et al. (2001)
Mordacia praecox Australian brook lamprey	20	124			474 ± 7	326–675		699 ± 18		Hughes and Potter (1969)
Tetrapleurodon geminis Mexican brook lamprey	3	130	108–148	5.5	2,222 ± 1,233	990–3,456	386 ± 138			Álvarez del Villar (1966)

*95% confidence limit instead of 1 SD

[a] Excluding individuals identified as not yet mature

[b] TL at or approaching maturity in October; TL at freshwater entry 655 mm

[c] TL at or approaching maturity (Oct 2013); TL at freshwater entry 506 mm in Aug 2012 (Cindy F. Baker, National Institute of Water and Atmospheric Research, Hamilton, NZ, personal communication 2018)

[d] Autumn migrants

[e] Spring migrants

[f] From Gambicki and Steinhart (2017)

[g] Transformers

[h] Means for 19 populations

[i] Total eggs released while mating in laboratory

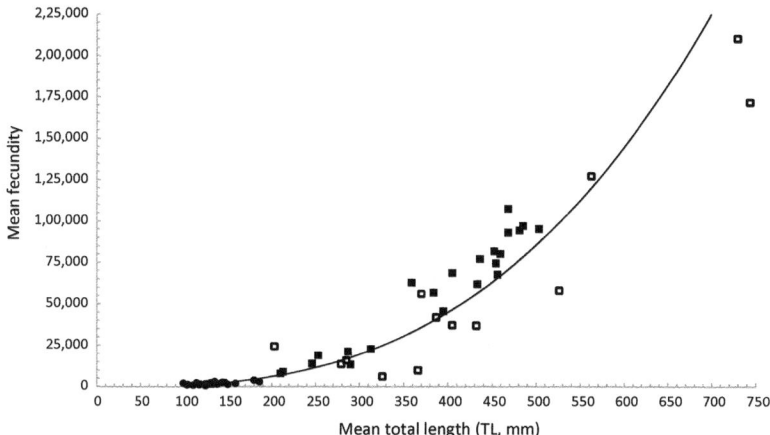

Fig. 1.13 Relationship between mean absolute fecundity (total number of eggs in adults) and total body length (TL) in 11 parasitic and 13 non-parasitic lamprey species; values are taken from Table 1.11 where both mean fecundity and mean TL were available (n = 64). *Solid circles* represent non-parasitic species; *solid squares* represent freshwater parasitic species or populations; *open squares* represent anadromous parasitic lampreys. A power function (*line*) describes the relationship between fecundity (y) and TL (x) as $y = 0.0014x^{2.8896}$ ($R^2 = 0.9375$, $p < 0.0001$)

tic lamprey is similarly variable, given the wide range of adult body sizes seen among and within populations (Table 1.11). However, it appears that some interspecific differences in fecundity are not strictly due to size differences. Short-headed lamprey display low fecundities (5,992–9,794) for their body length (mean TL 326–366 mm) relative to their Northern Hemisphere counterparts (Hughes and Potter 1969). Even the relatively small-bodied western river lamprey (mean TL 203) has higher fecundity (mean 24,343, albeit measured in only two specimens; Vladykov and Follett 1958).

Freshwater-resident parasitic lampreys are also smaller at maturity than the largest anadromous species and have correspondingly lower fecundities, although the landlocked sea lamprey in the Great Lakes, Cayuga Lake, and Lake Champlain are larger and more fecund than many of the smaller-bodied anadromous lampreys (Table 1.11). At maturity, female landlocked sea lamprey range in size from ~300 to 550 mm TL and have fecundities less than half those of their anadromous counterpart. The lowest mean fecundity reported (45,598 eggs) was for sea lamprey from Cayuga Lake (mean TL 395 mm; Wigley 1959), and the highest mean (97,016 eggs) was reported in sea lamprey from the Green Bay region of Lake Michigan (mean TL 485 mm; Johnson et al. 1982). Individual variation is high, ranging from about 14,000 to over 146,000, with an overall average of ~70,000 eggs per female. Fecundities of other freshwater parasitic species are well below those for sea lamprey, in accord with their smaller sizes (Table 1.11). For example, Nursall and Buchwald (1972) reported an average of 21,415 eggs for a freshwater-resident population of Arctic lamprey in Great Slave Lake in the Northwest Territories (~170–300 mm TL). Fecundity in chestnut and silver lampreys averages ~14,000 (mean TL 246 mm; Schuldt et al. 1987) and ~19,000 (mean TL 253–313 mm; Vladykov 1951; Schuldt et al. 1987), respectively, and the smallest freshwater-resident parasitic lamprey, the Miller Lake lamprey (~70–95 mm TL), has a mean fecundity of only ~600 eggs (Kan and Bond 1981).

Fecundity (and body size) is less variable in the non-parasitic brook lampreys. Of the 16 species listed in Table 1.11 (i.e., where both fecundity and TL were provided or could be approximated), TL at maturity typically ranges from ~100 to 150 mm, although individuals as small as 89 mm (Kern brook lamprey; Lapierre and Renaud 2015) and as large as 230 mm (Siberian brook lamprey *Lethenteron kessleri*; Yamazaki et al. 2001) have been reported. Mean fecundity in brook lampreys ranges from 474 in the Australian brook lamprey (Hughes and Potter 1969) to 3,787 in the American brook lamprey (Kott 1971). Interestingly, the Australian brook lamprey is not particularly small (mean TL 124 mm). Therefore, in general, it appears that all three Southern Hemisphere species examined to date have lower fecundities than Northern Hemisphere species of the same size. Although egg counts in a few individuals of other brook lamprey species are within the range observed in the Australian brook lamprey (326–675), fecundity for most brook lamprey species ranges from ~1,400 to 2,500, and the overall mean from the 28 studies included in Table 1.11 was 1,773.

Despite the exceptions noted above, there is a clear relationship between fecundity and female body size across all species. Using all studies for which mean TL and fecundity were available (Table 1.11), the relationship is described by the equation:

$$F_m = 0.0014 \, TL_m^{2.8896} \left(n = 64, \; R^2 = 0.9375, \; p < 0.001 \right), \text{ or} \qquad (1.1)$$

$$\text{Log } F_m = 2.8896 \log TL_m - 2.867 \left(n = 64, \; R^2 = 0.9375, \; p < 0.001 \right) \quad (1.2)$$

where F_m is the mean fecundity for each species in each reported investigation and TL_m is the corresponding mean total length (mm). The number of individuals representing each species ranged from 1 to 310 and averaged 22.6. Each record of mean fecundity was entered separately and not combined with other records for the same species. Therefore, across populations and species, the total number of eggs in lampreys increases approximately with the cubic power of TL (Fig. 1.13). In broad terms, with the doubling of TL, fecundity increases by almost an order of magnitude regardless of species. Hardisty (1964) reasoned that the relationship between egg number and TL should be similar regardless of species, because all lampreys exhibit essentially similar body forms and there is relatively little variation in egg size among species (see Sect. 1.6.1).

This general relationship between fecundity and TL also appears to apply within most species (Table 1.12). Fecundity was related to the cubic power (2.77, 2.82, 3.10, 3.21, 3.46) of TL in five brook lamprey species, although the relationship was less consistent in the parasitic species. In five sea lamprey populations (one anadromous and four landlocked populations), the exponent was found to be 2.11–3.21, but it was only 0.69–1.71 in European river lamprey and two other landlocked sea lamprey populations. The nature of the relationship between fecundity and TL may differ in some of these studies because of small size ranges or measurement and sample size limitations. Alternatively, there may be subtle differences in the relationship between fecundity and TL between non-parasitic and parasitic species (see below).

Although TL is generally a more convenient measure of body size than weight (particularly in the field), Smith and Marsden (2007) found body weight to be a better descriptor of size for estimating fecundity of sea lamprey from Lake Champlain and other freshwater-resident populations. They examined a number of morphometric indices including weight, TL, GSI, and other morphological measures (i.e., to test if one body segment is a more consistent predictor of fecundity than TL if loss of length during the spawning run occurs disproportionately over different segments). The best model combined both female wet weight and GSI, but support was also strong for a model based on wet weight alone. In the Lake Champlain population, wet weight alone explained 68% of the variation in fecundity, and wet weight and GSI combined explained 72% of the variability. Therefore, in situations where collection of GSI data is not convenient or possible (e.g., when non-lethal sampling is desirable), wet weight was sufficient. A model based on TL alone was not well supported.

Across all species, where mean fecundity (F_m) and mean weight (W_m in g) were available (Table 1.11), the relationship was described by the equation:

$$\text{Log } F_m = 0.9637 \log W_m + 2.6509 \left(n = 40, \ R^2 = 0.9572, \ p < 0.0001\right)$$
(1.3)

Thus, there is an almost 1:1 increase in egg number with increases in female body weight (e.g., 2,111 eggs in a 5-g American brook lamprey and 19,418 eggs in a 50-g silver lamprey). Given the large range in body weight observed across all species (e.g., from 3.8 g in northern brook lamprey to 842 g in anadromous sea lamprey), log transformation of the data was required. Within species, a linear relationship using untransformed data was the best descriptor (Table 1.12), although there was little difference in the fit quality with log-transformed data, and the same general relationship between fecundity and weight appears to apply within most species. Details of the relationship (i.e., slope and intercept) varied among species, but there was still an approximate doubling of fecundity with doubling of female body weight. For the nine data sets where both TL and weight were available, R^2 values were higher in six cases using weight data.

That female weight is, in general, a better predictor of fecundity is not surprising, given that the eggs can constitute ~25–35% of the female's weight at maturity (see Sect. 1.5.3). Thus, it is not merely a matter of expecting heavier females to be able to produce or mature more eggs, but more fecund females will generally be heavier as a result of these eggs. Using the eight data sets that provided ovary (or total egg) weight, we found that R^2 values were consistently (although only marginally) higher when fecundity was regressed on total body weight (0.188–0.817) rather than ovary-free body weight (0.124–0.771). Thus, although total body weight is a

Table 1.12 Relationship between absolute fecundity (F) and total length (TL, mm) and weight (W, g), and between relative fecundity (RF) and TL in two parasitic and four non-parasitic (brook) lamprey species, where b is the slope, a is the intercept, and R^2 is the coefficient of determination. For the relationships with length, TL, F and RF were log transformed. TL and W range for each population is given in Table 1.11

Species	n	log F = b log TL + a[a]			F = b W + a			Log RF = b log TL + a			References
		b	a	R^2	b	a	R^2	b	a	R^2	
PARASITIC SPECIES											
Lampetra fluviatilis European river lamprey (anadromous)	26[b]	0.976	2.026	0.128							Witkowski and Jęsior (2000)
Petromyzon marinus sea lamprey (anadromous) St Lawrence R, QC (20 May 1949)	10	2.879	−3.048	0.320	199	4.297	0.584*	−0.987	5.138	0.121	Vladykov (1951)
Petromyzon marinus sea lamprey (freshwater)											
Carp Cr and Ocqueoc R, MI (L Huron) 15 Apr–10 July 1947	70[c]	2.231	−1.106	0.677*	260	14.905	0.778*	−0.668	4.300	0.188*	Applegate (1949)
Thessalon R, ON (North Channel, L Huron) 02 June 1948	10	2.847	−2.618	0.710*	246	22.489	0.521*	−0.518	3.954	0.058	Vladykov (1951)
Hibbards Cr, WI (L Michigan) 09 June–07 July 1948	10	1.709	0.421	0.626*	285	26.677	0.775*	−1.418	6.338	0.447*	Vladykov (1951)
Cayuga Inlet, NY (Cayuga L) 30 Apr 1951	29	3.214	−0.513	0.696*	260	7,357	0.769*	−0.204	3.016	0.012*	Wigley (1959)

(continued)

Table 1.12 (continued)

Species	n	log F = b log TL + a[a]			F = b W + a			Log RF = b log TL + a			References
		b	a	R²	b	a	R²	b	a	R²	
Chocolay R, MI (L Superior)	29[d]	0.689	3.033	0.121	104	52,108	0.188*	−2.179	8.330	0.620*	Manion (1972)
26 May 1960	7	0.720	2.934	0.135	105	47,288	0.327	−1.879	7.500	0.655*	
16 June 1960	22	1.226	1.649	0.135	264	31,941	0.319*	−1.290	6.039	0.182*	
Tahquamenon and Betsey R, MI (L Superior) May, June 2011	30	2.110	−0.708	0.501*							Gambicki and Steinhart (2017)
BROOK LAMPREYS											
Ichthyomyzon fossor northern brook lamprey	9	3.104	−3.461	0.661*	350	20	0.786*	0.165	2.196	0.008	Vladykov (1951)
Lampetra aepyptera least brook lamprey	75[e]	3.460	−4.03	0.679*				0.56	1.25	0.076*	Docker and Beamish (1994)
Jay Cr, AL	13							−1.89	6.77	0.627*	
L Whippoorwill Cr, KY	14							−5.93	15.7	0.386*	
Lampetra planeri European brook lamprey	14	2.818	−2.835	0.742*							Hardisty (1964)
Lethenteron appendix American brook lamprey	10[f]	2.767	−2.606	0.869*	405	232	0.817*	0.183	2.258	0.019	Vladykov (1951)
	16	2.831	−5.361	0.829*	217	1,465	0.387*	−1.489	5.907	0.314*	Kott (1971)

*Statistically significant (p < 0.05)

[a]Sample calculation of F from TL: e.g., for northern brook lamprey where TL = 130 mm, Log F = 3.1014 and F = $10^{3.1014}$ or 1,263 eggs

[b]Autumn and spring migrants were pooled

[c]Predominantly (n = 58) from Carp Cr; n = 8, 1, and 3 from Ocqueoc R, Ocqueoc L, and Cheboygan R

[d]Migrants from 26 May and 16 June were pooled

[e]Data from eight populations were pooled

[f]Data from two populations were pooled

good predictor of an individual female's fecundity, ovary-free body weight (or other independent measures of size or nutritional status) may be better metrics when trying to explain the factors that determine the number of eggs that survive to maturity. The presumption is that individuals with a high nutritional status are better able to provide sufficient nourishment for developing oocytes and, conversely, that individuals with low nutritional levels are forced to provide this nourishment from the catabolism of a portion of their oocytes and other reserves. While such statements seem implicitly obvious, they lack explicit or supportive evidence. Further, there is currently no standard by which to appropriately evaluate "nutritional status." Condition factor ($W/TL^3 \times 10^6$, where W is weight in g and TL is total length in mm) is often used to infer a fish's nutritional condition (Ricker 1975) and is frequently used to measure "plumpness" of lampreys (e.g., Manzon et al. 2015). However, body weight of lampreys consists mostly of water (Lowe et al. 1973); as lipids and proteins are catabolized during the non-trophic spawning migration, they are replaced by water, thus further increasing its large contribution to body weight (Beamish et al. 1979; Bird and Potter 1983; Araújo et al. 2013). Proximate body composition of anadromous lampreys has been assessed at different stages of their life cycle (e.g., Beamish et al. 1979; Bird and Potter 1983; Araújo et al. 2013), but no comparisons have been made among individuals to determine if fecundity is correlated with these measures of nutritional status.

Nevertheless, deviations from the predictable relationship between body size and fecundity will help provide some initial insights into the factors that might "fine tune" this relationship, both among and within species. For example, all five species of brook lampreys studied to date showed that the total number of eggs increased with the cubic power of TL, leading us to conclude that the number of eggs brought to maturity in non-parasitic species approaches the physiological or anatomical limits imposed by body size. In contrast, the more pronounced variation seen among parasitic lampreys suggests species- or population-specific differences in the proportion of energy allocated to eggs. It is interesting that many of the anadromous lampreys have slightly lower fecundities than would be predicted based on the general power relationship with TL (Fig. 1.13). This appears not to be a function of their larger size alone, because the same pattern is not seen among the larger-bodied freshwater-resident lampreys, and it instead may be related to how much of a female lamprey's finite energy reserves are devoted to elaboration of the eggs relative to energy expended during the upstream spawning migration or other demands. Phylogenetic differences should also be explored; as pointed out above, the three of the Southern Hemisphere lamprey species examined to date have lower fecundities than their Northern Hemisphere counterparts at the same TL. Hardisty et al. (1986) suggested that the number of eggs in the pouched lamprey might be limited by its slender trunk, although its relatively long trunk (which, as a proportion of TL, increases during the spawning migration in females but not in males) appears to help compensate for the small body depth. The very different spawning behavior shown by pouched lamprey may also help explain its disproportionately low fecundity relative to its TL. Pouched lamprey eggs and embryos may suffer less mortality compared to Northern Hemisphere lampreys, because pouched lamprey eggs are attached to the underside of large boulders in cryptic nests, and the extended survival of the spawning adults may provide further protection (Baker et al. 2017). If early mortality rates are lower, selection may have favored lower fecundity but with more resources apportioned to the long upstream migration and post-spawning survival. Spawning behavior in the other Southern Hemisphere lamprey species has not been described to date.

Exploring intra-specific differences in the relationship between body size and fecundity will also be informative. For example, Manion (1972) found that fecundity ranged from 45,285 to 89,565 eggs in three females of identical TL (394 mm) that were all collected on the same day; understanding the factors that contribute to such individual differences would be valuable. Comparisons among sea lamprey populations (e.g., before and after initiation of sea lamprey control or among locations) would also be informative. Determining whether spatial and temporal differences in fecundity are merely reflective of differences in body size or whether they represent a proportionately greater or smaller allocation of resources to gonadal development will help us understand the effect of sea lamprey control measures on the reproductive potential of different populations (see Sect. 1.6.5).

As a final note in this section, because methodological differences employed when estimating potential fecundity appear to have resulted in large discrepancies among studies, it should be pointed out that methodological differences also exist among the various studies cited here. However, given the ease of counting mature eggs compared to larval oocytes, differences related to methodology are likely to be much smaller than when estimating potential fecundity. For the less fecund brook lampreys, many studies counted the total number of eggs per female (e.g., Docker and Beamish 1991; Lapierre and Renaud 2015), although others estimated the total number from subsamples and then extrapolated to estimate total fecundity by multiplying by the total weight of the ovary. For example, Vladykov (1951) compared fecundity estimated from a 1-g subsample of the ovary and total egg counts in nine American brook lamprey and nine northern brook lamprey, and he found that the values differed by an average of 0.4 and 1.1%, respectively, and never exceeded 5%. In Beamish et al. (1994), the eggs from each southern brook lamprey female were spread over a grid marked in 25-mm^2 squares, and the total number of eggs was estimated by multiplying the number of squares covered by the mean count per square. The proportion of eggs counted ranged from 20 to 50% of the total. Estimated numbers were compared against total numbers in 12 individuals, and the two values were never significantly different. Studies estimating fecundity in parasitic species likewise used various subsampling approaches: manually counting the number of eggs in known weight or known volume subsamples (e.g., Applegate 1949; Vladykov 1951; Potter et al. 1983; Schuldt et al. 1987), or using the "photocopy method" whereby individual eggs were spread along the bottom of a petri dish and copied and enlarged (200%) for ease of counting (Smith and Marsden 2007; Clemens et al. 2013). As with the above studies, the accuracy of the estimates was often assessed by performing total counts on a small number of individuals (e.g., Applegate 1949; Schuldt et al. 1987).

Virtually all the studies cited here counted eggs within the ovaries, although it is interesting to note that egg numbers recorded by Yamazaki and Koizumi (2017) are

from eggs that were released by the Siberian brook lamprey during mating experiments in the laboratory. The number of eggs shed (mean 1,895) is similar to total
egg counts for other brook lampreys and for this species in particular, suggesting
that the majority of eggs within the ovary are likely released, at least under normal
conditions. This is consistent with studies on other species (Applegate 1949; Manion and McLain 1971; Manion and Hanson 1980) that likewise concluded that the
proportion of unspawned eggs is generally low (see Sect. 1.5.1).

Although methodological differences related to the way in which eggs were
counted are thought to introduce only minor biases (if any), variation in the stage of
sexual maturation examined in the different studies may be more significant. Stage of
maturation likely had little effect on total egg counts, because studies to date suggest
that atresia is not significant during the final stages of maturation (e.g., Applegate
1949; Docker and Beamish 1991; see Sect. 1.5.1). However, because lampreys can
experience considerable shrinkage even during the final stages of maturation (see
Sect. 1.5.4), differences among (and within) studies in the timing of collection will
affect the relationship between fecundity and TL. All else being equal, the number of
eggs per unit length would be lower in individuals caught earlier in the spawning run.
Different preservation methods can also affect the relationship between fecundity and
TL. Because preservation in 5 and 10% formalin has been shown to cause ~2.8–3.0%
and 3.3–3.7% shrinkage, respectively, in larval TL (F. W. H. Beamish 1982; Neave
et al. 2007), studies that have used formalin-preserved specimens (e.g., Vladykov
1951; Vladykov and Follett 1958; Schuldt et al. 1987; Lapierre and Renaud 2015)
will have overestimated the number of eggs per unit fresh length if measurements
have not been corrected to those of live animals. Some studies correct for shrinkage
during maturation and as the result of preservation (e.g., F. W. H. Beamish 1982;
Docker and Beamish 1991; Beamish et al. 1994), but most do not. These differences should be kept in mind when comparing relationships between body size and
fecundity.

1.6.4 Relative Fecundity

Because absolute fecundity is positively associated with female body size, showing
an almost 1:1 increase in egg number with increases in body weight, relative fecundity—the number of eggs per gram of body weight—is a useful comparator among
and within species. In the 10 parasitic and seven non-parasitic species for which
relative fecundity was provided or could be estimated, mean relative fecundity typically falls between ~250 and 500 eggs/g, although considerable variation has been
reported and the precise stage of maturity is seldom provided (Table 1.11). However,
despite the attempt to use relative fecundity to standardize comparison across lampreys of different sizes, relative fecundity itself is significantly related to lamprey
body size, being negatively associated with both TL and weight. The lowest relative fecundity values reported are those for the anadromous sea lamprey, averaging
205–236 eggs/g (Vladykov 1951; Beamish and Potter 1975), and mean values <300

eggs/g have also been reported in pouched lamprey, European river lamprey, and one population of Arctic lamprey (Potter et al. 1983; Witkowski and Jęsior 2000; Kucheryavyi et al. 2007). However, higher mean relative fecundities (397–486 eggs/g) have been reported in Caspian and Pacific lampreys (Kan 1975; Nazari and Abdoli 2010; Clemens et al. 2013) and in Arctic lamprey (610–943 eggs/g) collected in the Amur River in Russia (Morozova 1956). Among the freshwater-resident parasitic species, mean relative fecundity values <300 eggs/g have been reported only in the silver lamprey (Schuldt et al. 1987); other mean values range from ~320 to 550 eggs/g, and this full range is evident within the landlocked sea lamprey (Table 1.11). Similarly, in the non-parasitic brook lampreys, relative fecundity generally ranges from 300 to 500 eggs/g. Across species, the relationship between mean relative fecundity (RF_m, eggs/g) and mean TL (TL_m) is described by the equation:

$$\text{Log } RF_m = -0.157 \log TL_m + 2.9374 \left(n = 46, \ R^2 = 0.1739, \ p = 0.0039 \right)$$
(1.4)

Equation 1.4 excludes western river lamprey (n = 2), but the relationship was still significantly negative when western river lamprey was included (b = –0.543, R^2 = 0.1751, p = 0.0039).

Within species (or at least within populations), there likewise appears to be a negative relationship between relative fecundity and TL (Table 1.12). In most cases, the relationship has been shown to be significant (e.g., Applegate 1949; Wigley 1959; Kott 1971; Manion 1972; Nazari and Abdoli 2010), although in some studies, significance was demonstrated only intermittently (e.g., Vladykov 1951, where n = 9–10 per species). Docker and Beamish (1991) similarly found a significant negative relationship between relative fecundity and TL in only two of eight populations of the least brook lamprey (n = 5–14 per population). Surprisingly, however, when all 75 least brook lamprey from these eight populations were pooled, there was a significant positive relationship between relative fecundity and TL. Furthermore, even studies showing that relative fecundity decreased significantly with increased female body size showed considerable differences in the nature of the relationship (Table 1.12). Relative fecundity in Cayuga Lake sea lamprey (Wigley 1959) decreased only modestly with increases in size (e.g., from 305 to 291 eggs/g in 400- and 500-mm females, respectively), while relative fecundity in sea lamprey from the Manion (1972) study decreased much more dramatically (from 457 to 281 eggs/g at 400 and 500 mm, respectively).

Therefore, the relationship in lampreys between relative fecundity and size is not yet clear. This uncertainty brings into question the idea that conversion from actual to relative fecundity removes the effect of lamprey size, and body size should still be kept in mind when comparing among species, populations, and individuals.

1.6.5   Temporal and Spatial Differences in Sea Lamprey Fecundity

Large variations in fecundity and body size have been reported in freshwater-resident sea lamprey, among locations and over the course of the population expansion and collapse seen in the Great Lakes prior to and following initiation of sea lamprey control (see Sect. 1.2.6; Chap. 5). For example, mean fecundity was only 45,602 in Cayuga Lake sea lamprey in 1951 where mean TL was 395 mm (Wigley 1959) but was more than double that in Lake Michigan (97,016) in 1980 and Lake Ontario (95,212) in 1998/1999 when mean TL was 485 and 468 mm, respectively (Johnson 1982; O'Connor 2001). Gambicki and Steinhart (2017) compared recent and historical data within lake basins and found that mean fecundity in Lake Superior increased by 17% between 1960 and 2011, over which time mean TL and weight increased by ~13 and 45%, respectively. In this section, we examine if these differences are merely a function of the effect of female body size on egg numbers or whether fecundity has varied spatially or temporally in a manner disproportionate to changes in size. Being able to predict the extent to which changes in female body size and condition affect sea lamprey fecundity is necessary to determine if gains achieved through sea lamprey control might be offset by increases in fecundity.

In brief, there were relatively few deviations from the general relationship between body size and fecundity, although the most notable exception was the Cayuga Lake sample (Fig. 1.14). Wigley (1959) suggested that the small body size of Cayuga Lake sea lamprey, relative to those from Seneca Lake, was related to the high sea lamprey-to-lake trout ratio in Cayuga Lake at this time. Nevertheless, absolute and relative fecundity in Cayuga Lake sea lamprey were still considerably lower than expected based on TL. Smith and Marsden (2007), examining historical fecundity data for sea lamprey in Lake Champlain and other landlocked populations, suggested that the availability of food resources might affect fecundity independently of size. They argued that the highest historical fecundity estimates in the upper Great Lakes were recorded where lake trout were still available or had just disappeared, and the lowest values were recorded in the North Channel of Lake Huron (in 1948) where lake trout had been absent for 5 years. However, most differences in fecundity in Great Lakes sea lamprey were largely attributable to differences in body size (Fig. 1.14). Likewise, lake trout stocks in Lake Superior were at an all-time low in 1960 (Heinrich et al. 1980), but absolute and relative fecundity in 1960 were, if anything, slightly higher than predicted based on TL alone. The increases in fecundity observed in Lakes Michigan and Huron between 1948 and 1980/1981 likewise paralleled that expected based on the observed increase in TL. In terms of spatial differences, sea lamprey in Lake Ontario and those from the Green Bay region of Lake Michigan produce a greater number of eggs per female relative to other areas. However, sea lamprey from these two regions tend to be larger than sea lamprey from the rest of the Great Lakes (Smith 1971; Johnson 1982), and, as with the temporal differences noted, the increase in fecundity is in proportion to their body size.

Fig. 1.14 Relationship between: **a** log mean fecundity (F, total number of eggs) and log total body length (TL in mm); and **b** log mean relative fecundity (RF, eggs/g) and TL in freshwater-resident sea lamprey *Petromyzon marinus*. *Solid circles* represent sea lamprey from Lake Superior, *open circles* Lake Michigan, *closed squares* Lake Huron, *open squares* Lake Ontario, *closed triangles* Lake Erie, *open triangles* Lake Champlain, and *open diamonds* Cayuga Lake. The linear relationships shown are represented by the equations: **a** Log F = 1.985 Log TL − 0.368 (n = 15, R^2 = 0.645, p = 0.0003) and **b** Log RF = −0.968 Log RF + 5.128 (n = 15, R^2 = 0.375, p = 0.0152). Data sources are given in Table 1.11

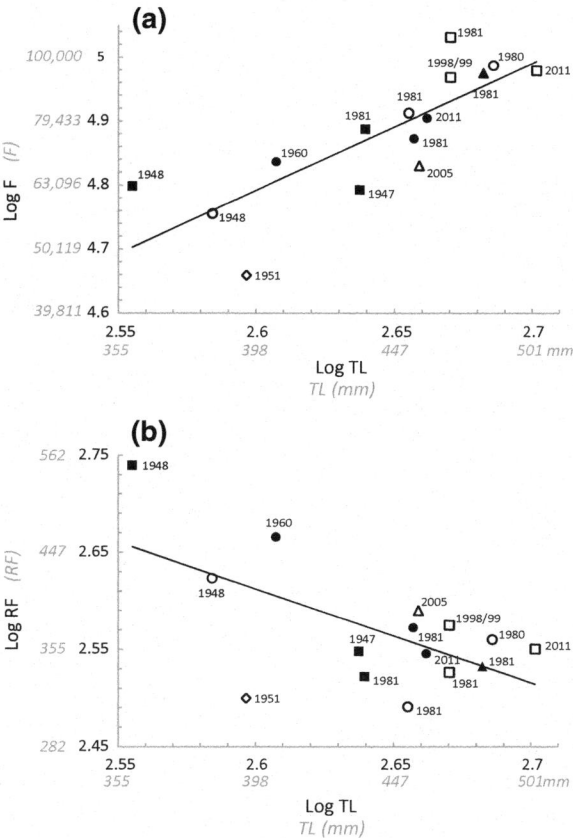

Therefore, factors that result in increases in sea lamprey body size (e.g., related to altered thermal regimes or density- and prey-dependent effects on growth; Cline et al. 2014: Gambicki and Steinhart 2017) are expected to increase the reproductive potential of individual sea lamprey. In this respect, the sea lamprey control program may be a "victim of its own success" to some extent, but larger females do not appear to be disproportionately more fecund. Likewise, although there are spatial differences among sea lamprey populations with respect to body size and fecundity, changes in fecundity are largely in proportion to changes in size. Thus, temporal and spatial differences in sea lamprey fecundity are largely predictable, and it does not appear that different fecundity estimators would be needed for different lake systems at different times.

1.7   Gonadal Steroids

All vertebrates have been shown to regulate reproduction through the hypothalamic-pituitary-gonadal (HPG) axis: the hypothalamus produces gonadotropin-releasing hormones (GnRHs), which stimulate the pituitary to produce one or more gonadotropins (GTHs), which in turn stimulate the gonads to produce steroid hormones[1] (see Sower 2015). The gonadal hormones in turn have myriad effects, which include controlling sex differentiation, maturation, reproductive behavior, and development of secondary sex characteristics (Norris and Carr 2013). The general organization of the HPG axis is common to all vertebrates, and much has been learned about the evolution of the vertebrate HPG axis by studying the lamprey HPG axis (Sower 2015, 2018). However, the hormones that coordinate the axis and regulate reproductive physiology are often different among vertebrate groups: lampreys have three unique GnRHs that have been characterized and a single GTH (or gonadotropic pituitary glycoprotein hormone, GpH) in comparison to most other vertebrates that have two GTHs (follicle stimulating hormone and luteinizing hormone, FSH and LH, respectively; Sower 2015, 2018). Likewise, the steroids produced by vertebrates are variable in structure and effect, and fishes in particular use a variety of "classical" steroids seen in later-evolving vertebrates and "non-classical" steroids observed only in fish (Kime 1993). Lampreys appear to use a mix of classical and non-classical steroids (Bryan et al. 2008), and the study of lamprey steroidogenesis and steroid receptors has helped contribute to our understanding of the evolution of steroid hormones as transcriptions factors in vertebrates (Thornton 2001; Baker 2004).

Furthermore, a better understanding of the gonadal steroids and their function in lampreys will have important management and conservation applications. Many of the non-pesticide control techniques aimed at the Great Lakes sea lamprey are designed to disrupt reproduction (Christie and Goddard 2003; Li et al. 2003), and a better understanding of lamprey reproductive physiology may make these techniques more effective or open the way to new techniques (Docker et al. 2003; Sower 2003). Similarly, a better understanding of the proximal controls on lamprey reproduction may aid in developing better conservation measures aimed at reproductive-stage lampreys and may also lead to better tools (such as better hormone assays) to understand the effect that conservation measures have on lamprey reproductive physiology (e.g., Mesa et al. 2010; Abedi et al. 2017).

In this section, we review the current state of research on the steroid synthetic pathways in lampreys, the classical and non-classical gonadal steroids detected in lampreys and their putative roles, and what is known to date regarding steroid receptors in lampreys. As in previous sections (e.g., Sects. 1.3.1 and 1.4.2.1) where future research in lampreys will be guided by knowledge gained to date in other vertebrates, we hope that this section will also serve as a primer for lamprey biologists not previously familiar with these topics. The hypothalamic and pituitary components of

---

[1] Where steroid refers to a molecular structure and hormone refers to a function; not all steroids are hormones (e.g., some are parts of synthetic pathways but do not function as hormones), and not all molecules that function as hormones (e.g., GnRH) are steroids.

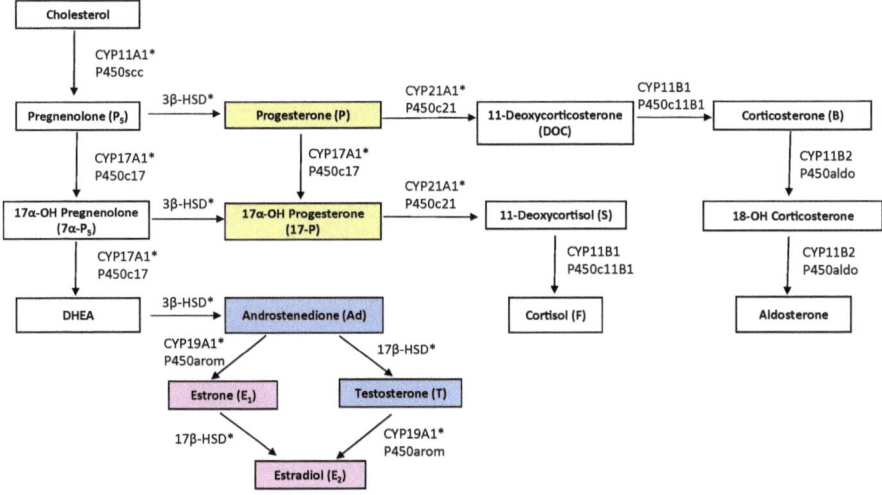

Fig. 1.15 Steroid biosynthesis pathway indicating enzymes involved in the synthesis of classical sex and adrenal steroids in vertebrates; non-classical steroids (e.g., the 15-hydroxylated steroids found in lampreys) are not included. The sex steroids include the progestagens (*yellow*), estrogens (*pink*), and androgens (*blue*); steroid abbreviations are the same as those in Table 1.13. The enzymes catalyzing each reaction are shown, where the CYP enzymes belong to the cytochrome P450 family (e.g., *CYP11A1* refers to CYP450 Family 11 Subfamily A Member 1); alternative names are also given (*P450scc* P450 cholesterol side chain cleavage; *P450c17* steroid 17-α hydroxylase; *P450arom* aromatase). 3β-HSD and 17β-HSD are hydroxysteroid dehydrogenases. *Asterisks* represent the steroidogenic enzymes that have been identified or inferred to date in lampreys (Adapted from Baker 2004.)

the lamprey HPG axis have been reviewed recently and thoroughly by Sower (2015, 2018).

1.7.1 Steroid Synthesis

Steroids in vertebrates are all derived from cholesterol, through a synthetic pathway that relies on enzymes from the cytochrome P450 family (CYP450s) and hydroxysteroid dehydrogenase enzymes (HSDs) (Fig. 1.15). CYP450s are ancient enzymes that evolved through gene duplication and divergence into a diverse protein family that metabolizes a wide variety of chemicals that function in both the synthesis of sterols and detoxification of xenobiotics (Baker 2004, 2011; Markov et al. 2009). HSDs have also undergone gene duplications and divergence, and they provide an important mechanism for regulating the actions of steroids. For example, 17β-HSD regulates the levels of active androgens and estrogens, and at least ten 17β-HSDs have been identified in mammals (Baker 2004).

In the first step of steroid synthesis, the C17 side chain is cleaved on cholesterol by CYP11A (i.e., CYP450 Family 11 Subfamily A, also known as P450 cholesterol side chain cleavage enzyme or P450scc) to form pregnenolone, which then serves as the precursor to all other steroids (Fig. 1.15). These other steroids include the sex steroids that are typically synthesized in the gonads (i.e., progestagens, which are C21 steroids; androgens, which are C19 steroids; and estrogens, which are C18 steroids), as well as the corticosteroids typically made in the adrenal cortex (i.e., glucocorticoids such as cortisol, and mineralocorticoids such as aldosterone). Pregnenolone can be metabolized by 3β-hydroxysteroid dehydrogenase (3β-HSD) to form progesterone (P), which can in turn be metabolized by the CYP17 enzyme to form a second progestagen, 17α-hydroxyprogesterone (17-P). In addition to functioning as sex hormones (see Sect. 1.7.2.1), both of these progestagens serve as precursors for the synthesis of corticosteroid hormones (Fig. 1.15); thus, progestagens are also produced in non-gonadal tissues (Norris and Carr 2013). Pregnenolone can also be converted to 17α-hydroxypregnenolone, which then serves as a precursor in the synthesis of androgens (e.g., testosterone, T) and estrogens (e.g., estrone, E_1, and estradiol, E_2) via the metabolic intermediates dehydroepiandrosterone (DHEA) and androstenedione (Ad). Ad can be metabolized by 17β-hydroxysteroid dehydrogenase-type 1 (17β-HSD-1) to T, or by CYP19 (also known as P450arom or aromatase) to E_1, and both T and E_1 can be further metabolized (by CYP19 and 17β-HSD-1, respectively) to E_2. In addition to these classical vertebrate steroids, a number of non-classical steroids have been observed in fish (Kime 1993; see Sect. 1.7.2).

Much of our initial knowledge regarding steroid synthesis in lampreys was inferred from studies that incubated radiolabeled precursors with gonadal or other tissue extracts (Table 1.13). The primary purpose of these studies was to identify the functional steroids in lampreys, but they also allowed researchers to deduce which enzymes must be present in lampreys to have converted the radiolabeled precursors into the detected products. Many of the studies were inconclusive because several of the products could not be identified through comparison to known steroid standards, but they did indicate that lampreys have many of the same steroidogenic enzymes as higher vertebrates, particularly those related to the synthesis of the sex steroids (Fig. 1.15). For example, production of small amounts of 17-P, 11-deoxycorticosterone (DOC), and 11-deoxycortisol from P provide evidence for CYP17 and CYP21 activity in lamprey testes and presumptive adrenocortical tissue (Weisbart and Youson 1975), and Weisbart et al. (1978) found evidence of weak 3β-HSD activity in lamprey testes. Callard et al. (1980) found that Ad was converted to E_1 and E_2 in sea lamprey ovary and testis, respectively, indicating that lampreys possess CYP19 (aromatase) and 17β-HSD, and recovery of 5α-Ad showed 5α-reductase activity. 5α-reductase also catalyzes the conversion of T to 5α-dihydrotestosterone (DHT), which is about 10$\times$ more potent than T (Baker et al. 2015), but DHT was not detected in sea lamprey tissues (Callard et al. 1980). Following incubation of testicular, ovarian, and presumed adrenocortical sea lamprey tissues with radiolabeled cholesterol, Weisbart et al. (1978) were unable to detect pregnenolone or any other identifiable products. However, they concluded that the absence of identifiable steroids was not due to lack of transformation of the cholesterol (i.e., by CYP11A), and recent phylogenetic analysis has shown that the gene encoding CYP11A is present in lampreys (see below). However, there is no evidence for the presence of CYP11B1 (or CYP11B2) in lampreys. Based on the presence of DOC and 11-deoxycortisol in sea lamprey but the absence of corticosterone and cortisol, Close et al. (2010) suggested that CYP11B1 was not present in early vertebrate evolution, and their inability to find CYP11B1 in the sea lamprey genome supports this conclusion.

The inability of researchers to identify the majority of the radiolabeled products in these precursor studies led them to hypothesize that lampreys used non-classical steroids as their main gonadal hormones (see Sect. 1.7.2). Subsequent steroid biosynthesis studies supported this hypothesis. In addition to the enzymes they share with other vertebrates, lamprey gonads (particularly testes) appear to show strong 15-hydroxylase activity. Kime and Rafter (1981) demonstrated that European river lamprey gonadal tissue extracts convert T and P to 15-hydroxylated forms (Table 1.13), and it was later shown that sea lamprey testis extracts convert Ad and T to 15α-hydroxytestosterone (15α-T; Kime and Callard 1982). These findings have been confirmed and further explored in sea lamprey by Lowartz et al. (2003, 2004) and Bryan et al. (2003, 2004), and 15α-hydroxyprogesterone (15α-P) was also shown to be a major steroid product of lamprey gonads. The presence of 15α-hydroxylase has also been confirmed in silver, chestnut, and American brook lampreys using the same methods (Bryan et al. 2006).

The steroidogenic enzymes present in lampreys have also been investigated recently from an evolutionary perspective, either by attempting to clone and sequence the various steroidogenic enzyme genes in lampreys and other chordates or by searching the genomes of these organisms for their orthologs (i.e., genes in different species that are evolved from a common ancestor; Baker 2004, 2011; Baker et al. 2015). Studies related to the origin of these enzymes are providing important insights into the origin of steroid hormone signaling in vertebrates (Baker 2004, 2011; Markov et al. 2009, 2017; Baker et al. 2007, 2015). In brief, orthologs of CYP11A, CYP17, CYP19, CYP21, 3β-HSD, and 17β-HSD-14 have been found or are inferred to exist in lampreys (Close et al. 2010; Baker et al. 2015), which is in agreement with the results from the precursor experiments. However, no clear ortholog has been found with close similarity to human 17β-HSD-1 in lampreys, which indicates that 17β-HSD-14 or another 17β-HSD is involved in lamprey estrogen synthesis (Baker et al. 2015). An ortholog of CYP27 (which catalyzes the synthesis of 27-hydroxycholesterol, a novel physiological estrogen in mammals, directly from cholesterol) was also found in lampreys (Baker et al. 2015). Of the orthologs found in lampreys, all but CYP21 were also found in amphioxus *Branchistoma* spp. (i.e., a non-vertebrate chordate), thus suggesting that CYP21, which is descended from a duplicated CYP17 gene, arose in the ancestor of vertebrates (Baker et al. 2015). As suggested by Close et al. (2010), a CYP11B ortholog was not found in lampreys, and phylogenetic analysis suggests that this gene first appeared in the ancestor to the jawed vertebrates (i.e., the gnathostomes), coinciding with the evolution of separate mineralocorticoid and glucocorticoid receptors (Baker et al. 2015; see Sect. 1.7.3).

Table 1.13 Summary of lamprey steroid biosynthesis studies (Reprinted and updated from Bryan et al. 2008.)

Study	Species	Tissue	Stage	Precursor	Products	Not produced
Weisbart and Youson (1975)	Sea lamprey	PAT	Larval, parasite	P	S, 17-P, Ad, UCs	F, E, B, T, DOC
		Testis	Parasite	P	DOC, UCs	F, E, B, S, T, 17-P, Ad
Weisbart and Youson (1977)	Sea lamprey	Intracardiac injection	Parasite	P	DOC, UCs	F, E, B, S, T, 17-P, Ad
Weisbart et al. (1978)	Sea lamprey	Testis, ovary, PAT	Adult	Cholesterol	UCs	F, E, B, S, T, DOC, Ad, P, P_5, 17-P, 7α-P_5
Callard et al. (1980)	Sea lamprey	Ovary, kidney	Adult	Ad	E_1, 5α-Ad, UCs	DHT
		Testis	Adult	Ad	E_2, E, 5α-Ad, UCs	DHT
Kime and Rafter (1981)	European river lamprey	Ovary	Adult	P, T	15α-P, Ad, 15β-T	T, 17-P, E_2
		Testis	Adult	P, T	15α-P, Ad, 15β-T	T, 17-P, E_2
Kime and Callard (1982)	Sea lamprey	Testis	Adult	Ad	15α-T, 15α-Ad	15β-T
		Brain, liver, kidney, ovary	Adult	Ad	15α-Ad	
Bryan et al. (2003)	Sea lamprey	Testis	Adult	T	15α-T	15β-T
Lowartz et al. (2003)	Sea lamprey	Testis	Adult	P_5, 17-P, Ad	15α-T, 15α-P, small amount of E_2	T, P
		Ovary		P_5, 17-P, Ad	15α-estrogens, small amount of E_2	T, P

(continued)

Table 1.13 (continued)

Study	Species	Tissue	Stage	Precursor	Products	Not produced
Bryan et al. (2004)	Sea lamprey	Testis	Adult	P	15α-P, UCs	
Lowartz et al. (2004)	Sea lamprey	Ovary, testis	Larval, metamorphosis, parasite	P_5, P, Ad	7α-P_5, 15α-P, 15α-Ad, 15α-T, 15α-E_2, small amount of E_2, UCs	T, 15β-steroids
Bryan et al. (2006)	Silver lamprey Chestnut lamprey American brook lamprey	Ovary, testis	Adult	P, T	15α-P, 15α-T, UCs	
Bryan et al. (2007)	Sea lamprey	Testis	Adult	P	Ad (in tissue extracts only), 15α-P, UCs	
Bryan et al. (unpublished)	European river lamprey	Testis	Adult	P, T	15α-P, 15α-T, UCs	

PAT presumptive adrenocorticol tissue, *P* progesterone, *P_5* pregnenolone, *7α-P_5* 7α-hydroxypregnenolone, *17-P* 17-hydroxyprogesterone, *Ad* androstenedione, *5α-Ad* 5α-reduced androstenedione, *5α-Ad* 15α-hydroxyandrostenedione, *T* testosterone, *S* 11-deoxycortisol, *DOC* 11-deoxycorticosterone, *F* cortisol, *E* cortisone, *B* corticosterone, *E_1* estrone, *15α-T* 15α-hydroxytestosterone, *15β-T* 15β-hydroxytestosterone, *15α-Ad* 15α-hydroxyandrostenedione, *15α-P* 15α-hydroxyprogesterone, *UCs* unidentified compounds

1.7.2 Sex Steroids in Lampreys

Fishes use a variety of both classical steroids and non-classical steroids (Kime 1993). For example, T, E_2, and P are the main steroid hormones in some fish species, but others use 11-ketotestosterone (11-KT) or 17α,20β-dihydroxyprogesterone (17α,20β-P), and it is not well understood why the steroid hormones of fishes vary so much among species. Lampreys likewise appear to use a mix of classical and non-classical steroids, including steroids that are different from those of other vertebrates by possessing an additional hydroxyl group at the C15 position (Bryan et al. 2008). It is possible that these 15-hydroxylated steroids evolved as functional hormones in lampreys as a response to parasitism (i.e., so that the parasitic lamprey would be less susceptible to the influence of the reproductive hormones in its host's blood), or they are simply a "primitive" form of steroid hormone (Docker 2006).

1.7.2.1 Progestagens

In other vertebrates, progestagens (also known as progestogens or progestins) have been shown to regulate key physiological activities for reproduction in both sexes (Norris and Carr 2013). P, 17α-P, and $17\alpha,20\beta$-P are the most studied progestagens in other fish species (Kime 1993). In lampreys, P and the non-classical 15α-hydroxyprogesterone (15α-P) are the most commonly studied (Table 1.14).

Baseline levels of P have been detected in lamprey plasma by immunoassay (Table 1.14). Levels are generally low (<1 ng/mL), particularly in pre-ovulating and pre-spermiating lampreys (Sower et al. 1987, 1993; Sower 1989; Sower and Larsen 1991; Bolduc and Sower 1992; Deragon and Sower 1994; Gazourian et al. 2000; Farrokhnejad et al. 2014; see Sects. 1.5.1 and 1.5.2). Higher levels have been occasionally reported: for example, P levels up to 10, 4, and 3.2 ng/mL have been detected in European river lamprey, adult Pacific lamprey, and pre-ovulatory Caspian lamprey, respectively (Barannikova et al. 1995; Mesa et al. 2010; Ahmadi et al. 2011). However, plasma P levels often differ between the sexes (e.g., sexually mature males generally have higher P than females; Linville et al. 1987; Mesa et al. 2010; Farrokhnejad et al. 2014), increase with maturity (Mesa et al. 2010; Farrokhnejad et al. 2014), increase after 2–5 injections of GnRH (Sower et al. 1987; Deragon and Sower 1994; Gazourian et al. 2000), and decrease after hypophysectomy (Sower and Larsen 1991). As a result of these changes in P with stage of maturity and in response to GnRH, Sower (1990) suggested that P is a functional hormone in lampreys.

Studies examining 15α-P levels in lampreys show that baseline levels are similarly low but increase even more dramatically (i.e., up to 36 and 100 ng/L in pre-spermiating males) in response to GnRH injections (Bryan et al. 2004; Young et al. 2007) or pituitary extracts containing GTH (Young et al. 2007). Furthermore, 15α-P is also the only steroid that appears to respond to GnRH or pituitary extract in a dose-dependent fashion (Young et al. 2007). As with P, 15α-P levels are higher in males than in females, although they are higher in mature females relative to immature females (Bryan et al. 2004). The plasma concentrations of 15α-P combined with the response to upstream stimulation makes it likely that 15α-P is an active hormone in lampreys. In support of this, Bryan et al. (2015) found that pre-spermiating male sea lamprey given time-release implants of P reached spermiation faster, and they had higher plasma concentrations of 15α-P and the lamprey sex pheromone 3-keto-petromyzonal sulfate (3kPZS; see Johnson et al. 2015b). Although a high proportion of P was converted in vivo to 15α-P, it was not possible to determine which of the two progestagens had a stronger biological activity, but the results indicated that progestagens likely play a role in both gonadal maturation and pheromone production in male sea lamprey. However, a receptor for 15α-P (see Sect. 1.7.3) has yet to be identified, and the specific physiological role of 15α-P is still unknown.

Table 1.14 Studies using steroid immunoassays in lampreys, showing plasma level of different steroids in males and females at different stages of maturity and in response to injection with GnRH; differences between sexes or stages are shown when compared (Reprinted and updated from Bryan et al. 2008.)

Study	Stage and sex	Steroid	Range or mean in plasma (ng/mL)	Difference between sexes?	Difference between reproductive stages?	Response to GnRH (ng/mL)
Caspiomyzon wagneri Caspian lamprey						
Ahmadi et al. (2011)	POF, PSM (fall and spring migrants)	E_2	0.03–0.04 in POF 0.09–0.11 in PSM	M > F	Fall M > spring M	
		T	1.5–2.0 in POF 8.5–9.0 in PSM	M > F	No	
		P	1.0–3.2 in POF 1.5–2.0 in PSM	F > M in spring	Fall M > spring M, but spring F > fall F	
Farrokhnejad et al. (2014)	POF, OF, PSM, SM	E_2	1.27 in POF 1.00 in OF 0.83 in PSM 1.38 in SM	POF > PSM, but SM > OF	SM > PSM, but POF > OF	
		T	0.46 in POF 0.54 in OF 0.38 in PSM 0.44 in SM	No	No	
		P	0.18 in POF 0.20 in OF 0.15 in PSM 0.23 in SM	SM > OF		
Abedi et al. (2017)	PSM, POF	E_2	1.25–2.75			
		17α-P	3.0		Up to 11 ng/mL after HCG injection	
Entosphenus tridentatus Pacific lamprey						
Mesa et al. (2010)	Adults	P	0–4	M > F	Seasonal changes	
		15α-T	0.25–1	M > F		
		E_2	0.5–4	M > F		
Lampetra fluviatilis European river lamprey						
Kime and Larsen (1987)	POF, PSM	T	0.1	No	Up to 1.2 ng/mL after gonadectomy	
		E_2	1	No	Up to 2 ng/mL after gonadectomy	
Barannikova et al. (1995)	POF, OF, PSM, SM	P	1–10		No	
		E_2	0.5–3.5		Decreased near ovulation	
Mewes et al. (2002)	POF, PSM, OF, SM	E_2	0.01–3.2	M > F	No	
Lampetra planeri European brook lamprey						
Seiler et al. (1985)	Adults	P_5	2–3		No	
		Ad	0–2.5		No	
		T	2.5–17		Increased with maturity	

(continued)

Table 1.14 (continued)

Study	Stage and sex	Steroid	Range or mean in plasma (ng/mL)	Difference between sexes?	Difference between reproductive stages?	Response to GnRH (ng/mL)
Lethenteron camtschaticum Arctic lamprey						
Fukayama and Takahashi (1985)	POF, PSM, OF, SM	T	0			
		E_2	0.4–3.22 in POF 1.0 in OF 0.8–4.5 in PSM 2.7 in SM	M > F near spawning	Increased with maturity in M; increased in F with vitellogenesis, decreased at spawning	
Petromyzon marinus sea lamprey						
Weisbart et al. (1980)	POF and PSM	$17\alpha,20\beta$-P	1.6–3.1			
		T	4.1			
Katz et al. (1982)	OF, SM	P	0–1.25	No	Rose after stress	
		DHT	0	No		
		T	0	No		
		Ad	1.05–5.58	No	Rose after stress	
		E	0.74–7.77			
		E_2	0.51–3.14	No		
Sower et al. (1983)	POF, OF	E_2	3–5		POF > OF	Up to 12
Sower et al. (1985a)	POF, OF, PSM, SM	T	0.1–0.2	No	No	
		E_2	0.5–3.0	SM > OF	At spawning, rose in M, dropped in F	
Sower et al. (1985b)	POF, OF, PSM, SM	T	0.10–0.18	No	No	No effect
		E_2	2	No	In F, decreased with maturity	Up to 6.5
Linville et al. (1987)	POF, OF, PSM, SM	T	0.005–0.170	M > F	No	
		P	0.1–2.8	M > F		
		E_2	0.6–2.3	M > F	At spawning, dropped in F	
Sower et al. (1987)	POF, OF	P	<1			Increased
		E_2	1.5			Up to 5.5
Sower (1989)	PSM	P	0.25			Up to 3
		E_2	1.5			Up to 3
Sower and Larsen (1991)	POF	P	0.3		Decreased after hypophy-sectomy	
		E_2	1.91		Decreased after hypophy-sectomy	
Bolduc and Sower (1992)	POF, OF	P	0.1–0.6		Fluctuated or slowly increased	
		E_2	0.25–3		Increased through spawning season, then decreased suddenly	

(continued)

Table 1.14 (continued)

Study	Stage and sex	Steroid	Range or mean in plasma (ng/mL)	Difference between sexes?	Difference between reproductive stages?	Response to GnRH (ng/mL)
Sower et al. (1993)	POF	P	0.52			Up to 0.71
		E_2	0.64			Up to 2.06
Deragon and Sower (1994)	PSM	P	0.2			Up to 2
		E_2	1.2			Up to 2.4
Gazourian et al. (1997)	POF	P	2			Up to 12
		E_2	0.2			Up to 0.5
Gazourian et al. (2000)	PSM	E_2	2			Up to 8
		P	0.2			Up to 2.6
Rinchard et al. (2000)	SM	T	0.03–0.15			
		E_2	1–2			
		P	0.4–1.2			
Bryan et al. (2003)	POF, OF, PSM, SM	15α-T	<1	M > F	POF > OF, PSM > SM	
Bryan et al. (2004)	POF, OF, PSM, SM	15α-P	<1–2.48	M > F	Increased with maturity	Up to 36 in PSM
Young et al. (2004a)	PSM	15α-T	<0.5			Up to 3
Young et al. (2004b)	PSM	15α-T	0.3			Up to 0.6
		E_2	1			Up to 3.5
Young et al. (2007)	POF, PSM	15α-T	0.15 in POF 0.2–0.4 in PSM		Up to 0.6 ng/mL in PSM after injection with pituitary extract	Up to 0.7 in PSM Up to 0.3 in POF
		15α-P	1.2–2.0 in PSM 0.12 in POF	M > F	Up to 25 ng/mL in PSM after injection with pituitary extract	Up to 100 in PSM Up to 0.44 in POF
Bryan et al. (2007)	PSM	Ad	< 1			Up to 1.9
Sower et al. (2011)	POF, OF, PSM, SM	E_2	0.01–0.65 in POF 0.65 in OF 1.4 in PSM 0.9 in SM	M > F	Increased with maturity	

POF pre-ovulating females, *OF* ovulating females, *PSM* pre-spermiating males, *SM* spermiating males, *HCG* human chorionic gonadotropin, *P* progesterone, P_5 pregnenolone, *$17\alpha,20\beta$-P* $17\alpha,20\beta$-dihydroxyprogesterone, *T* testosterone, *Ad* androstenedione, *DHT* 5α-dihydrotestosterone, E_1 estrone, E_2 estradiol, *15α-T* 15α-hydroxytestosterone, *15α-P* 15α-hydroxyprogesterone

Plasma concentrations of 17α-P and $17\alpha,20\beta$-P, two progestagens studied in other fishes, have been measured in Caspian lamprey (Abedi et al. 2017) and sea lamprey (Weisbart et al. 1980), respectively. In pre-spermiating and pre-ovulatory Caspian lamprey, baseline 17α-P levels were 3.0 ng/mL but increased to 11 ng/mL following injection with human chorionic gonadotropin (Abedi et al. 2017). $17\alpha,20\beta$-P has been detected in pre-spermiating and pre-ovulatory sea lamprey, measuring 1.6–3.1 ng/mL (Weisbart et al. 1980).

1.7.2.2 Androgens

Androgens have been linked to the development of male reproductive tissues, male secondary sex characteristics, and male reproductive behavior in other fishes (Knapp and Carlisle 2011), but there is relatively little information regarding the functional androgens in lampreys (Bryan et al. 2008). Classical androgens, most notably T, have not been produced in radiolabeling experiments (see Sect. 1.7.1; Table 1.13), and circulating T levels have generally been shown to be undetectable or very low (<1 ng/mL) (e.g., Fukayama and Takahashi 1985; Kime and Larsen 1987; Sower et al. 1985a, b; Linville et al. 1987; Farrokhnejad et al. 2014; Table 1.14). Higher levels of T have been reported (up to 9–17 ng/mL) in European brook lamprey and Caspian lamprey (Seiler et al. 1985; Ahmadi et al. 2011), but such reports are rare. Similarly, although some studies reported that T levels were significantly higher in males than females (Linville et al. 1987; Ahmadi et al. 2011) and increased with maturity (Seiler et al. 1985), most found no differences between the sexes (e.g., Katz et al. 1982; Sower et al. 1985a, b; Kime and Larsen 1987; Farrokhnejad et al. 2014) or reproductive stages (e.g., Sower et al. 1985a, b; Linville et al. 1987; Ahmadi et al. 2011; Farrokhnejad et al. 2014), and T levels did not change following GnRH stimulation (Sower et al. 1985b).

Androstenedione, which is the direct precursor to T in the steroid synthesis pathway, was shown by Bryan et al. (2007) to have androgenic effects in sea lamprey. Implants of time-release Ad pellets accelerated maturation in male sea lamprey and caused an increase in size of the dorsal "rope" tissue, a secondary sex characteristic unique to mature male sea lamprey (see Johnson et al. 2015b). Plasma concentrations of Ad appear to be low (<1 ng/mL; Bryan et al. 2007), although Katz et al. (1982) and Seiler et al. (1985) reported levels as high as 5.6 and 2.5 ng/mL in sea and European brook lampreys, respectively. Bryan et al. (2007) found that concentrations of Ad (but not of T) in sea lamprey plasma and testis increased (up to 1.9 ng/mL) following GnRH injection, and they also reported the existence of a receptor for Ad. The capacity and high affinity of this receptor means that much of the Ad is bound in the testis (rather than circulating in the plasma), which can cause high local concentrations of Ad in the testis, despite low circulating levels. Thus, Ad does appear to act as an androgenic hormone in sea lamprey, but the mechanism by which this happens has yet to be identified.

15α-T is the other androgen that has been identified in lamprey plasma (Bryan et al. 2003; Young et al. 2004a, b, 2007; Mesa et al. 2010). Plasma concentrations of 15α-T are generally low, but there are differences between the sexes and maturational states (Bryan et al. 2003; Mesa et al. 2010) and small but significant changes in 15α-T in response to hypothalamic and pituitary hormones (Young et al. 2004a, b, 2007). However, no hormonal role or receptor has been found yet for 15α-T.

1.7.2.3   Estrogens

Estrogens are synthesized in all vertebrates. They are involved in controlling the function of female reproductive organs and processes, and they play roles in some male-specific processes such as sperm maturation (Eick and Thornton 2011; Bondesson et al. 2015). E_2 is the most studied classical steroid in lampreys (Table 1.14), and there is strong evidence that it is a functional hormone in lampreys (Sower 1990). Baseline plasma levels up to ~3–5 ng/mL have been reported (Sower et al. 1983; Fukayama and Takahashi 1985; Barannikova et al. 1995; Mewes et al. 2002; Mesa et al. 2010; Abedi et al. 2017), and E_2 levels have been shown to vary with reproductive stage (Sower et al. 1985a; Linville et al. 1987; Bolduc and Sower 1992) and in response to heterologous and lamprey GnRH stimulation (Sower et al. 1983, 1985b, 1987, 1993; Sower 1989; Derragon and Sower 1994; Gazourian et al. 1997, 2000). Interestingly, however, circulating E_2 levels are often higher in males than in females (e.g., Fukayama and Takahashi 1985; Linville et al. 1987; Mewes et al. 2002; Mesa et al. 2010; Ahmadi et al. 2011), and it appears that E_2 also plays a major role in the reproductive physiology of male lampreys (Bryan et al. 2008). In European river lamprey and Arctic lamprey, plasma E_2 levels have been associated with vitellogenesis in females, but they were also shown to increase at spawning in males (Fukayama and Takahashi 1985; Barannikova 1995; Mewes et al. 2002). However, the increased concentration of E_2 in European river lamprey plasma after gonadectomy suggests that this steroid may be an inactive precursor synthesized in extra-gonadal endocrine tissues (Kime and Larsen 1987).

   Production of 15α-E_2 in lamprey gonads has been inferred from studies using radiolabeled precursors (Lowartz et al. 2004; see Sect. 1.7.1), and molecular modeling experiments have shown that the lamprey estrogen receptor (ER) may bind to it (Baker et al. 2009). However, 15α-E_2 is present in the plasma in levels lower than E_2, and the levels do not change after injection of GnRH (Mara B. Bryan, unpublished data), suggesting that it is likely not a hormone in lampreys. There is clearly still much to be learned about the gonadal steroid hormones in lampreys.

1.7.3   Steroid Receptors

To act as a hormone, a steroid must bind to a receptor, and this action must result in a physiological effect. Vertebrate steroids have been shown to act through two different types of receptors: (1) nuclear receptors, which act as transcription factors when the hormone is bound to the receptor, thus up- or down-regulating expression of particular genes (Baker 1997; Eick and Thornton 2011); and (2) membrane-bound G-protein coupled receptors (GPCRs), which trigger non-genomic changes in the function of the cell by the process of transduction (Norman et al. 2004). GPCRs result in a much quicker response than nuclear receptors (Norman et al. 2004; Freamat and Sower 2013). The interaction between steroids and their receptors thus induces complex genomic and non-genomic effects within the cell, triggering direct activation of transcription and activation of other signaling pathways (Freamat and Sower 2013).

Lampreys are an important model for understanding the evolution of vertebrate adrenal and sex steroid receptors. Nuclear receptors can be subdivided into estrogen receptors (ERs) and 3-ketosteroid receptors such as the glucocorticoid receptor (GR), mineralocorticoid receptor (MR), progesterone receptor (PR), and androgen receptor (AR) (Bridgham et al. 2010; Baker et al. 2015). Ho et al. (1987) identified an ER in the testis of sea lamprey, and orthologs of the ER, PR, and corticosteroid receptor (CR) have been found in sea lamprey and Arctic lamprey. However, there is no evidence of an AR, GR, or MR in lampreys (Thornton 2001; Baker et al. 2015). Thornton (2001) established that the ER is the ancestral steroid receptor. An ER has been cloned from amphioxus, and this non-vertebrate chordate also possesses a steroid receptor (SR) which shares a common ancestor with the AR, GR, MR, and PR of jawed vertebrates. Surprisingly, however, amphioxus ER does not bind E_2 (Paris et al. 2008), but it appears to be a transcriptional activator of amphioxus SR which, also surprisingly, does not bind 3-keto-steroids (Katsu et al. 2010). Thornton and colleagues therefore suggested that the cephalochordate ER lost its response to ligands while the SR retained the response to E_2 (Bridgham et al. 2008; Eick and Thornton 2011). Lampreys (and other vertebrates) therefore inherited the ER from their non-vertebrate ancestor, and a CR and PR (both present in lampreys and hagfishes) subsequently evolved in early vertebrates. The remaining sex and adrenal steroid receptors (i.e., an AR and a separate GR and MR) evolved in the jawed vertebrates (Baker et al. 2015).

The lamprey ER has been heterologously expressed (i.e., with the lamprey ligand-binding domain cloned into a vector) so that binding experiments could be performed, and a reporter assay determined that the ligand for the receptor was indeed E_2 (Paris et al. 2008). Binding assays using radiolabeled E_2 previously found binding activity in the lamprey testes (Ho et al. 1987). Two distinct ERs (ERα and ERβ) have been identified in amphibians, reptiles, birds, and mammals (Kuiper et al. 1997), and orthologs of two ERs have also been reported in sea and Arctic lampreys (Baker et al. 2015). However, Katsu et al. (2016) indicated that these two lamprey ERs (Esr1a and Esr1b) are the result of a lineage-specific gene duplication within the jawless fishes, different from the duplication event in the jawed vertebrates. In the Arctic lamprey, Esr1a showed both constitutive transcription (i.e., at a relatively constant rate) and estrogen-dependent activation of gene transcription. Esr1a displayed strong expression in the gut and liver in both sexes and stronger expression in the heart and gonad of females compared to males. In comparison, Esrb1 showed strong expression in female heart, liver, and gut and in male heart and gut. However, Esrb1 did not bind E_2 in Arctic lamprey and was not stimulated by other estrogens, androgens, or corticosteroids (Katsu et al. 2016). Using a 3D model, Katsu et al. (2016) concluded that, although E_2 fits into the steroid binding site of Esrb1, the lack of stabilizing contacts between the ligand and the receptor side chains appears to prevent E_2 binding activity. Therefore, Esr1a appears to be the functional ER in lampreys.

Binding experiments have not yet been performed with the lamprey PR and CR, but Bryan et al. (2015) investigated changes in the expression of the PR gene in putative target tissues in male sea lamprey at different stages of maturation. Messenger RNA (mRNA) transcript levels in the testis were significantly higher in spermiating males than in any other group, next highest in pre-spermiating males, and lowest in small and large parasitic-phase individuals (i.e., sexually immature juveniles). Levels of PR gene expression increased in spermiating males after injection of lamprey GnRH-I, but they did not change in pre-spermiating males in response to GnRH stimulation. In brain, gills, and liver, levels of PR gene expression were likewise highest in spermiating males. However, injections of GnRH-I and -III resulted in significantly higher gene expression only in brains of pre-spermiating males and in livers of spermiating males (although there was a trend toward an increase in the gills of spermiating males). Therefore, a nuclear PR is present in male sea lamprey, and the location and gene expression levels are consistent with some of the known male reproductive functions (e.g., gonadal maturation, reproductive behavior, and sex pheromone production).

As mentioned above, there is no evidence of an AR in lampreys. Orthologs of the gnathostome AR have not been found in the sea lamprey genome (Baker et al. 2015), nor could an AR resembling the gnathostome AR be amplified using PCR (Thornton 2001). This is consistent with suggestions from previous studies that lampreys lack functional androgens (see Sect. 1.7.2.2). However, Baker et al. (2015) noted that a novel nuclear receptor may mediate responses to androgens in lampreys, and the binding assays performed by Bryan et al. (2007) led to the discovery of a binding moiety in lamprey tissues that appears to function as a steroid receptor. The Ad binding moiety, which bound Ad with high affinity, was found in nuclear and cytosolic extracts of various tissues (but was highest in testes), and the Ad-moiety complex bound to DNA (Bryan et al. 2007). However, the protein that binds Ad has not yet been purified and identified, so it is unclear how it is related to the nuclear steroid receptor family.

It should be noted that much of the work to date on steroid receptors has been done using cytosolic, nuclear, or membrane extracts and radiolabeled ligands (e.g., Ho et al. 1987; Close et al. 2010; Bryan et al. 2015). Thus, because commercially available 15α-hydroxylated radiolabeled steroids are not available, binding experiments to detect receptors for lamprey-specific 15α-hydroxylated steroids have not been performed. Because all previous work has relied on using the native 15α-hydroxylase in lamprey testes to make label using tritiated precursors (e.g., Bryan et al. 2003, 2004), the specific activity of the radiolabeled 15α-hydroxylated steroids is unknown, and the radiolabeled compounds used in these experiments are likely contaminated with endogenous steroids of the same type. Based on the patterns of 15α-P plasma concentrations (see Sect. 1.7.2.1), it seems likely that it is a hormone in lampreys, but a receptor is needed for confirmation. Because binding experiments would need custom radiolabeled 15α-P, it may be most expedient to do this with a reporter assay as in the ER research by Paris et al. (2008).

As a final note, steroidal effects in vertebrates can also be mediated by membrane-bound steroid receptors (Thomas et al. 2006). Two families of such membrane-bound receptors were identified that have no relationship to each other and no relationship to the traditional nuclear receptors (Thomas et al. 2007). One of these families, known as

GPR30, is involved in mediating estrogen responses (Filardo and Thomas 2005), and the other, known as the membrane progestin receptor (mPR), is involved in mediating progesterone responses (Thomas et al. 2006, 2007). Bryan et al. (2015) identified the membrane receptor adaptor protein "progesterone receptor membrane component 1" (*pgrmc1*) gene in male sea lamprey, and they studied mRNA expression levels between different life stages and tissues and in response to lamprey GnRH-I and GnRH-III. Expression of *pgrmc1* in testes was highest in pre-spermiating and spermiating males (compared to sexually immature individuals in the parasitic feeding phase), and it increased in spermiating males following GnRH stimulation. In the brain, expression levels were not significantly different among stages, but GnRH injection resulted in higher expression levels in pre-spermiating (but not spermiating) males. In gills and liver, *pgrmc1* expression was highest in spermiating males. After GnRH injection, gene expression in the gills increased in both groups of maturing males; in the liver, it increased in spermiating males but decreased in pre-spermiating males. As with the nuclear PR, the location and expression levels of *pgrmc1* are consistent with a putative role in known male reproductive functions. However, although pgrmc1 was accepted as a membrane receptor at the time that Bryan et al. (2015) performed their study, it has since been concluded that this "adaptor" protein has only moderate specificity for P and may have higher affinity for T and cortisol (Thomas et al. 2014). Functionality of the estrogen membrane receptor (and probably membrane receptors for many other compounds) is also dependent on the presence of this protein, and thus increased *pgrmc1* expression is not proof that this increase is specifically related to the activity of P (Bryan et al. 2015). The lamprey genome does contain DNA sequences similar to the "fast-acting" membrane progestin receptor (mPR), which has five variants in higher vertebrates (Pang and Thomas 2011; Pang et al. 2013), and these deserve further study. No membrane ERs have been found so far in lampreys, and the mechanism by which lamprey steroids in general act as intercellular signals is not understood yet. Whether they compete for nuclear receptor binding sites with classical steroids or whether they have their own cognate nuclear or membrane receptors still needs to be determined (Bryan et al. 2008).

1.8 Conclusions

Lamprey gonadal development is intriguing, but it continues to be challenging to study. Our understanding of many aspects of the reproductive biology of these fascinating jawless vertebrates has been complicated by their phylogenetic distinctiveness and complex life cycle, the long period during which their gonads remain histologically undifferentiated, and their unique sex steroids. We hope that the overview of topics given in this chapter has provided readers with a deeper understanding of what we know to date about lamprey gonadal development and an appreciation of the many remaining unanswered questions. We believe that collaboration between lamprey biologists and other researchers can now help answer many questions previously believed to be intractable. Integration of detailed field-based research and observations with new advances in laboratory-based techniques (e.g., genomics, molecular biology, microscopy) will be especially productive. Here, we briefly highlight some key outstanding research questions introduced in this chapter, with a particular emphasis on areas where synergies between lamprey biologists and researchers with expertise in other fields would be especially rewarding.

The complex lamprey life cycle has made stage-specific sex ratios difficult to interpret, which has made it challenging to confidently ascertain whether sex determination in lampreys is subject to environmental influence. In particular, there are many unanswered questions related to the dramatic but transient shift in adult sex ratios observed in sea lamprey in the three upper Great Lakes following initiation of the control program. For example, do female sea lamprey experience higher mortality during the parasitic feeding phase (particularly following the onset of vitellogenesis) due to the higher energetic demands of ovarian maturation relative to testicular maturation? If so, were mortality rates in females during this stage disproportionately higher than in males when sea lamprey were at their peak of abundance and prey abundance was at its lowest? Do female sea lamprey undergo metamorphosis at older ages than male sea lamprey—or did they prior to sea lamprey control? Has there been selection in females for earlier metamorphosis as a consequence of lampricide treatments in the Great Lakes basin? What features of Lakes Erie and Ontario explain the lack of sex ratio shifts in their sea lamprey populations compared to those in the upper Great Lakes? Answering these questions will require a deeper understanding of lamprey biology and the stream and lake systems in which the sea lamprey occur. Incorporating genomics technologies into such studies (e.g., identifying sex-determining genes or conclusively ruling out a genetic basis to sex determination) would be of great benefit. For example, if sex-associated loci can be identified that are generally, if not always, correlated with phenotypic sex when sex ratios are at parity, mismatches between genotypic and phenotypic sex under other circumstances would provide strong evidence for an environmental influence overriding GSD (e.g., Patil and Hinze 2008; Cavileer et al. 2015). Non-lethal sexing of lampreys (using genetic markers, acoustic microscopy, or gonadal biopsy methods) would allow researchers to more effectively test for sex-specific differences in growth and mortality.

Clear delineation of the critical sex differentiation period in lampreys has further complicated our understanding of the factors influencing sex determination and differentiation. We currently do not know if a histologically undifferentiated gonad is truly bipotential or if the germinal and somatic elements are already differentiated at a molecular level. Similarly, different paths of male differentiation have been proposed (e.g., direct and indirect development), but we do not know if there is an underlying genetic basis for the difference (e.g., if indirect male differentiation is a form of sex reversal in genotypic females). Regardless of whether the master switch at the top of the cascade is genetic or environmental, identification of the genes involved in the subsequent development of the gonad into an ovary or testis would greatly improve our ability to study the sex differentiation process. For example, upregulation of the gene encoding the steroidogenic enzyme aromatase has been observed in the future ovaries of several teleost fish species (Nakamura et al. 1998; D'Cotta et al. 2001;

Tao et al. 2013), and other genes often show male-specific expression patterns in advance of testicular differentiation (Geffroy et al. 2016). Developing similar tools for lampreys would aid researchers investigating the earliest stages of ovarian and testicular differentiation, help determine if germ cell differentiation is induced by the somatic tissue or vice versa, and help highlight some of the earliest developmental differences between parasitic and non-parasitic lampreys.

Sexual maturation in lampreys is somewhat better understood than sex determination and differentiation, although there is still much to be learned. For example, although it is well known that absolute fecundity increases with female body size, the extent to which other factors such as nutritional status or migratory distance affect the number of eggs that survive to maturity is not known. Likewise, the relationship between reproductive fitness and testis size or other male traits (e.g., sperm concentration, motility, life span) is unrecognized and warrants investigation. Furthermore, there are substantial knowledge gaps regarding the gonadal steroid hormones and receptors involved in sexual maturation and reproduction in lampreys, as well as factors that regulate the cessation of feeding and subsequent onset of sexual maturation. Many of the gonadal steroids thought to act as hormones still have not had functions clearly defined. This is particularly true of the 15α-hydroxylated steroids which are different than all other studied vertebrate steroids. In addition, the androstenedione receptor has not yet been identified, which means that the earliest mechanism of androgen action in vertebrates remains unresolved. Future research into all aspects of lamprey gonadal development will be very rewarding.

Acknowledgements We gratefully acknowledge Kristian Sattelberger for helping to prepare the figures and the reference list, Drs. Colin J. Garroway and Alison E. Wright for their input and discussions related to lamprey sex determination, and Drs. John B. Hume and Murray D. Wiegand for providing insightful comments on an earlier draft of this chapter.

References

Abdoli A, Farokhnejad MR, Nazari H, Hassanzadeh Kiabi B (2017) The role of temperature and daytime as indicators for the spawning migration of the Caspian lamprey *Caspiomyzon wagneri* Kessler, 1870. J Appl Ichthyol 33:553–557

Abedi M, Mojazi Amiri B, Abdoli A et al (2017) Effect of human chorionic gonadotropin on sexual maturation, sex steroids and thyroid hormone levels in Caspian lamprey (*Caspiomyzon wagneri* Kessler, 1870). Caspian J Environ Sci 15:237–247

Abou-Seedo FS, Potter IC (1979) Estuarine phase in the spawning run of the river lamprey *Lampetra fluviatilis*. J Zool 188:5–25

Adolfi MC, Carreira AC, Jesus LW et al (2015) Molecular cloning and expression analysis of dmrt1 and sox9 during gonad development and male reproductive cycle in the lambari fish, *Astyanax altiparanae*. Reprod Biol Endocrinol 13:1–15

Adolfi MC, Nakajima RT, Nóbrega RH, Schartl M (2019) Intersex, hermaphroditism, and gonadal plasticity in vertebrates: evolution of the Müllerian duct and Amh/Amhr2 signaling. Annu Rev Anim Biosci 7:7.1–7.24

Afzelius BA, Nicander L, Sjödén I (1968) Fine structure of egg envelopes and the activation changes of cortical alveoli in the river lamprey, *Lampetra fluviatilis*. Development 19:311–318

Ahmadi M, Amiri BM, Abdoli A, Fakharzade SME, Hoseinifar SH (2011) Sex steroids, gonadal histology and biological indices of fall and spring Caspian lamprey (*Caspiomyzon wagneri*) spawning migrants in the Shirud River, Southern Caspian Sea. Environ Biol Fishes 92:229–235

Ajmani N (2017) Transcriptomic analysis of ovarian development in parasitic *Ichthyomyzon castaneus* (chestnut lamprey) and non-parasitic *Ichthyomyzon fossor* (northern brook lamprey). MSc thesis, University of Manitoba, Winnipeg, MB

Álvarez del Villar J (1966) Ictiologia michoacana, IV. Contribución al conocimiento biológico y sistemático de las lampreas de Jacona, Mich., México. An Esc Nac Cienc Biol Méx 13:107–144

Applegate VC (1949) Sea lamprey investigations: II Egg development, maturity, egg production, and percentage of unspawned eggs of sea lampreys, *Petromyzon marinus*, captured in several Lake Huron tributaries. Pap Mich Acad Sci Arts and Lett 35:71–90

Applegate VC (1950) Natural history of the sea lamprey (*Petromyzon marinus*) in Michigan. US Fish Wildl Serv Spec Sci Rep Fish 55:1–237

Applegate VC, Thomas MLH (1965) Sex ratios and sexual dimorphism among recently transformed sea lampreys, *Petromyzon marinus* Linnaeus. J Fish Res Board Can 22:695–711

Araújo MJ, Ozório RO, Bessa RJ et al (2013) Nutritional status of adult sea lamprey (*Petromyzon marinus* Linnaeus, 1758) during spawning migration in the Minho River, NW Iberian Peninsula. J Appl Ichthyol 29:808–814

Arntzen EV, Mueller RP (2017) Video-based electroshocking platform to identify lamprey ammocoete habitats: field validation and new discoveries in the Columbia River basin. N Am J Fish Manag 37:676–681

Bahamonde PA, Munkittrick KR, Martyniuk CJ (2013) Intersex in teleost fish: are we distinguishing endocrine disruption from natural phenomena? Gen Comp Endocrinol 192:25–35

Baker CF, Jellyman DJ, Reeve K et al (2017) First observations of spawning nests in the pouched lamprey (*Geotria australis*). Can J Fish Aquat Sci 74:1603–1611

Baker ME (1997) Steroid receptor phylogeny and vertebrate origins. Mol Cell Endocrinol 135:101–107

Baker ME (2004) Co-evolution of steroidogenic and steroid-inactivating enzymes and adrenal and sex steroid receptors. Mol Cell Endocrinol 215:55–62

Baker ME (2011) Origin and diversification of steroids: co-evolution of enzymes and nuclear receptors. Mol Cell Endocrinol 334:14–20

Baker ME, Chandsawangbhuwana C, Ollikainen N (2007) Structural analysis of the evolution of steroid specificity in the mineralocorticoid and glucocorticoid receptors. BMC Evol Biol 7:24

Baker ME, Chang DJ, Chandsawangbhuwana C (2009) 3D model of lamprey estrogen receptor with estradiol and 15α-hydroxy-estradiol. PLoS ONE 4:e6038

Baker ME, Nelson DR, Studer RA (2015) Origin of the response to adrenal and sex steroids: roles of promiscuity and co-evolution of enzymes and steroid receptors. J Steroid Biochem Mol Biol 151:12–24

Barannikova IA, Boev AA, Arshavskaya SV, Dyubin VP (1995) Role of hormonal effects in maturation stimulation of river lamprey females (*Lampetra fluviatilis*) from different populations. Vop Ikhtiol 36:260–267 [in Russian]

Barker LA, Beamish FWH (2000) Gonadogenesis in landlocked and anadromous forms of the sea lamprey, *Petromyzon marinus*. Environ Biol Fishes 59:229–234

Barker LA, Morrison BJ, Wicks BJ, Beamish FWH (1998) Potential fecundity of landlocked sea lamprey larvae, *Petromyzon marinus*, with typical and atypical gonads. Copeia 1998:1070–1075

Baron D, Cocquet J, Xia X et al (2004) An evolutionary and functional analysis of *FoxL2* in rainbow trout gonad differentiation. J Mol Endocrinol 33:705–715

Baron D, Houlgatte R, Fostier A et al (2005) Large-scale temporal gene expression profiling during gonadal differentiation and early gametogenesis in rainbow trout. Biol Reprod 73:959–966

Barrionuevo FJ, Hurtado A, Kim GJ et al (2016) *Sox9* and *Sox8* protect the adult testis from male-to-female genetic reprogramming and complete degeneration. eLIFE 5:1–23

Barton MC, Shapiro DJ (1988) Transient administration of estradiol-17β establishes an autoregulatory loop permanently inducing estrogen receptor mRNA. Proc Natl Acad Sci USA 85:7119–7123

Battle HI, Hayashida K (1965) Comparative study of the intraperitoneal alimentary tract of parasitic and nonparasitic lampreys from the Great Lakes region. J Fish Res Board Can 22:289–306

Baumann H, Conover DO (2011) Adaptation to climate change: contrasting patterns of thermal-reaction-norm evolution in Pacific versus Atlantic silversides. Proc R Soc B 278:2265–2273

Beamish FWH (1980) Biology of the North American anadromous sea lamprey, *Petromyzon marinus*. Can J Fish Aquat Sci 37:1924–1943

Beamish FWH (1982) Biology of the southern brook lamprey, *Ichthyomyzon gagei*. Environ Biol Fishes 7:305–320

Beamish FWH (1993) Environmental sex determination in southern brook lamprey, *Ichthyomyzon gagei*. Can J Fish Aquat Sci 50:1299–1307

Beamish FWH, Barker LA (2002) An examination of the plasticity of gonad development in sea lampreys, *Petromyzon marinus*: observations through gonadal biopsies. J Great Lakes Res 28:315–323

Beamish FWH, Griffiths RW (2001) Compensatory mechanisms in sea lamprey. Great Lakes Fisheries Commission Completion Report, Ann Arbor, MI

Beamish FWH, Medland TE (1988) Metamorphosis of the mountain brook lamprey *Ichthyomyzon greeleyi*. Environ Biol Fishes 23:45–54

Beamish FWH, Potter IC (1975) The biology of the anadromous sea lamprey (*Petromyzon marinus*) in New Brunswick. J Zool 177:57–72

Beamish FWH, Thomas EJ (1983) Potential and actual fecundity of the "paired" lampreys. *Ichthyomyzon gagei* and *I. castaneus*. Copeia 1983:367–374

Beamish FWH, Thomas EJ (1984) Metamorphosis of the southern brook lamprey, *Ichthyomyzon gagei*. Copeia 1984:502–515

Beamish FWH, Potter IC, Thomas E (1979) Proximate composition of the adult anadromous sea lamprey, *Petromyzon marinus*, in relation to feeding, migration and reproduction. J Anim Ecol 48:1–19

Beamish FWH, Ott PK, Roe SL (1994) Interpopulational variation in fecundity and egg size in southern brook lamprey, *Ichthyomyzon gagei*. Copeia 1994:718–725

Beamish RJ (1980) Adult biology of the river lamprey (*Lampetra ayresi*) and the Pacific lamprey (*Lampetra tridentata*) from the Pacific coast of Canada. Can J Fish Aquat Sci 37:1906–1923

Beamish RJ (1982) *Lampetra macrostoma*, a new species of freshwater parasitic lamprey from the west coast of Canada. Can J Fish Aquat Sci 39:736–747

Beamish RJ (1985) Freshwater parasitic lamprey on Vancouver Island and a theory of the evolution of the freshwater parasitic and nonparasitic life history types. In: Foreman RE, Gorbman A, Dodd JM, Olsson R (eds) Evolutionary biology of primitive fishes. Plenum Press, New York, pp 123–140

Beamish R, Withler R, Wade J, Beacham T (2016) A nonparasitic lamprey produces a parasitic life history type: the Morrison Creek lamprey enigma. In: Orlov A, Beamish R (eds) Jawless fishes of the world, vol 1. Cambridge Scholars Publishing, Newcastle upon Tyne, pp 191–230

Beard J (1893) Notes on lampreys and hags (*Myxine*). Anat Anz 8:59–60

Beaulaton L, Taverny C, Castelnaud G (2008) Fishing, abundance and life history traits of the anadromous sea lamprey (*Petromyzon marinus*) in Europe. Fish Res 92:90–101

Beaumont HM, Mandl AM (1962) Quantitative and cytological study of oogonia and oocytes in foetal and neonatal rat. Proc R Soc B 155:557–579

Becker P, Roland H, Reinboth R (1975) An unusual approach to experimental sex inversion in the teleost fish, *Betta* and *Macropodus*. In: Reinboth R (ed) Intersexuality in the animal kingdom. Springer, Berlin, pp 236–242

Berg LS (1948) Freshwater fishes of the USSR and adjacent countries. Guide to the Fauna of the USSR No. 27, vol 1:1–504

Berbejillo J, Martinez-Bengochea A, Bedo G et al (2012) Expression and phylogeny of candidate genes for sex differentiation in a primitive fish species, the Siberian sturgeon, *Acipenser baerii*. Mol Reprod Dev 79:504–516

Bergstedt RA, Twohey MB (2007) Research to support sterile-male-release and genetic alteration techniques for sea lamprey control. J Great Lakes Res 33:48–69

Beullens K, Eding E, Gilson P, Ollevier F (1997a) Gonadal differentiation, intersexuality and sex ratios of European eel (*Anguilla anguilla* L.) maintained in captivity. Aquaculture 153:135–150

Beullens K, Eding E, Ollevier F, Komen J, Richter C (1997b) Sex differentiation, changes in length, weight and eye size before and after metamorphosis of European eel (*Anguilla anguilla* L.) maintained in captivity. Aquaculture 153:151–162

Biederman NK, Nelson MM, Asalone KC et al (2018) Discovery of the first germline-restricted gene by subtractive transcriptomic analysis in the zebra finch, *Taenopygia guttata*. Curr Biol 28:1620–1627

Bird DJ, Potter IC (1979) Metamorphosis in the paired species of lampreys, *Lampetra fluviatilis* (L.) and *Lampetra planeri* (Bloch). 2. Quantitative data for body proportions, weights, lengths and sex ratios. Zool J Linn Soc 65:145–160

Bird DJ, Potter IC (1983) Proximate composition at various stages of adult life in the Southern Hemisphere lamprey, *Geotria australis* Gray. Comp Biochem Physiol 74A:623–633

Bird DJ, Ellis DJ, Potter IC (1993) Comparisons between the fatty acid compositions of the muscle and ovary of the nonparasitic lamprey *Lampetra planeri* (Bloch) and their counterparts in the anadromous and parasitic *Lampetra fluviatilis* (L.). Comp Biochem Physiol 105B:327–332

Bolduc TG, Sower SA (1992) Changes in brain gonadotropin-releasing hormone, plasma estradiol 17-ß, and progesterone during the final reproductive cycle of the female sea lamprey, *Petromyzon marinus*. J Exp Zool 264:55–63

Bondesson M, Hao R, Lin CY, Williams C, Gustafsson JÅ (2015) Estrogen receptor signaling during vertebrate development. Biochim Biophys Acta Gene Regul Mech 1849:142–151

Boney SE, Shelton WL, Yang SL, Wilken LO (1984) Sex reversal and breeding of grass carp. Trans Am Fish Soc 113:348–353

Bowles J, Koopman P (2001) New clues to the puzzle of mammalian sex determination. Genom Biol 2:1

Bravener G, Twohey M (2016) Evaluation of a sterile-male release technique: a case study of invasive sea lamprey control in a tributary of the Laurentian Great Lakes. N Am J Fish Manag 36:1125–1138

Brennan J, Capel B (2004) One tissue, two fates: molecular genetic events that underlie testis versus ovary development. Nat Rev Gen 5:509

Bridgham JT, Brown JE, Rodriguez-Mari A, Catchen JM, Thornton JW (2008) Evolution of a new function by degenerative mutation in cephalochordate steroid receptors. PLoS Genet 4:e1000191

Bridgham JT, Eick GN, Larroux C et al (2010) Protein evolution by molecular tinkering: diversification of the nuclear receptor superfamily from a ligand-dependent ancestor. PLoS Biol 8:e1000497

Brown EE, Baumann H, Conover DO (2014) Temperature and photoperiod effects on sex determination in a fish. J Exp Mar Biol Ecol 461:39–43

Browne R, Kaurova S, Uteshev V et al (2015) Sperm motility of externally fertilizing fish and amphibians. Theriogenology 83:1–13

Bryan MB, Scott A, Cerný I, Yun S-S, Li W (2003) 15α-Hydroxytestosterone produced in vitro and in vivo in the sea lamprey, *Petromyzon marinus*. Gen Comp Endocrinol 132:418–426

Bryan MB, Scott AP, Cerný I, Young BA, Li W (2004) 15α-Hydroxyprogesterone in male sea lampreys, *Petromyzon marinus* L. Steroids 69:473–481

Bryan MB, Young BA, Close DA et al (2006) Comparison of synthesis of 15α-hydroxylated steroids in males of four North American lamprey species. Gen Comp Endocrinol 146:149–156

Bryan MB, Scott AP, Li W (2007) The sea lamprey (*Petromyzon marinus*) has a receptor for androstenedione. Biol Reprod 77:688–696

Bryan MB, Scott AP, Li W (2008) Sex steroids and their receptors in lampreys. Steroids 73:1–12

Bryan MB, Chung-Davidson YW, Ren J et al (2015) Evidence that progestins play an important role in spermiation and pheromone production in male sea lamprey (*Petromyzon marinus*). Gen Comp Endocrinol 212:17–27

Bryant SA, Herdy JR, Amemiya CT, Smith JJ (2016) Characterization of somatically-eliminated genes during development of the sea lamprey (*Petromyzon marinus*). Mol Biol Evol 33:2337–2344

Bulmer M (1987) Sex determination in fish. Nature 326:440

Bulun SE, Sebastian S, Takayama K et al (2003) The human *CYP19* (aromatase P450) gene: update on physiologic roles and genomic organization of promoters. J Steroid Biochem Mol Biol 86:219–224

Busson-Mabillot S (1967) Gonadogenèse, différentiation sexuelle et structure de l'ovaire larvaire chez la lamproie de planer, *Lampetra planeri* (Bloch). Arch Zool Exp Gen 108:293–318

Callard GV, Petro Z, Ryan KJ (1980) Aromatization and 5α-reduction in brain and nonneural tissues of a cyclostome, *Petromyzon marinus*. Gen Comp Endocrinol 42:155–159

Capel B (2017) Vertebrate sex determination: evolutionary plasticity of a fundamental switch. Nat Rev Genet 18:675

Castro-Santos T, Shi X, Haro A (2017) Migratory behavior of adult sea lamprey and cumulative passage performance through four fishways. Can J Fish Aquat Sci 74:790–800

Cavileer TD, Hunter SS, Olsen J, Wenburg J, Nagler JJ (2015) A sex-determining gene (*sdY*) assay shows discordance between phenotypic and genotypic sex in wild populations of Chinook Salmon. Trans Am Fish Soc 144:423–430

Chambers R (1997) Environmental influences on egg and propagule sizes in marine fishes. In: Chambers RC, Trippel EA (eds) Early life history and recruitment in fish populations. Chapman and Hall, NY, pp 63–102

Charlesworth D, Charlesworth B, Marais G (2005) Steps in the evolution of heteromorphic sex chromosomes. Heredity 95:118–128

Charnov EL, Bull JJ (1977) When is sex environmentally determined? Nature 266:828–830

Chen S, Zhang G, Shao C et al (2014) Whole-genome sequence of a flatfish provides insights into ZW sex chromosome evolution and adaptation to a benthic lifestyle. Nat Genet 46:253–260

Chen TR, Reisman HM (1970) A comparative chromosome study of the North American species of sticklebacks (Teleostei: Gasterosteidae). Cytogenet Genom Res 9:321–332

Chen Y, Dong Y, Xiang X, Zhang X, Zhu B (2008) Sex determination of *Microtus mandarinus mandarinus* is independent of *Sry* gene. Mamm Genom 19:61–68

Christie GC, Goddard CI (2003) Sea Lamprey International Symposium (SLIS II): advances in the integrated management of sea lamprey in the Great Lakes. J Great Lakes Res 29(Suppl 1):1–14

Christie WJ, Kolenosky DP (1980) Parasitic phase of the sea lamprey (*Petromyzon marinus*) in Lake Ontario. Can J Fish Aquat Sci 37:2021–2038

Churchill W (1945) The brook lamprey in the Brule River. Trans Wisc Acad Sci Arts Lett 37:337–346

Ciereszko A, Glogowski J, Dabrowski K (2000) Fertilization in landlocked sea lamprey: storage of gametes, optimal sperm : egg ratio, and methods of assessing fertilization success. J Fish Biol 56:495–505

Ciereszko A, Dabrowski K, Toth GP, Christ SA, Glogowski J (2002) Factors affecting motility characteristics and fertilizing ability of sea lamprey spermatozoa. Trans Am Fish Soc 131:193–202

Clemens BJ, van de Wetering S, Kaufman J, Holt RA, Schreck CB (2009) Do summer temperatures trigger spring maturation in Pacific lamprey, *Entosphenus tridentatus*? Ecol Freshw Fish 18:418–426

Clemens BJ, Mesa MG, Magie RJ, Young DA, Schreck CB (2012) Pre-spawning migration of adult Pacific lamprey, *Entosphenus tridentatus*, in the Willamette River, Oregon, USA. Environ Biol Fish 93:245–254

Clemens BJ, van de Wetering S, Sower SA, Schreck CB (2013) Maturation characteristics and life-history strategies of the Pacific lamprey, *Entosphenus tridentatus*. Can J Zool 91:775–788

Clemens B, Schreck C, Sower S, van de Wetering S (2016) The potential roles of river environments in selecting for stream- and ocean-maturing Pacific lamprey, *Entosphenus tridentatus* (Gairdner, 1836). In: Orlov A, Beamish R (eds) Jawless fishes of the world, vol 1. Cambridge Scholars Publishing, Newcastle upon Tyne, pp 299–322

Cline TJ, Kitchell JF, Bennington V et al (2014) Climate impacts on landlocked sea lamprey: implications for host-parasite interactions and invasive species management. Ecosphere 5:68

Close DA, Fitzpatrick MS, Li HW (2002) The ecological and cultural importance of a species at risk of extinction, Pacific lamprey. Fisheries 27:19–25

Close DA, Yun SS, McCormick SD, Wildbill AJ, Li WM (2010) 11-Deoxycortisol is a corticosteroid hormone in the lamprey. Proc Natl Acad Sci USA 107:13942–13947

Cochran PA (2009) A comparison of native and exotic hosts for the silver lamprey. In: Brown LR, Chase SD, Mesa MG, Beamish RJ, Moyle PB (eds) Biology, management, and conservation of lampreys in North America. Am Fish Soc Symp 72:165–172

Colombo G, Grandi G (1996) Histological study of the development and sex differentiation of the gonad in the European eel. J Fish Biol 48:493–512

Colombo RE, Wills PS, Garvey JE (2004) Use of ultrasound imaging to determine sex of shovelnose sturgeon. N Am J Fish Manag 24:322–326

Conover DO (1984) Adaptive significance of temperature-dependent sex determination in a fish. Am Nat 123:297–313

Conover DO (2004) Temperature-dependent sex determination in fishes. In: Valenzuela N, Lance V (eds) Temperature-dependent sex determination. Smithsonian Institution Press, Washington, DC, pp 11–20

Conover DO, Fleisher MH (1986) Temperature-sensitive period of sex determination in the Atlantic silverside, *Menidia menidia*. Can J Fish Aquat Sci 43:514–520

Conover DO, Kynard BE (1981) Environmental sex determination: interaction of temperature and genotype in a fish. Science 213:577–579

Conover DO, Van Voorhees DA, Ehtisham A (1992) Sex ratio selection and the evolution of environmental sex determination in laboratory populations of *Menidia menidia*. Evolution 46:1722–1730

Corkum LD, Sapota MR, Skora KE (2004) The round goby, *Neogobius melanostomus*, a fish invader on both sides of the Atlantic Ocean. Biol Invasion 6:173–181

Cotton S, Wedekind C (2007) Introduction of Trojan sex chromosomes to boost population growth. J Theor Biol 249:153–161

Cutting A, Chue J, Smith CA (2013) Just how conserved is vertebrate sex determination? Dev Dyn 242:380–387

Dantzer B, Swanson EM (2012) Mediation of vertebrate life histories via insulin-like growth factor-1. Biol Rev Camb Philos Soc 87:414–429

Davis KB, Simco BA, Goudie CA et al (1990) Hormonal sex manipulation and evidence for female homogamety in channel catfish. Gen Comp Endocrinol 78:218–223

Davis RM (1967) Parasitism by newly-transformed anadromous sea lampreys on landlocked salmon and other fishes in a coastal Maine lake. Trans Am Fish Soc 96:11–16

Dawson HA, Quintella BR, Almeida PR, Treble AJ, Jolley JC (2015) The ecology of larval and metamorphosing lampreys. In: Docker MF (ed) Lampreys: biology, conservation and control, vol 1. Springer, Dordrecht, pp 75–137

Daye PG, Glebe BD (1984) Fertilization success and sperm motility of Atlantic salmon (*Salmo salar* L.) in acidified water. Aquaculture 43:307–312

D'Cotta H, Fostier A, Guiguen Y, Govoroun M, Baroiller JF (2001) Aromatase plays a key role during normal and temperature-induced sex differentiation of tilapia *Oreochromis niloticus*. Mol Reprod Dev 59:265–276

de Kretser DM, Loveland KL, Meinhardt A, Simorangkir D, Wreford N (1998) Spermatogenesis. Hum Reprod 13:1–8

de Rooij DG, Russell LD (2000) All you wanted to know about spermatogonia but were afraid to ask. J Androl 21:776–798

deVlaming V, Grossman G, Chapman F (1982) On the use of the gonadosomatic index. Comp Biochem Physiol 73A:31–39

Dean B, Sumner FB (1898) Notes on the spawning habits of the brook lamprey (*Petromyzon wilderi*). Trans N Y Acad Sci 16:321–324

Dendy JS, Scott DC (1953) Distribution, life history, and morphological variations of the southern brook lamprey, *Ichthyomyzon gagei*. Copeia 1953:152–161

Deragon KL, Sower SA (1994) Effects of lamprey gonadotropin-releasing hormone-III on steroido-genesis and spermiation in male sea lampreys. Gen Comp Endocrinol 95:363–367

Devlin RH, Nagahama Y (2002) Sex determination and sex differentiation in fish: an overview of genetic, physiological, and environmental influences. Aquaculture 208:191–364

Docker MF (1992) Labile sex determination in lampreys: the effect of larval density and sex steroids on gonadal differentiation. PhD thesis, University of Guelph, Guelph, ON

Docker MF (2006) Bill Beamish's contributions to lamprey research and recent advances in the field. Guelph Ichthyol Rev 7:1–52

Docker MF (2009) A review of the evolution of nonparasitism in lampreys and an update of the paired species concept. In: Brown LR, Chase SD, Mesa MG, Beamish RJ, Moyle PB (eds) Biology, management, and conservation of lampreys in North America. Am Fish Soc Symp 72:71–114

Docker MF, Beamish FWH (1991) Growth, fecundity, and egg size of least brook lamprey, *Lampetra aepyptera*. Environ Biol Fishes 31:219–227

Docker MF, Beamish FWH (1994) Age, growth, and sex ratio among populations of least brook lamprey, *Lampetra aepyptera*, larvae: an argument for environmental sex determination. Environ Biol Fishes 41:191–205

Docker MF, Sower SA, Youson JH, Beamish FWH (2003) Future sea lamprey control through regulation of metamorphosis and reproduction: a report from the SLIS II new science and control workgroup. J Great Lakes Res 29(Suppl 1):801–807

Docker MF, Hume JB, Clemens BJ (2015) Introduction: a surfeit of lampreys. In: Docker MF (ed) Lampreys: biology, conservation and control, vol 1. Springer, Dordrecht, pp 1–34

Dockray GJ, Pickering AD (1972) The influence of the gonad on the degeneration of the intestine in migrating river lampreys: *Lampetra fluviatilis* L. (Cyclostomata). Comp Biochem Physiol 43A:279–286

Dodd JM, Wiebe JP (1968) Endocrine influences on spermatogenesis in cold-blooded vertebrates. Arch Anat Histol Embryol 51:155–174

Dodd J, Evennett P, Goddard C (1960) Reproductive endocrinology in cyclostomes and elasmo-branchs. Symp Zool Soc Lond 1:77–103

Donaldson EM, Hunter GA (1982) Sex control in fish with particular reference to salmonids. Can J Fish Aquat Sci 39:99–110

Duffy TA, Hice LA, Conover DO (2015) Pattern and scale of geographic variation in environmental sex determination in the Atlantic silverside, *Menidia menidia*. Evolution 69:2187–2195

Dziewulska K, Domagała J (2009) Ripening of the oocyte of the river lamprey (*Lampetra fluviatilis* L.) after river entry. J Appl Ichthyol 25:752–756

Eick GN, Thornton JW (2011) Evolution of steroid receptors from an estrogen-sensitive ancestral receptor. Mol Cell Endocrinol 334:31–38

Ellegren H (2010) Evolutionary stasis: the stable chromosomes of birds. Trends Ecol Evol 25:283–291

Eshenroder RL (2014) The role of the Champlain Canal and Erie Canal as putative corridors for colonization of Lake Champlain and Lake Ontario by Sea Lampreys. Trans Am Fish Soc 143:634–649

Evans TM, Janvier P, Docker MF (2018) The evolution of lamprey (Petromyzontida) life history and the origin of metamorphosis. Rev Fish Biol Fish 28:825–838

Evennett PJ, Dodd JM (1963) Endocrinology of reproduction in the river lamprey. Nature 197:715–716

Fahien CM, Sower SA (1990) Relationship between brain gonadotropin-releasing hormone and final reproductive period of the adult male sea lamprey, *Petromyzon marinus*. Gen Comp Endocrinol 80:427–437

Farrokhnejad MR, Amiri BM, Abdoli A, Nazari H (2014) Gonadal histology and plasma sex steroid concentrations in maturing and mature spring migrants of Caspian lamprey *Caspiomyzon wagneri* in the Shirud River, southern Caspian Sea, Iran. Ichthyol Res 61:42–48

Feist G, Schreck CB, Fitzpatrick MS, Redding JM (1990) Sex steroid profiles of coho salmon (*Oncorhynchus kisutch*) during early development and sexual differentiation. Gen Comp Endocrinol 80:299–313

Ferguson MW, Joanen T (1982) Temperature of egg incubation determines sex in *Alligator mississippiensis*. Nature 296:850–853

Filardo EJ, Thomas P (2005) GPR30: a seven-transmembrane-spanning estrogen receptor that triggers EGF release. Trends Endocrinol Metab 16:362–367

Fleming IA (1998) Pattern and variability in the breeding system of Atlantic salmon (*Salmo salar*), with comparisons to other salmonids. Can J Fish Aquat Sci 55:59–76

Fodale MF, Bronte CR, Bergstedt RA, Cuddy DW, Adams JV (2003) Classification of lentic habitat for sea lamprey (*Petromyzon marinus*) larvae using a remote seabed classification device. J Great Lakes Res 29(Suppl 1):190–203

Forconi M, Canapa A, Barucca M et al (2013) Characterization of sex determination and sex differentiation genes in *Latimeria*. PLoS ONE 8:e56006

Freamat M, Sower SA (2013) Integrative neuro-endocrine pathways in the control of reproduction in lamprey: a brief review. Front Endocrinol 4:151

Frimpong EA, Henebry ML (2012) Short-term effects of formalin and ethanol fixation and preservation techniques on weight and size of fish eggs. Trans Am Fish Soc 141:1472–1479

Fukayama S, Takahashi H (1982) Sex differentiation and development of the gonad in the Japanese river lamprey, *Lampetra japonica*. Bull Fac Fish Hokkaido Univ 33:206–216

Fukayama S, Takahashi H (1983) Sex differentiation and development of the gonad in the sand lamprey, *Lampetra reissneri*. Bull Fac Fish Hokkaido Univ 34:279–290

Fukayama S, Takahashi H (1985) Changes in serum levels of estradiol-17ß and testosterone in the Japanese river lamprey, *Lampetra japonica,* in the course of sexual maturation. Bull Fac Fish Hokkaido Univ 36:163–169

Fukayama S, Takahashi H, Matsubara T, Hara A (1986) Profiles of the female-specific serum protein in the Japanese river lamprey, *Lampetra japonica* (Martens), and the sand lamprey, *Lampetra reissneri* (Dybowski), in relation to sexual maturation. Comp Biochem Physiol 84A:45–48

Gage MJ, Stockley P, Parker GA (1995) Effects of alternative male mating strategies on characteristics of sperm production in the Atlantic salmon (*Salmo salar*): theoretical and empirical investigations. Philos Trans R Soc Lond B 350:391–399

Gambicki S, Steinhart GB (2017) Changes in sea lamprey size and fecundity through time in the Great Lakes. J Great Lakes Res 43:209–214

Gazourian L, Deragon KL, Chase CF et al (1997) Characteristics of GnRH binding in the gonads and effects of lamprey GnRH-I and -III on reproduction in the adult sea lamprey. Gen Comp Endocrinol 108:327–339

Gazourian L, Evans EL, Hanson L, Chase CF, Sower SA (2000) The effects of lamprey GnRH-I, -III and analogs on steroidogenesis in the sea lamprey (*Petromyzon marinus*). Aquaculture 188:147–165

Geffroy B, Guiguen Y, Fostier A, Bardonnet A (2013) New insights regarding gonad development in European eel: evidence for a direct ovarian differentiation. Fish Physiol Biochem 39:1129–1140

Geffroy B, Guilbaud F, Amilhat E et al (2016) Sexually dimorphic gene expressions in eels: useful markers for early sex assessment in a conservation context. Sci Rep 6:34041

Gilbert SF (2000) Developmental biology. Sinauer Associates, Sunderland, MA

GLFC (1960) Annual report to the Great Lakes Fishery Commission 1960, Ann Arbor, MI

GLFC (1961) Annual report to the Great Lakes Fishery Commission 1961, Ann Arbor, MI

GLFC (1962) Annual report to the Great Lakes Fishery Commission 1962, Ann Arbor, MI

GLFC (1963) Annual report to the Great Lakes Fishery Commission 1963, Ann Arbor, MI

GLFC (1964) Annual report to the Great Lakes Fishery Commission 1964, Ann Arbor, MI

GLFC (1965) Annual report to the Great Lakes Fishery Commission 1965, Ann Arbor, MI

GLFC (1966) Annual report to the Great Lakes Fishery Commission 1966, Ann Arbor, MI

GLFC (1967) Annual report to the Great Lakes Fishery Commission 1967, Ann Arbor, MI

GLFC (1968) Annual report to the Great Lakes Fishery Commission 1968, Ann Arbor, MI

GLFC (1969) Annual report to the Great Lakes Fishery Commission 1969, Ann Arbor, MI
GLFC (1970) Annual report to the Great Lakes Fishery Commission 1970, Ann Arbor, MI
GLFC (1971) Annual report to the Great Lakes Fishery Commission 1971, Ann Arbor, MI
GLFC (1972) Annual report to the Great Lakes Fishery Commission 1972, Ann Arbor, MI
GLFC (1973) Annual report to the Great Lakes Fishery Commission 1973, Ann Arbor, MI
GLFC (1974) Annual report to the Great Lakes Fishery Commission 1974, Ann Arbor, MI
GLFC (1975) Annual report to the Great Lakes Fishery Commission 1975, Ann Arbor, MI
GLFC (1976) Annual report to the Great Lakes Fishery Commission 1976, Ann Arbor, MI
GLFC (1977) Annual report to the Great Lakes Fishery Commission 1977, Ann Arbor, MI
GLFC (1978) Annual report to the Great Lakes Fishery Commission 1978, Ann Arbor, MI
GLFC (1996) Integrated management of sea lampreys in the Great Lakes 1995. Annual report to
    the Great Lakes Fishery Commission, Ann Arbor, MI
GLFC (1997) Integrated management of sea lampreys in the Great Lakes 1996. Annual report to
    the Great Lakes Fishery Commission, Ann Arbor, MI
GLFC (1998) Integrated management of sea lampreys in the Great Lakes 1997. Annual report to
    the Great Lakes Fishery Commission, Ann Arbor, MI
GLFC (1999) Integrated management of sea lampreys in the Great Lakes 1998. Annual report to
    the Great Lakes Fishery Commission, Ann Arbor, MI
GLFC (2000) Integrated management of sea lampreys in the Great Lakes 1999. Annual report to
    the Great Lakes Fishery Commission, Ann Arbor, MI
GLFC (2001) Integrated management of sea lampreys in the Great Lakes 2000. Annual report to
    the Great Lakes Fishery Commission, Ann Arbor, MI
GLFC (2002) Integrated management of sea lampreys in the Great Lakes 2001. Annual report to
    the Great Lakes Fishery Commission, Ann Arbor, MI
GLFC (2003) Integrated management of sea lampreys in the Great Lakes 2002. Annual report to
    the Great Lakes Fishery Commission, Ann Arbor, MI
GLFC (2004) Integrated management of sea lampreys in the Great Lakes 2003. Annual report to
    the Great Lakes Fishery Commission, Ann Arbor, MI
GLFC (2005) Integrated management of sea lampreys in the Great Lakes 2004. Annual report to
    the Great Lakes Fishery Commission, Ann Arbor, MI
GLFC (2006) Integrated management of sea lampreys in the Great Lakes 2005. Annual report to
    the Great Lakes Fishery Commission, Ann Arbor, MI
GLFC (2007) Integrated management of sea lampreys in the Great Lakes 2006. Annual report to
    the Great Lakes Fishery Commission, Ann Arbor, MI
GLFC (2008) Integrated management of sea lampreys in the Great Lakes 2007. Annual report to
    the Great Lakes Fishery Commission, Ann Arbor, MI
GLFC (2009) Integrated management of sea lampreys in the Great Lakes 2008. Annual report to
    the Great Lakes Fishery Commission, Ann Arbor, MI
GLFC (2010) Integrated management of sea lampreys in the Great Lakes 2009. Annual report to
    the Great Lakes Fishery Commission, Ann Arbor, MI
GLFC (2011) Integrated management of sea lampreys in the Great Lakes 2010. Annual report to
    the Great Lakes Fishery Commission, Ann Arbor, MI
GLFC (2012) Sea lamprey control in the Great Lakes 2011. Annual report to the Great Lakes
    Fishery Commission, Ann Arbor, MI
GLFC (2013) Sea lamprey control in the Great Lakes 2012. Annual report to the Great Lakes
    Fishery Commission, Ann Arbor, MI
GLFC (2014) Sea lamprey control in the Great Lakes 2013. Annual report to the Great Lakes
    Fishery Commission, Ann Arbor, MI
GLFC (2015) Sea lamprey control in the Great Lakes 2014. Annual report to the Great Lakes
    Fishery Commission, Ann Arbor, MI
GLFC (2016) Sea lamprey control in the Great Lakes 2015. Annual report to the Great Lakes
    Fishery Commission, Ann Arbor, MI

GLFC (2017) Sea lamprey control in the Great Lakes 2016. Annual report to the Great Lakes Fishery Commission, Ann Arbor, MI

Goldstein JL, Wilson JD (1975) Genetic and hormonal control of male sexual differentiation. J Cell Physiol 85:365–377

Gorbman A (1983) Reproduction in cyclostome fishes and its regulation. In: Hoar WS, Randall DJ, Donaldson EM (eds) Fish physiology, vol 9. part A. Academic Press, New York, pp 1–29

Goudie CA, Redner BD, Simco BA, Davis KB (1983) Feminization of channel catfish by oral administration of steroid sex hormones. Trans Am Fish Soc 112:670–672

Gözler AM, Engin S, Turan D, Sahin C (2011) Age, growth and fecundity of the Turkish Brook Lamprey, *Eudontomyzon lanceolata* (Kux & Steiner, 1972), in north-eastern Turkey (Petromyzontiformes: Petromyzontidae). Zool Middle East 52:57–61

Graves JA (2013) How to evolve new vertebrate sex determining genes. Dev Dyn 242:354–359

Graves JA, Peichel CL (2010) Are homologies in vertebrate sex determination due to shared ancestry or to limited options? Genom Biol 11:205

Gutierrez JB, Teem JL (2006) A model describing the effect of sex-reversed YY fish in an established wild population: the use of a Trojan Y chromosome to cause extinction of an introduced exotic species. J Theor Biol 241:333–341

Gutowsky LF, Fox MG (2011) Occupation, body size and sex ratio of round goby (*Neogobius melanostomus*) in established and newly invaded areas of an Ontario river. Hydrobiologia 671:27–37

Hackmann E, Reinboth R (1974) Delimitation of the critical stage of hormone-influenced sex differentiation in *Hemihaplochromis multicolor* (Hilgendorf) (Cichlidae). Gen Comp Endocrinol 22:42–53

Hagelin, Steffner N (1958) Notes on the spawning habits of the river lamprey (*Petromyzon fluviatilis*). Oikos 9:221–238

Hale MC, Xu P, Scardina J et al (2011) Differential gene expression in male and female rainbow trout embryos prior to the onset of gross morphological differentiation of the gonads. BMC Genom 12:404

Hansen MJ, Hayne DW (1962) Sea lamprey larvae in Ogontz Bay and Ogontz River, Michigan. J Wild Manag 26:237–247

Hardisty MW (1960a) Sex ratio of ammocoetes. Nature 186:988–989

Hardisty MW (1960b) Development of the gonads in parasitic and nonparasitic lampreys. Nature 187:341–342

Hardisty MW (1961a) Studies on an isolated spawning population of brook lampreys. J Anim Ecol 30:339–355

Hardisty MW (1961b) Sex composition of lamprey populations. Nature 191:116–117

Hardisty MW (1961c) Oocyte numbers as a diagnostic character for the identification of ammocoete species. Nature 191:1215–1216

Hardisty MW (1964) The fecundity of lampreys. Arch Hydrobiol 60:340–357

Hardisty MW (1965a) Sex differentiation and gonadogenesis in lampreys. Part I. The ammocoete gonads of the brook lamprey, *Lampetra planeri*. J Zool 146:305–345

Hardisty MW (1965b) Sex differentiation and gonadogenesis in lampreys. Part II. The ammocoete gonads of the landlocked sea lamprey, *Petromyzon marinus*. J Zool 146:346–387

Hardisty MW (1969) A comparison of gonadal development in the ammocoetes of the landlocked and anadromous forms of the sea lamprey *Petromyzon marinus* L. J Fish Biol 2:153–166

Hardisty MW (1970) The relationship of gonadal development to the life cycles of the paired species of lamprey, *Lampetra fluviatilis* (L.) and *Lampetra planeri* (Bloch). J Fish Biol 2:173–181

Hardisty MW (1971) Gonadogenesis, sex differentiation and gametogenesis. In: Hardisty MW, Potter IC (eds) The biology of lampreys, vol 1. Academic Press, New York, pp 295–360

Hardisty MW (1986) *Lampetra fluviatilis* (Linnaeus, 1758). In: Holčík J (ed) The freshwater fishes of Europe: vol 1, Part I: Petromyzontiformes. AULA-Verlag Wiesbaden, pp 249–278

Hardisty MW (2006) Lampreys: life without jaws. Forrest Text, Ceredigion, UK

Hardisty MW, Taylor BJ (1965) The effect of sex hormones on the ammocoete larva. Life Sci 4:743–747

Hardisty MW, Potter IC, Sturge R (1970) A comparison of the metamorphosing and macrophthalmia stages of the lampreys *Lampetra fluviatilis* and *L. planeri*. J Zool 162:383–400

Hardisty MW, Potter IC, Hilliard RW (1986) Gonadogenesis and sex differentiation in the southern hemisphere lamprey, *Geotria australis* Gray. J Zool 209:477–499

Hardisty MW, Potter IC, Koehn JD (1992) Gonadogenesis and sex differentiation in the Southern Hemisphere lamprey *Mordacia mordax*. J Zool 226:491–516

Hattori RS, Murai Y, Oura M et al (2012) A Y-linked anti-Müllerian hormone duplication takes over a critical role in sex determination. Proc Natl Acad Sci USA 109:2955–2959

Heinrich JW, Weise JG, Smith BR (1980) Changes in biological characteristics of the sea lamprey (*Petromyzon marinus*) as related to lamprey abundance, prey abundance, and sea lamprey control. Can J Fish Aquat Sci 37:1861–1871

Hess JE, Campbell NR, Docker MF et al (2015) Use of genotyping by sequencing data to develop a high-throughput and multifunctional SNP panel for conservation applications in Pacific lamprey. Mol Ecol 15:187–202

Hewitt LM, Tremblay L, Van Der Kraak GJ, Solomon KR, Servos MR (1998a) Identification of the lampricide 3-trifluoromethyl-4-nitrophenol as an agonist for the rainbow trout estrogen receptor. Environ Toxicol Chem 17:425–432

Hewitt LM, Munkittrick KR, Van Der Kraak GJ et al (1998b) Hepatic mixed function oxygenase activity and vitellogenin induction in fish following a treatment of the lampricide 3-trifluoromethyl-4-nitrophenol (TFM). Can J Fish Aquat Sci 55:2078–2086

Ho SM, Press D, Liang LC, Sower S (1987) Identification of an estrogen receptor in the testis of the sea lamprey, *Petromyzon marinus*. Gen Comp Endocrinol 67:119–125

Holčík J (1986a) *Lethenteron japonicum* (Martens, 1868). In: Holčík J (ed) The freshwater fishes of Europe: vol 1, Part I: Petromyzontiformes. AULA-Verlag Wiesbaden, pp 198–219

Holčík J (1986b) *Caspiomyzon wagneri* (Kessler, 1870). In: Holčík J (ed) The freshwater fishes of Europe: vol 1, Part I: Petromyzontiformes. AULA-Verlag, Wiesbaden, pp 119–142

Holčík J (1986c) *Lethenteron kessleri* (Anikin, 1905). In: Holčík J (ed) The freshwater fishes of Europe: vol1, Part I: Petromyzontiformes. AULA-Verlag Wiesbaden, pp 200–236

Holčík J, Delić A (2000) New discovery of the Ukrainian brook lamprey in Croatia. J Fish Biol 56:73–86

Holleley CE, O'Meally D, Sarre SD et al (2015) Sex reversal triggers the rapid transition from genetic to temperature-dependent sex. Nature 523:79

Holleley CE, Sarre SD, O'Meally D, Georges A (2016) Sex reversal in reptiles: reproductive oddity or powerful driver of evolutionary change? Sex Dev 10:279–287

Houston KA, Kelso JRM (1991) Relation of sea lamprey size and sex ratio to salmonid availability in three Great Lakes. J Great Lakes Res 17:270–280

Heule C, Salzburger W, Böhne A (2014) Genetics of sexual development: an evolutionary playground for fish. Genetics 196:579–591

Hsu L, Chen H, Lin Y, Chu W, Lin M (2008) DAZAP1, an hnRNP protein, is required for normal growth and spermatogenesis in mice. RNA 14:1814–1822

Hubert TD (2003) Environmental fate and effects of the lampricide TFM: a review. J Great Lakes Res 29(Suppl 1):456–474

Hughes RL, Potter IC (1969) Studies on gametogenesis and fecundity in the lampreys *Mordacia praecox* and *M. mordax* (Petromyzonidae). Aust J Zool 17:447–464

Hunter GA, Donaldson EM, Goetz FW, Edgell PR (1982) Production of all-female and sterile groups of coho salmon (*Oncorhynchus kisutch*) and experimental evidence for male heterogamety. Trans Am Fish Soc 111:367–372

Hunter GA, Donaldson EM, Stoss J, Baker I (1983) Production of monosex female groups of Chinook salmon (*Oncorhynchus tshawytscha*) by the fertilization of normal ova with sperm from sex-reversed females. Aquaculture 33:355–364

Ijiri S, Kaneko H, Kobayashi T et al (2008) Sexual dimorphic expression of genes in gonads during early differentiation of a teleost fish, the Nile tilapia *Oreochromis niloticus*. Biol Reprod 78:333–341

Ishijima J, Uno Y, Nunome M et al (2017) Molecular cytogenetic characterization of chromosome site-specific repetitive sequences in the Arctic lamprey (*Lethenteron camtschaticum*, Petromyzontidae). DNA Res 24:93–101

Jang M-H, Lucas MC (2005) Reproductive ecology of the river lamprey. J Fish Biol 66:499–512

Janzen FJ, Krenz JG (2004) Phylogenetics: which was first, TSD or GSD? In: Valenzuela N, Lance VA (eds) Temperature-dependent sex determination in vertebrates. Smithsonian Institution Press, Washington, DC, pp 121–130

Janzen F, Phillips P (2006) Exploring the evolution of environmental sex determination, especially in reptiles. J Evol Biol 19:1775–1784

Josso N, Clemente N, Goue L (2001) Anti-Müllerian hormone and its receptors. Mol Cell Endocrinol 179:25–32

Johnson BGH (1969) Some statistics of the populations of parasitic phase sea lampreys in Canadian waters of the Great Lakes. In: Proceedings of the 12th conference on Great Lakes research. University of Michigan, Ann Arbor, MI, pp 45–52

Johnson BGH, Anderson WC (1980) Predatory-phase sea lampreys (*Petromyzon marinus*) in the Great Lakes. Can J Fish Aquat Sci 37:2007–2020

Johnson NS, Tix JA, Hlina BL et al (2015a) A sea lamprey (*Petromyzon marinus*) sex pheromone mixture increases trap catch relative to a single synthesized component in specific environments. J Chem Ecol 41:311–321

Johnson NS, Buchinger TJ, Li W (2015b) Reproductive ecology of lampreys. In: Docker MF (ed) Lampreys: biology, conservation and control, vol 1. Springer, Dordrecht, pp 265–303

Johnson NS, Swink WD, Brenden TO (2017) Field study suggests that sex determination in sea lamprey is directly influenced by larval growth rate. Proc R Soc B 284:20170262

Johnson W (1982) Body lengths, body weight and fecundity of sea lampreys (*Petromyzon marinus*) from Green Bay, Lake Michigan. Trans Wisc Acad Sci Arts Lett 70:73–77

Jones ML, Bergstedt RA, Twohey MB et al (2003) Compensatory mechanisms in Great Lakes sea lamprey populations: implications for alternative control strategies. J Great Lakes Res 29(Suppl 1):113–129

Kamiya T, Kai W, Tasumi S et al (2012) A trans-species missense SNP in *Amhr2* is associated with sex determination in the tiger pufferfish, *Takifugu rubripes* (fugu). PLoS Genet 8:e1002798

Kamler E (2005) Parent–egg–progeny relationships in teleost fishes: an energetics perspective. Rev Fish Biol Fish 15:399–421

Kan TT (1975) Systematics, variation, distribution, and biology of lampreys of the genus *Lampetra* in Oregon. PhD thesis, Oregon State University, Corvallis, OR

Kan TT, Bond CE (1981) Notes on the biology of the Miller Lake lamprey *Lampetra (Entosphenus) minima*. Northwest Sci 55:70–74

Kato T, Esaki M, Matsuzawa A, Ikeda Y (2012) NR5A1 is required for functional maturation of Sertoli cells during postnatal development. Reproduction 143:663–672

Katsu Y, Kubokawa K, Urushitani H, Iguchi T (2010) Estrogen-dependent transactivation of amphioxus steroid hormone receptor via both estrogen and androgen response elements. Endocrinology 151:639–648

Katsu Y, Cziko PA, Chandsawangbhuwana C et al (2016) A second estrogen receptor from Japanese lamprey (*Lethenteron japonicum*) does not have activities for estrogen binding and transcription. Gen Comp Endocrinol 236:105–114

Katz Y, Dashow L, Epple A (1982) Circulating steroid hormones of anadromous sea lampreys under various experimental conditions. Gen Comp Endocrinol 48:261–268

Kettlewell JR, Raymond CS, Zarkower D (2000) Temperature-dependent expression of turtle *Dmrt1* prior to sexual differentiation. Genesis 26:174–178

Khan A (2017) Investigation of candidate sex determination and sex differentiation genes in sea lamprey, *Petromyzon marinus*, and Pacific lamprey, *Entosphenus tridentatus*. MSc thesis, University of Manitoba, Winnipeg, MB

Kikuchi K, Hamaguchi S (2013) Novel sex-determining genes in fish and sex chromosome evolution. Dev Dyn 242:339–353

Kille RA (1960) Fertilization of the lamprey egg. Exp Cell Res 20:12–27

Kime DE (1993) Classical and non-classical reproductive steroids in fish. Rev Fish Biol Fish 3:160–180

Kime DE, Callard GV (1982) Formation of 15α-hydroxylated androgens by the testis and other tissues of the sea lamprey, *Petromyzon marinus*, in vitro. Gen Comp Endocrinol 46:267–270

Kime DE, Larsen LO (1987) Effect of gonadectomy and hypophysectomy on plasma steroid levels in male and female lampreys (*Lampetra fluviatilis*, L.). Gen Comp Endocrinol 68:189–196

Kime DE, Rafter JJ (1981) Biosynthesis of 15-hydroxylated steroids by gonads of the river lamprey, *Lampetra fluviatilis*, in vitro. Gen Comp Endocrinol 44:69–76

Knapp R, Carlisle (2011) Testicular function and hormonal regulation in fish. In: Norris DO, Lopez KH (eds) Hormones and reproduction of vertebrates, vol 1. Academic Press, Amsterdam, pp 43–58

Knowles FGW (1939) The influence of anterior-pituitary and testicular hormones on the sexual maturation of lampreys. J Exp Biol 16:535–548

Kobayashi T, Matsuda M, Kajiura-Kobayashi H et al (2004) Two DM domain genes, *DMY* and *DMRT1*, involved in testicular differentiation and development in the medaka, *Oryzias latipes*. Dev Dyn 231:518–526

Kobayashi W (1993) Effect of osmolality on the motility of sperm from the lamprey, *Lampetra japonica*. Zool Sci 10:281–285

Kobayashi W, Yamamoto TS (1994) Fertilization of the lamprey (*Lampetra japonica*) eggs: implication of the presence of fast and permanent blocks against polyspermy. J Exp Zool 269:166–176

Komen J, Lodder PAJ, Huskens F, Richter CJJ, Huisman EA (1989) Effects of oral administration of 17α-methyltestosterone and 17β-estradiol on gonadal development in common carp, *Cyprinus carpio* L. Aquaculture 78:349–363

Kopp CR (2017) An historical comparison of Lake Superior sea lamprey fecundity and egg size. Senior Capstone in Natural Resources, Northland College, Ashland, WI

Kott E (1971) Characteristics of pre-spawning American brook lampreys from Big Creek, Ontario. Can Field-Nat 85:235–240

Kovtun I (1979) The significance of sex ratio in the spawning population of round goby *Neogobius melanostomus* Pallas for the reproduction of generations of the Sea of Azov. Vopr Ikhtiol 17:642–649

Kucheryavyi AV, Savvaitova KA, Pavlov DS et al (2007) Variations of life history strategy of the Arctic lamprey *Lethenteron camtschaticum* from the Utkholok River (Western Kamchatka). J Ichthyol 47:37–52

Kuiper GG, Carlsson BO, Grandien KAJ et al (1997) Comparison of the ligand binding specificity and transcript tissue distribution of estrogen receptors α and β. Endocrinology 138:863–870

Kuznetsov Y, Mosyagina M, Zelennikov O (2016) The formation of fecundity in ontogeny of lampreys. In: Orlov A, Beamish R (eds) Jawless fishes of the world, vol 1. Cambridge Scholars Publishing, Newcastle upon Tyne, pp 323–345

Lack D (1954) The natural regulation of animal numbers. The Clarendon Press, Oxford

Lagomarsino IV, Conover DO (1993) Variation in environmental and genotypic sex-determining mechanisms across a latitudinal gradient in the fish, Menidia menidia. Evolution 47:487–494

Lampman RT (2011) Passage, migration behavior, and autoecology of adult Pacific lamprey at Winchester Dam and within the North Umpqua River Basin, Oregon, USA. MS thesis, Oregon State University, Corvallis, OR

Lapierre K, Renaud CB (2015) Fecundity of the lampreys *Lampetra hubbsi* and *Eudontomyzon hellenicus*. Environ Biol Fish 98:2315–2320

Labrie F, Lin S, Claude L, Simard J (1997) The key role of 17β-hydroxysteroid dehydrogenases in sex steroid biology. Steroids 62:148–158

Larsen LO (1969a) Effects of hypophysectomy before and during sexual maturation in the cyclostome *Lampetra fluviatilis* (L.) Gray. Gen Comp Endocrinol 12:200–208

Larsen LO (1969b) Effects of gonadectomy in the cyclostome *Lampetra fluviatilis*. Gen Comp Endocrinol 13:516–517

Larsen LO (1970) The lamprey egg at ovulation (*Lampetra fluviatilis* L. Gray). Biol Reprod 2:37–47

Larsen LO (1972) Endocrine control of intestinal atrophy in normal lampreys and of intestinal hypertrophy in gonadectomized lampreys, *Lampetra fluviatilis*. Gen Comp Endocrinol 18:602

Larsen LO (1973) Development in adult, freshwater river lampreys and its hormonal control. Starvation, sexual maturation and natural death. PhD thesis, University of Copenhagen, Denmark

Larsen LO (1974) Effects of testosterone and oestradiol on gonadectomized and intact male and female river lampreys (*Lampetra fluviatilis* (L.) Gray). Gen Comp Endocrinol 24:305–313

Larsen LO (1980) Physiology of adult lampreys, with special regard to natural starvation, reproduction, and death after spawning. Can J Fish Aquat Sci 37:1762–1779

Larsen LO (1987) The role of hormones in initiation of sexual maturation in male river lampreys (*Lampetra fluviatilis*, L.): gonadotropin and testosterone. Gen Comp Endocrinol 68:197–201

Larson GL, Christie GC, Johnson DA et al (2003) The history of sea lamprey control in Lake Ontario and updated estimates of suppression targets. J Great Lakes Res 29(Suppl 1):637–654

Leach WJ (1940) Occurrence and life history of the northern brook lamprey, *Ichthyomyzon fossor*, in Indiana. Copeia 1940:21–34

Lewis JC, McMillan DB (1965) The development of the ovary of the sea lamprey (*Petromyzon marinus* L.). J Morphol 117:425–466

Li M, Shen Q, Xu H et al (2011) Differential conservation and divergence of fertility genes *boule* and *dazl* in the rainbow trout. PLoS ONE 6:e15910

Li MH, Yang HH, Li MR et al (2013) Antagonistic roles of Dmrt1 and Foxl2 in sex differentiation via estrogen production in tilapia as demonstrated by TALENs. Endocrinology 154:4814–4825

Li W, Siefkes MJ, Scott AP, Teeter JH (2003) Sex pheromone communication in the sea lamprey: implications for integrated management. J Great Lakes Res 29(Suppl 1):85–94

Li W, Deng F, Wang H, Zhen Y (2006) Germ cell-less expression in zebrafish embryos. Dev Growth Differ 48:333–338

Li W, Twohey M, Jones M, Wagner M (2007) Research to guide use of pheromones to control sea lamprey. J Great Lakes Res 33:70–86

Liew WC, Orban L (2014) Zebrafish sex: a complicated affair. Brief Funct Genom 13:172–187

Linville JE, Hanson LH, Sower SA (1987) Endocrine events associated with spawning behavior in the sea lamprey (*Petromyzon marinus*). Horm Behav 21:105–117

Lowartz SM, Beamish FWH (2000) Novel perspectives in sexual lability through gonadal biopsy in larval sea lampreys. J Fish Biol 56:743–757

Lowartz SM, Holmberg DL, Ferguson HW, Beamish FWH (1999) Healing of abdominal incisions in sea lamprey larvae: a comparison of three wound-closure techniques. J Fish Biol 54:616–626

Lowartz S, Petkam R, Renaud R et al (2003) Blood steroid profile and in vitro steroidogenesis by ovarian follicles and testis fragments of adult sea lamprey, *Petromyzon marinus*. Comp Biochem Physiol 134A:365–376

Lowartz S, Renaud R, Beamish FWH, Leatherland JF (2004) Evidence for 15α-and 7α-hydroxylase activity in gonadal tissue of the early-life stages of sea lampreys, *Petromyzon marinus*. Comp Biochem Physiol 138B:119–127

Lowe DR (1972) Variations in body composition of pre-adult landlocked sea lamprey, *Petromyzon marinus* L. in relation to size and season. MSc thesis, University of Guelph, Guelph, ON

Lowe DR, Beamish FWH, Potter IC (1973) Changes in the proximate body composition of the landlocked sea lamprey *Petromyzon marinus* (L.) during larval life and metamorphosis. J Fish Biol 5:673–682

Lowerre-Barbieri SK, Brown-Peterson NJ, Murua H et al (2011) Emerging issues and methodological advances in fisheries reproductive biology. Mar Coast Fish 3:32–51

Maekawa M, Ito C, Toyama Y et al (2004) Stage-specific expression of mouse germ cell-less-1 (mGCL-1), and multiple deformations during *mgcl-1* deficient spermatogenesis leading to reduced fertility. Arch Histol Cytol 67:335–347

Maeva E, Bruno I, Zielinski BS et al (2004) The use of pulse-echo acoustic microscopy to non-invasively determine sex of living larval sea lamprey, *Petromyzon marinus*. J Fish Biol 65:148–156

Mair GC, Scott AG, Penman DJ, Beardmore JA, Skibinski DOF (1991a) Sex determination in the genus *Oreochromis*. 1. Sex reversal, gynogenesis and triploidy in *Oreochromis niloticus* (L). Theor Appl Genet 82:144–152

Mair GC, Scott AG, Penman DJ, Skibinski DOF, Beardmore JA (1991b) Sex determination in the genus *Oreochromis*. 2. Sex reversal, hybridization, gynogenesis and triploidy in *Oreochromis aureus* Steindachner. Theor Appl Genet 82:153–160

Maitland PS, Morris KH, East K et al (1984) The estuarine biology of the river lamprey, *Lampetra fluviatilis*, in the Firth of Forth, Scotland, with particular reference to size composition and feeding. J Zool 203:211–225

Maitland PS, Morris KH, East K (1994) The ecology of lampreys (Petromyzonidae) in the Loch Lomond area. Hydrobiologia 290:105–120

Maitland PS, Renaud CB, Quintella BR, Close DA, Docker MF (2015) Conservation of native lampreys. In: Docker MF (ed) Lampreys: biology, conservation and control, vol 1. Springer, Dordrecht, pp 375–428

Malmqvist B (1978) Population structure and biometry of *Lampetra planeri* (Bloch) from three different watersheds in south Sweden. Arch Hydrobiol 84:65–86

Malmqvist B (1980) The spawning migration of the brook lamprey, *Lampetra planeri* Bloch, in a south Swedish stream. J Fish Biol 16:105–114

Malmqvist B (1986) Reproductive ecology of lampreys. In: Uyeno T, Arai R, Taniuchi T, Matsuura K (eds) Indo-Pacific fish biology: Proceedings of the Second International Conference on Indo-Pacific Fishes. Ichthyological Society of Japan, Tokyo, pp 20–30

Manion PJ (1972) Fecundity of the sea lamprey (*Petromyzon marinus*) in Lake Superior. Trans Am Fish Soc 101:718–720

Manion PJ, Hanson LH (1980) Spawning behavior and fecundity of lampreys from the upper three Great Lakes. Can J Fish Aquat Sci 37:1635–1640

Manion PJ, McLain AL (1971) Biology of larval sea lampreys (*Petromyzon marinus*) of the 1960 year class, isolated in the Big Garlic River, Michigan, 1960-65. Great Lakes Fish Comm Tech Rep 16:1–35

Manion PJ, Smith BR (1978) Biology of larval and metamorphosing sea lampreys, *Petromyzon marinus*, of the 1960 year class in the Big Garlic River, Michigan, Part II, 1966–72. Great Lakes Fish Comm Tech Rep 30:1–35

Manzon RG, Youson JH, Holmes JA (2015) Lamprey metamorphosis. In: Docker MF (ed) Lampreys: biology, conservation and control, vol 1. Springer, Dordrecht, pp 139–214

Markov GV, Tavares R, Dauphin-Villemant C et al (2009) Independent elaboration of steroid hormone signaling pathways in metazoans. Proc Natl Acad Sci USA 106:11913–11918

Markov GV, Gutierrez-Mazariegos J, Pitrat D et al (2017) Origin of an ancient hormone/receptor couple revealed by resurrection of an ancestral estrogen. Sci Adv 3:e1601778

Marsh E (1984) Egg size variation in central Texas populations of *Etheostoma spectabile* (Pisces: Percidae). Copeia 1984:291–301

Marsh E (1986) Effects of egg size on offspring fitness and maternal fecundity in the orangethroat darter, *Etheostoma spectabile* (Pisces: Percidae). Copeia 1986:18–30

Martin RW, Myers J, Sower SA, Phillips DJ, McAuley C (1983) Ultrasonic imaging, a potential tool for sex determination of live fish. N Am J Fish Manag 3:258–264

Mateus CS, Stange M, Berner D et al (2013) Strong genome-wide divergence between sympatric European river and brook lampreys. Curr Biol 23:R649–R650

Matsuda M, Nagahama Y, Shinomiya A et al (2002) *DMY* is a Y-specific DM-domain gene required for male development in the medaka fish. Nature 417:559–563

Mawaribuchi S, Musashijima M, Wada M et al (2017) Molecular evolution of two distinct *dmrt1* promoters for germ and somatic cells in vertebrate gonads. Mol Biol Evol 34:724–733

McCauley DW, Docker MF, Whyard S, Li W (2015) Lampreys as diverse model organisms in the genomics era. BioScience 65:1046–1056

McConville MB, Hubert TD, Remucal CK (2016) Direct photolysis rates and transformation pathways of the lampricides TFM and niclosamide in simulated sunlight. Environ Sci Tech 50:9998–10006

McIlraith BJ, Caudill CC, Kennedy BP, Peery CA, Keefer ML (2015) Seasonal migration behaviors and distribution of adult Pacific Lampreys in unimpounded reaches of the Snake River Basin. N Am J Fish Manag 35:123–134

Meek SE (1889) The fishes of the Cayuga Lake basin. Ann N Y Acad Sci 4:297–316

Mesa MG, Bayer JM, Bryan MB, Sower SA (2010) Annual sex steroid and other physiological profiles of Pacific lampreys (*Entosphenus tridentatus*). Comp Biochem Physiol 155A:56–63

Mewes KR, Latz M, Golla H, Fischer A (2002) Vitellogenin from female and estradiol-stimulated male river lampreys (*Lampetra fluviatilis* L.). J Exp Zool 292:52–72

Morais da Silva S, Hacker A, Harley V et al (1996) *Sox9* expression during gonadal development implies a conserved role for the gene in testis differentiation in mammals and birds. Nat Genet 14:62–68

Morinaga C, Saito D, Nakamura S et al (2007) The *hotei* mutation of medaka in the anti-Müllerian hormone receptor causes the dysregulation of germ cell and sexual development. Proc Natl Acad Sci USA 104:9691–9696

Mork L, Czerwinski M, Capel B (2014) Predetermination of sexual fate in a turtle with temperature-dependent sex determination. Dev Biol 386:264–271

Morkert SB, Swink WD, Seelye JG (1998) Evidence for early metamorphosis of sea lampreys in the Chippewa River, Michigan. N Am J Fish Manag 18:966–971

Morozova T (1956) Contribution to the biology and systematics of the Pacific [Arctic] lamprey. Vopr Ikhtiol 7:149–157 [in Russian]

Morris K (1989) A multivariate morphometric and meristic description of a population of freshwater-feeding river lampreys *Lampetra fluviatilis* (L.), from Loch Lomond, Scotland. Zool J Linn Soc 96:357–371

Moser ML, Almeida PR, Kemp PS, Sorensen PW (2015) Lamprey spawning migration. In: Docker MF (ed) Lampreys: biology, conservation and control, vol 1. Springer, Dordrecht, pp 215–263

Mullett KM, Heinrich JW, Adams JV et al (2003) Estimating lake-wide abundance of spawning-phase sea lampreys (*Petromyzon marinus*) in the Great Lakes: extrapolating from sampled streams using regression models. J Great Lakes Res 29(Suppl 1):240–252

Munkittrick KR, Servos MR, Parrott JL et al (1994) Identification of lampricide formulations as a potent inducer of MFO activity in fish. J Great Lakes Res 20:355–365

Murauskas JG, Orlov AM, Siwicke KA (2013) Relationships between the abundance of Pacific Lamprey in the Columbia River and their common hosts in the marine environment. Trans Am Fish Soc 142:143–155

Murdoch SP, Docker MF, Beamish FWH (1992) Effect of density and individual variation on growth of sea lamprey (*Petromyzon marinus*) larvae in the laboratory. Can J Zool 70:184–188

Myosho T, Otake H, Masuyama H et al (2012) Tracing the emergence of a novel sex-determining gene in medaka, *Oryzias luzonensis*. Genetics 191:163–170

Nakamoto M, Matsuda M, Wang DS, Nagahama Y, Shibata N (2006) Molecular cloning and analysis of gonadal expression of *Foxl2* in the medaka, *Oryzias latipes*. Biochem Biophys Res Commun 344:353–361

Nakamura M, Kobayashi T, Chang XT, Nagahama Y (1998) Gonadal sex differentiation in teleost fish. J Exp Zool 281:362–372

Nanda I, Kondo M, Hornung U et al (2002) A duplicated copy of *DMRT1* in the sex-determining region of the Y chromosome of the medaka, *Oryzias latipes*. Proc Natl Acad Sci USA 99:11778–11783

Navarro-Martin L, Vinas J, Ribas L et al (2011) DNA methylation of the gonadal aromatase (*cyp19a*) promoter is involved in temperature-dependent sex ratio shifts in the European sea bass. PLoS Genet 7:e1002447

Nazari H, Abdoli A (2010) Some reproductive characteristics of endangered Caspian Lamprey (*Caspiomyzon wagneri* Kessler, 1870) in the Shirud River southern Caspian Sea, Iran. Environ Biol Fish 88:87–96

Neave FB, Mandrak NE, Docker MF, Noakes DL (2007) An attempt to differentiate sympatric *Ichthyomyzon* ammocoetes using meristic, morphological, pigmentation, and gonad analyses. Can J Zool 85:549–560

Nicander L, Afzelius BA, Sjödén I (1968) Fine structure and early fertilization changes of the animal pole in eggs of the river lamprey, *Lampetra fluviatilis*. Development 19:319–326

Norman AW, Mizwicki MT, Norman DP (2004) Steroid-hormone rapid actions, membrane receptors and a conformational ensemble model. Nat Rev Drug Dis 3:27–41

Norris DO, Carr JA (2013) Vertebrate endocrinology, 5th edn. Academic Press, Amsterdam

Nursall JR, Buchwald D (1972) Life history and distribution of the Arctic lamprey (*Lethenteron japonicum* [Martens]) of Great Slave Lake, N.W.T. Fish Res Board Can Tech Rep 304:1–28

O'Connor LM (2001) Spawning success of introduced sea lampreys (*Petromyzon marinus*) in two streams tributary to Lake Ontario. MSc thesis, University of Guelph, Guelph, ON

Ohno S (1974) Animal cytogenetics. Borntraeger, Berlin

Okkelberg P (1921) The early history of the germ cells in the brook lamprey, *Entosphenus wilderi* (Gage), up to and including the period of sex differentiation. J Morphol 35:1–151

Ospina-Álvarez N, Piferrer F (2008) Temperature-dependent sex determination in fish revisited: prevalence, a single sex ratio response pattern, and possible effects of climate change. PLoS ONE 3:e2837

Owens IPF, Short RV (1995) Hormonal basis of sexual dimorphism in birds: implications for new theories of sexual selection. Trends Ecol Evol 10:44–47

Palaiokostas C, Bekaert M, Taggart JB et al (2015) A new SNP-based vision of the genetics of sex determination in European sea bass (*Dicentrarchus labrax*). Genet Sel Evol 47:68

Pang Y, Thomas P (2011) Progesterone signals through membrane progesterone receptors (mPRs) in MDA-MB-468 and mPR-transfected MDA-MB-231 breast cancer cells which lack full-length and N-terminally truncated isoforms of the nuclear progesterone receptor. Steroids 76:921–928

Pang Y, Dong J, Thomas P (2013) Characterization, neurosteroid binding and brain distribution of human membrane progesterone receptors δ and ϵ (mPRδ and mPRϵ) and mPRδ involvement in neurosteroid inhibition of apoptosis. Endocrinology 154:283–295

Papaioannou MD, Nef S (2010) microRNAs in the testis: building up male fertility. J Androl 31:26–33

Paris M, Pettersson K, Schubert M et al (2008) An amphioxus orthologue of the estrogen receptor that does not bind estradiol: insights into estrogen receptor evolution. BMC Evol Biol 8:219

Parker KA (2018) Evidence for the genetic basis and inheritance of ocean and river-maturing ecotypes of Pacific lamprey (*Entosphenus tridentatus*) in the Klamath River, California. MS thesis, Humboldt State University, Arcata, CA

Parma P, Radi O (2012) Molecular mechanisms of sexual development. Sex Dev 6:7–17

Patil JG, Hinze SJ (2008) Simplex PCR assay for positive identification of genetic sex in the Japanese medaka, *Oryzias latipes*. Mar Biotechnol 10:641–644

Patiño R, Takashima F (1995) Gonads. In: Takahashi F, Hibiya T (eds) An atlas of fish histology: normal and pathological features. Gustav Fisher Verlag, Tokyo, pp 128–153

Peichel CL, Ross JA, Matson CK et al (2004) The master sex-determination locus in threespine sticklebacks is on a nascent Y chromosome. Curr Biol 14:1416–1424

Peng JX, Xie JL, Zhou L, Hong YH, Gui JF (2009) Evolutionary conservation of *Dazl* genomic organization and its continuous and dynamic distribution throughout germline development in gynogenetic gibel carp. J Exp Zool B Mol Dev Evol 312:855–871

Piavis GW (1971) Embrylogy. In: Hardisty MW, Potter IC (eds) The biology of lampreys, vol 1. Academic Press, London, pp 361–400

Pickering AD (1976a) Effects of gonadectomy, oestradiol and testosterone on the migrating river lamprey, *Lampetra fluviatilis* (L). Gen Comp Endocrinol 28:473–480

Pickering AD (1976b) Stimulation of intestinal degeneration by oestradiol and testosterone implantation in the migrating river lamprey, *Lampetra fluviatilis* L. Gen Comp Endocrinol 30:340–346

Pickering AD, Dockray GJ (1972) The effects of gonadectomy on osmoregulation in the migrating river lamprey: *Lampetra fluviatilis* L. Comp Biochem Physiol 41A:139–147

Piferrer F (2001) Endocrine sex control strategies for the feminization of teleost fish. Aquaculture 197:229–281

Piferrer F, Donaldson EM (1989) Gonadal differentiation in coho salmon, *Oncorhynchus kisutch*, after a single treatment with androgen or estrogen at different stages during ontogenesis. Aquaculture 77:251–262

Piferrer F, Guiguen Y (2008) Fish gonadogenesis. Part II: Molecular biology and genomics of sex differentiation. Rev Fish Sci 16:35–55

Piferrer F, Blázquez M, Navarro L, González A (2005) Genetic, endocrine, and environmental components of sex determination and differentiation in the European sea bass (*Dicentrarchus labrax* L.). Gen Comp Endocrinol 142:102–110

Piferrer F, Ribas L, Díaz N (2012) Genomic approaches to study genetic and environmental influences on fish sex determination and differentiation. Mar Biotechnol 14:591–604

Pigozzi MI, Solari AJ (1998) Germ cell restriction and regular transmission of an accessory chromosome that mimics a sex body in the zebra finch, *Taeniopygia guttata*. Chromosome Res 6:105–113

Pigozzi MI, Solari AJ (2005) The germ-line-cell restricted chromosome in the zebra finch: recombination in females and elimination in males. Chromosoma 114:403–409

Pintus E, Ros-Santaella JL, Garde JJ (2015) Beyond testis size: links between spermatogenesis and sperm traits in a seasonal breeding mammal. PLoS ONE 10:e0139240

Pletcher FT (1963) The life history and distribution of lampreys in the Salmon and certain other rivers in British Columbia, Canada. PhD thesis, University of British Columbia, Vancouver

Pokorná MJ, Kratochvíl L (2016) What was the ancestral sex-determining mechanism in amniote vertebrates? Biol Rev 91:1–12

Porter LL, Galbreath PF, McIlraith BJ, Hess JE (2017) Sex ratio and maturation characteristics of adult Pacific Lamprey at Willamette Falls, Oregon. Columbia River Inter-Tribal Fish Comm Tech Rep 17–01:1–33

Potter IC, Beamish FWH (1977) Freshwater biology of adult anadromous sea lampreys *Petromyzon marinus*. J Zool 181:113–130

Potter IC, Robinson ES (1991) Development of the testis during post-metamorphic life in the Southern Hemisphere lamprey *Geotria australis* Gray. Acta Zool 72:113–119

Potter IC, Rothwell B (1970) The mitotic chromosomes of the lamprey, *Petromyzon marinus* L. Experientia 26:429–430

Potter IC, Lanzing WJR, Strahan R (1968) Morphometric and meristic studies on populations of Australian lampreys of the genus *Mordacia*. J Linn Soc Lond Zool 47:533–546

Potter IC, Beamish FWH, Johnson BGH (1974) Sex ratios and lengths of adult sea lampreys, *Petromyzon marinus*, from a Lake Ontario tributary. J Fish Res Board Can 31:122–124

Potter IC, Hilliard RW, Bird DJ (1980) Metamorphosis in the southern hemisphere lamprey, *Geotria australis*. J Zool 190:405–430

Potter IC, Hilliard RW, Bird DJ, Macey DJ (1983) Quantitative data on morphology and organ weights during the protracted spawning-run period of the Southern Hemisphere lamprey *Geotria australis*. J Zool 200:1–20

Potter IC, Gill HS, Renaud CB, Haoucher D (2015) The taxonomy, phylogeny, and distribution of lampreys. In: Docker MF (ed) Lampreys: biology, conservation and control, vol 1. Springer, Dordrecht, pp 35–73

Purvis HA (1970) Growth, age at metamorphosis, and sex ratio of northern brook lamprey in a tributary of southern Lake Superior. Copeia 1970:326–332

Purvis HA (1979) Variations in growth, age at transformation, and sex ratio of sea lampreys reestablished in chemically treated tributaries of the upper Great Lakes. Great Lakes Fish Comm Tech Report 35:1–49

Quinn AE, Georges A, Sarre SD et al (2007) Temperature sex reversal implies sex gene dosage in a reptile. Science 316:411–411

Quinn TP, Hendry AP, Wetzel LA (1995) The influence of life history trade-offs and the size of incubation gravels on egg size variation in sockeye salmon (*Oncorhynchus nerka*). Oikos 74:425–438

Ramm SA, Schärer L (2014) The evolutionary ecology of testicular function: size isn't everything. Biol Rev 89:874–888

Radder RS, Pike DA, Quinn AE, Shine R (2009) Offspring sex in a lizard depends on egg size. Curr Biol 19:1102–1105

Reading BJ, Sullivan CV, Schilling (2017) Vitellogenesis in fishes. In: Ferrell AP (ed) Encyclopedia of fish physiology: from genome to environment. Elsevier, Maryland Heights, MO, pp 635–646

Reichwald K, Petzold A, Koch P et al (2015) Insights into sex chromosome evolution and aging from the genome of a short-lived fish. Cell 163:1527–1538

Renaud CB (1982) Revision of the lamprey genus *Eudontomyzon* Regan, 1911. MSc thesis, University of Ottawa, Ottawa, ON

Renaud CB (1986) *Eudontomyzon hellenicus* Vladykov, Renaud, Kott, and Economidis, 1982. In: Holčík J (ed) The freshwater fishes of Europe, vol 1, Part I: Petromyzoniformes. AULA-Verlag, Weisbaden, pp 186–195

Renaud CB (2011) Lampreys of the world. An annotated and illustrated catalogue of lamprey species known to date. FAO Species Cat Fish Purp 5, FAO, Rome

Renaud CB, Holčík J (1986) *Eudontomyzon danfordi* Regan, 1911. In: Holčík J (ed) The freshwater fishes of Europe, vol 1, Part I: Petromyzoniformes. AULA-Verlag, Weisbaden, pp 146–164

Renaud CB, Naseka AM (2015) Redescription of the Far Eastern brook lamprey *Lethenteron reissneri* (Dybowski, 1869) (Petromyzontidae). ZooKeys 506:75–93

Ribas L, Robledo D, Gómez-Tato A et al (2016) Comprehensive transcriptomic analysis of the process of gonadal sex differentiation in the turbot (*Scophthalmus maximus*). Mol Cell Endocrinol 422:132–149

Ricker WE (1975) Computation and interpretation of biological statistics of fish populations. Bull Fish Res Board Can 191:1–382

Rinchard J, Ciereszko A, Dabrowski K, Ottobre J (2000) Effects of gossypol on sperm viability and plasma sex steroid hormones in male sea lamprey, *Petromyzon marinus*. Toxicol Lett 111:189–198

Rinchard J, Dabrowski K, Garcia-Abiado MA (2006) High efficiency of meiotic gynogenesis in sea lamprey *Petromyzon marinus*. J Exp Zool 306:521–527

Robinson TC, Sorensen PW, Bayer JM, Seelye JG (2009) Olfactory sensitivity of Pacific lampreys to lamprey bile acids. Trans Am Fish Soc 138:144–152

Rohde FC, Arndt RG, Wang JCS (1976) Life history of the freshwater lampreys, *Okkelbergia aepyptera* and *Lampetra lamottenii* (Pisces: Petromyzonidae), on the Delmarva Peninsula (East Coast, United States). Bull South Calif Acad Sci 75:99–111

Rothbard S, Moav B, Yaron Z (1987) Changes in steroid concentrations during sexual ontogeny in tilapia. Aquaculture 61:59–74

Sadovy Y, Colin PL (1995) Sexual development and sexuality in the Nassau grouper. J Fish Biol 46:961–976

Saito D, Morinaga C, Aoki Y et al (2007) Proliferation of germ cells during gonadal sex differentiation in medaka: insights from germ cell-depleted mutant *zenzai*. Dev Biol 310:280–290

Sandra GE, Norma MM (2010) Sexual determination and differentiation in teleost fish. Rev Fish Biol Fish 20:101–121

Sarre SD, Georges A, Quinn A (2004) The ends of a continuum: genetic and temperature-dependent sex determination in reptiles. BioEssays 26:639–645

Sawyer PJ (1960) A new geographic record for the American brook lamprey, *Lampetra lamottei*. Copeia 1960:136–137

Schepers G, Wilson M, Wilhelm D, Koopman P (2003) SOX8 is expressed during testis differentiation in mice and synergizes with SF1 to activate the *Amh* promoter *in vitro*. J Biol Chem 278:28101–28108

Schill DJ, Heindel JA, Campbell MR, Meyer KA, Mamer ER (2016) Production of a YY male brook trout broodstock for potential eradication of undesired brook trout populations. N Am J Aquacult 78:72–83

Schleen LP, Christie GC, Heinrich RA et al (2003) Development and implementation of an integrated program for control of sea lamprey in the St Marys River. J Great Lakes Res (Suppl 1):677–693

Schuldt RJ, Heinrich JW, Fodale MF, Johnson WJ (1987) Prespawning characteristics of lampreys native to Lake Michigan. J Great Lakes Res 13:264–271

Schulz RW, De França LR, Lareyre JJ et al (2010) Spermatogenesis in fish. Gen Comp Endocrinol 165:390–411

Schwanz LE, Cordero GA, Charnov EL, Janzen FJ (2016) Sex-specific survival to maturity and the evolution of environmental sex determination. Evolution 70:329–341

Seagle HH, Nagel JW (1982) Life cycle and fecundity of the American brook lamprey, *Lampetra appendix*, in Tennessee. Copeia 1982:362–366

Seiler K, Seiler R, Ackermann W, Claus R (1985) Estimation of steroids and steroid metabolism in a lower vertebrate (*Lampetra planeri* Bloch). Acta Zool 66:145–150

Sekido R, Lovell-Badge R (2008) Sex determination involves synergistic action of SRY and SF1 on a specific *Sox9* enhancer. Nature 453:930–934

Senior AM, Lokman P, Closs GP, Nakagawa S (2015) Ecological and evolutionary applications for environmental sex reversal of fish. Q Rev Biol 90:23–44

Seversmith HF (1953) Distribution, morphology and life history of *Lampetra aepyptera*, a brook lamprey, in Maryland. Copeia 1953:225–232

Shetty S, Kirby P, Zarkower D, Graves JM (2002) DMRT1 in a ratite bird: evidence for a role in sex determination and discovery of a putative regulatory element. Cytogenet Genom Res 99:245–251

Shirood Mirzaie FS, Renaud CB, Ghorbani R (2017) Biological characteristics of autumn and spring runs of *Caspiomyzon wagneri* into the Shirood River, Iran. Environ Resour Res 5:111–122

Siegfried K (2010) In search of determinants: gene expression during gonadal sex differentiation. J Fish Biol 76:1879–1902

Silva S, Barca S, Cobo F (2016) Advances in the study of sea lamprey *Petromyzon marinus* Linnaeus, 1758 in the NW of the Iberian Peninsula. In: Orlov A, Beamish R (eds) Jawless fishes of the world, vol 1. Cambridge Scholars Publishing, Newcastle upon Tyne, pp 346–385

Slade JW, Adams JV, Christie GC et al (2003) Techniques and methods for estimating abundance of larval and metamorphosed sea lampreys in Great Lakes tributaries, 1995 to 2001. J Great Lakes Res 29(Suppl 1):137–151

Smith BR (1971) Sea lampreys in the Great Lakes of North America. In: Hardisty MW, Potter IC (eds) The biology of lampreys, vol 1. Academic Press, London, pp 207–247

Smith BR, Tibbles JJ (1980) Sea lamprey (*Petromyzon marinus*) in Lakes Huron, Michigan, and Superior—history of invasion and control, 1936–78. Can J Fish Aquat Sci 37:1780–1801

Smith CA, McClive PJ, Western PS, Reed KJ, Sinclair AH (1999) Evolution: conservation of a sex-determining gene. Nature 402:601–602

Smith CC, Fretwell SD (1974) The optimal balance between size and number of offspring. Am Nat 108:499–506

Smith JJ, Antonacci F, Eichler EE, Amemiya CT (2009) Programmed loss of millions of base pairs from a vertebrate genome. Proc Natl Acad Sci USA 106:11212–11217

Smith JJ, Kuraku S, Holt C et al (2013) Sequencing of the sea lamprey (*Petromyzon marinus*) genome provides insights into vertebrate evolution. Nat Genet 45:415–421

Smith JJ, Timoshevskaya N, Ye C et al (2018) The sea lamprey germline genome provides insights into programmed genome rearrangement and vertebrate evolution. Nat Genet 50:270–277

Smith SJ, Marsden JE (2007) Predictive morphometric relationships for estimating fecundity of sea lampreys from Lake Champlain and other landlocked populations. Trans Am Fish Soc 136:979–987

Smith SJ, Marsden JE (2009) Factors affecting sea lamprey egg survival. N Am J Fish Manag 29:859–868

Solar I, I Donaldson EM, Douville D (1991) A bibliography of gynogenesis and androgenesis in fish 1913–1989. Can Tech Rep Fish Aquat Sci 1788:1–41

Sower SA (1989) Effects of lamprey gonadotropin-releasing hormone and analogs on steroidogenesis and spermiation in male sea lampreys. Fish Physiol Biochem 7:101–107

Sower SA (1990) Neuroendocrine control of reproduction in lampreys. Fish Physiol Biochem 8:365–374

Sower SA (2003) The endocrinology of reproduction in lampreys and applications for male lamprey sterilization. J Great Lakes Res 29(Suppl 1):50–65

Sower SA (2015) The reproductive hypothalamic-pituitary axis in lampreys. In: Docker MF (ed) Lampreys: biology, conservation and control, vol 1. Springer, Dordrecht, pp 305–373

Sower SA (2018) Landmark discoveries in elucidating the origins of the hypothalamic-pituitary system from the perspective of a basal vertebrate, sea lamprey. Gen Comp Endocrinol 264:3–15

Sower SA, Larsen LO (1991) Plasma estradiol and progesterone after hypophysectomy and substitution with pituitary in female sea lampreys (Petromyzon marinus). Gen Comp Endocrinol 81:93–96

Sower SA, Dickhoff WW, Gorbman A, Rivier JE, Vale WW (1983) Ovulatory and steroidal responses in the lamprey following administration of salmon gonadotropin and agonistic and antagonistic analogues of gonadotropin-releasing hormone. Can J Zool 61:2653–2659

Sower SA, Dickhoff WW, Flagg TA, Mighell JL, Mahnken CV (1984) Effects of estradiol and diethylstilbesterol on sex reversal and mortality in Atlantic salmon (Salmo salar). Aquaculture 43:75–81

Sower SA, Plisetskaya E, Gorbman A (1985a) Changes in plasma steroid and thyroid hormones and insulin during final maturation and spawning of the sea lamprey, Petromyzon marinus. Gen Comp Endocrinol 58:259–269

Sower SA, Plisetskaya E, Gorbman A (1985b) Steroid and thyroid hormone profiles following a single injection of partly purified salmon gonadotropin or GnRH analogues in male and female lamprey. J Exp Zool 235:403–408

Sower SA, King JA, Millar RP, Sherwood NM, Marshak DR (1987) Comparative biological properties of lamprey gonadotropin-releasing hormone in vertebrates. Endocrinology 120:773–779

Sower SA, Chiang YC, Lovas S, Conlon JM (1993) Primary structure and biological activity of a third gonadotropin-releasing hormone from lamprey brain. Endocrinology 132:1125–1131

Sower SA, Balz E, Aquilina-Beck A, Kavanaugh SI (2011) Seasonal changes of brain GnRH-I,-II, and-III during the final reproductive period in adult male and female sea lamprey. Gen Comp Endocrinol 170:276–282

Spice EK (2013) Ovarian differentiation in an ancient vertebrate: timing, candidate gene expression, and global gene expression in parasitic and non-parasitic lampreys. MSc thesis, University of Manitoba, Winnipeg, MB

Spice EK, Docker MF (2014) Reduced fecundity in non-parasitic lampreys may not be due to heterochronic shift in ovarian differentiation. J Zool 294:49–57

Spice EK, Whyard S, Docker MF (2014) Gene expression during ovarian differentiation in parasitic and non-parasitic lampreys: implications for fecundity and life history types. Gen Comp Endocrinol 208:116–125

Stearns SC (1989) Trade-offs in life-history evolution. Func Ecol 3:259–268

Steeves TB, Slade JW, Fodale MF, Cuddy DW, Jones ML (2003) Effectiveness of using backpack electrofishing gear for collecting sea lamprey (Petromyzon marinus) larvae in Great Lakes tributaries. J Great Lakes Res 29(Suppl 1):161–173

Stier K, Kynard B (1986) Abundance, size, and sex ratio of adult sea-run sea lampreys, Petromyzon marinus, in the Connecticut River. US Natl Mar Fish Serv Fish Bull 84:476–548

Sullivan WP, Christie GC, Cornelius FC et al (2003) The sea lamprey in Lake Erie: a case history. J Great Lakes Res 29(Suppl 1):615–636

Surface HA (1899) Removal of lampreys from the interior waters of New York. In: Fourth annual report of the state of New York Commissioners of Fisheries. Game and Forests, Albany, NY, pp 191–245

Suzuki DG, Grillner S (2018) The stepwise development of the lamprey visual system and its evolutionary implications. Biol Rev 93:1461–1477

Takada S, Mano H, Koopman P (2005) Regulation of *Amh* during sex determination in chickens: *Sox* gene expression in male and female gonads. Cell Mol Life Sci 62:2140–2146

Takahashi H (1975) Process of functional sex reversal of the gonad in the female guppy, *Poecilia reticulata*, treated with androgen before birth. Dev Growth Differ 17:167–175

Takehana Y, Matsuda M, Myosho T et al (2014) Co-option of *Sox3* as the male-determining factor on the Y chromosome in the fish *Oryzias dancena*. Nat Commun 5:4157

Tammaru T, Haukioja E (1996) Capital breeders and income breeders among Lepidoptera: consequences to population dynamics. Oikos 77:561–564

Tao W, Yuan J, Zhou L et al (2013) Characterization of gonadal transcriptomes from Nile tilapia (*Oreochromis niloticus*) reveals differentially expressed genes. PLoS ONE 8:e63604

Thomas P, Dressing G, Pang Y et al (2006) Progestin, estrogen and androgen G-protein coupled receptors in fish gonads. Steroids 71:310–316

Thomas P, Pang Y, Dong J et al (2007) Steroid and G protein binding characteristics of the seatrout and human progestin membrane receptor α subtypes and their evolutionary origins. Endocrinology 148:705–718

Thomas P, Pang Y, Dong J (2014) Enhancement of cell surface expression and receptor functions of membrane progestin receptor α (mPRα) by progesterone receptor membrane component 1 (PGRMC1): evidence for a role of PGRMC1 as an adaptor protein for steroid receptors. Endocrinology 155:1107–1119

Thomaz D, Beall E, Burke T (1997) Alternative reproductive tactics in Atlantic salmon: factors affecting mature parr success. Proc R Soc B 264:219–226

Thorgaard GH (1977) Heteromorphic sex chromosomes in male rainbow trout. Science 196:900–902

Thornton JW (2001) Evolution of vertebrate steroid receptors from an ancestral estrogen receptor by ligand exploitation and serial genome expansions. Proc Natl Acad Sci USA 98:567–5676

Thresher RE, Hayes K, Bax NJ et al (2014) Genetic control of invasive fish: technological options and its role in integrated pest management. Biol Invasion 16:1201–1216

Thresher RE, Jones M, Drake DAR (2019) Evaluating active genetic options for the control of Sea Lampreys (*Petromyzon marinus*) in the Laurentian Great Lakes. Can J Fish Aquat Sci (in press)

Tong SK, Hsu HJ, Chung BC (2010) Zebrafish monosex population reveals female dominance in sex determination and earliest events of gonad differentiation. Dev Biol 344:849–856

Torblaa RL, Westman RW (1980) Ecological impacts of lampricide treatments on sea lamprey (*Petromyzon marinus*) ammocoetes and metamorphosed individuals. Can J Fish Aquat Sci 37:1835–1850

Tsimbalov IA, Kucheryavyi AV, Veselov AE, Pavlov DS (2015) Description of the European river lamprey *Lampetra fluviatilis* (L., 1758) from the Lososinka River (Onega Lake basin). Dokl Biol Sci 462:124–127

Tsuneki K (1976) Effects of estradiol and testosterone in the hagfish *Eptatretus burgeri*. Acta Zool 57:137–146

Twohey MB, Heinrich JW, Seelye JG et al (2003) The sterile-male-release technique in Great Lakes sea lamprey management. J Great Lakes Res 29(Suppl 1):410–423

Uller T, Pen I, Wapstra E, Beukeboom LW, Komdeur J (2007) The evolution of sex ratios and sex-determining systems. Trends Ecol Evol 22:292–297

van den Hurk R, Lambert JGD, Peute J (1982) Steroidogenesis in the gonads of rainbow-trout fry (*Salmo gairdneri*) before and after the onset of gonadal sex differentiation. Reprod Nutr Dev 22:413–425

Vandeputte M, Dupont-Nivet M, Chavanne H, Chatain B (2007) A polygenic hypothesis for sex determination in the European sea bass *Dicentrarchus labrax*. Genetics 176:1049–1057

Vasil'eva N (1961) Functional variability in intestinal epithelium in Baltic river lamprey *Lampetra fluviatilis* (L.). Arh Anat Gistol Embriol 41:96–100

Vidal VP, Chaboissier MC, de Rooij DG, Schedl A (2001) *Sox9* induces testis development in XX transgenic mice. Nature 28:216–217

Vizziano D, Randuineau G, Baron D, Cauty C, Guiguen Y (2007) Characterization of early molecular sex differentiation in rainbow trout, *Oncorhynchus mykiss*. Dev Dyn 236:2198–2206

Vladykov VD (1951) Fecundity of Quebec lampreys. Can Fish Cult 10:1–14

Vladykov VD (1973) A female sea lamprey (*Petromyzon marinus*) with a true anal fin, and the question of the presence of an anal fin in Petromyzonidae. Can J Zool 51:221–224

Vladykov VD, Follett WI (1958) Redescription of *Lampetra ayresii* (Günther) of western North America, a species of lamprey (Petromyzontidae) distinct from *Lampetra fluviatilis* (Linnaeus) of Europe. J Fish Res Board Can 15:47–77

Vladykov VD, Kott E (1976) A new nonparasitic species of lamprey of the genus *Entosphenus* Gill, 1862 (Petromyzonidae) from south central California. Bull South Calif Acad Sci 75:60–67

Vladykov VD, Kott E (1978) A new nonparasitic species of the holarctic lamprey genus *Lethenteron* Creaser and Hubbs, 1922, (Petromyzontidae) from northwestern North America with notes on other species of the same genus. Biol Pap Univ Alsk 19:1–74

Vladykov VD, Mukerji GN (1961) Order of succession of different types of infraoral lamina in landlocked sea lamprey (*Petromyzon marinus*). J Fish Board Can 18:1125–1143

Wade J, Dealy L, MacConnachie S (2018) First record of nest building, spawning and sexual dimorphism in the threatened Cowichan Lake lamprey *Entosphenus macrostomus*. Endanger Species Res 35:39–45

Wagner T, Wirth J, Meyer J et al (1994) Autosomal sex reversal and campomelic dysplasia are caused by mutations in and around the *SRY*-related gene *SOX9*. Cell 79:1111–1120

Wagner WC, Stauffer TM (1962) Sea lamprey larvae in lentic environments. Trans Am Fish Soc 91:384–387

Waldman J, Daniels R, Hickerson M, Wirgin I (2009) Mitochondrial DNA analysis indicates sea lampreys are indigenous to Lake Ontario: response to comment. Trans Am Fish Soc 138:1190–1197

Wallis MC, Waters PD, Graves JAM (2008) Sex determination in mammals—before and after the evolution of *SRY*. Cell Mol Life Sci 65:3182

Wang XG, Bartfai R, Sleptsova-Freidrich I, Orban L (2007) The timing and extent of 'juvenile ovary' phase are highly variable during zebrafish testis differentiation. J Fish Biol 70:33–44

Warner DA (2011) Sex determination in reptiles. In: Norris DO, Lopez KH (eds) Hormones and reproduction of vertebrates, vol 1. Academic Press, Amsterdam, pp 1–38

Weisbart M, Youson JH (1975) Steroid formation in the larval and parasitic adult sea lamprey, *Petromyzon marinus* L. Gen Comp Endocrinol 27:517–526

Weisbart M, Youson JH (1977) *In vivo* formation of steroids from [1,2,6,7-^{3}H]-progesterone by the sea lamprey, *Petromyzon marinus* L. J Steroid Biochem 8:1249–1252

Weisbart M, Youson JH, Wiebe JP (1978) Biochemical, histochemical, and ultrastructural analyses of presumed steroid-producing tissues in the sexually mature sea lamprey, *Petromyzon marinus* L. Gen Comp Endocrinol 34:26–37

Weisbart M, Dickhoff WW, Gorbman A, Idler DR (1980) The presence of steroids in the sera of the Pacific hagfish, *Eptatretus stouti*, and the sea lamprey, *Petromyzon marinus*. Gen Comp Endocrinol 41:506–519

Weissenberg R (1927) Beiträge zur Kenntnis der Biologie und Morphologie der Neunaugen. ii. Dar Reifewachstum der Gonaden bei *Lampetra fluviatilis* und *planeri*. Z Mikrosk Anat Forsch 8:193–249

Whitlock SL, Schultz LD, Schreck CB, Hess JE (2017) Using genetic pedigree reconstruction to estimate effective spawner abundance from redd surveys: an example involving Pacific lamprey (*Entosphenus tridentatus*). Can J Fish Aquat Sci 74:1646–1653

Wicks BJ, Morrison BJ, Barker LA, Beamish FWH (1998a) Unusual sex ratios in larval sea lamprey, *Petromyzon marinus*, from Great Lakes tributaries. Great Lakes Fisheries Commission Completion Report, Ann Arbor, MI

Wicks BJ, Barker LA, Morrison BJ, Beamish WH (1998b) Gonadal variation in Great Lakes sea lamprey, *Petromyzon marinus*, larvae. J Great Lakes Res 24:962–968

Wiegand MD (1982) Vitellogenesis in fishes. In: Richter CJJ, Goos HJT (eds) Reproductiv physiology of fish. Pudoc, Wageningen, The Netherlands, pp 136–146

Wigley RL (1959) Life history of the sea lamprey of Cayuga Lake, New York. US Fish Wildl Serv Fish Bull 59:559–617

Wilhelm D, Palmer S, Koopman P (2007) Sex determination and gonadal development in mammals. Physiol Rev 87:1–28

Witkowski A, Jęsior M (2000) Fecundity of river lamprey *Lampetra fluviatilis* (L.) in Drweca River (Vistula Basin, northern Poland). Arch Ryb Pol 8:225–232

Woolgar L, Trocini S, Mitchell N (2013) Key parameters describing temperature-dependent sex determination in the southernmost population of loggerhead sea turtles. J Exp Mar Biol Ecol 449:77–84

Wright AE, Dean R, Zimmer F, Mank JE (2016) How to make a sex chromosome. Nat Comm 7:12087

Wu C, Yan L, Depre C et al (2009) Cytochrome *c* oxidase III as a mechanism for apoptosis in heart failure following myocardial infarction. Am J Physiol Cell Physiol 297:C928–C934

Xu H, Li M, Gui J, Hong Y (2007) Cloning and expression of medaka *dazl* during embryogenesis and gametogenesis. Gene Expr Patterns 7:332–338

Yamamoto T (1969) Sex differentiation. In: Hoar WS, Randall DJ (eds) Fish physiology, vol 3. Academic Press, New York, pp 117–175

Yamamoto Y, Zhang Y, Sarida M, Hattori RS, Strüssmann CA (2014) Coexistence of genotypic and temperature-dependent sex determination in pejerrey *Odontesthes bonariensis*. PLoS ONE 9:e102574

Yamazaki C, Koizumi I (2017) High frequency of mating without egg release in highly promiscuous nonparasitic lamprey *Lethenteron kessleri*. J Ethol 35:237–243

Yamazaki F (1983) Sex control and manipulation in fish. Aquaculture 33:329–354

Yamazaki Y, Konno S, Goto A (2001) Interspecific differences in egg size and fecundity among Japanese lampreys. Fish Sci 67:375–377

Yano A, Guyomard R, Nicol B et al (2012) An immune-related gene evolved into the master sex-determining gene in rainbow trout, *Oncorhynchus mykiss*. Curr Biol 22:1423–1428

Yano A, Nicol B, Jouanno E et al (2013) The sexually dimorphic on the Y-chromosome gene (*sdY*) is a conserved male-specific Y-chromosome sequence in many salmonids. Evol Appl 6:486–496

Yano A, Nicol B, Jouanno E, Guiguen Y (2014) Heritable targeted inactivation of the rainbow trout (*Oncorhynchus mykiss*) master sex-determining gene using zinc-finger nucleases. Mar Biotech 16:243–250

Yaron A, Sivan B (2005) Reproduction. In: Evans DH, Claiborne JB (eds) The physiology of fishes. CRC, Boca Raton, FL, pp 343–385

Yen PH (2004) Putative biological functions of the DAZ family. Int J Androl 27:125–129

Yoshida S (2016) From cyst to tubule: innovations in vertebrate spermatogenesis. Wiley Interdiscip Rev Dev Biol 5:119–131

Yoshikawa H, Oguri M (1978) Sex differentiation in a cichlid, *Tilapia zillii*. Bull Jpn Soc Sci Fish 44:1093–1097

Young BA, Bryan MB, Sower SA, Scott AP, Li W (2004a) 15α-hydroxytestosterone induction by GnRH I and GnRH III in Atlantic and Great Lakes sea lamprey (*Petromyzon marinus* L.). Gen Comp Endocrinol 136:276–281

Young BA, Bryan MB, Sower SA, Li W (2004b) The effect of chemosterilization on sex steroid production in male sea lampreys. Trans Am Fish Soc 133:1270–1276

Young BA, Bryan MB, Glenn JR et al (2007) Dose-response relationship of 15α-hydroxylated sex steroids to gonadotropin-releasing hormones and pituitary extract in male sea lampreys (*Petromyzon marinus*). Gen Comp Endocrinol 151:108–115

Young JA, Marentette JR, Gross C et al (2010) Demography and substrate affinity of the round goby (*Neogobius melanostomus*) in Hamilton Harbour. J Great Lakes Res 36:115–122

Young RT, Cole LJ (1900) On the nesting habits of the brook lamprey (*L. wilderi*). Am Nat 34:617–620

Youson JH, Beamish RJ (1991) Comparison of the internal morphology of adults of a population of lampreys that contains a nonparasitic life-history type, *Lampetra richardsoni*, and a potentially parasitic form, *L. richardsoni* var. *marifuga*. Can J Zool 69:628–637

Zanandrea G (1951) Rilievi e confronti biometrici e biologici sul *Petromyzon* (*Lampetra*) *planeri* Bloch nelle acque della Marca Trevigiana. Boll Pesca Piscicolt Idrobiol 6:53–78

Zanandrea G (1959) *Lampetra fluviatilis* catturata in mare nel golfo di Gaeta. Pubb Staz Zool Napoli 31:265–307

Zanandrea G (1961) Studies on European lampreys. Evolution 15:523–534

Zerrenner A, Marsden JE (2005) Influence of larval sea lamprey density on transformer life history characteristics in Lewis Creek, Vermont. Trans Am Fish Soc 134:687–696

Zeyl JN, Love OP, Higgs DM (2014) Evaluating gonadosomatic index as an estimator of reproductive condition in the invasive round goby, *Neogobius melanostomus*. J Great Lakes Res 40:164–171

Chapter 2
Lamprey Reproduction and Early Life History: Insights from Artificial Propagation

Mary L. Moser, John B. Hume, Kimmo K. Aronsuu, Ralph T. Lampman and Aaron D. Jackson

Abstract Artificial propagation of lampreys was first developed to produce specimens for the study of evolutionary development in vertebrates. In recent years, artificially propagated larvae have been used to improve identification methods for native lamprey species, to study invasive sea lamprey *Petromyzon marinus* in the Laurentian Great Lakes and to provide animals for genomic studies, and for restoration and conservation research. In the course of developing methods for lamprey cultivation, insights into lamprey behavior, biology, genetics, and early life history have been gained. Broodstock holding has indicated that adult lampreys can be kept at extremely high densities when provided with cold, oxygenated water. Sexual maturation is controlled primarily by temperature, but may be affected by photoperiod, the presence of other lampreys, and suitable substrate. Fertilization and incubation experiments have revealed that gamete contact times are very short and that embryos are resilient to low flow, poor water quality, or variable substrates. Early larvae are also resilient to these factors and can tolerate abrupt changes in temperature and

M. L. Moser (✉)
Northwest Fisheries Science Center, National Marine Fisheries Service, National Oceanic and Atmospheric Administration, 2725 Montlake Boulevard East, Seattle, WA 98112, USA
e-mail: mary.moser@noaa.gov

J. B. Hume
Department of Fisheries and Wildlife, Michigan State University, 480 Wilson Road, East Lansing, MI 48824, USA
e-mail: jhume@msu.edu

K. K. Aronsuu
Centre for Economic Development, Transport, Environment for North Ostrobothnia, PO Box 86, 90101 Oulu, Finland
e-mail: kimmo.aronsuu@ely-keskus.fi

R. T. Lampman
Fisheries Resources Management Program, Department of Natural Resources, Yakama Nation, 401 Fort Road, Toppenish, WA 98948, USA
e-mail: lamr@yakamafish-nsn.gov

A. D. Jackson
Fisheries Program, Department of Natural Resources, Confederated Tribes of the Umatilla Indian Reservation, 46411 Timine Way, Pendleton, OR 97801, USA
e-mail: aaronjackson@ctuir.org

© Springer Nature B.V. 2019   187
M. F. Docker (ed.), *Lampreys: Biology, Conservation and Control*,
Fish & Fisheries Series 38, https://doi.org/10.1007/978-94-024-1684-8_2

extended periods of starvation. However, they cannot survive sudden changes in water quality, excessive disturbance, and lack of adequate burrowing media. These observations have resulted in more efficient and effective lamprey propagation and have yielded important information about the early life stage requirements of lampreys in the wild. Further study is needed on a broader array of species to allow inter specific comparisons of early life history. However, information from lampreys receiving the most attention to date (European river lamprey *Lampetra fluviatilis*, sea lamprey, and Pacific lamprey *Entosphenus tridentatus*) indicates that culture and environmental requirements of the early life stages are remarkably similar, allowing generalization across species.

Keywords Broodstock · Culture · Development · Fertilization · Incubation · Propagation · Rearing · Spawning

2.1   Introduction

Artificial propagation of lampreys was first developed to produce specimens for the study of evolutionary development in vertebrates. While this continues to be an important purpose of artificial propagation (see Chap. 6), other uses have recently come to the fore. As a result, there has been increased awareness of lampreys as both model organisms and as critical components of ecological systems. Moreover, lampreys are of significant cultural importance in many parts of the world (Docker et al. 2015). Restoring lamprey populations to levels that allow for sustainable harvest is a goal of fisheries managers in Finland (Vikström 2002), Japan (Hokkaido Fish Hatchery 2008), and in the northwestern United States (Close et al. 2002). At the same time, the proliferation of invasive sea lamprey *Petromyzon marinus* populations in the Laurentian Great Lakes has increased the demand for information on factors that limit lamprey production.

The increasing need for lamprey propagation tools for research, conservation, and control has led to a proliferation of studies designed to perfect artificial production methods. These efforts have increased our knowledge of lamprey genetics, physiology, and behavior. While the focus of this research has typically been to improve culture technology, many of the lessons learned may be applicable to lamprey biology in the wild. In contrast to the many decades of developmental studies using lampreys, we know comparatively little about the early life history of lampreys in their natural environments.

Hence, the aim of this chapter is to provide a comprehensive review of research that has stemmed from artificial propagation of lampreys and to compare this body of knowledge with what is known regarding lamprey early life history in the wild. The spawning requirements and mating behavior of lampreys in the wild and the ecology of larval lampreys are reviewed by Johnson et al. (2015) and Dawson et al. (2015), respectively; sexual maturation in lampreys is reviewed in Chap. 1. Embryos of a few model lamprey species have long been generated and studied in the laboratory to better understand the origin and development of the vertebrate body plan (see Sects. 2.1.1 and 2.1.3; Chap. 6), but other research and management needs are now requiring culture of additional species, to later stages and in larger numbers than attempted previously. The current chapter focuses on these recent advances in artificial propagation and laboratory or hatchery rearing of lampreys. The transfer of findings from natural studies to the laboratory has benefitted artificial rearing, allowing sufficient production for restoration and mitigation programs. Conversely, artificial rearing studies have and continue to shed light on early lamprey life history in nature. Reviewing this information has helped to identify critical knowledge gaps and provide directions for future research that will inform not only lamprey biology, but also their conservation and control.

2.1.1   Lampreys as Model Organisms

Hagfishes (Myxinidae) and lampreys represent the most ancient vertebrate groups alive today (see Docker et al. 2015). Their pedigree extends back a minimum of 395 million years (Janvier and Lund 1983; Gess et al. 2006), and they may have remained functionally unchanged for as long as 125 million years (Chang et al. 2014; see Chap. 4). As cyclostomes, lampreys have a primitive appearance, perhaps exemplified most clearly by their lack of jaws (Kuratani et al. 2002; Kuratani 2005), and they have captured the attention of generations of biologists seeking insight into the evolutionary development of vertebrates (e.g., Richardson et al. 2010; Shimeld and Donoghue 2012; McCauley et al. 2015; see Docker et al. 2015). Lampreys are well regarded as model organisms in fields such as embryonic development, organ differentiation, and phylogenetics. As such, they have provided deep insight into the evolution of vertebrate nervous, endocrine, and immune systems (Johnels 1956; Kusakabi and Kuratani 2005; Nikitina et al. 2009; Richardson et al. 2010; Kuratani 2012; Shimeld and Donoghue 2012; Green and Bronner 2014; Sower 2015; Xu et al. 2016). Such lines of inquiry have yielded an astonishing understanding of the origin and subsequent evolution of the vertebrate lineage (see Chap. 6).

Vertebrates can be loosely characterized by their possession of a complex cranial region bearing paired sensory organs linked to a well-developed brain-neural network, along with a hinged jaw for processing food (Kuratani 2012). Although lampreys lack some of these anatomical features, they, along with all other vertebrates, possess a neural crest during embryonic development, the region of tissue largely responsible for cranial development (Shimeld and Donoghue 2012; Green and Bronner 2014; see Chap. 6). A landmark development in the evolution of vertebrates was the acquisition of articulating jaws that enabled more active predatory foraging strategies (Gans and Northcutt 1983; Kuratani 2012). There is remarkable similarity between lampreys and gnathostomes (jawed vertebrates) in the expression of transcription factor genes during pharyngeal patterning (Horigome et al. 1999; Kuratani et al. 1999; Neidert et al. 2001; Shigetani et al. 2002). A change in the interaction between the neural crest cells and pharyngeal tissue may have led to the development of jaws in the gnathostome ancestor (Shigetani et al. 2002; Kuratani and Ota 2008). Key differences in transcription factor expression in the first pharyngeal arch between lampreys and gnathostomes (Cerny et al. 2010; Kuraku et al. 2010) suggest that the gnathostome lineage focused pharyngeal patterning towards development of a joint crucial to the hinged jaw arrangement (Shimeld and Donoghue 2012). However, the precise evolutionary steps leading to the acquisition of jaws in the gnathostome ancestor remain a subject of debate (Kuratani 2012; see Chap. 6).

Lampreys have a relatively simple nervous system with large neurons, which have enabled extensive investigations of neural pathways (Khonsari et al. 2009; Murakami and Watanabe 2009). The lamprey brain comprises five distinct regions, from anterior to posterior: forebrain, diencephalon, midbrain, cerebellum (or cerebellum-like structure), and medulla (Murakami and Kuratani 2008). Brain development studies in lampreys may help inform vertebrate developmental pathways in general, such as the discovery that Sonic Hedgehog/Hedgehog (Shh/Hh) signaling in the lamprey embryonic midline is responsible for vertebrate forebrain development (Rétaux and Kano 2010). Our understanding of the lamprey brain-neural network is so complete that it has even enabled the development of model robotic lampreys capable of complex swimming motions and response to visually detected objects (Kamali et al. 2013). The eyes of larval lampreys, both in their developmental mechanisms and neural function, are representative of an evolutionarily primitive state in the acquisition of "camera-style" eyes (Lamb et al. 2007; Suzuki et al. 2015). As larvae, lampreys do possess an eye, although it is covered by skin, and the lens is not fully developed (Kleerekoper 1972), so it functions simply as a light detector (Suzuki et al. 2015). Following metamorphosis, the retinotectal projection—the part of the brain responsible for visual reflexes—is arranged in a manner similar to that of gnathostomes (Jones et al. 2009), and adult lamprey eyes are considered fully functioning camera-style eyes (Villar-Cerviño et al. 2006; Collin 2010).

The point at which the adaptive immune system of vertebrates first appeared has long captured the attention of researchers in the field of immunology (Amemiya et al. 2007; Shimeld and Donoghue 2012). Early investigators demonstrated that exposure to antigens such as anthrax resulted in antibody (agglutinin) production in lampreys (Fujii et al. 1979). However, lampreys lack immune receptors common to other vertebrates (T-cell receptors, B-cell receptors, and major histocompatibility complex). Hence, the lamprey adaptive immune system remained obscure for many more years (Ardavin and Zapata 1988; Cooper and Alder 2006). Lampreys and hagfishes were found to have a similar but different set of lymphocyte cells relative to other vertebrates, which provide the same function of adaptive immunity (Shintani et al. 2000; Pancer et al. 2004; Amemiya et al. 2007). Where exactly lymphocytes are produced in lampreys, and therefore how immunity is conferred, remains uncertain. One potential region of production is the typhlosole, an intestinal fold common to lampreys and several other chordates (Shintani et al. 2000; Bajoghli et al. 2011). Recognition of an alternative autoimmune system in lampreys has sparked renewed interest in the group as a model species in immunology, particularly in the age of genomic investigations (Amemiya et al. 2007; see Chap. 6).

2.1.2   Anatomy and Developmental Staging

As introduced above, lamprey anatomy exemplifies their primitive condition and basal positioning among vertebrates. An excellent example of this unique position is evidenced by the transformation of the lamprey endostyle into a thyroid gland during metamorphosis. In non-vertebrate chordate groups (tunicates and cephalochordates) as well as in larval lampreys, the endostyle is an organ that produces mucus for feeding (Olsson 1963). In contrast, non-lamprey vertebrates directly develop a thyroid gland, which shares with the endostyle a common embryonic origin and a partial overlap in enzyme production and gene expression (McCauley and Bronner-Fraser 2002; Kluge et al. 2005). Lampreys therefore represent a prime example of organ evolution recapitulated in the ontogeny of a single organism (Wright and Youson 1976).

Initial reports of lamprey artificial propagation can be traced back to university and hatchery reports from Japan between 1893 and 1951 using Arctic lamprey *Lethenteron camtschaticum* (Hatta 1893, 1907; Isahaya 1934) or Far Eastern brook lamprey *Lethenteron reissneri* (Yamada 1951). Investigations of lamprey embryonic development also became a focal point resulting from the pressing need to control the spread of sea lamprey within the Great Lakes basin (Piavis 1961; see Chap. 5). Along with later work on Far Eastern brook lamprey by Tahara (1988), the work of Piavis (1961, 1971) remains the foundation of developmental staging for all other lamprey species today (Nikitina et al. 2009; Richardson et al. 2010). Species whose embryology has been investigated thus far include: chestnut lamprey *Ichthyomyzon castaneus* and silver lamprey *Ichthyomyzon unicuspis* (Smith et al. 1968), sea lamprey (Piavis 1961; Langille and Hall 1988; Richardson and Wright 2003), Pacific lamprey *Entosphenus tridentatus* (Yamazaki et al. 2003; Meeuwig et al. 2006), American brook lamprey *Lethenteron appendix* and northern brook lamprey *Ichthyomyzon fossor* (Smith et al. 1968), western brook lamprey *Lampetra richardsoni* (Meeuwig et al. 2006), Far Eastern brook lamprey (Fujimoto and Takaoka 1960; Tahara 1988), European brook lamprey *Lampetra planeri* (Damas 1944; Horigome et al. 1999), and most recently Korean lamprey *Eudontomyzon morii* (Feng et al. 2018) (Table 2.1). In each of these species, embryology is similar, with developmental rate largely responsible for any interspecific differences. A comprehensive review of the embryonic developmental stages of lampreys can be found in Richardson et al. (2010), but the major pattern of appearance is briefly summarized here.

We follow nomenclature established by Piavis (1961) and consider an embryo the developing lamprey that has not yet hatched and a prolarva the stage after hatching but prior to the onset of exogenous feeding. Once the yolk sac has been consumed and exogenous feeding begins, the lamprey is termed a larva. At metamorphosis, it is considered a juvenile until sexual maturation to the adult form. The developmental process may be subdivided into 18 discrete stages, beginning with the fertilized ovum or zygote (stage 1) and ending with the larval stage (stage 18) at the onset of exogenous feeding; stage 14 (i.e., hatching) marks the beginning of the prolarval stage and stage 17 begins when prolarvae begin burrowing (Piavis 1961). The earliest investigations of embryonic development included observations of the external

Table 2.1 Summary of studies that describe lamprey in vitro propagation; maximum duration refers to maximum for all studies

Species	Maximum duration or stage of rearing	References
Entosphenus tridentatus Pacific lamprey	5 years (through metamorphosis)	Close et al. (2002), Yamazaki et al. (2003), Meeuwig et al. (2005, 2006), Lampman et al. (2016), Moser et al. (2016), Maine et al. (2017, 2018)
Eudontomyzon morii Korean lamprey	25 days (larvae)	Feng et al. (2018)
Geotria australis pouched lamprey	1 year (larvae)	Cindy F. Baker, National Institute of Water and Atmospheric Research, Hamilton, NZ, personal communication, 2018
Ichthyomyzon castaneus chestnut lamprey	Stage 17 (burrowing prolarvae)	Smith et al. (1968), Piavis et al. (1970)
Ichthyomyzon fossor northern brook lamprey	28 days (larvae)	Piavis et al. (1970), Neave et al. (2019)
Ichthyomyzon unicuspis silver lamprey	Stage 17 (burrowing prolarvae)	Smith et al. (1968), Piavis et al. (1970)
Lampetra richardsoni western brook lamprey	1 year (larvae)	Meeuwig et al. (2005, 2006)
Lampetra fluviatilis European river lamprey	72 days (larvae)	Damas (1944), Kainua et al. (1983), Kainua and Ojutkangas (1984), Ojutkangas and Laukkanen (1985), Törrönen et al. (1988), Ryapolova and Mitans (1991), Myllynen et al. (1997), Vikström (2002), Aronsuu and Virkkala (2014), Rougemont et al. (2015), Kujawa et al. (2017), Tsimbalov et al. (2018)
Lampetra planeri European brook lamprey	Stage 17 (burrowing prolarvae)	Hume et al. (2013), Tsimbalov et al. (2018)
Lethenteron appendix American brook lamprey	Stage 17 (burrowing prolarvae)	Piavis et al. (1970)

(continued)

Table 2.1 (continued)

Species	Maximum duration or stage of rearing	References
Lethenteron camtschaticum Arctic lamprey	4 years (through metamorphosis)	Hatta (1893, 1907), Isahaya (1934), Hosoya et al. (1979), Kataoka et al. (1980b), Kobayashi (1993), Yamazaki and Goto (1997), Fukutomi et al. (2002), Hokkaido Fish Hatchery (2008)
Lethenteron reissneri Far Eastern brook lamprey	28 days	Yamada (1951), Fujimoto and Takaoka (1960)
Petromyzon marinus sea lamprey	98 days (larvae)	Lennon (1955), Piavis (1961), Piavis and Howell (1969), Hanson et al. (1974), Langille and Hall (1988), Fredricks and Seelye (1995), Ciereszko et al. (2000, 2002), Rodríguez-Muñoz et al. (2001), Rodríguez-Muñoz and Ojanguren (2002), Smith and Marsden (2009)

appearance of the blastopore and neural groove imposed atop the ridge-like neural plate (Shipley 1887; Hatta 1900) (Fig. 2.1).

Embryonic lampreys then take on a characteristically curved "comma-shape" as they elongate dorsally, beginning with the definition of the head region from the mass of yolk and followed by the trunk bending around the yolk itself (Hatta 1923; Veit 1939; Damas 1944) (Fig. 2.1). The anterior portion expresses some swelling, as tissues that will later form the oral region and pharyngeal pouch undertake a period of expansion and migration (Damas 1944; Tahara 1988; Richardson and Wright 2003). After the embryo hatches, the heart initiates pumping, and the upper lip, mouth, and nasohypophyseal openings rapidly approach their final positions and appearance (Scott 1887) (Fig. 2.2). The prolarval stage is complete when the branchiopores open and the digestive tract connects with the esophagus and anus (Richardson et al. 2010).

2.1.3 Artificial Propagation for Evo-Devo Research

The comparative ease of obtaining and rearing lampreys, compared to hagfishes, is a primary reason for their attractiveness in studies of evolutionary development (Nikitina et al. 2009; Lampman et al. 2016; see Chap. 6). Various methodologies have been used to investigate lamprey development, including embryonic manipulation and a burgeoning number of gene expression studies (Shimeld and Donoghue

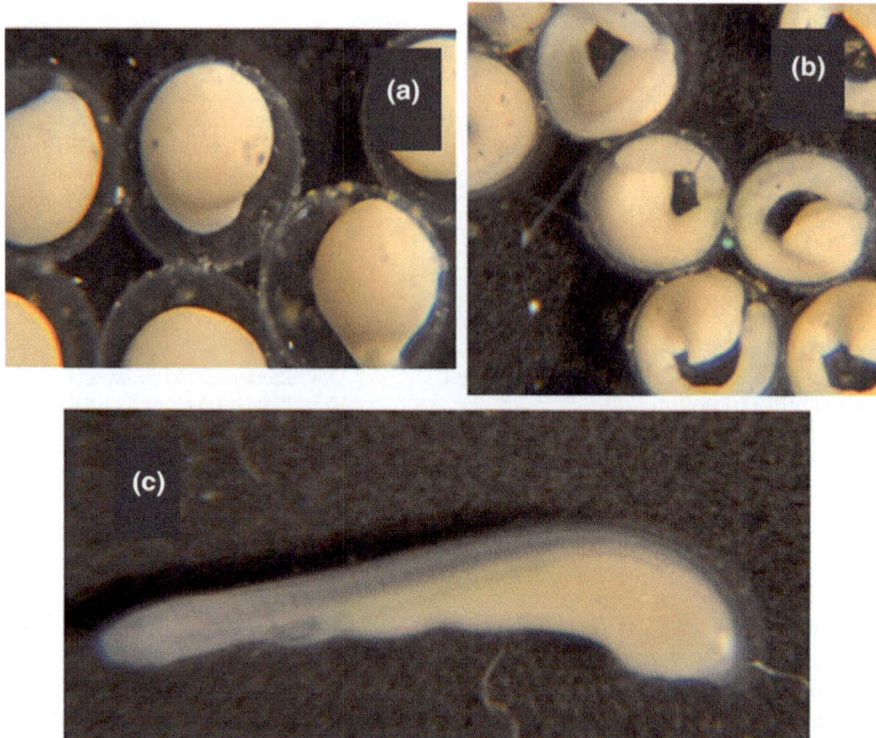

Fig. 2.1 Larval development of Pacific lamprey: **a** early neurula stage 11; **b** pre-hatching stage 13; and **c** hatching stage 14 (*Photo* © Mary L. Moser)

Fig. 2.2 Pacific lamprey prolarva (stage 15; see Sect. 2.1.2) (*Photo* © Mary L. Moser)

2012; McCauley et al. 2015). Lineage tracing using dye markers and the removal of embryonic tissue has been used to examine development in a more classical manner (Langille and Hall 1988; Shigetani et al. 2002; McCauley and Bronner-Fraser 2006).

Lampreys reared in captivity have also been used in more technologically advanced gene expression studies, such as those using messenger RNA (mRNA) and microRNA (miRNA) visualization (Ogasawara et al. 2000; Murakami et al. 2001; Neidert et al. 2001; Boorman and Shimeld 2002; Derobert et al. 2002; Ota et al. 2007; Pierce et al. 2008; Nikitina et al. 2009), gene knockdown (McCauley and Bronner-Fraser 2006; Sauka-Spengler et al. 2007), and even transgenesis (Kusakabe et al. 2003). In Arctic lamprey, pharmacological methods have also been used to investigate the developing lamprey brain by inhibiting signaling of Hedgehog (Hh) and fibroblast growth factor (FBF) (Murakami et al. 2004; Sugahara et al. 2011).

The ability to obtain lamprey embryos and parental tissues with relative ease has opened the door for fascinating new investigations of the lamprey genome. Amazingly, it has been discovered that during embryonic development, a large portion of the lamprey genome (~20%) is naturally eliminated from cells (Smith et al. 2012). This process could shed light on the mechanisms responsible for the distribution of chromosomes into daughter cells and the subsequent maintenance of genomes during cell division (Timoshevskiy et al. 2016). Such programmed genome rearrangements likely act to silence genes to prevent their incorrect expression during embryonic development, and may even protect against the formation of certain cancer cells in developing embryos (Bryant et al. 2016).

2.1.4 Artificial Production for Identification and Restoration

In recent years, lamprey production for use in field identification and restoration has increased. The necessity for accurate identification of larvae belonging to lamprey populations of conservation concern has fueled greater interest in examining early life stages collected in streams (e.g., Meeuwig et al. 2006; Goodman et al. 2009). Historically, small larvae were not routinely collected by electrofishing gear and were overlooked (Churchill 1945; McLain and Dahl 1968). However, advances in sampling gear and in our understanding of its effectiveness have improved collection of small larvae, which are difficult to identify (Bowen et al. 2003; Steeves et al. 2003; Moser et al. 2007; Dunham et al. 2013). Although individuals of less than 35 mm have been captured by standard electrofishing techniques (Derosier 2001; Lasne et al. 2010a; Dunham et al. 2013; Silva et al. 2014a), these techniques do not typically produce large sample sizes, and may cause an unknown degree of mortality or sublethal effects. Much smaller larvae (<10 mm), and even fertilized eggs, are collected by plankton nets set within the water column to intercept downstream drift after spawning (Manion 1968; Derosier 2001; Laroche et al. 2004; Brumo 2006; Pavlov et al. 2014; Zvezdin et al. 2016) or by dredging the sediment after settlement (Derosier 2001; Lasne et al. 2010a; Whitlock et al. 2017).

Identification of larvae based on the presence of spawning adults is not reliable. It was previously believed that young-of-the year (YOY, 0+) lampreys remained close to the nests in which they were deposited (Okkelberg 1922). However, in natural streams and rivers, prolarvae emerge from interstitial spaces of the substrate and become displaced downstream (Piavis 1961; Manion and Smith 1978; Malmqvist 1983; Beamish and Lowartz 1996; White and Harvey 2003; Derosier et al. 2007; Kirillova et al. 2011; Pavlov et al. 2014). The rate and extent to which they are displaced is a function of the stream gradient and other hydrographic features, such as velocity, depth, temperature, and substrate particle size (Applegate 1961; Hardisty 1961a; Hardisty and Potter 1971; Manion and McLain 1971; Manion and Smith 1978; Malmqvist 1980; Morman et al. 1980; Potter 1980; Kelso and Todd 1993; see Dawson et al. 2015). Few studies of displacement have been made, but estimates range from less than 1 km to more than 3 km in sea and American brook lampreys (Thomas 1962; Derosier 2001; Derosier et al. 2007). Furthermore, mixed-species spawning associations have been observed where two or more species occupy the same nest (Huggins and Thompson 1970; Manion and Hanson 1980; Brumo 2006; Lasne et al. 2010b; see Johnson et al. 2015). Genetic studies have been used to develop or refine keys to identification (Goodman et al. 2009; Hess et al. 2015; Docker et al. 2016; see Chap. 7), although it should be noted that diagnostic genetic markers are not yet available to distinguish between most "paired" species (i.e., closely related parasitic and non-parasitic lampreys; see Sect. 2.1.6; Chap. 4) and rearing lampreys in the laboratory until they reach stages that can be definitively identified remains the best way to verify species identification (Richards et al. 1982; Meeuwig et al. 2006).

Artificial propagation programs have also been proposed and erected to halt the decline of species broadly distributed across the Northern Hemisphere. These include the European river (Kainua et al. 1983; Kainua and Ojutkangas 1984; Ojutkangas and Laukkanen 1985; Törrönen et al. 1988; Aronsuu 2015), Arctic (Hokkaido Fish Hatchery 2008) and Pacific (Close et al. 2002; Moyle et al. 2009; Luzier et al. 2011; Lampman et al. 2016) lampreys. Artificial propagation programs provide animals for research, or as broodstock for refuge sites should populations become extirpated. This uptick in practical management concern has driven lamprey early life history biology forwards once again, and there are now well characterized methodologies for large-scale production of larval lampreys (Lampman et al. 2016). Developments in this area could also aid research to control invasive sea lamprey in the Great Lakes region (Sect. 2.1.5).

For species of conservation concern, producing sufficient numbers of embryos to mitigate population decline requires a more industrial or mass-production approach compared with experimental studies for research. Standardization of laboratory methods will allow replication when scaled up. Pacific lamprey has been the subject of numerous such methodological investigations, and today, millions of its prolarvae can be produced each year (Lampman et al. 2016). Early trials indicated that McDonald jars could accommodate thousands of Pacific lamprey eggs in suspension (Meeuwig et al. 2005). Similar approaches using upwelling jars were used for mass rearing of European river lamprey; 10-L upwelling jars accommodated 200,000 eggs and circulated the developing embryos to prevent clumping (Vikström 2002).

For Pacific lamprey, emphasis has recently shifted from embryo production (Lampman et al. 2016) to larval growth following the onset of exogenous feeding (Barron et al. 2015). Dietary studies using very small larvae (<15 mm) have shown promise in developing methods to promote and maintain growth in the laboratory (see Sects. 2.6.5 and 2.6.7). The goal of this research is to produce healthy larvae of a size large enough to escape early mortality when outplanted and to provide animals for fish passage research without the need to mine wild stocks (Lampman et al. 2016; Barron et al. 2017; Maine et al. 2017; Moser et al. 2017a, b).

2.1.5 Artificial Production in Support of Research Related to Sea Lamprey Control

Artificial propagation of sea lamprey dates back to the 1950s, where relatively crude methodologies were employed. Lennon (1955) documents hand-stripping gametes from mature adults into glass jars containing water from Lake Huron that was refreshed frequently. Despite the success in development through to hatching, the authors were unable to induce prolarvae to burrow into sediment, or to feed. Piavis and Howell (1969), however, were able to induce 50–71% of prolarvae to burrow into sediment following development in distilled water, but did not report overall mortality or growth rates.

Other early attempts to rear sea lamprey larvae in the laboratory were equally underwhelming. In an unpublished study by Hanson and colleagues (cited in Hanson et al. 1974), only 1.75% of 17,500 prolarvae survived after 4 months in aquaria supplied with fully exchanged stream water, even when experimenting with various diatom cultures for food. Hanson et al. (1974) achieved greater success following the addition of yeast cakes (11.6–36.5% survivorship through year 1), even at extremely high densities (>600 larvae per m^2), and survival averaged 13.7% by the end of year 2. Growth rates during the first year of life when provisioned with this feed were as high as 0.11 mm/day (Hanson et al. 1974), and average length of 1- and 2-year-old larvae was 29.7 and 48.7 mm, respectively.

More recently, as a consequence of the success of an integrated program to control the sea lamprey (see Chap. 5), there is now a limited availability of particular life stages for research. Given that sea lamprey control aims to kill larvae before they transform into parasitic juveniles, metamorphosing larvae and outmigrating juveniles can be hard to collect in large numbers, and parasites and sexually mature males are also limited. At face value, this appears to be a "good problem" to have, yet this lack of specimens hinders further progress towards a more efficient and effective control program. The Great Lakes Fishery Commission developed a rearing facility for sea lamprey in the 1990s (Mike Steeves, Fisheries and Oceans Canada, Sea Lamprey Control Centre, Sault Ste. Marie, ON, personal communication, 2018). It was initially designed to provide juveniles for mark-recapture studies to estimate overall population size, but some animals were also made available for basic research.

Because of difficulties providing adequate nutrition in the laboratory and potential density-dependent effects on metamorphosis (Dawson et al. 2015), a retrofitted outdoor raceway facility on a former fish farm was developed. The new rearing facility allowed for the diversion of stream water to provide both cool fresh water and nutrients to large larvae that could be held until metamorphosis. However, management of the stream provisioning this rearing facility, coupled with a reduced flow rate within the raceway, resulted in unsuitably high water temperatures as well as significant macrophyte growth and subsequent biological oxygen demand from decaying organic matter. Furthermore, there was some indication that predators were getting into the raceways and consuming larval sea lamprey. Furthermore, obtaining sufficient numbers of large larvae to stock such facilities is still a major hurdle, and it is exceedingly difficult to collect several thousand larval sea lamprey each year as they approach the size (i.e., length ≥ 120 mm and weight ≥ 3.0 g) at which they are expected to metamorphose (Manzon et al. 2015). If smaller larvae are collected, more time is required until they will undergo metamorphosis. The U.S. Geological Survey's Hammond Bay Biological Station in Michigan currently maintains several thousand larval sea lamprey for the extraction of larval odors used in research (e.g., Meckley et al. 2014). These animals are maintained in large (1,000 L) outdoor tanks provisioned with cool (5–10 °C) water drawn from Lake Huron. Therefore, it is certainly feasible to hold these larvae for many months with minimal mortality, but growth rates are unlikely to be high given low temperatures and the traditional artificial diet of brewer's yeast.

Future strategies to rear larval sea lamprey in significant numbers may include the use of closed ponds capable of producing their own food supply, requiring little maintenance until metamorphosis (Nicholas S. Johnson, U.S. Geological Survey, Hammond Bay Biological Station, MI, personal communication, 2018). This has previously been suggested as a means to restore Pacific lamprey populations on the west coast of North America, where larvae have been found to colonize fish farm abatement ponds in high densities (Nelson and Nelle 2007). Perhaps when it comes to production of larval lampreys, these nature-like environments may be the most successful (Kataoka 1985; see Sect. 2.6.8).

2.1.6 Artificial Production for Other Experimental Purposes

Despite the significant problems that must be overcome, the ability to successfully and consistently rear larval lampreys through metamorphosis will represent a major breakthrough for researchers. In particular, development of optimal egg fertilization and rearing methods across a variety of species could help resolve the "paired species problem" in lampreys (see Docker 2009; Chap. 4). Many sympatrically occurring paired species can be observed spawning in the same nests (e.g., Manion and Hanson 1980; Lasne et al. 2010b; Rougemont et al. 2015, 2016) and evidence of contemporary gene flow (Docker et al. 2012; Rougemont et al. 2015, 2016) suggests that they are capable of successfully hybridizing at least to some extent. Moderate larval survivorship has been achieved in vitro in a variety of species pairs: European river and brook lampreys (Staponkus and Kesminas 2014; Hume et al. 2013), silver and northern brook lampreys (Piavis et al. 1970), and western river and brook lampreys (Beamish and Neville 1992), but hybrids have rarely been reared beyond the burrowing prolarval stage (Table 2.1). Beamish and Neville (1992) reared the western river and western brook lamprey hybrid larvae for 2.5 years, but it is still unknown what happens at metamorphosis (i.e., when the two life history types diverge) or at maturity. Rougemont et al. (2017) used genomic markers to identify first-generation (F1) European river and brook lamprey hybrids, but a virtual absence of later-generation hybrids suggests reduced hybrid survival or fertility. Hume et al. (2018) speculated that the extent of hybrization ebbs and flows with relative abundance on the spawning grounds. Testing for intrinsic postzygotic barriers in individuals of known parentage would require robust animal husbandry methods that must be maintained for multiple years as genetic incompatibilities are generally best revealed in F2 hybrids or when F1 hybrids backcross with one of the parental species (see Chap. 4). Difficulty rearing lampreys from fertilization through metamorphosis (but see Sect. 2.6.8) has also frustrated attempts at determining if feeding type in paired species is heritable or plastic (i.e., environmentally determined). Neave et al. (2019) attempted common garden and reciprocal transplant experiments with progeny of silver and northern brook lampreys to see if the feeding type of offspring was always the same as that of their parents, but mass mortality of developing larvae resulted in inconclusive findings.

Elucidating the genetic basis of sex determination in lamprey has also been hampered by the challenges associated with maintaining larvae in the laboratory for prolonged periods of time while trying to adequately mirror natural conditions. Observed correlations between sex ratio and larval density or growth rate, for example, have led to suggestions that lamprey sex determination may be influenced by environmental conditions (e.g., Docker and Beamish 1994; Johnson et al. 2017; see Chap. 1). However, attempts to test the effect of density on sex ratio under controlled conditions were inconclusive because survival and growth rates were low, and it was not possible to exclude differential mortality between the sexes (Docker 1992).

2.2 Artificial Propagation: Broodstock Holding

In parasitic lampreys, there is high intra- and interspecific variation in the duration of pre-spawning maturation in rivers (Applegate 1950; Clemens 2011; Aronsuu et al. 2015; see Moser et al. 2015). Whereas sea lamprey of both the anadromous and Great Lakes populations spend only 1–2 months in rivers before spawning (Applegate 1950; Almeida et al. 2002; Clemens et al. 2010), many populations of European river and Pacific lampreys overwinter in fresh water prior to reproduction (Masters et al. 2006; Clemens et al. 2012; Starcevich et al. 2014; Aronsuu et al. 2015; see Chap. 1). Overwintering individuals cease upstream migration when water temperature drops following the autumn season; they become passive and hide in refuges from predators and light (Robinson and Bayer 2005; Lampman 2011; Clemens et al. 2012; Starcevich et al. 2014; Aronsuu et al. 2015). European river lamprey enter into an energy-saving hypometabolic state during these winter months, resuming higher levels of activity with increased river discharge in spring (Abou-Seedo and Potter 1979; Gamper and Savina 2000). Hence, adult lampreys used for artificial propagation (broodstock) need to be provided with conditions that allow energy conservation during this extended period, in addition to cues necessary for successful final maturation.

2.2.1 Broodstock Density

When housed in aquaria for artificial propagation purposes, pre-spawning lampreys can be maintained in high densities, if they have adequately cool, clean, and well-oxygenated water. In Finland, hatcheries maintain densities as high as 2,000 European river lamprey adults/1,000 L of water (Vikström 2002). These lamprey can also survive through the winter beneath river ice when held in 200-L barrels provided with small-diameter holes for water exchange. Under these conditions, densities of up to 20 kg per barrel were successfully held (Jukka Pakkala, Centre for Economic Development, Transport and the Environment for South Ostrobothnia, Kokkola, Finland, personal communication, 2017).

Experience housing adult pre-spawning Pacific lamprey indicates that this species can also tolerate unnatural conditions (i.e., tanks without substrate) and high densities. For translocation and artificial propagation programs, Pacific lamprey broodstock are held at densities up to 60 kg/1,000 L or ~150 individuals/1,000 L. However, much higher densities are often observed in fishways and at winter aggregation areas below dams (Fig. 2.3). Winter temperatures during broodstock holding range from 2.8 to 15.5 °C, and mortality and disease incidence is very low under these conditions. However, as Pacific lamprey reach final sexual maturation and temperatures increase, they become more susceptible to *Aeromonas salmonicida* infection. Up to 21% of adults sacrificed following use for artificial propagation tested positive for this bacterium (Moser et al. 2016).

2.2.2 Broodstock Environmental Conditions

Providing adult lampreys with a sufficient flow of clean, oxygenated water is critical during winter holding. High flow rates of either oxygenated well water (with complete water turnover in 25–30 min) or natural spring water (19 L/min) have been used to maintain the high densities of Pacific lamprey described in Sect. 2.2.1. In Finland, European river lamprey adults were housed at water flows of 100 L/min (~1 L/min/kg lamprey; Vikström 2002).

Low pH and high metal concentration can deteriorate the quality of eggs during the wintering period of adults (Mäenpää et al. 2001). In Finland, European river lamprey

Fig. 2.3 Dense aggregation of pre-spawning Pacific lamprey at a Columbia River fishway (*Photo* © Donald Larsen)

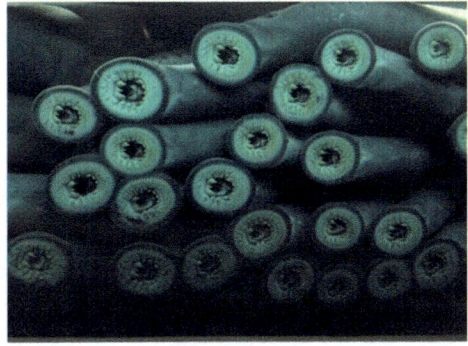

were held overwinter in upriver sites with high water quality (pH >5.5, aluminium concentration ~3 mg/L) and downriver sites where water quality was degraded (pH 5.2–5.5 and aluminium concentrations up to 4 mg/L). Egg fertilization was 85% in the upriver sites but only 55% in the downriver sites. Low fertilization rates were noted even when there were no obvious ill effects on the adults (Mäenpää et al. 2001).

In natural conditions, pre-spawning lampreys overwinter in darkened sites that are sheltered from direct water flow (Lampman 2011; Clemens et al. 2012; Baker et al. 2017). At high latitudes, they can overwinter beneath ice covered with snow. In Finland, European river lamprey housed through winter for artificial propagation increased their activity and restlessness when exposed to bright light during the day when no shelter was provided. This was presumed to increase stress levels, as lamprey housed in uncovered tanks had significantly higher mortality rates than those in dark, covered tanks (Juha Iivari, Natural Resource Institute, Keminmaa, Finland, personal communication, 2016). Langille and Hall (1988) also observed that sea lamprey exposed to a cycle of 16 h light:8 h dark were more agitated than those kept in dim light or complete darkness. These authors recommended maintaining pre-spawning lampreys in low or no-light conditions to reduce motor activity and decrease associated mortality.

However, lamprey broodstock held without any environmental cues can fail to mature, lack synchrony, and may even die without releasing eggs (Lampman et al. 2016). Piavis (1961) recommended that sea lamprey broodstock be held in complete darkness, but did not report on the percentage of fish that achieved full maturation. Of Pacific lamprey held in dark, coolwater tanks through winter and up to the time of spawning without provision of rocky substrate, only about half matured, even when both sexes were held together (Aaron D. Jackson, unpublished data). Hence, other cues may be required to stimulate final maturation (e.g., presence of substrate, mates, temperature fluctuation, or appropriate photoperiod; see Johnson et al. 2015). Johnson et al. (2012) concluded that the presence of male mating pheromones is likely an important trigger that synchronizes maturation. Further studies are needed to investigate the effect of pheromone presence or absence on the timing of final maturation.

2.3 Artificial Propagation: Broodstock Maturation

Development of species-specific secondary sexual characteristics are signs that maturation in lampreys is complete, including changes to the shape, position, and size of dorsal fins; shape of abdomen, and extension of the urogenital papilla (Hagelin and Steffner 1958; Kataoka et al. 1980a; Larsen 1980; Mesa et al. 2010; Johnson et al. 2015). In pouched lamprey *Geotria australis*, this includes the elaboration of the gular pouch in males and a raised dorsal ridge in females (Baker et al. 2017). In Pacific and European river lampreys, the closing of the gap between the two previously separated dorsal fins is also a good indicator of reproductive readiness (Vikström 2002; Clemens et al. 2009; Lampman et al. 2016). In artificial propagation programs, these characteristics are used for monitoring the maturation process and for segregating by sex or maturity level. This eases operational workflow during the fertilization process and prevents potential volitional spawning of broodstock in holding tanks (Vikström 2002; Lampman et al. 2016).

2.3.1 Broodstock Substrate

Lampreys rarely spawn in bare holding tanks, even when both sexes are present (Juha Iivari, Natural Resource Institute, Keminmaa, Finland, personal communication, 2016). Many studies indicate that unidirectional flow and a gravel substrate are required for lampreys to spawn in captivity (Hagelin 1959; Fredricks and Seelye 1995; Kusuda 2012; Aronsuu and Tertsunen 2015). Sea lamprey spawning was induced in static thermal conditions at 18 ± 2 °C by providing adults with 3–6 cm diameter substrate and a circulating water velocity of 0.2–0.3 m/s (Fredricks and Seelye 1995). Lack of these environmental factors may be one reason why lampreys do not readily spawn in holding tanks.

All lampreys are semelparous and most die shortly after spawning (Johnson et al. 2015; but see Baker et al. 2017). According to Hagelin (1959), wild lampreys may fail to spawn if they are unable to locate suitable spawning ground. Vikström (2002) observed that European river lamprey died in captivity within 48 h of completing maturation if they were not hand-stripped of their gametes. Pacific lamprey also will die without spawning if suitable substrate is not provided; thus, regular assessment of maturation state is critical for artificial propagation of this species (Lampman et al. 2016).

2.3.2 Broodstock Temperature

Rising water temperature in spring apparently triggers final maturation and spawning in both the laboratory (Vikström 2002; Clemens et al. 2009; Moser et al. 2018) and in the field (Larsen 1980; Binder and McDonald 2008a; Binder et al. 2010; Cochran et al. 2012; see Johnson et al. 2015). In the wild, lampreys often begin final maturation and spawn when temperatures approach 10–14 °C (Applegate 1950; Hagelin and Steffner 1958; Kan 1975). However, Larsen (1980) reported that European river lamprey held in the laboratory can mature even when kept at a stable temperature of 6 °C. At temperatures >7 °C, only 70% of European river lamprey males matured (Cejko et al. 2016). European brook lamprey typically will not start spawning activities until water temperature reaches at least 10–11 °C (Hardisty 1961b). For sea lamprey, the temperature threshold is higher, at ~15 °C, and spawning occurs closer to 20 °C (Applegate 1950; Gardner et al. 2012). Peak spawning of Pacific lamprey is typically observed at ~13–15 °C (Brumo 2006; Starcevich et al. 2014) and pouched lamprey was observed to spawn when stream temperatures were 10–13 °C (Baker et al. 2017).

Under natural conditions, the spawning period of European river lamprey has been reported to last for several weeks (Jang and Lucas 2005), as has the spawning period for sea lamprey in the Great Lakes (Applegate 1950). In Pacific lamprey, the natural spawning period may extend over 2 months (Brumo 2006). There is a tendency for the spawning season to be shortest at high latitudes and when water temperatures are steady and high; spawning periods are longer when temperatures are low and variable (Hardisty and Potter 1971; Johnson et al. 2015). Aquaculture of European river lamprey demonstrated that when water temperature continues to rise after exceeding 10 °C in spring, almost all overwintered lamprey mature within 1–3 days (Vikström 2002; Juha Iivari, Natural Resource Institute, Keminmaa, Finland, personal communication, 2016) and are ready for hand spawning shortly thereafter (Fig. 2.4). However, if temperature dropped near or below 10 °C, there was asynchrony in timing of maturation, and maturation could cease completely for weeks (Vikström 2002). For Pacific lamprey held under identical tank conditions, maturation rate tends to vary considerably among individuals (Lampman et al. 2016).

Temperature regimes that adults experience well before spawning can also influence maturation. In Finnish rivers, European river lamprey overwinter at close to 0 °C, and temperature during winter fluctuates very little. In contrast, the freshwater pre-spawning period for Pacific lamprey and pouched lamprey may last more than a year (Moser et al. 2015; Baker et al. 2017; see Chap. 1). Pacific lamprey broodstock are typically maintained at higher temperatures (2.8–15.5 °C) than European river lamprey. Clemens et al. (2009) showed that holding temperature during summer has a pronounced effect on maturation the following spring. In their experiments, Pacific lamprey held during summer at 13.6 °C experienced less weight loss and later maturation than those held during summer at 21.8 °C. In addition, all Pacific lamprey held at higher summer temperatures matured in spring, while only 53% of those held at lower summer temperatures matured (Clemens et al. 2009).

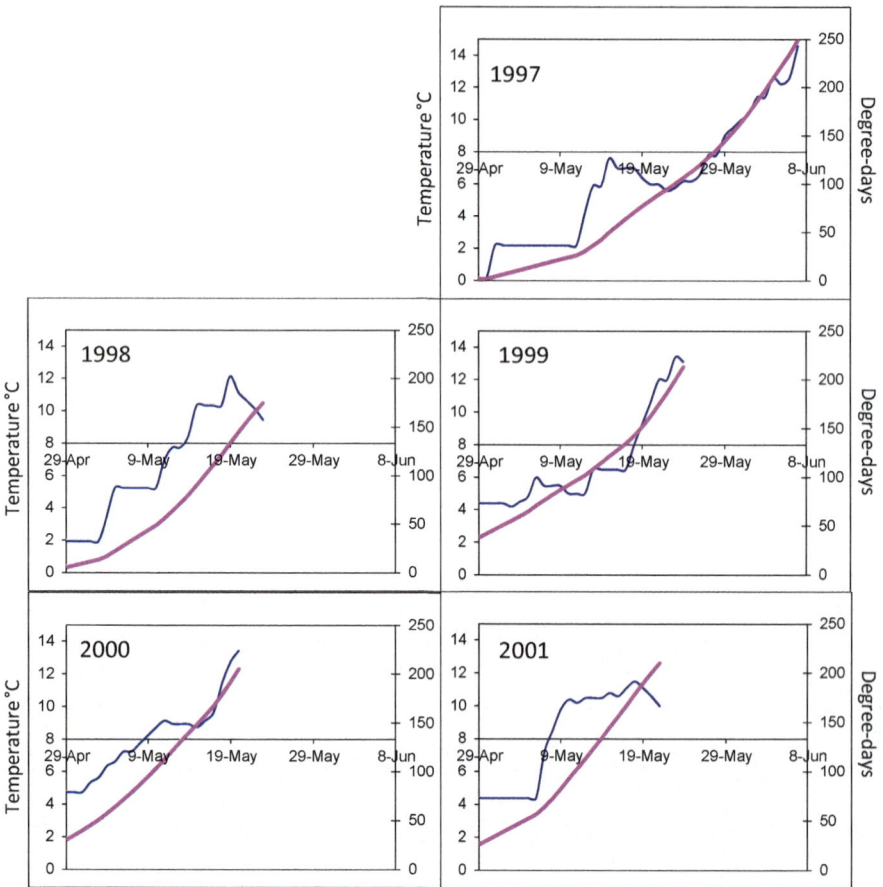

Fig. 2.4 Water temperature (°C, *blue*) and the numbers of degrees-days for female European river lamprey to reach maturity (*pink*) in 1997–2001 (This figure was originally published in Vikström (2002) and reproduced with permission of R. Vikström.)

2.3.3 Broodstock Photoperiod

Pre-spawning lampreys are best held in low or no-light conditions (see Sect. 2.2.2). When fully mature, however, lampreys lose their negative phototactic response (Sjöberg 1977; Binder and McDonald 2008b), which may indicate that photoperiod has an effect on maturation and therefore spawn timing. European river lamprey, when maintained in captivity under low light levels, matured later than those exposed to brighter lights and an ambient photoperiod (Vikström 2002). However, many studies show that light is not an important factor controlling final maturation, as lampreys have completed maturation even when maintained in complete darkness

Fig. 2.5 Proportion of sexually mature male (*dark blue*), mature female (*light red*), and immature (*gray*) Pacific lamprey held under complete darkness, artificial light (12:12), and natural lighting (01 March–30 June 2017) (This figure was originally published in Moser et al. (2018) and reproduced with permission of the authors.)

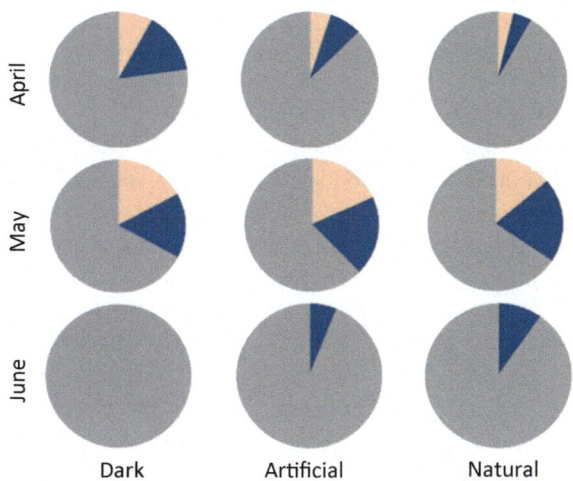

(Larsen 1980; Langille and Hall 1988) or have matured early as a consequence of increased temperatures (Brumo 2006; Cochran et al. 2012).

To assess this further, adult Pacific lamprey broodstock were held under three light treatments: natural lighting, a 12:12 artificial light regime, and in complete darkness during final maturation in March–June (Moser et al. 2018). Three replicate tanks containing 300 individuals were used for each treatment, and maturation state of each lamprey was assessed monthly. Lamprey held in complete darkness matured earlier than those exposed to either artificial lighting or natural light cycles (Fig. 2.5). However, there was no significant difference among treatments in the overall proportion of fish that matured (darkness = 44%, artificial light = 46%, natural light = 38%). These results suggest that photoperiod mediates the timing of maturation somewhat, but does not appear to trigger this process.

2.4   Artificial Propagation: Fertilization Methods

In vitro fertilization has been conducted in locations worldwide using a variety of large-bodied parasitic lampreys: Arctic lamprey in Japan (Kobayashi 1993; Yamazaki and Goto 1997; Fukutomi et al. 2002; Hokkaido Fish Hatchery 2008), Pacific lamprey in Japan and the Pacific Northwest (Yamazaki et al. 2003; Lampman et al. 2016), sea lamprey in the Great Lakes region (Langille and Hall 1988; Fredricks and Seelye 1995; Ciereszko et al. 2000, 2002) and in Spain (Rodríguez-Muñoz et al. 2001; Rodríguez-Muñoz and Ojanguren 2002), and European river lamprey in Russia (Ryapolova and Mitans 1991) and Finland (Vikström 2002) (Table 2.1). In addition, in vitro fertilization has also been conducted using several species of non-parasitic (brook) lampreys: Far Eastern brook lamprey in Japan (Fujimoto and Takaoka 1960), western brook lamprey in the Pacific Northwest (Ralph T. Lampman, unpublished data), and European brook lamprey in Scotland (Hume et al. 2013). The most salient difference among species is the much greater number of eggs produced by the large-bodied lampreys since fecundity increases approximately with the cubic power of length (see Chap. 1). Number of eggs, for example, averages more than 140,000 and 170,000 in Pacific and sea lampreys, respectively (Kan 1975; Beamish and Potter 1975) versus the smaller-bodied parasitic (e.g., ~37,000 in European river lamprey; Witkowski and Jęsior 2000) and non-parasitic species (e.g., ~1,500 in northern brook lamprey; Vladykov 1951). Otherwise, in vitro fertilization methods have been similar across these species, and researchers have been able to build on earlier work conducted, even if it was with a different species (Lampman et al. 2016).

2.4.1   Number of Parents

Lampreys in the wild visit multiple nests and contribute to multiple clutches (Cochran et al. 2008; Johnson et al. 2015). This observation has been confirmed by parentage analysis paired with nest mapping for wild Pacific lamprey (Whitlock et al. 2017) and Great Lakes sea lamprey (Scribner and Jones 2002). Whitlock et al. (2017) reported that the same parents contributed to progeny in nests that were up to 815 m apart. For smaller-bodied lamprey species (e.g., chestnut lamprey, Arctic lamprey, and European river lamprey), communal spawning is typical, and dozens of individuals have been counted in one spawning excavation (Case 1970; Savvaitova and Maksimov 1979; Jang and Lucas 2005; Lasne et al. 2010b). The larger-bodied sea and Pacific lampreys have been described as monogamous tending toward polygynous, although they appear to show variation in their mating systems (Johnson et al. 2015; Baker et al. 2017) and recent genetic evidence indicates that in Pacific lamprey, both polygyny and polyandry may be more common than previously believed (Whitlock et al. 2017). Thus, polygynandry (i.e., multiple males mating with multiple females) appears to be the most prevalent mating system in lampreys (Johnson et al. 2015). For this reason and to maximize genetic diversity, Pacific lamprey propagation protocols emphasize use of multiple males to fertilize eggs from multiple females (Lampman et al. 2016).

There is some limited evidence for lack of sperm dominance in lampreys. Parentage analysis in studies with both propagated Pacific lamprey (Hess et al. 2015) and wild sea lamprey (Scribner and Jones 2002) have successfully assigned progeny to known parents at very high rates (>95%), with lack of assignment likely owing to poor DNA preservation quality. In a common garden experiment with propagated Pacific lamprey, the number of progeny assigned to two females and three males in the family were roughly in proportion to the quantity of gametes contributed by each parent (Hess et al. 2015). This result hints that sperm competition may be similar across males of this species. Moreover, these results indicate that parentage assignment can help elucidate mating systems for lampreys and allow estimation of the numbers of successful wild spawners (Hess et al. 2015; Whitlock et al. 2017).

2.4.2   Fertilization Timing

To maximize the quality and quantity of gametes obtained for artificial propagation, timing of gamete harvest is critical. In sea and Pacific lampreys, forcefully stripping gametes can result in premature adult mortality, damaged gametes, and unsuccessful egg development (Langille and Hall 1988; Lampman et al. 2016). For these reasons, it is important that during gamete harvest the adults are at a high plane of anesthesia and that the gametes are allowed to flow with minimal pressure (Fig. 2.6). Surgical removal of Pacific lamprey eggs when females were not quite ready (based on secondary sexual characteristics) resulted in lower mean fertilization success (63.4%) than surgical removal of eggs when the female was fully ripe (90%, Moser et al. 2016).

There is only slight interspecific variation in behavior among the Northern Hemisphere lampreys (family Petromyzontidae) during the spawning act (Johnson et al. 2015). Spawning begins when the female attaches to a large rock or stone and orients her body with the water flow. The male approaches the female from behind, attaches to the female's head, and wraps the lower half of his body around the female, forming a loose coil around her trunk. This tail-loop is then tightened and both male and female raise their branchial region up from their anchor point at an acute angle and violently vibrate and thrash their tails for several seconds. This results in the expulsion of ova and milt into a gravel depression, which is rapidly covered in sand and small gravel. Eggs typically adhere to the downstream ridge of the nest (Applegate 1950; Hagelin 1959). Only a portion of eggs, if any, is released during a single spawning (Huggins and Thompson 1970; Yamazaki and Koizumi 2017). Thus, spawning in the wild can last several days for each individual, and superimposed spawning is common (Manion and Hanson 1980; Jang and Lucas 2005; Brumo 2006). Pouched lamprey nests and post-spawning behavior were recently described by Baker et al. (2017), but spawning behavior has not yet been reported in any of the four Southern Hemisphere species (families Geotridae and Mordaciidae).

Artificial propagation programs have taken advantage of the fact that reproductively mature lampreys can be successfully spawned multiple times, over the course of several days in females to >1 week in males (Hagelin and Steffner 1958; Langille

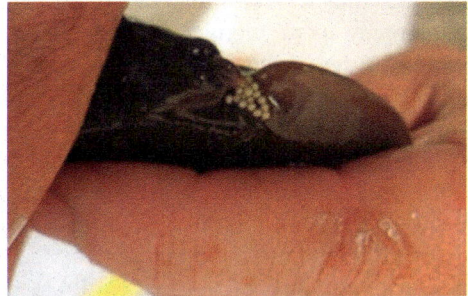

Fig. 2.6 Expressing eggs from a fully anesthetized female Pacific lamprey (*Photo* © Ralph T. Lampman)

and Hall 1988). This has allowed a greater number of pairings when lamprey brood-stock are limited (Lampman et al. 2016). Close synchrony between male and female reproductive readiness is key and is likely mediated by a number of cues: environmental (temperature and photoperiod), physiological (neuroendocrine activity and pheromone production), and behavioral (presence of mates and nest-building activity) (see Johnson et al. 2015; Sower 2015).

2.4.3 Gamete Viability and Contact Time

Lamprey gametes are generally viable for much longer than those of other fishes (Johnson et al. 2015; see Chap. 1) and are resilient to environmental changes, which allows for flexibility in artificial propagation programs. Eggs from freshly dead lampreys can still be viable, and there is good evidence that both eggs and milt are still viable at environmental temperature after several hours (Lampman et al. 2016; see Chap. 6). For Pacific lamprey gametes held at 4 °C, fertilization success was lower after 24 h for eggs and after 3 days for milt (Moser et al. 2016). In these studies, sperm motility could be extended for another day with provision of oxygen, and there was some evidence that cryopreservation methods might be successful (Lampman et al. 2016). Similarly, >95% viability was observed in sea lamprey eggs after 24 h storage at 15 °C, but viability decreased to <20% after 3 days; >60% fertilization was achieved with milt stored at 1 °C for 2 days (Ciereszko et al. 2000).

In production of lampreys for conservation purposes, particularly non-parasitic species that produce only ~1,000–2,000 eggs (see Sect. 2.4; Chap. 1), artificial propagation methods need to yield maximal fertilization success while minimizing egg loss from damage or adhesion. Lampman et al. (2016) found that a 2–5% solution of milt from one or more males mixed directly with ova, followed by the immediate addition of culture water at a volume representing $1-1.5 \times$ the egg weight, maximized fertilization. They recommended very short gamete contact and holding times (30 s each) before a thorough rinsing of the eggs with culture water before installation in incubation chambers. Recommendations for European river lamprey fertilization are quite similar (Jääskä 2002). Sea lamprey eggs could be fertilized for up to 1 h after contact with fresh water (Ciereszko et al. 2000), but the ability of sperm to fertilize them was only 27% just 2 min after activation (Ciereszko et al. 2002). Interestingly, sea lamprey sperm motility could be increased slightly by incubation in water containing 4% female coelomic fluid or water that had contained eggs (Ciereszko et al. 2002). Very short gamete contact times and low levels of egg tumbling after fertilization in the laboratory are consistent with conditions in the wild. Spawning occurs in flowing streams, so gametes are likely in contact for very short periods of time. Since eggs are adhesive immediately upon water hardening (see Sect. 2.4.4), both fertilized and unfertilized eggs can quickly acquire a coating of sand or silt particles. This may result in retention on or near the nest excavation and protection from excessive tumbling (Silva et al. 2014b).

2.4.4 Egg Adhesion

Lamprey eggs are highly adhesive (Yorke and McMillan 1979) and this characteristic has important implications for survival both in the wild and in the laboratory. In the wild, lamprey eggs adhere to a small amount of sand that helps to embed them in the interstices of gravel substrates (Applegate 1950). In addition, small sand particles may separate eggs from one another, functioning to prevent mortality from fungus (Smith and Marsden 2009). The lamprey egg coating includes both an amorphous apical tuft over the animal pole and a heavily textured coating over most of the rest of the egg (Yorke and McMillan 1979). These coatings allow lamprey eggs to stick to rocks and help to anchor them in the relatively benign nest environment.

For artificial propagation, egg adhesion can cause loss or clumping of eggs that leads to increased incidence of fungal infection (Piavis 1961; Vikström 2002; Lampman et al. 2016). Yorke and McMillan (1979) found that egg adhesiveness could be diminished by exposure to various proteins and sulphydryl-blocking agents. Lampman et al. (2016) reported that immersion of newly fertilized eggs in a 1% solution of fresh pineapple juice for 1–2 min could completely inhibit the adhesive capacity of the egg coating without affecting egg viability. Gentle rinsing of fertilized eggs in culture water also helps to reduce clumping and spread eggs more evenly in incubation chambers (Vikström 2002; Lampman et al. 2016).

While egg adhesion has important consequences for lamprey culture operations, investigation of the role of egg adhesion in the wild might have equally important ramifications for lamprey conservation or control. Lampreys do not always properly cover the eggs after spawning, and eggs deposited in the excavation are easily flushed out of the depression by ongoing spawning activity (Huggins and Thompson 1970) or water flow (Silva et al. 2014b). Consequently, it has been hypothesized that most fertilized eggs drift downstream from the excavation during the spawning act or soon thereafter and incubate somewhere below the nest (Manion and Hanson 1980; Smith and Marsden 2009; Silva et al. 2014b). Silva et al. (2014b) proposed that the nests of European river lamprey may function as egg dispersal structures rather than as egg shelter structures. In contrast, pouched lamprey egg masses adhere to the underside of a boulder and are thereby protected from water currents and predators (Baker et al. 2017). The male has been observed to "groom" the eggs with his gular pouch, which may reduce the incidence of fungal infection. The pouched lamprey larvae have adhesive tails that allow them to remain adhered to the nest boulder for at least 2 weeks after hatching. Understanding the role of egg/larval adhesion may provide insights into mechanisms of early embryo mortality such as susceptibility to predation, protection of incubation habitats, and the role of nest building for various species.

2.5 Artificial Propagation: Incubation Methods

Lampreys spawn in fast-flowing parts of rivers, such as pool tailouts, glides, and deep riffles, where substrate consists of gravel often mixed with sand and cobbles (Jang and Lucas 2005; Brumo 2006; Gunckel et al. 2009; Nika and Virbickas 2010; see Johnson et al. 2015). There is a tendency for larger species to spawn in deeper sites with higher water velocities and coarser substrate than smaller species (e.g. Applegate 1950; Sokolov et al. 1992; Takayama 2002; Gunckel et al. 2009; Nika and Virbickas 2010). Such differences in spawning habitats indicate that incubation conditions for eggs could differ among species. Consequently, there may be variation among lamprey species in optimal methods of artificial incubation.

Identifying the environmental conditions required for successful egg incubation is fundamental to any artificial propagation program. For conservation and restoration of lamprey species, incubation success is particularly important. Broodstock of depleted populations may be difficult to obtain, and production of fertilized eggs can be limited by both synchrony in adult maturation and gamete viability. Hence, maximizing incubation success has been a primary objective of many native lamprey propagation efforts (Rodríguez-Muñoz et al. 2001; Vikström 2002; Hokkaido Fish Hatchery 2008; Lampman et al. 2016).

2.5.1 Incubation Temperature

Lamprey eggs typically hatch in 1–4 weeks, and temperature has a profound effect on incubation timing and, ultimately, the survival and success of larvae (Potter 1980; Dawson et al. 2015). In sea lamprey cultured at 18.4 °C, Piavis (1961) observed hatching at 10–13 days post-fertilization. With increasing temperature, Rodríguez-Muñoz et al. (2001) found that mean time to 50% hatch decreased from 27 days at 11 °C to 7.5 days at 23 °C, but exposure to temperatures above 19 °C resulted in mortality of larvae. Field observations of incubation temperature for sea lamprey in the River Stella in northern Spain indicated that eggs incubate at 11–20 °C in the wild (Rodríguez-Muñoz et al. 2001). In a similar field study, most Pacific lamprey egg incubation was found to occur when stream temperature ranged from 9 to 16 °C (Fig. 2.7; Aaron D. Jackson, unpublished data). However, in the laboratory, Pacific lamprey eggs were successfully incubated at 20 °C (Alexa N. Maine, Confederated Tribes of the Umatilla Indian Reservation, Pendleton, OR, personal communication, 2018).

In the laboratory, Pacific lamprey typically require 184–294 cumulative degree-days for incubation (Yamazaki et al. 2003; Meeuwig et al. 2005; Lampman et al. 2016). Arctic lamprey hatched in 234 degree-days (18 days at 11.8–12.9 °C; Hosoya et al. 1979). Experience from the artificial propagation of European river lamprey indicates that egg incubation takes 1–3 weeks, depending on water temperature. Usually, hatching starts after 190–220 degree-days, and all eggs will have hatched after

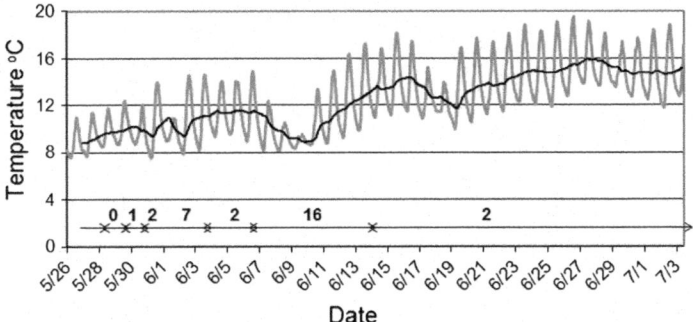

Fig. 2.7 Hourly water temperature (*gray line*) and moving average of 24 h (*black line*) in an Umatilla River tributary (Meacham Creek, OR). The number of new viable Pacific lamprey nests observed (*numbers above arrows*) is shown for a 0.9 km reach (Aaron D. Jackson, unpublished data)

an additional 50 degree-days. However, if temperature rises soon after fertilization (up to 20 °C), all hatching can occur within just 150 degree-days (Vikström 2002).

Fertilized lamprey eggs are very resilient to periods of high temperature and to abrupt changes in temperature. Upper temperature limits during embryonic development appear similar among lamprey species. Both Pacific and western brook lampreys suffer highest mortality rates at temperatures >22°C (Meeuwig et al. 2005). Sea lamprey embryos can survive at temperatures >21.1 °C (Piavis 1961) and perhaps as high as 23 °C (Piavis 1971; Rodríguez-Muñoz et al. 2001). With respect to tolerance for rapid temperature changes, Pacific lamprey embryos have been successfully held at 5 °C for 24 h during transport and returned immediately to the initial incubation temperature (13 °C) with no appreciable mortality. This is not surprising, as lamprey eggs in the wild are likely exposed to rapid and substantial changes in temperature during spring freshets and periods of intense solar radiation. For example, in the Umatilla River drainage in northeastern Oregon, temperature can vary by more than 6 °C in a day (Fig. 2.7), yet 60–100% of the eggs in Pacific lamprey nests were typically viable (Fig. 2.8; Aaron D. Jackson, unpublished data).

2.5.2 Photoperiod and Water Quality

Although temperature is clearly a factor that controls the timing of Pacific lamprey embryonic development, considerable variation in hatch timing (15–23 d) has been observed between years or among individuals even when temperature was nearly constant (Fig. 2.9). In 2015, embryos from the same female held in replicate chambers (n = 15) with no flow in a 14 °C water bath were checked daily for developmental changes. Three water sources were tested (n = 5 replicates per treatment): natural creek water; de-chlorinated, UV-irradiated city water; and the city water source

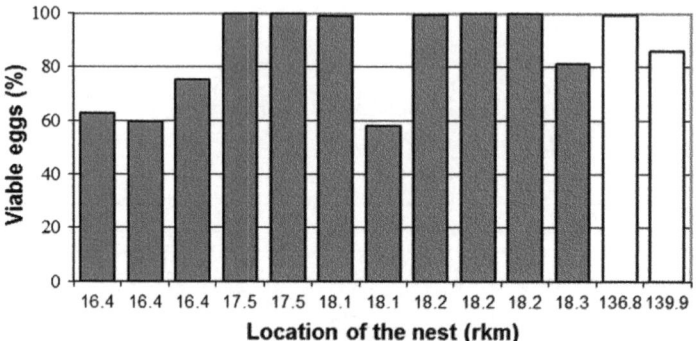

Fig. 2.8 The percentage of viable Pacific lamprey eggs at stages 12–14 in 11 nests from Meacham Creek (*gray bars*), and two nests in the mainstem Umatilla River (*white bars*); rkm is river kilometers from the creek or river mouth (Aaron D. Jackson, unpublished data)

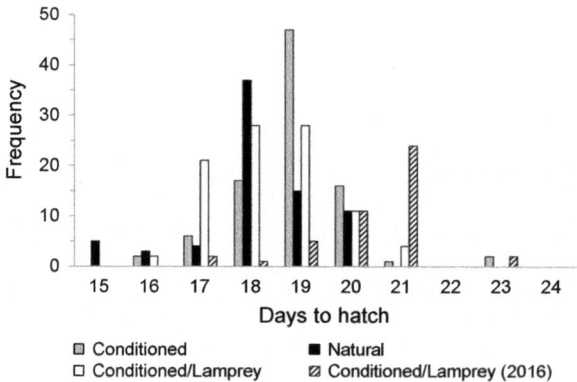

Fig. 2.9 Frequency distribution of individual hatch dates for Pacific lamprey held in static conditions at 14 °C. Eggs from one female in 2015 were incubated in natural creek water (*black*), conditioned city water with no conspecifics (*gray*), or conditioned city water with larval lamprey (*white*); eggs from one female in 2016 (*hatched*) were incubated in conditioned city water with larval lamprey (This figure was originally published in Maine et al. (2017) and reproduced with permission of the authors.)

with conspecific larvae present. The three water treatments did not affect survival to hatching or median incubation period. However, in the following year, median hatch times were shifted by several days. The only differences between study years were the parents used and a slight change in natural photoperiod; spawn dates were 21 April 2015 and 05 May 2016 (i.e., 2 weeks earlier in 2015) and day-length on the spawning dates was 13.9 h in 2015 and 14.6 h in 2016 (Maine et al. 2017). Piavis (1961) and Kataoka et al. (1980b) recommended incubation of lamprey eggs in darkness to synchronize hatching.

Piavis and Howell (1969) reported that sea lamprey embryos could be incubated in distilled water, thereby potentially reducing the potential for fungal infection. Alter-

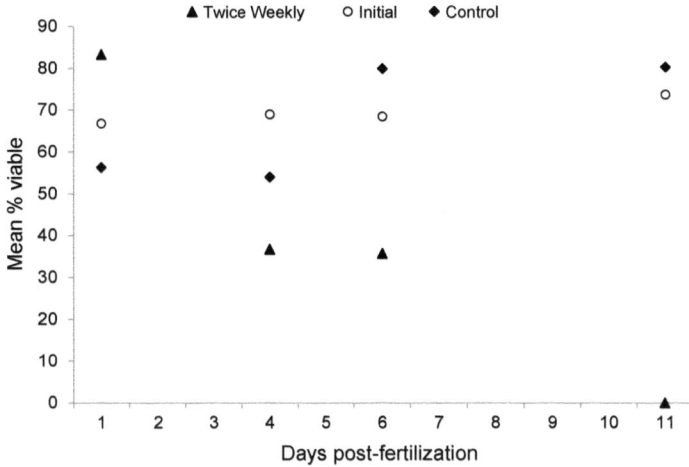

Fig. 2.10 Percentage of viable Pacific lamprey eggs from subsamples examined at 1, 4, 6, and 11 days post-fertilization following multiple disinfection treatments at 3-day intervals (i.e., twice weekly; *black triangles*), those exposed to a single initial disinfection immediately after fertilization (*open circles*), and for non-disinfected controls (*black diamonds*) (This figure was originally published in Moser and Jackson (2013) and reproduced with permission of the authors.)

natively, Pacific lamprey embryos up to 4 days old can be safely disinfected by 10-min immersion in a 100-parts per million (ppm) buffered iodophor bath (Fig. 2.10). Formalin was also successfully used to disinfect fertilized Pacific lamprey eggs up to 14 days after fertilization when used at a concentration of 0.8 mg/L (1:1,250) for continuous exposure (Maine et al. 2017) and at a dilution of 1.7 mg/L (1:600) for embryos 3–10 days after fertilization (Lampman et al. 2016).

These results indicate that early lamprey embryos are generally resilient to water quality insults. However, they become more sensitive to water quality as they near hatching (Lampman et al. 2016). Myllynen et al. (1997) showed that incubation of European river lamprey embryos in water with low pH (5–6), high aluminum (0.45–0.6 mg/L), and/or high iron (1.5–3 mg/L) concentrations caused reduced hatching rates and low larval survival. Controls held in low pH without heavy metals were unaffected.

2.5.3 Water Flow and Substrate

Lamprey eggs can be incubated successfully in both flowing and static water conditions. In experiments where eggs from the same Pacific lamprey female were held under flowing (2 L/min) and static conditions with UV-irradiated water, there was no difference in survival to hatching (Fig. 2.11). This is not a new development, as the eggs of many lamprey species have been cultured in static conditions for embryolog-

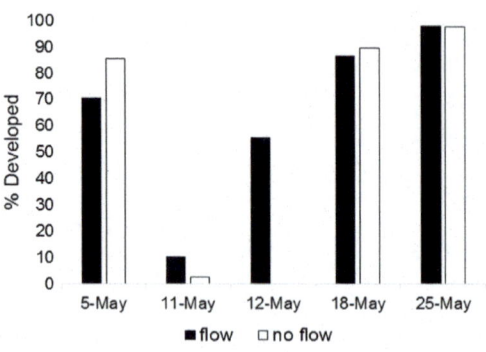

Fig. 2.11 The percentage of Pacific lamprey embryos that survived to hatching (% developed; stage 14) for each spawn date in flowing (2 L/min, *black bars*) or static (*white bars*, photo inset) conditions (This figure was originally published in Maine et al. (2017) and reproduced with permission of the authors.)

ical studies (e.g., Damas 1944; Piavis 1961; Kuratani et al. 1997; Yamazaki et al. 2003; Hokkaido Fish Hatchery 2008). However, the experiment was interesting in light of the fact that, in the wild, adult lampreys typically build nests in flowing water (Johnson et al. 2015), where the eggs would potentially be exposed to some degree of turbulence and hyporheic flow (Fixler 2017).

Laboratory observations of spawning activity by both European river and Pacific lampreys indicate that adults seek out areas of relatively high flow for egg fertilization and incubation. Aronsuu and Tertsunen (2015) observed that regardless of substrate provided, European river lamprey constructed nests near tank walls, where current velocity was lowest. Pacific lamprey also selected areas near water inlets for nest construction (Alexa N. Maine, Confederated Tribes of the Umatilla Indian Reservation, personal communication, 2017). Yet current velocity measurement of nests constructed by European river lamprey in the wild indicated that water velocities near the nest bottom (2 cm from the substrate) can be less than half those at a height of 10 cm above the substrate (Aronsuu and Tertsunen 2015). For Pacific lamprey nests measured in the Umatilla River basin (n = 18), water velocity 5 cm above the substrate inside the nest averaged 15.3 cm/s and ranged from 0 to 41 cm/s (Aaron D. Jackson, unpublished data). Therefore, eggs that adhere to the bottom of the nest or become wedged in substrate may experience very low flows during incubation in the wild.

Smith and Marsden (2009) incubated sea lamprey eggs on a variety of substrates and found that they were relatively insensitive to suffocation: survival to at least stage 12 (Piavis 1961; see Sect. 2.1.2) was not significantly different for eggs incubated in sand and silt treatments. In the laboratory, proliferation of fungus can occur when lamprey eggs are incubated in low to no-flow conditions (Piavis 1961; Lampman et al. 2016). This may be the reason that wild lampreys spawn in areas where eggs will be exposed to enough current velocity to protect them from fungal infestation.

Eggs that become covered with a thin layer of particles may be somewhat protected from fungal infection (Smith and Marsden 2009). Thus, eggs that are inadvertently flushed from the nest (up to 86% of a clutch; Manion 1968; Manion and Hanson 1980) are susceptible to predation (Applegate 1950; Manion 1968), but may not suffer from deposition on silty substrate (Smith and Marsden 2009).

2.5.4 Incubation Mortality

Survival of eggs to hatching in the laboratory can be 100% under ideal conditions (Fig. 2.11), but stage-specific mortality rates of embryos in the wild are difficult to assess. Kujawa et al. (2017) noted hatching rates of only 10% in wild European river lamprey. In a study in the Umatilla River drainage, eggs were collected from freshly constructed nests of Pacific lamprey (n = 16, Fig. 2.8; Aaron D. Jackson, unpublished data). After eggs had developed to at least stage 12 (Piavis 1961), a sample of 200 eggs was taken from each nest, fixed in 10% formalin, and assessed for viability under a dissecting microscope. Eggs were classified as unviable if covered with fungus or deformed. Several nests had 100% viable embryos. Of the dead embryos, 75% were infested with fungus and 25% had developmental deformities. These conditions have also been described for eggs incubated under controlled laboratory conditions (Piavis 1961; Lampman et al. 2016). Moreover, 19% of the nests did not contain enough eggs for an adequate sample. Similarly, Whitlock et al. (2017) reported that over half of the Pacific lamprey nests they sampled in a western Oregon stream did not contain any eggs. Whether these empty nests represent test digging, failed spawning attempts, scouring, or losses from disease or predation is unknown.

2.6 Artificial Propagation: Rearing Early Larvae

Developing methods to rear early stages of fish at the production level is notoriously difficult (Sifa and Mathias 1987; Kujawa et al. 2017). Many fish species exhibit a critical period between hatching and first feeding, as hypothesized by Hjort (1914, 1926). As fish switch from endogenous (yolk sac) to exogenous sources of nutrition, high specific mortality rates often occur (Sifa and Mathias 1987). This is coincident with profound changes in larval morphology, physiology, and ecology (Dabrowski 1984). Switching to exogenous feeding often involves changes in body function that must be precisely synchronized, such as sensory organ development for food capture or collection, muscular elaboration for manipulation of prey, or gut development for processing of new foods. Imperfect synchrony or underdevelopment of crucial systems can combine with mechanical constraints to retard efficient feeding and can ultimately result in larval starvation (China and Holtzman 2014). These problems occur in lamprey culture and demand a more thorough understanding of both larval physiology and feeding behavior.

2.6.1   Timing of Exogenous Feeding

While their feeding morphology is relatively simple, larval lampreys must orchestrate a complicated switch from yolk-sac feeding in the nest or on the sediment surface to active burrowing and collection and processing of relatively nutrient-poor food particles from the environment (Manion 1968; Moore and Beamish 1973; Sutton and Bowen 1994; Yap and Bowen 2003). The end of endogenous feeding (i.e., the transition from prolarva to larva) is signaled by completion of the digestive tract and connection to the anus (Fig. 2.12). At this time, lampreys can begin to supplement yolk-sac feeding with collection of exogenous food particles. In the wild, these particles are typically microalgae and detritus (Manion 1967; Sutton and Bowen 1994; Yap and Bowen 2003; see Sect. 2.6.5). Exogenous feeding is accomplished with a mucus-lined pharynx that delivers particles to the simple, straight gut tract via peristalsis (Mallatt 1981). Elaboration of the oral hood and completion of the gut tract accompany a dramatic change in behavior, from resting on the substrate surface to seeking and actively burrowing into substrate of the appropriate particle size (Lampman 2016; Lampman et al. 2016).

In a culture situation, facilitating the switch to exogenous feeding in larval lampreys requires identification of the appropriate time to start providing appropriate feed for a given life stage (Barron et al. 2016). Artificial propagation of lampreys provides a unique opportunity to study the timing of this shift from endogenous to exogenous feeding, as this stage is rarely encountered in the field (Manion 1968; Brumo 2006; Schultz et al. 2014). Barron et al. (2016) found that growth of Pacific lamprey larvae was maximized when feed was provided coincident with the onset of first feeding or slightly earlier (16–24 days after hatching). Individual hatch times for a single spawning event can vary over 7 days in Pacific lamprey (Fig. 2.9), so early initiation of feeding ensures that all larvae are accommodated. There is very little information on variation in larval development times in wild lampreys. However, Whitlock et al. (2017) noted that the ages of embryos collected from wild Pacific lamprey nests were all within 5 days of each other.

For lampreys in culture, the length of time from hatching to first feeding is similar among species, but varies substantially with temperature (Piavis and Howell 1969; Langille and Hall 1988; Fredricks and Seelye 1995; Vikström 2002; Richardson and Wright 2003; Hokkaido Fish Hatchery 2008). Completion of gut tract and eye formation were observed 32 days post-fertilization (14 days post-hatch) in Arctic

Fig. 2.12 Pacific lamprey larva (stage 17). Note the completed connection of the digestive tract to the anus, signaling the start of exogenous feeding (*Photo* © Alexa N. Maine)

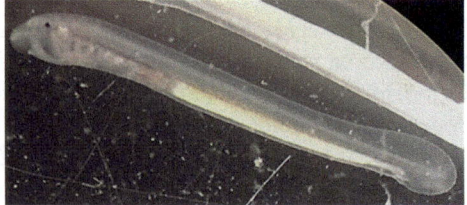

lamprey held at 11.8–12.9 °C (Hosoya et al. 1979). Lampman et al. (2016) reported that burrowing in Pacific lamprey larvae started at 26–33 days after egg fertilization, and that this corresponded to 369–469 cumulative degree-days. Similar times to first burrowing (17–33 days after fertilization) were reported for sea lamprey in culture (at 18 °C; Piavis 1961) and Pacific and sea lampreys in the wild (Manion 1968; Brumo 2006). Piavis (1961) observed transition from prolarva to larva (i.e., when the gut was fully differentiated) at 33–40 days after fertilization at 18 °C, and Richardson and Wright (2003) observed that gut formation in this species was completed at 23–36 days. Rodríguez-Muñoz et al. (2001) found that burrowing in sea lamprey can occur before the yolk is fully depleted in embryos held at temperatures >19 °C, but embryos incubated at 15 °C only started burrowing when their yolk was nearly exhausted. The body mass at first feeding also increased with incubation temperature (Rodríguez-Muñoz et al. 2001).

Determination of the optimal time to start feeding larval lampreys is critical to the success of aquaculture operations and can provide insight into both the timing of first feeding in the wild and larval capacity for starvation. Experiments were conducted with first-feeding larvae to assess the consequences to survival of delayed feed provision. Barron et al. (2016) found that delaying the onset of first feeding by only a few days could have profound effects on growth in larval Pacific lamprey. However, these larvae were also surprisingly resilient to starvation and have been known to survive for up to a month without substantial food inputs (Lampman et al. 2016). It is likely that these larvae were able to subsist on micro-organisms that persist in culture even when no food is added. Given the low metabolic rate of larval lampreys (Hill and Potter 1970; Potter and Rogers 1972) and the ability of metamorphosing and adult lampreys to survive extended periods of fasting during these non-trophic stages (Clemens et al. 2010; Manzon et al. 2015; Moser et al. 2015), it is not surprising that larval lampreys exhibit high tolerance to starvation relative to larval teleosts (see Sect. 2.6.5).

2.6.2 Feeding and Sheltering Behavior

Lampreys are thought to passively filter particles from the seston (Yap and Bowen 2003). However, laboratory experiments with Pacific lamprey suggest that they can also feed on particles from substrate pore water (Alexa N. Maine, Confederated Tribes of the Umatilla Indian Reservation, personal communication, 2014). In these experiments, 1-L static beakers with 3 cm of either fine (<149 μm) or coarse (149–595 μm) sand were prepared and placed in a 14.4 °C water bath. Immediately before experimentation, a mixture of commercially prepared (Reed Mariculture) concentrated (3–8 billion cells/mL) marine algae cells (0.5 mL *Nannochloropsis* and 0.5 mL *Pavlova*) was injected into the sediment. The 85-day-old larvae used in the experiments were not fed for a week prior to experimentation, and were gently introduced into the chambers individually on the same day (10 larvae/L). After 3 days, larvae were examined under a dissecting microscope for the presence of algal cells in the gut. An average of 96 and 98% of larvae had algal cells in their guts in the coarse and fine sediment treatments, respectively. These data suggest that lampreys can obtain food particles from within the sediment pore water as a deposit feeder, and that this feeding mechanism might contribute to their nutrition. This is consistent with lamprey isotope studies (Limm and Power 2011; Evans and Bauer 2015, 2016) which all point to the importance of substrate and deposited organic matter in nutritional uptake by larval lampreys. Substrate is also very important in the hatchery environment to allow for normal feeding and development of cultured larval lampreys (Lampman et al. 2016; Sect. 2.6.3).

Behavioral observations of cultured larvae indicate that they are mobile at just a few days after hatching and capable of moving vertically into flowing currents at night (Moser and Jackson 2013; Lampman et al. 2016). Hence, very small mesh size (<300 μm) and complete tank seals are necessary to keep very young larvae from escaping (see Chap. 6). The downstream drift of wild YOY European river lamprey, Great Lakes sea lamprey, Pacific lamprey, and Arctic lamprey also takes place during hours of darkness (Manion and McLain 1971; Bennett and Ross 1995; Derosier 2001; White and Harvey 2003; Brumo 2006; Kirillova et al. 2011; Pavlov et al. 2014; Zvezdin et al. 2016, 2017). Derosier (2001) found that sea lamprey prolarvae (i.e., after hatching but prior to the onset of exogenous feeding) emerge from the nest during the darkest hours of the night (1200–0300 h), and that the emergence period is short, with 80% of prolarvae emerging after 8–14 days on average. Such diel timing is likely a common strategy in other species (Potter 1980; Dawson et al. 2015).

The reliance of larval lampreys on optimal substrate, depth, and flow conditions in the field has been intensively studied for a broad range of species (e.g., Morman et al. 1980; Potter et al. 1986; Sugiyama and Goto 2002; Torgersen and Close 2004; Nazarov et al. 2016; see Sect. 2.6.3). However, settlement mechanisms are poorly understood. Presumably, wild larval drift slows down in areas of silty substrate, allowing lampreys to passively settle in areas with appropriate depth, particle size, and flow (Applegate 1950; Bennett and Ross 1995; Derosier 2001). Thus, settlement of prolarvae could be entirely passive, occurring when current strength weakens or when individuals find themselves in a backwater or pool environment. However, it is also possible that settlement is non-random and that they use olfactory cues from other larval lampreys to identify and potentially reject rearing habitat (Zielinski 1996). Active substrate selection by subyearling larvae has been studied only in the laboratory; European river lamprey prolarvae selected sieved gravel in which to shelter and started to select for fine-grain substrates at just 8 mm in length (Aronsuu and Virkkala 2014).

2.6.3 Substrate

A key aspect of lamprey culture is the provision of sufficient substrate for functional burrowing (Kelso 1993), and substrate characteristics must be closely coordinated with ontogeny. Immediately after hatching, prolarval lampreys are unable to burrow

(Piavis 1961), but require areas to shelter. They actively select substrate with interstitial spaces available (Aronsuu and Virkkala 2014), a sheltering behavior that likely evolved to increase survival after they leave the nest. Hence, in situations where lampreys are transplanted into the wild shortly after hatch, Aronsuu and Virkkala (2014) recommended that European river lamprey <8 mm total length should be outplanted in areas where substrate provides interstitial spaces, and indicated that gravel areas with low or moderate currents may offer the best option. Shelter is likewise important when rearing lampreys under laboratory or hatchery conditions, and the switch from the relatively sterile, clean hatchery tanks used for egg incubation to substrates that allow prolarvae to shelter and burrow needs to be carefully timed. Provision of a fiber mat or other material to shade the substrate is recommended to reduce prolarval activity and stress prior to the burrowing stage (Lampman et al. 2016).

Burrowing capabilities are developed by the last prolarval stage (Piavis 1961), and wild YOY larvae are typically found in fine silt and sand. Hence, when outplanting subyearling European river lamprey larvae >8 mm, Aronsuu and Virkkala (2014) recommended fine sediment with a high proportion of particles <125 μm. As larvae grow, they start to select slightly coarser material for burrowing. Numerous studies have shown that smaller substrate particle sizes are selected by the youngest larvae, while older larvae are able to occupy a broader range of sediment grain sizes (Morman et al. 1980; Sugiyama and Goto 2002; Quintella et al. 2007; Aronsuu and Virkkala 2014; Dawson et al. 2015; Alexa N. Maine, Confederated Tribes of the Umatilla Indian Reservation, personal communication, 2017).

Differences in habitat preference with body size may be related to burrowing abilities. Quintella et al. (2007) found that smaller sea lamprey larvae showed poorer burrowing performance than larger individuals across all substrate types tested, but particularly so in coarser substrates where, if particles are too large, they can impair burrowing. Similarly, in experiments with 85-day-old cultured larval Pacific lamprey, time to complete burrowing was significantly faster (66 s) in sand <149 μm in diameter than in coarser material (146 s) where particle size was 149–595 μm (Alexa N. Maine, personal communication, 2017). If young larvae are not provided with adequate substrate, they do not grow and can suffer increased mortality rates (Lampman et al. 2016). Kujawa et al. (2017) found that the survival rates and growth of subyearling European river lamprey larvae were much higher in tanks with sand substrate than without it (see Sect. 2.6.5).

2.6.4 Flow

Food delivery is an essential aspect of lamprey culture. Unlike other fish species, lampreys probably do not actively intercept food particles (Mallatt 1981; Malmqvist and Brönmark 1982). Hence, their culture is analogous to rearing of sessile invertebrates, such as mussels (Kamermans et al. 2013), abalone (Bouma 2007), or oysters (Jacob et al. 1993). The density of food particles, flow rate through tanks, and length of time that lampreys are exposed to food are important considerations. In the wild, lampreys likely have nearly continuous exposure to low levels of microalgae and detritus with occasional spikes in feeding after freshets or spates (Malmqvist and Brönmark 1982). Indeed, European brook lamprey respond to low food concentrations by increasing their filtration rate (Malmqvist and Brönmark 1982). However, if particle concentrations are too high (85–330 mg/L), the filtration apparatus of sea lamprey can become clogged (Mallatt 1981).

Larval lampreys have been cultured in completely flow-through systems as well as in recirculating and static flow systems. In flow-through systems, flow is often kept at a minimum or shut off during feeding times to give the food time to settle and to allow lampreys the opportunity to feed before it is swept away (Hanson et al. 1974; Mallatt 1983; Swink 1995; Barron et al. 2015, 2016). This method has resulted in very rapid growth in Pacific lamprey larvae (e.g., 34 mm at 71 days after hatching, Lampman 2017; 45 mm at 163 days after hatching, Barron et al. 2016). However, feeding rates reported for larvae reared in some recirculating and static systems are slower, perhaps because of the food quality, food delivery system, water quality, and/or larval lamprey density (Mallatt 1983; Murdoch et al. 1991, 1992; Rodríguez-Muñoz et al. 2003; see Sects. 2.6.5, 2.6.6 and 2.6.7).

2.6.5   Feed

Larval lampreys feed by trapping small, water-borne particles in mucus within the pharynx (Mallatt 1983), and the majority of the ingested materials are typically organic detritus (Mundahl et al. 2005). Lampreys can survive from this seemingly low quality food source primarily due to their high assimilation efficiency (Bowen 1993; Yap and Bowen 2003) and extremely low metabolic rates (Moore and Mallatt 1980; Sutton and Bowen 1994). Although organic matter/detritus is typically abundant in lamprey-bearing streams, lamprey growth is generally reduced when density is high (see Sect. 2.6.7). This is also true in the laboratory environment (Murdoch et al. 1992), although higher feeding rates can compensate for density effects to some extent (MacDonald 1963; Hanson et al. 1974; Moore and Potter 1976; Griffiths et al. 2001; Lampman et al. 2016; Kujawa et al. 2017; Schultz et al. 2017).

The key constituents of the larval lamprey diet has been a topic of interest and debate for decades (e.g., Applegate 1950; Potter et al. 1986), and a variety of studies have investigated this question (see reviews by Hardisty 2006; Aronsuu et al. 2015; Dawson et al. 2015). Many of these studies have described the importance of organic matter as substratum and habitat (Applegate 1950; Hardisty and Potter 1971; Potter et al. 1986; Beamish and Lowartz 1996), and some studies have gone further to describe the importance of organic matter as a food source (Hardisty and Potter 1971; Beamish and Jebbink 1994; Sutton and Bowen 1994; Shirakawa et al. 2009; Sutton and Bowen 2009; Smith et al. 2011). Others have highlighted the seasonal importance of other food ingredients, such as algae including diatoms and desmids (Potter et al. 1986; Quintella 2000) and microbes including biofilm (Bowen 1993; Yap and Bowen 2003). While some studies suggest that larval lampreys are not capable of digesting diatoms and bacteria efficiently enough to make them a primary food ingredient (Moore and Beamish 1973; Rogers et al. 1980), it is likely that these ingredients are important when they are available (Yap and Bowen 2003).

Our ability to readily recognize a variety of microorganisms (with diverse decomposition rates in streams and lamprey guts) are certainly not equal and this affects our ability to accurately identify key constituents of the larval lamprey diet (Hardisty 2006). In addition, detritus, organic matter, and biofilm can originate from, form alongside, and/or contain a wide variety of microorganisms simultaneously (e.g., bacteria, archaea, protozoa, phytoplankton, and fungi). Organic matter/detritus can also originate from both autochthonous and allochtonous sources, further complicating the elucidation of the larval lamprey diet.

As detritivores, larval lampreys live off organic matter breakdown, including the detrital fraction, fungi, and the myriad other microorganisms that exist within the detritus/biofilm complex (Moore and Beamish 1973; Sutton and Bowen 1994; Mundahl et al. 2005). A primary dietary criterion appears to be particle size. Particles in the range 5–340 μm are common in the guts of both small and large wild lamprey larvae (Moore and Mallatt 1980). Brewer's yeast and active dry yeast (cells of which are 5–10 μm in diameter) have been used successfully for feeding larval lampreys in the laboratory since at least the 1950s (e.g., Schroll 1959), even for prolonged periods of time (Hanson et al. 1974; Mallatt 1983; Rodríguez-Muñoz et al. 2003).

Development of optimal feeds for early larvae in the laboratory can provide a wealth of information on early larval feeding in wild lampreys. Larvae of many other fish species exhibit selection for preferred prey very early in their development (Robert et al. 2014). Although larval lampreys likely have less control than teleost fishes over the particles they ingest, laboratory investigations have indicated that there may be some selection that occurs on the basis of particle size and shape. Pacific lamprey larvae not yet feeding exogenously were provided with a diet of 80% yeast and 20% dry larval fish feed (Otohime A1) in static chambers held at 14 °C (Moser et al. 2017a). As soon as they started to feed, growth (in length) was apparent, and larvae provided with the smallest particle sizes (<50 μm) showed an early growth advantage relative to those provided with particles 50–150 μm (Fig. 2.13). In contrast, wild sea lamprey larvae showed no relationship between particle size and lamprey length (Moore and Mallatt 1980).

Cultured lampreys exhibited great variation in individual growth rates within treatment groups (Fig. 2.13), even when chambers were small (1 L) and variation in food encounter rates was minimized (Moser et al. 2017a). This suggests individual variation in filtering rates or metabolism. Evidence for high variation in individual growth has been observed for older dye-marked or PIT-tagged larvae in culture (Murdoch et al. 1992; Moser et al. 2017b), and in the wide range of larval sizes resulting from a single spawning event in the wild (Hess et al. 2015). Further study is needed to evaluate the mechanisms behind such variable growth.

Potential ontogenetic changes in lamprey nutrition has been hypothesized (Evans 2012), and recent artificial propagation research has also indicated that nutritional requirements of larval lampreys change as lamprey grow. Using Pacific lamprey, Barron et al. (2015) found that at 51 days post-hatch, artificially propagated larvae

Fig. 2.13 Mean length (mm) over time for larval Pacific lamprey spawned on 05 May 2016 that were fed starting on 21 June 2017 with three food particle size treatments (*diamonds* = small <50 μm; *squares* = medium 50–100 μm; *triangles* = large 100–150 μm). Error bars denote standard deviation (This figure was originally published in Moser et al. (2017a) and reproduced with permission of the authors.)

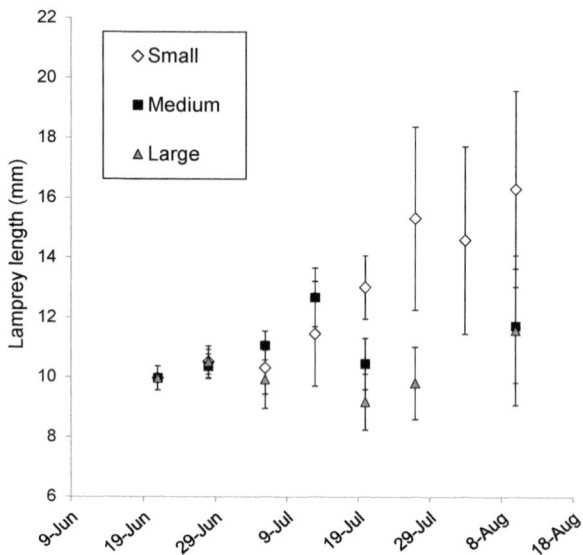

grew fastest and had the highest lipid retention when fed a diet of yeast supplemented with larval fish food (Otohime A1). In contrast, larvae in these experiments that were fed microalgae had relatively slow growth, even though algae and detritus are commonly found in the gut of wild specimens. Mallatt (1983) reared Pacific lamprey in the laboratory for >1 year on yeast alone and found that adding vitamins or switching to a commercial fish food did not improve growth or survival. Jolley et al. (2015) also experimented with larger wild-caught Pacific lamprey larvae (59–120 mm) and found that growth was highest under diets of algae wafers or salmon carcass analog pellets and that growth was lowest for larvae fed allochthonous detritus or yeast.

The contribution of marine-derived nutrients may be important for some lamprey populations. Many lamprey populations worldwide have experienced dramatic declines, often as the result of habitat degradation and the construction of dams that are barriers to migration (Maitland et al. 2015). In these areas, declines in co-occurring anadromous salmonids and sturgeons have also been observed (Jolley et al. 2015). Semelparous Pacific salmon are especially important sources of marine-derived nutrients (Naiman et al. 2002), and the loss of naturally occurring carcasses from these species may be further impacting native lampreys by reducing larval growth rates in these areas (Kucheryavyi et al. 2007; Jolley et al. 2015).

There appear to be interspecific differences in optimal feeds. Stable isotope studies of wild larval sea lamprey larvae indicated that they depend heavily on autochthonous sources of nutrition (i.e., algae, nutrition from aquatic sediments), with terrestrial plants being less important (Evans and Bauer 2016). In contrast, American brook lamprey larvae subsist almost exclusively on detritus (Mundahl et al. 2005), and cultured Arctic lamprey YOY similarly grew better (6.8 mm/30 days) on heated and sifted willow leaves than they did (<1 mm/30 days) on salmon carcass and control diets (Arakawa 2018). However, Pacific lamprey larvae provided with this same heated and sifted willow leaf feed exhibited negative growth (–1.5 mm/30 days; Ralph T. Lampman, unpublished data). Kujawa et al. (2017) achieved a very high growth rate (15 mm/30 days) in European river lamprey fed a mixture of live *Artemia salina* nauplii and dry feed (Hikari Plankton). Pacific lamprey YOY fed a mixture of yeast, wheat flour, and alfalfa pellets also attained a very high growth rate (14 mm/30 days; Lampman 2018).

Providing a sufficient ration is critical to achieving rapid growth in larval lamprey culture (Lampman et al. 2016). Encounter rates with food particles are also undoubtedly an important factor in wild populations. Perhaps this is best illustrated by the tight relationship between larval density in the field and very specific flow, depth, and substrate conditions (Morman et al. 1980; Potter et al. 1986; Sugiyama and Goto 2002; Torgersen and Close 2004; Nazarov et al. 2016). For lamprey culture work, ration is typically based on larval lamprey weight (Mallatt 1983). Lampman et al. (2016) demonstrated a strong logarithmic correlation (r = 0.881) between ration and growth rate. In these experiments, active dry yeast was the primary feed, constituting ~50% of the overall feed. A mixed feed ration of 10–20 g/week/fish weight (g) resulted in growth rates of 7–12 mm per month between late July and late September. In this same study, a positive linear relationship (r = 0.706) was also observed between ration per surface area and growth rate: a mixed feed of 400–700 g/m^2 resulted in growth rates ranging from 7.5 to 12 mm/month. Wild larval lamprey density is typically limited (0–45 g/m^2) even in preferred habitats (Silva et al. 2014a; Dawson et al. 2015; Beals and Lampman 2018); however, rapid growth of cultured lampreys held at high densities (100–217 g/m^2; see Sect. 2.6.7) can often be achieved by providing a high ration of yeast and supplemental feeds (Barron et al. 2015; Lampman 2017, 2018). These supplemental feeds (e.g., Otohime A1, wheat flour, brown rice flour, and alfalfa pellets) likely help promote the complex of other nutrients and microbes available in natural organic matter, detritus, and/or biofilm beyond those provided by yeast.

Larval lampreys can tolerate near starvation for extended periods. Although growth rates were negative, McGree et al. (2008) reported a high survival rate (~96%) in large larval Pacific lamprey (~2 g each) that were not fed anything but unfiltered creek water over the course of 3.5 months. Using dechlorinated tap water, death occurred after 7–8 months without food in Pacific lamprey larvae of a similar size (Mallatt 1983). However, in very young sea lamprey larvae, Hanson et al. (1974) found that feeding too little or not at all appeared to kill most of the larvae within 2 months. With Pacific lamprey, 5-month-old larvae reared using only well water without additional feed began dying after 60 days. Overall survival rate after 73 days was 51% in a tank with sand substrate and 73% in a tank with organic rich, fine substrate (Lampman et al. 2016). Despite the lack of supplemental feed, total length and weight still increased slightly (by 2.2–3.7 mm and 0–6 mg) in the surviving larvae. This was most likely due to organic content and bacteria available within the fine substrate provided (Nevejan et al. 2017).

2.6.6 Water Quality

Along with the need for adequate food and optimal physical habitat attributes, larval lampreys in the wild and in the laboratory show survival and growth effects related to water quality. Understanding these relationships is important for lamprey conservation in the face of increased human population growth and development, climate change, and the ever-changing field of environmental contaminants (Holmes 2011; Maitland et al. 2015; Nilsen et al. 2015). This is exemplified by recent studies to assess the effects of climate change on lamprey habitat, both in light of needs for lamprey conservation and to avoid unwanted consequences related to control of invasive sea lamprey (Macey and Potter 1978; reviewed in Griffiths et al. 2001 and Meeuwig et al. 2005; see Chap. 5). The ability to test growth and survival in propagated lamprey larvae allows assessment of these effects over a broad range of life history stages (Piavis 1961).

Lamprey larvae appear remarkably tolerant of high water temperatures. Potter and Beamish (1975) reported that upper incipient lethal temperatures for four lamprey species (Great Lakes sea lamprey and northern, American, and European brook lampreys) ranged from 27 to 31.4 °C depending on the season, acclimation state, and species. Pouched lamprey larvae were shown to survive temperatures up to 28.3 °C (Macey and Potter 1978), and incipient lethal temperature for Arctic lamprey larvae was estimated to be 29.3 °C (Arakawa 2018). The upper incipient lethal temperature for wild-caught Pacific lamprey larvae was 28.5 °C, and they were able to live indefinitely at 27 °C upon immediate transfer from water of 20 °C (Christina Uh, U.S. Fish and Wildlife Service, Vancouver, WA, personal communication, 2017). In addition, testing with younger Pacific lamprey larvae (<20 days post-hatch) indicated that they too were capable of surviving rapid thermal shocks (immediate transfer from 13 to 20 °C) and survived at 20 °C for >24 h (Moser et al. 2018). This has important implications for hatchery management, as lamprey larvae can survive power outages and short-term water quality changes better than salmonid hatchery residents.

Larval lampreys are generally also tolerant of degraded water quality, both in the wild and in laboratory environments (Bettaso and Goodman 2010; Linley et al. 2016; Moser et al. 2017a). Larval sea and Pacific lampreys, for example, have been reported in lagoons contaminated with untreated municipal sewage or pollution abatement ponds (Morman et al. 1980; Nelson and Nelle 2007). Both wild-caught Pacific lamprey larvae older than 1 year and artificially propagated larvae younger than 30 days post-hatch were able to tolerate abrupt, short-term exposures to salinities below 14 parts per thousand (ppt) and disinfectant concentrations of formalin at 15 mg/L (Silver 2015; Maine et al. 2018). Older larvae were also able to survive for up to 12 h in full-strength sea water (35 ppt), not entirely surprising in light of the fact that lampreys in tidally dominated rivers and streams may be exposed to oscillating salinity regimes throughout the rearing period (Silver 2015). On the other hand, excessive eutrophication and other forms of pollution can have negative effects on larval lampreys (see Maitland et al. 2015), and water of low pH (<5) combined with high metal concentrations has been shown to increase mortality of subyearling European river lamprey larvae (Myllynen et al. 1997).

Sensitivity to water quality can change dramatically with growth and development (Rodríguez-Muñoz et al. 2001). For example, Pacific lamprey prolarvae (2, 5, and 11 days post-hatch) transported without aeration for 7- and 24-h periods exhibited nearly 100% survival if they were maintained in the water supply used for transport. However, nearly all died when they were transitioned to a new water supply after transport (Lampman et al. 2016). In contrast, a similar change in the water supply did not appear to affect younger (developing eggs) or older (>60 days old) stages (Alexa N. Maine, Confederated Tribes of the Umatilla Indian Reservation, personal communication, 2018). This suggests that prior to and during the transition to exogenous feeding, lamprey larvae may be sensitive to water quality changes.

However, these same stages of first feeding lampreys are resilient to stable but low dissolved oxygen (<2 mg/L) and high un-ionized ammonia (Galloway et al. 1987; Barron et al. 2017; Mary L. Moser, unpublished data). Oxygen consumption rates by mountain brook lamprey *Ichthyomyzon greeleyi* larvae (Hill and Potter 1970) and Pacific lamprey prolarvae (Mary L. Moser, unpublished data) were lower than those of larval teleosts (Winberg 1956). Potter et al. (1970) observed that larval lampreys can tolerate oxygen tensions as low as 7–10 mmHg at 5 °C, 12–16 mmHg at 15.5 °C, and 13–21 mmHg at 22.5 °C for up to 4 days. This may be an adaptation needed for proliferation of dense lamprey beds in low-flow pool habitats and silt banks (Hill and Potter 1970).

2.6.7 *Larval Culture Density, Growth, and Survival*

There is large variation in larval densities reported during field investigations, ranging from hundreds to thousands of larvae per m^2 to <1 individual per m^2 (e.g., Churchill 1945; Kainua and Voltanten 1980; Kelso and Todd 1993; Griffiths et al. 2001; Jellyman and Glova 2002; see Dawson et al. 2015). However, this is likely due to a combination of the spatial scale measured, gear selectivity, environmental variation, and larval supply and size. Larval lampreys can exploit even very small patches of suitable habitat (Thomas 1962; Malmqvist 1980; Nazarov et al. 2016) and, in optimal habitats, particularly as YOY, densities of up to 2,000 larvae/m^2 have been reported (e.g., Churchill 1945; Tuunainen et al. 1980). Evidence for a negative effect of density on larval growth rate in the wild is inconsistent (e.g., Morman 1987; Zerrenner 2004), but in the laboratory, lampreys exhibit density-dependence under a variety of holding conditions. Growth suppression has been observed in sea lamprey larvae held in the laboratory for 8 months at high densities, and larvae even shrank in length at the highest density (~345 larvae or 500 g/m^2; Murdoch et al. 1992). Interestingly, Rodríguez-Muñoz et al. (2003) also observed reduced growth rates in sea lamprey larvae exposed to water from high-density tanks. In re-circulating systems, Pacific lamprey prolarvae cultured at densities above 800 larvae/m^2 (~2 g/m^2) exhibited reduced growth and survival (Maine et al. 2017). However, density effects have also been observed in flow-through systems where lamprey were held at densities higher than 100–130 g/m^2 (Lampman et al. 2016; Lampman 2017). Recently, Bowen and Yap (2018) demonstrated, using field surveys and in situ cage studies, that food utilization (e.g., feeding rate and assimilation efficiency) of northern brook lamprey larvae decreased with increasing larval density. These authors suggested that crowding results in physical disturbance of the sediment, thus interfering with the larvae's efficient utilization of available food. This would explain growth depression observed at high densities even when food is thought not to be limiting (e.g., Murdoch et al. 1992); its implications to artificial propagation warrant further study.

Understanding the limits to lamprey growth under high density is important for production-level culture, but also for assessment of habitat carrying capacity in lamprey conservation and control (Griffiths et al. 2001; Zerrenner 2004; Johnson et al. 2014). Where examined, the growth rate of recently hatched larval lampreys in streams during their first year appears similar across species, ranging from 0.06 to 0.18 mm/day in sea and chestnut lampreys, as well as in American brook and northern brook lampreys (Purvis 1970; Holt and Durkee 1983; Griffiths et al. 2001; Evans 2017), and even in the more distantly related pouched lamprey (Todd and Kelso 1993). In contrast, larval Pacific lamprey in culture can be reared at growth rates ranging from 0.25 to 0.55 mm/day (Barron et al. 2015; Lampman 2017; Fig. 2.12). In the wild, specific growth rates of larval lampreys are most rapid during the first 2 years of life and slow when larvae approach ~80 mm in length (Hardisty 1961a; Kan 1975; Purvis 1979; Potter 1980; Morman 1987; Murdoch et al. 1991; Weise and Pajos 1998; Griffiths et al. 2001; Quintella et al. 2003; see Dawson et al. 2015). A similar pattern has been observed with artificially propagated larvae (Lampman et al. 2016).

In nature, lampreys can suffer high rates of mortality during early development (see Dawson et al. 2015). Once again, few empirical data are available for comparison among field-based studies, but hatching success rates from natural nests of sea lamprey in the Great Lakes were found to be as low as 0.4–1.1% (Applegate 1950) and 5.3–7.8% (Manion 1968). Predation on fertilized eggs and prolarvae by a range of organisms is also likely to be high (Schultz 1930; Dendy and Scott 1953; Hardisty 1961a, b; Heard 1966; Manion 1968; Potter 1980; Derosier 2001). Kujawa et al. (2017) noted that only 1% of European river lamprey likely survive to metamorphosis in the wild. In artificially propagated Pacific lamprey, a bottleneck to survival is observed at first feeding (Lampman et al. 2016). Hence, examination of the factors affecting early survival has been a top priority in recent Pacific lamprey culture research (Lampman et al. 2016; Barron et al. 2017; Lampman 2017; Maine et al. 2017).

Despite this early survival bottleneck and their apparent vulnerability to predation, larval lampreys have exceedingly high survival rates after 2 months of age. Survival rates of YOY after initiation of exogenous feeding have been estimated at 0.64–0.81 in western brook lamprey (Schultz et al. 2017), 0.44–0.95 in sea lamprey from the Great Lakes (Jones et al. 2009, 2015; Robinson et al. 2013; Johnson et al. 2014), and 0.81–0.95 in sea lamprey from Lake Champlain (Zerrenner 2004; Howe et al. 2012).

Similarly high survival rates have been observed for artificially propagated Pacific lamprey larvae in these age classes (Lampman et al. 2016).

2.6.8 Metamorphosis

Arctic lamprey have been successfully reared from embryos to metamorphosis after only 2–4 years in culture (Kataoka 1985). Researchers in Niigata, Japan, reared Arctic lamprey to metamorphosis by releasing artificially propagated larvae into a large concrete rearing pond (Kataoka et al. 1980a, b; Kataoka and Hoshino 1983; Kataoka 1985) that received ambient river water ranging in temperature from 2 °C in winter up to 24 °C in summer (Kataoka and Hoshino 1983). Of the 8,400 first-feeding larvae released in 1980 (density of 518 larvae/m^2), 7.2% remained after 315 days (Kataoka et al. 1980b). The first larvae showing early signs of metamorphosis (i.e., eyes beginning to appear behind the epidermis) were observed after 25 months, and metamorphosed juveniles were collected in year 3 and 4 as well, totaling 399 (Kataoka and Hoshino 1983; Kataoka 1985). These researchers estimated that 10–20% of the larvae transformed at ~2.3 years old, 30–40% at 3.3 years old, and 40–50% at 4.3 years old. Larvae consistently metamorphosed in August and September each year, in a usually very narrow span of time (Kataoka 1985). This is consistent with the highly synchronized metamorphosis observed in natural populations (see Manzon et al. 2015). Newly metamorphosed juveniles were ≥150 mm in length and all larvae >180 mm underwent metamorphosis, which is consistent with size at metamorphosis in natural populations of this species (see Docker 2009). At the beginning of year 4, all lamprey were transferred to smaller aquaria receiving well water at 11–13 °C. A high rate of metamorphosis occurred in year 4, indicating that water temperature fluctuation was not a key factor in triggering metamorphosis (see Manzon et al. 2015). A modest amount of freshwater eel feed (e.g., containing fish meal, pregelatinized starch, wheat flour) was fed in this new tank setting due to the lack of river water. Most of the resulting juveniles successfully parasitized salmonids and cyprinids introduced into the freshwater tanks.

More recently, artificially propagated Pacific lamprey have been observed to metamorphose at the Prosser Fish Hatchery in Prosser, Washington (Ralph T. Lampman, unpublished data). To date, metamorphosis has been observed in ~25 individuals. The youngest juvenile recorded (1.3 years old) was 108 mm in length. Size at metamorphosis in wild Pacific lamprey is typically slightly larger than this (e.g., 108–136 mm and 112–135 mm; Pletcher 1963 and Beamish and Levings 1991, respectively), and wild Pacific lamprey are known to metamorphose in as little as 3.3 years. For example, juveniles were collected from Indian Creek in western Washington in summer 2016, after adults first spawned there in spring 2013 following removal of the Elwha Dam (Moser and Paradis 2017). Genetic parentage analysis indicated that the offspring of translocated Pacific lamprey outmigrated 3–9 years after the adults were released into natural streams (Jon E. Hess, Columbia River Inter-Tribal Fish Commission, Portland, OR, personal communication, 2017). A similar wide range of ages at metamorphosis was observed in a single year class of sea lamprey larvae isolated above a barrier (Manion and Smith 1978), but such individual variation generally goes unnoticed when known-age populations cannot be examined (see Chap. 7). Through a combination of field and artificial propagation research and an assortment of new technologies, we anticipate that our understanding of lamprey metamorphosis and the juvenile life stage will expand considerably in the near future.

2.7 Conclusions

We have made great strides in understanding lamprey biology as well as early development in vertebrates as a result of increased efficiency in artificial fertilization, observation of embryonic development, and larval rearing of different species in the laboratory. Substantial contributions to our understanding of early life history have come from recent studies of lamprey genetic programming (Bryant et al. 2016; Timoshevskiy et al. 2016) and of anatomical features throughout their ontogeny (Kusakabe and Kuratani 2005; Amemiya et al. 2007; Khonsari et al. 2009; Richardson et al. 2010; Kuratani 2012; Green and Bronner 2014; Suzuki et al. 2015; see Chap. 6). Thus, it is rightly the case that lampreys are considered a model organism in biology (Nikitina et al. 2009; Shimeld and Donoghue 2012; Xu et al. 2016; see Docker et al. 2015).

Knowledge gained from the development of artificial propagation methods has provided tools for restoration of imperiled lamprey species and potentially for efforts to control invasive sea lamprey in the Laurentian Great Lakes. In many parts of the world, lampreys are declining or have been extirpated from their native range (Maitland et al. 2015). The ability to produce larvae from viable donor stocks could potentially lead to outplanting of artificially reared lampreys in streams where they have been extirpated (Ward et al. 2012; Clemens 2017). These efforts are already underway in Finland, where production of European river lamprey larvae was initiated in the 1980s as mitigation for losses in lamprey recruitment due to dam operations (Aronsuu 2015). However, this decades-long research program indicates that lamprey culture is costly and may not return the same benefits as habitat improvements and/or aids to adult passage (Aronsuu 2015). Control of sea lamprey in the Great Lakes has traditionally relied on use of pesticides; however, there is increasing pressure to investigate alternative control methods (see Chap. 5). Should development of lamprey embryos prove manipulable and ethically tolerable (e.g., gene knockdown, sex ratio distortion), these techniques may become important tools in modern control efforts (McCauley et al. 2015; see Chap. 7).

While great advances have been made in obtaining reliable methodologies for gamete collection, incubation, and fertilization, there is still much to learn regarding the requirements for large-scale rearing of early larvae. One of the most challenging aspects of lamprey production is the necessity for provision of burrowing substrate and the attendant problems of space required for long-term lamprey culture. This two-dimensional aspect makes rearing even a single cohort challenging (Lampman et al. 2016). Research is needed to investigate the potential for polyculture, water re-use and tank stacking, or other methods to make more efficient use of space.

The need for sediment in larval lamprey rearing also brings problems associated with culture cleanliness and disease prevention. Lampreys can apparently survive relatively poor water quality, and frequent disturbance from tank cleaning can reduce growth (Barron et al. 2017). Understanding the interplay among these factors is critical if lamprey culture operations are to be scaled up. Moreover, very little is known about the microbial requirements of larval lampreys. This has become an important area of interest in finfish aquaculture (Ringø and Song 2016), and may be of even more importance for lampreys. Due to their simple intestinal tract, apparently passive feeding mode, and continuous contact with sediment, lampreys are likely very sensitive to the microbial environment both in culture and in the wild. Studies on the gut microbiome of lampreys may provide important clues to improving growth and survival of artificially reared larvae (Tetlock et al. 2012; Zuo et al. 2017; Arakawa 2018; Alexa N. Maine, Confederated Tribes of the Umatilla Indian Reservation, Pendleton, OR, personal communication, 2018).

Very little is known of the disease organisms specific to lampreys (Maitland et al. 2015; Moser et al. 2016). While lampreys appear relatively insensitive to the pathogens typically associated with salmonid culture, little is known about their vulnerabilities to other pathogens. Widespread disease screening of lampreys has not been conducted, and methods for assessing lamprey diseases and parasites have not been standardized or orchestrated (Moser et al. 2016). Those studies that examine parasites or other pathogens in lampreys generally focus on their potential role in disease transmission or impact to human health rather than their effect on the lampreys themselves (e.g., Gadd et al. 2010; Bao et al. 2013). Greater attention to this topic will be required as efforts to hold lampreys in dense cultures are undertaken.

Holding lampreys in the laboratory provides a rare opportunity to document many aspects of basic biology, such as feeding, growth, and survival rates, as well as to examine fine-scale patterns of larval physiology, behavior, and genetic control. Insights gained from these observations can be used to inform management of lampreys in the wild. For example, a better understanding of the relationship between food quality and growth or production may help to delineate habitat characteristics that are most important for lamprey conservation or control. The role of temperature and potential effects of climate change on lamprey populations can be assessed using lamprey larvae held under controlled conditions. Comparisons of wild and artificially produced lamprey performance may afford insight into underlying recruitment mechanisms. Such comparisons will also help to elucidate cases where artificially produced lampreys are suitable surrogates for wild lampreys or where extrapolations to natural populations must be made cautiously.

Finally, large-scale artificial propagation and long-term rearing methods have been developed for only a few species, notably in the large-bodied Pacific lamprey, European river lamprey, sea lamprey, and Arctic lamprey. Information gathered thus far indicates that there is room for broad generalization across species, but increasing the scope of lamprey propagation to include other species will undoubtedly provide valuable information on the differences among species and perhaps also on the determinants of sex, growth, metamorphosis, and/or feeding and migratory type in lampreys.

Acknowledgements This review would not have been possible without the support of the follow-ing agencies and personnel: Chelan Public Utility District (S. Hemstrom), the Bonneville Power Administration (D. Docherty), and the U.S. Bureau of Reclamation (S. Camp). We thank D. Dey, J. Butzerin, K. Frick, and J. Jolley for their comments on a draft manuscript.

References

Abou-Seedo FS, Potter IC (1979) The estuarine phase in the spawning run of the river lamprey *Lampetra fluviatilis*. J Zool 188:5–25

Almeida PR, Quintella BR, Dias NM (2002) Movement of radio-tagged anadromous sea lamprey during the spawning migration in the River Mondego (Portugal). Hydrobiologia 483:1–8

Amemiya CT, Saha NR, Zapata A (2007) Evolution and development of immunological structures in the lamprey. Curr Opin Immunol 19:535–541

Applegate VC (1950) Natural history of the sea lamprey (*Petromyzon marinus*) in Michigan. US Fish Wildl Serv Spec Sci Rep Fish 55:1–237

Applegate VC (1961) Downstream movement of lampreys and fishes in the Carp Lake River, Michigan. US Fish Wildl Serv Spec Sci Rep Fish 387:1–71

Arakawa H (2018) Investigating causes for the decline and examining conservation methods of the endangered species Arctic Lamprey (*Lethenteron japonicum*) inhabiting the Noto Peninsula. MS thesis, Ishikawa Prefectural University, Ishikawa, Japan [in Japanese]

Ardavin CF, Zapata A (1988) The pharyngeal lymphoid tissue of lampreys. A morpho-functional equivalent of the vertebrate thymus? Thymus 11:59–65

Aronsuu K (2015) Lotic life stages of the European river lamprey (*Lampetra fluviatilis*): anthropogenic detriment and rehabilitation. PhD thesis, University of Jyväskylä, Jyväskylä, Finland

Aronsuu K, Tertsunen J (2015) Selection of spawning substratum by European river lampreys (*Lampetra fluviatilis*) in experimental tanks. Mar Freshw Behav Physiol 48:41–50

Aronsuu K, Virkkala P (2014) Substrate selection by subyearling European river lampreys *(Lampetra fluviatilis)* and older larvae (*Lampetra* spp). Ecol Freshw Fish 23:644–655

Aronsuu K, Marjomäki TJ, Tuohino J et al (2015) Migratory behaviour and holding habitats of adult river lampreys (*Lampetra fluviatilis*) in two Finnish rivers. Boreal Environ Res 20:120–144

Bajoghli B, Guo P, Aghaallaei N et al (2011) A thymus candidate in lampreys. Nature 470:90–94

Baker CF, Jellyman DJ, Reeve K et al (2017) First observations of spawning nests in the pouched lamprey (*Geotria australis*). Can J Fish Aquat Sci 74:1603–1611

Bao M, Garci ME, Antonio JM, Pascual S (2013) First report of *Anisakis simplex* (Nematoda, Anisakidae) in the sea lamprey (*Petromyzon marinus*). Food Control 33:81–86

Barron JM, Twibell RG, Hill HA, Hanson KC, Gannam AL (2015) Development of diets for the intensive culture of Pacific lamprey. Aquacult Res 2015:1–8

Barron J, Twibell R, Gannam A, Hanson K (2016) Development of artificial propagation methods for production of juvenile Pacific lamprey (*Entosphenus tridentatus*) for the use in research associated with Section 4.2.3 of the Rocky Reach Pacific Lamprey Management Plan. Annual Report for Contract No. 15-105 to the Chelan County Public Utility District, Wenatchee, WA

Barron J, Hawke K, Twibell R, Gannam A (2017) Development of artificial propagation methods for production of juvenile Pacific lamprey (*Entosphenus tridentatus*) for the use in research associated with Section 4.2.3 of the Rocky Reach Pacific Lamprey Management Plan. Annual Report for Contract No. 15-105 to the Chelan County Public Utility District, Wenatchee, WA

Beals T, Lampman R (2018) Intensive monitoring of larval lamprey and associated habitat in Columbia River tributaries within Yakama Nation Ceded Lands in Washington State. Appendix in

2018 Annual Progress Report from the Yakama Nation Pacific Lamprey Project to the Bonneville Power Administration (Project No. 2008-470-00), Portland, OR

Beamish FWH, Jebbink J (1994) Abundance of lamprey larvae and physical habitat. Environ Biol Fish 39:209–214

Beamish FWH, Lowartz S (1996) Larval habitat of American brook lamprey. Can J Fish Aquat Sci 53:693–700

Beamish RJ, Neville CE (1992) The importance of size as an isolating mechanism in lampreys. Copeia 1992:191–196

Beamish FWH, Potter IC (1975) The biology of the anadromous sea lamprey *Petromyzon marinus* in New Brunswick. J Zool Lond 177:52–72

Beamish RJ, Levings CD (1991) Abundance and freshwater migrations of the anadromous parasitic lamprey, *Lampetra tridentata*, in a tributary of the Fraser River, British Columbia. Can J Fish Aquat Sci 48:1250–1263

Bennett RM, Ross RM (1995) Stream drift of newly hatched sea lampreys in the upper Delaware River. J Pa Acad Sci 69:7–9

Bettaso JB, Goodman DH (2010) A comparison of mercury contamination in mussel and ammocoete filter feeders. J Fish Wildl Manag 1:142–145

Binder TR, McDonald DG (2008a) The role of temperature in controlling diel activity in upstream migrant sea lampreys (*Petromyzon marinus*). Can J Fish Aquat Sci 65:113–1121

Binder TR, McDonald DG (2008b) The role of dermal photoreceptors during the sea lamprey (*Petromyzon marinus*) spawning migration. J Comp Physiol 194A:921–928

Binder TR, McLaughlin RL, McDonald DG (2010) Relative importance of water temperature, water level, and lunar cycle to migratory activity in spawning-phase sea lampreys in Lake Ontario. Trans Am Fish Soc 139:700–712

Boorman CJ, Shimeld SM (2002) Cloning and expression of a Pitx homeobox gene from the lamprey, a jawless vertebrate. Dev Genes Evol 212:349–353

Bouma JV (2007) Early life history dynamics of pinto abalone (*Haliotis kamtschatkana*) and implications for recovery in the San Juan archipelago, Washington State. MS thesis, University of Washington, Seattle, WA

Bowen AK (1993) Identification of the nutritional resource supporting growth in larvae of two lamprey species in the Great Lakes Basin. Great Lakes Fishery Commission Completion Report, Ann Arbor, MI

Bowen AK, Weisser JW, Bergstedt RA, Famove F (2003) Response of larval sea lampreys (*Petromyzon marinus*) to pulsed DC electrical stimuli in laboratory experiments. J Great Lakes Res 29(Suppl 1):174–182

Bowen S, Yap MR (2018) Crowding reduces feeding rate, effectiveness of diet selection, and efficiency of digestion by northern brook lamprey ammocoetes. Environ Biol Fish 101:1385–1394

Brumo AF (2006) Spawning, larval recruitment, and early life survival of Pacific lampreys in the South Fork Coquille River, Oregon. MS thesis, Oregon State University, Corvallis, OR

Bryant SA, Herdy JR, Amemiya CT, Smith JJ (2016) Characterization of somatically-eliminated genes during development of the sea lamprey (*Petromyzon marinus*). Mol Biol Evol 33:2337–2344

Case B (1970) Spawning behaviour of chestnut lamprey (*Ichthyomyzon castaneus*). J Fish Res Board Can 27:1872–1874

Cejko BI, Judycka S, Kujawa R (2016) The effect of different ambient temperatures on river lamprey (*Lampetra fluviatilis*) egg and sperm production under controlled conditions. J Thermal Biol 62:70–85

Cerny R, Cattell M, Sauka-Spengler T et al (2010) Evidence for the prepattern/cooption model of vertebrate jaw evolution. Proc Natl Acad Sci USA 107:17262–17267

Chang M-M, Wu F, Miao D, Zhang J (2014) Discovery of fossil lamprey larva from the Lower Cretaceous reveals its three-phased life cycle. Proc Natl Acad Sci USA 111:15486–15490

China V, Holtzman R (2014) Hydrodynamic starvation in first-feeding larval fishes. Proc Natl Acad Sci USA 11:8083–8088

Churchill WS (1945) The brook lamprey in the Brule River. Trans Wis Acad Sci Arts Lett 37:337–346

Ciereszko A, Glogowski J, Dabrowski K (2000) Fertilization in landlocked sea lamprey: storage of gametes, optimal sperm: egg ratio, and methods of assessing fertilization success. J Fish Biol 56:495–505

Ciereszko A, Dabrowski K, Toth GP, Christ SA, Glogowski J (2002) Factors affecting motility characteristics and fertilizing ability of sea lamprey spermatozoa. Trans Am Fish Soc 131:193–202

Clemens BJ (2011) The physiological ecology and run diversity of adult Pacific lamprey, *Entosphenus tridentatus*, during the freshwater spawning migration. PhD thesis, Oregon State University, Corvallis, OR

Clemens BJ (2017) Progress report: Miller Lake Lamprey. 2017 progress report, Oregon Department of Fish and Wildlife, Corvallis, OR

Clemens B, van de Wetering S, Kaufman J, Holt RA, Schreck CB (2009) Do summer temperatures trigger spring maturation in Pacific Lamprey, *Entosphenus tridentatus*? Ecol Freshw Fish 18:418–426

Clemens BJ, Binder TR, Docker MF, Moser ML, Sower SA (2010) Similarities, differences, and unknowns in biology and management of three parasitic lampreys of North America. Fisheries 35:580–594

Clemens BJ, Mesa MG, Magie RJ, Young DA, Schreck CB (2012) Pre-spawning migration of adult Pacific lamprey, *Entosphenus tridentatus*, in the Willamette River, Oregon, U.S.A. Environ Biol Fish 93:245–254

Close DA, Fitzpatrick MS, Li HW (2002) The ecological and cultural importance of a species at risk of extinction, Pacific lamprey. Fisheries 27:19–25

Cochran PA, Bloom DD, Wagner RJ (2008) Alternative reproductive behaviors in lampreys and their significance. J Freshw Ecol 23:437–444

Cochran PA, Ross MA, Walker TS, Biederman T (2012) Early spawning by the American brook lamprey (*Lethenteron appendix*) in southeastern Minnesota. Can Field-Nat 126:204–209

Collin SP (2010) Evolution and ecology of retinal photoreception in early vertebrates. Brain Behav Evol 75:174–185

Cooper MD, Alder MN (2006) The evolution of adaptive immune systems. Cell 124:815–822

Dabrowski K (1984) The feeding of fish larvae: present "state of the art" and perspectives. Reprod Nutr Dev 24:807–833

Damas H (1944) Recherches sur le dévelopment de *Lampetra fluviatilis* L. Contribution à l'étude de la céphalogenèse des vertébrés. Arch Biol 55:5–284

Dawson HA, Quintella BR, Almeida PR, Treble AJ, Jolley JC (2015) The ecology of larval and metamorphosing lampreys. In: Docker MF (ed) Lampreys: biology, conservation and control, vol 1. Springer, Dordrecht, pp 75–137

Dendy JS, Scott DC (1953) Distribution, life history, and morphological variations of the southern brook lamprey, *Ichthyomyzon gagei*. Copeia 1953:152–162

Derobert Y, Baratte B, Lepage M, Mazan S (2002) *Pax6* expression patterns in *Lampetra fluviatilis* and *Scyliorhinus canicula* embryos suggest highly conserved roles in the early regionalization of the vertebrate brain. Brain Res Bull 57:277–280

Derosier AL (2001) Early life history of sea lamprey: emergence, movements, and dispersal. MS thesis, Michigan State University, East Lansing, MI

Derosier AL, Jones ML, Scribner KT (2007) Dispersal of sea lamprey larvae during early life: relevance for recruitment dynamics. Environ Biol Fish 78:271–284

Docker MF (1992) Labile sex determination in lampreys: the effect of larval density and sex steroids on gonadal differentiation. PhD thesis, University of Guelph, Guelph, ON

Docker MF (2009) A review of the evolution of nonparasitism in lampreys and an update of the paired species concept. In: Brown LR, Chase SD, Mesa MG, Beamish RJ, Moyle PB (eds) Biology, management, and conservation of lampreys in North America. Am Fish Soc Symp 72:71–114

Docker MF, Beamish FWH (1994) Age, growth, and sex ratio among populations of least brook lamprey, *Lampetra aepyptera*, larvae: an argument for environmental sex determination. Environ Biol Fishes 41:191–205

Docker MF, Mandrak NE, Heath DD (2012) Contemporary gene flow between "paired" silver (*Ichthyomyzon unicuspis*) and northern brook (*I. fossor*) lampreys: implications for conservation. Conserv Genet 134:823–835

Docker MF, Hume JB, Clemens BJ (2015) Introduction: a surfeit of lampreys. In: Docker MF (ed) Lampreys: biology, conservation and control, vol 1. Springer, Dordrecht, pp 1–34

Docker MF, Silver GS, Jolley JC, Spice EK (2016) Simple genetic assay distinguishes lamprey genera *Entosphenus* and *Lampetra*: comparison with existing genetic and morphological identification methods. N Am J Fish Manag 36:780–787

Dunham JB, Chelgren ND, Heck MP, Clark SM (2013) Comparison of electrofishing techniques to detect larval lampreys in wadeable streams in the Pacific Northwest. N Am J Fish Manag 33:1149–1155

Evans TM (2012) Assessing food and nutritional resources of native and invasive lamprey larvae using natural abundance isotopes. MS thesis, The Ohio State University, Columbus, OH

Evans TM (2017) Measuring the growth rate in three populations of larval lampreys with mark-recapture techniques. Trans Am Fish Soc 146:147–159

Evans TM, Bauer JE (2015) Using stable isotopes and C:N ratios to examine the life-history strategies and nutritional sources of larval lampreys. J Fish Biol 88:638–654

Evans TM, Bauer JE (2016) Identification of the nutritional resources of larval sea lamprey in two Great Lakes tributaries using stable isotopes. J Great Lakes Res 42:99–107

Feng B, Zhang T, Wu F, Chen S, Xu A (2018) Artificial propagation and embryonic development of Yalu River lamprey, *Lampetra morii*. Acta Biochim Biophys Sin 50:828–830

Fixler S (2017) Hyporheic flow possibilities within lamprey (*Petromyzon marinus*) redds on the Blackledge River in Marlborrough, Connecticut. Honors thesis, Connecticut College, New London, CT

Fredricks KT, Seelye JG (1995) Flowing recirculated-water system for inducing laboratory spawning of sea lampreys. Prog Fish-Cult 57:297–301

Fujimoto T, Takaoka K (1960) Artificial insemination and the early development of the Planeri Lamprey. Okayama University Medical School, Department of Anatomy, Okayama, Japan [in Japanese]

Fujii T, Nakagawa H, Murakawa S (1979) Immunity in lamprey II. Antigen-binding responses to sheep erythrocytes and hapten in the ammocoete. Dev Comp Immunol 3:609–620

Fukutomi N, Nakamura T, Doi T, Takeda K, Oda N (2002) Records of *Entosphenus tridentatus* from the Naka River system, central Japan: physical characteristics of possible spawning redds and spawning behavior in the aquarium. Jap J Ichthyol 49:53–59 [in Japanese]

Gadd T, Jakava-Viljanen M, Einer-Jensen K et al (2010) Viral haemorrhagic septicaemia virus (VHSV) genotype II isolated from European river lamprey *Lampetra fluviatilis* in Finland during surveillance from 1999 to 2008. Dis Aquat Org 88:189–198

Galloway R, Potter IC, Macey DJ, Hilliard RW (1987) Oxygen consumption and responses to hypoxia of ammocoetes of the southern hemisphere lamprey *Geotria australis*. Fish Physiol Biochem 4:63–72

Gamper N, Savina MV (2000) Reversible metabolic depression of hepatocytes in lamprey (*Lampetra fluviatilis*) during pre-spawning: regulation by substrate availability. Comp Biochem Physiol 127B:147–154

Gans C, Northcutt RG (1983) Neural crest and the origin of vertebrates: a new head. Science 220:268–273

Gardner C, Coghlan SM Jr, Zydlewski J (2012) Distribution and abundance of anadromous Sea Lamprey spawners in a fragmented stream: current status and potential range expansion following barrier removal. Northeast Nat 19:99–110

Gess RW, Coates MI, Rubidge BS (2006) A lamprey from the Devonian period of South America. Nature 443:981–984

Goodman DH, Kinziger AP, Reid SB, Docker MF (2009) Morphological diagnosis of *Entosphenus* and *Lampetra* ammocoetes (Petromyzontidae) in Washington, Oregon, and California. In: Brown LR, Chase SD, Mesa MG, Beamish RJ, Moyle PB (eds) Biology, management, and conservation of lampreys in North America. Am Fish Soc Symp 72:223–232

Green SA, Bronner ME (2014) The lamprey: a jawless vertebrate model system for examining origin of the neural crest and other vertebrate traits. Differentiation 87:44–51

Griffiths RW, Beamish FWH, Morrison BJ, Barker LA (2001) Factors affecting larval sea lamprey growth and length at metamorphosis in lampricide-treated streams. Trans Am Fish Soc 130:289–306

Gunckel SL, Jones KK, Jacobs SE (2009) Spawning distribution and habitat use of adult Pacific and western brook lampreys in Smith River, Oregon. In: Brown LR, Chase SD, Mesa MG, Beamish RJ, Moyle PB (eds) Biology, management, and conservation of lampreys in North America. Am Fish Soc Symp 72:173–189

Hagelin L-O (1959) Further aquarium observations on the spawning habits of the river lamprey (*Petromyzon fluviatilis*). Oikos 10:50–64

Hagelin L-O, Steffner (1958) Notes on the spawning habits of the river lamprey (*Petromyzon fluviatilis*). Oikos 9:221–238

Hanson LH, King EL Jr, Howell JH, Smith AJ (1974) A culture method for sea lamprey larvae. Prog Fish-Cult 36:122–128

Hardisty MW (1961a) The growth of larval lampreys. J Anim Ecol 30:357–371

Hardisty MW (1961b) Studies on an isolated spawning population of the brook lamprey (*Lampetra planeri*). J Anim Ecol 30:339–355

Hardisty MW (2006) Life without jaws. Forrest Text, Ceredigion, UK

Hardisty MW, Potter IC (1971) The behaviour, ecology and growth of larval lampreys. In: Hardisty MW, Potter IC (eds) The biology of lampreys, vol 1. Academic Press, London, pp 85–125

Hatta S (1893) On the formation of the germinal layers in Petromyzon. J Coll Sci Imp Univ Tokyo 5:129–148

Hatta S (1900) Contributions to the morphology of Cyclostomata. II: The development of pronephros and segmental duct in Petromyzon. J Coll Sci Imp Univ Tokyo 13:311–425

Hatta S (1907) On the gastrulation in Petromyzon. J Coll Sci Imp Univ Tokyo 21:3–44

Hatta S (1923) Über die Entwicklung des Gefäßsystems des Neunauges, *Lampetra mitsukurii* Hatta. Zool Jahr Anat 4:1–264

Heard WR (1966) Observations on lampreys in the Naknek River system of southwest Alaska. Copeia 1966:332–339

Hess JE, Campbell NR, Docker MF et al (2015) Use of genotyping-by-sequencing data to develop a high-throughput and multi-functional SNP panel for conservation applications in Pacific lamprey. Mol Ecol Resour 15:187–202

Hill BJ, Potter IC (1970) Oxygen consumption in ammocoetes of the lamprey *Ichthyomyzon hubbsi* Raney. J Exp Biol 53:47–57

Hjort J (1914) Fluctuations in the great fisheries of northern Europe. Rapp P-v Réun Cons Int Explor Mer 20:1–228

Hjort J (1926) Fluctuations in the year classes of important food fishes. J Cons Perm Int Explor Mer 1:5–38

Hokkaido Fish Hatchery (2008) Manual for artificial hatching of Arctic lamprey. Inland Water Resources Department, Hokkaido Fish Hatchery, Hokkaido, Japan

Holmes JA (2011) Sea lamprey as an early responder to climate change in the Great Lakes basin. Trans Am Fish Soc 119:292–300

Holt CS, Durkee PA (1983) The distribution, ecology and growth of the chestnut lamprey, *Ichthyomyzon castaneus*, in the Clearwater River, Minnesota. J Minn Acad Sci 49:35–38

Horigome N, Myojin M, Ueki T et al (1999) Development of the cephalic neural crest cells in embryos of *Lampetra japonica*, with special reference to the evolution of the jaw. Dev Biol 207:287–308

Hosoya H, Kojima T, Kataoka T, Kaneko F (1979) Ecology of larval Arctic Lamprey *Entosphenus japonicas* (short communications). Research Report No. 7 from Niigata Prefecture Inland Fisheries Experimental Station, Japan [in Japanese]

Howe EA, Marsden JE, Donovan TM, Lamberson RH (2012) A life cycle approach to modeling sea lamprey population dynamics in the Lake Champlain basin to evaluate alternative control strategies. J Great Lakes Res 38(Suppl 1):101–114

Huggins RJ, Thompson A (1970) Communal spawning of brook and river lampreys, *Lampetra planeri* Bloch and *Lampetra fluviatilis* L. J Fish Biol 2:53–54

Hume JB, Adams CE, Mable B, Bean C (2013) Post-zygotic hybrid viability in sympatric species pairs: a case study from European lampreys. Biol J Linn Soc 108:378–383

Hume JB, Recknagel H, Bean CW, Adams CE, Mable BK (2018) RADseq and mate choice assays reveal unidirectional gene flow among three lamprey ecotypes despite weak assortative mating: insights into the formation and stability of multiple ecotypes in sympatry. Mol Ecol 27:4572–4590

Isahaya T (1934) Lamprey behavior and methods of artificial propagation. Seasonal Reports of Hokkaido Fisheries Experiment [in Japanese]

Jääskä T (2002) The effect of fertilization time and amount of water on river lamprey's sperm (*Lampetra fluviatilis*) fertilization capability, and the effect of rinsing time on spawn surviving under conditions of artificial fertilization. BSc thesis, Turku University of Applied Science, Turku, Finland [in Finnish]

Jacob GS, Pruder GD, Wang JK (1993) Growth trial with the American oyster *Crassostrea virginica* using shrimp pond water as feed. J World Aquacult Soc 24:344–351

Jang MH, Lucas MC (2005) Reproductive ecology of the river lamprey. J Fish Biol 66:499–512

Janvier P, Lund R (1983) *Hardistiella montanensis* N. gen. et sp. (Petromyzontida) from Lower Carboniferous of Montana, with remarks on the affinities of the lampreys. J Vert Paleontol 2:407–413

Jellyman DH, Glova GJ (2002) Habitat use by juvenile lampreys (*Geotria australis*) in a large New Zealand river. N Z J Mar Fresh Res 36:503–510

Jolley JC, Uh CT, Silver GS, Whitesel TA (2015) Feeding and growth of larval Pacific Lamprey reared in captivity. N Am J Aquacult 77:449–459

Johnels AG (1956) On the peripheral autonomic nervous system of the trunk region of *Lampetra planeri*. Acta Zool 1956:251–283

Johnson NS, Yun SS, Buchinger TJ, Li W (2012) Multiple functions of a multi-component mating pheromone in sea lamprey *Petromyzon marinus*. J Fish Biol 80:538–554

Johnson NS, Swink WD, Brenden TO et al (2014) Survival and metamorphosis of low-density populations of larval sea lampreys (*Petromyzon marinus*) in streams following lampricide treatment. J Great Lakes Res 40:155–163

Johnson NS, Buchinger TJ, Li W (2015) Reproductive ecology of lampreys. In: Docker MF (ed) Lampreys: biology, conservation and control, vol 1. Springer, Dordrecht, pp 265–303

Johnson NS, Swink WD, Brenden TO (2017) Field study suggests that sex determination in sea lamprey is directly influenced by larval growth rate. Proc Biol Sci B 284:20170262

Jones ML, Brenden TO, Irwin BJ (2015) Re-examination of sea lamprey control policies for the St. Marys River: completion of an adaptive management cycle. Can J Fish Aquat Sci 72:1538–1551

Jones MR, Grillner S, Robertson B (2009) Selective projection patterns from subtypes of retinal ganglion cells to tectum and pretectum: distribution and relation to behavior. J Comp Neurol 517:257–275

Kainua K, Ojutkangas E (1984) Management of European river lamprey populations in regulated rivers. Report of the (1983) investigations. Bothnian Bay Research Station Handout, University of Oulu, Finland [in Finnish]

Kainua K, Valtonen T (1980) Distribution and abundance of European river lamprey (*Lampetra fluviatilis*) larvae in three rivers running into Bothnian Bay, Finland. Can J Fish Aquat Sci 37:1960–1966

Kainua K, Rintamäki P, Valtonen T (1983) Report on growing lamprey larvae at the Montta fish farm in 1982. Bothnian Bay Research Station Handout, University of Oulu, Finland [in Finnish]

Kamali SI, Kozlov A, Harischandra N, Grillner S, Ekeberg Ö (2013) A computational model of visually guided locomotion in lamprey. Biol Cybern 107:497–512

Kamermans P, Galley T, Boudry P et al (2013) Blue mussel hatchery technology in Europe. Adv Aquacult Hatch Tech 242:339–373

Kan TT (1975) Systematics, variation, distribution, and biology of lampreys of the genus *Lampetra* in Oregon. PhD thesis, Oregon State University, Corvallis, OR

Kataoka T (1985) Studies on propagation of lamprey, *Entosphenus japonicas*—IV: On metamorphosis and parasitism of the keeping larvae. Research Report No. 12 from Niigata Prefecture Inland Fisheries Experimental Station, Japan [in Japanese]

Kataoka T, Hoshino M (1983) Studies on propagation of lamprey, *Entosphenus japonicas*—III: On artificial incubation and keeping fry and ecology in river. Research Report No. 10 from Niigata Prefecture Inland Fisheries Experimental Station, Japan [in Japanese]

Kataoka T, Hosoya H, Emura K (1980a) Study on the propagation of lampreys, *Entosphenus japonicas* (Martens)—I: On the adult lampreys. Research Report No. 8 from Niigata Prefecture Inland Fisheries Experimental Station, Japan [in Japanese]

Kataoka T, Hosoya H, Emura K (1980b) Study on propagation of lamprey, *Entosphenus japonicas* (Martens)—II: On maturing of lamprey and artificial propagation. Research Report No. 9 from Niigata Prefecture Inland Fisheries Experimental Station, Japan [in Japanese]

Kelso JRM (1993) Substrate selection by *Geotria australis* ammocoetes in the laboratory. Ecol Freshw Fish 2:116–120

Kelso JRM, Todd PR (1993) Instream size segregation and density of *Geotria australis* ammocoetes in two New Zealand streams. Ecol Freshw Fish 2:108–115

Khonsari RH, Li B, Vernier P, Northcutt RG, Janvier P (2009) Agnathan brain anatomy and craniate phylogeny. Acta Zool 90(Suppl 1):52–68

Kleerekoper H (1972) The sense organs. In: Hardisty MW, Potter IC (eds) The biology of lampreys, vol 2. Academic Press, New York, pp 373–404

Kluge B, Renault N, Rohr KB (2005) Anatomical and molecular reinvestigation of lamprey endostyle development provides new insight into thyroid gland evolution. Dev Genes Evol 215:32–40

Kirillova EA, Kirillov PI, Kucheryavyi AV, Pavlov DS (2011) Downstream migration in ammocoetes of the Arctic lamprey *Lethenteron camtschaticum* in some Kamchatka rivers. J Ichthyol 51:1117–1125

Kobayashi W (1993) Effect of osmolality on the motility of sperm from lamprey, *Lampetra japonica*. Zool Sci 10:281–285

Kucheryavyi AV, Savvaitova KA, Pavlov DS et al (2007) Variations of life history strategy of the Arctic lamprey *Lethenteron camtschaticum* from the Utkholok River (western Kamchatka). J Ichthyol 47:37–52

Kujawa R, Fopp-Bayat D, Cejko BI et al (2017) Rearing river lamprey *Lampetra fluviatilis* (L.) larvae under controlled conditions as a tool for restitution of endangered populations. Aquacult Int 26:27–36

Kuraku S, Takio Y, Sugahara F, Takeuchi M, Kuratani S (2010) Evolution of oropharyngeal patterning mechanisms involving Dlx and endothelins in vertebrates. Dev Biol 341:315–323

Kuratani S (2005) Developmental studies of the lamprey and hierarchical evolutionary steps towards the acquisition of the jaw. J Anat 207:489–499

Kuratani S (2012) Evolution of the vertebrate jaw from developmental perspectives. Evol Dev 14:76–92

Kuratani S, Ota GK (2008) The primitive versus derived traits in the developmental program of the vertebrate head: views from cyclostome developmental studies. J Exp Biol B Mol Dev Biol 310:294–314

Kuratani S, Ueki T, Aizawa S, Hirano S (1997) Peripheral development of cranial nerves in a cyclostome, *Lampetra japonica*: morphological distribution of nerve branches and the vertebrate body plan. J Comp Neurol 384:483–500

Kuratani S, Horigome N, Hirano S (1999) Development morphology of the cephalic mesoderm and reevaluation of segmental theories of the vertebrate head: evidence from the embryos of an agnathan vertebrate, *Lampetra japonica*. Dev Biol 210:381–400

Kuratani S, Kuraku S, Murakami Y (2002) Lamprey as an evo-devo model: lessons from comparative embryology and molecular phylogenetics. Genesis 34:175–195

Kusakabe R, Kuratani S (2005) Evolution and development patterning of the vertebrate skeletal muscles: perspectives from the lamprey. Dev Dynam 234:824–834

Kusakabe R, Tochinai S, Kuratani S (2003) Expression of foreign genes in lamprey embryos: an approach to study evolutionary changes in gene regulation. J Exp Zool B Mol Dev Evol 296:87–97

Kusuda S (2012) Lamprey—a god sent gift: participating in the International Lamprey Restoration and Artificial Propagation Forum. Uo to Mizu 48:7–10 [in Japanese]

Lamb TD, Collin SP, Pugh EN (2007) Evolution on the vertebrate eye: opsins, photoreceptors, retina and eye cup. Nature 8:960–975

Lampman R (2011) Passage, migration behavior, and autoecology of adult Pacific lamprey at Winchester Dam and within the North Umpqua River Basin, Oregon, USA. MS thesis, Oregon State University, Corvallis, OR

Lampman R (2016) Evaluation of Pacific lamprey (*Entosphenus tridentatus*) life stage transition from prolarva to larva and timing of first feeding. Appendix J1 in 2016 Annual Progress Report from the Yakama Nation Pacific Lamprey Project to the Bonneville Power Administration (Project No. 2008-470-00), Portland, OR

Lampman R (2017) Development of artificial propagation methods for production of juvenile Pacific lamprey (*Entosphenus tridentatus*) for the use in research associated with Section 4.2.3 of the Rocky Reach Pacific Lamprey Management Plan. 2016 Annual Report from the Yakama Nation Fisheries to the Public Utility District No. 1 of Chelan County, Wenatchee, WA

Lampman R (2018) Development of artificial propagation methods for production of juvenile Pacific lamprey (*Entosphenus tridentatus*) for the use in research associated with Section 4.2.3 of the Rocky Reach Pacific Lamprey Management Plan. 2017 Annual Report from the Yakama Nation Fisheries to the Public Utility District No. 1 of Chelan County, Wenatchee, WA

Lampman R, Moser ML, Jackson AD et al (2016) Developing techniques for artificial propagation and early rearing of Pacific Lamprey (*Entosphenus tridentatus*) for species recovery and restoration. In: Orlov A, Beamish R (eds) Jawless fishes of the world, vol 2. Cambridge Scholars Publishing, Newcastle upon Tyne, pp 160–195

Langille RM, Hall BK (1988) Artificial fertilization, rearing and timing of stages of embryonic development of the anadromous sea lamprey, *Petromyzon marinus* L. Can J Zool 66:549–554

Laroche W, Martin CD, Wimmer HP (2004) Exploratory study of dismantling sea lamprey nests to reduce egg and larval production in two Lake Champlain basin tributaries. Lake Champlain Basin Program, Technical Report No. 44, Grand Isle, VT

Larsen LO (1980) Physiology of adult lampreys with special regard to natural starvation, reproduction and death after spawning. Can J Fish Aquat Sci 37:1762–1779

Lasne E, Sabatié M-R, Tremblay J et al (2010a) A new sampling technique for larval lamprey population assessment in small river catchments. Fish Res 106:22–26

Lasne E, Sabatié M-R, Evanno G (2010b) Communal spawning of brook and river lampreys (*Lampetra planeri* and *L. fluviatilis*) is common in the Oir River (France). Ecol Freshw Fish 19:323–325

Lennon RE (1955) Artificial propagation of the sea lamprey, *Petromyzon marinus*. Copeia 1955:235–236

Limm MP, Power ME (2011) Effect of the western pearlshell mussel *Margaritifera falcata* on Pacific lamprey *Lampetra tridentata* and ecosystem processes. Oikos 120:1076–1082

Linley T, Krogstad E, Mueller R, Gill G, Lasorsa B (2016) Mercury concentrations in Pacific Lamprey (*Entosphenus tridentatus*) and sediments in the Columbia River basin. Environ Toxicol Chem 35:2571–2576

Luzier CW, Schaller HA, Brostrom JK et al (2011) Pacific lamprey (*Entosphenus tridentatus*) assessment and template for conservation measures. U.S. Fish and Wildlife Service, Portland, OR

MacDonald TH (1963) Rates of growth in British ammocoetes. Am Midl Nat 69:198–204

Macey DJ, Potter IC (1978) Lethal temperatures of ammocoetes of the Southern Hemisphere lamprey, *Geotria australis* Gray. Environ Biol Fish 3:241–243

Mäenpää E, Myllynen K, Pakkala J, Aronsuu K, Koskenniemi E (2001) The effect of water quality during the wintering period of mature lampreys (*Lampetra fluviatilis*): the physiological state and egg fertilization rate. West Finland Regional Environment Center, Kokkola, Finland [in Finnish]

Maine AN, Moser ML, Jackson AD (2017) Development of artificial propagation methods for Pacific Lamprey (*Entosphenus tridentatus*) 2016. Report for Bureau of Reclamation, Boise, ID

Maine AN, Moser ML, Jackson AD (2018) Development of artificial propagation methods for Pacific Lamprey (*Entosphenus tridentatus*) 2017. Report for Bureau of Reclamation, Boise, ID

Maitland PS, Renaud CB, Quintella BR, Close DA, Docker MF (2015) Conservation of native lampreys. In: Docker MF (ed) Lampreys: biology, conservation and control, vol 1. Springer, Dordrecht, pp 375–428

Mallatt J (1981) The suspension feeding mechanism of the larval lamprey *Petromyzon marinus*. J Zool 194:103–142

Mallatt J (1983) Laboratory growth of larval lampreys (*Lampetra (Entosphenus) tridentata* Richardson) at different food concentrations and animal densities. J Fish Biol 22:293–301

Malmqvist B (1980) Habitat selection of larval brook lampreys (*Lampetra planeri*, Bloch) in a south Swedish stream. Oecologia 45:35–38

Malmqvist B (1983) Breeding behaviour of brook lampreys *Lampetra planeri*: experiments on mate choice. Oikos 43:43–48

Malmqvist B, Brönmark C (1982) Filter feeding in larval *Lampetra planeri*: effects of size, temperature and particle concentration. Oikos 38:40–46

Manion PJ (1967) Diatoms as food of larval sea lampreys in a small tributary of northern Lake Michigan. Trans Am Fish Soc 96:224–226

Manion PJ (1968) Production of sea lamprey larvae from nests in two Lake Superior streams. Trans Am Fish Soc 97:484–485

Manion PJ, Hanson LH (1980) Spawning behavior and fecundity of lampreys from the upper three Great Lakes. Can J Fish Aquat Sci 37:1635–1640

Manion PJ, McLain AL (1971) Biology of larval sea lampreys (*Petromyzon marinus*) of the 1960 year class, isolated in the Big Garlic River, Michigan, 1960–1965. Great Lakes Fish Comm Tech Rep 16:1–35

Manion PJ, Smith BR (1978) Biology of larval and metamorphosing sea lampreys, *Petromyzon marinus*, of the 1960 year class in the Big Garlic River, Michigan, Part II, 1966–72. Great Lakes Fish Comm Tech Rep 30:1–35

Manzon RG, Youson JH, Holmes JA (2015) Lamprey metamorphosis. In: Docker MF (ed) Lampreys: biology, conservation and control, vol 1. Springer, Dordrecht, pp 139–214

Masters JEG, Jang M-H, Ha K et al (2006) The commercial exploitation of a protected anadromous species, the river lamprey (*Lampetra fluviatilis* (L.)), in the tidal River Ouse, north-east England. Aquat Conserv 16:77–92

McCauley DW, Bronner-Fraser M (2002) Conservation of Pax gene expression in ectodermal placodes of the lamprey. Gene 287:129–139

McCauley DW, Bronner-Fraser M (2006) Importance of SoxE in neural crest development and the evolution of the pharynx. Nature 441:750–752

McCauley DW, Docker MF, Whyard S, Li W (2015) Lampreys as diverse model organisms in the genomics era. BioScience 65:1046–1056

McGree M, Whitesel TA, Stone J (2008) Larval metamorphosis of individual Pacific lampreys reared in captivity. Trans Am Fish Soc 137:1866–1878

McLain AL, Dahl FH (1968) An electric beam trawl for the capture of larval lampreys. Trans Am Fish Soc 97:289–293

Meckley TD, Wagner CM, Gurarie E (2014) Coastal movements of migrating sea lamprey (*Petromyzon marinus*) in response to a partial pheromone added to river water: implications for management of invasive populations. Can J Fish Aquat Sci 71:533–544

Meeuwig MH, Bayer JM, Seelye JG (2005) Effects of temperature on the survival and development of early life stage Pacific and western brook lampreys. Trans Am Fish Soc 134:19–27

Meeuwig MH, Bayer JM, Reiche RA (2006) Morphometric discrimination of early life stage *Lampetra tridentata* and *L. richardsoni* (Petromyzonidae) from the Columbia River Basin. J Morph 267:623–633

Mesa MG, Bayer JM, Bryan MB, Sower SA (2010) Annual sex steroid and other physiological profiles of Pacific lampreys (*Entosphenus tridentatus*). Comp Biochem Physiol 155A:56–63

Moore JW, Beamish FW (1973) Food of larval sea lamprey (*Petromyzon marinus*) and American brook lamprey (*Lampetra lamottei*). J Fish Res Board Can 30:7–15

Moore JW, Mallatt JM (1980) Feeding of larval lamprey. Can J Fish Aquat Sci 37:1658–1664

Moore JW, Potter IC (1976) A laboratory study on the feeding of larvae of the brook lamprey *Lampetra planeri* (Bloch). J Anim Ecol 45:81–90

Morman RH (1987) Relationship of density to growth and metamorphosis of caged larval sea lampreys, *Petromyzon marinus* Linnaeus, in Michigan streams. J Fish Biol 30:173–181

Morman RH, Cuddy DW, Rugen PC (1980) Factors influencing the distribution of sea lamprey (*Petromyzon marinus*) in the Great Lakes. Can J Fish Aquat Sci 37:1811–1826

Moser ML, Jackson AD (2013) Development of artificial propagation methods for Pacific Lamprey (*Entosphenus tridentatus*) 2012. Report for Bureau of Reclamation, Boise, ID

Moser ML, Paradis RL (2017) Pacific lamprey restoration in the Elwha River drainage following dam removals. Am Curr 42:3–8

Moser ML, Butzerin JM, Dey DB (2007) Capture and collection of lampreys: the state of the science. Rev Fish Biol Fish 17:45–56

Moser ML, Almeida PR, Kemp PS, Sorenson PW (2015) Lamprey spawning migration. In: Docker MF (ed) Lampreys: biology, conservation and control, vol 1. Springer, Dordrecht, pp 215–263

Moser ML, Jackson AD, Maine AN (2016) Development of artificial propagation methods for Pacific Lamprey (*Entosphenus tridentatus*) 2015. Report for Bureau of Reclamation, Boise, ID

Moser ML, Maine AN, Jackson AD (2017a) Development of artificial propagation methods for production of juvenile Pacific Lamprey (*Entosphenus tridentatus*) for the use in research associated with Section 4.2.3 of the Rocky Reach Pacific Lamprey Management Plan. 2016 Annual Report from NOAA Fisheries to the Public Utility District No. 1 of Chelan County, Wenatchee, WA

Moser ML, Jackson AD, Mueller RP, Maine AN, Davisson M (2017b) Effects of passive integrated transponder (PIT) implantation on Pacific lamprey ammocoetes. Anim Biotelem 5:1

Moser ML, Maine AN, Jackson AD (2018) Development of artificial propagation methods for production of juvenile Pacific Lamprey (*Entosphenus tridentatus*) for the use in research associated with Section 4.2.3 of the Rocky Reach Pacific Lamprey Management Plan. 2017 Annual Report from NOAA Fisheries to the Public Utility District No. 1 of Chelan County, Wenatchee, WA

Moyle PB, Brown LB, Chase SD, Quiñones RM (2009) Status and conservation of lampreys in California. In: Brown LR, Chase SD, Mesa MG, Beamish RJ, Moyle PB (eds) Biology, management, and conservation of lampreys in North America. Am Fish Soc Symp 72:279–293

Mundahl ND, Erickson C, Johnston MR, Sayeed GA, Taubel S (2005) Diet, feeding rate, and assimilation efficiency of American brook lamprey larvae. Environ Biol Fish 72:67–72

Murakami Y, Kuratani S (2008) Brain segmentation and trigeminal projections in the lamprey; with reference to vertebrate brain evolution. Brain Res Bull 75:218–224

Murakami Y, Watanabe A (2009) Development of the central and peripheral nervous systems in the lamprey. Dev Growth Diff 51:197–205

Murakami Y, Ogasawara M, Sugahara F et al (2001) Identification and expression of the lamprey Pax6 gene: evolutionary origin of the segmented brain of vertebrates. Development 128:3521–3531

Murakami Y, Pasqualetti M, Takio Y et al (2004) Segmental development of reticulospinal and branchiomotor neurons in the lamprey: insights into the evolution of the vertebrate hindbrain. Development 131:983–995

Murdoch SP, Beamish FWH, Docker MF (1991) Laboratory study of growth and interspecific competition in larval lampreys. Trans Am Fish Soc 120:653–656

Murdoch SP, Docker MF, Beamish FWH (1992) Effect of density and individual variation on growth of sea lamprey *(Petromyzon marinus)* larvae in the laboratory. Can J Zool 70:184–188

Myllynen K, Ojutkangas E, Nikinmaa M (1997) River water with high iron concentration and low pH causes mortality of lamprey roe and newly hatched larvae. Ecotoxicol Environ Saf 36:43–48

Naiman RJ, Bilby RE, Schindler DE, Helfield JM (2002) Pacific salmon, nutrients, and the dynamics of freshwater ecosystems. Ecosystems 5:399–417

Nazarov D, Kucheryavyi A, Pavlov D (2016) Distribution and habitat types of the lamprey larvae in rivers across Eurasia. In: Orlov A, Beamish R (eds) Jawless fishes of the world, vol 1. Cambridge Scholars Publishing, Newcastle upon Tyne, pp 280–298

Neave FB, Steeves TB, Pratt TC et al (2019) Stream characteristics associated with feeding type in silver *(Ichthyomyzon unicuspis)* and northern brook *(I. fossor)* lampreys and tests for phenotypic plasticity. Environ Biol Fish (in press)

Neidert AH, Virupnannavar V, Hooker GW, Langeland JA (2001) Lamprey Dlx genes and early vertebrate evolution. Proc Natl Acad Sci USA 98:1665–1670

Nelson MC, Nelle RD (2007) Juvenile Pacific lamprey use of a pollution abatement pond on the Entiat National Fish Hatchery. Final Report, U. S. Fish and Wildlife Service, Leavenworth, WA, US

Nevejan N, De Schryver P, Wille M et al (2017) Bacteria as food in aquaculture: do they make a difference? Rev Aquacult 10:180–212

Nika N, Virbickas T (2010) Brown trout *Salmo trutta* redd superimposition by spawning *Lampetra* species in a lowland stream. J Fish Biol 77:2358–2372

Nikitina N, Bronner-Fraser M, Sauka-Spengler T (2009) The sea lamprey *Petromyzon marinus*: a model for evolutionary and developmental biology. Emerging model organisms: a laboratory manual. CSHL, Cold Spring Harbor, NY, pp 405–429

Nilsen EB, Hapke WB, McIlraith B, Markovchick D (2015) Reconnaissance of contaminants in larval Pacific lamprey *(Entosphenus tridentatus)* tissues and habitats in the Columbia River Basin, Oregon and Washington, USA. Environ Pollut 201:121–130

Ogasawara M, Shigetani Y, Hirano S, Satoh N, Kuratani S (2000) *Pax1/Pax9*-related genes in an agnathan vertebrate, *Lampetra japonica*: expression pattern of *LjPax9* implies sequential evolutionary events toward the gnathostome body plan. Dev Biol 223:399–410

Okkelberg P (1922) Notes on the life history of the brook lamprey, *Ichthyomyzon unicolor*. Occas Pap Mus Zool Univ Mich 125:1–14

Olsson R (1963) Endostyles and endostylar secretions: a comparative histochemical study. Acta Zool 44:299–328

Ota KG, Kuraku S, Kuratani S (2007) Hagfish embryology with reference to the evolution of the neural crest. Nature 446:672–675

Ojutkangas E, Laukkanen E (1985) Management of lamprey population in regulated rivers. Report on trials carried out in the field hatchery by the river Lestijoki. Kokkola water district, Kokkola, Finland [in Finnish]

Pancer Z, Amemiva CT, Ehrhardt GR et al (2004) Somatic diversification of variable lymphocyte receptors in the agnathan sea lamprey. Nature 430:174–180

Pavlov DS, Nazarov DY, Zvezdin AO, Kucheryavyi AV (2014) Downstream migration of early larvae of the European river lamprey *Lampetra fluviatilis*. Dokl Biol Sci 459:344–347

Piavis GW (1961) Embryological stages in the sea lamprey and effects of temperature on development. US Fish Wildl Serv Fish Bull 61:111–143

Piavis GW (1971) Embryology. In: Hardisty MW, Potter IC (eds) The biology of lampreys, vol 1. Academic Press, London, pp 361–400

Piavis GW, Howell JH (1969) Rearing of sea lamprey, *Petromyzon marinus,* embryos in distilled water. Copeia 1969:204–205

Piavis GW, Howell JH, Smith AJ (1970) Experimental hybridization among five species of lampreys from the Great Lakes. Copeia 1970:29–37

Pierce ML, Weston MD, Fritzch B et al (2008) MicroRNA-183 family conservation and ciliated neurosensory organ expression. Evol Dev 10:106–113

Pletcher FT (1963) The life history and distribution of lampreys in the Salmon and certain other rivers in British Columbia, Canada. PhD thesis, University of British Columbia, Vancouver

Potter IC (1980) Ecology of larval and metamorphosing lampreys. Can J Fish Aquat Sci 37:1641–1657

Potter IC, Beamish FWH (1975) Lethal temperatures in ammocoetes of four species of lampreys. Acta Zool 56:85–91

Potter IC, Rogers MJ (1972) Oxygen consumption in burrowed and unburrowed ammocoetes of *Lampetra planeri*. Comp Biochem Physiol 41A:427–432

Potter IC, Hilliard RW, Bradley JS, McKay RJ (1986) The influence of environmental variables on the density of larval lampreys in different seasons. Oecologia 70:433–440

Potter IC, Hill BJ, Gentleman S (1970) Survival and behaviour of ammocoetes at low oxygen tensions. J Exp Biol 53:59–73

Purvis HA (1970) Growth, age at metamorphosis, and sex ratio of northern brook lamprey in a tributary of southern Lake Superior. Copeia 1970:326–332

Purvis HA (1979) Variations in growth, age at transformation, and sex ratio of sea lampreys reestablished in chemically treated tributaries of the upper Great Lakes. Great Lakes Fish Comm Tech Rep 35:1–36

Quintella BR (2000) Ecology of the sea lamprey (*Petromyzon marinus* L.) larval phase in the River Mondego. BSc thesis, University of Lisbon, Portugal

Quintella BR, Andrade NO, Almeida PR (2003) Distribution, larval stage duration and growth of the sea lamprey ammocoetes, *Petromyzon marinus* L., in a highly modified river basin. Ecol Freshw Fish 12:286–293

Quintella BR, Andrade NO, Dias NM, Almeida PR (2007) Laboratory assessment of sea lamprey larvae burrowing performance. Ecol Freshw Fish 16:177–182

Rétaux S, Kano S (2010) Midline signaling and evolution of the forebrain in chordates: a focus on the lamprey Hedgehog case. Integr Comp Biol 50:98–109

Richards JE, Beamish RJ, Beamish FWH (1982) Descriptions and keys for ammocoetes of lampreys from British Columbia, Canada. Can J Fish Aquat Sci 39:1484–1495

Richardson MK, Wright GM (2003) Developmental transformations in a normal series of embryos of the sea lamprey *Petromyzon marinus* (Linnaeus). J Morphol 257:348–363

Richardson MK, Admiraal J, Wright GM (2010) Developmental anatomy of lampreys. Biol Rev 85:1–33

Ringø E, Song SK (2016) Applications of dietary supplements (synbiotics and probiotics in combination with plant products and β-glucans) in aquaculture. Aquacult Nutr 22:4–24

Robert D, Murphy HM, Jenkins GP et al (2014) Poor taxonomical knowledge of larval fish prey preference is impeding our ability to assess the existence of a "critical period" driving year-class strength. ICES J Mar Sci 71:2042–2052

Robinson JM, Wilberg MJ, Adams JV, Jones ML (2013) A spatial age-structured model for describing sea lamprey (*Petromyzon marinus*) population dynamics. Can J Fish Aquat Sci 70:1709–1722

Robinson TC, Bayer JM (2005) Upstream migration of Pacific lampreys in the John Day River, Oregon: behavior, timing, and habitat use. Northwest Sci 79:106–119

Rodríguez-Muñoz R, Ojanguren AF (2002) Effect of short-term preservation of sea lamprey gametes on fertilisation rate and embryo survival. J Appl Ichthyol 18:127–128

Rodríguez-Muñoz R, Nicieza AG, Braña F (2001) Effects of temperature on developmental performance, survival and growth of sea lamprey embryos. J Fish Biol 58:475–486

Rodríguez-Muñoz R, Nicieza AG, Braña F (2003) Density-dependent growth of sea lamprey larvae: evidence for chemical interference. Funct Ecol 17:403–408

Rogers PA, Glenn AR, Potter IC (1980) The bacterial flora of the gut contents and environment of larval lamprey. Acta Zool 61:23–27

Rougemont Q, Gaigher A, Lasne E et al (2015) Low reproductive isolation and highly variable levels of gene flow reveal limited progress towards speciation between European river and brook lampreys. J Evol Biol 28:2248–2263

Rougemont Q, Roux C, Neuenschwander S et al (2016) Reconstructing the demographic history of divergence between European river and brook lampreys using approximate Bayesian computations. PeerJ 4:e1910

Rougemont Q, Gagnaire PA, Perrier C et al (2017) Inferring the demographic history underlying parallel genomic divergence among pairs of parasitic and nonparasitic lamprey ecotypes. Mol Ecol 26:142–162

Ryapolova NI, Mitans AR (1991) Biological background and biotechnical principles of river lamprey (*Lampetra fluviatilis*) hatchery reproduction. Aquaculture at the Baltics. Avots, Riga, pp 84–99 [in Russian]

Savvaitova KA, Maksimov VA (1979) Spawning of the Arctic lamprey of the genus *Lampetra* in connection with the problem of the taxonomic status of small forms. J Ichthyol 18:555–560

Sauka-Spengler T, Meulemans D, Jones M, Bronner-Fraser M (2007) Ancient evolutionary origin of the neural crest gene regulatory network. Dev Cell 13:405–420

Schroll F (1959) Zur Ernährungsbiologie der Steirischen Ammocöten *Lampetra planeri* und *Eudontomyzon danfordi*. Int Rev Ges Hydrobiol 44:395–429

Schultz LD, Mayfield MP, Wyss LA et al (2014) The distribution and relative abundance of spawning and larval Pacific Lamprey in the Willamette River basin. Final Report to the Columbia River Inter-Tribal Fish Commission, Portland, OR

Schultz LD, Chasco SL, Whitlock MH et al (2017) Estimating growth and mortality of larval lamprey: implications of ageing error in fitting models. J Fish Biol 90:1305–1320

Schultz LP (1930) The life history of *L. planeri* Bloch, with a statistical analysis of the rate of growth of the larvae from western Washington. Occas Pap Mus Zool Univ Mich 221:1–35

Scott WB (1887) Notes on the development of *Petromyzon*. J Morphol 1:253–310

Scribner KT, Jones ML (2002) Genetic assignment of larval parentage as a means of assessing mechanisms underlying adult reproductive success and larval dispersal. Great Lakes Fishery Commission Completion Report, Ann Arbor, MI

Shigetani Y, Sugahara F, Kawakami Y et al (2002) Heterotopic shift of epithelial-mesenchymal interactions in vertebrate jaw evolution. Science 296:1316–1319

Shimeld SM, Donoghue PCJ (2012) Evolutionary crossroads in developmental biology: cyclostomes (lamprey and hagfish). Development 139:2091–2099

Shipley AE (1887) On some points in the development of *Petromyzon fluviatilis*. Q J Microsc Sci 27:1–46

Shintani S, Terzic J, Sato A et al (2000) Do lampreys have lymphocytes? The Spi evidence. Proc Natl Acad Sci USA 97:7417–7422

Shirakawa H, Yanai S, Kouchi K (2009) Habitat selection of fluvial lamprey larvae *Lethenteron japonicum* change with growth stage. Ecol Civil Eng 12:87–98

Sifa L, Mathias JA (1987) The critical period of high mortality of larvae fish—a discussion based on current research. Chin J Oceanol Limnol 5:80–96

Silva S, Viera-Lanero R, Barca S et al (2014a) Single pass electrofishing method for assessment and monitoring of larval lamprey populations. Limnetica 33:217–226

Silva S, Gooderham A, Forty M, Morland B, Lucas MC (2014b) Egg drift and hatching success in European river lamprey *Lampetra fluviatilis*: is egg deposition in gravel vital to spawning success? Aquat Conserv Mar Freshw Ecosyst 25:534–543

Silver GS (2015) Investigations of larval Pacific lamprey *Entosphenus tridentatus* osmotic stress tolerance and occurrence in a tidally-influenced estuarine stream. MS thesis, Portland State University, Portland, OR

Sjöberg K (1977) Locomotor activity of river lamprey *Lampetra fluviatilis* (L.) during the spawning season. Hydrobiologia 55:265–270

Smith AJ, Howell JH, Piavis GW (1968) Comparative embryology of five species of lamprey of the upper Great Lakes. Copeia 1968:461–469

Smith DM, Welsh SA, Turk PJ (2011) Selection and preference of benthic habitat by small and large ammocoetes of the least brook lamprey (*Lampetra aepyptera*). Environ Biol Fish 91:421–428

Smith JJ, Baker C, Eichler EE et al (2012) Genetic consequences of programmed genome rearrangement. Curr Biol 22:1524–1529

Smith SJ, Marsden JE (2009) Factors affecting sea lamprey egg survival. N Am J Fish Manag 29:859–868

Sokolov LI, Tsepkin YA, Barabanova YR (1992) Reproductive ecology of the western brook lamprey, *Lampetra planeri* (Petromyzontidae). J Ichthyol 32:145–150

Sower SA (2015) The reproductive hypothalamic-pituitary axis in lampreys. In: Docker MF (ed) Lampreys: biology, conservation and control, vol 1. Springer, Dordrecht, pp 305–373

Staponkus R, Kesminas V (2014) Confirmation of hybridisation between river lamprey (*L. fluviatilis*) and brook lamprey (*L. planeri*) from *in situ* experiments. Pol J Nat Sci 29:49–54

Starcevich SJ, Gunckel SL, Jacobs SE (2014) Movements, habitat use, and population characteristics of adult Pacific lamprey in a coastal river. Environ Biol Fish 97:939–953

Steeves TB, Slade JW, Fodale MF, Cuddy DW, Jones ML (2003) Effectiveness of using backpack electrofishing gear for collecting sea lamprey (*Petromyzon marinus*) larvae in Great Lakes tributaries. J Great Lakes Res 29 (Suppl 1):161–173

Sugahara F, Aota S, Kuraku S et al (2011) Involvement of Hedgehog and FGF signaling in the lamprey telencephalon: evolution of regionalization and dorsoventral patterning of the vertebrate forebrain. Development 138:1217–1226

Sugiyama H, Goto A (2002) Habitat selection by larvae of a fluvial lamprey, *Lethenteron reissneri*, in a small stream and an experimental aquarium. Ichthyol Res 49:62–68

Sutton TM, Bowen SH (1994) Significance of organic detritus in the diet of larval lampreys in the Great Lakes basin. Can J Fish Aquat Sci 51:2380–2387

Sutton TM, Bowen SH (2009) Diel feeding by larval northern brook lampreys in two northern Michigan streams. In: Brown LR, Chase SD, Mesa MG, Beamish RJ, Moyle PB (eds) Biology, management, and conservation of lampreys in North America. Am Fish Soc Symp 72:153–164

Suzuki DG, Murakami Y, Escriva H, Wada H (2015) A comparative examination of neural circuit and brain patterning between the lamprey and amphioxus reveals the evolutionary origin of the vertebrate visual center. J Comp Neurol 523:251–261

Swink WD (1995) Effect of larval sea lamprey density on growth. Great Lakes Fishery Commission Completion Report, Ann Arbor, MI

Tahara Y (1988) Normal stages of development in the lamprey, *Lampetra reissneri* (Dybowski). Zool Sci 55:109–118

Takayama M (2002) Spawning activities and physical characteristics of the spawning ground of *Lethenteron reissneri* at the headstream of the Himekawa River, central Japan. Ichthyol Res 49:165–170

Tetlock A, Yost CK, Stavrinides J, Manzon RG (2012) Changes in the gut microbiome of the sea lamprey during metamorphosis. Appl Environ Microbiol 78:7638–7644

Thomas MLH (1962) Observations on the ecology of ammocoetes of *Petromyzon marinus* L. and *Entosphenus* (*Lampetra lamottei* LeSueur) in the Great Lakes watershed. MSc thesis, University of Toronto, Toronto, ON

Timoshevskiy VA, Herdy JR, Keinath MC et al (2016) Cellular and molecular features of developmentally programmed genome rearrangement in a vertebrate (sea lamprey: *Petromyzon marinus*). PLoS Genet 12:e1006103

Todd PR, Kelso JRM (1993) Distribution, growth and transformation timing of larval *Geotria australis* in New Zealand. Ecol Freshw Fish 1993:99–107

Törrönen J, Kokko H, Päivänen K (1988) Trials on lamprey farming and outplanting in Kymi province in 1985–1987. Handout series of the water and environment administration 75, Kouvola, Finland [in Finnish]

Torgersen CE, Close DA (2004) Influence of habitat heterogeneity on the distribution of larval Pacific lamprey (*Lampetra tridentata*) at two spatial scales. Freshw Biol 49:614–630

Tsimbalov IA, Kucheryuavyi AV, Pavlov DS (2018) Results of hybridization between anadromous and resident forms of European river lamprey Lampetra fluviatilis. J Ichthyol 58:122–125

Tuunainen P, Ikonen E, Auvinen H (1980) Lampreys and lamprey fisheries in Finland. Can J Fish Aquat Sci 37:1953–1959

Veit O (1939) Contributions to the insight into the head of vertebrates. Morph Jahrb 84:86–107 [in German]

Vikström R (2002) Notes on lamprey propagation. Regional Environment Publications 252. Ykkös-Offset, Vaasa, Finland [in Finnish]

Villar-Cerviño V, Abalo XM, Villar-Cheda B et al (2006) Presence of glutamate, glycine, and y-aminobutyric acid in the retina of the larval sea lamprey: comparative immunohistochemical study of classical neurotransmitters in larval and postmetamorphic retinas. J Comp Neurol 499:810–827

Vladykov VD (1951) Fecundity of Quebec lampreys. Can Fish Cult 10:1–14

Ward DL, Clemens BJ, Clugston D et al (2012) Translocating adult Pacific lamprey within the Columbia River Basin: state of the science. Fisheries 37:351–361

Weise JG, Pajos TA (1998) Intraspecific competition between larval sea lamprey year-classes as Salem Creek was recolonized, 1990–1994, after a lampricide application. N Am J Fish Manag 18:561–568

Whitlock S, Schultz LD, Schreck CB, Hess JE (2017) Using pedigree reconstruction to estimate effective spawner abundance from redd surveys: an example involving Pacific lamprey. Can J Fish Aquat Sci 74:1646–1653

White JL, Harvey BC (2003) Basin-scale patterns in the drift of embryonic and larval fishes and lamprey ammocoetes in two coastal rivers. Environ Biol Fish 67:369–378

Winberg GG (1956) Rate of metabolism and food requirements of fish. Fish Res Board Can Transl Ser 194:1–253

Witkowski A, Jęsior M (2000) Fecundity of river lamprey Lampetra fluviatilis (L.) in Drwęca River (Vistula basin, northern Poland). Arch Ryb Pol 8:225–232

Wright GM, Youson H (1976) Transformation of the endostyle of the anadromous sea lamprey, Petromyzon marinus L., during metamorphosis. Light microscopy and autoradiography with 125I. Gen Comp Endocrinol 30:243–257

Xu Y, Zhu S-W, Li Q-W (2016) Lamprey: a model for vertebrate evolutionary research. Zool Res 37:263–269

Yap MR, Bowen SH (2003) Feeding by northern brook lamprey (Ichthyomyzon fossor) on sestonic biofilm fragments: habitat selection results in ingestion of a higher quality diet. J Great Lakes Res 29(Suppl 1):15–25

Yamada H (1951) Life history of Lethenteron reissneri. Collect Breed 13:121–127 [in Japanese]

Yamazaki Y, Fukutomi N, Takeda K, Iwata A (2003) Embryonic development of the Pacific lamprey, Entosphenus tridentatus. Zool Sci 20:1095–1098

Yamazaki Y, Goto A (1997) Morphometric and meristic characteristics of two groups of Lethenteron reissneri. Ichthyol Res 44:15–25

Yamazaki C, Koizumi I (2017) High frequency of mating without egg release in highly promiscuous nonparasitic lamprey Lethenteron kessleri. J Ethol 35:237–243

Yorke MA, McMillan DB (1979) Nature and cellular origin of the adhesive coats of the lamprey egg (Petromyzon marinus). J Morphol 162:313–325

Zerrenner A (2004) Effect of density and age on larval sea lamprey growth and survival in three Lake Champlain streams. J Freshw Ecol 19:515–519

Zielinski B (1996) Sea lamprey olfactory activity: prolarvae and larvae. Great Lakes Fishery Commission Completion Report, Ann Arbor, MI

Zuo Y, Xie W, Pang Y et al (2017) Bacterial community composition in the gut content of Lampetra japonica revealed by 16S rRNA gene pyrosequencing. PLoS ONE 12:e0188919

Zvezdin AO, Pavlov DS, Nazarov DY, Kucheryavyi AV (2016) Photopreferendum of migratory and nonmigratory larvae of European river lamprey Lampetra fluviatilis. J Ichthyol 56:171–173

Zvezdin AO, Pavlov DS, Kucheryavyi AV, Tsimbalov IA (2017) Experimental study of the European river lamprey, *Lampetra fluviatilis* (L.), migratory behavior in the period of initial dispersion of juveniles. Inland Water Biol 10:209–218

Chapter 3
Post-metamorphic Feeding in Lampreys

Claude B. Renaud and Philip A. Cochran

Abstract Eighteen of 41 lamprey species worldwide feed post metamorphosis; nine in either marine waters or fresh waters and nine exclusively in fresh waters. Four feeding modes have been identified: blood feeding, flesh feeding, blood and flesh feeding, and carrion feeding. Adaptations to these feeding modes are associated with characteristics of the dentition of the oral disc and tongue-like piston, the oral papillae and fimbriae, the velar tentacles, and the buccal glands. The duration of the adult feeding phase varies from a few months to 4 years and during this time the various species grow either slightly or up to nearly eight times the length that they reached as larvae. The post-metamorphic diet consists usually of fishes but in some cases may include marine mammals. Feeding behavior is complex and highly variable and differs between the two major modes of blood feeding and flesh feeding. Blood feeders tend to selectively attack larger hosts and tend to attach ventrally to them in deep water but dorsally in shallower habitats. Flesh feeders tend to attach dorso-laterally to schooling fishes, and their hosts may be relatively small compared to those used by blood feeders.

Keywords Feeding behavior · Feeding modes · Hosts · Parasitic phase

3.1 Introduction

Eighteen lamprey species feed following metamorphosis (Potter et al. 2015). The most noticeable change observed at metamorphosis is the development of the oral disc and its associated structures that together comprise the post-metamorphic feeding apparatus (see Sect. 3.2). The other important change is the switch from flow-

C. B. Renaud (✉)
Research and Collections, Canadian Museum of Nature, P.O. Box 3443, Station D, Ottawa, ON
K1P 6P4, Canada
e-mail: crenaud@nature.ca

P. A. Cochran
Biology Department, Saint Mary's University of Minnesota, 700 Terrace Heights, Winona, MN
55987, USA

© Springer Nature B.V. 2019
M. F. Docker (ed.), *Lampreys: Biology, Conservation and Control*,
Fish & Fisheries Series 38, https://doi.org/10.1007/978-94-024-1684-8_3

through gill ventilation in the larva (Dawson et al. 2015) to tidal ventilation in the post-metamorphic individual (Manzon et al. 2015). The change of ventilation is critical because it enables the metamorphosed lamprey to simultaneously carry out the functions of feeding and breathing while attached to a host.

The mechanisms of attachment and subsequent feeding by parasitic lampreys have been described previously (e.g., Reynolds 1931; Lanzing 1958; Gradwell 1972; Farmer 1980; Kawasaki and Rovainen 1988). In brief, parasitic lampreys attach to the host by their oral disc and penetrate the skin through the action of their toothed tongue-like piston. Secretions having both cytolytic and anticoagulatory properties issue from their buccal glands and into the wound to break down tissues and keep the blood free-flowing. As a result of this parasitic feeding, post-metamorphic lampreys exceed the maximum lengths reached as larvae, in some cases by a factor of two or more (Table 3.1).

Previous reviews of post-metamorphic feeding in lampreys by Farmer (1980) and Swink (2003) have primarily focused on the feeding mechanisms, host selection, energy uptake, growth rate, and host-lamprey interactions in sea lamprey *Petromyzon marinus* from the Laurentian Great Lakes. Much of what has been revealed about feeding by parasitic lampreys has been spurred by efforts to understand and manage the sea lamprey in this freshwater ecosystem. When access by this species to the upper Great Lakes was facilitated by human activity, a scenario was created by which a non-native lamprey was established in a system with host species with which it had not recently co-evolved. Moreover, compared to its anadromous populations, sea lamprey in the Great Lakes had access to relatively extensive, high quality spawning habitat. It is therefore not surprising that the sea lamprey contributed to dramatic population declines in several fish species in the Great Lakes (see Chap. 5). Because of the value of these stocks, some of the most intensive studies of lamprey-host interactions have involved the sea lamprey and Great Lakes hosts, including lake trout *Salvelinus namaycush* (Moore and Lychwick 1980; Pycha 1980; Swanson and Swedberg 1980; Wells 1980) and lake whitefish *Coregonus clupeaformis* (Spangler et al. 1980). However, although some aspects of sea lamprey feeding biology can be generalized, the sheer volume of published research on resident sea lamprey from the Great Lakes has contributed to a lack of appreciation for the diversity in feeding displayed by other lampreys.

In this review, we examine the relationship between structure of the oral apparatus and diet, the duration of the feeding phase, and growth during the feeding phase, and we compare various aspects of feeding behavior among lamprey species. We present the results of both field and laboratory observations and cover, as much as possible, research published since the reviews of Farmer (1980) and Swink (2003). Theoretical and quantitative models of lamprey feeding and interactions with hosts at the level of the individual or population (Bence et al. 2003; Madenjian et al. 2003, 2008) are beyond the scope of this review. However, these models have provided useful insights into potential relationships among variables that are relatively easy to measure, such as lamprey growth or marking rates on hosts, and others more difficult to assess directly, such as the intake of energy by a lamprey from its host or the impact of a lamprey population on a host population (see Bence et al. 2003 for a review). They

Table 3.1 Duration of the parasitic phase in lampreys and absolute growth during this phase, as inferred from the difference between the maximum larval length and the maximum adult length

Parasitic lamprey species	Duration of parasitic phase (months)	Maximum larval length (mm)	Maximum adult length (mm)	Difference in length (mm)	Maximum adult length/maximum larval length	Source
Caspiomyzon wagneri Caspian lamprey	?	* 127	553	426	4.35	Berg (1931), Ginzburg (1970)
Entosphenus macrostomus Vancouver lamprey	≤12	170	273	103	1.60	Beamish (1982, 1987a)
Entosphenus minimus Miller Lake lamprey	A few	141	145	4	1.03	Kan and Bond (1981), Lorion et al. (2000), Vladykov and Kott (1976)
Entosphenus similis Klamath lamprey	?	?	269	?	?	Vladykov and Kott (1979)
Entosphenus tridentatus Pacific lamprey	20–42	158	850	692	5.38	Beamish (1980), Chase (2001), Kan (1975), Orlov et al. (2008)
Eudontomyzon danfordi Carpathian lamprey	7–9	216	300	84	1.39	Kux (1965), Kux and Weisz (1960), Vladykov (1925, 1931)
Eudontomyzon morii Korean lamprey	?	?	290	?	?	Ma and Yu (1959)
Geotria australis Pouched lamprey	≥15	111	788	677	7.10	de Castelnau (1872), Potter et al. (1983), Renaud (2011)
Ichthyomyzon bdellium Ohio lamprey	?	169	279	100	1.65	Hubbs and Trautman (1937), Lanteigne (1981, 1988)
Ichthyomyzon castaneus Chestnut lamprey	7–12	165	363	198	2.20	Cochran et al. (2003b), Hall (1963), Lanteigne (1981, 1988), Moore and Kernodle (1965)

(continued)

Table 3.1 (continued)

Parasitic lamprey species	Duration of parasitic phase (months)	Maximum larval length (mm)	Maximum adult length (mm)	Difference in length (mm)	Maximum adult length/maximum larval length	Source
Ichthyomyzon unicuspis Silver lamprey	?	138	415	277	3.01	Cochran (2004), Lanteigne (1981, 1988)
Lampetra ayresii Western river lamprey	3–4	172	311	139	1.81	Beamish (1980), Vladykov and Follett (1958)
Lampetra fluviatilis European river lamprey	3–24	~129	492	363	3.81	Abou-Seedo and Potter (1979), Bahr (1933), Berg (1948), Goodwin et al. (2006), Inger et al. (2010), Lanzing (1959), Maitland (1980). Maitland et al. (1984, 1994), Potter and Osborne (1975), Zanandrea (1959)
Lethenteron camtschaticum Arctic lamprey	24–48	~220	790	570	3.59	Nikolskii (1956), Nursall and Buchwald (1972), Orlov et al. (2014)
Mordacia lapicida Chilean lamprey	?	154	540	386	3.51	Neira (1984), Neira et al. (1988)
Mordacia mordax Short-headed lamprey	23	160	~500	340	3.12	Allen (1989), Potter et al. (1968), Strahan (1960)
Petromyzon marinus Sea lamprey:						
Landlocked	12–20	179	645	466	3.60	Applegate (1950), Bergstedt and Swink (1995), Manion and Smith (1978), Swink and Johnson (2014)
Anadromous	10.5–14.5, 23–28	153	1,200	1,047	7.84	Beamish (1980b), Halliday (1991), Hardisty (1969), Oliva (1953), Silva et al. (2013c)
Tetrapleurodon spadiceus Chapala lamprey	24	?	310	?	?	Àlvarez del Villar (1966)

? = unknown

may also be used to identify questionable assumptions (Cochran et al. 2003a) or critical aspects of lamprey behavior that warrant further investigation (Cochran and Kitchell 1986).

The diversity and complexity of life histories employed by lampreys (see Chap. 4) challenge our ability to neatly categorize them with simple words. We therefore follow several conventions as a matter of convenience. Thus, we use the terms "parasites" and "hosts" to refer respectively to lampreys that feed after metamorphosis and to the organisms they feed upon (rather than "predators" and "prey"), even though, in many cases, the latter are quickly killed as a result of the encounter. Furthermore, although recent molecular studies have revealed that the inter-specific distinction between parasitic lampreys and their non-parasitic derivatives may be more blurred than morphological differences suggest (Docker 2009 and references therein), we use the traditional species designations for parasitic and non-parasitic forms. Finally, we use the term "adult" to refer to post-metamorphic lampreys, including those that are actively feeding, although some workers have restricted this term to lampreys that have reached sexual maturity and use the term "juvenile" to refer to the sexually-immature feeding phase (e.g., Beamish 1980a, b; Docker et al. 2015).

3.2   Functional Morphology of Feeding

Of the 18 lamprey species that feed following the completion of metamorphosis (Table 3.1), nine are anadromous (four, possibly five of these also possess permanent freshwater-resident populations) and nine live exclusively in fresh water (Potter et al. 2015; Chap. 4). Collectively, adult lampreys possess four different modes of feeding (Renaud et al. 2009), a characteristic reflected at the generic level. There are blood feeders (*Ichthyomyzon*, *Mordacia*, *Petromyzon*), flesh feeders (*Eudontomyzon*, *Geotria*, *Lampetra*, *Lethenteron*), blood and flesh feeders (*Entosphenus*, *Tetrapleurodon*), and a presumed carrion feeder (*Caspiomyzon*). Alternative names for these feeding types are respectively, parasites, predators, intermediates, and scavengers.

Potter and Hilliard (1987) were the first to propose functional relationships among the dentition of the oral disc and tongue-like piston, the size of the buccal glands, and the diets of the various lamprey species that feed as adults. Their landmark study was expanded on, with the inclusion of more species, by Renaud et al. (2009). Furthermore, to ensure that the proper assignment of feeding types was made, Renaud et al. (2009) conducted a microscopic examination of the intestinal contents and tested these with Hema-Screen, an assay for detecting blood. The Northern Hemisphere blood feeders (*Ichthyomyzon* and *Petromyzon*) have labial teeth entirely covering all fields of the oral disc, a narrow supraoral lamina, a w-shaped transverse lingual lamina with uniformly-sized cusps, and large buccal glands. Although the Southern Hemisphere blood feeders (*Mordacia*) also have their labial teeth entirely covering all fields of the oral disc and a w-shaped transverse lingual lamina, they differ from the Northern Hemisphere blood feeders in having two triangular supraoral laminae, the transverse lingual lamina having slightly enlarged median and subterminal cusps, and their buccal glands being relatively small. The w-shaped transverse lingual lamina with its uniformly or mostly uniformly-sized cusps in blood feeders is particularly well suited for rasping a hole in the host's skin. The larger buccal glands in Northern Hemisphere blood feeders is related to their need to produce sufficient quantities of lamphredin, with its anticoagulatory properties, to keep the blood meal flowing. The smaller size of the three buccal glands in *Mordacia* is compensated for by their location at the entrance of the oral aperture, combined with their secretion of a unique mucus-lamphredin mixture that adheres to the host tissue, making their action particularly effective. In all other lampreys, there are only two buccal glands and these are positioned further posteriorly and deeply embedded in the basilaris muscles. The Northern Hemisphere flesh feeders (*Eudontomyzon, Lampetra, Lethenteron*) do not have their labial teeth fully covering the available space on the oral disc, but possess a wide supraoral lamina, a u-shaped transverse lingual lamina with a prominent central cusp, and small buccal glands. While the single Southern Hemisphere flesh feeder *Geotria* is different from the Northern Hemisphere flesh feeders in having its labial teeth entirely cover all fields of the oral disc, the other characteristics are the same. The presence of a stout central cusp on the u-shaped transverse lingual lamina in the flesh feeders is an adaptation for removing large chunks of flesh from the prey through gouging. The relatively undigested condition of the tightly-packed and clearly identifiable muscle chunks found in the intestine of the flesh feeders (Renaud et al. 2009), would indicate that large quantities of cytolytically-active lamphredin are not required for effective flesh feeding. Those lampreys that feed on both blood and flesh show intermediate conditions to those of the above two feeding modes, with the transverse lingual lamina being slightly w-shaped in *Entosphenus* and slightly u-shaped in *Tetrapleurodon*, and in both genera, with a slightly enlarged central cusp. Even though *Caspiomyzon* has remarkably blunt dentition on the oral disc and tongue-like piston, it is suspected to be a carrion feeder due to the potential compensatory action of its moderately large buccal glands that presumably secrete lamphredin. This requires confirmation as no blood or flesh has ever been found in its intestine.

Khidir and Renaud (2003) examined the number of oral papillae and oral fimbriae in the different lamprey feeding types. Oral papillae and oral fimbriae are fleshy appendages that lie very close to each other at the perimeter of the oral disc, with the papillae arranged in a circle just outside of the fimbriae. Blood feeders have high numbers of oral papillae and fimbriae (although the Southern Hemisphere *Mordacia* lacks oral fimbriae). In contrast, the flesh feeders have fewer oral papillae and fimbriae. The blood-flesh feeders exhibit a mixture of these characters, with fewer oral papillae than blood feeders but more oral fimbriae than flesh feeders. Oral papillae are innervated (Borri 1922) and are believed to have a sensory function (Lethbridge and Potter 1981; Khidir and Renaud 2003). Their higher number in blood feeders is presumed to be linked to the greater requirement by these species to find attachment sites on their hosts where adequate sources of blood are available. The higher number of oral fimbriae in blood feeders is presumed to be linked to the greater need by these species for a tight seal with the host's body surface to prevent any loss of the blood meal. The absence of oral fimbriae in the blood-feeding *Mordacia* may be compensated for by the presence of its dual triangular supraoral laminae, mentioned above, and numerous elongate and multicuspid circumoral laminae that would help in securing a strong attachment, in conjunction with a well-developed marginal membrane that would help in creating a good seal. Since intermediates feed on a mixture of blood and flesh, the numbers of oral papillae and fimbriae that they possess represent a compromise between their need to find the most beneficial site on the host for feeding, and their need to achieve an effective seal at the oral disc-host surface interface, to prevent the loss of food.

Renaud et al. (2009) studied the role played by the tentacles of the velar apparatus in the feeding process. The velar apparatus lies at the junction of the dorsally-positioned esophagus and the ventrally-positioned, blind-ending water tube (=branchial tube), which connects directly to the seven pairs of gill pouches (Randall 1972). It bears tentacles that project anteriorly into the pharynx. The blood feeders have short and few velar tentacles. In contrast, the flesh feeders have longer and more numerous velar tentacles. Velar tentacles guard the entrance to the water tube and deflect food upwards into the esophagus. The longer and more numerous velar tentacles in the flesh feeders compared to the blood feeders are linked to the need of the former to prevent solid material from entering the branchial pouches via the water tube and potentially clogging the gills, thus interfering with respiration. This is not as critical a requirement in blood feeders because of the liquid nature of their diet, as long as the anticoagulant component of lamphredin is secreted in sufficient quantity to perform its function.

Although the four modes of feeding described broadly explain the feeding adaptations exhibited by lampreys, they are not exclusive categories as carrion feeding has been observed in species that belong to each of the other three modes of feeding. Thus, the blood-feeding chestnut lamprey *Ichthyomyzon castaneus* [reported as *Petromyzon concolor*, re-identified in Hubbs and Trautman (1937: 73)], was attached to a dead sucker *Catostomus* sp. in Wilder Creek, Kalamazoo River basin, Michigan (Bollman 1890). The blood-flesh feeding Miller Lake lamprey *Entosphenus minimus* was observed feeding in the field on dead tui chub *Siphateles bicolor*, and even dead conspecifics, until all soft tissue was removed (Kan and Bond 1981). The flesh-feeding Carpathian lamprey *Eudontomyzon danfordi* will feed in the field on recently dead fishes and on the remains of birds and mammals from slaughterhouses (Grossu et al. 1962; Bănărescu 1969).

Superimposing the mode of feeding onto the morphologically-based cladogram of parasitic lampreys (Gill et al. 2003) suggests that in the Northern Hemisphere lampreys (Petromyzontidae), blood feeding is the ancestral condition, flesh feeding is the derived condition and the blood-flesh feeders represent a transitional stage in the evolution of feeding adaptations (Renaud et al. 2009). However, the unresolved trichotomy between the two Southern Hemisphere lamprey families (Geotriidae and Mordaciidae) and the Petromyzontidae (Potter et al. 2015) does not permit the determination of the ancestral condition for the order Petromyzontiformes.

3.3   Duration of and Growth During the Parasitic Phase

Numerous authors make statements regarding the length of the adult period (i.e., between the end of metamorphosis and death following spawning), but few give precise indications of the duration of the parasitic phase within this period. The parasitic phase follows metamorphosis and begins during or at the end of the downstream or feeding migration to a larger river, lake or to the sea, and ends near the beginning of the upstream or spawning migration (Farmer 1980; Larsen 1980; Bird et al. 1994; Dawson et al. 2015; Moser et al. 2015).

Among the 12 parasitic species for which the duration of the parasitic phase is either known or has been estimated (Table 3.1), this period varies considerably, ranging from a few months in the Miller Lake lamprey up to 42 months (i.e., 3.5 years) in anadromous Pacific lamprey *Entosphenus tridentatus* (Kan and Bond 1981; Beamish 1980) and even 48 months (i.e., 4 years) in anadromous Arctic lamprey *Lethenteron camtschaticum* (Orlov et al. 2014). However, the marine feeding phase in Pacific lamprey may be as short as 20 months (Kan 1975) and in Arctic lamprey as short as 24 months (Nikolskii 1956). Nikolskii (1956) examined 22 feeding-phase Arctic lamprey measuring 147–293 mm total length and collected off the northwest shore of Sakhalin Island, Russia. He suggested that the sample represented three groups, the shortest (147 mm) having just entered the sea, the intermediate one (170–270 mm) having spent 1 year at sea, and the longest (280–293 mm) having spent 2 years feeding at sea. Orlov et al. (2014) collected 472 feeding-phase Arctic lamprey widely distributed between the Sea of Japan and the Bering Sea and determined that four year classes were involved: 150–320, 330–530, 540–650, and 660–790 mm total length. The marine trophic phase of the anadromous pouched lamprey *Geotria australis* has been estimated to last more than 1 year (Potter et al. 1979). Adults of anadromous short-headed lamprey *Mordacia mordax* spend about 5 months feeding in the variably saline Gippsland Lakes, Victoria, Australia, and another 18 months feeding at sea (i.e., about 2 years in total) before returning to fresh water to embark on their spawning run (Potter et al. 1968). On the other hand, in the anadromous western river lamprey *Lampetra ayresii*, the length of time spent feeding at sea appears to be very short, perhaps only 3–4 months, between June and September (Beamish 1980). Similarly, in the European river lamprey *Lampetra fluviatilis*, the length of the feeding phase at sea seems to be about 3 months, from the end of July to October (Bahr 1933). However, the various populations of the European river lamprey exhibit wide variation in the length of their trophic phase, even within a single river. In their monitoring study of the early upstream-migrating European river lamprey into the River Severn estuary in England, Abou-Seedo and Potter (1979) determined that two anadromous forms of the species occurred; a larger typical form and a smaller praecox form (sensu Berg 1931, 1948). Abou-Seedo and Potter (1979) estimated that the typical form spent 18 months feeding at sea, whereas the praecox form fed at sea for 12 months only. On the other hand, Berg (1948) suggested that the praecox form of the European river lamprey that enters the Neva River, Russia, in autumn had spent only one summer feeding in the sea. Zanandrea (1959) examining European river lamprey collected from the Gulf of Gaeta, off the west coast of Italy, estimated that these spent 12–24 months feeding at sea. A population of European river lamprey in which most individuals are permanent freshwater residents, but a few may go to sea, feeds in Loch Lomond, Scotland, only during the months of June to October, based on the incidence of fresh lamprey-produced wounds on powan *Coregonus clupeoides* (=*C. lavaretus*), being restricted to that period only (Maitland 1980; Maitland et al. 1994), whereas a population in the estuarine waters of the Firth of Forth, Scotland, feeds on clupeids, and occasionally on gadids, from June to November (Maitland et al. 1984). Another population of European river lamprey in which few if any individuals are anadromous is said to actively feed from May to October in Lough Neagh, Northern Ireland (Goodwin et al. 2006). However, fresh lamprey-produced wounds on Irish pollan *Coregonus pollan* (=*C. autumnalis*) were observed only between 9 April and 27 August, although Goodwin et al. (2006) suggested that later in the year the lamprey may switch to other uncollected prey or quickly kill Irish pollan, thus preventing the capture of that host. The first of these possibilities is supported by the stable isotope carbon ratio study of Inger et al. (2010) which indicated that freshwater bream *Abramis brama* was the main diet item of European river lamprey in Lough Neagh between June and November. In summary, we can conclude from the above studies that European river lamprey has a feeding phase that ranges widely between 3 and 24 months, whether it feeds in fresh water or at sea. Based on the presence of fresh wounds on prey throughout the year, Beamish (1982) suggested that the freshwater Vancouver lamprey *Entosphenus macrostomus* feeds for 12 months. Beamish (1987a) further specified that the Vancouver lamprey begins feeding heavily on 1- and 2-year-old coho salmon *Oncorhynchus kisutch* in the spring following metamorphosis and that feeding continues uninterrupted into the winter, with spawning believed to occur the following year. According to Kux (1965), the adult trophic phase of the freshwater Carpathian lamprey begins in March or April and extends to October or November of the same year, giving a range of 7–9 months. Very little is known about the duration of the feeding phase in the parasitic species of the exclusively freshwater genus *Ichthyomyzon*. Hall (1963) reported that in the Manistee River, Michigan, the chestnut lamprey attacked fish hosts during a 7-month period, from April through October, and was largely inactive from November through April. However, Cochran et al. (2003b) reported parasitic attachments by chestnut lamprey to host fishes during winter in Wisconsin (more data were available for silver lamprey *I. unicuspis*, which gained significant mass between October and March). Therefore, the duration of the feeding phase of the chestnut lamprey is tentatively inferred to extend for the entire year. Álvarez del Villar (1966) suggested a duration of the feeding phase in the freshwater Chapala lamprey, *Tetrapleurodon spadiceus*, of 2 years.

There is an intraspecific difference in the duration of the parasitic phase in the landlocked versus anadromous forms of sea lamprey. In the former, the duration is 12–20 months (Applegate 1950; Bergstedt and Swink 1995), whereas in the latter it is 23–28 months (Beamish 1980b; Halliday 1991). This difference is reflected in the size attained by the two; <650 mm in the landlocked form (Applegate 1950; Bergstedt and Swink 1995) and >800 mm in the anadromous form (Grinyuk 1970; Beamish 1980b; Halliday 1991; Holčík et al. 2004; see Chap. 4). However, Silva et al. (2013c) reported the case of a young anadromous sea lamprey captured feeding on a golden grey mullet *Liza aurata* in the River Ulla estuary in northern Spain that was tagged and recaptured on its upstream migration only 13.5 months later. Taking into account that at capture the lamprey may have been feeding for some time and upon recapture had already ceased feeding for some time, these authors estimated a marine feeding phase of 10.5–14.5 months. This case notwithstanding, based on the information on sea lamprey presented above, one could infer that the duration of the feeding phase in anadromous lampreys is longer than in those species restricted to fresh water. While this general statement holds true in a number of cases, that is, in the anadromous Pacific lamprey, pouched lamprey, Arctic lamprey, and short-headed lamprey versus the freshwater Vancouver lamprey, Miller Lake lamprey, Carpathian lamprey, and chestnut lamprey, an exception to the rule is the anadromous western river lamprey versus the freshwater Chapala lamprey (Table 3.1). The case of European river lamprey is difficult to assess because the duration of its feeding phase is highly variable (Table 3.1) and the habitat in which the feeding occurs, whether in marine or fresh waters, is not always clear from the literature.

Growth achieved during the post-metamorphic feeding phase may be roughly estimated by comparing the maximum total length attained by the feeding adult with the maximum total length attained by the ammocoete larva and expressing these as a ratio (Table 3.1). Those data are available for 15 species and exhibit wide variation from 1.03 in Miller Lake lamprey to 7.84 in the anadromous sea lamprey. In the landlocked sea lamprey, the ratio is only 3.60, a reflection of the shorter duration of its parasitic phase relative to that of the anadromous form (Table 3.1). Silva et al. (2013c) provide a unique direct measure of growth rate during the marine feeding phase in anadromous sea lamprey: a feeding individual tagged at a size of 218 mm total length and 20 g wet weight was recaptured 13.5 months later at a size of 895 mm and 1,218 g. The authors point out that this growth is probably underestimated because at the time of recapture the individual was on its upstream spawning migration and not feeding.

3.4 Trends in Feeding Behavior

Lamprey behavior is quite variable both among and within species. Although they display several trends or patterns with respect to parasitic feeding, rarely do lampreys display all-or-none responses with respect to the aspects of behavior discussed below. Although variability in feeding might be dismissed as random or suboptimal behavior, it might also be viewed as evidence of adaptive flexibility (sensu Dill 1983). This variability may also provide challenges to researchers attempting to design powerful experiments or effective sampling protocols.

For lampreys that feed on blood, a conceptual model of a feeding bout may help organize consideration of the various aspects of feeding that will be considered in this review. Note that subsequent to a feeding event, a lamprey may undergo an interval of non-feeding that may include spending time in a sheltered location as well as searching for a new host. Very little is known about this non-feeding interval in general or the search process in particular, although Cochran (2014) has found parasitic-phase chestnut lamprey in crevices beneath boulders or other cover objects in the same areas where other lamprey were attached to fish. After a blood-feeding lamprey has attached to a host, it typically penetrates the host's skin and scale layer to gain access to the greater blood supply beneath. To consider this process analogous to "food handling" in general ecological models of foraging may be justified by Farmer et al.'s (1975) estimate that the sea lamprey obtains less than 2% of its food intake as the result of tissue cytolysis, but some host species may have lipid-rich layers associated with the skin that might provide considerable energy to a lamprey (see Sects. 3.4.3, 3.4.4, and 3.4.6). Once a lamprey begins feeding on its host's blood, some simplifying assumptions are that the percentage of host blood volume removed per day is constant and proportional to lamprey mass and inversely proportional to host mass, that host blood "quality" is maintained in the face of this ongoing removal for some time (e.g., through mobilization of reserves in the spleen) before declining over a period that ends with host death or termination of the feeding bout by the lamprey, and that both the length of time that host blood quality can be maintained and the length of time that a host can survive an attack are negatively related to the percentage of host blood volume removed per day (Cochran and Kitchell 1986, 1989; Cochran 1994).

To measure the consumption of a fluid such as blood is logistically challenging. However, experiments in which lake trout and rainbow trout *Oncorhynchus mykiss* blood was tagged with radioactive chromium (^{51}Cr) permitted estimation of the amount of blood removed by sea lamprey (Farmer 1974; Farmer et al. 1975). Their data have been used to justify the assumptions about lamprey feeding listed above. For example, the percentage of host blood volume removed daily (V) was negatively related to the mass of the host (F) and positively related to the mass of the lamprey (L): $V/100 = 2.54(L/F)$. Assuming that the blood volume of the host is 4.7% of its wet mass, the mean daily ration of the lamprey in terms of wet mass is a constant percentage (11.9%) of the lamprey's mass. However, there was substantial variability among individual sea lamprey in the rate at which they removed blood from their hosts, so that the range of individual estimates of daily ration was 3–30% (Farmer et al. 1975). Their data were also used to construct a quantifiable and testable energetics-based model of sea lamprey feeding and subsequent growth (Kitchell and Breck 1980; Cochran and Kitchell 1986, 1989); a model subsequently refined to include the effects of water temperature (Cochran et al. 1999) and the change in energy density of sea lamprey tissue with increasing body size (Cochran et al. 2003a).

There are few measurements available for host blood energy density. Some attempt has been made to measure the constituents of fish blood in the context of lamprey attacks (e.g., Kinnunen and Johnson 1985; Edsall 1999; Swink and Fredricks 2000; Edsall and Swink 2001). By using energy equivalents for the various blood constituents, it might be possible to assess which components contribute most to variation in overall energy density.

Although the general lamprey feeding model described above has been used to make theoretical predictions about lamprey feeding behavior (Cochran 1994), it has been quantified to date only for sea lamprey feeding on lake trout and rainbow trout. Nevertheless, the sea lamprey model has been applied to some other situations. For example, a comparison of observed growth by sea lamprey feeding on burbot *Lota lota* during 55 feeding bouts (Swink and Fredricks 2000) to model predictions of growth by sea lamprey feeding on lake trout and rainbow trout of identical sizes for identical lengths of time suggested that the two salmonids and burbot were equivalent as food resources for sea lamprey (Cochran et al. 2003a). As new technologies are developed to facilitate the analysis of blood and as interest in lamprey-host interactions broadens from a narrow focus on sea lamprey and salmonids [e.g., studies on lake sturgeon *Acipenser fulvescens* by Sepúlveda et al. (2012a, b)], it is hoped that the lamprey feeding model can be quantified for more combinations of lamprey and host species.

Aspects of lamprey feeding behavior to be covered in this section include the diel timing of attacks; selectivity by lampreys with respect to host species, host size, and site of attachment on the host; multiple attachments on hosts and patterns of distribution of attacks among hosts; and the duration of feeding bouts and lethality of attacks.

3.4.1 Daily Timing of Attacks

Some parasitic lampreys attack significantly more often at night than by day. Evidence reviewed by Cochran (1986a) was especially strong for chestnut lamprey and silver lamprey, and included both field and laboratory data. Evidence for the sea lamprey was equivocal, but it was based on laboratory studies not specifically designed to test for differences between nocturnal and diurnal attack rates. Typically these experiments are monitored at 12-h intervals, at the beginning of each light and each dark period. One potential bias in this sort of study arises when trials are routinely started at the same time of day (i.e., in the morning at the beginning of the light period), because a disproportionate number of lampreys may initiate attachments during the first time period in which hosts are available. The bias may be exaggerated if new hosts are routinely added to tanks (e.g., to replace hosts that have died) at the same time of day, because addition of new fish to a tank may stimulate attachments. These biases may be especially important when overall sample size is low (e.g., in studies in which lampreys are allowed to remain attached to hosts until they voluntarily detach).

Cochran (1986a) suggested an experimental design that would increase sample size and minimize the effect of the initial exposure of lampreys to new hosts. For lampreys on a 12 h:12 h light:dark cycle, this would entail gently detaching lampreys from their hosts at intervals of 1.5 days and allowing them to attack during the intervening intervals, until a sufficient balanced number of light and dark intervals had passed. Cochran and Lyons (2004) used this design with 13 silver lamprey allowed to attack common carp *Cyprinus carpio* over a 9-day experiment, allowing for three 36-h periods each to be initiated by light and dark periods. With this design, a total of 39 attachments were recorded. Significantly more attachments occurred at night than during the day (29 vs. 10).

The timing of lamprey attacks during the day is related to the sensory modes used to detect hosts. Nocturnal foraging is consistent with the documented ability of lampreys to orient to the sources of host odors (Kleerekoper and Mogensen 1963; Kleerekoper 1972), and both silver and sea lampreys have been shown to possess electroreceptors (Bodznick and Preston 1983) that may also contribute to detection of hosts. Cochran (2014) suspended an all-glass tank containing a common carp in a larger tank as a way of providing visual cues to chestnut lamprey in the absence of other sensory information. Eight individual lamprey were sequentially tested. None showed any apparent response to the visual stimulus provided by the carp, but after each lamprey was transferred to an identical tank with an unconfined carp, it had attached to the carp by the next morning. Moreover, when the same eight lamprey were individually released into a tank with a carp in the absence of light (in a room originally used as a photographic darkroom), each had attached by the end of a 12-h dark period. It would appear, therefore, that visual cues are not necessary for successful attack.

3.4.2 Host Species Selectivity

Parasitic lampreys collectively attack a wide range of sizes and diversity of host species (Table 3.2); from small darters (Cochran and Jenkins 1994) to large whales (e.g., Pike 1951; Nichols and Tscherter 2011). Hosts for four species, two from the Northern Hemisphere (Caspian lamprey and Klamath lamprey) and two from the Southern Hemisphere (pouched lamprey and Chilean lamprey) have yet to be identified. All of these, except Klamath lamprey, are anadromous and probably feed exclusively in the marine environment over widely dispersed areas, making the study of their hosts difficult. There are additional host species not listed in Table 3.2 that are known to have been parasitized by lampreys, but from the evidence available it was not possible to assign the lampreys to host species. For example, Hubley (1961) reported lamprey scars on bowfin *Amia calva*, river carpsucker *Carpiodes carpio*, bigmouth buffalo *Ictiobus cyprinellus*, yellow bullhead *Ameiurus natalis*, flathead catfish *Pylodictis olivaris*, and black crappie *Pomoxis nigromaculatus* in the upper Mississippi River, where it was possible for either chestnut or silver lampreys to have been responsible. There are also two marine mammals, the pygmy sperm whale *Kogia breviceps* and the dwarf sperm whale *K. sima* reported by McAlpine (2002) to bear lamprey marks, but given that these whales range worldwide in temperate and tropical waters of the Atlantic, Pacific and Indian oceans, a number of anadromous lamprey candidates are possible. An additional report by Heyning (2002) of marks on Cuvier's beaked whale *Ziphius cavirostris* was dismissed because they were attributed either to lampreys or cookie cutter sharks *Isistius* spp.

Individual lamprey species are also known to attack a wide range of hosts. For example, Renaud (2002) provided a list of 20 fish species known to be attacked by the

Table 3.2 Parasitic lamprey species and their native and non-native hosts under natural conditions. To qualify, a host must either possess a lamprey-induced mark on its body, or be identifiable in the lamprey gut contents, or have a parasitic lamprey attached to a part of its body where nourishment to the lamprey is possible. Scientific names of fish hosts follow Page et al. (2013) for North America and FishBase (www.fishbase.org) elsewhere, and for marine mammal hosts follow Wilson and Reeder (2005). Non-native hosts are indicated with an asterisk

Parasitic lamprey species	Host	Source
Caspiomyzon wagneri Caspian lamprey	None yet positively identified	This study
Entosphenus macrostomus Vancouver lamprey	Freshwater fishes: *Oncorhynchus clarkii, O. kisutch, Salvelinus malma*	Beamish (1982, 2001)
Entosphenus minimus Miller Lake lamprey	Freshwater fishes: *Rhinichthys osculus klamathensis, Salvelinus fontinalis*[*], *Salmo trutta*[*], *Siphateles bicolor*	Bond and Kan (1973), Lorion et al. (2000)
Entosphenus similis Klamath lamprey	None yet positively identified	This study
Entosphenus tridentatus Pacific lamprey	Freshwater fishes: *Catostomus rimiculus, Oncorhynchus mykiss, O. nerka* Marine fishes: *Anoplopoma fimbria, Atheresthes evermanni, A. stomias, Clidoderma asperrimum, Clupea pallasii, Gadus chalcogrammus, G. macrocephalus, Hippoglossoides elassodon, Hippoglossus stenolepis, Merluccius productus, Oncorhynchus gorbuscha, O. keta, O. kisutch, O. mykiss, O. nerka, O. tshawytscha, Ophiodon elongatus, Platichthys stellatus, Pleurogrammus monopterygius, Reinhardtius hippoglossoides, Sebastes aleutianus, S. alutus, S. borealis, S. reedi* Marine mammals: *Balaenoptera borealis, B. physalus, Megaptera novaeangliae, Physeter catodon*	Abakumov (1964), Beamish (1980), Coots (1955), Kucheryavyy et al. (2016), Murauskas et al. (2013), Novikov (1963), Nemoto (1955), Orlov (2016), Orlov et al. (2008, 2009), Pelenev et al. (2008), Pike (1951), Simpson and Wallace (1978), Wade and Beamish (2016)
Eudontomyzon danfordi Carpathian lamprey	Freshwater fishes: *Alburnoides bipunctatus, Barbatula barbatula, Barbus barbus, B. carpathicus, Chondrostoma nasus, Cottus gobio, C. poecilopus, Gobio carpathicus, Phoxinus phoxinus, Salmo sp., Squalius cephalus*	Bănărescu (1969), Grossu et al. (1962), Holčík (1963), Talabishka et al. (2012)
Eudontomyzon morii Korean lamprey	Freshwater fishes: *Barbatula toni, Carassius auratus, Cobitis taenia, Pseudogobio esocinus, Rhynchocypris percnurus, Sarcocheilichthys soldatovi*	Ma and Yu (1959)

(continued)

Table 3.2 (continued)

Parasitic lamprey species	Host	Source
Geotria australis Pouched lamprey	None yet positively identified	This study
Ichthyomyzon bdellium Ohio lamprey	Freshwater fishes: *Catostomus commersonii, Cyprinus carpio* *, Etheostoma vulneratum, Hypentelium nigricans, Micropterus dolomieu, Moxostoma carinatum, M. erythrurum, Nocomis micropogon, Noturus flavus, Sander vitreus*	Cochran and Jenkins (1994), Cooper (1983), Hubbs and Trautman (1937), Jenkins and Burkhead (1993), Smith (1985), Trautman (1981)
Ichthyomyzon castaneus Chestnut lamprey	Freshwater fishes: *Acipenser fulvescens, Ambloplites constellatus, Ameiurus nebulosus, Catostomus commersonii, Cyprinus carpio* *, Esox lucius, E. niger, Hiodon alosoides, Hypentelium nigricans, Hypophthalmichthys nobilis*, Ictalurus furcatus, I. punctatus, Ictiobus bubalus, Lepisosteus osseus, Lepomis cyanellus, L. macrochirus, Lota lota, Micropterus dolomieu, M. punctulatus, M. salmoides, Minytrema melanops, Moxostoma anisurum, M. carinatum, M. erythrurum, M. macrolepidotum, M. pisolabrum, M. poecilurum, Oncorhynchus mykiss *, Polyodon spathula, Rhinichthys atratulus, Salvelinus fontinalis, Salmo trutta *, Sander canadensis, S. vitreus, Semotilus atromaculatus*	Becker (1983), Cochran (unpubl. data), Cochran (2009), Cochran and Jenkins (1994), Cross and Metcalf (1963), Flammang and Olson (2010), Gudger (1930), Hall (1960, 1963), Hall and Moore (1954), Hubbs and Trautman (1937), Keleher (1952), Knapp (1951), Massé (pers. comm. 2010), Mayden et al. (1989), Moore and Kernodle (1965), Pflieger (1997), Renaud and de Ville (2000), Salinger (2019), Stewart and Watkinson (2004), Thompson (1898), Watkinson (pers. comm. 2010)
Ichthyomyzon unicuspis Silver lamprey	Freshwater fishes: *Acipenser fulvescens, A. oxyrinchus, Ambloplites rupestris, Ameiurus nebulosus, Aplodinotus grunniens, Carassius auratus *, Catostomus catostomus, C. commersonii, Coregonus clupeaformis, Cyprinus carpio *, Esox lucius, E. masquinongy, Ictalurus punctatus, Ictiobus niger, Lepisosteus osseus, Micropterus dolomieu, Morone chrysops, M. saxatilis, Polyodon spathula, Salvelinus fontinalis, S. namaycush, Sander vitreus*	Allin (1951), Becker (1983), Bensley (1915), Choudhury and Dick (1993), George (1985), Hubbs and Trautman (1937), Hubley (1961), Lyons (1993), Lyons (unpubl. data), Renaud (2002), Roy (1973), Runstrom et al. (2001), Vladykov (1949, 1985), Wagner (1904, 1908)
Lampetra ayresii Western river lamprey	Freshwater fishes: *Oncorhynchus kisutch, O. nerka* Marine fishes: *Alosa sapidissima *, Clupea pallasii, Engraulis mordax, Oncorhynchus nerka, O. tshawytscha*	Beamish (1980), Bond et al. (1983), Roos et al. (1973), Vladykov and Follett (1958), Weitkamp et al. (2015)

(continued)

Table 3.2 (continued)

Parasitic lamprey species	Host	Source
Lampetra fluviatilis European river lamprey	Freshwater fishes: *Abramis brama*[*], *Coregonus albula, C. clupeoides, C. pollan, Osmerus eperlanus, Perca fluviatilis, Rutilus rutilus*[a], *Salmo salar, S. trutta, Sander lucioperca* Marine fishes: *Clupea harengus, Gadus morhua, Pollachius virens, Sprattus sprattus*	Bahr (1933), Berg (1948), Eglite (1958), Goodwin et al. (2006), Inger et al. (2010), Maitland (1980), Tambs-Lyche (1963), Tuunainen et al. (1980)
Lethenteron camtschaticum Arctic lamprey	Freshwater fishes: *Catostomus catostomus, Coregonus artedi, C. clupeaformis, Gasterosteus aculeatus, Lota lota, Prosopium coulterii, Oncorhynchus gorbuscha, O. keta, O. mykiss, O. nerka, Rutilus rutilus, Salvelinus namaycush, Stenodus leucichthys* Marine fishes: *Clupea harengus, C. pallasii, Coregonus autumnalis, C. laurettae, C. nasus, C. sardinella, Eleginus gracilis, Oncorhynchus gorbuscha, O. keta, O. kisutch, O. masou, O. tshawytscha, Osmerus dentex, Platichthys stellatus, Salvelinus alpinus, S. malma, Salmo salar, Stenodus leucichthys*	Birman (1950)[b], Heard (1966), Gritsenko (1968), Holcík (1986), Hunter (1981), Ioganzen (1935), McPhail and Lindsey (1970), Nikolskii (1956), Novomodnyy and Belyaev (2002), Nursall and Buchwald (1972), Riske (1960), Sandstrom et al. (1997)
Mordacia lapicida Chilean lamprey	None yet positively identified	This study
Mordacia mordax Short-headed lamprey	Freshwater/brackish water fishes: *Acanthopagrus butcheri, Aldrichetta forsteri, Salmo trutta*[*] Marine fish: *Thyrsites atun*	Potter et al. (1968)

(continued)

Table 3.2 (continued)

Parasitic lamprey species	Host	Source
Petromyzon marinus Sea lamprey	Freshwater fishes: *Abramis brama, Acipenser fulvescens, Alosa aestivalis, A. alosa, A. fallax, A. pseudoharengus, A. sapidissima, Ambloplites rupestris, Ameiurus melas, A. nebulosus, Amia calva, Anguilla rostrata, Aplodinotus grunniens, Catostomus catostomus, C. commersonii, Coregonus artedi, C. clupeaformis, C. johannae, C. nigripinnis, C. zenithicus, Couesius plumbeus, Cyprinus carpio*, Erimyzon sucetta, Esox lucius, E. masquinongy, E. niger, Ictalurus punctatus, Lepisosteus osseus, Lepomis gibbosus, Leuciscus leuciscus, Lota lota, Luciobarbus bocagei, Micropterus dolomieu, M. salmoides, Morone americana, M. chrysops, Moxostoma erythrurum, M. macrolepidotum, Notemigonus crysoleucas, Oncorhynchus gorbuscha*, O. kisutch*, O. mykiss*, O. tshawytscha*, Osmerus mordax, Perca flavescens, Prosopium cylindraceum, Pseudochondrostoma duriense, Rutilus rutilus, Salmo salar, S. trutta, Salvelinus fontinalis, S. namaycush, Sander canadensis, S. vitreus*	Applegate (1950), Araújo et al. (2013), Beamish (1980b), Beamish and Potter (1975), Berst and Wainio (1967), Bigelow and Schroeder (1948), Davis (1967), Gallant et al. (2006), Halliday (1991), Japha (1910), Jensen and Schwartz (1994), Jensen et al. (1998), Lennon (1954), Maitland (1980), Mansueti (1962), Miller et al. (1989), Nichols and Hamilton (2004), Nichols and Tscherter (2011), Noltie (1987), Penczak (1964), Pearce et al. (1980), Potter and Beamish (1977), Samarra et al. (2012), Silva et al. (2013a, b, 2014), Smith (1971), Smith and Tibbles (1980), Surface (1898, 1899), Tambs-Lyche (1963), van Utrecht (1959), Wigley (1959)
	Marine fishes: *Acipenser sp., Alosa aestivalis, A. alosa, A. pseudoharengus, A. sapidissima, Anarhichas lupus, Anguilla rostrata, Belone belone, Boops boops, Brevoortia tyrannus, Carcharhinus obscurus, C. plumbeus, Cetorhinus maximus, Clupea harengus, Cynoscion regalis, Gadus morhua, Galeocerdo cuvier, Liza aurata, L. ramada, Melanogrammus aeglefinus, Merluccius merluccius, Morone saxatilis, Myliobatis aquila, Pollachius virens, Pomatomus saltatrix, Prionace glauca, Reinhardtius hippoglossoides, Salmo salar, Scomber scombrus, Sebastes mentella, Solea solea, Somniosus microcephalus, Thunnus thynnus, Trachurus trachurus, Trichiurus lepturus, Urophycis sp., Xiphias gladius*	
	Marine mammals: *Balaenoptera acutorostrata, B. borealis, B. physalus, Eubalaena glacialis, Grampus griseus, Mesoplodon bidens, Orcinus orca, Phocoena phocoena, Physeter catodon*	

(continued)

Table 3.2 (continued)

Parasitic lamprey species	Host	Source
Tetrapleurodon spadiceus Chapala lamprey	Freshwater fishes: *Chirostoma* sp., *Ictalurus* sp., *Moxostoma austrinum*, *Oncorhynchus mykiss*[*], and either *Algansea popoche*, *Yuriria alta* or *Y. chapalae* Freshwater mammal: *Trichechus manatus*[*]	Álvarez del Villar (1966), Cochran and Jenkins (1994), Cochran et al. (1996)

[a]The roach is parasitized by *Lampetra fluviatilis*, both as a native (Maitland 1980) and a non-native host (Inger et al. 2010)

[b]The Pacific lamprey that Birman (1950) refers to is *Lethenteron camtschaticum* and not *Entosphenus tridentatus* as erroneously supposed by various authors, including Beamish (1980). The confusion stems from the fact that the common name Pacific lamprey was until recently used for the former species in Russia, but has always been used for the latter species in North America

silver lamprey. Further evidence of the non-specificity of the lamprey diet is provided by the ability of the sea lamprey to use the native fish species it encountered after invading the upper Great Lakes, and in particular, burbot, lake whitefish, the ciscoes (*Coregonus artedi*, *C. johannae*, *C. nigripinnis*, and *C. zenithicus*), and lake trout, and by the apparent readiness of native lamprey species to include non-native species in their diets (Cochran 1994). Nine of the 14 lamprey species for which hosts have been identified (Miller Lake lamprey, Ohio lamprey, chestnut lamprey, silver lamprey, western river lamprey, European river lamprey, short-headed lamprey, sea lamprey, and Chapala lamprey) are known to have parasitized non-native hosts (Table 3.2). In one case, reports by commercial fishers indicated that the Chapala lamprey attacked West Indian manatees *Trichechus manatus* when the latter were introduced into Lake Chapala, Mexico, in an attempt to control water hyacinth (Cochran et al. 1996). Using stable isotope analyses coupled with Bayesian mixing models, as well as direct observation of lamprey-induced scars, Inger et al. (2010) indicated that, in addition to feeding on native Irish pollan (see Sect. 3.3), European river lamprey in Lough Neagh also fed on native brown trout *Salmo trutta* and European perch *Perca fluviatilis*, and on non-native freshwater bream and roach *Rutilus rutilus*. Hume et al. (2013) suggested that European river lamprey in Loch Lomond has switched from feeding on native powan (see Sect. 3.3), which have dramatically declined in numbers, to feeding on the abundant non-native ruffe *Gymnocephalus cernua*. The new non-native host proposed in the last study, however, requires confirmation, as it is based on indirect evidence.

Even though a lamprey species may be observed feeding on many different host species throughout its geographic range, it may nevertheless concentrate its attacks on one or a few species at any particular locality, and it may be more abundant where certain host species are also in abundance. Multiple lampreys have been observed attached to single individuals of apparently preferred host species. For example, Vladykov (1985) reported 61 silver lamprey attached to a lake sturgeon in Quebec, whereas Wagner (1904, 1908) and Becker (1983) reported as many as 10–27 silver and chestnut lampreys attached to individual lake sturgeon and paddlefish *Polyodon spathula* in Wisconsin. Typically, host species attacked by blood feeders are among the largest fish species available, but there is a dearth of studies that have assessed host species selectivity while adequately controlling for the effect of host body size. Fish species that are reported as hosts for blood-feeding lampreys also tend to have naked skin (e.g., paddlefish) or small scales (e.g., salmonids) and tend to form schools or otherwise aggregate. Flesh-feeding lampreys, such as western and European river lampreys, which typically kill their hosts quickly as compared to blood feeders, tend to feed on smaller schooling fishes (e.g., clupeids and young salmon), at least in marine environments (Bahr 1933; Beamish 1980; Beamish and Neville 1995).

Cochran (1994) noted that host species used by lampreys tended to coincide with the commercially and recreationally important fishes preferred by humans. Fishes preferred by humans tend to be large and have high muscle mass, and those valued for their fighting ability have high aerobic capacity. These traits are correlated with large blood volume, a quality that may provide a lamprey with an extended period of feeding. The piscivorous fishes often preferred by humans tend to have reduced layers of skin and scales, which may be associated with a reduced investment of energy and handling time for a lamprey that penetrates to underlying blood vessels. Commercially important fishes tend to be concentrated in shoals, a condition that would lead to reduced search times for lampreys seeking new hosts.

Selection of hosts may be related to habitat. The majority of lamprey species feed on both benthic and pelagic hosts, indicating plasticity in the foraging habitat that they utilize (Table 3.2). In separate cases of European river lamprey feeding in large lakes, Inger et al. (2010) and Hume et al. (2013) suggested that a switch from feeding pelagically to feeding benthically occurred as the result of new host introductions into the system that altered the trophic dynamics between parasite and host and between the historical host and the new one.

Much remains to be learned about host species selectivity. In particular, it is not known how a lamprey's previous feeding experience affects its selection of subsequent hosts. Moreover, although diversity among parasitic lampreys with respect to feeding adaptations has been recognized (Potter and Hilliard 1987; Renaud et al. 2009), diversity among host species with respect to qualities that enhance or inhibit the lamprey feeding process has not been investigated in depth. Cochran (2009), for example, noted that silver lamprey feeding on paddlefish may benefit from high concentrations of lipids in their skin. Wilkie et al. (2004) noted that sea lamprey feeding on basking shark *Cetorhinus maximus* must be able to penetrate the shark's dermal denticles and rapidly excrete the high urea content of its body fluids. It would be of interest to know whether these challenges are outweighed by the advantages of feeding on a host that is large relative to a host such as Atlantic cod *Gadus morhua*.

3.4.3 Host Size Selectivity

Perhaps the most consistent aspect of parasitic lamprey feeding behavior is a tendency to attach selectively to larger hosts (Farmer and Beamish 1973; Cochran 1985; Swink 1991). Evidence used to evaluate size selectivity may result from direct observation of lampreys attached to hosts in the field or the laboratory or from marking rates derived from field samples of host populations. We use the term "mark" as a general term to include any evidence of a prior attachment, including surface abrasions, wounds (typically, for blood feeders, with a central puncture through the skin/scale layer to the underlying musculature), and scars (=healed wounds). Attempts have been made to standardize the classification and reporting of sea lamprey marks (King and Edsall 1979; King 1980; Ebener et al. 2003, 2006; Patrick et al. 2007), including stages in their healing, but the various terms have not been consistently applied in the literature. As will be seen, some marks provide better information about size selectivity than others.

Field data that resulted in significant correlations between host size and marking rates or significant differences in marking rates between hosts in different size categories were reported in many papers cited by Cochran (1985) and Swink (1991). Additional examples include sea lamprey attacking white sucker *Catostomus com-*

mersonii (Henderson 1986), pink salmon *Oncorhynchus gorbuscha* (Noltie 1987), and lake trout (Sitar et al. 1997), and silver lamprey attacking muskellunge *Esox masquinongy* (Renaud 2002), lake sturgeon (Cochran et al. 2003b), and paddlefish (Cochran and Lyons 2010). However, field evidence in support of size selectivity must be interpreted with caution, and in fact it would be surprising if marking rates were not correlated with host size even if lampreys were attaching to hosts randomly with respect to host size. To the extent that larger hosts tend to be older, they have had more time to accumulate lamprey marks. For this reason, some analyses have been restricted to fresh wounds and have excluded healed scars (e.g., Sitar et al. 1997). Larger hosts are also more likely to survive lamprey attacks and show up as marked individuals in field samples. Finally, it is possible that larger hosts swim faster and farther and are therefore more likely to encounter foraging lampreys.

Two field studies have provided evidence for size selective feeding above and beyond the typical assessment of marking rates: (1) Nuhfer (1993) caged brown trout of different size classes in the Upper Manistee River in Michigan, where they were exposed to attacks by chestnut lamprey. Although he did not analyze the resulting data in the manner proposed by Cochran (1985), he observed that higher percentages of large trout (26–41 cm) were attacked, that the number of lamprey marks on an individual host was positively correlated with trout size, and that the probability of a host being attacked was more accurately predicted with logistic regression by using trout surface area rather than trout length or mass. (2) Schneider et al. (1996) used bottom trawls to recover lake trout recently killed by sea lamprey in Lake Ontario. A comparison of the size distributions of dead fish with those of living fish with and without lamprey wounds revealed that the majority of recent wounds on living trout were on large trout ($\geq$600 mm), with smaller trout ($\leq$400 mm) typically not wounded when larger individuals were available. The conclusion that the sea lamprey was size selective was strengthened by the absence of smaller individuals from the size distribution of trout killed by the lamprey.

Laboratory assessments of size selectivity may allow potentially confounding factors to be controlled. Even in the laboratory, however, it is important to consider the appropriate null hypothesis to be tested and the power of the experimental design to detect frequencies of attack that are significantly different from random. Cochran (1985) recommended that the "numbers-dependent null hypothesis" (that lamprey attacking randomly should attach to hosts of different sizes in frequencies proportional to their relative abundances) be replaced by the more conservative "surface area-dependent null hypothesis" (since a lamprey attacks by attaching to a surface, then lamprey attacking randomly should attach to hosts of different sizes in frequencies proportional to their relative surface areas). According to the numbers-dependent null hypothesis, if lamprey are provided with hosts of two size classes that are equally abundant, then the expected frequencies of attack for those two size classes should be equal. However, according to the surface area-dependent null hypothesis, expected attack frequency for the larger size class should be greater because larger hosts provide greater potential surface area for attachments. Cochran (1985) noted that it is easier to detect significant departure from the area-dependent null hypothesis when the numbers of small hosts are increased relative to large hosts, so that their total surface areas are more equal than in traditional experimental designs, where equal numbers of hosts of different sizes are employed.

Size selectivity by parasitic lampreys may depend on the size range of hosts available. The sea lamprey, for example, displayed evidence for size selectivity in laboratory trials when all hosts were very small (47–95 g) and individual lamprey were exposed to host pairs that were closely matched in size (Cochran and Jenkins 1994), but not when all hosts were very large (>615 mm in total length) and not closely matched in size (Swink 1991). It may be that any host that exceeds a minimum size threshold provides the maximum possible benefit to a feeding lamprey in terms of net rate of energy intake. Swink (1991, 2003) suggested that host size selection by Great Lakes sea lamprey resulted mostly from avoidance of lake trout shorter than 600 mm in total length when larger hosts were available. Cochran (1985) suggested that, because the time to death of a host is inversely related to the proportion of its blood removed daily (Farmer et al. 1975), the most important benefit of size selectivity may be that a lamprey is assured of a longer period of feeding on a larger host. When Swink (1990) subjected lake trout of three size classes to single sea lamprey attacks, he recorded significantly greater mortality in the smallest size class (469–557 mm in total length) but no difference in mortality between the two largest size classes (559–643 mm and 660–799 mm).

Swink (1991, 2003) assessed host size selectivity by sea lamprey of three size classes (<50 g, 50–100 g, and >100 g). Because the relative frequencies of attacks on hosts of different sizes were similar for the three lamprey size classes, it was concluded that the pattern of size selection does not change with sea lamprey size. However, newly transformed sea lamprey have sometimes been reported to feed on relatively small hosts. In some cases, these represent species encountered as the lamprey move downstream through rivers and estuaries toward the ocean (Mansueti 1962; Davis 1967; Silva et al. 2013a, b); in addition to being relatively abundant, these small hosts may have scale and skin layers easier for small lamprey to penetrate, and they may pose less risk as potential predators. In the Great Lakes, newly transformed sea lamprey that enter the lakes in fall, winter, or early spring move into deep water and feed on ciscoes. This apparent preference for relatively small prey species may actually represent a temperature preference during a period when the warmest temperature is available in deep water (Johnson and Anderson 1980).

3.4.4   Site Selectivity on Hosts

Site selectivity by lampreys attaching to hosts was reviewed by Cochran (1986b), who identified several trends: (1) Flesh-feeders tend to attack dorsally or dorso-laterally where host muscle mass is greatest. (2) Blood-feeders in deep waters tend to attach ventrally, often just posterior to the paired fins (especially the pectoral fins). (3) Blood-feeders in shallow habitats tend to attach dorsally. (4) Catostomids, compared to other host species, tend to be attacked more often on their relatively large heads and on the upper surfaces of their relatively large paired fins (although it has not been assessed whether attachments in these areas are greater than would be expected based on relative surface area). (5) Captive lampreys in small tanks tend to attach dorsally, even if they typically attach ventrally in nature. Subsequent reports were generally consistent with these trends. Captive sea lamprey attached to lake trout in 151-L tanks in approximately equal numbers above and below the lateral line (Swink and Hanson 1986, 1989) and to rainbow trout primarily dorsally (Swink and Hanson 1989). Noltie (1987) reported that almost all attacks by sea lamprey on pink salmon in Lake Superior occurred below the lateral line, and Bergstedt et al. (2001) reported that most attacks by sea lamprey on lake trout in Lake Ontario were ventral (and they were especially common behind the pectoral fins). Similarly, Cochran and Lyons (2004) observed that captive silver lamprey attached to common carp dorsally more often than ventrally, even though they were collected at a site in the Wisconsin River where attachments to paddlefish in deep water were significantly more often ventral rather than dorsal (Cochran and Lyons 2010). Finally, Nuhfer (1993) reported that chestnut lamprey attached primarily dorsally to brown trout in the Upper Manistee River in Michigan at depths of 25–50 cm.

Cochran (1986b) discussed a combination of factors that might explain patterns of site selection by blood-feeding lampreys. Ventral attachments in areas with thinner skin and scale layers (i.e., behind the pectoral fins) may be associated with reduced handling costs and handling time prior to feeding (Farmer and Beamish 1973; Christie and Kolenosky 1980). It is also possible that lampreys achieve greater rates of blood removal from blood vessels accessed through ventral attachments. However, although Farmer (1974) reported slightly greater rates of blood removal for sea lamprey attached in the pectoral region, differences among sites were not statistically significant. Attachments to the dorsal surfaces of hosts in shallow lotic habitats, including to the upper surfaces of the paired fins of catostomids, could minimize the likelihood of detachment or injury to the lamprey through abrasion against rough substrate. Aquaria of the sizes typically used in laboratory experiments may elicit behavior similar to that exhibited by lampreys feeding in shallow streams and rivers. Renaud (2002) suggested that ventral attachments by silver lamprey to muskellunge occurred less often than expected because this surface would be less available when the host was hovering motionless in heavily vegetated habitat. Novikov (1963) and Abakumov (1964) noted that Pacific lamprey marks were observed almost exclusively on the blind (i.e., ventral) side of Greenland halibut *Reinhardtius hippoglossoides*, but it is unknown whether this represents the result of selective behavior by the lamprey.

Several studies have provided additional commentary with respect to host site selection by lampreys. Swink and Hanson (1989) speculated that attachments by captive sea lamprey to the head region of lake trout, including two inside the mouth, resulted from attempts at predation by the lake trout. Bergstedt et al. (2001) compared locations of healed sea lamprey marks on living lake trout from Lake Ontario with locations of marks on dead fish. The lack of a significant difference implied that lamprey-induced mortality did not vary among attachment sites. Bergstedt et al. (2001) also suggested that a tendency for lamprey attachments to be concentrated on the anterior half of the host's body may be related to the lower amplitude of lateral swimming undulations relative to the posterior half. A tendency for anterior attachment was reported by Patrick et al. (2009) in a study of parasitism by captive sea lamprey on lake sturgeon. Many attachments occurred near the insertion of the pectoral fins or on the lower part of the head. Cochran (2009) compared skin density of paddlefish in terms of dry mass (g/cm^2) for samples taken from ventral, lateral, and dorsal locations. Unexpectedly, skin density from ventral locations, where silver lamprey tend to attach in the field, was significantly greater than in locations higher on the body, and even though paddlefish skin is scaleless, it was comparable to skin density of common carp. However, dried samples of paddlefish skin contained a substantial lipid residue, and skin density in terms of ash weight was much less than that of carp. It is possible that silver lamprey attaching ventrally may benefit from easy access to high energy lipids. Cochran and Lyons (2010) compared observed frequencies of attachment by silver lamprey to paddlefish in various body regions to frequencies expected on the basis of their relative surface areas. As expected from previous anecdotal observations, ventral attachments on the body were significantly more common than dorsal attachments. In addition, attachments to the rostrum occurred significantly less often than expected on the basis of its surface area, and, unlike on the body proper, attachments to its dorsal surface were significantly more common than to its ventral surface. Previous anecdotal accounts reported paddlefish collected with lamprey "attached in the gill" region (Thomas Say in Keating 1824; Becker 1983), and Cochran and Lyons (2010) noted attachments to the isthmus at the base of the gills within the branchial cavity significantly more often than expected on the basis of its relatively small surface area. As many as four lamprey were observed within the branchial cavity of a single paddlefish, and they were sometimes completely obscured from external view by the gular flap. Attaching within the branchial cavity may protect silver lamprey from detachment when paddlefish breach, and this location may also provide easy access to blood under pressure in the ventral aorta.

3.4.5 Multiple Attachments and Distribution of Lampreys Among Individual Hosts

Individual hosts may sometimes be parasitized by more than one lamprey at a time. As noted previously, paddlefish and lake sturgeon are known for multiple attachments by silver or chestnut lampreys (e.g., Becker 1983; Vladykov 1985), but individuals of smaller host species, including stream salmonids, catostomids, and esocids, may also be subject to multiple attacks by these and other lampreys (Nuhfer 1993; Renaud 2002; Hume et al. 2013; Philip A. Cochran unpublished data). For example, when Nuhfer (1993) caged groups of brown trout for 20–21 days in a Michigan stream where they were accessible to chestnut lamprey, 43 of 128 trout sustained multiple (2–20) wounds, whereas 46 trout received no wounds at all. Multiple attachments to the same host might occur just by chance. If lampreys attach to individual hosts independently of each other, then the probability that a particular host will suffer two attacks during a given period is the square of the probability of suffering a single attack.

Because Lennon (1954) reported that hosts struggling to dislodge sea lamprey attracted additional lampreys, it might be expected that the distribution of lamprey among hosts would be clumped rather than random. However, Farmer and Beamish (1973) reported that relative numbers of white sucker with single and multiple attacks by captive sea lamprey did not differ from those expected under a binomial distribution, an indication that an attachment by one lamprey did not attract additional attacks. Beamish (1980b) reached a similar conclusion based on frequencies of anadromous sea lamprey scars on Atlantic salmon, *Salmo salar*. Frequencies of paddlefish in the Wisconsin River with different numbers of lampreys attached were not significantly different from those expected if the lampreys were distributed among paddlefish randomly (i.e., according to a Poisson distribution) (Cochran and Lyons 2016). Some models of sea lamprey wounding have assumed that the number of wounds per fish of a given length follow a Poisson distribution (Bence et al. 2003; Rutter and Bence 2003).

3.4.6 Duration of Attachments to Individual Hosts and Lethality of Attacks

How long a lamprey stays attached to a host is of great theoretical and practical interest (Cochran and Kitchell 1986, 1989; Bence et al. 2003), but it is difficult to ascertain under natural conditions. In the Namekagon River of Wisconsin, where it is possible to observe chestnut lamprey attached to hosts from bridge crossings, Philip A. Cochran (unpublished data) has observed lamprey apparently attached to the same hosts for minimum periods of 1–12 days. These are minimum estimates because neither attachments nor detachments were observed.

More accurate measurements of attachment durations are possible in laboratory experiments, but even under controlled conditions the interpretation of attachment duration is not straightforward. Whereas lampreys sometimes detach voluntarily from living hosts, in other cases attachments are terminated when the hosts die. We consider the latter scenario first. Experiments in which lake and rainbow trout blood was tagged with radioactive chromium (^{51}Cr) permitted estimation of the amount of blood removed by sea lamprey (Farmer 1974). The time in days that it takes a host to die from a sea lamprey attack (D) is negatively related to the percentage of its blood volume (V) removed daily (Farmer et al. 1975), as expressed by Cochran and Kitchell (1989) in the following equation: $\ln(D) = 8.03 - 1.63(\ln V)$.

Cochran and Kitchell (1989) recorded attachment times for captive sea lamprey on rainbow trout in the presence and absence of an alternative host of the same species. Individual lamprey were exposed to both treatments, with half initially exposed to one trout and half initially exposed to two. Overall, 15 of 28 attacks resulted in the death of the host, with deaths almost equally split between treatments. Five lamprey killed their hosts in both treatments, but individual lamprey that killed one host were not more likely to kill the second. Neither the attachment times nor the latency periods prior to attack for individual lamprey were correlated between the two experimental treatments, evidence that variation in feeding behavior was not due to consistent differences among individuals. Attachment times that resulted in host death averaged 12.7 days (range = 1–40 days), whereas those that left the host alive averaged 13.1 days (range = 1–70 days). Although it might be expected that attacks interrupted by the death of the hosts would be shorter, the means were not significantly different. Neither were mean attachment times in the presence (10.1 days) and absence (15.4 days) of an alternative host significantly different, although the difference was in the direction predicted by optimal foraging theory (Cochran and Kitchell 1989). However, the first attachment by a lamprey tended to be of longer duration than its second, regardless of whether an alternative host was present. Also, attachment duration was significantly and positively correlated with the latency period prior to attack.

Swink's (2003) review of extensive laboratory experiments at the Hammond Bay Biological Station allowed for a regression analysis of attachment times for sea lamprey confined with single hosts (Bence et al. 2003). Some trends were consistent with those suggested by Cochran and Kitchell (1989). Attachment time was positively related to the ratio of host weight to sea lamprey weight, presumably because this ratio is inversely related to the proportion of the host's blood volume that can be removed daily. Attachment time was also positively related to the previous latency to attack. Swink's (2003) data also indicated that water temperature was negatively related to attachment time, that host species affected the relationship between attachment time and host/sea lamprey weight ratio, and that trends were not qualitatively different when non-lethal attacks were considered separately.

Cochran and Kitchell (1986) noted that attachment times recorded for captive lampreys were quite variable. For example, a sea lamprey that did not feed for 62 days subsequently attached to a rainbow trout for 70 days and more than doubled in weight, but it did not kill its host. Similarly, non-lethal attachment times of >35 days have been recorded for chestnut lamprey (Cochran and Kitchell 1986). Hall (1960) determined that chestnut lamprey attachment to hatchery trout, either rainbow trout or brook trout *Salvelinus fontinalis*, varied from 0.6 to 18.3 days, and 61% of these attachments (11 of 18) resulted in the death of the host. The 39% of attachments in which the trout survived likewise lasted between 0.6 and 18.2 days. Variability in attachment times presumably reflects variability in rates of blood removal. As noted previously, variability in feeding rates among feeding bouts can be inferred from the data collected by Farmer (1974) and Farmer et al. (1975), and Cochran and Kitchell (1986) considered the potential adaptive benefit of adjusting rate of blood removal along with attachment time in response to changes in host population density. No attempt to assess variation in feeding rate within a feeding bout has been reported (i.e., to determine whether a lamprey feeds discontinuously while it is attached to a host). However, in a neurophysiological study, Kawasaki and Rovainen (1988) concluded that feeding behavior of parasitic phase silver lamprey under laboratory conditions was more labile and more complex than respiratory behavior with respect to sensory regulation, motor output, pattern generation, and sensitivity to stress and higher centers.

It might seem that some attachment durations would be limited by the death of the host, and it would be typical for a lamprey researcher to remove dead hosts from aquaria promptly to limit the possibility of fungal or other diseases infecting lampreys or other hosts. However, lampreys can sometimes be found attached to dead hosts in the field (see Sect. 3.2), and captive lampreys may sometimes remain attached to dead hosts (Beamish 1980) for up to several days (Philip A. Cochran unpublished observations). Whether a lamprey is able to obtain some nutrition from a dead fish, while perhaps waiting for a subsequent host, or whether this behavior is anomalous or maladaptive is open to question, but it is likely that feeding on carrion is part of the evolutionary history of the parasitic phase since it has occasionally been observed in representatives of all three other recognized feeding modes (see Sect. 3.2). In any case, the death of the host most often coincides closely with the termination of an attachment.

Kawasaki and Rovainen (1988) noted that virtually nothing is known of appetite and satiation in lampreys. An interesting question is whether the high lipid content of some host species, such as paddlefish in the Wisconsin River (Cochran 2009) and siscowet, a deepwater, fatty form of lake trout in Lake Superior, leads to more rapid satiation of parasitic lampreys and shorter durations of attachment. This might explain the high wounding rates coupled with apparently high survival observed in both cases (Cochran et al. 2003b; Moody et al. 2011).

Finally, little information is available about the extent to which host behavior in nature may contribute to the involuntary termination of lamprey attachments. Anecdotal evidence suggests that lampreys may be dislodged when large hosts breach (Cochran and Lyons 2010). In addition, it may be easier for hosts to scrape off lampreys in natural habitats with boulders and other rough surfaces than in aquaria or raceways with smooth surfaces.

3.5 Facultative Parasitism

There are two, and perhaps, three cases of facultative parasitism among the 23 otherwise non-parasitic species of lampreys (Potter et al. 2015). Eight "giant" adults of American brook lamprey *Lethenteron appendix* measuring 260–354 mm in total length have been reported from Lake Huron and Lake Michigan basins (Manion and Purvis 1971; Cochran 2008). These adults exceed the maximum total length of 240 mm reported for the larvae of the species (Mundahl et al. 2005) and, hence, must have fed post-metamorphosis. It has been suggested that the otherwise non-parasitic American brook lamprey feeds facultatively in the adult stage, either parasitically (Manion and Purvis 1971) or on fish eggs or organic detritus (Vladykov and Kott 1980). Cochran (2008) found acanthocephalans in the guts of two of the giants, an indication that these parasites were acquired through predatory feeding or carrion feeding.

Balon and Holčík (1964) reported a lamprey-induced wound on the side of a European chub *Squalius cephalus* (reported as *Leuciscus cephalus*) in Jelešná Brook, Orava Valley Reservoir basin, Slovakia. However, all 590 lamprey specimens, collected from 10 localities throughout the basin (Holčík et al. 1965), belonged to the non-parasitic Ukrainian brook lamprey *Eudontomyzon mariae* (reported as *Lampetra vladykovi*). Such an instance of post-metamorphic feeding is probably rare in the species as Abakumov (1966) found that Ukrainian brook lamprey adults from the Kuban' River basin, Russia, never develop a completely patent foregut, thus effectively precluding their feeding.

Perhaps, American and Ukrainian brook lampreys have only relatively recently diverged from their parasitic ancestors and on occasion will exhibit atavistic behavior (Renaud 1982; Docker 2009; see Chap. 4). An indication of that recent divergence, at least in American brook lamprey, is the discovery of chloride cells in recently metamorphosed individuals of this species (Bartels et al. 2011). This adaptation for dealing with hypertonic marine conditions was apparently inherited from the Arctic lamprey, its anadromous and parasitic ancestor, but is no longer required. However, a study of the prevalence of chloride cells among all lampreys in relation to their taxonomic relationships is needed to properly evaluate this link.

A third case of facultative parasitism may be that of the enigmatic population of western brook lamprey in Morrison Creek, British Columbia, the taxonomically unrecognized *marifuga* variety, in which feeding was observed in adults under laboratory conditions, but not in the field (Beamish and Withler 1986; Beamish 1987b). Renaud (1997) suggested an alternative hypothesis, that this population is in fact a permanent freshwater form of the morphologically-similar, parasitic and usually anadromous western river lamprey. Permanent freshwater forms of other anadromous parasitic species are well known in the Northern Hemisphere Petromyzontidae, occurring in the congeneric European river lamprey, as well as in Pacific lamprey, Arctic lamprey, and sea lamprey (Renaud 1997; see Chap. 4). Further research is needed to conclusively establish whether the Morrison Creek population feeds as an adult in the natural environment and also whether it possesses chloride cells.

3.6 Conclusions

Post-metamorphic feeding behavior in lampreys is complex and highly variable in terms of daily timing of attacks, host species, size and site selectivity, the duration of attachments and the lethality of the attacks. While the functional morphology and the main behavioral characteristics of the two principal modes of feeding (i.e., blood feeding and flesh feeding) are well established, much remains unknown. For example, the hosts for four out of the 18 parasitic species have yet to be identified and the duration of the feeding phase in six species has not been determined. Future studies should seek to clarify these lacunae.

Acknowledgements We thank the many colleagues and students who have assisted us in the field and in the laboratory. D. A. Watkinson and H. Massé provided positive identification for three additional species of fish hosts for *Ichthyomyzon castaneus* and J. Lyons did the same for two additional fish hosts for *I. unicuspis*. Nina Bogutskaya translated Ioganzen (1935) and Gritsenko (1968).

References

Abakumov VA (1964) On the marine phase of the Pacific three-toothed lamprey, *Entosphenus tridentatus* (Richardson). Inst Morsk Rybn Khoz Okeanogr 49:253–256 [in Russian]

Abakumov VA (1966) Systematics and ecology of the Ukrainian lamprey (*Lampetra mariae* Berg). Vop Ikhtiol 6:609–618 [in Russian]

Abou-Seedo FS, Potter IC (1979) The estuarine phase in the spawning run of the river lamprey *Lampetra fluviatilis*. J Zool 188:5–25

Allen GR (1989) Freshwater fishes of Australia. TFH Publications, Neptune City, NJ

Allin AE (1951) Records of the sea lamprey and the silver lamprey from Canadian waters of the western end of Lake Superior. Can Field-Nat 65:184–185

Álvarez del Villar J (1966) Ictiologia michoacana, IV. Contribución al conocimiento biológico y sistemático de las lampreas de Jacona, Mich., México. An Esc Nac Cienc Biol Méx 13:107–144

Applegate VC (1950) Natural history of the sea lamprey (*Petromyzon marinus*) in Michigan. US Fish Wildl Serv Spec Sci Rep Fish 55:1–237

Araújo MJ, Novais D, Antunes C (2013) Record of a newly metamorphosed sea lamprey (*Petromyzon marinus* Linnaeus, 1758) feeding on a freshwater fish. J Appl Ichthyol 29:1380–1381

Bahr K (1933) Das Flussneunauge (*Lampetra fluviatilis*) als Urheber von Fischverletzungen. Mitt Dtsch Seefisch-Ver 49:3–8

Balon EK, Holčík J (1964) Kilka nowych dla Polski form krągłoustych i ryb dorzecza Dunaju (Czarna Orawa). Fragm Faunist 11:189–206

Bănărescu P (1969) Cyclostomata şi Chondrichthyes (ciclostomi şi selacieni). Fauna Repub Social România 12:5–54

Bartels H, Fazekas U, Youson JH, Potter IC (2011) Changes in the cellular composition of the gill epithelium during the life cycle of a nonparasitic lamprey: functional and evolutionary implications. Can J Zool 89:538–545

Beamish FWH (1980a) Osmoregulation in juvenile and adult lampreys. Can J Fish Aquat Sci 37:1739–1750

Beamish FWH (1980b) Biology of the North American anadromous sea lamprey, *Petromyzon marinus*. Can J Fish Aquat Sci 37:1924–1943

Beamish FWH, Potter IC (1975) The biology of the anadromous sea lamprey (*Petromyzon marinus*) in New Brunswick. J Zool 177:57–72

Beamish RJ (1980) Adult biology of the river lamprey (*Lampetra ayresi*) and the Pacific lamprey (*Lampetra tridentata*) from the Pacific coast of Canada. Can J Fish Aquat Sci 37:1906–1923

Beamish RJ (1982) *Lampetra macrostoma*, a new species of freshwater parasitic lamprey from the west coast of Canada. Can J Fish Aquat Sci 39:736–747

Beamish RJ (1987a) Status of the lake lamprey, *Lampetra macrostoma*, in Canada. Can Field-Nat 101:186–189

Beamish RJ (1987b) Evidence that parasitic and nonparasitic life history types are produced by one population of lamprey. Can J Fish Aquat Sci 44:1779–1782

Beamish RJ (2001) Updated status of the Vancouver Island Lake Lamprey, *Lampetra macrostoma*, in Canada. Can Field-Nat 115:127–130

Beamish RJ, Neville C-EM (1995) Pacific salmon and Pacific herring mortalities in the Fraser River plume caused by river lamprey (*Lampetra ayresi*). Can J Fish Aquat Sci 52:644–650

Beamish RJ, Withler RE (1986) A polymorphic population of lampreys that may produce parasitic and a nonparasitic varieties. In: Uyeno T, Arai R, Taniuchi T, Matsuura K (eds) Indo-Pacific fish biology: proceedings of the second international conference on Indo-Pacific fishes. Ichthyological Society of Japan, Tokyo, pp 31–49

Becker GC (1983) Fishes of Wisconsin. University of Wisconsin Press, Madison, WI

Bence JR, Bergstedt RA, Christie GC et al (2003) Sea lamprey (*Petromyzon marinus*) parasite-host interactions in the Great Lakes. J Great Lakes Res 29(Suppl 1):253–282

Bensley BA (1915) The fishes of Georgian Bay. Contributions to Canadian Biology Sessional Paper 39b Fasc II:1–51 + 6 figs + 2 pls

Berg LS (1931) A review of the lampreys of the northern hemisphere. Annuaire du Musée Zoologique de l'Académie des Sciences de l'URSS 32:87–116 + 8 pls

Berg LS (1948) Freshwater fishes of the USSR and adjacent countries. Guide to the Fauna of the USSR No. 27, vol 1:1–504

Bergstedt RA, Swink WD (1995) Seasonal growth and duration of the parasitic life stage of the landlocked sea lamprey (*Petromyzon marinus*). Can J Fish Aquat Sci 52:1257–1264

Bergstedt RA, Schneider CP, O'Gorman R (2001) Lethality of lamprey attacks on lake trout in relation to location on the body surface. Trans Am Fish Soc 130:336–340

Berst AH, Wainio AA (1967) Lamprey parasitism of rainbow trout in southern Georgian Bay. J Fish Res Board Can 24:2539–2548

Bigelow HB, Schroeder WC (1948) Cyclostomes. Memoir of Sears Foundation for Marine Research No. 1, Part 1:29–58

Bird DJ, Potter IC, Hardisty MW, Baker BI (1994) Morphology, body size and behaviour of recently-metamorphosed sea lampreys, *Petromyzon marinus*, from the lower River Severn, and their relevance to the onset of parasitic feeding. J Fish Biol 44:67–74

Birman IB (1950) Parasitism of salmon of the genus *Oncorhynchus* by the Pacific lamprey. Izvestiia TINRO 32:158–160 [in Russian]

Bodznick D, Preston DG (1983) Physiological characterization of electroreceptors in the lampreys *Ichthyomyzon unicuspis* and *Petromyzon marinus*. J Comp Physiol A 152:209–217

Bollman CH (1890) A report upon the fishes of Kalamazoo, Calhoun, and Antrim counties, Michigan. Bull US Fish Comm 8:219–225

Bond CE, Kan TT (1973) *Lampetra (Entosphenus) minima* n. sp., a dwarfed parasitic lamprey from Oregon. Copeia 1973:568–574

Bond CE, Kan TT, Myers KW (1983) Notes on the marine life of the river lamprey, *Lampetra ayresi*, in Yaquina Bay, Oregon, and the Columbia River estuary. Fish Bull 81:165–167

Borri C (1922) L'apparecchio labiale dei petromyzonti. Atti Soc Toscana Sci Nat 34:249–316

Chase SD (2001) Contributions to the life history of adult Pacific lamprey (*Lampetra tridentata*) in the Santa Clara River of southern California. Bull South Calif Acad Sci 100:74–85

Choudhury A, Dick TA (1993) Parasites of lake sturgeon, *Acipenser fulvescens* (Chondrostei: Acipenseridae), from central Canada. J Fish Biol 42:571–584

Christie WJ, Kolenosky DP (1980) Parasitic phase of the sea lamprey (*Petromyzon marinus*) in Lake Ontario. Can J Fish Aquat Sci 37:2021–2038

Cochran PA (1985) Size-selective attack by parasitic lampreys: consideration of alternate null hypotheses. Oecologia 67:137–141

Cochran PA (1986a) The daily timing of lamprey attacks. Environ Biol Fish 16:325–329

Cochran PA (1986b) Attachment sites of parasitic lampreys: comparisons among species. Environ Biol Fish 17:71–79

Cochran PA (1994) Why lampreys and humans "compete" (and when they don't): toward a theory of host species selection by parasitic lampreys. In: Stouder DJ, Fresh KL, Feller RJ (eds) Theory and application in fish feeding ecology. Belle W. Baruch Library in Marine Science, No 18, University of South Carolina Press, Columbia, SC, pp 329–345

Cochran PA (2004) Historical notes on lampreys in Wisconsin. Am Curr 30:4–8

Cochran PA (2008) Observations on giant American brook lampreys (*Lampetra appendix*). J Freshw Ecol 23:161–164

Cochran PA (2009) A comparison of native and exotic hosts for the silver lamprey (*Ichthyomyzon unicuspis*). In: Brown LR, Chase SD, Mesa MG, Beamish RJ, Moyle PB (eds) Biology, management and conservation of lampreys in North America. Am Fish Soc Sym 72:165–172

Cochran PA (2014) Field and laboratory observations on the ecology and behavior of the chestnut lamprey (*Ichthyomyzon castaneus*). J Freshw Ecol 29:491–505

Cochran PA, Jenkins RE (1994) Small fishes as hosts for parasitic lampreys. Copeia 1994:499–504

Cochran PA, Kitchell JF (1986) Use of modeling to investigate potential feeding strategies of parasitic lampreys. Environ Biol Fish 16:219–223

Cochran PA, Kitchell JF (1989) A model of feeding by parasitic lampreys. Can J Fish Aquat Sci 46:1845–1852

Cochran PA, Lyons J (2004) Field and laboratory observations on the ecology and behavior of the silver lamprey (*Ichthyomyzon unicuspis*) in Wisconsin. J Freshw Ecol 19:245–253

Cochran PA, Lyons J (2010) Attachments by parasitic lampreys within the branchial cavities of their hosts. Environ Biol Fish 88:343–348

Cochran PA, Lyons J (2016) The silver lamprey and the paddlefish. In: Orlov A, Beamish RJ (eds) Jawless fishes of the world, vol 2. Cambridge Scholars Publishing, Newcastle upon Tyne, pp 214–233

Cochran PA, Lyons J, Merino-Nambo E (1996) Notes on the biology of the Mexican lampreys, *Lampetra spadicea* and *L. geminis* (Agnatha: Petromyzontidae). Ichthyol Explor Freshw 7:173–180

Cochran PA, Swink WD, Kinziger AP (1999) Testing and extension of a sea lamprey feeding model. Trans Am Fish Soc 128:403–413

Cochran PA, Hodgson JY, Kinziger AP (2003a) Change in the energy density of the sea lamprey (*Petromyzon marinus*) during its parasitic phase: implications for modeling food consumption and growth. J Great Lakes Res 29(Suppl 1):297–306

Cochran PA, Lyons J, Gehl MR (2003b) Parasitic attachments by overwintering silver lampreys, *Ichthyomyzon unicuspis*, and chestnut lampreys, *Ichthyomyzon castaneus*. Environ Biol Fish 68:65–71

Cooper EL (1983) Fishes of Pennsylvania and the northeastern United States. The Pennsylvania State University Press, University Park, PA

Coots M (1955) The Pacific lamprey, *Entosphenus tridentatus*, above Copco Dam, Siskiyou County, California. Calif Fish Game 41:118–119

Cross FB, Metcalf AL (1963) Records of three lampreys (*Ichthyomyzon*) from the Missouri River system. Copeia 1963:187

Davis RM (1967) Parasitism by newly-transformed anadromous sea lampreys on landlocked salmon and other fishes in a coastal Maine lake. Trans Am Fish Soc 96:11–16

Dawson HA, Quintella BR, Almeida PR, Treble AJ, Jolley JC (2015) The ecology of larval and metamorphosing lampreys. In: Docker MF (ed) Lampreys: biology, conservation and control, vol 1. Springer, Dordrecht, pp 75–137

de Castelnau F (1872) Contribution to the ichthyology of Australia. No. I. The Melbourne fish market. Proc Zool Acclim Soc Vic 1:29–242

Dill LM (1983) Adaptive flexibility in the foraging behavior of fishes. Can J Fish Aquat Sci 40:398–408

Docker MF (2009) A review of the evolution of nonparasitism in lampreys and an update of the paired species concept. In: Brown LR, Chase SD, Mesa MG, Beamish RJ, Moyle PB (eds) Biology, management, and conservation of lampreys in North America. Am Fish Soc Sym 72:71–114

Docker MF, Hume JB, Clemens BJ (2015) Introduction: a surfeit of lampreys. In: Docker MF (ed) Lampreys: biology, conservation and control, vol 1. Springer, Dordrecht, pp 1–34

Ebener MP, Bence JR, Bergstedt RA, Mullet KM (2003) Classifying sea lamprey marks on Great Lakes lake trout: observer agreement, evidence on healing time between classes, and recommendations for reporting of marking statistics. J Great Lakes Res 29(Suppl 1):283–296

Ebener MP, King EL Jr, Edsall TA (2006) Application of a dichotomous key to the classification of sea lamprey marks on Great Lakes fish. Great Lakes Fish Comm Misc Publ 2006–02:1–21

Edsall CC (1999) A blood chemistry profile for lake trout. J Aquat Anim Health 11:81–86

Edsall CC, Swink WD (2001) Effects of nonlethal sea lamprey attack on the blood chemistry of lake trout. J Aquat Anim Health 13:51–55

Eglite RM (1958) Feeding habits of *Lampetra fluviatilis* L. in the sea. Zool Zhurnal 37:1509–1514 [in Russian]

Farmer GJ (1974) Food consumption, growth and host preferences of the sea lamprey, *Petromyzon marinus* L. PhD thesis, University of Guelph, Guelph, ON

Farmer GJ (1980) Biology and physiology of feeding in adult lampreys. Can J Fish Aquat Sci 37:1751–1761

Farmer GJ, Beamish FWH (1973) Sea lamprey (*Petromyzon marinus*) predation on freshwater teleosts. J Fish Res Board Can 30:601–605

Farmer GJ, Beamish FWH, Robinson GA (1975) Food consumption of the adult landlocked sea lamprey, *Petromyzon marinus* L. Comp Biochem Physiol A: Mol Integr Physiol 50:753–757

Flammang MK, Olson JR (2010) The occurrence of chestnut lamprey (*Ichthyomyzon castaneus*; Pisces: Petromyzontidae) in the Chariton River in south-central Iowa. J Iowa Acad Sci 117:1–3

Gallant J, Harvey-Clark C, Myers RA, Stokesbury MJW (2006) Sea lamprey attached to a Greenland shark in the St. Lawrence estuary, Canada. Northeast Nat 13:35–38

George CJ (1985) Occurrence of the silver lamprey in the Stillwater sector of the Hudson River. N Y Fish Game J 32:95

Gill HS, Renaud CB, Chapleau F, Mayden RL, Potter IC (2003) Phylogeny of living parasitic lampreys (Petromyzontiformes) based on morphological data. Copeia 2003:687–703

Ginzburg YI (1970) Reproduction of the lamprey [*Caspiomyzon wagneri* (Kessler)] below the Volgograd dam and the development of its larvae. J Ichthyol 10:485–493

Goodwin CE, Griffiths D, Dick JTA, Elwood RW (2006) A freshwater-feeding *Lampetra fluviatilis* L. population in Lough Neagh, Northern Ireland. J Fish Biol 68:628–633

Gradwell N (1972) Hydrostatic pressure and movements of the lamprey, *Petromyzon*, during suction, olfaction, and gill ventilation. Can J Zool 50:1215–1223

Grinyuk IN (1970) The sea lamprey [*Petromyzon marinus* (L.)] caught off the Murmansk coast. J Ichthyol 10:135–137

Gritsenko OF (1968) On the question of an ecological parallelism between lamprey and salmons. Inst Ryb Khoz Okeanogr 65:157–169 [in Russian]

Grossu A, Homei, V, Barbu P, Popescu A (1962) Contribution à l'étude des pétromyzonides de la République Populaire Roumaine. Trav Mus Hist Nat "Gr. Antipa" 3:253–279

Gudger EW (1930) *Ichthyomyzon concolor*, in the Coosa River at Rome, Georgia, with notes on other lampreys in our South Atlantic and Gulf drainages. Copeia 1930:145–146

Hall GE, Moore GA (1954) Oklahoma lampreys: their characterization and distribution. Copeia 1954:127–135

Hall JD (1960) Preliminary studies on the biology of native Michigan lampreys. MS thesis, University of Michigan, Ann Arbor, MI

Hall JD (1963) An ecological study of the chestnut lamprey, *Ichthyomyzon castaneus* Girard, in the Manistee River, Michigan. PhD thesis, University of Michigan, Ann Arbor, MI

Halliday RG (1991) Marine distribution of the sea lamprey (*Petromyzon marinus*) in the northwest Atlantic. Can J Fish Aquat Sci 48:832–842

Hardisty MW (1969) Information on the growth of the ammocoete larva of the anadromous sea lamprey, *Petromyzon marinus* in British rivers. J Zool 159:139–144

Heard WR (1966) Observations on lampreys in the Naknek River system of southwest Alaska. Copeia 1966:332–339

Henderson BA (1986) Effect of sea lamprey (*Petromyzon marinus*) parasitism on the abundance of white suckers (*Catostomus commersoni*) in South Bay, Lake Huron. J Anim Ecol 23:381–389

Heyning JE (2002) Cuvier's beaked whale *Ziphius cavirostris*. In: Perrin WF, Würsig B, Thewissen JGM (eds) Encyclopedia of marine mammals. Academic Press, London, pp 305–307

Holčík J (1963) Notes on the Czechoslovakian lampreys with redescription of *Lampetra* (*Eudontomyzon*) *vladykovi* (Oliva and Zanandrea), 1959. Věstník Československé Společnosti Zoologické 27:51–61 + figs 3, 6–8

Holčík J (1986) *Lethenteron japonicum* (Martens, 1868). In: Holčík J (ed) The freshwater fishes of Europe, vol 1, Part I. Petromyzontiformes. AULA-Verlag, Wiesbaden, Germany, pp 198–219

Holčík J, Mišík V, Bastl I, Kirka A (1965) Ichtyologický výskum Karpatského Oblúka. 3. Ichtyofauna povodia Oravskej Priehrady a jej prítokov. Acta Rer Nat Mus Natur Slov 11:93–139

Holčík J, Delić A, Kučinić M, Bukvić V, Vater M (2004) Distribution and morphology of the sea lamprey from the Balkan coast of the Adriatic Sea. J Fish Biol 64:514–527

Hubbs CL, Trautman MB (1937) A revision of the lamprey genus *Ichthyomyzon*. Misc Publ Mus Zool Univ Mich 35:7–109 + 2 pls

Hubley RC Jr (1961) Incidence of lamprey scarring on fish in the upper Mississippi River, 1956–58. Trans Am Fish Soc 90:83–85

Hume JB, Adams CE, Bean CW, Maitland PS (2013) Evidence of a recent decline in river lamprey *Lampetra fluviatilis* parasitism of a nationally rare whitefish *Coregonus lavaretus*: is there a diamond in the ruffe *Gymnocephalus cernuus*? J Fish Biol 82:1708–1716

Hunter JG (1981) Distribution and abundance of fishes of the south eastern Beaufort Sea. Beaufort Sea Technical Report No. 7, Beaufort Sea Project, Department of Fisheries and Environment, Victoria, BC

Inger R, McDonald RA, Rogowski D et al (2010) Do non-native invasive fish support elevated lamprey populations? J Appl Ecol 47:121–129

Ioganzen [Johansen] BG (1935) Morphobiological features of cyclostomes of Siberia. Zool Zhurnal 14:353–370 [in Russian]

Japha A (1910) Weitere Beiträge zur Kenntnis der Walhaut. Zool Jahrb Suppl 12:711–718

Jenkins RE, Burkhead NM (1993) Freshwater fishes of Virginia. American Fisheries Society, Bethesda, MD

Jensen C, Schwartz FJ (1994) Atlantic Ocean occurrences of the sea lamprey, *Petromyzon marinus* (Petromyzontiformes: Petromyzontidae), parasitizing sandbar, *Carcharhinus plumbeus*, and dusky, *C. obscurus* (Carcharhiniformes: Carcharhinidae), sharks off North and South Carolina. Brimleyana 21:69–72

Jensen C, Schwartz FJ, Hopkins G (1998) A sea lamprey (*Petromyzon marinus*)-tiger shark (*Galeocerdo cuvier*) parasitic relationship off North Carolina. J Elisha Mitchell Sci Soc 114:72–73

Johnson BGH, Anderson WC (1980) Predatory-phase sea lampreys (*Petromyzon marinus*) in the Great Lakes. Can J Fish Aquat Sci 37:2007–2020

Kan TT (1975) Systematics, variation, distribution, and biology of lampreys of the genus *Lampetra* in Oregon. PhD thesis, Oregon State University, Corvallis, OR

Kan TT, Bond CE (1981) Notes on the biology of the Miller Lake lamprey *Lampetra* (*Entosphenus*) *minima*. Northwest Sci 55:70–74

Kawasaki R, Rovainen CM (1988) Feeding behavior by parasitic phase lampreys, *Ichthyomyzon unicuspis*. Brain Behav Evol 32:317–329

Keating WH (1824) Narrative of an expedition to the source of St. Peter's River, Lake Winnepeek, Lake of the Woods, &c, &c. performed in the year 1823, by order of the Hon. J. C. Calhoun, Secretary of War, under the command of Stephen H. Long, Major U.S.T.E. compiled from the notes of Major Long, Messers. Say, Keating, and Colhoun [sic], vol II. HC Carey & I Lea, Philadelphia

Keleher JJ (1952) Notes on fishes collected from Lake Winnipeg region. Can Field-Nat 66:170–173

Khidir KT, Renaud CB (2003) Oral fimbriae and papillae in parasitic lampreys (Petromyzontiformes). Environ Biol Fish 66:271–278

King EL Jr (1980) Classification of sea lamprey (*Petromyzon marinus*) attack marks on Great Lakes lake trout (*Salvelinus namaycush*). Can J Fish Aquat Sci 37:1989–2006

King EL Jr, Edsall TA (1979) Illustrated field guide for the classification of sea lamprey attack marks on Great Lakes lake trout. Great Lakes Fish Comm Spec Publ 79–1:1–42

Kinnunen RE, Johnson HE (1985) Impact of sea lamprey parasitism on the blood features and hemopoietic tissues of rainbow trout. Great Lakes Fish Comm Tech Rep 46:1–17

Kitchell JF, Breck JE (1980) Bioenergetics model and foraging hypothesis for sea lamprey (*Petromyzon marinus*). Can J Fish Aquat Sci 37:2159–2168

Kleerekoper H (1972) The sense organs. In: Hardisty MW, Potter IC (eds) The biology of lampreys, vol 2. Academic Press, London, pp 373–404

Kleerekoper H, Mogensen J (1963) Role of olfaction in the orientation of *Petromyzon marinus*. I. Response to a single amine in prey's body odor. Physiol Zool 36:347–360

Knapp FT (1951) Additional reports of lampreys from Texas. Copeia 1951:87

Kucheryavyy A, Tsimbalov I, Kirillova E, Nazarov D, Pavlov D (2016) The need for a new taxonomy for lampreys. In: Orlov A, Beamish RJ (eds) Jawless fishes of the world, vol 1. Cambridge Scholars Publishing, Newcastle upon Tyne, pp 251–277

Kux Z (1965) *Lampetra gracilis*, nový neparasitický druh mihule z východního slovenska. Acta Mus Morav 50:293–302

Kux Z, Weisz T (1960) Příspěvek k poznání ichtyofauny Dunajce, Propradu, Váhu a Hronu. Acta Mus Morav 45:203–240

Lanteigne J (1981) The taxonomy and distribution of the North American lamprey genus *Ichthyomyzon*. MSc thesis, University of Ottawa, Ottawa, ON

Lanteigne J (1988) Identification of lamprey larvae of the genus *Ichthyomyzon* (Petromyzontidae). Environ Biol Fish 23:55–63

Lanzing W (1958) Structure and function of the suction apparatus of the lamprey. Proc K Ned Akad Wet Ser C Biol Med Sci 61:300–307

Lanzing WJR (1959) Studies on the river lamprey, *Lampetra fluviatilis*, during its anadromous migration. Uitgeversmaatschappij Neerlandia 1959:11–82

Larsen LO (1980) Physiology of adult lampreys, with special regard to natural starvation, reproduction, and death after spawning. Can J Fish Aquat Sci 37:1762–1779

Lennon RE (1954) Feeding mechanism of the sea lamprey and its effect on host fishes. US Fish Wildl Serv Fish Bull 56:247–293

Lethbridge RC, Potter IC (1981) The skin. In: Hardisty MW, Potter IC (eds) The biology of lampreys, vol 3. Academic Press, London, pp 377–448

Lorion CM, Markle DF, Reid SB, Docker MF (2000) Redescription of the presumed-extinct Miller Lake Lamprey, *Lampetra minima*. Copeia 2000:1019–1028

Lyons J (1993) Status and biology of paddlefish (*Polyodon spathula*) in the lower Wisconsin River. Trans Wisc Acad Sci Arts Lett 81:123–135

Ma CF, Yu CL (1959) Preliminary observations on *L. morii*. Chin J Zool 3:115–117 [in Chinese]

Madenjian CP, Cochran PA, Bergstedt RA (2003) Seasonal patterns in growth, blood consumption, and effects on hosts by parasitic-phase sea lampreys in the Great Lakes: an individual-based model approach. J Great Lakes Res 29(Suppl 1):332–346

Madenjian CP, Chipman BD, Marsden JE (2008) New estimates of lethality of sea lamprey (*Petromyzon marinus*) attacks on lake trout (*Salvelinus namaycush*): implications for fisheries management. Can J Fish Aquat Sci 65:535–542

Maitland PS (1980) Scarring of whitefish (*Coregonus lavaretus*) by European river lamprey (*Lampetra fluviatilis*) in Loch Lomond, Scotland. Can J Fish Aquat Sci 37:1981–1988

Maitland PS, Morris KH, East K et al (1984) The estuarine biology of the river lamprey, *Lampetra fluviatilis*, in the Firth of Forth, Scotland, with particular reference to size composition and feeding. J Zool 203:211–225

Maitland PS, Morris KH, East K (1994) The ecology of lampreys (Petromyzonidae) in the Loch Lomond area. Hydrobiologia 290:105–120

Manion PJ, Purvis HA (1971) Giant American brook lampreys, *Lampetra lamottei*, in the upper Great Lakes. J Fish Res Board Can 28:616–620

Manion PJ, Smith BR (1978) Biology of larval and metamorphosing sea lampreys, *Petromyzon marinus*, of the 1960 year class in the Big Garlic River, Michigan, Part II, 1966–72. Great Lakes Fish Comm Tech Rep 30:1–35

Mansueti RJ (1962) Distribution of small, newly metamorphosed sea lampreys, *Petromyzon marinus*, and their parasitism on menhaden, *Brevoortia tyrannus*, in mid-Chesapeake Bay during winter months. Chesapeake Sci 3:137–139

Manzon RG, Youson JH, Holmes JA (2015) Lamprey metamorphosis. In: Docker MF (ed) Lampreys: biology, conservation, and control, vol 1. Springer, Dordrecht, pp 139–214

Mayden RL, Matson RH, Kuhajda BR, Pierson JM, Mettee MF, Frazer KS (1989) The chestnut lamprey, *Ichthyomyzon castaneus* Girard, in the Mobile basin. Southeast Fish Counc Proc 20:10–13

McAlpine DF (2002) Pygmy and dwarf sperm whales, *Kogia breviceps* and *K. sima*. In: Perrin WF, Würsig B, Thewissen JGM (eds) Encyclopedia of marine mammals. Academic Press, London, pp 1007–1009

McPhail JD, Lindsey CC (1970) Freshwater fishes of northwestern Canada and Alaska. Fish Res Board Can Bull 173:1–381

Miller RR, Williams JD, Williams JE (1989) Extinctions of North American fishes during the past century. Fisheries 14:22–38

Moody EK, Weidel BC, Ahrenstorff TD, Mattes WP, Kitchell JF (2011) Evaluating the growth potential of sea lampreys (*Petromyzon marinus*) feeding on siscowet lake trout (*Salvelinus namaycush*) in Lake Superior. J Great Lakes Res 37:343–348

Moore GA, Kernodle M (1965) A new size record for the chestnut lamprey, *Ichthyomyzon castaneus* Girard in Oklahoma. Proc Okla Acad Sci 45:68–69

Moore JD, Lychwick TJ (1980) Changes in mortality of lake trout (*Salvelinus namaycush*) in relation to increased sea lamprey (*Petromyzon marinus*) abundance in Green Bay, 1974–78. Can J Fish Aquat Sci 37:2052–2056

Moser ML, Almeida PR, Kemp PS, Sorenson PW (2015) Lamprey spawning migration. In: Docker MF (ed) Lampreys: biology, conservation and control, vol 1. Springer, Dordrecht, pp 215–263

Mundahl ND, Erickson C, Johnston MR, Sayeed GA, Taubel S (2005) Diet, feeding rate, and assimilation efficiency of American brook lamprey larvae. Environ Biol Fish 72:67–72

Murauskas JG, Orlov AM, Siwicke KA (2013) Relationships between the abundance of Pacific lamprey in the Columbia River and their common hosts in the marine environment. Trans Am Fish Soc 142:143–155

Neira FJ (1984) Biomorfologia de las lampreas parasitas chilenas *Geotria australis* Gray, 1851 y *Mordacia lapicida* (Gray, 1851) (Petromyzoniformes). Gayana Zool 48:3–40

Neira FJ, Bradley JS, Potter IC, Hilliard RW (1988) Morphological variation among widely dispersed larval populations of anadromous southern hemisphere lampreys (Geotriidae and Mordaciidae). Zool J Linn Soc 92:383–408

Nemoto T (1955) White scars on whales (I) Lamprey marks. Sci Rep Whales Res Inst Bull 10:69–77

Nichols OC, Hamilton PK (2004) Occurrence of the parasitic sea lamprey, *Petromyzon marinus*, on western North Atlantic right whales, *Eubalaena glacialis*. Environ Biol Fishes 71:413–417

Nichols OC, Tscherter UT (2011) Feeding of sea lampreys *Petromyzon marinus* on minke whales *Balaenoptera acutorostrata* in the St Lawrence Estuary, Canada. J Fish Biol 78:338–343

Nikolskii GV (1956) Some data on marine period of life of Pacific lamprey *Lampetra japonica* (Martens). Zoologicheskii Zhurnal 35:588–591 [in Russian]

Noltie DB (1987) Incidence and effects of sea lamprey (*Petromyzon marinus*) parasitism on breeding pink salmon (*Oncorhynchus gorbuscha*) from the Carp River, eastern Lake Superior. Can J Fish Aquat Sci 44:1562–1567

Novikov NP (1963) Cases of the Pacific three-toothed lamprey *Entosphenus tridentatus* (Gairdner) attacking halibut and other fish in the Bering Sea. Vopr Ikhtiol 3:567–569 [in Russian]

Novomodnyy GV, Belyaev VA (2002) Predation by lamprey smolts *Lampetra japonica* as the main cause of Amur chum salmon and pink salmon mortality in the early sea period of life. N Pac Anadromous Fish Comm Tech Rep 4:81–82

Nuhfer AJ (1993) Chestnut lamprey predation on caged, and free-living brown trout in the Upper Manistee River, Michigan. Michigan Department of Natural Resources, Fisheries Division, Fisheries Research Report 1986, Lansing, MI

Nursall JR, Buchwald D (1972) Life history and distribution of the Arctic lamprey (*Lethenteron japonicum* (Martens)) of Great Slave Lake, N.W.T. Fish Res Board Canada Tech Rep 304:1–28

Oliva O (1953) Příspěvek k přehledu našich mihulí (Petromyzones Berg 1940). Věstník Královské české společnosti nauk 9:1–19 + 2 pls

Orlov A (2016) Relationships between Pacific lamprey and their prey. In: Orlov A, Beamish R (eds) Jawless fishes of the world, vol 2. Cambridge Scholars Publishing, Newcastle upon Tyne, pp 234–285

Orlov AM, Savinyh VF, Pelenev D (2008) Features of the spatial distribution and size structure of the Pacific lamprey *Lampetra tridentata* in the North Pacific. Russ J Mar Biol 34:276–287

Orlov AM, Beamish RJ, Vinnikov AV, Pelenev DV (2009) Feeding and prey of Pacific lamprey in coastal waters of the western North Pacific. In: Haro AJ, Smith KL, Rulifson RA et al (eds) Challenges for diadromous fishes in a dynamic global environment. Am Fish Soc Sym 69:875–877

Orlov AM, Baitalyuk AA, Pelenev DV (2014) Distribution and size composition of the Arctic lamprey *Lethenteron camtschaticum* in the North Pacific. Oceanology 54:180–194

Page LM, Espinosa-Pérez H, Findley LT et al (2013) Common and scientific names of fishes from the United States, Canada, and Mexico, 7th edn. American Fisheries Society Special Publication 34, Bethesda, MD

Patrick HK, Sutton TM, Swink WD (2007) Application of a dichotomous key to the classification of sea lamprey *Petromyzon marinus* marks on lake sturgeon *Acipenser fulvescens*. Great Lakes Fish Comm Misc Publ 2007–02:1–25

Patrick HK, Sutton TM, Swink WD (2009) Lethality of sea lamprey parasitism on lake sturgeon. Trans Am Fish Soc 138:1065–1075

Pearce WA, Braem RA, Dustin SM et al (1980) Sea lamprey (*Petromyzon marinus*) in the lower Great Lakes. Can J Fish Aquat Sci 37:1802–1810

Pelenev D, Orlov A, Klovach N (2008) Predator-prey relations between the Pacific lamprey *Lampetra tridentatus* and Pacific salmon (*Oncorhynchus* spp.). North Pacific Anadromous Fish Commission Document 1097, Russian Federal Research Institute of Fisheries and Oceanography (VNIRO), Moscow

Penczak T (1964) Report on catching *Petromyzon marinus* L. in the river of Pilica. Naturwissenschaften 51:322

Pflieger WL (1997) The fishes of Missouri. Conservation Commission of the State of Missouri, Jefferson City, MO

Philippi RA (1865) Ueber die chilenische Anguilla. Arch Naturgeschichte 31:107–109

Pike GC (1951) Lamprey marks on whales. J Fish Res Board Can 8:275–280

Potter IC, Beamish FWH (1977) The freshwater biology of adult anadromous sea lampreys *Petromyzon marinus*. J Zool 181:113–130

Potter IC, Hilliard RW (1987) A proposal for the functional and phylogenetic significance of differences in the dentition of lampreys (Agnatha: Petromyzontiformes). J Zool 212:713–737

Potter IC, Osborne TS (1975) The systematics of British larval lampreys. J Zool 176:311–329

Potter IC, Lanzing WJR, Strahan R (1968) Morphometric and meristic studies on populations of Australian lampreys of the genus *Mordacia*. J Linn Soc Lond Zool 47:533–546

Potter IC, Prince PA, Croxall JP (1979) Data on the adult marine and migratory phases in the life cycle of the southern hemisphere lamprey, *Geotria australis* Gray. Environ Biol Fish 4:65–69

Potter IC, Hilliard RW, Bird DJ, Macey DJ (1983) Quantitative data on morphology and organ weights during the protracted spawning-run period of the southern hemisphere lamprey *Geotria australis*. J Zool Soc 200:1–20

Potter IC, Gill HS, Renaud CB, Haoucher D (2015) The taxonomy, phylogeny, and distribution of lampreys. In: Docker MF (ed) Lampreys: biology, conservation and control, vol 1. Springer, Dordrecht, pp 35–73

Pycha RL (1980) Changes in mortality of lake trout (*Salvelinus namaycush*) in Michigan waters of Lake Superior in relation to sea lamprey (*Petromyzon marinus*) predation, 1968–78. Can J Fish Aquat Sci 37:2063–2073

Randall DJ (1972) Respiration. In: Hardisty MW, Potter IC (eds) The biology of lampreys, vol 2. Academic Press, London, pp 287–306

Renaud CB (1982) Revision of the lamprey genus *Eudontomyzon* Regan, 1911. MSc thesis, University of Ottawa, ON

Renaud CB (1997) Conservation status of northern hemisphere lampreys (Petromyzontidae). J Appl Ichthyol 13:143–148

Renaud CB (2002) The muskellunge, *Esox masquinongy*, as a host for the silver lamprey, *Ichthyomyzon unicuspis*, in the Ottawa River, Ontario/Québec. Can Field-Nat 116:433–440

Renaud CB (2011) Lampreys of the world, an annotated and illustrated catalogue of the lamprey species known to date. FAO Species Cat Fish Purp 5:1–109

Renaud CB, de Ville N (2000) Three records of the chestnut lamprey, *Ichthyomyzon castaneus*, new to Québec. Can Field-Nat 114:333–335

Renaud CB, Gill HS, Potter IC (2009) Relationships between the diets and characteristics of the dentition, buccal glands and velar tentacles of the adults of the parasitic species of lamprey. J Zool 278:231–242

Reynolds FE (1931) Hydrostatics of the suctorial mouth of the lamprey. Univ Calif Publ Zool 37:15–34

Riske ME (1960) A comparative study of north Pacific and Canadian Arctic herring (*Clupea*). MSc thesis, University of Alberta, Edmonton, AB

Roos JF, Gilhousen P, Killick SR et al (1973) Parasitism on juvenile Pacific salmon (*Oncorhynchus*) and Pacific herring (*Clupea harengus pallasi*) in the Strait of Georgia by the river lamprey (*Lampetra ayresi*). J Fish Res Board Can 30:565–568

Roy J-M (1973) Croissance, comportement et alimentation de la lamproie du nord (*Ichthyomyzon unicuspis*, Hubbs et Trautman) en captivité. Trav Pêcheries Québec 41:3–144

Runstrom AL, Vondracek B, Jennings CA (2001) Population statistics for paddlefish in the Wisconsin River. Trans Am Fish Soc 130:546–556

Rutter MA, Bence JR (2003) An improved method to estimate sea lamprey wounding rate on hosts with application to lake trout in Lake Huron. J Great Lakes Res 29(Suppl 1):320–331

Salinger JM (2019) Host usage and evidence of Chestnut Lamprey distribution in selected Arkansas streams. Southeast Nat (in press)

Samarra FI, Fennell A, Aoki K, Deecke VB, Miller PJO (2012) Persistence of skin marks on killer whales (*Orcinus orca*) caused by the parasitic sea lamprey (*Petromyzon marinus*) in Iceland. Mar Mamm Sci 28:395–401

Sandstrom SJ, Lemieux PJ, Reist JD (1997) Enumeration and biological data from the upstream migration of Dolly Varden charr (*Salvelinus malma*) (W.), from the Babbage River, Yukon North Slope, 1990 to 1992. Can Data Rep Fish Aquat Sci 1018:1–132

Schneider CP, Owens RW, Bergstedt RA, O'Gorman R (1996) Predation by sea lamprey (*Petromyzon marinus*) on lake trout (*Salvelinus namaycush*) in southern Lake Ontario, 1982–1992. Can J Fish Aquat Sci 53:1921–1932

Sepúlveda MS, Patrick HK, Sutton TM (2012a) A single sea lamprey attack causes acute anemia and mortality in lake sturgeon. J Aquat Anim Health 24:91–99

Sepúlveda MS, Sutton TM, Patrick HK, Amberg JJ (2012b) Blood chemistry values for shovelnose and lake sturgeon. J Aquat Anim Health 24:135–140

Silva S, Servia MJ, Vieira-Lanero R, Cobo F (2013a) Downstream migration and hematophagous feeding of newly metamorphosed sea lampreys (*Petromyzon marinus* Linnaeus, 1758). Hydrobiologia 700:277–286

Silva S, Servia MJ, Vieira-Lanero R, Nachón DJ, Cobo F (2013b) Haematophagous feeding of newly metamorphosed European sea lampreys *Petromyzon marinus* on strictly freshwater species. J Fish Biol 82:1739–1745

Silva S, Servia MJ, Vieira-Lanero R, Barca S, Cobo F (2013c) Life cycle of the sea lamprey *Petromyzon marinus*: duration of and growth in the marine life stage. Aquat Biol 18:59–62

Silva S, Araújo MJ, Bao M, Mucientes G, Cobo F (2014) The haematophagous feeding stage of anadromous populations of sea lamprey *Petromyzon marinus*: low host selectivity and wide range of habitats. Hydrobiologia 734:187–199

Simpson JC, Wallace RL (1978) Fishes of Idaho. University Press of Idaho, Moscow, ID

Sitar SP, Bence JR, Johnson JE, Taylor WW (1997) Sea lamprey wounding rates on lake trout in Lake Huron, 1984–1994. Mich Acad 29:21–27

Smith BR (1971) Sea lampreys in the Great Lakes of North America. In: Hardisty MW, Potter IC (eds) The biology of lampreys, vol 1. Academic Press, London, pp 207–247

Smith BR, Tibbles JJ (1980) Sea lamprey (*Petromyzon marinus*) in lakes Huron, Michigan, and Superior: history of invasion and control, 1936–78. Can J Fish Aquat Sci 37:1780–1801

Smith CL (1985) The inland fishes of New York State. The New York State Department of Environmental Conservation, Albany, NY

Spangler GR, Robson DS, Regier HA (1980) Estimates of lamprey-induced mortality in whitefish, *Coregonus clupeaformis*. Can J Fish Aquat Sci 37:2146–2150

Stewart KW, Watkinson DA (2004) The freshwater fishes of Manitoba. University of Manitoba Press, Winnipeg, MB

Strahan R (1960) A comparison of the ammocoete and macrophthalmia stages of *Mordacia mordax* and *Geotria australis* (Petromyzonidae). Pac Sci 14:416–420

Surface HA (1898) The lampreys of central New York. Bull US Fish Comm 17:209–215

Surface HA (1899) Removal of lampreys from the interior waters of New York. In: Fourth annual report of the State of New York Commissioners of Fisheries, Game and Forests. Albany, NY, pp 191–245

Swanson BL, Swedberg DV (1980) Decline and recovery of the Lake Superior Gull Island Reef lake trout (*Salvelinus namaycush*) population and the role of sea lamprey (*Petromyzon marinus*) predation. Can J Fish Aquat Sci 37:2074–2080

Swink WD (1990) Effect of lake trout size on survival after a single lamprey attack. Trans Am Fish Soc 119:996–1002

Swink WD (1991) Host-size selection by parasitic sea lampreys. Trans Am Fish Soc 120:637–643

Swink WD (2003) Host selection and lethality of attacks by sea lampreys (*Petromyzon marinus*) in laboratory studies. J Great Lakes Res 29(Suppl 1):307–319

Swink WD, Fredricks KT (2000) Mortality of burbot from sea lamprey attack and initial analyses of fish blood. In: Paragamian VL, Willis D (eds) Burbot biology, ecology, and management. American Fisheries Society, Fisheries Management Section, Publication 1, Spokane, WA, pp 147–154

Swink WD, Hanson LH (1986) Survival from sea lamprey (*Petromyzon marinus*) predation by two strains of lake trout (*Salvelinus namaycush*). Can J Fish Aquat Sci 43:2528–2531

Swink WD, Hanson LH (1989) Survival of rainbow trout and lake trout after sea lamprey attack. N Am J Fish Manag 9:35–40

Swink WD, Johnson NS (2014) Growth and survival of sea lampreys from metamorphosis to spawning in Lake Huron. Trans Am Fish Soc 143:380–386

Talabishka EM, Bogutskaya NG, Naseka AM (2012) Local migration and feeding habits of Carpathian lamprey *Eudontomyzon danfordi* (Petromyzontes: Petromyzontidae) in Tisza River system (Danube drainage, Ukraine). Proc Zool Inst RAS 316:361–368

Tambs-Lyche H (1963) Norwegian Petromyzontidae. Sarsia 11:21–24

Thompson ES (1898) A list of the fishes known to occur in Manitoba. For Str 51:214

Trautman MB (1981) The fishes of Ohio with illustrated keys. Ohio State University Press, Columbus, OH

Tuunainen P, Ikonen E, Auvinen H (1980) Lampreys and lamprey fisheries in Finland. Can J Fish Aquat Sci 37:1953–1959

van Utrecht WL (1959) Wounds and scars in the skin of the common porpoise, *Phocaena phocaena* (L.). Mammalia 23:100–122

Vladykov VD (1925) Über einige neue Fische aus der Tschechoslowakei (Karpathorußland). Zool Anz 64:248–252

Vladykov VD (1931) Poissons de la Russie sous-carpathique (Tchécoslovaquie). Mém Soc Zool Fr 29:217–374

Vladykov VD (1949) Quebec lampreys (Petromyzonidae). I-List of species and their economical importance. Contrib Dép Pêcheries Québec 26:7–67

Vladykov VD (1985) Record of 61 parasitic lampreys (*Ichthyomyzon unicuspis*) on a single sturgeon (*Acipenser fulvescens*) netted in the St. Lawrence River (Québec). Nat Can 112:435–436

Vladykov VD, Follett WI (1958) Redescription of *Lampetra ayresii* (Günther) of western North America, a species of lamprey (Petromyzontidae) distinct from *Lampetra fluviatilis* (Linnaeus) of Europe. J Fish Res Board Can 15:47–77 + 15 figs

Vladykov VD, Kott E (1976) A second nonparasitic species of *Entosphenus* Gill, 1862 (Petromyzonidae) from Klamath River system, California. Can J Zool 54:974–989

Vladykov VD, Kott E (1979) A new parasitic species of the Holarctic lamprey genus *Entosphenus* Gill, 1862 (Petromyzonidae) from Klamath River, in California and Oregon. Can J Zool 57:808–823

Vladykov VD, Kott E (1980) Description and key to metamorphosed specimens and ammocoetes of Petromyzonidae found in the Great Lakes region. Can J Fish Aquat Sci 37:1616–1625

Wade J, Beamish R (2016) Trends in the catches of river and Pacific lampreys in the Strait of Georgia. In: Orlov A, Beamish R (eds) Jawless fishes of the world, vol 2. Cambridge Scholars Publishing, Newcastle upon Tyne, pp 57–72

Wagner G (1904) Notes on *Polyodon* I. Science 19:554–555

Wagner G (1908) Notes on the fish fauna of Lake Pepin. Trans Wisc Acad Sci Arts Lett 16:23–37

Weitkamp LA, Hinton SA, Bentley PJ (2015) Seasonal abundance, size, and host selection of western river (*Lampetra ayresii*) and Pacific (*Entosphenus tridentatus*) lampreys in the Columbia River estuary. Fish Bull 113:213–226

Wells L (1980) Lake trout (*Salvelinus namaycush*) and sea lamprey (*Petromyzon marinus*) populations in Lake Michigan, 1971–78. Can J Fish Aquat Sci 37:2047–2051

Wigley RL (1959) Life history of the sea lamprey of Cayuga Lake, New York. US Fish Wildl Serv Fish Bull 59:561–617

Wilkie MP, Turnbull S, Bird J et al (2004) Lamprey parasitism of sharks and teleosts: high capacity urea excretion in an extant vertebrate relic. Comp Biochem Physiol A: Mol Integr Physiol 138:485–492

Wilson DE, Reeder DM (eds) (2005) Mammal species of the world. A taxonomic and geographic reference. Johns Hopkins University Press, Baltimore, MD

Zanandrea G (1959) *Lampetra fluviatilis* catturata in mare nel golfo di Gaeta. Pubbl Stn Zool Napoli 31:265–307

Chapter 4
Life History Evolution in Lampreys: Alternative Migratory and Feeding Types

Margaret F. Docker and Ian C. Potter

Abstract Despite their highly conserved body plan and larval stage, adult life history type in lampreys diverges on two main axes related to migration and feeding. Of the 41–45 recognized lamprey species, 18 species feed parasitically after metamorphosis and their juvenile (sexually immature) feeding phase lasts from 3–4 months to 2–4 years. Nine of these species are exclusively freshwater resident; five are exclusively or almost exclusively anadromous, and four (sea lamprey, European river lamprey, Arctic lamprey, and, to a lesser extent, Pacific lamprey) are largely anadromous but with established freshwater populations. The other 23–27 described species are non-parasitic "brook" lampreys which remain within their natal streams. They initiate sexual maturation during metamorphosis, and, because the non-trophic periods of metamorphosis and sexual maturation are superimposed, the parasitic feeding phase is eliminated; this makes them the only vertebrates known to have non-trophic adults. Body size at maturity varies dramatically among life history types, ranging from ~110 to 150 mm total length (TL) in non-parasitic species to 800–900 mm TL in the anadromous sea lamprey. Freshwater forms are typically intermediate in size, although those that inhabit small systems may be no larger than non-parasitic lampreys and others (particularly the Great Lakes sea lamprey) are quite large. Some anadromous species (most notably European river lamprey, Pacific lamprey, and Arctic lamprey) show considerable intraspecific variation, consisting of typical large-bodied forms and dwarf or "praecox" forms that appear to feed at sea for a reduced period of time. Establishment in fresh water is more common in species that are consistently small-bodied or those with praecox forms. The only exceptions are the very small-bodied western river lamprey (mean TL at maturity ~200 mm), which does not produce freshwater parasitic forms (although it has given rise to innumerable non-parasitic freshwater populations), and the sea lamprey

M. F. Docker (✉)
Department of Biological Sciences, University of Manitoba, 50 Sifton Road,
Winnipeg, MB R3T 2N2, Canada
e-mail: margaret.docker@umanitoba.ca

I. C. Potter
School of Veterinary and Life Sciences, Centre for Fish and Fisheries Research, Murdoch
University, Perth, WA 6150, Australia
e-mail: i.potter@murdoch.edu.au

© Springer Nature B.V. 2019
M. F. Docker (ed.), *Lampreys: Biology, Conservation and Control*,
Fish & Fisheries Series 38, https://doi.org/10.1007/978-94-024-1684-8_4

which, despite its very large size, has successfully colonized the Great Lakes. Abundant prey of a suitable size range is critical for establishment of freshwater parasitic populations. However, even with abundant prey, abandonment of anadromy is expected only under circumstances where decreases in mortality and the costs associated with migration make the reduction in size at maturity, and the accompanying reduction in fecundity, worthwhile. Pacific lamprey generally fail to establish when isolated above recently constructed barriers, likely because the reservoirs in which they have been isolated are relatively small and because they appear to osmoregulate poorly in fresh water. However, because colonization of fresh water appears to select for individuals "pre-adapted" to feed and grow to maturity in fresh water (i.e., relying on existing genetic variation within the source population), probability of establishment would likely increase with the number of founders. The existence of three closely related freshwater parasitic species suggests that Pacific lamprey successfully colonized fresh water in the past. Whether sea lamprey colonized Lake Ontario and Lake Champlain post-glacially or in historic times is debated. At present, the "invasion-by-canal" hypothesis appears to be the most convincing, but definitive resolution should be possible with genome-level analyses. Given the decimation of the Great Lakes ecosystem by sea lamprey, it is critical to be able to predict the potential for anadromous lampreys to become invasive in other freshwater systems. Migratory type is rarely considered a species-specific character unless it is accompanied by identifiable morphological differences. In contrast, variability in feeding type has long been considered a species-specific character because size-assortative mating was thought to result in reproductive isolation between parasitic and non-parasitic forms. However, not all parasitic and non-parasitic forms appear to be reproductively isolated, and different species show different degrees of divergence from their presumed parasitic ancestor. "Paired" non-parasitic species are defined as those that are morphologically similar to a particular parasitic species in all aspects other than body size, and "relict" brook lampreys are those that cannot be obviously paired with extant parasitic forms. However, molecular analyses have: (1) identified the closest extant parasitic relative to these relict species (although a few "orphan" species still remain, where identification of the closest living relative still sheds little light on the identity of the parasitic ancestor); (2) shown that the distinction between paired and relict species is sometimes unclear; and (3) demonstrated that there is also considerable variation among paired species in the degree to which they are morphologically and genetically differentiated from their parasitic ancestors. We review this "speciation continuum," particularly in European river and brook lamprey populations where recent genetic and genomic studies show significant gene flow between these species where they co-occur (i.e., refuting assumptions of complete reproductive isolation resulting from size-assortative mating) while also showing that there are genome-level differences between the feeding types (i.e., refuting the hypothesis of phenotypic plasticity). In sympatry, European river and brook lampreys appear to be partially reproductively isolated ecotypes that nevertheless maintain distinct phenotypes, because regions of the genome involved in reproductive isolation and local adaptation resist the homogenizing effect of introgression. Interestingly, the results of analyses used to reconstruct the demographic history of divergence in this species pair are inconsistent with recent and rapid divergence in sympatry following the recent glacial retreat; rather, they support divergence in allopatry ~200,000–250,000 years ago and re-establishment of secondary contact ~90,000 years ago. Some loci have been identified that differ between the forms (e.g., the vasotocin and gonadotrophin-releasing hormone 2 precursor genes), but an understanding of the genetic basis of life history evolution in lampreys remains elusive. There appear to be strong parallels between factors that promote or constrain loss of anadromy in parasitic species (and reduction in duration of the feeding phase) and those that lead to or limit the evolution of non-parasitism (i.e., total elimination of the feeding phase). Smaller-bodied parasitic species have been far more prolific in producing non-parasitic derivatives than others; western river lamprey are already so small at maturity that "skipping" right to a non-parasitic form represents a more profitable trade-off between mortality and fecundity than freshwater parasitism. In contrast, there is a conspicuous absence of brook lamprey derivatives from the large-bodied sea, pouched, and Caspian lampreys. Our comparisons suggest that 300 mm TL (with a 10–12× reduction in fecundity) is the cut-off above which shifts to non-parasitism would not be beneficial. Therefore, we predict that Great Lakes sea lamprey is not an "intermediate" freshwater parasitic form that will give rise to a non-parasitic derivative, because complete elimination of the parasitic feeding phase would represent too large (~40×) a reduction in fecundity.

Keyword Ancestral life history type · Contemporary gene flow · Ecotypes · Ecological constraints · Evolution of metamorphosis · Fecundity · Genetic basis of life history type · Freshwater colonization · Invasive species · Life history trade-offs · Non-parasitism · Non-trophic adults · Osmoregulation · Parasitism · Partial anadromy · Reproductive isolation · Speciation · Species pairs

4.1   Introduction

Lampreys (order Petromyzontiformes) are one of the two surviving groups of jawless vertebrates. This small remnant group consists of three extant families and 10 genera. A total of 41 species were recognized by Potter et al. (2015); 37 species in the Northern Hemisphere are assigned to the family Petromyzontidae, and three and one species of Southern Hemisphere lampreys are allocated to Mordaciidae and Geotriidae, respectively. All lampreys have a protracted microphagous larval phase (generally lasting ~3–7 years) which is spent in fresh water (see Dawson et al. 2015) and culminates in a dramatic metamorphosis. During metamorphosis, all lampreys develop a suctorial oral disc and tongue-like piston (both of which bear teeth) and fully formed eyes, and they undergo a range of other anatomical, physiological, and biochemical changes (see Manzon et al. 2015; Potter et al. 2015).

Despite their highly conserved body plan and larval stage, lamprey life history type diverges at metamorphosis on two main axes related to migration and feeding. At the completion of metamorphosis, 18 species remain sexually immature and enter a trophic phase (the juvenile or parasitic feeding phase) in which they feed on the blood or tissue of predominantly actinopterygian fish hosts (see Chap. 3). The parasitic phase lasts between a few months and 2 or more years, with the duration varying among, and sometimes within, species. After the juvenile feeding phase, they embark on a non-trophic upstream migration (see Moser et al. 2015), undergo sexual maturation (see Chap. 1), spawn, and die (see Johnson et al. 2015). Nine of the parasitic species remain in fresh water, usually feeding in either large lakes or rivers (see Potter et al. 2015; Chap. 3). The remaining nine parasitic species are predominantly anadromous, that is, feeding in marine environments (see Chap. 3) before returning to fresh water to spawn. Mean size at maturity may be as small as ~125 mm total length (TL) in freshwater species like the Miller Lake lamprey *Entosphenus minimus* or as large as ~700 and ~900 mm in anadromous Pacific lamprey *En. tridentatus* and sea lamprey *Petromyzon marinus*, respectively.

Within parasitic species, there is also variation with respect to migratory type and duration of the feeding phase, and this intraspecific variation is often underappreciated. Migratory type is rarely considered a species-specific character unless it is accompanied by identifiable morphological differences (Potter et al. 2015). Three of the nine anadromous species (sea lamprey, Arctic lamprey *Lethenteron camtschaticum*, and European river lamprey *Lampetra fluviatilis*) have given rise to permanent freshwater-resident populations, and Pacific lamprey may have as well, although the propensity of this species to establish in fresh water appears more limited (Wallace and Ball 1978; Beamish and Northcote 1989). In another two species (western river lamprey *Lampetra ayresii* and short-headed lamprey *Mordacia mordax*), there are rare reports of at least some individuals remaining in fresh water throughout their life cycle. Even greater life history diversity within species is apparent when anadromous praecox forms are included. Praecox literally means "very early" (or premature or early onset) and is used to refer to lampreys that feed at sea, but presumably for a reduced period of time. Anadromous sea lamprey and pouched lamprey *Geotria australis* appear to be consistently large (i.e., without any known praecox forms), but Arctic, European river, and Pacific lampreys occur as both large "typical" forms and smaller praecox forms (e.g., Abou-Seedo and Potter 1979; Kucheryavyi et al. 2007). Within these species, taxonomic distinctions are rarely made between typical and praecox forms, and, in many cases, there is not even a clear dimorphism between the forms. In some rivers, bimodal distribution in the size of upstream-migrating European river lamprey can be used to distinguish typical and praecox forms (e.g., Abou-Seedo and Potter 1979), but often only a range of sizes (generally with differences among geographic regions) is evident. These differences are thought to be partly attributable to intraspecific differences in the duration of the marine feeding phase and differences in abundance and size of prey. These inter- and intraspecific differences are generally referred to as variation in migratory type, but where lampreys feed and for how long overlaps with the second axis of divergence, feeding.

Variability in feeding type per se, specifically parasitic versus non-parasitic types, has generally commanded more attention and appreciation than variation in migratory type. This "parting of the ways" observed at metamorphosis (Hardisty 2006) is more obviously dimorphic, and it has long commanded the interest of biologists (e.g., Loman 1912). Non-parasitic "brook" lampreys spend their entire life in fresh water. Like all lampreys, they enter a non-trophic metamorphosis at the end of the larval phase; however, unlike parasitic lampreys, they initiate sexual maturation during metamorphosis. As a result, the non-trophic periods of metamorphosis and sexual maturation are superimposed, and the juvenile (parasitic) feeding phase is eliminated. Lampreys are the only vertebrates known to have a non-trophic adult (Hendler and Dojiri 2009), which is likely another reason variability in feeding type has generally commanded more interest than variability in migratory type. In addition, conspicuous morphological differences distinguish non-parasitic adults from parasitic forms, with the most notable difference being adult body size. Adult brook lampreys will be smaller than the largest larvae, generally measuring ~110–150 mm TL at maturity (see Docker 2009). The morphological similarity between several pairs of non-parasitic and parasitic lampreys, and their often overlapping geographic distributions, led to suggestions that particular brook lamprey species evolved from a form similar to that of the extant parasitic lamprey (e.g., Hubbs 1925; Zanandrea 1959). Because it is generally thought that size-assortative mating would result in reproductive isolation between parasitic and non-parasitic forms (e.g., Hardisty and Potter 1971a; Beamish and Neville 1992), most lamprey taxonomists recognize feeding type as a species-specific character. There is past (e.g., Enequist 1937) and continuing (e.g., Artamonova et al. 2011) debate on this subject (i.e., are paired species "real" species?), but there is likely not a simple "one size fits all" answer to this question.

There is also lack of agreement regarding whether geographically disjunct non-parasitic derivatives of the same presumed ancestor constitute one or multiple species. Thus, although there is little or no dissent regarding the number of recognizable parasitic species (18), different taxonomies often vary in the number of non-parasitic species recognized. Many of the past debates have largely been resolved (e.g., that Pacific brook lamprey *Lampetra pacifica* is distinct from western brook lamprey *La. richardsoni*), and Potter et al. (2015) recognized 23 species of non-parasitic lampreys. However, there is continuing discussion whether three brook lamprey populations from Portugal are distinct species (the Nabão lamprey *Lampetra auremensis*, Costa de Prata lamprey *La. alavariensis*, and Sado lamprey *La. lusitanica*; Mateus et al. 2013a) or whether they are synonymous with European brook lamprey *Lampetra planeri* (Potter et al. 2015). Another new brook lamprey species named *Lampetra soljani*, which appears to be related to, but distinct from, the Po brook lamprey *Lampetra zanandreai*, was recently described from the southern Adriatic Sea basin (Tutman et al. 2017). Also, several genetically distinct populations may represent new species that have not yet been formally described (e.g., Yamazaki and Goto 1996, 1998; Yamazaki et al. 2006; Boguski et al. 2012). We do not attempt to definitively answer the question "exactly how many brook lamprey species are there?" We agree with previous authors (Potter et al. 2015; Tuniyev et al. 2016) that, for the sake of stability, lamprey taxonomy should not be hastily revised without full systematic examination, and, for the most part, we follow the taxonomy of Potter et al. (2015). However, we recognize the strengths and limitations of different species concepts (see Docker et al. 2015) and that the transition from parasitic ancestor to non-parasitic forms represents a continuum (Docker 2009) that is difficult to objectively partition. Thus, we mention these other newly described and as-yet-undescribed species here in an attempt to provide a fuller discussion regarding the transition process itself. Such discussion should help inform future decisions regarding species delimitation in lampreys (see Chap. 7).

In this chapter, we synthesize the available information regarding lamprey life history divergence on both migratory and feeding axes, and, by so doing, we attempt to offer some novel insights into life history evolution in these ancient vertebrates. We discuss the putative life history type of the ancestral lamprey and then attempt to provide a greater appreciation for the breadth of life history diversity in extant lampreys. We give an in-depth review of inter- and intraspecific variation with respect to migratory type within parasitic species, which we feel has been underappreciated in the past. We also continue earlier discussions (e.g., Hardisty and Potter 1971a; Potter 1980; Salewski 2003; Hardisty 2006; Docker 2009) on the evolution of non-parasitism in lampreys, particularly with respect to insights provided from recent genomic studies (e.g., Mateus et al. 2013b; Rougemont et al. 2017). In doing this, we intend to provide a broader view of life history diversity in lampreys than has been presented thus far and to move away from categorization of lampreys along two independent axes: anadromous or freshwater-resident and parasitic or non-parasitic. We extend the argument proposed by previous authors (e.g., Beamish 1985; Salewski 2003; Hardisty 2006) that variation in migratory type among and within parasitic species is the "jumping-off point" for the evolution of non-parasitism. The common factor in the transition from anadromy to freshwater residency and from parasitism to non-parasitism is a reduction in size at maturity and fecundity. Thus, strong parallels appear to exist between factors that promote or constrain loss of anadromy in parasitic species (and generally reduction in duration of the feeding phase) and those that lead to or limit the evolution of non-parasitism (i.e., total elimination of the feeding phase).

4.2 Life History of the Ancestral Lamprey

Among modern lampreys, parasitism is clearly the ancestral life history type. Modern non-parasitic species retain teeth on their oral disc (albeit generally reduced) and buccal glands that produce an anticoagulant necessary for parasitic feeding (see Docker 2009). However, this should not be interpreted as meaning that the earliest lampreys were necessarily parasitic (presumably parasitism originated after the evolution of fishes upon which they could feed) or that the only other alternative is that the ancestral lamprey life cycle was similar to that of modern non-parasitic lampreys (i.e., with an extended filter-feeding larval stage and an entirely non-trophic adult stage). Likewise, although anadromy is considered ancestral among modern lampreys in that anadromous species (most notably the sea lamprey) are known to colonize fresh water (see Sect. 4.3.3), this does not mean that the ancestral lamprey was anadromous or that the only other alternative is that it was entirely freshwater resident (i.e., the two options among extant lampreys). Furthermore, although all modern lampreys share a highly conserved larval stage, parting ways only at metamorphosis, this does not mean that the ancestral lamprey life history type included this larval stage. Here, we use the life history and morphology of extant non-vertebrate chordates, the limited lamprey fossil record, and interpretations of the external environment of the earliest fishes to make inferences regarding the evolution of metamorphosis and the characteristic larval ("ammocoete") stage in lampreys and to deduce the feeding and migratory type of early lampreys.

4.2.1   Evolution of Metamorphosis and the Prolonged Larval Stage

Many metazoan phyla undergo metamorphosis (i.e., indirect development), during which they undergo dramatic physiological, molecular, behavioral, and ecological changes as they transition from a larva to a morphologically distinct juvenile (Bishop et al. 2006; Paris and Laudet 2008; Laudet 2011). Metazoans that undergo metamorphosis include invertebrate taxa (e.g., insects, echinoderms), as well as the non-vertebrate chordates (cephalochordates and urochordates) and some vertebrate chordates (e.g., lampreys, eels, flatfishes, amphibians). Among all metazoans, there is clearly more than one origin of metamorphosis, but there is still debate whether metamorphosis evolved independently in those chordate lineages with it or whether it was an ancestral feature of all chordates. Considerable morphological diversity exhibited during metamorphosis in different chordate lineages has been used to support independent origins (Sly et al. 2003; Heyland et al. 2005). In contrast, Paris and Laudet (2008) suggested that the common role of a thyroid hormone-producing gland (the endostyle or thyroid gland) in the metamorphosis of all chordates—although sometimes in apparently different ways and by mechanisms not fully understood (see Manzon et al. 2015)—suggests an ancestral origin.

Whether the earliest vertebrates exhibited metamorphosis is also unknown. Some authors (e.g., Northcutt and Gans 1983; Mallatt 1984, 1985) suggested that the earliest vertebrates showed metamorphosis and that, like most extant non-vertebrate chordates, they had a pelagic larval stage and benthic adult stage. Northcutt and Gans (1983) suggested that the larvae were pelagic suspension feeders and the adults were benthic predators; in contrast, Mallatt (1984, 1985) suggested that the larvae were pelagic "raptorial" feeders (i.e., taking individual food particles from the water column) and the adults were benthic suspension feeders. However, hagfishes, the other extant ancient vertebrate lineage, are direct developers, meaning either: (1) indirect development was not a trait shared by the last common ancestor of hagfishes and lampreys, and lampreys since acquired the trait; or (2) hagfishes secondarily abandoned metamorphosis. It is now generally accepted that hagfishes and lampreys are each other's closest living relatives (rather than lampreys sharing an ancestor more recently with the gnathostomes, the jawed vertebrates; see Docker et al. 2015). However, they have still been separated for long periods of evolutionary time, having diverged ~486–444 million years ago (Ma; Kuraku and Kuratani 2006), and, despite retaining many ancestral vertebrate characteristics, both have become specialized in their own ways. Many hagfish features once thought to be primitive (e.g., their degenerate eyes) represent secondary losses associated with their deepsea habitat (see Docker et al. 2015). Thus, cyclostome monophyly alone does not allow us to distinguish between these two scenarios.

In this section, we address the questions: (1) did metamorphosis evolve in lampreys, or was it inherited from its early chordate or vertebrate ancestor? and (2) what was the body form of the earliest lamprey (i.e., which came first: the ammocoete or the adult)? Given the similarities in the body plans of modern lamprey larvae and cephalochordates (i.e., lancelets or amphioxi), lamprey larvae are often taken as representing the primitive early vertebrate bauplan (see Hardisty et al. 1989; Evans et al. 2018). Because extant cephalochordates undergo subtle metamorphosis, where the pelagic asymmetric larvae transform into benthic symmetric juveniles (Paris and Laudet 2008), one might assume that the earliest lampreys underwent a similar metamorphosis. However, several authors have concluded that early lampreys did not, in fact, metamorphose (e.g., Youson and Sower 2001; Chang et al. 2014), although there is a lack of agreement regarding the body form of lampreys prior to the evolution of metamorphosis. Youson and Sower (2001), whose argument has been termed the "larval-first" hypothesis by Evans et al. (2018), suggested that early lampreys were marine and probably resembled the larvae from which the urochordate larvaceans were derived. These authors proposed that metamorphosis (giving rise to a sedentary benthic adult) appeared later after entry into fresh water, where the iodide-concentrating efficiency of the endostyle was a critical factor in the evolution of metamorphosis (Youson and Sower 2001; Youson 2004). Diogo and Ziermann (2015), based on the anatomy and development of chordate cephalic muscles, concluded that the inferred adult muscles of the last common ancestor of vertebrates are strikingly similar to the condition that is present in the lamprey larva, and likewise support the suggestion that the adult lamprey phenotype is derived. However, the assumption that a blind protochordate-like stage is the ancestral lamprey body form is inconsistent with our understanding that the earliest vertebrates were characterized by a suite of advancements that included a cranium and pronounced cephalization and a set of highly specialized paired sense organs (including image-forming eyes and a lateral line; see Docker et al. 2015).

Also, the larval-first hypothesis is not concordant with the fossil evidence. To date, lamprey fossils mostly resemble very small modern lamprey juveniles or adults (i.e., following metamorphosis, when the lamprey is sexually immature or mature, respectively); evidence of animals resembling modern larval lampreys is not known before 125 Ma—although one that may have been a larva dates back to ~320 Ma (Chang et al. 2014). From the oldest fossils, which had not yet been discovered when Youson and Sower (2001) proposed that the earliest lampreys were likely larva-like, to the most recent, the known fossils of lampreys can briefly be described as follows:

- *Priscomyzon riniensis*, from upper Devonian marine or estuarine deposits (~360 Ma) in South Africa, had clearly developed eyes, a large oral disc, and circumoral teeth, but its TL was only 42 mm (Gess et al. 2006). Apart from its small size and differences in body proportions (e.g., an oral disc proportionately larger than in living lampreys), it looked astonishingly like modern juvenile or adult lampreys.

- *Hardistiella montanensis* from lower Carboniferous deposits (~320 Ma) in Montana resembled modern juvenile or adult lampreys less clearly (Janvier and Lund 1983). There was no evidence of an oral sucker, and instead, the mouth may have been surrounded by a simple oral hood similar to that of metamorphosing lampreys. The holotype measured ~115 mm TL. A 50-mm lamprey fossil from this same locality could be a larval lamprey, but poor preservation has prevented definitive identification (Lund and Janvier 1986). Another specimen (<100 mm TL) reported by Janvier et al. (2004) also showed no trace of preserved cranial cartilages (e.g., no piston or annular cartilages), but evidence of a "large, globulous" snout was taken as support for the presence of a sucking device.

- *Mayomyzon pickoensis* from upper Carboniferous (~280 Ma) deposits in Illinois (Bardack and Zangerl 1968, 1971) clearly possessed many of the morphological and anatomical characters of the adults of extant lampreys. Although an oral disc and circumoral dentition were not evident, an annular cartilage (which maintains the structural integrity of the oral disc), a piston cartilage (which implies the presence of a rasping tongue), and dorsolateral eyes were apparent. Again, however, the specimens were small; the holotype (which was a presumed adult) measured only 48 mm TL and presumed juvenile specimens measured 33–61 mm TL.

- *Mesomyzon mengae* from lower Cretaceous (~125 Ma) freshwater shale deposits in China had a well-developed oral disc and a long snout; it possessed ~80 myomeres, but was still relatively small (~85 mm TL). However, Chang et al. (2014) subsequently discovered well-preserved fossils of larval (40–67 mm TL) and metamorphosing (82–94 mm TL) *M. mengae*. The larval specimens looked "surprisingly modern," exhibiting tiny eyes, an oral hood and lower lip, and detritus in the gut. The fossils presumed to be metamorphosing lampreys had enlarged eyes and a thickened oral hood or pointed snout; an oral disc was not evident, but it was assumed that these individuals represented early stages of metamorphosis (see Manzon et al. 2015).

Therefore, the larval form and "three-phased" life cycle (i.e., with larval, metamorphosing, and adult stages) appear to be derived characters in lampreys. Chang et al. (2014) proposed that lampreys initially evolved without (or with at most a limited) larval period, and relied on the juvenile/adult form for all or the majority of their lives; they suggested that introduction (and subsequent lengthening) of the larval stage and metamorphosis came later. Evans et al. (2018) referred to this as the "juvenile-first" hypothesis. Hardisty et al. (1989) similarly indicated that, even if a larval phase had been present in early lampreys, it would have been of only short duration. The absence of clear larval lamprey fossils prior to 125 Ma, of course, does not rule out a considerably earlier origin, particularly given the fossil of a possible larval *H. montanensis* specimen from ~320 Ma. The fact that all extant lampreys share this highly similar triphasic life cycle suggests an origin that predates the separation of Northern and Southern Hemisphere lampreys. Using molecular data, Kuraku and Kuratani (2006) placed the divergence between the families Petromyzontidae and Geotriidae at 280–220 Ma, and the split between the two Southern Hemisphere families (Geotriidae and Mordaciidae) is assumed to have occurred at approximately the same time (Gill et al. 2003; Potter et al. 2015). This means that the modern lamprey life cycle had evolved by 280–220 Ma; otherwise, we would have to accept that it evolved independently in each lineage, which is not likely considering that the features of this life cycle are so highly conserved among all extant lampreys.

A third hypothesis has been proposed by Evans et al. (2018). Similar to the larval-first and juvenile-first hypotheses, Evans et al. (2018) suggested that the earliest lampreys were without a distinctive metamorphosis and only underwent gradual ontogenetic changes during development. However, Evans et al. (2018) suggested that initially early lampreys had a body form somewhat intermediate between that of modern larvae and modern juveniles/adults. In fossil specimens, the external (more obvious) features resemble the juvenile/adult form, but the position of the otic capsules and other features (more subtly) resemble modern lamprey larvae. They further proposed that, during the evolution of lampreys, the "larval" characters became segregated in the beginning of the life cycle and appearance of the "juvenile" characters was delayed; eventually, development of the juvenile characters was condensed into (and accelerated during) the distinct phase of metamorphosis following a progressively longer larval stage. Elongation of the larval stage and reactivation of development at metamorphosis is evident in modern lampreys in such processes as gonadal differentiation (e.g., with testicular differentiation delayed until the onset of metamorphosis; see Chap. 1) and eye development (which appears to "pause" after reaching a very immature stage before resuming near the end of the larval stage and at metamorphosis; Suzuki and Grillner 2018). Evans et al. (2018) termed their hypothesis the "condensation" hypothesis, in line with terminology used to describe the evolution of metamorphosis in other organisms (e.g., Schoch and Fröbisch 2006). Here, to contrast this hypothesis more explicitly to the "larval-first" and "juvenile-first" hypotheses, we refer to this hypothesis as the "segregation and specialization" hypothesis. Early lampreys were neither larva-like nor juvenile-like, but, over time, the larval and juvenile characters became segregated during development, which allowed each form to become more highly specialized. Colonization of fresh water (see Sect. 4.2.4) and exploitation of new trophic niches (see Sect. 4.2.2) may have selected for increasingly specialized larval and juvenile forms, respectively. A radical metamorphosis was required to effect the transition between these now distinctive periods, and the modern lamprey life history appeared. With the specialized larval and juvenile forms, each well-adapted to their respective environments, the growth potential of each period was maximized, enabling the large body size that now characterizes modern parasitic lampreys. As outlined above, evolution of this dramatic metamorphosis would have occurred at least 280–220 Ma.

As a final point when discussing the various hypotheses regarding the evolution of metamorphosis in lampreys, it should be noted that there is little or no support for a fourth hypothesis, the so-called "larval transfer hypothesis" (Williamson 2012). This hypothesis contends that lampreys and hagfishes had no larvae until an ancestor of modern lampreys acquired larvae by hybridizing with a cephalochordate; hagfishes, which never crossed with a cephalochordate, retained their direct development. Williamson's "long-cherished hypothesis" of a hybrid origin of other organisms with complex life cycles (e.g., Williamson 2001, 2009) has largely been discredited (e.g., Hart and Grosberg 2009; Minelli 2010).

Evans et al. (2018) reviewed the evolution of metamorphosis in two other groups of animals with complex life histories, insects and amphibians, and there are considerable similarities between the "condensation" or "segregation and specialization" hypothesis proposed for lampreys and the sequence of events that has been proposed for these other taxa. Early insect lineages did not undergo metamorphosis; instead, a continuous progression from egg to embryo to adult occurred (Truman and Riddiford 1999). Complete metamorphosis in insects evolved ~280–350 Ma and has largely been credited with fueling their dramatic radiation, because it presumably enabled stage-specific specializations to different habitats (Truman and Riddiford 1999; McMahon and Hayward 2016). Metamorphosis permitted the extreme adaptation of one stage for a particular role, such as dispersal, and allowed structures (e.g., wings) to be delayed in their appearance until needed (Truman and Riddiford 1999; Haug et al. 2016). Insect phylogeny shows a progression from groups that are ametabolous (no metamorphosis) to those that are hemimetabolous (partial metamorphosis, where there are more subtle differences from the younger stages to the adult, but not requiring a radical reshaping of the body at any time) to the most derived groups that are holometabolous (complete metamorphosis, with a radical change from a larva to the juvenile/adult) (Engel 2015). If we were to use the same terminology in lampreys, early lampreys appear to have been ametabolous and all modern lampreys are holometabolous; by the suggested "segregation and specialization" hypothesis, hemimetabolous might be an appropriate term to describe lampreys during the initial stages of segregation and specialization.

Early amphibians were also direct developers (Schoch 2009), and metamorphosis is thought to have evolved by ~300 Ma (Schoch and Fröbisch 2006). The development of juvenile characters was initially delayed but was then accelerated and condensed into the distinct phase of metamorphosis (Schoch 2009). As with insects and lampreys, this dramatic metamorphosis allows amphibian larvae and adults to efficiently exploit different resources, with amphibian larvae using suction feeding in an aquatic habitat and adults generally becoming adapted to the capture of terrestrial insects with tongue-supported feeding (Schoch 2009, 2014). However, unlike extant lampreys, some amphibians have since reduced or eliminated metamorphosis; some direct-developing frogs and salamanders show only vestiges of metamorphosis during early development, and other salamanders (e.g., axolotls, mudpuppies, some tiger salamanders) have eliminated metamorphosis completely, and retain larval characteristics and remain in aquatic habitats as adults (Johnson and Voss 2013).

Thus, although lampreys are often used as model ancient vertebrates, the life history type of extant lampreys appears not be representative of the earliest vertebrates. However, we have been able to make inferences only about the life cycle of early lampreys, but not necessarily the earliest lampreys. The lamprey and hagfish lineages are thought to have diverged ~486–444 Ma (Kuraku and Kuratani 2006), although this only indicates when they last shared a common ancestor and not necessarily the origin of lampreys per se. Considering the stability of lamprey morphology in the past 360 Ma, Janvier (2008) indicated that it would not be surprising if recognizable fossil lampreys "turned up" 50 or 100 Ma earlier. However, the above arguments do suggest that the dramatic metamorphosis that characterizes all modern lampreys was not present in the earliest known lampreys (~360 Ma), but had evolved by at least 280–220 Ma.

4.2.2   Origin of Parasitism

The earliest known fossil lamprey, *Priscomyzon riniensis* from ~360 Ma, had a large oral disc and circumoral teeth (Gess et al. 2006), but it was very small (42 mm TL). No oral disc was evident in *Mayomyzon pickoensis* from ~280 Ma, although the annular cartilage that supports the oral disc in extant lampreys is evident (Bardack and Zangerl 1968, 1971). Hardisty et al. (1989) suggested that the small size of *M. pickoensis* (48 mm TL in the adult holotype) and apparent lack of circumoral teeth made it unlikely that this species fed parasitically, and there was no evidence of parasitic feeding on other vertebrates in the deposit in which it was found. However, Hardisty et al. (1989) indicated that the presence of a piston cartilage suggests that it might have fed on carrion (as some extant lampreys do; see Chap. 3) or even browsed on surface algal films. As reviewed above, it appears that metamorphosis in lampreys had evolved by ~280–320 Ma, which permitted subsequent specialization and elongation of both the larval and juvenile forms (see Sect. 4.2.1). Thereafter, the juvenile form became fully specialized to take advantage of the newly diversifying jawed fish fauna, while the larval form became specialized to take advantage of newly hospitable freshwater environments (see Sect. 4.2.4).

   We do not know on what the earliest parasitic lamprey would have fed, but, based on known hosts of extant lampreys (see Chap. 3), we would assume that they would have a general preference for fishes with few or small scales but that their tastes otherwise would have been rather catholic (meaning, in this context, "all-embracing" or "including a wide variety of things"). Some groups of armored jawless fishes are evident in the fossil record 488–443 Ma (Janvier 2007) and were diverse and abundant during the Devonian "Age of Fishes" (~419–359 Ma; Janvier 1996), but the armor likely limited lamprey feeding opportunities. Fossils of scales and dermal denticles indicate that cartilaginous fishes date back to ~455 Ma (Janvier 1996), but, notwithstanding the observation that sea lamprey may sometimes feed on sharks (Wilkie et al. 2004), cartilaginous fishes are not common hosts of modern parasitic lampreys. Thus, evolution of the ray-finned fishes ~439–383 Ma (class Actinopterygii, which includes sturgeons, paddlefishes, gars, bowfins, and teleosts) and diversification of some of the earliest extant teleost lineages ~333–286 Ma (Near et al. 2012) may have presented the first substantial parasitic feeding opportunities for juvenile lampreys. The salmoniform/esociform lineage (e.g., giving rise to salmonids and pikes) and clupeomorphs (e.g., herrings) arose and began diversifying by ~201–145 Ma and

252–201 Ma, respectively (Near et al. 2012). Clupeomorphs had a broad distribution in freshwater, marine, and brackish environments by 145–66 Ma (Vernygora et al. 2016). Hardisty et al. (1989) likewise concluded that suitable thin-scaled fishes similar to those used by modern lampreys to sustain rapid growth and attain large body size would not have been available before the end of the Carboniferous period (i.e., before ~299 Ma). Presumably, extension of the juvenile parasitic feeding phase occurred gradually, as growth opportunities increased, and perhaps with the evolution of anadromy, since larger body size might have been required to withstand the distance and rigor of upstream migrations (see Sect. 4.2.4). It is entirely feasible that this gradual extension of the juvenile feeding phase occurred independently in each lineage following separation of Northern and Southern Hemisphere lampreys, once the fundamental aspects of the modern lamprey life cycle that provided the capacity for growth were established.

With respect to the mode of feeding of the earliest parasitic lampreys, among modern parasitic lampreys, Potter and Hilliard (1987) concluded that blood feeding is ancestral. They argued that blood feeding would be less detrimental (to the hosts and thus, ultimately, to the lampreys which depend on them) in less productive waters. Blood is a renewable resource if the rate of feeding is not excessive relative to the size of the host, and the wounds would be smaller and less likely to be fatal. They concluded that flesh feeding came later, with access to smaller but more plentiful coastal fishes (e.g., herrings and other clupeids) or where lampreys travel farther offshore to feed in very productive waters (see Chap. 3; Sect. 4.4.2).

4.2.3 Origin of Non-parasitism

Even if the earliest lampreys were not parasitic (see Sect. 4.2.2), they did not resemble modern non-parasitic lampreys. Modern non-parasitic lampreys, which retain clear vestiges of the parasitic feeding mode, are derived from parasitic lampreys through a subsequent abandonment of the parasitic feeding phase (see Docker 2009). Non-parasitic derivatives of parasitic species are known in both hemispheres in two of the three extant families, Petromyzontidae and Mordaciidae. However, unlike the evolution of metamorphosis and the modern triphasic life cycle, the complexity and similarity of which would necessitate that it evolved once and was inherited in each of the extant families from their common ancestor, elimination of parasitic feeding can occur independently. In fact, even within the Petromyzontidae, non-parasitism has evolved independently in six of the eight genera and within genera as well (see Docker 2009; Sect. 4.6). In general, it is easier to lose complex traits than it is to acquire them because, from a strictly genetic point of view, most mutations are more likely to be degenerative than constructive (see Strathmann and Eernisse 1994; Gompel and Prud'homme 2009). Of course, evolution of non-parasitism in lampreys did not just involve a mutation that prevented feeding following metamorphosis; such mutations are unlikely to have been adaptive. The evolution of non-parasitism appears to have required a heterochronic shift in development that accelerated sexual maturation relative to metamorphosis (Docker 2009). We still know nothing regarding the genetic basis for this acceleration of sexual maturation or the extent to which the genetic changes are parallel among different independently derived species (see Sects. 4.6.3.3 and 4.6.3.4). However, for the most part, regardless of the genetic mechanism, we would expect acceleration of sexual maturation relative to metamorphosis to produce similar phenotypic results even when occurring independently. The non-trophic period of metamorphosis would merge with the non-trophic period of sexual maturation, resulting in elimination of the intervening juvenile feeding phase and eventual degeneration of the teeth and other structures associated with feeding (Docker 2009).

Nevertheless, apart from saying that non-parasitism did not need to evolve in the ancestor to all modern lampreys (i.e., before 280–220 Ma), we cannot say when the feeding phase was first eliminated. On the basis of mitochondrial DNA (mtDNA) sequence data, some of the oldest extant non-parasitic species are estimated to have diverged from any known parasitic species a few to several million years ago (e.g., least brook lamprey *Lampetra aepyptera* at least 2 Ma, and the Macedonia and Epirus brook lampreys *Eudontomyzon hellenicus* and *Eu. graecus* at least 5.5 Ma; see Sect. 4.6.2). However, given the increasing appreciation of the "non-clock-like" nature of mtDNA (see Galtier et al. 2009) and general concerns regarding the precision of molecular timescales (Graur and Martin 2004), these divergence times are very likely underestimates. In the least brook lamprey, Martin and White (2008) suggested that a vicariance event during the Pliocene (~5.3–2.6 Ma) produced strong phylogeographic structuring (see Sect. 4.6.2.3), which suggests that the species itself is considerably more than 2 million years old. Furthermore, these estimates do not represent the age of these species per se, only the approximate time since divergence from other extant species and certainly should not be interpreted as representing the first non-parasitic species. It has been suggested that non-parasitic species are more prone to extinction than parasitic species, because they typically show a more limited distribution and smaller populations (Spice et al. 2019). Older non-parasitic species will have become extinct but new ones will have evolved.

Despite the elimination of post-metamorphic feeding, all lampreys still undergo metamorphosis, unlike paedomorphic salamanders where elimination of metamorphosis has evolved rapidly and independently multiple times (Page et al. 2010; Johnson and Voss 2013). Metamorphosis in lampreys is energetically costly, requiring extensive remodeling of the body and subsequent maturation of the gonads while the lamprey is not feeding for 6–10 months (Docker 2009). Therefore, since it has been retained in all lampreys, it appears that the changes associated with metamorphosis are required, not just for parasitic feeding during the juvenile phase, but also for reproduction (Manzon et al. 2015). The oral disc that is a key feature of parasitic lampreys is also used by brook lampreys to attach to rocks during the short upstream migration, during nest building, and for attachment to mates during spawning (see Johnson et al. 2015). Other aspects of mate choice and reproduction may rely on adult sensory capabilities (Johnson et al. 2015). Youson and Sower (2001) suggested that a complex interplay between the thyroid and reproductive axes evolved during and subsequent to the evolution of metamorphosis in lampreys, and the two may not be easy to disentangle. There has been one report of apparent paedomorphism or neoteny in lampreys (Zanandrea 1957), but the general view is that true paedomorphism is not present in any extant lamprey species (Vladykov 1985). Zanandrea (1957) reported finding 12 Po brook lamprey larvae that showed well-developed ovaries with eggs that were in an advanced state of maturity. One larva had well-developed secondary sex characteristics (e.g., enlargement of the dorsal fins and urogenital papilla) and a transparent body wall through which the eggs could be seen. The endostyle and other larval features were still visible, and there was no evidence of any post-metamorphic features except for the well-developed ovary. However, artificial fertilization of the eggs was not attempted, so it is unknown if they were viable. The Po brook lamprey is considered to be an older "relict" non-parasitic species (see Sect. 4.6.2), so it is possible that the changes associated with the heterochronic shift in the timing of sexual maturation relative to metamorphosis produced rare individuals that began gonadal maturation in the absence of metamorphosis. However, there is no evidence that it could have successfully reproduced without completing a full metamorphosis.

4.2.4 Marine or Freshwater Origin of Lampreys

For decades, biologists and paleontologists have debated whether the vertebrates originated in the sea or in fresh water. This question is not trivial; differences between these two habitats would have involved much more than just the salt content of the water (i.e., it is not just a question of the osmoregulatory abilities of the earliest vertebrates), but also has profound ecological and evolutionary implications. When vertebrates arose (~500 Ma), the seas possessed a rich and diverse fauna while freshwater rivers and lakes are believed to have lacked multicellular animals and were rather unproductive (Halstead 1985). This would mean that, if the vertebrates originated in the sea, they would have evolved under conditions of diverse and abundant food supplies but with intense predation and competition. In fresh water, on the other hand, there would have been few predators or competitors, but food sources would have been limited to algae and other unicellular organisms (Griffith 1987, 1994). Arguments for a freshwater origin were based on some geological evidence suggesting that, at least in North America, early vertebrate fossils appeared to originate from freshwater localities, as well as the "naïve" suggestion that the streamlined shape of fish was a direct response to the flow of running waters and the "more sophisticated views" of Smith (1953) that the glomerular kidney evolved as a means of combating the problem of osmosis in fresh water (Halstead 1985). In fact, Romer (1955) concluded "I see no reason for serious consideration of a marine history for the early vertebrates." However, hagfishes also have a glomerular kidney, and it was subsequently concluded that the evolution of the glomerular kidney did not depend on a freshwater environment for its initial development (see Halstead 1985). For the most part, the argument has since been "conclusively settled in favor of a marine origin for the vertebrates," because all the non-vertebrate chordates (and hagfishes) are exclusively marine and the earliest vertebrate fossils come from marine environments (see Holland and Chen 2001; Janvier 2007). A third argument proposes that the early vertebrates were anadromous. Griffith (1987, 1994) suggested that marine "pre-vertebrates" invaded food-rich estuaries where, given the large fluctuations in salinity in these environments, they evolved osmoregulatory features that enabled reproduction in fresh water. Since fewer competitors would have been present in fresh water, it provided a safe haven from predators, although at the expense of growth. Slow-growing filter-feeding larvae developed in fresh water, and adults subsequently returned to sea to feed. Griffith (1987, 1994) argued that many of the characteristic vertebrate features (cephalization, paired sensory organs, complex endocrine system) could have evolved as a response to the demands of anadromous migrations and seasonal spawning. However, these are the same arguments given for a somewhat later invasion of fresh water by marine vertebrates (see below), and it is more likely that anadromy was a later addition to the repertoire of many fishes; evidence suggests that anadromy was acquired secondarily and independently within multiple lineages (McDowall 1988, 1993; Hardisty et al. 1989; Dodson et al. 2009).

Whether the earliest lampreys were marine, freshwater, or anadromous involves similar arguments. The earliest known fossil lamprey from ~360 Ma was recovered in marine or estuarine deposits (Gess et al. 2006). Fossils from ~320 Ma were also found in marine deposits (Janvier and Lund 1983), although a third specimen reported by Janvier et al. (2004) was found in an area that would have shown wide fluctuations in salinity ranging from brackish (slightly salty) to hypersaline conditions. *Mayomyzon pieckoensis* from ~280 Ma was recovered in a diverse collection of ~300 predominantly marine species, which suggests a coastal deltaic area of fluctuating marine and fresh waters (Bardack and Zangerl 1971). The fact that the lampreys were well preserved suggests a rapid death and burial in the area of deposition, but it cannot be determined with certainty whether they lived in fresh or marine waters. *Mesomyzon mengae* from ~125 Ma represented the first lamprey fossil from unambiguous freshwater deposits; it was found with other freshwater or terrestrial animals, and there was no indication that the area had any connection with the sea since the Triassic period (~250–200 Ma; Chang et al. 2006). The first (and only) definitive larval lamprey fossils were also recovered from these deposits (Chang et al. 2014; see Sect. 4.2.1).

However, Lutz (1975) suggested that lampreys evolved in fresh or brackish water based on their relatively low serum osmolality relative to marine fish species. The very low blood concentrations of Na^+ and Cl^- in larval lampreys in particular (and juveniles to a lesser extent) was used as support for a relatively long history of life in fresh water; Na^+ and Cl^- concentrations were similarly low in lungfishes and polypteroids (bichirs), which are "presumed never to have left freshwater" (Hardisty et al. 1989). It has also been argued that the early life stages of fish species survive best in the type of osmoregulatory environment ancestral to the group (see McDowall 1993), likewise suggesting a freshwater origin of lampreys since modern-day lamprey larvae are generally unable to osmoregulate in water with salinities higher than ~28% sea water (i.e., ~10 parts per thousand, ppt; see Dawson et al. 2015). Furthermore, Hardisty et al. (1989) argued that the small size of the earliest lampreys (see Sect. 4.2.1) would likely have imposed a greater osmotic stress in full-strength sea water, leading to the conclusion that early lampreys lived in a stable and predominantly freshwater or brackish environment.

Nevertheless, the above evidence and arguments are not entirely contradictory; the overlap appears to be the suggestion that the earliest lampreys occupied brackish environments. Janvier (2007) stressed that, during the Devonian (419–360 Ma), the continental margins were occupied by vast deltas and tidal flats, and that the environment of most fishes during this time (and probably earlier) was most likely comparable to present-day major tropical deltas and mangroves. Coastal deltaic areas would have been supplied by numerous slow flowing streams, and they would have provided diverse habitats with respect to salinity and depth (Hardisty et al. 1989). Furthermore, the reasoning made by Lutz (1975) cannot distinguish between evolution in fresh (or brackish) water and a subsequent long history of life in fresh water.

Thus, it appears that the earliest lampreys were likely marine, but they probably lived in coastal areas of fluctuating salinity that "prepared" them for subsequent invasion of fresh water following the development of terrestrial flora and more hospitable inland areas. The first true plants emerged onto land 470–425 Ma (Gibling and Davies 2012), and the earliest known trees date back to 385 Ma (Stein et al. 2007). The evolution of terrestrial plants led to the development of soils and terrestrial animal assemblages (Gibling and Davies 2012). Before this, inland streams and rivers would have been unproductive and, without stabilizing vegetation, extremely "flashy" (i.e., where water levels rise very quickly, making rivers prone to flooding). This timing is consistent with the suggestion that the larval stage and metamorphosis had evolved in lampreys by 280–220 Ma and may have been evident by ~320 Ma.

Early freshwater environments were relatively unproductive, but they also supported few predators, providing a "safe haven" for reproduction and rearing of early developmental stages. As suggested by Hardisty et al. (1989) and Evans et al. (2018), the development of the larval stage in the lamprey life cycle likely coincided with the invasion of fresh water, followed by subsequent specialization to the new environment. A pelagic larval stage would have been maladaptive in flowing water, for example, selecting for the evolution of the benthic (burrowing) larval stage. Subsequent elongation of the larval period would have taken advantage of the safe but relatively unproductive fresh waters, and size at metamorphosis approached that of at least some modern lampreys by ~125 Ma (see Sect. 4.2.1). Thus, metamorphosis in lampreys, insects, and amphibians appears to have evolved at similar times to take advantage of the opportunities presented by newly hospitable terrestrial or inland habitats.

The above discussion still does not address the question of whether, following colonization of fresh water for reproduction, early lampreys restricted their freshwater use to reproduction and rearing of the filter-feeding larval stage (i.e., were anadromous, returning to more productive marine coastal areas to feed as adults) or were entirely freshwater resident. Growth opportunities for the juvenile stage would clearly be better in the marine or estuarine environment (particularly given the especially depauperate nature of fresh waters at that time), but duration of the juvenile stage and size achieved during feeding at sea likely increased only gradually over

time. Long feeding and spawning migrations characterized by some of the largest
extant anadromous lamprey species were likely not a feature of the earliest anadromous lampreys. Based on plasma osmolalities of juvenile lampreys, Lutz (1975)
and Hardisty et al. (1989) concluded that there has been a relatively recent origin of
the marine feeding phase in lampreys. The freshwater lamprey genus *Ichthyomyzon*
either evolved in fresh water or has been freshwater resident for long periods of
evolutionary time (Bartels et al. 2012; see Sect. 4.3.2.4), but again this observation
only allows us to determine that they have been in fresh water for a "relatively" long
period. Regardless, the life history of the earliest lampreys that invaded fresh water
was still not entirely similar to either extant anadromous or freshwater-resident parasitic lampreys. Despite their highly conserved body plan—which allows 360-Ma
fossils to be immediately recognizable as lampreys—the above discussion indicates
that there was likely a gradual transition in terms of life history from the earliest
lampreys to those seen in modern forms and subsequent diversification in migratory
and feeding types.

4.3   Variation in Migratory Type

Parasitic lampreys exhibit two basic migratory types, anadromous and freshwater-
resident, and anadromous lampreys are sometimes divided into the large-bodied
forms ("forma typica") and the smaller-bodied "forma praecox." Praecox lampreys
are assumed to have a reduced marine phase relative to the typical form (e.g., Abou-
Seedo and Potter 1979), although differences in the quantity or quality of available
host fishes cannot be ruled out. The term praecox (not to be confused with *Mordacia praecox*, the non-parasitic precocious lamprey or Australian brook lamprey)
has sometimes also been used to refer to freshwater-resident parasitic forms that
are also generally smaller than anadromous forms (e.g., Hardisty 1986a; Maitland
et al. 1994) or without distinction between small-bodied anadromous and freshwater forms (e.g., Berg 1948). Here, however, we use the term praecox to refer only
to the smaller-bodied anadromous form. Some anadromous lampreys also appear
to show considerable intraspecific variation with respect to migration timing (also
known as run timing). Although some species show little intraspecific variation in
the onset and duration of their upstream migration (e.g., sea lamprey and pouched
lamprey appear to consistently enter fresh water 1–2 and 15–16 months, respectively,
prior to spawning), others (e.g., Caspian and European river lampreys and, to some
extent, Pacific lamprey) show variation among and within populations (see Moser
et al. 2015; Chap. 1). For the non-anadromous lampreys, we use the term freshwater-
resident to describe those that spend their entire life cycle within fresh water (i.e.,
excluding anadromous species that feed prior to or during outmigration to sea; see
Sect. 4.3.4.4). Freshwater-resident populations (including all non-parasitic "brook"
lampreys; see Sect. 4.6) are also sometimes referred to as potamodromous (i.e.,
showing directed movement within fresh water; Moser et al. 2015). The term "landlocked" is also frequently used, although many of these populations retain access to

the sea, and, in some locations, freshwater and anadromous lampreys co-occur in a lake basin (e.g., European river lamprey in Lake Ladoga and Arctic lamprey in the Naknek River system; Sects. 4.3.4.2 and 4.3.4.3). In most cases, however, the majority of freshwater species or populations (e.g., sea lamprey in the Great Lakes, Arctic lamprey in Great Slave Lake) appear to be permanently freshwater resident without any individuals that go to sea. Whether or not the former cases represent polymorphism within a single population or spatially overlapping (but reproductively isolated and genetically determined) ecotypes has not yet been tested.

Of the 18 extant parasitic lamprey species, nine are exclusively freshwater resident and nine are largely anadromous, but with at least some reports of freshwater-resident populations in all but two of these species. Some of the freshwater-resident species and populations are clearly derived from anadromous species (e.g., the three freshwater-resident parasitic species in the genus *Entosphenus* and the Great Lakes sea lamprey); in constrast, others (e.g., species in the genus *Ichthyomyzon*) appear to have evolved in or been freshwater resident for long periods of evolutionary time (see Sect. 4.3.2.4). This section reviews these different migratory types in lampreys, with an emphasis on intraspecific variation in migratory type and features of the species or environments where the different types are found.

4.3.1 Exclusively Anadromous

Five lamprey species are exclusively or almost exclusively anadromous (Fig. 4.1). Rare or unconfirmed freshwater populations have been reported in three of these species (see Sect. 4.3.3), leaving only the pouched lamprey and Chilean lamprey *Mordacia lapicida* with no reports of freshwater-resident or praecox forms. The length ranges of fully grown individuals of these species are thus relatively narrow. At the commencement of their upstream migration, the majority of pouched lamprey typically measure 530–740 mm TL in Australia (Potter et al. 1983) and 445–570 mm in Chile (Neira 1984). Adults of the Chilean lamprey are reported to range in size from 278 to 313 mm (Neira 1984). Duration of the feeding phase is unknown in both species, but it is thought to be quite long in the pouched lamprey given its size (Renaud 2011; see Chap. 3).

4.3.2 Exclusively Freshwater

Nine parasitic lamprey species are exclusively freshwater resident, including: three species in the genus *Entosphenus*, which all appear to be recent derivatives of the anadromous Pacific lamprey; the Korean lamprey *Eudontomyzon morii* which, despite its current placement in the genus *Eudontomyzon*, may be an older freshwater derivative of the anadromous Arctic lamprey or an Arctic lamprey-like ancestor (see Sect. 4.3.2.2); and five species (Carpathian lamprey *Eu. danfordi*, silver lamprey

PARASITIC		NON-PARASITIC		
		Recent "Paired"		
Presumed Ancestor	Recent Derivatives	Species	"Relict" Species	"Orphan" Species

Caspiomyzon
Caspian lamprey *C. wagneri* [A (AP, FW?)]

         ??   10.5 – 10.7%   Epirus brook lamprey *Eu. graecus* (1)
                             Macedonia brook lamprey *Eu. hellenicus* (2)
                                              0.4%

Entosphenus
Pacific lamprey *E. tridentatus* [A, AP (FW?)]

     0 – 0.2%   Vancouver lamprey *E. macrostomus* [FW]
     0.3 – 0.4%   Klamath lamprey *E. similis* [FW]
     0.4 – 0.6%   Miller Lake lamprey *E. minimus* [FW]

               0 – 0.4%   Pit-Klamath brook lamprey *E. lethophagus* (3)
               0.5%   N California brook lamprey *E. folletti* (4)
                                        0.8%

Eudontomyzon
Carpathian lamprey *Eu. danfordi* [FW]

          2.2 – 3.7%   Ukrainian brook lamprey *Eu. mariae* (5)
          2.5 – 2.7%   Drin brook lamprey *Eu. stankokaramani* (6)
                                        2.5 – 3.2%

Ichthyomyzon
Chestnut lamprey *I. castaneus* [FW]
          0.1%   Southern brook lamprey *I. gagei* (7)

Ohio lamprey *I. bdellium* [FW]
          0.3%   Mountain brook lamprey *I. greeleyi* (8)

Silver lamprey *I. unicuspis* [FW]
          0%   Northern brook lamprey *I. fossor* (9)

Geotria
Pouched lamprey *G. australis* [A]

***Lampetra* (Atlantic Basin)**
European river lamprey
La. fluviatilis [A, AP, FW]

          0 – 0.6%   European brook lamprey *La. planeri* (10)
          0.4 – 0.8%   Nabão lamprey *La. auremensis* (11)
          0.5 – 0.8%   Costa de Prata lamprey *La. alavariensis* (12)
          0.8 – 1.1%   Sado lamprey *La. lusitanica* (13)
                                 0.5 – 1.1%
      2.8 – 3.5%   *La. soljani* (14)
      2.8 – 3.1%   Po brook lamprey *La. zanandreai* (15)
                           0.7 – 3.5%
      3.2 – 4.2%   Turkish brook lamprey *La. lanceolata* (16)
      3.8 – 4.2%   Western Transcaucasian brook lamprey *Le. ninae* (17)
                                             0.2 – 0.9%
      4.3 – 4.8%   Least brook lamprey *La. aepyptera* (18)

***Lampetra* (Pacific Basin)**
Western river lamprey
La. ayresii [A (FW?)]

          0 – 2.3%   Western brook lamprey *La. richardsoni* (19)
      2.3 – 3.5%   Pacific brook lamprey *La. pacifica* (20)
      2.3 – 4.0%   Kern brook lamprey *La. hubbsi* (21)
      2.5 – 3.4%   *Lampetra* sp. Fourmile OR
      2.2 – 3.3%   *Lampetra* sp. Siuslaw OR   2.6 – 6.5%
      4.3 – 4.9%   *Lampetra* sp. Mark West CA
      5.7 – 6.5%   *Lampetra* sp. Kelsey CA

Lethenteron
Arctic lamprey
Le. camtschaticum [A, AP, FW]

          0 – 0.4%   Alaskan brook lamprey *Le. alaskense* (22)
          0 – 0.5%   Siberian brook lamprey *Le. kessleri* (23)
          0.1 – 0.5%   Far Eastern brook lamprey *Le. reissneri* (24)
          0.2 – 1.1%   American brook lamprey *Le. appendix* (25)
      3.5 – 3.9%   *Lethenteron* sp. N (26)
      ??   11.7%   *Lethenteron* sp. S (27)

    ??   1.3%   Korean lamprey *Eu. morii* [FW]

Mordacia
Short-headed lamprey *M. mordax* [A (FW?)]   0.4%   Precocious lamprey *M. praecox* (28)

Chilean lamprey *M. lapicida* [A]

Petromyzon
Sea lamprey *P. marinus* [A, FW]

Tetrapleurodon
Mexican lamprey *T. spadiceus* [FW]   ?   Mexican brook lamprey *T. geminus*

◄**Fig. 4.1** Postulated relationships between freshwater-resident parasitic and non-parasitic species and their presumed ancestor in 10 genera of extant lampreys. The 41 species recognized by Potter et al. (2015) are shown in *black*; taxa not recognized by Potter et al. (2015) and new or tentative species that have not been formally described are given in *gray* (see Yamazaki et al. 2006; Boguski et al. 2012; Mateus et al. 2013a; Tutman et al. 2017). Presumed parasitic ancestors (as represented by 14 species in the contemporary fauna) are shown with recent exclusively freshwater parasitic derivatives (4 species) and recent ("paired") and older ("relict") non-parasitic (brook lamprey) derivatives (although, in some cases, the distinction between "paired" and "relict" species is not entirely clear; see Sect. 4.6.2.1; Fig. 4.2). Relict species whose parasitic counterparts in the contemporary fauna are particularly difficult to identify but whose affinities might be inferred from morphology (as indicated by current generic placement) or molecular data are also shown. Genetic divergence (Kimura's two-parameter distance, K2P) between presumed parasitic ancestors and derivative species (and between multiple derivatives of the same presumed ancestor) is based on >300 cytochrome b gene sequences (>1,131 base pairs, bp) from GenBank, except for Northern California brook lamprey (where only 384 bp were available; Margaret F. Docker, unpublished data) and *Lampetra soljani* (where only cytochrome oxidase I sequence data was available, but where divergence at cytochrome oxidase I in these lamprey species was 0.75 times that of cytochrome b and adjusted accordingly); no sequence data was available for the Mexican lamprey. Migratory type (anadromous, anadromous praecox, and freshwater resident) is indicated (*A, AP, FW*) for parasitic species (see Sect. 4.3); numbers in parentheses after the name of each non-parasitic species are those referred to in Fig. 4.2

Ichthyomyzon unicuspis, chestnut lamprey *I. castaneus*, Ohio lamprey *I. bdellium*, and Mexican lamprey *Tetrapleurodon spadiceus*) from exclusively freshwater genera (Fig. 4.1). The duration of the parasitic feeding phase (and consequently body size) is reduced in these species relative to most anadromous forms.

4.3.2.1 Freshwater Parasitic *Entosphenus* Species

Mean and maximum TL of the three freshwater parasitic species of the genus *Entosphenus* are: 125 and 145 mm for Miller Lake lamprey (Bond and Kan 1973; Lorion et al. 2000) and 231 and 269 mm for Klamath lamprey *Entosphenus similis* (Vladykov and Kott 1979a), the two species found in the Klamath basin of Oregon and California; and 174 and 273 mm for the Vancouver or Cowichan lamprey *En. macrostomus* (Beamish 1982), which is endemic to the Cowichan Lake system on southeastern Vancouver Island, British Columbia. The particularly small size of the Miller Lake lamprey (presumably corresponding with the relatively small size of Miller Lake, 2.3 km^2) suggests a very brief post-metamorphic feeding period, and there have even been suggestions that some individuals do not feed at all (Bond and Kan 1973). Little is known about the feeding habits of the Klamath lamprey, but, given its size, it presumably feeds parasitically for no more than 1 year. The Vancouver lamprey also likely feeds for ≤1 year (see Chap. 3). All three species have been described as distinct taxa based on diagnostic morphological differences (see Renaud 2011; Potter et al. 2015), although low levels of genetic differentiation between these species and anadromous Pacific lamprey (Docker et al. 1999; Lorion et al. 2000; Lang et al. 2009; see Fig. 4.1) suggest that all are recent freshwater derivatives of the latter species. Kan and Bond (1981) proposed rapid speciation of the Miller Lake lamprey from Pacific lamprey or a Pacific lamprey-type ancestor following the eruption of Mount Mazama ~6,600 years ago that isolated Miller Lake from the Williamson drainage. Similarly, the Vancouver lamprey was likely established post-glacially with the formation of Cowichan Lake <15,000 years ago (see Taylor et al. 2012).

The Vancouver lamprey, endemic to Cowichan Lake (62 km^2), Mesachie Lake (0.6 km^2), and the interconnecting Bear Lake, is the best studied of these three species. Beamish (1982) described the Vancouver lamprey as a distinct taxon due to its smaller size and differences in body proportions, pigmentation, physiology, and spawning time and location. Most notably, the Vancouver lamprey is distinguished by its larger oral disc (which produces relatively large wounds on hosts that can be mistaken for attacks by Pacific lamprey), its ability to survive following metamorphosis in both fresh and salt water, and the fact that it spawns in the lake rather than tributary streams. During the parasitic feeding phase, Vancouver lamprey prey on a number of salmonid species found within the Cowichan system, including cutthroat trout *Oncorhynchus clarkii*, rainbow trout *O. mykiss*, coho salmon *O. kisutch*, Dolly Varden *Salvelinus malma*, and non-anadromous sockeye salmon or kokanee *O. nerka* (Beamish 1982, 2001; COSEWIC 2008). This species is not landlocked, but appears to be parapatric rather than sympatric with anadromous Pacific lamprey. Pacific lamprey are found in the Cowichan River downstream of Skutz Falls. A fishway at the falls was constructed in the 1950s to facilitate upstream passage of salmonids, but these falls were probably not a complete barrier to upstream movement by Pacific lamprey and other anadromous fishes (Taylor et al. 2012). However, there is no evidence that Pacific lamprey enter Cowichan Lake to spawn in its inlet tributaries, and, despite being indistinguishable in their cytochrome b gene sequences (Docker et al. 1999; Fig. 4.1), recent analysis using microsatellite DNA loci showed that Vancouver lamprey and Pacific lamprey in the Cowichan system represent distinct gene pools (Taylor et al. 2012).

Other potential freshwater parasitic *Entosphenus* forms have been reported that are sometimes referred to as non-anadromous Pacific lamprey populations but which may represent distinct but undescribed freshwater species. Hubbs (1925) suggested that the population in Goose Lake, a shallow alkaline lake in Oregon and California, represented a separate but unnamed race, and Moyle (2002) felt that this form should be considered a subspecies of Pacific lamprey. A non-migratory form of Pacific lamprey also occurs in the Sprague River and Upper Klamath Lake in the Klamath basin in Oregon (Hamilton et al. 2005), and it appears to be morphologically (e.g., in the number and structure of velar tentacles; Renaud 2011) and genetically (Lorion et al. 2000) distinct from other *Entosphenus* species. Both populations are presumably post-Pleistocene Pacific lamprey derivatives and not recently landlocked forms. Goose Lake formed from precipitation and melting glaciers at the end of the Pleistocene, and Upper Klamath Lake was also formed around this time when the much larger Lake Modoc receded and disappeared (Dicken and Dicken 1985). Non-anadromous Pacific lamprey-like populations have also been reported in three disjunct locations in southwestern British Columbia (Beamish 2001; see Sect. 4.3.4.1).

4.3.2.2 Korean Lamprey

The Korean lamprey from the Yalu River drainage in China and North Korea feeds in fresh water on small cypriniform fishes such as goldfish *Carassius auratus*, spine loach *Cobitis taenia*, and lake minnow *Rhynchocypris percnurus*. Maximum reported TL for this species is 279 mm, although adult size (i.e., following shrinkage prior to spawning; see Chap. 1) is typically ~150–200 mm (see Renaud 2011; Chap. 3). All other described species in the genus *Eudontomyzon* are freshwater resident (see Sect. 4.3.2.3), although there is debate regarding the generic placement (and hence the ancestor) of this species. Morphological phylogenies retain this species as sister to the Carpathian lamprey (Gill et al. 2003), but analyses using cytochrome b sequence data—albeit using a single metamorphosing individual—place Korean lamprey in the clade with *Lethenteron* (Lang et al. 2009). Given the disjunct distribution of the genus, Berg (1931) similarly suggested that the Korean lamprey evolved from the Arctic lamprey while the *Eudontomyzon* species in the Black Sea basin were derived from the European river lamprey. Cytochrome b gene sequences in Korean and Arctic lampreys differed by 1.3% (Kimura two-parameter, K2P, values; Lang et al. 2009; Li 2014). In comparison, Pacific lamprey and its three described freshwater derivatives differ by 0–0.6% in cytochrome b DNA sequence (Fig. 4.1), suggesting that the Korean lamprey is a somewhat older freshwater derivative. Although the precision of molecular clocks based on mtDNA sequence divergence has been debated (see Galtier et al. 2009), sequence data are useful for comparing relative divergence times between species.

4.3.2.3 Carpathian Lamprey

Carpathian lamprey feed for only 7–9 months (see Chap. 3). The maximum TL recorded during the feeding phase is 300 mm, although the largest mature adult recorded was only 207 mm and mean TL in different rivers ranged from 141 to 199 mm (Renaud and Holčík 1986). The Carpathian lamprey is an exclusively freshwater parasitic species (Potter et al. 2015), but the origin of this species and its presumed non-parasitic counterpart, the Ukrainian brook lamprey *Eudontomyzon mariae*, has been questioned. An extinct *Eudontomyzon* sp. nov. "migratory" lamprey from the Dniester, Dnieper and Don drainages in the Black Sea basin are included on the IUCN Red List of Threatened Species. Adults of this undescribed form were the target of fisheries during their autumn and spring migrations, but it is not known whether this lamprey fed at sea or entirely within fresh water (Kottelat et al. 2005; Freyhof and Kottelat 2008a). The absence of seawater-type mitochondria-rich cells (SW-MRCs, formerly known as chloride cells) in the gills of Ukrainian brook lamprey strongly suggests that this species did not evolve recently from an anadromous ancestor (Bartels et al. 2017), because non-parasitic (European brook lamprey, western brook lamprey, and American brook lamprey *Lethenteron appendix*) and freshwater parasitic (Great Lakes sea lamprey) species that are derived post-glacially from anadromous ancestors still develop SW-MRCs in their gills during metamorphosis (Youson and Freeman 1976; Youson and Beamish 1991; Bartels et al. 2011, 2015). However, further studies are needed to determine whether the Carpathian lamprey also lacks SW-MRCs, particularly since the Carpathian lamprey and Ukrainian brook lamprey are one of the very few lamprey species pairs studied that do not share cytochrome b haplotypes (e.g., Docker 2009; Lang et al. 2009; Fig. 4.1) and thus may not share the same recent evolutionary history as most paired species (see Sect. 4.6.3.1). Nevertheless, the cytochrome b network analysis performed by Bartels et al. (2017) suggests that the hypothetical ancestor of both Carpathian and Ukrainian brook lampreys was probably freshwater resident. Cytochrome b gene sequence in the Carpathian lamprey differed by 4.5–5.2% (K2P) from the European river lamprey, its closest extant anadromous relative.

4.3.2.4   Freshwater Parasitic *Ichthyomyzon* Species

Like other freshwater lampreys, adults of the three parasitic *Ichthyomyzon* species, the chestnut, Ohio, and silver lampreys, are smaller than most anadromous lampreys. Mean TL in the adults examined by Hubbs and Trautman (1937) were 189 mm for Ohio lamprey, 216 mm for chestnut lamprey, and 224 mm for silver lamprey, although maximum lengths of 279, 363, and 415 mm, respectively, have been reported during the feeding phase (see Chap. 3). Like other freshwater lampreys, the trophic phase is likely no more than ~1 year (Vladykov and Roy 1948; Hall 1963; Cochran et al. 2003). Access to large-bodied hosts (e.g., native and introduced salmonids, lake sturgeon *Acipenser fulvescens*, northern pike *Esox lucius*, muskellunge *E. masquinongy*, and American paddlefish *Polyodon spathula*) may permit the large size observed in some chestnut and silver lampreys relative to smaller-bodied freshwater lampreys such as the Miller Lake lamprey, Korean lamprey, and Carpathian lamprey (see Chap. 3). However, unlike most of the other freshwater lampreys discussed in this section, lampreys in the genus *Ichthyomyzon* likely evolved in fresh water or have been confined to fresh water for long periods of evolutionary time. Unlike lamprey species that were presumably derived post-glacially from an anadromous ancestor (see Sect. 4.3.2.3), silver and chestnut lampreys lack SW-MRCs in their gills (Bartels et al. 2012, 2015). This genus is confined to river systems and lakes in central and eastern North America at a considerable distance from the sea (Potter et al. 2015) and co-occurs in the large Mississippi River system with a number of ancient actinopterygian fishes also found only in fresh water (Hubbs and Potter 1971; Bartels et al. 2012). Phylogenetic analysis places *Ichthyomyzon* at or near the base of the phylogenetic tree of Northern Hemisphere lampreys, sister to *Petromyzon* with its anadromous and freshwater-resident sea lamprey (Gill et al. 2003; Lang et al. 2009). However, these two genera are not close relatives; cytochrome b gene sequences differ by ~14–15%. It appears that the genus *Ichthyomyzon* either evolved from a freshwater species that never possessed SW-MRCs, or SW-MRCs became lost in *Ichthyomyzon* during a long period of separation from an anadromous ancestor (Bartels et al. 2012, 2015).

4.3.2.5 Mexican Lamprey

Very little is known regarding the biology of the Mexican or Chapala lamprey. It is critically endangered but probably not yet extinct (Snoeks et al. 2009; Maitland et al. 2015). The reported range for adult TL (229–286 mm; Lyons et al. 1996) is comparable to that of most other freshwater parasitic species, although a parasitic feeding period of 2 years rather than ≤1 year has been suggested (see Chap. 3). Molecular studies (using cytochrome b sequence data from the presumably closely related Mexican brook lamprey *T. geminis*) place *Tetrapleurodon* as sister to *Entosphenus* (Lang et al. 2009); Mexican brook lamprey and anadromous Pacific lamprey cytochrome b gene sequences differ by 3.8%.

4.3.3 Anadromous Species with Rare Praecox or Freshwater Populations

Rare or unconfirmed freshwater populations have been reported in three of the 18 parasitic lamprey species, the Caspian lamprey *Caspiomyzon wagneri*, western river lamprey, and short-headed lamprey.

4.3.3.1 Caspian Lamprey

Berg (1948) reported two forms of the Caspian lamprey in the Volga Delta: a typical anadromous form where average TL ranged from 370 to 410 mm, with a maximum recorded TL of 553 mm, and smaller praecox individuals measuring 191–290 mm TL. The typical form underwent its spawning migration in November to March, and the praecox form was found migrating from mid-September to March. In contrast, in the European river lamprey, the typical form shows an earlier and more protracted migration (see Sect. 4.3.4.2). Berg (1948) also reported numerous praecox Caspian lamprey (250–370 mm) in the Sura River basin, a tributary to the Volga River. Freyhof and Kottelat (2008b) indicated that Caspian lamprey may have formed landlocked populations in reservoirs in the lower Volga River. The Volgograd and Saratov reservoirs (~3,120 and 1,830 km², respectively), constructed in 1958–1961 and 1955–1956, support dozens of fish species, including bream *Abramis brama*, European cisco or vendace *Coregonus albula*, burbot *Lota lota*, European perch *Perca fluviatilis*, and roach *Rutilus rutilus* (Ermolin 2010). Although no hosts of Caspian lamprey have been positively identified yet, these fishes are all preyed upon by other lamprey species (see Chap. 3). However, in the Saratov Reservoir, it appears that Caspian lamprey may now be extirpated. Ermolin (2010) reported that this species was found at low abundance in 1955–1967, decreased to very low abundance in 1969–1985, was represented by a solitary specimen in 1986–1995, and was not observed at all in 1996–2007. Construction of the Mingechaur Reservoir (605 km^2) on the Kura River in Azerbaijan in 1953 likewise appears to have resulted in loss of Caspian lamprey above the dam (Nazari et al. 2017), and this species also appears to have been extirpated from the Sura River following construction of the Cheboksary Reservoir (2,190 km^2) at its mouth in 1968–1986 (Ruchin et al. 2012). Since the Caspian Sea has a salinity approximately one-third that of most sea water, it would seem that an inability to osmoregulate at low salinities during the feeding phase would not hinder Caspian lamprey from establishing freshwater-resident populations (see Sect. 4.4.3). Most recent studies on the Caspian lamprey focus on those spawning in rivers in Iran, particularly in the Shirud River where TL ranges from 271 to 492 mm and in the Talar River where TL ranges from 295 to 428 mm (Nazari et al. 2017).

Differences in migration timing have been reported in Caspian lamprey from the Shirud River, with autumn migrants initiating upstream migration from mid-September to late October and spring migrants entering the river in mid-March to late April (Ahmadi et al. 2011; see Chap. 1). However, fall and spring migrants do not differ in size (Ahmadi et al. 2011), making it appear that the "premature migration" strategy shown by the fall migrants does not significantly cut short their growth opportunities at sea (see Quinn et al. 2016; Sect. 4.4.4).

4.3.3.2   Western River Lamprey

The western river lamprey or North American river lamprey is, on average, the smallest of the anadromous lampreys. Although a 324-mm individual was reported recently in the Columbia River estuary (Weitkamp et al. 2015), adult TL typically ranges from 168 to 236 mm, with a very narrow mean of 196–198 mm (Vladykov and Follett 1958; Beamish and Neville 1992; Weitkamp et al. 2015). The western river lamprey is considerably smaller than typical anadromous European river lamprey, with which it was considered conspecific until 1958 (although it is not as closely related as once thought; Docker et al. 1999; Lang et al. 2009; Li 2014), and it is even smaller than most praecox forms of the latter species (see Sect. 4.3.4.2). It could thus be argued that all western river lamprey correspond to the anadromous praecox type. Apparently, this species occurs only as widely separated populations, generally associated with larger estuarine systems (Moyle 2002; Boguski et al. 2012), and it feeds at sea for a single summer (3–4 months; R. J. Beamish 1980), compared with ~12 and 18 months for praecox and typical European river lamprey, respectively. Even "at sea" (e.g., in the Strait of Georgia), western river lamprey tend to remain in surface waters and are concentrated in the general vicinity of the larger rivers where salinity is reduced.

Despite what would seem as a "predisposition" for freshwater residency, western river lamprey—unlike the European river lamprey—appear to rarely, if ever, form freshwater-resident parasitic populations. Nevertheless, the western river lamprey or a western river lamprey-like ancestor has given rise to numerous freshwater non-parasitic derivatives (see Sects. 4.6.3.1 and 4.7). Some authors (e.g., Renaud 2011; Potter et al. 2014) have suggested that parasitic individuals produced within a western brook lamprey population on Vancouver Island may represent freshwater-resident western river lamprey (Beamish 1987; Beamish et al. 2016). However, unlike the freshwater-resident Vancouver lamprey which has retained its ability to osmoregulate in salt water (Beamish 1982), these parasitic variants (the so-called marifuga variety) are unable to osmoregulate in salt water (Beamish 1987). This and other features of these individuals suggest instead that they are western brook lamprey that show an "atavistic" reversal to the parasitic feeding type (see Docker 2009).

Beamish and Youson (1987) found that only 3% of western river lamprey held in the laboratory were able to feed and spawn entirely in fresh water, and there are only anecdotal suggestions of freshwater-resident populations of this species. Adult western river lamprey have been recorded in Lake Sammamish (19.8 km^2) and Lake Washington (88 km^2) in October and December, respectively, and the lamprey from Lake Washington was attached to and possibly feeding on a kokanee (Vladykov and Follett 1958). However, each record is represented by a single specimen (311 and 279 mm, respectively) and could represent individuals returning to fresh water to spawn following a marine feeding phase. Nevertheless, there are current reports of a healthy freshwater-resident western river lamprey population in Lake Washington, which may have become non-anadromous when the lake lost its natural connection with Puget Sound following construction of the Lake Washington Ship Canal in 1916 (Molly Hallock, Washington Department of Fish and Wildlife, Lacey, WA, personal communication, 2012). There are dozens of fish species in this lake, including coastal cutthroat trout, juvenile sockeye salmon, longfin smelt *Spirinchus thaleichthys*, and northern pikeminnow *Ptychocheilus oregonensis* (Quinn et al. 2012). There are also reports (in 1931 and 1959) of western river lamprey in Lake Cushman (United States Fish and Wildlife Service 2004), a lake that was expanded into a reservoir (16.2 km^2) after dam construction in 1924–1926. The 1931 specimen (Burke Museum Ichthyology Collection Catalog Number 1509) was collected in June by a commercial fisherman and could represent a remnant of the original anadromous population. Details of the 1959 specimen require verification; the only western river lamprey collected in 1959 in the Burke Museum Ichthyology Collection (Catalog Number 15726) was collected in March in the Yakima River in the Columbia River basin, below Easton Dam. A few transformed western river lamprey are collected each year at the Chandler Juvenile Fish Monitoring Facility in the lower Yakima River, but adult western river lamprey have never been observed here (Ralph Lampman, Yakama Nation, Fisheries Resources Management Program, Toppenish, WA, personal communication, 2018). Similarly, a single putative juvenile western river lamprey was recently reported above the John Day Dam, the third upriver mainstem dam on the Columbia River, but there is no evidence of an established freshwater population above the dam (Jolley et al. 2016). In both cases, it appears that these individuals (presumably captured during their outmigration to sea) have arisen from within the normally non-parasitic western brook lamprey populations upstream (see Sect. 4.6.3.3).

4.3.3.3   Short-Headed Lamprey

The short-headed lamprey is an anadromous species found in drainages and coastal waters in southeastern Australia. Individuals as large as ~500 mm have been reported during the feeding phase, and TL for adults at maturity is ~280–420 mm (Potter et al. 1968; see Chap. 3). In the Gippsland Lakes region of Victoria, this species is known to feed in fresh water on introduced brown trout *Salmo trutta,* black bream *Acanthopagrus butcheri,* and yellow-eyed mullet *Aldrichetta forsteri* during the summer and fall before going to sea; at sea, they appear to spend another 18 months feeding before returning to fresh water. However, there are also indications that a small population remains throughout the feeding phase in Lake Wellington, the largest (147 km^2) and least saline of the three lakes (Potter et al. 1968). Commercial fishermen report catching lamprey attached to yellow-eyed mullet in Lake Wellington throughout the year, but lamprey attached to black bream are reported in Lake Victoria (1.4 km^2) only during the summer months. Since the lakes have been completely cut off from the sea at times, Potter et al. (1968) considered it feasible that conditions existing in the Gippsland Lakes in the past led to the evolution of a form of short-headed lamprey which restricts its feeding to fresh or brackish water.

4.3.4   Anadromous Species with Established Praecox or Freshwater Populations

Praecox or freshwater populations appear to be more common and are certainly better known in the remaining four parasitic lamprey species: the Pacific, European river, Arctic, and sea lampreys. Compared to the other three species, there are few known freshwater-resident populations of Pacific lamprey, but this may be partly the result of many populations that are known having been described as distinct freshwater species (see Sect. 4.3.2.1). There are also relatively few freshwater-resident populations of sea lamprey (e.g., none are known from Europe), but, of course, where they do occur (i.e., in the Great Lakes, Finger Lakes, and Lake Champlain in North America), they have become very abundant and well established (see Chap. 5). Anadromous praecox (or dwarf) Pacific lamprey have been described from several locations, but no small-bodied anadromous sea lamprey are known. Praecox and freshwater-resident European river and Arctic lampreys are common.

4.3.4.1   Pacific Lamprey

The Pacific lamprey is typically large-bodied, although considerable geographic variation in body size has been reported. Over a broad scale, TL of spawning adults appears to be positively correlated with latitude and distance from the sea. Feeding phase individuals up to 850 mm TL have been captured in the Bering Sea (Orlov et al. 2008), and relatively large adults have been reported from large interior rivers within the Skeena River drainage in northern British Columbia and the Columbia River basin. In the Skeena River drainage, adult Pacific lamprey measuring 550–670 and 410–590 mm TL were reported in Babine Lake and Babine Creek, respectively, and adults ranged from 410 to 720 mm in the Bulkley River (R. J. Beamish 1980). In the Columbia River basin, Clemens et al. (2012) reported upstream migrants measuring 560–710 mm at Willamette Falls (235 river kilometers, rkm, upstream from the ocean), and migrants measuring ~710–810 mm TL have been observed in the Methow River, one of the most upstream spawning sites in the Columbia River drainage and accessible only to lamprey that are able to pass nine dams (John Crandall, Wild Fish Conservancy, Duvall, WA, personal communication, 2011). In the Umpqua River, on the Pacific coast, TL of upstream migrants has been recorded to be 415–644 mm (Lampman 2011), and the southernmost record of Pacific lamprey (from Revillagigedo Archipelago, Mexico) is that of a pre-spawning female measuring only 420 mm TL (Renaud 2008).

In contrast, Pacific lamprey appear to have smaller body sizes in small coastal streams or those draining into the Strait of Georgia (also known as the Salish Sea). Spawning migrants in four rivers on Vancouver Island draining east into the Strait of Georgia (Bonsall Creek and Qualicum, Chemainus, and Quinsam rivers) measured 130–380 mm TL. Those from three rivers on Vancouver Island draining west into the Pacific Ocean (Robertson Creek, Stamp River, and a tributary of Kennedy Lake) measured 220–510 mm TL (R. J. Beamish 1980). Smaller-bodied Pacific lamprey are also thought to occur in Duckabush River, which drains into Puget Sound. Similarly, Pacific lamprey in mainland rivers draining into the Strait of Georgia (i.e., the Salmon River, in the Lower Fraser Valley, and the Nicola River, a tributary of the Fraser River) appear to be considerably smaller than those in the Skeena and Columbia river systems; upstream migrants measuring only 193–214 mm and 273–453 mm were reported by Pletcher (1963) and Beamish and Levings (1991), respectively. Although not specifically described as praecox forms, these smaller-bodied anadromous Pacific lamprey are thought to spend less time feeding at sea. R. J. Beamish (1980) assumed that moderate to large lamprey spend up to 3.5 years feeding in salt water; minimum duration of the feeding phase has been estimated at 20 months (1.7 years). Given the range of sizes, there is not a clear distinction (i.e., based on a bimodal size distribution) between the typical and praecox forms; however, based on Pacific lamprey adults found in the Coquille River on the Oregon coast, Kostow (2002) considered mature individuals <370 mm TL to be the dwarf type and those >550 mm to constitute the typical form.

In addition to this variation in size at maturity, Pacific lamprey—at least in the Klamath River in California—show variation in migration timing. Although most populations appear to represent the typical "river-maturing" form (i.e., entering fresh water during the summer prior to spawning), Clemens et al. (2013) described an "ocean-maturing" form entering this river. These individuals are more sexually mature when they enter fresh water in late winter and likely spawn within weeks or months (see Chap. 1). However, unlike the European river lamprey in the Severn River, there is no apparent size difference associated with run timing (e.g., mean TL 609–625 mm for individuals entering in June–September and 612–618 mm for those entering in March–May; Parker 2018). Interestingly, Parker (2018) found that there was a genetic basis for these different ecotypes (see Sect. 4.4.3).

The wide variation in Pacific lamprey body size among rivers has often been used, reasonably so, as evidence of homing and local adaptation (e.g., Beamish and Withler 1986). There is now overwhelming evidence that migratory lampreys do not home to their natal streams (see Moser et al. 2015), but this does not mean that Pacific lamprey constitute a single panmictic population. Weak but significant genetic variation has been detected among widely separated populations, and evidence of isolation-by-distance suggests that limitations to their dispersal at sea prevent formation of an entirely homogenous population (Spice et al. 2012; Hess et al. 2013). Greater genetic differentiation among locations with dwarf or praecox forms suggests even more limited dispersal by smaller-bodied Pacific lamprey. This suggests that the praecox form may remain in estuarine or coastal areas, without the need or opportunity (i.e., because of a rich and/or less mobile prey base) to disperse more widely. Recent genomic studies have also indicated that there is a genetic basis for body size (Hess et al. 2013, 2014), and these individuals may represent a "jumping-off point" for the evolution of freshwater-resident parasitic and non-parasitic forms (see Hardisty 2006; Docker 2009; Sect. 4.7).

There have been several reports of freshwater-resident or landlocked Pacific lamprey, although it should be noted that some of the earlier reports of lacustrine forms once considered to be non-anadromous races of Pacific lamprey have since been elevated to species status (Bond and Kan 1973; Vladykov and Kott 1979a; Beamish 1982; see Sect. 4.3.2.1). These non-anadromous forms are thought to have arisen within the past 6,600–15,000 years, post-glacially in the case of the Vancouver lamprey and following the eruption of Mount Mazama in the case of the Miller Lake lamprey (see Sect. 4.3.2.1). Kan (1975) postulated that interior forms in Oregon and northern California were less affected by Pleistocene glaciation, but that non-anadromous forms arose and survived in large lakes when anadromous migrations were blocked off by the Cascade and Klamath mountain building. It is thought that some other non-anadromous populations (e.g., in Goose Lake and Upper Klamath Lake) likewise deserve recognition as distinct taxa.

More recently, there have been reports of a Pacific lamprey-like form or forms feeding parasitically in three disjunct lake systems in southwestern British Columbia: West Lake on Nelson Island, Village Bay Lake on Quadra Island, and two adjacent lakes (Ruby and Sakinaw lakes) on the Sechelt Peninsula (Beamish 2001; COSEWIC 2008). The freshwater lamprey in Ruby and Sakinaw lakes (~5 and 7 km^2, respectively) are the best studied. Although they shared cytochrome b DNA sequences with anadromous Pacific lamprey and Vancouver lamprey from the Cowichan lake system, analysis using microsatellite DNA markers showed that the lamprey in Ruby and Sakinaw lakes was distinct from the Vancouver lamprey and even more distinct from Pacific lamprey (Taylor et al. 2012). Although the taxonomic status of this population is still unresolved, there is good evidence that they feed in fresh water, including feeding on Sakinaw sockeye salmon, a stock that has been assessed as Endangered by the Committee on the Status of Wildlife in Canada, and perhaps also on small populations of coho salmon and chum salmon *Oncorhynchus keta* (COSEWIC 2016). Although much of the observed scarring in upstream-migrating and adult Sakinaw Lake sockeye salmon may be the result of parasitism by western river lamprey in the Strait of Georgia, there is also evidence of lamprey parasitism on sockeye fry and smolts within the lake (COSEWIC 2016). Other non-anadromous derivatives have been reported in British Columbia over the years (e.g., in the Columbia River, the Fraser River at Prince George, and Cultus Lake; McPhail and Lindsey 1970; Vladykov and Kott 1979a), but little is known about such lamprey and whether they, in fact, represent permanent freshwater-resident populations. There are also reports of Pacific lamprey in Grosvenor Lake and at the mouth of an unnamed creek in Brooks Lake in the Naknek River system in Alaska, but these reports were not substantiated by the extensive collections made by Heard (1966) when studying Arctic lamprey in this system. Lamprey remains have been reported in scat samples recovered from resident harbor seal *Phoca vitulina* in Iliamna Lake in Alaska. Although both Pacific and Arctic lampreys have been reported in the Kvichak River drainage (Hauser et al. 2008), these remains are likely from Arctic lamprey (see Sect. 4.3.4.3).

The freshwater derivatives of Pacific lamprey indicate that this species is capable of establishing in fresh water, although several reports of extirpations above dams suggest that not all populations or individuals are capable of doing so. Pacific lamprey were confined in Dworshak Reservoir (69 km^2) in the Columbia River basin in Idaho when dam construction was completed in September 1971 (Wallace and Ball 1978). Rainbow trout and kokanee were stocked into the impoundment in spring 1972 and 1973; during creel surveys in May 1973, 5% of the rainbow trout and kokanee were found to bear at least one lamprey scar, and this proportion increased to 16% by November 1973. However, incidence of scarring decreased to 4% in 1974 and <1% in 1975 and 1976, and no lamprey were directly observed after 1973. Beamish and Northcote (1989) reported a similar situation in Elsie Lake (6.7 km^2) on central Vancouver Island, following construction of five dams on the outlet of the lake in 1957–1959. Resident cutthroat and rainbow trout collected in the lake started showing evidence of lamprey scarring immediately after dam construction was complete in 1959. In this year, 74.5% of the trout had fresh wounds and 2.1% exhibited older wounds that were healing (Pletcher 1963). However, the small size of the wounds indicated that the lamprey probably grew very little during this time period, and, although the proportion of old scars increased from 1960 to 1963, there was a progressive decline in the proportion of fresh scars (to 0% in 1969, 1981, and 1987).

Likewise, Pacific lamprey appear to have been extirpated above barrier dams in the Willamette Basin in Oregon (Doug Larson and Matt Helstab, U.S. Forest Service, Middle Fork Ranger District, Westfir, OR, personal communication, 2017) after construction of two high-head flood control dams in the Middle Fork Willamette River in 1953 and 1954. Fish species in the resulting reservoirs (Dexter Reservoir and Lookout Point Reservoir, 4.2 and 17.6 km^2, respectively) included Pacific salmon *Oncorhynchus* spp., northern pikeminnow, and largescale sucker *Catostomus macrocheilus* (see Keefer et al. 2013). In 2001, however, six Pacific lamprey (two recently metamorphosed juveniles and four larvae) were captured during an

Environmental Protection Agency survey 4.8 km upstream of the uppermost dam. This would suggest that a small number of Pacific lamprey survived and reproduced upstream of the dams more than 45 years after their construction or that some escapement of anadromous lamprey occurred past the dams. Recent surveys have to date failed to detect additional Pacific lamprey at these sites, but investigation of this potential landlocked population is ongoing (Doug Larson and Matt Helstab, personal communication, 2017).

Extirpation of Pacific lamprey following confinement to fresh water is consistent with observations that post-metamorphic individuals held in fresh water in the laboratory fed poorly or not at all, and all ultimately died prior to maturation (Richards and Beamish 1981; Clarke and Beamish 1988). The ability to maintain blood sodium concentration in fresh water varied among lamprey from different source populations, with performance in fresh water (from best to worst) ranked as follows: Chemainus River > Puntledge River > Big Qualicum River > Kanaka Creek > Somass River > Babine River (Clarke and Beamish 1988). Chemainus, Puntledge, Big Qualicum, and Somass rivers are located on Vancouver Island; Kanaka Creek is a tributary of the Fraser River on the lower mainland of British Columbia, and the Babine River is located in northern British Columbia. Pacific lamprey from the Chemainus River survived in fresh water until July, but no Babine lamprey survived beyond mid-February. Survival in the laboratory was good in salt water (Beamish 1982). Thus, Clarke and Beamish (1988) concluded that confinement of Pacific lamprey in fresh water does not easily result in the formation of landlocked populations. Beamish and Northcote (1989) suggested that barriers to migration might select for a few individuals that genetically would be able to feed and grow to maturity in fresh water and that the chance of this happening would be higher if the size of the population was large (or if the founding population was naturally small-bodied or otherwise "predisposed" to freshwater residency; see Sect. 4.4.3). Beamish (1985) suggested that the genetic change required for Pacific lamprey to survive in fresh water is either extremely rare or requires a series of changes that are unlikely to occur immediately when faced with sudden barriers such as dams.

4.3.4.2 European River Lamprey

A wide range of sizes has been reported for anadromous European river lamprey, and there is evidence of geographical variation in size. Females at or near spawning in the Drwêca River in the Vistula River basin in northern Poland averaged 405 mm TL (337–462 mm; Witkowski and Jęsior 2000). Large size has also been reported among upstream-migrating European river lamprey in the River Meuse (>400 mm TL; Lanzing 1959) and in the Nemunas River in Lithuania (356–408 mm; Gaygalas and Matskevichyus 1968). This species seems to be somewhat smaller (mean TL for males and females was 310–320 mm and 320–340 mm, respectively; Berg 1948) in the River Neva, which flows into the Gulf of Finland, and even smaller in the U.K. In the Severn Estuary, mean TL in early upstream migrants was 290–306 mm in males and 301–318 in females (Hardisty and Huggins 1973), and average TL was ~300 mm near the end of the spawning run (Abou-Seedo and Potter 1979). In Scottish waters, TL of the anadromous form at or near spawning is ~300–350 mm TL (Hume 2013). Abou-Seedo and Potter (1979) suggested that the consistently larger size of European river lamprey in the Baltic and North seas, compared with those of the River Severn, indicates that feeding conditions in the former areas are probably better than they are in the region off the west coast of England. As with the western river lamprey, the European river lamprey may prefer water of reduced salinity (Bahr 1952), resulting in a more estuarine, coastal distribution, and more genetic variation among regions than wide-ranging, large-bodied lamprey species (Mateus et al. 2016; see Sect. 4.7.5).

The estimated duration of the marine trophic phase in typical European river lamprey is 18 months, from the spring of one year to the autumn of the next (Hardisty and Potter 1971b). In this form, the entire post-metamorphic period is thought to be ~2.5 years, including the non-trophic period following metamorphosis (i.e., migrating downstream in the fall, with growth over the first winter being minimal) and during upstream migration (starting in the fall prior to spawning; Abou-Seedo and Potter 1979). In addition to this typical form, a smaller praecox form with a presumably reduced marine trophic period has been identified in the River Neva (mean TL 225 mm) and Severn River (mean 240 mm) where, interestingly, the typical form is already smaller than elsewhere (Berg 1948; Abou-Seedo and Potter 1979). In both rivers, the praecox individuals formed a distinctly smaller size class alongside the typical form during upstream migration. Praecox European river lamprey (mean TL 249 mm) have also been reported in the River Bladnoch, which drains into the Solway Firth in southwestern Scotland (Hume 2013). In the Severn Estuary, the praecox form was less common than the typical form (comprising ~25% of all individuals), and it appeared to show differences in run timing, with a later and more contracted upstream migration. Typical anadromous individuals were occasionally found in the estuary as early as July and as late as April, with peak abundance generally in November; in contrast, the praecox form was present mainly between January and March. Abou-Seedo and Potter (1979) thus concluded that the praecox form spent ~12 months at sea, but, because it delayed its upstream migration until the winter or spring prior to spawning, its non-trophic period following metamorphosis lasts only 1.5 years. Therefore, the praecox form appears to reduce its post-metamorphic period by 1 year relative to the typical anadromous form but reduces the duration of feeding by only 6 months. In the River Neva, appreciable numbers of the praecox form were found in the delta in October and November, and Berg (1948) concluded that the praecox form began its upstream migration after feeding in the Gulf of Finland for <6 months. Other small-bodied European river lamprey (mean TL 220 and 240 mm for males and females, respectively) have been reported in the Narew River in western Belarus and northeastern Poland (Hardisty 1986a). These individuals were not identified specifically as the praecox form, but they likely would have had a reduced marine trophic phase relative to the large-bodied individuals in other tributaries of the Vistula River.

There are also several reports of small-bodied freshwater-resident European river lamprey populations, particularly in large lakes. Berg (1948) reported both large (up to 362 mm TL) and small (250–330 mm) European river lamprey in Lake Ladoga (~18,100 km^2) and Lake Onega (~9,700 km^2), the largest and second largest lakes in Europe, respectively. Lake Onega is connected to Lake Ladoga via the River Svir, and both lakes drain into the Gulf of Finland via the Neva River. Therefore, as with many other freshwater-resident populations, these European river lamprey were not landlocked; the larger form was presumably anadromous, and the smaller form was likely a freshwater-resident (lacustrine) form. However, it should be noted that the specimens examined by Berg (1948) were collected prior to construction of the first dam on the Svir River (1936), and now two dams present impassable barriers to anadromous river lamprey migrating upstream from the Baltic Sea and Lake Ladoga to Lake Onega. Nevertheless, the freshwater form still persists in these lakes. Tsimbalov et al. (2015) reported spawning-phase male and female lamprey measuring 202–241 mm and 200–247 mm TL, respectively, in the Lososinka River in the Lake Onega basin, and Kuznetsov et al. (2016) observed mature female lamprey measuring 264–348 mm TL in Lake Ladoga. Both lakes, particularly Lake Ladoga, have rich fish fauna, including freshwater Atlantic salmon *Salmo salar* populations (Ozerov et al. 2010) and other fishes (e.g., European perch, burbot, European cisco, and roach; Berezina and Strelnikova 2010). Berg (1948) found a large number of lamprey attached to European cisco. Lake Ladoga even supports a resident ringed seal population *Phoca hispida ladogensis* (Kunnasranta et al. 2001).

Small freshwater-resident European river lamprey (mean 200–225 mm TL) have also been reported in Lake Mjøsa in southern Norway (Berg 1948). Lake Mjøsa is the largest lake in Norway (365 km^2), as well as one of the deepest lakes in Europe, with at least 20 species of fish, including a fast-growing brown trout morph that can reach weights of >15 kg, northern pike, European perch, burbot, European smelt *Osmerus eperlanus*, European whitefish *Coregonus lavaretus*, and European cisco (Sandlund et al. 1987; Taugbol 1994; Mariussen et al. 2008). Freshwater-resident European river lamprey populations have also been reported in large lakes in the Vuoksi, Kymijoki, and Kokemaenjoki river drainages in Finland (Tuuainen et al. 1980).

Freshwater residency has also been demonstrated in European river lamprey inhabiting Lough Neagh in Northern Ireland (Goodwin et al. 2006). With a surface area of 390 km^2, it is the largest lake in the British Isles. Fishes within Lough Neagh have access to the sea, but all information suggests that all or most of the Lough Neagh river lamprey are non-anadromous. River lamprey are rarely caught in the River Bann downstream of the Lough Neagh outflow, and actively feeding lamprey (i.e., attached to pollan *Coregonus autumnalis* and with full guts) have been captured in the lough almost year-round (February to October). During this time, TL increased from 118 mm to a maximum of 391 mm. Scarring on pollan was first observed (at low incidence) in April and May and increased in late June and early July, although no fresh scars were evident in September or October. Multiple size classes were not present during the feeding phase, suggesting that this population feeds for ≤1 year in fresh water. Nevertheless, the size achieved by the end of the feeding phase indicates rapid growth during this year, making these individuals (mean 318 mm TL at maturity; Hume 2013) as large or larger than some of the praecox or even smaller-bodied typical anadromous European river lamprey that feed at sea for 1 or 1.5 years. Inger et al. (2010), using stable isotope analysis, subsequently confirmed that none of the 71 Lough Neagh river lamprey examined had fed in the marine environment. However, these authors identified brown trout and non-native bream as the main items in the river lamprey diet, with pollan representing the main food source only between May and July. This finding and the scarring data from Goodwin et al. (2006) suggest that river lamprey shift to larger fish species later in the year (see Chap. 3). Therefore, access to a range of prey sizes may be important in permitting these lamprey to reach a large size exclusively in fresh water.

The best-studied freshwater-resident European river lamprey is the population found in Loch Lomond (71 km^2) in Scotland. Loch Lomond contains the greatest number of fish species of any lake in Scotland. In addition to European river lamprey and European brook lamprey, and the occasional report of sea lamprey (see Sect. 4.3.4.4), Loch Lomond contains at least 12 other native fish species, including Atlantic salmon, brown trout, European whitefish (locally known as powan and sometimes recognized as *Coregonus clupeoides*), northern pike, roach, and European perch, and six non-native species (Maitland 1980; Hume et al. 2013a). Maitland (1980) found that brown trout, roach, and especially powan bore European river lamprey scars (with scars on 6, 5, and 45% of all individuals captured, respectively) and that 55% of the scarred powan had 2–8 scars each. Fresh wounds were recorded in May–November and were especially prevalent in July and August (Morris 1989; Maitland et al. 1994), that is, after they would normally outmigrate if they were anadromous. Maitland (1980) indicated that feeding river lamprey must be relatively common in most parts of the loch, because wounding rates were high on powan in all locations. However, it should be noted that Hume et al. (2013a) found that scarring on powan has become greatly reduced in recent years. In 2010, only 6% of powan had river lamprey scars, possibly as a result of declines in powan numbers following introduction of the ruffe *Gymnocephalus cernuus*.

The freshwater-resident European river lamprey in Loch Lomond, sometimes called a dwarf (Morris 1989) or praecox (Maitland et al. 1994) form, is not landlocked. Anadromous river lamprey can access the loch via the River Leven, which connects Loch Lomond to the Firth of Clyde via the River Clyde, and both forms spawn in the River Endrick, which flows into the eastern end of the loch (Morris 1989). The freshwater-resident form measures 155–257 mm TL during its upstream migration (Adams et al. 2008) and 164–197 mm (mean 185 mm) at spawning (Morris 1989), compared to 269–338 mm (mean 327 mm) in the anadromous form (Morris 1989; Adams et al. 2008; Hume 2013). Adams et al. (2008) used stable isotope analysis to show that the small and large lamprey size classes from the River Endrick corresponded to those feeding in fresh- and saltwater, respectively. Morris (1989) indicated that the freshwater form appears to feed for a few months compared to 15–18 months at sea for the typical anadromous form.

Morphological differences have been reported between the freshwater-resident and anadromous European river lamprey in Loch Lomond. The freshwater form has a bigger oral disc and eye, longer snout, and darker pigmentation (Morris 1989). Morris (1989) indicated that the freshwater form in Loch Lomond is in some ways intermediate between the normal forms of river and brook lampreys and suggested that it might represent an intermediate stage between these two species (see also Beamish 1985; Docker 2009; Sect. 4.7.3). However, there is evidence of temporal separation between forms, and some measurements and counts of the freshwater parasitic form were more extreme than either the anadromous river lamprey or brook lamprey; these observations suggest that the freshwater form is not a hybrid (Morris 1989). Furthermore, recent studies show that the freshwater form is more genetically differentiated from the anadromous and non-parasitic forms than either of these latter forms are from each other (Bracken et al. 2015; Hume et al. 2018), although there is evidence for ongoing gene flow among all three forms (Bracken et al. 2015).

4.3.4.3   Arctic Lamprey

Like the Pacific and European river lampreys, size at maturity in anadromous Arctic lamprey varies widely, and there appear to be geographical differences. Maximum TL during the feeding phase (790 mm) has been reported in the North Pacific Ocean (Orlov et al. 2014), and the largest size at, or approaching, maturation has been reported in southeastern Russia. Maximum TL at maturity was 625 mm (mean 505 mm) and 566 mm (mean 456 mm) in the Partizanskaya River (formerly known as the Suchan River) and the lower reaches of the Amur River, respectively (Bogaevskii 1949). In Japan, Yamazaki et al. (2001) reported maximum TL ranging from 400 to 442 mm at or near spawning, and in western Kamchatka, Russia, maximum TL of the typical anadromous form was 350 and 330 mm in males and females (with means of 293 and 279 mm, respectively; Kucheryavyi et al. 2007). Heard (1966) found that maximum TL of mature anadromous Arctic lamprey in southwestern Alaska was 311 mm TL (mean 253 mm), and they were even smaller (maximum and mean TL of 200 and 166 mm, respectively) on the southern Kuril Islands in the Western Pacific Ocean between Japan and the Kamchatka Peninsula (Sidorov and Pichugin 2005).

This latter population from the southern Kuril Islands could likely be described as a praecox anadromous form, although, given the range of sizes observed, the distinction between the typical and praecox forms is not entirely clear. Berg (1948) gave a brief description of a spent dwarf male Arctic lamprey (224 mm TL) from the mouth of the Kukhtui River, near the Sea of Okhotsk, and Kucheryavyi et al. (2007) explicitly described a praecox form in western Kamchatka (maximum and mean TL of 220 and 190 mm, respectively) that occurred alongside the typical anadromous form and a freshwater-resident non-parasitic form. The praecox form in western Kamchatka was less common (3% of all individuals) than the typical anadromous form (14%), and 92% of the praecox individuals were males, lending credence to the suggestion that life history transitions may occur more readily in males (see Docker 2009).

Duration of the feeding phase in this species can only be extrapolated from body size (see Chap. 3). However, given the range of sizes observed, there appears to be more intraspecific variation than in most other parasitic species, with the possible exception of the Pacific lamprey (see Sect. 4.3.4.1). Orlov et al. (2014) identified four size classes (150–320, 330–530, 540–650, and 660–800 mm) in Arctic lamprey captured in the North Pacific Ocean during the feeding phase, suggesting that individuals in this population feed at sea for up to 4 years. Presumably, populations with smaller sizes at maturity feed at sea for shorter periods of time, with those specifically identified as the praecox form having a marine trophic phase lasting for as little as several months to 1 year (Kucheryavyi et al. 2007). Nevertheless, given that we know relatively little about the marine feeding phase of lampreys, we cannot exclude the possibility that size differences are the result of differences in the quantity or quality of available host fishes. The anadromous sea lamprey can reach 700–800 mm TL in ≤2 years of feeding at sea (see Sect. 4.3.4.4); therefore, size at maturity (at least across species and potentially within species) may not be a reliable indicator of the duration of the feeding phase.

Freshwater-resident populations of Arctic lamprey have been reported in Asia and North America, but they are not known from Europe (Holčík 1986), and some freshwater-resident populations attributed to this species are apparently non-parasitic forms (Iwata and Hamada 1986; Kucheryavyi et al. 2007; Yamazaki et al. 1998, 2011; see Sect. 4.6.3.2). The best-studied populations where freshwater parasitism and permanent freshwater residency have been confirmed are in North America. The Naknek River system on the Alaska Peninsula consists of seven interconnected lakes, including Naknek Lake, which is the fifth largest lake in Alaska (584 km²). Heard (1966) reported that both anadromous and freshwater-resident parasitic Arctic lamprey occurred in this system and that the freshwater-resident form was more common than the anadromous form. Lamprey were found feeding on sockeye salmon, rainbow trout, pygmy whitefish *Prosopium coulterii*, and threespine stickleback *Gasterosteus aculeatus* in July, August, and September. They measured 115–226 mm (mean 167 mm) TL during the feeding phase and 117–188 mm (mean 155 mm) at spawning. Vladykov and Kott (1978) subsequently re-examined 47 specimens (122–172 mm) from Heard (1966), as well as specimens obtained from other areas, and described these lamprey as a new non-parasitic species, the Alaskan brook lamprey *Lethenteron alaskense*. However, Heard (1966) did not find any recently metamorphosed individuals with maturing gonads and presented clear evidence of feeding in fresh water throughout the summer and fall. He suggested that individuals within this system might exhibit flexibility in feeding, remaining in the lake if there is an adequate source of prey or, alternatively, starting to feed on salmon smolts in fresh water and being carried to sea when they outmigrated. Arctic lamprey were observed attached to upstream-migrating rainbow trout and sockeye salmon congregating at the outlet to one of the lakes in this system in July and August, and Heard (1966) suggested that these lamprey had probably started to migrate from the lake but remained at the outlet upon finding a large concentration of prey. This would suggest that Arctic lamprey in this system are opportunistically freshwater resident, rather than representing a discrete freshwater population, and this population may represent one of the few cases of partial migration in lampreys. Partial migration refers to the resident–migratory dimorphism seen in many populations of other fish species and birds, for example, where some individuals in the population migrate and others remain resident (Olsson et al. 2006; Chapman et al. 2012). Other freshwater-resident lampreys, even those that co-occur with the anadromous form (e.g., European river lamprey in Loch Lomond; see Sect. 4.3.4.2), appear instead to represent distinct gene pools rather than belonging to a single polymorphic population.

Unlike Arctic lamprey in the Naknek system, the freshwater-resident Arctic lamprey population in Great Slave Lake (27,200 km^2) in the Northwest Territories in Canada is thought to represent a permanent freshwater population. Although not land-locked, anadromous lamprey in this system would need to migrate up the Mackenzie River, the longest river in Canada (1,740 km), to reach the lake. Arctic lamprey in this system have been found to prey on lake whitefish *Coregonus clupeaformis*, cisco *Coregonus* spp., inconnu *Stenodus leucichthys*, lake trout *Salvelinus namaycush*, longnose sucker *Catostomus catostomus*, and burbot (Nursall and Buchwald 1972). Specific size information was not given for the 112 feeding or spawning-phase lamprey collected, but body size in this population appears somewhat larger than in the Naknek River system; feeding phase juveniles collected in the north arm of Great Slave Lake in late August and September often exceeded 300 mm in TL, and size at spawning included a spent female measuring ~168 mm and a ripe male measuring 226 mm TL. No juvenile lamprey were collected in the winter fishery, leading Nursall and Buchwald (1972) to conclude that this population likely fed following metamorphosis for <6 months (i.e., from downstream migration in the spring to early fall).

Other freshwater-resident Arctic lamprey have been reported in Tatlmain Lake, a tributary to the Pelly River in the Yukon Territory, Canada, and in Iliamna Lake in southwest Alaska (McPhail and Lindsey 1970). Iliamna Lake is the largest lake in Alaska and the eighth largest lake in the U.S. (2,600 km^2) and has abundant salmonids (e.g., sockeye salmon, Arctic char *Salvelinus alpinus*, Dolly Varden) and other fishes presumably capable of supporting Arctic lamprey during the feeding phase (Foote and Brown 1998; Hauser et al. 2008; May-McNally et al. 2015). This lake drains into Bristol Bay via the Kvichak River, with no barriers preventing access to the sea, and the abundant prey base supports even a resident harbor seal population (Hauser et al. 2008). Lamprey remains identified as either Arctic lamprey or Pacific lamprey (the two species that have been reported in the Kvichak River drainage) were found in 27% of scat samples collected from these seal in July–August 2001, 2005, and 2006 (Hauser et al. 2008), indicating that lamprey were present in the lake after the time at which they would normally outmigrate. McPhail and Lindsey (1970) reported that lamprey up to 254 mm TL were recorded in the lake. Hauser et al. (2008) reported that they did not observe any sockeye salmon in the lake with lamprey attached, but directed studies targeting Arctic lamprey specifically are needed.

4.3.4.4   Sea Lamprey

The anadromous sea lamprey is the largest of all extant lampreys, and there are no known anadromous praecox populations. If there is geographic variation in size at maturity, it is more subtle than what has been observed in Pacific, European river, and Arctic lampreys. Given that TL has been estimated to shrink by up to 24.3% in females and 18.6% in males during the course of the spawning migration (F. W. H. Beamish 1980a; see Chap. 1), observed differences could be the result of individuals being measured at different points on their migration. Maximum TL during the feeding phase is recorded at 1,200 mm (see Chap. 3), and spawning migrants at or approaching 900 mm TL have been reported in Europe (e.g., Hardisty 1986b; Silva et al. 2013a; Rooney et al. 2015). Adults entering the Mulkear River in Ireland, for example, measured 616–913 mm TL (mean 760 mm; Rooney et al. 2015). In North America, upstream migrants averaged 743 and 729 mm TL in the St. Lawrence River in Quebec and St. John River in New Brunswick (ranges 666–841 and ~640–840), respectively (Vladykov 1951; Beamish and Potter 1975), and sea lamprey captured on the spawning grounds in the Terra Nova River in Newfoundland averaged 601 mm (530–687 mm; Dempson and Porter 1993). It is thought that sea lamprey feed at sea for 23–28 months (F. W. H. Beamish 1980a). Halliday (1991) found two size classes during the feeding phase in the northwest Atlantic (120–380 and 560–840 mm TL), which would likewise be consistent with a marine feeding phase lasting at least 2 years. However, Halliday (1991) could not rule out a 1.5-year juvenile feeding period, and a recent study conclusively showed that some individuals can reach large size in considerably less than 2 years. Silva et al. (2013a) captured a 895-mm upstream migrant in northwestern Spain that had been tagged on its downstream migration a mere 13.5 months previously, suggesting that it spent as little as 10.5 months feeding at sea.

Freshwater-resident sea lamprey are well known in the Laurentian Great Lakes, Lake Champlain, the Finger Lakes (Cayuga and Seneca lakes), and Oneida Lake (see Chap. 5). These lamprey grow larger than any other freshwater-resident parasitic lampreys and larger than many anadromous lampreys. MacKay and MacGillivray (1949) reported that the majority of upstream migrants in the Little Thessalon River (in the Lake Huron basin) in 1946 were ~610 mm TL, with a maximum TL of 762 mm. However, mean TL at maturity typically falls within the 395–500 mm range (e.g., Applegate 1950; Wigley 1959; Manion 1972; Johnson 1982; O'Connor 2001; Smith and Marsden 2007), and the maximum TL recorded in recent decades is ~570 mm (Johnson 1982). Duration of the parasitic feeding phase has been established at 12–20 months (Applegate 1950; Bergstedt and Swink 1995; see Sect. 4.5.3).

Although these sea lamprey populations are frequently referred to as "landlocked," until recently there were no physical barriers preventing movement between the Atlantic Ocean and Lake Ontario and the Finger Lakes (i.e., through the St. Lawrence River), and Lake Champlain likewise remains accessible from the Atlantic Ocean via the Richelieu River (Marsden and Langdon 2012; Eshenroder 2014). However, large-bodied sea lamprey of presumably marine origin have not been observed in Lake Ontario, and, with the strong downstream current, movement into Lake Ontario is thought to be unlikely (Eshenroder 2014). Construction of the Moses Saunders Power Dam on the St. Lawrence River in 1954–1958 would now further impede upstream migration of anadromous sea lamprey. Sea lamprey larvae are absent from any upper St. Lawrence tributaries (i.e., in the 160-km downstream portion from Lake Ontario to the Moses Saunders Power Dam), although parasitic-phase sea lamprey are sometimes attached to fish in the St. Lawrence River itself (Pearce et al. 1980). Sea lamprey populations in the lower St. Lawrence River are presumably anadromous, but possibly derive from Lake Champlain (Eshenroder 2014). Sea lamprey in Lake Erie and the upper Great Lakes are essentially landlocked. Although colonization past Niagara Falls was initially permitted through the Welland Ship Canal (see Chap. 5), present-day movement through the Niagara River or the Welland Canal is thought to be very limited, although not impossible (Larson et al. 2003; Kim and Mandrak 2016).

Thus, like the Arctic lamprey in Great Slave Lake, the sea lamprey in the Great Lakes, Lake Champlain, and Cayuga, Seneca and Oneida lakes represent permanent freshwater-resident populations. These lakes are also "great" in size and have a rich prey base (e.g., lake trout, lake whitefish, Pacific salmonids; see Chaps. 3 and 5). Surface area of the Great Lakes ranges from ~19,000 km^2 for Lake Ontario to >82,000 km^2 for Lake Superior; the surface area of Lake Champlain is ~1,300 km^2, and Oneida, Cayuga, and Seneca lakes range in area from 172 to 207 km^2. There is debate regarding the origin of sea lamprey in these lakes (i.e., whether they have been present in Lake Ontario and Lake Champlain since the last glacial retreat or whether they invaded via canals less than 200 years ago; see Sect. 4.5). Resolution of this debate will help determine whether colonization of fresh water required adaptation over time or whether access to abundant and large-bodied prey is the main determinant of whether sea lamprey can survive and indeed flourish entirely in fresh water (see Sect. 4.4).

Despite their rapid spread from Lake Ontario into Lake Erie and the upper Great Lakes in the 1920s and 1930s (see Chap. 5), sea lamprey do not appear to have spread farther into smaller inland lakes. However, a recent study by Johnson et al. (2016) indicates that, should they gain access, establishment might be possible if a sufficient prey base is available. These authors present evidence that a small number (<200) of sea lamprey may complete their life cycle in the Cheboygan River system, upstream of a dam intended to prevent spawning-phase sea lamprey access from Lake Huron. Despite this dam, the watershed remains infested with sea lamprey. A navigational lock on the dam was generally thought to permit escapement past the dam, but Holbrook et al. (2014) showed little or no escapement through the Cheboygan lock and dam during the 2011 spawning migration. There are four lakes upstream of the dam, the two largest of which—Mullet Lake (70 km^2) and Burt Lake (69 km^2)—contain fishable populations of northern pike, smallmouth bass *Micropterus dolomieu*, walleye *Sander vitreus*, yellow perch *Perca flavescens*, rainbow and brown trout, and a threatened population of lake sturgeon. Lamprey wounds on fish have been reported previously by local fishermen (Applegate 1950), and Johnson et al. (2016) confirmed the presence of lamprey wounds (the majority of which were classified as sea lamprey wounds rather than native silver lamprey wounds) on northern pike and rainbow trout in these lakes. Parasitic-phase sea lamprey captured in August (a few of which were attached to rainbow trout) measured 330–440 mm TL, indicating that sea lamprey feed substantially in these lakes long after they would normally outmigrate to Lake Huron. Johnson et al. (2016) also found adult sea lamprey in the upper river before the first lock opening in the spring, so unless they passed through the dam from Lake Huron by an unknown route, these individuals would have overwintered in the Cheboygan River system. Six unmarked spawning-phase sea lamprey captured in the upper river (and assumed to be from the inland population) averaged 459 mm TL, compared to 493 mm in marked sea lamprey of Lake Huron origin captured in the lower river. The difference in size suggests slightly lower growth rates within the Cheboygan River system, either due to differences in prey availability—although Johnson et al. (2016) indicated that these lakes are more productive than Lake Huron and likely have higher prey density—or due to more rapid cooling of Mullett and Burt lakes in the fall relative to Lake Huron.

No other freshwater-resident sea lamprey populations are known throughout the species' range. There are a growing number of accounts of freshwater feeding by anadromous sea lamprey, but all appear to represent transitory freshwater feeding prior to or during outmigration. However, these reports are worthy of discussion because at least a few individuals have been shown to reach appreciable sizes (250–410 mm TL) feeding either in rivers while en route to the sea or in lakes prior to outmigration. F. W. H. Beamish (1980a), for example, reported newly metamorphosed sea lamprey attached to alewife *Alosa pseudoharengus*, shad *A. sapidissima*, and white sucker *Catostomus commersonii* in the St. John River. These individuals were still reasonably small (mean 132–136 mm TL in mid- to late May, respectively), but condition factor (CF = W/TL^3 × 10^6, where W is weight in g and TL is in mm) increased during this time (from 1.1 to 1.4 from mid- to late May). Furthermore, six individuals averaged 211 mm TL by mid-May, and two even larger sea lamprey (242–292 mm) were caught near the mouth of the river in early June. Feeding sea lamprey as large as 400 mm TL have been reported in rivers attached to anadromous fishes (e.g., Atlantic salmon, alewife, shad), but their occurrence is assumed to result from attacks initiated in marine or estuarine waters; sea lamprey juveniles found attached to the occasional non-anadromous fish species (lake trout, brook trout *Salvelinus fontinalis*) were always smaller (100–250 mm TL). Silva et al. (2013b) confirmed that feeding was initiated in fresh water when they observed outmigrating sea lamprey in two Spanish rivers attached to an exclusively freshwater cyprinid (northern straight-mouth nase *Pseudochondrostoma duriense*); they estimated that ~6% of this shoal-forming cyprinid had lamprey attached, and attachments or scarring were observed up to ~20 km from the upstream tidal limit. Silva et al. (2013c) suggested that 10–30% of all sea lamprey start feeding in the river, particularly on large anadromous fish species, and reported one individual attached to an Atlantic salmon measuring 315 mm TL. The other five sea lamprey that they found attached to anadromous brown trout or twaite shad *Alosa fallux* (20–40 km from the river mouth) measured 149–199 mm TL. Other downstream migrants (i.e., not attached to fish) measured 132–205 mm TL, suggesting that those at the upper end of the range had started feeding in fresh water as well. For comparison, newly metamorphosed sea lamprey feeding in the estuary measured 145–338 mm (mean 217 mm TL; Silva et al. 2013c). Recent observations by Baer et al. (2018) suggest a much lower incidence of in-river feeding by sea lamprey in the Rhine River in Germany, but indicate that those few individuals that do start feeding during outmigration can grow large. Only 28 of the 18,610 downstream migrants (0.15%) showed evidence of substantial feeding, but they measured 250–370 mm (mean 280 mm TL) compared to 90–190 mm for the remaining outmigrants. Over a dozen other fish species, including roach, bream, European perch, and zander *Sander lucioperca* occurred in this portion of the Rhine River. Unlike the previous studies, however, where most observations of in-river feeding were made near the river mouth, Baer et al. (2018) observed evidence of feeding 600 km upstream. These authors suggest that feeding in the river might increase in incidence as the lamprey move downstream.

Lacustrine feeding by anadromous sea lamprey prior to outmigration has also been observed. Davis (1967) reported sea lamprey feeding in Love Lake (2.7 km^2), Maine, where at least 17 teleost fish species occur. Virtually all newly metamorphosed sea lamprey outmigrated in November and December (98.4% in 1960–1961 when downstream movement was monitored between September and May), although a few downstream migrants were captured in February–May. Mean TL of outmigrants was 160 mm; the majority measured 140–175 mm TL, but 16.3% were larger than 175 mm, and the maximum reported was 234 mm TL. In 1960–1964, 85% of the landlocked Atlantic salmon captured had been attacked by newly transformed sea lamprey, and these salmon bore an average of 2.4 and a maximum of 8 wounds per fish. The incidence of fresh lamprey marks suggested that feeding occurred mainly in May–June and October–February. Only 5.1% of brook trout were observed with lamprey marks; marks were also observed on white perch *Morone americana*, white sucker, chain pickerel *Esox niger*, pumpkinseed *Lepomis gibbosus*, yellow perch, alewife, rainbow smelt *Osmerus mordax*, and other sea lamprey, but these marks were not necessarily indicative of successful feeding. Davis (1967) found no evidence of freshwater parasitism by sea lamprey in two other lakes in the East Machias River system downstream of Love Lake (4.8 and 1.2 km^2), but only Love Lake had a well-established population of landlocked Atlantic salmon. In Loch Lomond (71 km^2), Scotland, Atlantic salmon and brown trout occasionally carry lamprey scars that, based on their size and location, are likely caused by sea lamprey, and there are a few records of sea lamprey (up to 200 mm TL) attached to Atlantic salmon in this lake (Maitland 1980; Maitland et al. 1994). Two feeding sea lamprey (350 and 185 mm TL) were recovered on trout 9 and 18 years after construction of the Llandegefedd Reservoir in Wales in 1963 (1.7 km^2; Maitland et al. 1994). These individuals likely represented remnants of the original population, and there was no evidence of an established resident population once the offspring of the last anadromous sea lamprey died out. Likewise, a limited number of sea lamprey were observed feeding in two large reservoirs (Lakes Carrigadrohid and Iniscarra, 5.8 and 4.9 km^2, respectively) on the River Lee in Ireland in 1959–1965 following the construction of the reservoirs in 1957 (Kelly and King 2001). Subsequent monitoring in Iniscarra Reservoir has failed to show any additional evidence of sea lamprey feeding (King and O'Gorman 2018). Likewise, anadromous sea lamprey accidentally introduced into a reservoir above the Portodemouros Dam in northwestern Spain have been unable to form a self-sustaining population in fresh water (Silva et al. 2014). The reservoir, which has an area of 11 km^2, was formed in 1967 when the dam was built. In 2008, brown trout bearing sea lamprey wounds (including one with a 211-mm juvenile sea lamprey attached) were captured by anglers. Of 14 locations sampled across six tributaries of the reservoir, sea lamprey larvae were collected at only two locations 300 m downstream of a restaurant where groups of live adult lamprey showing reproductive behavior were held in nets within the river. There have been no other reports of sea lamprey in the reservoir, further suggesting that these captive individuals were the source of the parasites rather than a residual population that had been confined to fresh water since the building of the dam 40 years earlier.

Persistent freshwater feeding by sea lamprey has been reported in several other Irish lakes (i.e., year after year and over extended periods of the year). The largest individual recorded was 410 mm TL, but, again, there is no evidence that any individuals complete their life cycle in fresh water (King and O'Gorman 2018). Juvenile sea lamprey were reported feeding in Loughs Corrib (178 km^2), Conn (57 km^2), and Leane (19 km^2) in 1959–1965 and in Lough Derg (8.9 km^2) in 1996, and their continued presence in these lakes has been confirmed over the past decade. King and O'Gorman (2018) further reported feeding sea lamprey in two new locations, Muckross Lake (2.7 km^2) in 2009–2011 and Lough Gill (12.8 km^2) in 2011 and 2018. The majority of parasitic-phase individuals were collected during the annual brown trout angling season in May to early June, particularly in Loughs Derg and Conn, but juvenile feeding sea lamprey were collected by anglers in many months and a small number of sea lamprey that attached to open-water swimmers in Muckross Lake were collected in July–September. Sea lamprey attached to adult Atlantic salmon were observed in Loughs Conn and Leane, and attachments to pike and bream were also documented. Although many of the 79 juvenile sea lamprey collected when attached to fish were similar in size to the downstream migrants captured at the outflow from Lough Derg in January (~130–180 mm, mode 160 mm TL), 26.6% were >180 mm (up to ~250–280 mm in May–June), and two of them measured 400 and 410 mm TL (presumably from Muckross Lake later in the summer). King and O'Gorman (2018) found no evidence of "dwarf" sea lamprey in any tributary streams, arguing against the existence of self-sustaining freshwater populations.

Nevertheless, although sea lamprey appear not to readily become established in fresh water, the results of Johnson et al. (2016) suggest that colonization from the Great Lakes proper into other inland lakes is not impossible. Thus, preventing secondary spread to other inland lakes is a high priority (see Chap. 5). Johnson et al. (2016) also demonstrated that barriers to adult migration may not always be able to extirpate sea lamprey populations from upstream reaches if they have access to a sufficient prey base above the barriers. Similarly, dam removal on large streams such as the Black Sturgeon River in Ontario (McLaughlin et al. 2013) could expose large inland lakes to sea lamprey infestation that could result in parasitic feeding within these lakes (see Sect. 4.8.3).

4.4 Factors that Promote or Constrain Freshwater Residency in Parasitic Lampreys

Freshwater parasitic lampreys are said to have a restricted distribution relative to anadromous parasitic and freshwater non-parasitic lampreys, and it has been suggested that they represent an evolutionarily unstable and transitory form between anadromous parasitic and freshwater non-parasitic forms (Beamish 1985; Salewski 2003). Based on the above review, however, it could be argued that the distribution of freshwater lampreys is not particularly restricted, especially in North America where several species (e.g., chestnut and silver lampreys, Great Lakes sea lamprey) are abundant and well established in fresh water (see Sects. 4.3.2.4 and 4.3.4.4). Nevertheless, even if the freshwater parasitic life history type is not inherently unstable, it is apparent that specific ecological conditions are required for its evolution and persistence (Taylor et al. 2012). A minimum level of host fish acting as a forage base is clearly critical for the persistence of parasitic lampreys in fresh water, although the required minimum level apparently varies considerably among species. The ability of anadromous lampreys to colonize fresh water appears to be inversely related to their size (i.e., successful establishment is more likely for small-bodied lampreys) and directly related to availability of suitable prey (both in terms of abundance and size). However, different lamprey species may also show different genetic "predispositions" (e.g., related to osmoregulatory ability) to successful confinement in fresh water, and the "incentive" to abandon anadromy (i.e., related to life history trade-offs) will likely also differ between species and environments. Factors that constrain or promote freshwater residency in different species are, of course, not mutually exclusive. Body size and life history traits presumably also have a genetic basis (certainly among, if not within, species). The genetic basis of physiological and life history traits in lampreys is poorly understood, but recent advances in genomics in other fishes are beginning to identify genomic and transcriptomic differences between anadromous and freshwater forms (e.g., Czesny et al. 2012; Hale et al. 2013; Hecht et al. 2013) and will continue to contribute to our understanding of the genetic underpinnings of life history evolution (see Sect. 4.8.1).

4.4.1 Body Size

As shown by the above accounts (Sect. 4.3), all else being equal, establishment in fresh water is more common in small-bodied anadromous lamprey species or in large-bodied species in regions where dwarf or praecox forms have been reported. For example, at least six freshwater-resident populations of the smaller European river lamprey are known from western Russia, Norway, Finland, and the British Isles (see Sect. 4.3.4.2). In contrast, the large anadromous sea lamprey—although it can be seen feeding in fresh water during or prior to outmigration—appears not to have established any freshwater-resident populations in Europe (see Sect. 4.3.4.4). In this species, colonization of fresh water has only been successful in the Great Lakes and other large productive lakes in North America. Furthermore, the regions where lacustrine populations of European river lamprey have become established seem to be characterized by particularly small-bodied anadromous forms. Similarly, although some populations of Arctic lamprey reach appreciable sizes (e.g., in southeastern Russia), the only known freshwater-resident populations are in regions of Asia and North America where size at maturity of the anadromous population appears to be smaller (see Sect. 4.3.4.3). Likewise, the freshwater-resident *Entosphenus* species assumed to be post-glacial derivatives of the Pacific lamprey (see Sect. 4.3.2.1) and the few known non-anadromous Pacific lamprey populations generally occur in regions where smaller-bodied anadromous Pacific lamprey are known (see Sect. 4.3.4.1). Praecox Pacific lamprey tend to show more genetic differentiation among locations (i.e., more limited dispersal) than the larger anadromous form (Spice et al. 2012; Hess et al. 2013), and there appears to be a genetic basis for body size in this species (Hess et al. 2013, 2014). Thus, praecox individuals—by virtue of their smaller body size or other traits—could be genetically "predisposed" for freshwater colonization (see Sect. 4.4.3).

The only apparent exception to this pattern is the very small-bodied western river lamprey. Western river lamprey feed parasitically for only a few months, in estuaries (e.g., the Strait of Georgia) rather than at sea, and it could be argued that all individuals of this species constitute the praecox form (see Sect. 4.3.3.2). Therefore, we would predict that western river lamprey would be able to easily abandon anadromy. Nevertheless, there is only one report of a "healthy" non-anadromous population (in Lake Washington). However, the western river lamprey or a western river lamprey-like ancestor has given rise to numerous freshwater-resident non-parasitic lamprey populations (from Alaska to California), and perhaps this species, given its very small body size, readily abandons both anadromy and the parasitic feeding phase (see Sect. 4.7).

4.4.2 Prey Availability and Other Ecological Factors

Access to a sufficient prey base is clearly critical for the evolution and persistence of freshwater parasitic lampreys, but what qualifies as sufficient varies considerably among species. Small-bodied species that are exclusively freshwater resident can inhabit relatively small systems with small-bodied prey (e.g., the Miller Lake lamprey; Sect. 4.3.2.1), although others (e.g., silver and chestnut lampreys; Sect. 4.3.2.4) can grow quite large. Both prey abundance and prey size seem to be important. The consumption rate of large-bodied lampreys is high, and the host size necessary to support this consumption rate—particularly as the lampreys increase in size—can be limiting in fresh water (see Chap. 3). European river lamprey in Lough Neagh are estimated to feed in fresh water for only 1 year, but individuals in this population appear to grow as large or larger than some anadromous conspecifics that feed at sea for 1 or 1.5 years (see Sect. 4.3.4.2). Goodwin et al. (2006) showed that river lamprey in this lake shift to larger host species as they grow, and access to a range of prey sizes (and not just absolute prey biomass) may be important in fresh water.

It is no coincidence, of course, that large anadromous lampreys establish in large lakes only, generally where at least a dozen species of fish are present. Lough Neagh is the largest lake in the British Isles, and other lacustrine populations are known from lakes ranging in area from the size of Loch Lomond (71 km^2) to Lakes Onega and Ladoga (>9,700 km^2). The largest anadromous species, sea lamprey, has success-fully established freshwater-resident populations in only the Laurentian Great Lakes (~19,000–82,000 km^2), Lake Champlain (~1,300 km^2), and Oneida, Cayuga, and Seneca lakes (172–207 km^2). It is not known if persistence in the smallest of these lakes is dependent on recruitment from or feeding opportunities in Lake Ontario. Likewise, sea lamprey appear to have become established upstream of a barrier in the Lake Huron basin, where they feed on prey in Mullet Lake (70 km^2) and Burt Lake (69 km^2) (Johnson et al. 2016; see Sect. 4.3.4.4). However, it is not known if the long-term persistence of this population depends on occasional recruitment from Lake Huron. Arctic lamprey, another relatively large-bodied anadromous species, has established freshwater-resident populations in large lakes (Great Slave Lake and Iliamna Lake, 27,200 km^2 and 2,600 km^2, respectively; see Sect. 4.3.4.3); in con-trast, the Arctic lamprey in the Naknek River system (with a total lake surface area of ~700 km^2) might not be a permanent freshwater-resident population (Heard 1966). In the Naknek River system, individuals may feed opportunistically in fresh water, and there may be ongoing recruitment from the anadromous population.

Species for which few or no non-anadromous populations have been reported are generally those large-bodied species without access to large lakes. This includes the Pacific lamprey and pouched lamprey, and, to a lesser extent, the smaller-bodied western river and short-headed lampreys that share the respective ranges of these species. Compared to the three species discussed above, there are relatively few large lakes within the range of Pacific and western river lampreys, and the same is true for the Southern Hemisphere species. The small number of lamprey species in the Southern Hemisphere has been attributed to the general paucity of freshwater systems for spawning and larval rearing (Potter et al. 2015), and the shortage of large lakes in particular will limit freshwater parasitism. In Australia, natural freshwater lakes are rare due to the general absence of recent glaciation. Lake Wellington (147 km^2) appears to support a non-anadromous population of the smaller short-headed lamprey (see Sect. 4.3.3.3), but it is apparently not sufficiently large for the pouched lamprey.

There are more, but still relatively few, large lakes on the west coast of North America within the range of the Pacific lamprey. One of the largest lakes in this species' range in the contiguous United States (i.e., excluding Alaska) is Lake Roo-sevelt, which is a reservoir created in 1941 when the Columbia River was impounded by the Grand Coulee Dam. With a surface area of ~330 km^2, this lake is very small in comparison to the Laurentian Great Lakes and Great Slave Lake. The reservoir con-tains rainbow trout, kokanee, walleye, burbot, and smallmouth bass (Baldwin et al. 2003; Polacek et al. 2006), but there are no reports of Pacific lamprey becoming estab-lished following isolation above the dam. It is thus not surprising that large-bodied anadromous Pacific lamprey also failed to establish in the much smaller Dworshak

Reservoir and Elsie Lake (69 and 6.7 km^2, respectively; see Sect. 4.3.4.1). Lacustrine Pacific lamprey-like populations are known from Goose Lake (380 km^2) and Upper Klamath Lake (250 km^2), but these populations are likely of post-Pleistocene origin (see Sect. 4.3.2.1) and it is worth noting that Upper Klamath Lake is a remnant of the much larger Pluvial Lake Modoc, which at its largest, covered >2,600 km^2 (Dicken and Dicken 1985).

The largest lake within this species' range in British Columbia is Babine Lake (~480 km^2). Upstream-migrating anadromous Pacific lamprey were first recorded in this lake's tributaries in 1963, with the number increasing steadily by 1971, and the presence of larval Pacific lamprey was confirmed in many of the streams by the early 1980s (Farlinger and Beamish 1984). Thus, it appears that large-bodied anadromous Pacific lamprey (~480–690 mm TL; R. J. Beamish 1980) have only started accessing the system relatively recently, perhaps following removal of a major rock slide 65 km downstream of the lake (Farlinger and Beamish 1984). However, despite an unconfirmed report in 1977 of a lamprey feeding on a fish in Babine Lake (R. J. Beamish 1980), there are no documented reports of Pacific lamprey parasitizing freshwater fishes anywhere in the Skeena River drainage (Farlinger and Beamish 1984). The potential prey base in Babine Lake includes sockeye and coho salmon, rainbow trout and steelhead (i.e., anadromous rainbow trout), lake trout, and lake whitefish (Shortreed and Morton 2000). Whether a freshwater lamprey population could become established in this lake is unknown and should be monitored.

The western river lamprey likewise has access to relatively few large lakes within its range. It might be supposed that, given its very small body size, large lakes are not required. However, given the relative paucity of large lakes, but the abundance of spawning and larval rearing habitat in coastal systems, it may be more advantageous for this small-bodied species to bypass the parasitic feeding phase altogether and mature as a non-parasitic brook lamprey (see Sect. 4.7).

Although the above argument suggests that large-bodied anadromous lampreys should generally be capable of establishing non-anadromous populations in large productive lakes, isolation above dams appears not to have resulted in successful freshwater colonization in Caspian lamprey, even in very large reservoirs (1,830–3,120 km^2; see Sect. 4.3.3.1). As with Pacific lamprey, perhaps not all anadromous individuals are equally "prepared" for freshwater residency, and isolation of a relatively small number of individuals is insufficient for selection of a freshwater-resident phenotype (see Sect. 4.4.3).

However, feeding mode may also affect the ability of anadromous lampreys, particularly large-bodied ones, to succeed in fresh water. Potter and Hilliard (1987) proposed that blood feeding would have a selective advantage in freshwater environments, where a relatively restricted host population would be less likely to be decimated, especially if hosts are large enough to permit replacement of the blood lost during feeding. Furthermore, because wounds inflicted by blood-feeding lampreys are typically smaller than those caused by flesh-feeding species, they are less likely to be fatal. Therefore, the ability of sea lamprey to become established in the Great Lakes may be permitted in part by their blood-feeding mode. Mortalities are still observed, even among blood feeders (especially with Great Lakes sea lamprey, which are far larger than any other freshwater-resident lamprey), but sea lamprey behave more like a parasite than a predator (see Chap. 3). Likewise, chestnut and silver lampreys are blood feeders and have co-existed with their freshwater hosts for long periods of evolutionary time (e.g., Cochran and Lyons 2016). Potter and Hilliard (1987) suggested that flesh feeding, in contrast, evolved in lamprey populations which had access to estuarine and marine hosts with large, widespread populations. This would particularly be the case for small-bodied hosts that would provide only a small amount of blood and require the lamprey to frequently seek out new hosts (i.e., making blood feeding less efficient), but, given their abundance, they would not be easily depleted. The western river lamprey is categorized as a flesh feeder (Chap. 3), and, Beamish and Neville (1995) estimated that this species kills ~40–50 million Pacific salmon smolts per year and ~150–200 million Pacific herring *Clupea pallasii* >100 mm TL during its brief parasitic feeding phase as it enters the Strait of Georgia from the Fraser River. The Pacific lamprey is classified as a blood and flesh feeder (see Chap. 3). Both western river and Pacific lampreys are thus more likely to deplete their prey base in fresh water. The combined need for a particularly large prey base, but general lack of access to large lakes with a large prey base, may therefore explain why these two species have established few freshwater parasitic populations in contemporary time. The pouched lamprey is likewise a flesh feeder that travels farther offshore to feed in very productive waters (Potter and Hilliard 1987). Carrion feeding (e.g., by the exclusively freshwater-resident Carpathian lamprey) might also have a selective advantage in fresh water by reducing the pressure on local teleost populations (Potter and Hilliard 1987). Nevertheless, it does not appear that this presumed mode of feeding by the Caspian lamprey (see Chap. 3) helped it survive landlocking. Furthermore, the European river lamprey has successfully established in fresh water despite being a flesh feeder, and the Vancouver lamprey feeds on blood and flesh. However, both species are smaller-bodied than most of the blood-feeding freshwater species. Furthermore, although they probably inflict high mortality on their hosts (Beamish 1982; Potter and Hilliard 1987), the presence of fishes with healed scars (Goodwin et al. 2006; COSEWIC 2008) indicates that not all hosts succumb to their wounds, and a certain amount of host mortality would presumably be tolerated where prey are abundant.

As a final note regarding ecological constraints, it is also worth noting that, among other anadromous fishes, there is a greater tendency for populations at lower latitudes to become freshwater resident because fresh waters are more productive at these latitudes than ocean waters (Gross et al. 1988). In contrast, anadromous species tend to predominate in temperate regions, where the reverse is true. In lampreys, the majority of established freshwater parasitic populations are found at higher latitudes (i.e., the opposite of what would be expected). However, to explicitly test for the effect of latitude, we would need to control for other factors (e.g., the relative availability of large inland bodies of water and the presence of barrier dams), and we would also need to include freshwater-resident non-parasitic lampreys. Including brook lampreys, the expected pattern of fewer anadromous populations at low latitudes appears to hold (see Sect. 4.6).

4.4.3 Osmoregulatory Ability and Genetic Factors

Physiological constraints related to osmoregulation may limit the ability of some anadromous lampreys to colonize fresh water. It has been suggested that estuarine species such as European river lamprey prefer water of reduced salinity (Bahr 1952), and this may facilitate the transition of European river lamprey to fresh water. However, western river lamprey has likewise been thought to inhabit waters of lower salinity (R. J. Beamish 1980), but only one freshwater parasitic population has been reported, and this species appears to do poorly in fresh water during its parasitic feeding phase (see Sect. 4.3.3.2). Similarly, despite apparent adaptation to the very low salinity of the Caspian Sea (~one-third that of most sea water), Caspian lamprey do not appear to survive landlocking (see Sect. 4.3.3.1). Nevertheless, the failure of Pacific lamprey populations to survive above dams that prevent access to the sea is often attributed to their inability to osmoregulate in fresh water during the parasitic feeding phase (Clarke and Beamish 1988; see Sect. 4.3.4.1) and has been used to suggest that this species is ill-adapted for colonization of fresh water (e.g., Farlinger and Beamish 1984). By adaptation, we typically mean evolutionary change in a trait with a heritable basis (i.e., over the course of generations). This is different than acclimation or acclimatization, which is the adjustment of physiological traits (i.e., phenotypic flexibility within a single organism) to ambient environmental conditions in the laboratory or in nature, respectively (Piersma and Drent 2003). It is "notoriously difficult" to differentiate between adaptation and environmentally induced flexibility in wild populations because phenotypic variation could be the product of either mechanism or a combination of both (Laporte et al. 2016).

Invasion of fresh water by anadromous lampreys likely involves both adaptation and phenotypic flexibility. In populations suddenly isolated above a barrier, the immediate need to osmoregulate in fresh water during the feeding phase is likely met through a combination of selection on existing variation within the anadromous population and phenotypic flexibility within individuals. Existing variation and phenotypic flexibility allow initial survival; subsequent adaptation leads to further "improvements" and evolution of other traits that increase fitness in fresh water. In Pacific lamprey, some individuals or populations may be better adapted to freshwater feeding than others (i.e., based on existing genetic variation); in turn, feeding in fresh water may improve osmoregulatory ability (i.e., conferring phenotypic flexibility). Clarke and Beamish (1988) found that Pacific lamprey from different populations throughout British Columbia appeared to have different abilities to feed and survive in fresh water, and the ability may be inversely related to body size at maturity. Pacific lamprey from the Chemainus River (mean TL at maturity 270 mm) survived in fresh water until July, whereas no individuals from the Babine River (mean TL 480–640 mm) survived beyond mid-February (R. J. Beamish 1980; Clarke and Beamish 1988). These results are consistent with the suggestion above (Sect. 4.4.1) that anadromous lampreys from smaller-bodied source populations are the progenitors of freshwater-resident populations and species. However, the Pacific lamprey in this study fed poorly in fresh water, and plasma sodium concentrations were correlated with condition factor, suggesting that osmoregulatory failure may have resulted from, or at least been compounded by, depletion of body energy reserves. Hardisty et al. (1989) indicated that, because the blood and tissue of the lamprey's host fishes are isotonic with the internal milieu of the lamprey, feeding would be expected to ease the osmotic load. Ferreira-Martins et al. (2016) demonstrated that feeding on an isomotic meal helped anadromous sea lamprey compensate for ion gains from the seawater environment, and, in fresh water, feeding would be a source of ions (Wood and Bucking 2011). Failure of lampreys to feed during the parasitic phase would therefore compromise their osmoregulatory abilities.

R. J. Beamish and colleagues have suggested that confinement to fresh water might select for a few Pacific lamprey individuals that are genetically predisposed to feed and grow to maturity in fresh water, and the probability of this happening would likely be higher if the size of the population was large (Farlinger and Beamish 1984; Beamish and Northcote 1989). In this respect, Pacific lamprey may be similar to pink salmon *Oncorhynchus gorbuscha* and coho salmon. Pink salmon were previously thought to require salt water for completion of their life cycle, but they have nevertheless become firmly established in the Great Lakes following an accidental introduction into Lake Superior in 1956 (Emery 1981; Gharrett and Thomason 1987). Establishment of a freshwater-resident pink salmon population presumably involved differential survival of those individuals best adapted to fresh water, leading to strong and rapid selection. Non-anadromous coho salmon are also now common in the Great Lakes following introduction in the 1960s (Sandercock 1991), and there is recent evidence of "freshwater residualism" in coho salmon in two small lakes that drain into the Skeena River basin in British Columbia (Parkinson et al. 2016). To date, outside of the Great Lakes, only 15 coho salmon (11 of which were male) have been reported to reach maturity in fresh water (Parkinson et al. 2016). Freshwater residency outside of large lake systems is apparently rare for coho salmon, but rare does not mean impossible. Several authors (e.g., Hardisty 1969; Mathers and Beamish 1974; Beamish et al. 1978; F. W. H. Beamish 1980b) have suggested that colonization of the Great Lakes by anadromous sea lamprey involved selection for small individuals already predisposed for life in fresh water. However, if a relatively small number of sea lamprey colonized Lake Ontario in historical times via canals, this would imply that the genetic ability to osmoregulate in fresh water existed at higher levels within the anadromous population or that the relatively few were either "pre-adapted" (Briski et al. 2018) or "lucky" (see Sect. 4.5.3). Colonization of fresh water also may have involved selection for anadromous sea lamprey with accelerated gonadal development and reduced potential fecundity (Hardisty 1969; see Chap. 1), or these characteristics associated with existing freshwater-resident sea lamprey evolved rapidly following invasion.

Research into the genetic basis of life history traits in lampreys is still in its infancy. To date, however, loci correlating with body size and run timing have been identified in Pacific lamprey (Hess et al. 2013, 2014; Parker 2018), and some of the loci that differ between European river lamprey and European brook lamprey appear to be related to osmoregulation (i.e., rather than feeding type; see Sect. 4.6.3.3). Hess et al. (2014) found that body morphology, primarily TL of upstream migrants, was strongly associated with genetic variation at three single nucleotide polymorphism (SNP) loci (identified as *Etr_1806*, *Etr_4281*, and *Etr_5317*; Hess et al. 2013). The genetic mechanisms associated with these loci are likely complex, but Hess et al. (2013, 2014) have made some inferences regarding the function of the genes to which these SNPs were localized (i.e., when mapped to the sea lamprey reference genome; Smith et al. 2013). *Etr_1806* does not appear to localize within any described genes, but *Etr_4281* aligns with the human protocadherin related 15 (*PCDH15*) gene, which has an essential role in the maintenance of normal retinal and cochlear function, and *Etr_5317* localizes to the dymeclin (*DYM*) gene, which encodes a protein necessary for normal skeletal development and brain function in humans (Hess et al. 2013). Parker (2018) found that there was a strong correlation between run timing (i.e., whether anadromous Pacific lamprey were of the river-maturing or ocean-maturing ecotype; see Sect. 4.3.4.1) and two groups of linked loci. Individuals that were homozygous for the "ocean-maturing" allele at both linkage groups almost always had well-developed ovaries at the onset of their freshwater migration, but individuals that had at least one river-maturing allele in either linkage group had small ovaries. This means that the river-maturing ecotype carries standing genetic variation capable of producing both ecotypes (i.e., both dominant and recessive alleles), while the ocean-maturing ecotype carries a single (recessive) allele. The specific genes associated with these loci (represented by SNP loci *Etr_2878* and *Etr_2791*) have yet to be determined, but continuing improvements to the assembly and annotation of the sea lamprey genome (e.g., Smith et al. 2018) and growing genomic resources for other lamprey species will aid in these efforts.

4.4.4 Life History Trade-Offs

The above discussions largely revolve around factors that limit colonization of fresh water by anadromous lampreys. However, it is also possible—and, in fact, likely—that, in addition to these factors that constrain establishment in fresh water, there will also be factors that differentially promote freshwater residency. What constitutes sufficient "incentive" for lampreys to abandon anadromy will presumably vary among species. There has been considerable discussion regarding life history trade-offs in other anadromous and freshwater-resident fishes (e.g., Gross et al. 1988; Jonsson and Jonsson 1993, 2006; Fleming 1996; Klemetsen et al. 2003) and in parasitic and non-parasitic lampreys (e.g., Hardisty 2006; Docker 2009; see Sect. 4.7). Specifically, the reduction in fecundity in freshwater-resident lampreys (that results from reduced size at maturity) would require a compensatory reduction in mortality (resulting from generally shorter feeding and spawning migrations, reduced osmoregulatory costs, and reduced exposure to predators during a generally shorter feeding phase). Most non-anadromous lampreys are not actually "landlocked" (see Sect. 4.3); therefore, most appear not to be making the "best of a bad situation" by enduring fresh water when they are unable to feed at sea. Presumably, lampreys become parasitic in fresh water when they fare better in the freshwater system than they would in marine environments. Where the costs of anadromy outweigh its benefits or, conversely, where the benefits of freshwater residency outweigh its costs (e.g., in terms of lost growth opportunities at sea), freshwater residency should be favored. Large productive lakes at a considerable distance from the ocean (e.g., Great Slake Lake, the Great Lakes) would presumably favor freshwater residency; they offer access to a large, diverse prey base without the longer spawning migrations (see Moser et al. 2015). However, for large-bodied anadromous species, it likely becomes increasingly difficult for freshwater systems to sufficiently compensate for the loss of growth opportunities at sea. Across all populations and species of lampreys, fecundity increases approximately with the cubic power of length; thus as TL doubles, the number of eggs will increase by approximately an order of magnitude (see Chap. 1). Therefore, we would expect to see abandonment of anadromy only under circumstances where reduction of mortality and costs associated with migration make this decreased reproductive output worthwhile. For example, in sea lamprey, mean TL and fecundity are ~840 mm and 171,600 eggs in the North American anadromous population and ~440 mm and 70,000 in the Great Lakes (see Chap. 1 and references therein). This amounts to a 60% reduction in fecundity; presumably, this is an "acceptable loss" since growth conditions in the Great Lakes are very good and mortality rates during the shorter parasitic feeding and migratory stages are assumed to be lower. However, anadromy would presumably remain advantageous compared to feeding in small lakes or reservoirs where growth conditions are poor and reduction in fecundity could be 90–95% (e.g., if mean TL was reduced to 225–285 mm; see Chap. 1).

In contrast, the smaller-bodied anadromous lampreys have "less to lose" by abandoning anadromy. For example, large anadromous European river lamprey (mean TL ~400 mm) have an average fecundity of ~37,200, and freshwater-resident females (mean TL ~225 mm) produce an average of ~10,135 eggs (see Chap. 1 and references therein); this represents a >70% reduction in fecundity. However, there is virtually no difference in fecundity when comparing anadromous praecox (mean TL ~210 mm and 10,000 eggs) and freshwater-resident European river lamprey. Thus, for already small-bodied European river lamprey, abandoning anadromy would presumably reduce the costs and risks associated with the marine feeding phase, without sacrificing reproductive output. Likewise, for the Arctic lamprey, where large anadromous forms (TL 451–500 mm) produce ~102,000 eggs and freshwater-resident forms (TL 170–300 mm) produce ~21,400 eggs, colonization of fresh water by the large forms could result in an almost 80% reduction in fecundity. In contrast, abandonment of anadromy could be "a step up" for anadromous praecox Arctic lamprey (TL 280 mm and ~13,700 eggs). This is consistent with the general observation that freshwater forms arise in regions where smaller body sizes have been reported for anadromous lampreys (see Sect. 4.4.1).

The above argument depends on the relationship between body size and fecundity in female lampreys; whether or not male reproductive success is also related to size is not known. Large anadromous males would presumably have higher reproductive success than small freshwater males if strong size-assortative mating (Hardisty and Potter 1971b) ensures they mate with large anadromous females. However, sneak mating tactics have been observed in some lamprey species (e.g., Hume et al. 2013b) which would allow small males to fertilize at least some of the eggs of large females. Kucheryavyi et al. (2007) found that virtually all of the praecox anadromous form of Arctic lamprey in western Kamschatka were male, and they did not find size-assortative mating. The praecox males were observed to spawn jointly with the larger anadromous form but also with the even smaller non-parasitic form. In some other anadromous fishes, the dwarf or resident forms are entirely or predominantly male (e.g., Dalley et al. 1983; Heath et al. 1991), and Docker (2009) suggested that life history transitions in lampreys may occur more readily in males. However, with the exception of the praecox Arctic lamprey in western Kamschatka, male and female lampreys appear to abandon anadromy more or less equally.

Because lampreys are semelparous, it is relatively easy to quantify their lifetime reproductive success, at least in females, and the relationship between size at maturity and fecundity is reasonably well studied (see Chap. 1). However, the other life history variables "in the equation" (e.g., the costs associated with migrating to and from different environments and those associated with osmoregulating and feeding in different environments and for different durations) are more difficult to quantify. Costs of upstream migration in different anadromous and freshwater-resident populations can be estimated using an energetics approach (e.g., by calculating shrinkage or change in proximate body composition during migration; Beamish et al. 1979; R. J. Beamish 1980; Beamish 1982; see Chap. 1) or by comparing the length and duration of the spawning migration (see Moser et al. 2015). It should be noted, however, that the length and duration of upstream migration are not always correlated, and the apparent "paradox of premature migration" seen in anadromous salmonids is also observed in many lamprey species (Quinn et al. 2016). Some lamprey species, populations, or individuals (e.g., all pouched lamprey, most Pacific lamprey, some European river lamprey) enter fresh water 8–16 months prior to spawning, thus appearing to reduce their growth opportunities at sea, while others (e.g., anadromous and freshwater sea lamprey, ocean-maturing Pacific lamprey, and some praecox European lamprey) delay freshwater entry until 1–2 months before spawning. How these different strategies influence the relative benefits and costs of migration under different marine and freshwater conditions is unknown. Likewise, age-specific mortality rates are also difficult to quantify. Lampreys outmigrating to sea can significantly contribute to the diet of predatory fishes, birds, and pinnipeds (see Docker et al. 2015), but predation on downstream migrants in fresh water has not been quantified. Furthermore, although predation on juveniles is thought to be relatively low during the parasitic feeding phase (because the adults are well dispersed), predation on lampreys will often go undetected (Cochran 2009), and nothing is known of the age- or size-specific mortality rates during this stage (i.e., to balance the costs and benefits of delaying maturation for another year). More precise estimates of the relative cost and benefits of anadromy versus freshwater residency in different environments would be very informative.

4.5   Origin of Sea Lamprey in Lake Ontario and Lake Champlain

It is clear that sea lamprey invaded Lake Erie and the upper Great Lakes following completion of the Welland Canal in 1829 (or, perhaps more likely, following subsequent modifications in the early 1900s that resulted in the current Welland Ship Canal) which allowed them to bypass Niagara Falls (see Chap. 5). However, whether they are invasive or native to the Lake Ontario drainage (including the Finger Lakes and Oneida Lake)—that is, whether they entered in historic times (within the past 200 years through manmade canals) or prehistorically (as a result of post-Pleistocene natural colonization)—is still debated. Investigators have argued both for invasive (e.g., Aron and Smith 1971; Mandrak and Crossman 1992; Eshenroder 2009) and native (e.g., Hubbs and Lagler 1947; Wigley 1959; Bailey and Smith 1981; Daniels 2001; Waldman et al. 2004; Bryan et al. 2005) status. The origin of sea lamprey in Lake Champlain is similarly controversial (see Waldman et al. 2006; D'Aloia et al. 2015). Eshenroder (2009, 2014) thoroughly reviews the evidence for the competing "invasion-by-canal" and "native-but-rare" hypotheses and, in the 2014 paper in particular, argues very convincingly for the former hypothesis—or at least that sea lamprey were not present in Lake Ontario and Lake Champlain much before the 19th and 20th centuries, respectively. Nevertheless, it is also possible that sea lamprey adapted to fresh water following glacial retreat somewhere adjacent to these lake basins and then gained access to the Lake Ontario and Champlain basins in historical times. At present, support appears strongest for the "invasion-by-canal" hypothesis, but modern population genomic analyses may be able to conclusively resolve the issue. Resolution of this long-standing question will help us understand how quickly anadromous lampreys can become invasive in fresh water (e.g., whether colonization of fresh water requires gradual genetic change or whether it can happen almost immediately following access to large productive lakes).

4.5.1   Colonization in Historical Times: Invasion-by-Canal Hypothesis

The first record of sea lamprey in Lake Ontario is frequently dated to 1835, based on a diary description by the naturalist Charles Fothergill of a single adult said to have been collected in a creek just east of Toronto, Ontario (Lark 1973). Eshenroder (2014) reviewed the fisheries literature related to the earliest records of this species and concluded that this 1835 record is suspect. Although Fothergill's morphological description fits that of the sea lamprey, description of its natural history appears conflated with that of the American eel *Anguilla rostrata*, and Fothergill may have merely recorded information that he received second-hand and supplemented it with information taken from a textbook (Eshenroder 2014). Eshenroder (2014) concluded that the first credible report of sea lamprey in Lake Ontario was in 1888 when a parasitic-phase sea lamprey was found attached to a boat (Dymond et al. 1929). Eshenroder (2014) indicated that there were at least 13 instances when experts had the opportunity to encounter and report sea lamprey in the Lake Ontario drainage before 1888 but, with the exception of Fothergill's report 53 years previously, did not. The rare, smaller silver lamprey was reported twice in Lake Ontario before 1888. However, Waldman et al. (2009) noted that walleye, burbot, and yellow perch were not mentioned in earlier accounts despite having been present. Nevertheless, if sea lamprey were native, they likely would have been conspicuous in shallow tributary streams during their spawning runs, especially in streams blocked by mill dams, and scarring on frequently encountered host fishes (e.g., Atlantic salmon, lake trout, and lake whitefish) would have been noticed and commented upon (Eshenroder 2014). Thus, it is indeed hard to imagine that sea lamprey could have been present in Lake Ontario prior to the mid- to late 1800s and not be recorded.

The first credible account of sea lamprey in Cayuga Lake was an adult collected in 1875. Seven more were reported the following year, and over 1,000 adults were reported by 1886 (Eshenroder 2014). Zoologists at Cornell University were studying the fishes of Cayuga Lake at this time (Meek 1889) and presumably would have noticed and recorded sea lamprey had they been present much before this (Eshenroder 2014). Sea lamprey were apparently abundant in Seneca Lake by 1893 (Gage 1893) and were in Oneida Lake "near" 1894 (Gage 1928).

Eshenroder (2014) therefore concluded that sea lamprey entered the Lake Ontario drainage no earlier than the 1860s and quickly reached pest levels of abundance. This timeline is consistent with the time (~10 years) required for sea lamprey to reach pest proportions in Lake Ontario and the upper Great Lakes following colonization (Eshenroder and Amatangelo 2002). Eshenroder (2014) concluded that invasion of the Lake Ontario drainage was accomplished through the Erie Canal. This canal ran from Albany, New York on the Hudson River to Buffalo, New York on Lake Erie, thus allowing navigation between the Atlantic Ocean and the Great Lakes, and it also had an extensive network of lateral canals connecting to the Oswego River drainage and Lake Ontario. Anadromous sea lamprey once migrated reasonably far upstream in the Hudson River, probably at least as far as the mouth of the Mohawk River (Bigelow and Schroeder 1948). Daniels (2001) and Waldman et al. (2004) concluded that these 19th-century navigation canals would have been inhospitable to sea lamprey and would have had many barriers to passage. However, Eshenroder (2014) reviewed details of the construction and operation of the canal and suggested that completion of a dam and sloop lock on the Hudson River in 1823 could have diverted sea lamprey into the Erie Canal. He reasoned that the eastern section of the canal, with its high-quality water and directional current provided by feeder canals, would have facilitated upstream migration by adult lamprey. Even more likely, however, sea lamprey may have gained access in 1863 when a tributary of the Susquehanna River was diverted into the Oneida Lake drainage, creating a watershed breach between the Lake Ontario and Hudson River/Atlantic Ocean drainages (Eshenroder 2014). Water from the Susquehanna River drainage would have spilled directly into the Erie Canal or, under high-water conditions, spilled into Limestone Creek and eventually Oneida Lake. Sea lamprey spawning runs were known from the Susquehanna River drainage until it was occluded by construction of Conowingo Dam near its mouth in 1928 (F. W. H. Beamish 1980a; Waldman et al. 2009), and adults were observed in the upper reaches of the Susquehanna River "only a few miles south of Cayuga Lake" (Gage 1893). Larval pheromones in these upper reaches would have been diverted into the Erie Canal and could have served as an attractant to upstream-migrating sea lamprey at the canal's Hudson River entrance (Eshenroder 2014). Previous discussions regarding possible invasion routes into Lake Ontario (e.g., Aron and Smith 1971; Eshenroder 2009) were unaware of the watershed breach between the Susquehanna River and Lake Ontario drainages, but this later timing agrees well with the first reports and subsequent proliferation of sea lamprey in the Lake Ontario drainage in 1875–1888.

The "invasion-by-canals" hypothesis is consistent with arguments based on zoogeography. Mandrak and Crossman (1992), for example, classified sea lamprey as non-native to Lake Ontario, arguing that this species, if it dispersed into the St. Lawrence River and Lake Ontario basin during the Champlain Sea inundation (11,800–9,700 years ago), would have had the opportunity to colonize the Ottawa River and inland waters of eastern Ontario. Likewise, if it dispersed in the glacial lakes in the Ontario basin through the Susquehanna outlet 13,000–11,800 years ago, it would have had the opportunity to disperse into all the Great Lakes, because Niagara Falls was not established as a barrier to dispersal until ~12,500 years ago. Eshenroder (2009) briefly discussed two alternative versions of the non-native hypothesis to explain the belated appearance of sea lamprey in the Lake Ontario drainage. First, he suggested that extreme weather events could have caused a watershed breach between the Susquehanna River and Oneida Lake drainages or between the Mohawk River and Oneida Lake above the impassable falls. The divide, in several locations, is less than 2 km wide. Unusually heavy rains in 1818 breached the divide between the Mohawk River and Oneida Lake; due to the presence of two impassable falls, sea lamprey would not have been able to take advantage of this particular breach, but a similar breach above the impassable falls could have provided access. A second alternative to invasion via canals was human transplantation. Smith (1985) stated that Native Americans were not known to transplant fishes, but European colonists may have. Although there is no record of transplantation, sea lamprey were a popular food fish in Europe (see Docker et al. 2015), and larvae were used as bait and sometimes shipped to anglers at different locations (Daniels 2001).

In Lake Champlain, the first credible report of sea lamprey appears to date to 1929 (Greeley 1930). The literature related to records of sea lamprey in this lake is "long and convoluted" and is complicated by frequent revisions in taxonomy, interchangeable common names, and confusion regarding separate classifications for larval, juvenile, and adult lamprey of the same species (Eshenroder 2014). However, reviewing the literature, Eshenroder (2014) concluded that previous assumptions that sea lamprey were present in Lake Champlain since at least 1841 (Bryan et al. 2005) or 1894 (Waldman et al. 2006) were incorrect. The earliest report from 1841 (Thompson 1842) appears to have been a silver lamprey, and, as with the Lake Ontario basin, it is unlikely that spawning sea lamprey or sea lamprey scarring would have gone unnoticed had this species been native to Lake Champlain.

Eshenroder (2014) therefore concluded that anadromous sea lamprey colonized Lake Champlain via the Champlain Barge Canal, which opened in 1916. The Champlain Canal system connects the south end of Lake Champlain to the Hudson River, diverging from the Erie Canal just north of West Troy, New York. Eshenroder (2014) argued that the earlier Champlain Canals (i.e., the original 1825 Champlain Canal and the improved 1863 Champlain Canal) would not have permitted access to Lake Champlain. Instead, sea lamprey would have stayed in the Erie Canal, entered the Champlain Canal and migrated northward in the Mohawk River impoundment, or they would have continued northward in an artificial cut, remaining in the Hudson River. However, when the Champlain Barge Canal opened, sea lamprey migrating upstream in the Hudson River could have entered an artificial cut at Fort Edward, New York. At Fort Edward, a dam without locks (which has since been removed) would have blocked sea lamprey from further upstream migration and encouraged their entry into another artificial cut, allowing them to either swim back to the Hudson River or south to Lake Champlain.

4.5.2   Post-glacial Colonization: Native-but-Rare Hypothesis

Several authors have suggested that sea lamprey colonized the Lake Ontario drainage sometime after the retreat of the Laurentide ice sheet (e.g., Hubbs and Lagler 1947; Wigley 1959; Bailey and Smith 1981) by one of at least three hypothesized zoogeographic pathways (Waldman et al. 2004). These include colonization via Lake Ontario's present outlet, the St. Lawrence River (which has persisted for ~9,000 years), via temporary glacial outlets such as the Delaware-Susquehanna drainage (13,000–11,800 years ago), or via the Hudson-Mohawk system (12,500–12,000 years ago) (Underhill 1986; Mandrak and Crossman 1992; Wall and LaFleur 1995; Waldman et al. 2004). Similarly, a post-Pleistocene origin for sea lamprey in Lake Champlain has been suggested. Modern-day Lake Champlain, as well as the St. Lawrence and Ottawa River valleys, were once encompassed by the Champlain Sea, a temporary inlet of the Atlantic Ocean that was created during deglaciation ~11,800–9,700 years ago (Mandrak and Crossman 1992). Post-Pleistocene colonization by other anadromous fishes is known for Lake Ontario (e.g., Atlantic salmon, American eel, rainbow smelt; Mandrak and Crossman 1992) and Lake Champlain (e.g., Atlantic salmon, American eel; Marsden and Langdon 2012).

However, post-glacial colonization of Lakes Ontario and Champlain is inconsistent with historical records, and it is hard to imagine that sea lamprey could have been present much before the late 1800s in Lake Ontario and the early 1900s in Lake Champlain without having been observed (see Sect. 4.5.1). Nevertheless, some authors have suggested that sea lamprey went undetected for so long in these lakes because ecological conditions kept their numbers low; this is the "native, but rare" hypothesis (Waldman et al. 2009). Sea lamprey distribution and abundance, for example, may have been limited by the cooling and low productivity associated with the

"Little Ice Age" that lasted from the late 1500s to ~1850 (Patterson 1998). Warming and ecological changes associated with European settlement—for example, loss of the forest canopy and siltation as land was converted to farming or subsequently, as mill dams on tributaries were removed or deteriorated, thus opening up sea lamprey spawning habitat—may have served as a release for the previously small population (Waldman et al. 2004, 2009). It has likewise been suggested that fishing pressures (i.e., on Atlantic salmon and lake trout) and similar ecological changes from anthropogenic impacts initially depressed the native sea lamprey in Lake Champlain and then subsequently allowed it to increase to noticeable levels (Waldman et al. 2006; D'Aloia et al. 2015). In some cases, however, it appears that the onset of large-scale agricultural and clear-cutting practices may have suppressed, rather than released, the population (D'Aloia et al. 2015; see below). Furthermore, Eshenroder (2009) reasoned that it is unlikely that sea lamprey in Lake Ontario prior to the mid-1880s were significantly constrained by low temperatures and low productivity, because three large tributaries discharging into Lake Ontario were fed from lakes that should have generated nearly ideal temperatures for sea lamprey spawning (or at least comparable to, or warmer than, Lake Superior rivers currently infested with sea lamprey). He also argued that present-day Lake Superior is more oligotrophic and likely less productive than pre-1800s Lake Ontario (but it clearly still supports a large sea lamprey population), and abundant and not-yet-overfished populations of lake trout and deepwater ciscoes (*Coregonus* spp.) would have supported a large population of sea lamprey prior to the mid-1800s.

Genetic studies from the early 2000s have been used as support for post-Pleistocene colonization in both Lakes Ontario and Champlain. Studies using both mtDNA sequence data (Waldman et al. 2004, 2006, 2009) and microsatellite loci (Bryan et al. 2005) showed evidence for long-term vicariance (i.e., separation) of the freshwater and anadromous populations. These studies will be reviewed here briefly. Although they used the genetic markers available at the time, they are relatively limited compared to those now available and are generally no longer considered conclusive evidence of native status. Nevertheless, readers are referred to these influential papers and the rebuttal by Eshenroder (2009). In brief, these studies found significant genetic differentiation between sea lamprey in Lake Ontario or Lake Champlain and the Atlantic Ocean population. For example, Waldman et al. (2009) indicated that four of six haplotypes in the Lake Ontario population were rare (e.g., haplotype B) or absent (haplotype P) in the Atlantic population, and they argued that this was unlikely to have happened by stochastic lineage change in less than 200 years. A shift in haplotype frequency in rare alleles could have resulted from a founder effect (i.e., the loss of genetic variation that occurs when a new population is established from a larger population by a very small number of individuals), but founder effects normally result in the loss of rare alleles and haplotypes, not common ones (Waldman et al. 2009). Likewise, Bryan et al. (2005) found an allele that was exclusive to the Lake Ontario population and one that was represented only in the Lake Ontario and Lake Champlain populations, again suggesting that sea lamprey in these lakes have been separate from the Atlantic population for considerably longer than 200 years (i.e., that these exclusive alleles most likely evolved in the lake over thousands of years due to mutations that occurred during isolation since post-glacial colonization). Bryan et al. (2005) found statistical support for genetic bottlenecks (i.e., sharp reductions in population size) in Lake Ontario (and Cayuga Lake) sea lamprey populations, but not in Lake Champlain. These authors, as expected, also found evidence of sequential population bottlenecks as sea lamprey expanded into Lakes Erie, Huron, Michigan, and Superior, but they interpreted the bottlenecks seen in Lake Ontario and Cayuga Lake as possibly having been caused by environmental degradation during human settlement. Bryan et al. (2005) suggested multiple invasions of Lake Ontario and, using coalescence analysis, showed that colonization via the St. Lawrence River was more likely than via the Champlain Sea. Eshenroder (2009) argued that genetic differences between sea lamprey in Lakes Ontario and Champlain and the Atlantic population were likely the result of a recent genetic bottleneck at founding (rather than long-term residence followed by a recent bottleneck) and that the absence of rare alleles in the Atlantic population was likely due to a sampling artifact or recent declines in the Atlantic population. Haplotype P, for example, could have become extinct in the Atlantic since invasion of Lake Ontario in the late 1800s, or it might be restricted to, or more common in, regions not sampled by Waldman et al. (2004). Waldman et al. (2009) analyzed an expanded mtDNA data set that included samples representing most or all of the range of sea lamprey in the western Atlantic. They discovered three new haplotypes in the Atlantic population which, according to Eshenroder (2014), indicates that not all alleles existing in the Atlantic population have been recovered to date, but still failed to find haplotype P. Nevertheless, Waldman et al. (2009) estimated that the probability of obtaining the mtDNA results seen among the Lake Ontario specimens in less than 500 years of separation from the Atlantic population was considerably less than 1%.

Recently, D'Aloia et al. (2015) used additional genetic models to estimate the historical demography of sea lamprey in Lake Champlain, although with data comparable to the previous genetic studies, that is, with independently derived mtDNA sequence data and the summary statistics (i.e., rather than the complete microsatellite data set) from Bryan et al. (2005). These authors concluded that their results were most consistent with a post-Pleistocene origin of Lake Champlain sea lamprey. They identified an initial decline in effective population size which would have preceded the proposed invasion-by-canal hypothesis and a subsequent very recent population expansion (within the last 50 years). However, there was considerable uncertainty in both the magnitude and timing of these demographic events. For example, they dated the initial decline to ~400 years ago using BEAST analysis of the mtDNA data, but coalescent modeling of the microsatellite data suggested that the decline occurred ~1,230 years ago. D'Aloia et al. (2015) suggested that the initial decline, if ~400 years ago, could have been associated with land use and fishing pressure changes following European settlement. The very recent population expansion may have been associated with implementation of Atlantic salmon and lake trout stocking in the 1970s, following extirpation of these species in the basin in the mid- to late 1800s. D'Aloia et al. (2015) considered the alternative interpretation that the decline in effective population size and loss of genetic diversity might be the result of a founder event in the early 20th century, but, as discussed above, a recent founder event should lead to a loss of rare, not dominant, haplotypes. These genetic studies have provided some new insights into the demographic history of sea lamprey in Lake Ontario and Lake Champlain; it is hoped that definitive resolution will be possible with modern genome-level analyses (see Rougemont et al. 2017; Veale et al. 2018; Hohenlohe et al. 2019; see Chap. 7).

4.5.3 Morphological, Physiological, and Life History Differences

Morphological, physiological, and life history differences have been described between freshwater-resident sea lamprey and anadromous sea lamprey. For example, Gage (1893) reported that sea lamprey from Cayuga and Seneca lakes had a larger dorsal ridge (i.e., in sexually mature males), closer dorsal fins, a tendency for a greater number of cusps on the infraoral lamina, and differences in pigmentation relative to the Atlantic population. They are also considerably smaller (see Sect. 4.3.4.4), and Gage (1893) considered body size to be completely effective for reproductive isolation (see Sect. 4.6.3). He suggested that they be considered different species and thought that sea lamprey in the Finger Lakes had been separated from the Atlantic population since the end of the Pleistocene. However, it is not known whether body size has a significant heritable component or whether it is largely a plastic response to the freshwater environment. The transition to fresh water has involved a reduction in the duration of the parasitic phase from approximately 23–28 months (F. W. H. Beamish 1980a) to 12–20 months (Applegate 1950; Bergstedt and Swink 1995), but we do not know whether cessation of the feeding phase is triggered earlier in fresh water by environmental or endogenous cues (e.g., related to prey availability or growth rate) or whether reduction of the parasitic phase involved selection at the level of the genome. Bergstedt and Swink (1995) speculated that the large size of a few sea lamprey (~400–525 mm TL) collected in northern Lake Huron in April–May indicates that a small proportion of the population may feed parasitically for 2 years, but there is no proof of this (i.e., rather than representing unusually fast-growing individuals or those that started feeding earlier). Comparing mean TL during the parasitic feeding phase in anadromous and Great Lakes sea lamprey, Halliday (1991) suggested that the growth patterns are similar in both forms. Mean TL in Great Lakes sea lamprey in November (i.e., 1 year after metamorphosis and ~7–8 months prior to spawning) is ~430–475 mm (Applegate 1950; Bergstedt and Swink 1995), roughly comparable to the TL extrapolated for anadromous sea lamprey after their first year of feeding; Halliday (1991) estimated that TL increased from ~450 to 800 mm during their second year in the marine environment. An unpublished study by Roger A. Bergstedt at Hammond Bay Biological Station in Michigan (see Eshenroder 2009) suggested that Great Lakes sea lamprey showed better growth than anadromous sea lamprey when both were held in fresh water. When fed on white sucker in the laboratory, 92% of the Great Lakes sea lamprey grew compared to only

64% of the anadromous sea lamprey. However, although this might imply that the Great Lakes sea lamprey is better adapted to feeding in fresh water, the results were considered inconclusive because the anadromous lamprey may have been suffering from handling stress associated with the long transport from the Atlantic Ocean. Their corresponding performance in sea water was not assessed.

Apart from differences in the duration of the feeding phase, and the resulting differences in size at maturation and fecundity, there appear to be other differences in life history traits between the Great Lakes and anadromous sea lamprey. The best-studied differences are those associated with gonadal development during the larval stage and age and size at metamorphosis. There appears to be an acceleration of ovarian differentiation in the Great Lakes sea lamprey (i.e., occurring at 2–3 years of age versus when larvae are ~4–5 years old in the anadromous population; Hardisty 1969; Barker and Beamish 2000) and a concomitant reduction in potential (larval) fecundity (33,000–165,000 and 182,000–328,000 oocytes, respectively; Hardisty 1964, 1969, 1971; Barker et al. 1998; see Chap. 1). This acceleration of ovarian differentiation and reduction in potential fecundity is consistent with the shift seen following the transition from parasitic to non-parasitic lampreys (i.e., also with a reduction in size at maturity). This change in the phasing of oogenesis is assumed to have a genetic component because it happens during the larval phase, that is, prior to the divergent environmental influences experienced in the freshwater versus marine feeding phases (Hardisty 1964). Thus, if sea lamprey invaded the Great Lakes from the Atlantic Ocean in historical times (e.g., when a tributary of the Susquehanna River was diverted into the Oneida Lake drainage in 1863; see Sect. 4.5.1), it appears that changes in the timing of gonadogenesis and potential fecundity can evolve quickly in lampreys. Alternatively, as suggested by Hardisty (1971), differentiation of the "landlocked race" of sea lamprey may have involved selection for individuals who already exhibited low potential fecundity, reduced body size, and perhaps reduced osmoregulatory performance in salt water (see below). A non-parasitic form of Arctic lamprey arose in Japan following construction of a dam ~90 years previously (Yamazaki et al. 2011; see Sect. 4.6.3.2), but it is not known if there were corresponding changes to potential fecundity and the phasing of gonadogenesis similar to that observed in Great Lakes sea lamprey.

Differences in the size (and presumably age) at metamorphosis have also been reported. On average, sea lamprey in the Great Lakes appear to enter metamorphosis at a larger size than the anadromous form (~140 and 130 mm, respectively; Potter et al. 1978; Dawson et al. 2015; Manzon et al. 2015). However, there is considerable variation among and within populations, largely, or at least partially, attributable to variation in growth conditions (Dawson et al. 2015). Thus, it is not known if there has been selection (i.e., adaptation with a genetic basis) for a longer larval stage in the Great Lakes sea lamprey or whether most or all of the observed differences are the result of environmentally induced plasticity.

Physiological differences, mostly related to the ability of Great Lakes sea lamprey to osmoregulate in salt water during the parasitic feeding phase, have also been reported. Mathers and Beamish (1974) found that sea lamprey juveniles from Lake Ontario, when exposed to increasing concentrations of salt water (2 ppt per day), were able to osmoregulate up to concentrations of 16 ppt (where full-strength sea water is ~35 ppt). However, within 10 days at 26 ppt, over half of the small sea lamprey (mean TL 181 mm) had died, but all larger lamprey (mean TL 250 mm) survived for 15 days and were able to maintain their serum osmotic and ionic levels throughout this period. At 34 ppt, moderately large sea lamprey (mean TL 220–250 mm) were able to maintain serum osmotic and ionic levels for 4 days, but all had died by the ninth day. Only the largest category tested (mean TL 289 mm) survived at 34 ppt for 15 days with no mortality. As has been observed in other fishes (Fontaine 1930; Parry 1960), the reduced surface area-to-volume ratio in larger sea lamprey was likely an important factor in lowering their osmotic stress in salt water, although the relationship between body size and osmoregulatory ability is complicated by the fact that the surface area of the gill increases allometrically (Jonathan M. Wilson, Wilfrid Laurier University, Waterloo, ON, personal communication, 2018). Furthermore, Beamish et al. (1978) did not see a similar size effect in anadromous sea lamprey in salt water. Anadromous sea lamprey juveniles of all sizes (>135 mm) were able to osmoregulate between 0 and 35 ppt without mortality. In addition to greater survival rates, anadromous sea lamprey were better able to regulate serum osmolality at the higher salinities (26 and 34 ppt) than were small and large Great Lakes sea lamprey. For example, in the small anadromous individuals, serum osmolality at 34 ppt increased by less than 10% relative to that in fresh water, but it increased by ~25% in small Great Lakes sea lamprey (Beamish et al. 1978). Anadromous sea lamprey showed lower serum osmolality than landlocked sea lamprey at all salinities and large individuals had lower serum osmolality than small ones (Beamish et al. 1978). Anadromous sea lamprey juveniles, regardless of size, were able to regulate serum sodium levels in salt water more precisely than sea lamprey from Lake Ontario. It should be noted, however, that a more recent study—while also finding detectable differences in the inherent physiological capacity of landlocked and anadromous sea lamprey to osmoregulate in salt water—found that these differences were much more subtle than previously reported. Sea lamprey transformers from three landlocked populations (from Lakes Superior, Huron, and Champlain) showed survival rates ranging from ~40 to 100% (compared to ~90% for anadromous sea lamprey transformers) when held at 30 and 35 ppt for 30 days (Jessica L. Norstog and Stephen D. McCormick, University of Massachusetts, Amherst, MA and S. O. Conte Anadromous Fish Research Center, U.S. Geological Survey, Turners Falls, MA, personal communication, 2018). Survival rates between the landlocked and anadromous populations were not significantly different at 30 ppt, and only the Lake Champlain population showed significantly lower survival at 35 ppt. Unlike the studies above, these results suggest that even very small juvenile landlocked sea lamprey have robust salinity tolerance. Further research is required to clarify the contrasting results.

However, differences in osmoregulatory abilities between the anadromous and Great Lakes populations are not sufficient to conclusively resolve whether colonization was in historical or prehistoric times. The Vancouver lamprey, despite its presumed post-glacial origin, still retains the ability to osmoregulate in salt water (Beamish 1982). Landlocked sea lamprey still possess chloride cells (SW-MRCs) in their gills for osmoregulation in salt water during the feeding phase (Youson and Freeman 1976; see Sects. 4.3.2.3 and 4.3.2.4), but the retention of these cells does not distinguish between freshwater colonization that happened a few hundred years ago and a few thousand years ago (Bartels et al. 2012, 2015). Most salmonid populations that have been landlocked for several thousand years (i.e., post-glacially) do show decreased osmoregulatory ability in salt water (e.g., Staurnes et al. 1992; Nilsen et al. 2003), but other populations have shown no apparent decrease in this ability (e.g., McCormick et al. 1985; Nilsen et al. 2007). Conversely, there are examples where salmonid populations isolated above recently constructed barriers soon showed reduced ability to osmoregulate in salt water and reduced rates of smoltification (e.g., Thrower and Joyce 2004; Holecek et al. 2012). Moreover, there are multiple examples of rapid freshwater evolution in other fishes. Several populations of threespine stickleback, for example, have shown substantial changes in body shape and lateral plate phenotype within decades of freshwater colonization (e.g., Bell et al. 2004; Vamosi 2006; Gelmond et al. 2009; Aguirre and Bell 2012; Lescak et al. 2015). In fact, the results of Lescak et al. (2015) support the "intriguing hypothesis that most stickleback evolution in fresh water occurs within the first few decades after invasion." In many cases, rapid adaptation to fresh water may be due to selection on pre-existing variation in the ancestral anadromous population (e.g., Colosimo et al. 2005; Barrett et al. 2008; Lescak et al. 2015; Nelson and Cresko 2018). Similarly in lampreys, several authors (e.g., Hardisty 1969; Mathers and Beamish 1974; Beamish et al. 1978; F. W. H. Beamish 1980b) have suggested that sea lamprey colonization of the Great Lakes might have involved selection for traits advantageous in fresh water (i.e., smaller body size, lower potential fecundity, and reduced osmoregulatory abilities in salt water).

4.5.4 Does It Matter?

The debate regarding the origin of sea lamprey in the Lake Ontario drainage (including the Finger Lakes and Oneida Lake) and Lake Champlain is often discussed in terms of the implications to the sea lamprey control program. For example, Waldman et al. (2004) suggested that sea lamprey control policies aimed toward intense suppression might need re-evaluation if sea lamprey are shown to be native to Lake Ontario. Determining whether there would be continued "social license" (i.e., acceptance within the local community and among stakeholders) for controlling a native species that is a significant pest would involve public consultation (see Chap. 7). Regardless, determining if the sea lamprey is native in these lake systems has other important management implications and will improve our understanding of life history evolution in lampreys. For example, it is important to understand how quickly sea lamprey (or other anadromous lampreys) can become invasive in fresh water and the genetic basis of this adaptation. Does adaptation to fresh water require gradual genetic change or can it happen rapidly? Clarifying the demographic history of colonization (e.g., identifying the initial number of founders) and the genetic changes associated with colonization also will be very informative. A review of the factors that promote and constrain freshwater residency in parasitic lampreys (see Sect. 4.4) emphasizes the need for a large prey base for establishment of sea lamprey in fresh water, but virtually nothing is known regarding the genetic basis of freshwater adaptation in lampreys. Did successful colonization of the Great Lakes depend on existing genetic variation within the anadromous population (i.e., "pre-selection" for individuals that already showed traits advantageous to survival in fresh water) or could any anadromous sea lamprey colonize fresh water if permitted access (see Sect. 4.8.3)?

4.6 Feeding Type Variation: Evolution of Non-parasitism

A non-trophic adult feeding phase is unknown in any group of vertebrates other than lampreys. Thus, the evolution of non-parasitism in lampreys and the relationship between closely related parasitic and non-parasitic forms have long interested biologists. Loman (1912), for example, recognized that European river and brook lampreys were morphologically similar, but he noted that the brook lamprey exhibited delayed metamorphosis and accelerated sexual maturation relative to the river lamprey. The morphological similarity between several other non-parasitic and parasitic lampreys was likewise recognized by Hubbs (1925), who suggested several cases in which a particular brook lamprey species had apparently evolved from a form similar to that of an extant parasitic lamprey. The term "paired species" was later coined by Zanandrea (1959). Vladykov and Kott (1979b) introduced the more general term "satellite species," because there are several cases in which more than one brook lamprey (satellite) species has apparently been derived from a single parasitic (stem) species (see Potter 1980; Docker 2009).

In addition to the non-parasitic species that are paired with a parasitic counterpart, several so-called "relict" species have also been identified. Relict brook lampreys are non-parasitic species that occur at or near the extreme southern limits of distribution of the Northern Hemisphere lampreys and are generally those that cannot be unambiguously paired with an extant parasitic species (see Hubbs and Potter 1971; Docker et al. 1999; Potter et al. 2015). Much of the previous ambiguity of "who begat whom" has largely been removed through molecular phylogenetic studies (e.g., Docker et al. 1999; Lang et al. 2009; see Fig. 4.1), although it is now delineation between paired and relict species that is somewhat ambiguous. However, a better understanding of the apparent continuum between recently derived paired species and older relict species will be very informative. Non-parasitism has arisen independently in seven of the 10 extant lamprey genera—and often multiple times within each genus—with different non-parasitic species evolving at different times and in different locations (Hubbs and Potter 1971; Vladykov and Kott 1979b; Potter 1980; Docker 2009). By comparing the phenotypic, molecular, and ecological differences in parasitic–non-parasitic pairs that have only recently diverged (or are still in the process of diverging) to traits in progressively more differentiated relict species (i.e., those further down the pathway to non-parasitism), we can better understand the recurrent elimination of the adult feeding phase that is unique to lampreys.

4.6.1 Non-trophic Adults Unique Among Vertebrates

The characteristic elimination of the adult feeding phase in non-parasitic lampreys is unheard of in any other vertebrate. It is rare in animals in general, but it has been reported in a number of disparate insect and other invertebrate taxa (Hendler and Dojiri 2009; Benesh et al. 2013). All involve taxa with complex life cycles.

Among insects, non-trophic adults have been reported in eight of the ~120 families in the order Lepidoptera (butterflies and moths), two of the ~100 families in the order Hymenoptera (e.g., wasps, bees, and ants), seven of the ~160 families in the order Coleoptera (beetles), and seven of the ~150 families in the order Diptera (flies) (see Hendler and Dojiri 2009; Benesh et al. 2013). In the lepidopteran family Saturniidae (e.g., luna moth *Actia luna* and polyphemus moth *Antheraea polyphemus*), adults have vestigial mouthparts, lack functional digestive tracts, and generally live for <1 week following emergence from the pupa (Janzen 1984). Some species of geometer moths (family Geometridae) are similarly non-trophic as adults, and it has been proposed that loss of adult feeding is correlated with the evolution of flightlessness in forest habitats (Snäll et al. 2007). It has been suggested that, under conditions where female mobility lost its adaptive value (e.g., due to abundance of host plants for the larvae but scarcity of adult food in late summer), loss of wings—although preventing adult foraging, growth, and dispersal—allowed females to increase fecundity beyond the point at which egg loads would reduce flight performance. In the order Hymenoptera, a large number of species are parasitoids as larvae, and some feed on nectar or pollen as adults while the adults of other species do not feed at all (Benesh et al. 2013). The best-known insect order with non-trophic adults is likely Ephemeroptera (mayflies). Aptly named, the adult stage is very short lived (as short as 37 min in one species; Lancaster and Downes 2013), and its primary function is reproduction. Non-feeding adults have also been reported in species of the orders Plectoptera (stoneflies), Megaloptera (e.g., alderflies and fishflies), and Trichoptera (caddisflies) (Lancaster and Downes 2013). In some insect taxa (e.g., orders Strepsiptera and Embiidina), only males have evolved to be non-feeding, and, in many cases, their mouthparts are modified into mating appendages or sensory structures (Benesh et al. 2013 and references therein).

Non-trophic adults have also been reported in some crustaceans, all of which are parasitic as larvae: copepod species in the families Thaumatopsyllidae and Monstrillidae, isopods in the family Gnathiidae, and species in the subclass Tantulocarida. Likewise, horsehair or Gordian worms (phylum Nematomorpha), nematodes from the family Mermithidae, and ticks from the family Argasidae have non-feeding adults. In some barnacles and rotifers, only males are non-feeding as adults (Hendler and Dojiri 2009; Benesh et al. 2013 and references therein).

It thus appears that non-parasitic lampreys resemble these other species with complex life cycles, where the relatively long-lived larval stage is specialized for feeding and growth and the adult stage is specialized for reproduction (Hendler and Dojiri 2009; see Sect. 4.2.1). Benesh et al. (2013) proposed that the "no-growth strategy" should be found where "massive larval size can make adult growth superfluous" and showed theoretically that this counterintuitive strategy would be favored when the optimal larval size is greater than or equal to the optimal adult size for reproduction. It has already been suggested that non-parasitic lampreys evolve under conditions providing good larval growth opportunities (e.g., Kucheryayvi et al. 2007; Docker 2009). However, given the increase in fecundity achieved in lampreys with the inclusion of a parasitic feeding phase, it would be inaccurate to say that adult growth in lampreys is superfluous, although the trade-off between reduced mortality and reduced fecundity apparently makes it unnecessary under some conditions (see Sect. 4.7.4). It should be pointed out, however, that the duration of the non-trophic period in brook lampreys is appreciably longer (6–10 months) than that of invertebrate taxa with "ephemeral" non-feeding adult stages, even accounting for the shorter overall life cycle of the latter. Nevertheless, similarities and differences between non-parasitic lampreys and invertebrates with non-feeding adults could shed light on the mechanisms and selective pressures associated with elimination of the adult feeding phase.

4.6.2   Relict Species

"Relict" brook lampreys have been defined as non-parasitic species which cannot be obviously paired with extant parasitic forms and which have an extreme southerly distribution that seems to reflect their status as relicts of groups with a previously more widespread distribution (Hubbs and Potter 1971). Potter et al. (2015) recognized six relict species: the Po brook lamprey, least brook lamprey, Kern brook lamprey *Lampetra hubbsi*, Western Transcaucasian brook lamprey *Lethenteron ninae*, and Macedonia and Epirus brook lampreys. In each case, based on morphology, the identity of a possible parasitic ancestor has indeed been problematic, as evidenced by past or current uncertainty regarding generic placement (see Fig. 4.1; Sect. 4.6.2.3). However, molecular studies are helping to clarify the evolutionary history of many of these species although, in other cases, they are adding to the confusion. Mitochondrial DNA sequencing, for example, suggests that the Macedonia and Epirus brook lampreys are not closely related to any extant species, certainly none within the genus *Eudontomyzon*, while suggesting that other relict species are not as obviously "unpaired" as previously thought.

4.6.2.1   Older and More Divergent Brook Lamprey Species

With the inclusion of molecular data, the delineation between relict and "paired" species (i.e., non-parasitic species that are morphologically similar to a particular parasitic species in all aspects other than body size; Potter et al. 2015) has also become somewhat "fuzzier." These studies remind us that non-parasitism in lampreys has evolved independently at different times and in different locations (Hubbs and Potter 1971; Vladykov and Kott 1979b). Therefore, it is not surprising that there are different degrees of morphological and genetic divergence between a presumed parasitic ancestor and various non-parasitic derivatives (Fig. 4.1), and different brook lamprey species presumably represent different stages in the speciation process (Docker 2009).

We have tried to represent the different degrees of divergence between each non-parasitic species and its presumed ancestor (as best represented in the contemporary fauna) using a combination of differentiation at the mitochondrial cytochrome b gene (using Kimura's two-parameter distance, K2P) and reduction in the number of trunk myomeres (see Sect. 4.6.2.2) as proxies of time since divergence and degree of morphological divergence, respectively (Fig. 4.2). The non-parasitic species examined included the 23 brook lamprey species recognized by Potter et al. (2015), the three Portuguese species described by Mateus et al. (2013a), the recently described *Lampetra soljani* (Tutman et al. 2017), and the undescribed *Lethenteron* sp. N and sp. S (Yamazaki et al. 2006). Using this approach, non-parasitic species were divided into three categories:

(1) Species (n = 8) showing little or no genetic divergence ($\leq$0.4% K2P) and no reduction in number of trunk myomeres (i.e., 0–1.5 more trunk myomeres) compared to the presumed parasitic ancestors (see Sect. 4.6.3.1).
(2) Species (n = 9) showing intermediate genetic and/or morphological divergence (i.e., 0.2–2.9% K2P and 1–8.5 fewer myomeres) compared to the presumed parasitic ancestors (see Sect. 4.6.3.1).
(3) Species (n = 11) showing both higher genetic ($\geq$2.9% K2P) and morphological (2.5–13 fewer myomeres) divergence when compared to the most closely related extant parasitic species. This category includes the six relict species identified by Potter et al. (2015) plus the Turkish brook lamprey *Lampetra lanceolata*, the Pacific brook lamprey, *Lampetra soljani*, and *Lethenteron* sp. N and sp. S.

This third category therefore loosely corresponds with the relict species, but it is somewhat more inclusive. Potter et al. (2015) omitted Turkish brook lamprey from their list of relict species because, based on morphology (and confirmed with sequence data), it is probably derived from European river lamprey or a European river lamprey-like ancestor. However, molecular evidence suggests that this species is closely related to (or even conspecific) with the Western Transcausian brook lamprey (Li 2014; Tuniyev et al. 2016). Whether the two species are conspecific is not within the purview of this chapter, but, given the close relationship between the two, we decided to consider both (or neither) as relict species.

Likewise, Pacific brook lamprey was not considered a relict species by Potter et al. (2015), and its placement on Fig. 4.2 was somewhat intermediate between categories 1 and 2. Based on morphology, the Pacific brook lamprey is clearly a derivative of the western river lamprey or a western river lamprey-like ancestor. It is characterized by a low myomere count relative to western river lamprey, and, until recently, there was debate regarding whether it was distinct from or synonymous with western brook lamprey (see Reid et al. 2011; Potter et al. 2015). It is not found at the extreme southern distribution of Northern Hemisphere lampreys, but it appears restricted to the Columbia River basin, presumably south of the glacial margin during the Pleistocene (Reid et al. 2011). Furthermore, because Pacific brook lamprey showed sequence divergence from western river lamprey comparable to that of other relict species and their presumed ancestors, we include it here as an older non-parasitic derivative. It may not be a relict species by the definition of Hubbs and

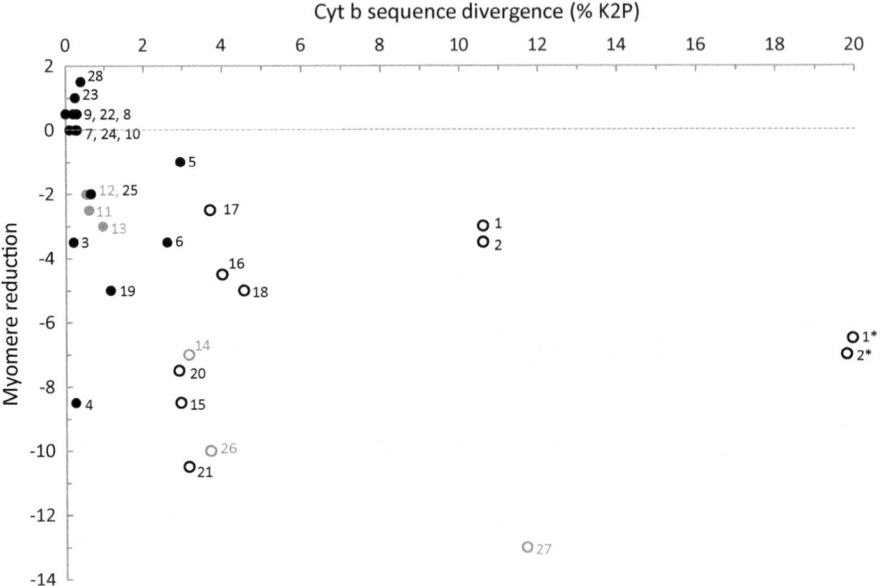

Fig. 4.2 Relationship between cytochrome b sequence divergence (Kimura's two-parameter distance, K2P) between non-parasitic lamprey species and their presumed parasitic ancestor and the extent to which the number of trunk myomeres has been reduced in the non-parasitic species. For sequence divergence, the mid-point in the known range is given (see Fig. 4.1); likewise, reduction in myomeres was calculated by subtracting the midpoint in the known myomere range for each non-parasitic species from that for the presumed ancestor. Myomere data were compiled from Seversmith (1953), Vladykov (1955), Vladykov and Kott (1976a, b, 1984), Yamazaki et al. (2006), Docker (2009), Naseka et al. (2009), Renaud and Economidis (2010), Reid et al. (2011), Renaud (2011), Mateus et al. (2013a), Renaud and Naseka (2015), Renaud et al. (2016), and Tutman et al. (2017); counts from larvae were used whenever possible. Numbers given as data labels apply to the non-parasitic lamprey species listed in Fig. 4.1; species recognized by Potter et al. (2015) are shown in *black*; other taxa or tentative species that have not been formally described are given in *gray*. *Closed circles* are those species considered here to be "paired" with their parasitic counterpart and *open circles* are considered "relict" species, although the distinction is not always clear (see Sect. 4.6.2.1). Species 1 and 2 (Macedonia and Epirus brook lampreys, *Eudontomyzon hellenicus* and *Eu. graecus*, respectively) are compared with both the Caspian lamprey *Caspiomyzon wagneri* with which they are most genetically similar and the Carpathian lamprey *Eu. danfordi* (*asterisk*) with which they are most similar morphologically

Potter (1971), but we feel that placing it in this category facilitates discussion related to the different stages in the evolution of non-parasitism. Boguski et al. (2012) found *Lampetra* sp. populations in Oregon and California that, based on cytochrome b gene sequence divergence from western river lamprey (or any known lamprey species), may also represent relict species. However, myomere counts (or other morphological characters) are not available for these populations.

Lethenteron sp. S has never been formally described (see Yamazaki et al. 2006), but its genetic distinctiveness and extreme reduction in number of trunk myomeres (relative to Arctic lamprey) suggest that it would also be considered a relict species. Yamazaki et al. (2006) suggested that it may be a descendent of the co-ancestor of *Lethenteron*, *Lampetra*, and *Entosphenus*. Based on its uncertain phylogenetic placement, we refer to it here as an "orphan" relict species to emphasize that it truly cannot be paired with an extant parasitic species. We likewise refer to the Macedonia and Epirus brook lampreys as "orphan" species because their putative parasitic ancestor cannot be identified (see Sect. 4.6.2.3).

Thus, whether we use the term "relict species" more loosely than originally defined or whether we merely refer to these as "older, more divergent brook lamprey species," category 3 includes brook lampreys with a relatively long separation from their parasitic ancestor. We recognize that this categorization is imperfect, since we are trying to divide a speciation continuum into discrete categories using only one gene and one morphological character as proxies of time since divergence. Caution needs to be exercised when relative divergence times are estimated from mtDNA sequence data alone, because genetic drift between isolated populations can obscure inferences (Galtier et al. 2009). Likewise, we also need to be aware that some species may show more rapid rates of morphological divergence due to drift or selection relative to morphologically conserved, but genetically divergent, populations. Furthermore, those species that are morphologically and/or genetically divergent from any known parasitic species may appear so due to extinction of their recent parasitic ancestor. However, these categories are useful for the purpose of trying to understand the sequence of changes involved in the evolution of non-parasitism. In particular, there appears to be a number of brook lamprey species within both North American and Eurasian *Lampetra* that show different degrees of genetic and morphological differentiation from the two (or a few closely related) parasitic ancestors (Fig. 4.1). In these two clades alone, different stages in the speciation process are represented within the contemporary fauna, and replication is provided among taxa.

4.6.2.2   Morphological Degeneracy in Relict Species

Despite their placement in different genera, relict brook lamprey species look rather similar. Over time, reduction in traits associated with parasitic feeding has occurred in the different species independently, allowing them to converge on a similar phenotype. Reduction in number of trunk myomeres (corresponding with smaller body size at maturity) is the easiest trait to quantify across taxa (hence its use in Fig. 4.2), but relict species are also characterized by more degenerate dentition and fewer velar tentacles (Hubbs and Potter 1971; Vladykov and Kott 1979b; Potter et al. 2015). Although Bond and Kan (Kan 1975; Bond and Kan 1986) suggested a latitudinal cline in trunk myomere counts, Goodman et al. (2009) and Reid et al. (2011) found no such pattern and concluded that any observed differences were taxonomic rather than latitudinal. The number of myomeres appears to remain unchanged or increase at speciation (i.e., in closely related non-parasitic species that are sympatric with their parasitic ancestor) but appears to decrease with time since divergence (Vladykov and Kott 1979b). Similarly, dentition in non-parasitic species appears to be more variable (e.g., in terms of number of teeth and cusps) in the initial stages of divergence but then becomes reduced over time (Hubbs and Potter 1971). The dentition of several of the relict species, most notably the least brook lamprey, is highly degenerate and contains only a few small, blunt teeth (Hubbs and Potter 1971; Potter et al. 2015). Thus, the long-standing difficulty in trying to pair the various relict species with extant parasitic forms based on morphology is not surprising.

The number, arrangement, and structure of papillae along the posterior margin of the gill pore in mature lampreys have recently been described as useful morphological characters for taxonomic discrimination in lampreys (Beamish 2010, 2016). There seems to be a general trend showing reduction in the number of marginal papillae in freshwater parasitic and non-parasitic species derived from an anadromous ancestor, and, as with number of myomeres and dentition, the extent of the loss appears correlated with presumed time since divergence. The trend is less obvious in the genus *Entosphenus*, where all derivative species are relatively recent, but is quite pronounced in North American *Lampetra*. In *Lampetra*, the number of marginal papillae decreased from 24–34 in western river lamprey to 27–29 in western brook lamprey, 8–12 in Pacific brook lamprey, and 6–13 in Kern brook lamprey (Beamish 2010). The function of these papillae is not known, but, because they increase in size at maturity, Beamish (2016) inferred that they may have some sensory function related to reproduction.

There may also be a progressive reduction in potential fecundity in non-parasitic species as they diverge from their parasitic ancestor. It has been suggested that recently derived non-parasitic species may still "atavistically" produce a large number of oocytes during the larval stage (i.e., in line with the high potential fecundity of their parasitic ancestor) and then reduce the number of oocytes prior to maturation (i.e., through atresia), so that fecundity at maturity is in line with their now-smaller adult body size (see Chap. 1). For example, Hardisty (1964) estimated that up to 90% of the larval oocytes in the recently derived European brook lamprey are lost to atresia. In contrast, the least brook lamprey is thought to experience little or no atresia, leading to the suggestion that natural selection in this relict species has had sufficient time to reduce the number of oocytes elaborated during the larval stage to a level that could reasonably be brought to maturity in the adult (Docker and Beamish 1991). However, there is debate whether extent of atresia is indeed correlated with time since divergence (see Docker 2009), likely due to the difficulties associated with accurately estimating potential fecundity and the relative dearth of such estimates (see Chap. 1).

Recent studies have started examining genome-level differences between the closely related European brook and river lampreys (Mateus et al. 2013b; Rougemont et al. 2017; Hume et al. 2018; see Sect. 4.6.3.3), but none have examined genomic differences between parasitic or recently derived non-parasitic species and the relict species. Such studies would help elucidate the genomic basis for the changes in morphological and life history characters observed at different stages in the transition from parasitism to non-parasitism.

4.6.2.3 Taxonomic Uncertainties in Relict Species

As indicated above, all the traditional relict species have presented— and sometimes continue to present—taxonomic uncertainties. Given their degenerate dentition, there has been uncertainty regarding generic placement as well as some confusion regarding species delimitation. For example, based on dentition, past authorities have sometimes placed the Po brook lamprey in the genus *Lethenteron* (see Potter et al. 2015); the least brook lamprey was once placed in the "provisional and noncommittal" genus *Okkelbergia* (Hubbs and Potter 1971), and the Kern brook lamprey was originally and until recently referred to the genus *Entosphenus* (see Docker et al. 1999; Potter et al. 2015). Molecular data have helped resolve many of these conflicts, identifying a putative parasitic ancestor, although it is interesting that the degree of divergence seen in the morphological characters is not always consistent with molecular distance. For example, the least brook lamprey is often seen as the "poster child" for degenerate dentition, but its genetic divergence from European river lamprey (4.3–4.8% K2P at the cytochrome b gene) is only moderately higher than that observed between the Turkish brook lamprey and European river lamprey (3.2–4.2%) and is considerably less than that inferred for the three "orphan" species (Figs. 4.1 and 4.2). Likewise, the Kern brook lamprey, despite its morphological distinctness from other *Lampetra* brook lampreys, particularly in terms of its dentition, was only moderately more divergent genetically from western river lamprey (2.3–4.0%) compared to the more morphologically conserved Pacific brook lamprey (2.3–3.5%).

The Western Transcaucasian brook lamprey, although originally and currently referred to the genus *Lethenteron*, likely belongs in the genus *Lampetra*. Based on re-examination of morphological characters, Tuniyev et al. (2016) concluded that features such as a tricuspid middle endolateral, low number of trunk myomeres, and absence of velar wings suggest that this species should be assigned to *Lampetra*. The Western Transcaucasian brook lamprey possesses a row of posterial teeth, which is typical for the genus *Lethenteron*, but Tuniyev et al. (2016) acknowledge that this characteristic may have evolved independently in *Lethenteron* and *Lampetra*. Mitochondrial DNA sequence data also place the Western Transcaucasian brook lamprey into the European *Lampetra* clade (Li 2014). However, we concur with the decision of Tuniyev et al. (2016) that reclassification should be delayed until a total evidence cladistic analysis (integrating both morphological and molecular characters) has been completed (see Chap. 7).

Cytochrome b gene sequences also show that the Western Transcaucasian brook lamprey is very closely related to the Turkish brook lamprey (Li 2014; 0.2–0.9% K2P). The two species are found in close geographic proximity to each other (the south and east Black Sea basins, respectively), and they are both no longer sympatric with European river lamprey (Naseka et al. 2009). We do not attempt to decide here whether these two species should be synonymized. What is relevant (i.e., what should be kept in mind when considering the recurrent evolution of non-parasitism in lampreys) is that these two brook lampreys may not represent independent non-parasitic derivatives of the European river lamprey or a European river lamprey-like ancestor. No satellite species is recognized as the ancestor of another satellite species (Vladykov and Kott 1979b), but, in cases like these, vicariance (i.e., geographical separation) following divergence from the common parasitic ancestor may be more likely than independent derivation from this ancestor. In a similar manner, *Lampetra soljani* from the southern Adriatic Sea basin appears closely related to the Po brook lamprey in terms of morphology (e.g., number of velar tentacles and trunk myomeres) and DNA sequence (0.7–3.5% K2P; Tutman et al. 2017). Several genetically divergent populations of *Lampetra* brook lampreys have been identified in Oregon and California (Boguski et al. 2012), but they have not been formally described. Again, we do not try to resolve here the taxonomic status of these species or populations, but we remind the reader that some now-distinct species or populations likely represent independent transitions to non-parasitism at different times and different locations but caution against assuming that all do without further study (see Sect. 4.6.3.4).

Vicariance leading to pronounced phylogeographic structure has been inferred in the least brook lamprey using mtDNA sequence data (i.e., suggesting evolution of non-parasitism followed by vicariance and not separate transitions to non-parasitism). Martin and White (2008) examined control region and NADH dehydrogenase subunit 3 (ND3) gene sequences in least brook lamprey from 21 populations in Maryland, Ohio, Missouri, North Carolina, Tennessee, Kentucky, Mississippi, and Alabama. They detected as many as 12 highly differentiated clades, differing in sequence by an average of 4.5% (range 2.5–9.0%), and their distribution corresponded to different drainages or parts of drainages. They detected distinct Atlantic coastal, Ohio River, and Obion-Mississippi drainage clades, but these three clades were embedded within the very diverse Gulf drainage clade. Given the lack of further resolution among the clades, Martin and White (2008) suggested that vicariance occurred over a relatively short time (e.g., as the result of rising sea levels during the Pliocene, 5.3–2.6 Ma). These authors suggested that the Obion-Mississippi drainage populations in Tennessee and Kentucky may represent an undescribed taxon, but this clade was not necessarily any more distinct than other clades. There has been debate over the years whether the least brook lamprey consists of a single or multiple species. Hubbs and Potter (1971) indicated that dentition in populations from the Atlantic Coastal Plain may be less degenerate than that from populations in the Gulf Coastal Plain. Vladykov et al. (1975) described some individuals from the Tennessee, Alabama, and Tombigbee river systems as a distinct species, *Lampetra meridionale* (separable from least brook lamprey within the same watersheds), although this species has since been synonymized with *Lampetra aepyptera* (Nelson et al. 2004).

Nevertheless, as many of the formerly ambiguous species (Kern brook lamprey, Po brook lamprey, Western Transcaucasian brook lamprey) are assigned to *Lampetra*, it is becoming increasing clear that the majority of brook lampreys appear to have originated from European and western river lampreys or now-extinct ancestors resembling these parasitic species. Characteristics of parasitic species that appear to promote evolution of non-parasitism are discussed in Sect. 4.7.

However, perhaps the most intriguing findings from molecular phylogenetic studies indicate that the Macedonia and Epirus brook lampreys are not closely related to any extant species within the genus *Eudontomyzon* and that the most closely related species is the Caspian lamprey (Lang et al. 2009). However, we agree with the conclusion of Renaud and Economidis (2010) that considering these two Greek non-parasitic species as sister to the Caspian lamprey is premature without a cladistic analysis that integrates multiple morphological and molecular characters. Furthermore, it is important to note that the molecular analysis suggests only that the Caspian lamprey is the closest living relative of these two species, not that it is a close relative and certainly not that it is the parasitic ancestor. The level of sequence divergence (10.5–10.7%) observed between these brook lampreys and the Caspian lamprey is substantial and comparable to some genus-level differences. For example, European river and Arctic lampreys (*Lampetra* and *Lethenteron*) differ at the cytochrome b gene by 8.2–9.7%, and Pacific and Arctic lampreys (*Entosphenus* and *Lethenteron*) differ by 9.9%. Note, however, that the Macedonia and Epirus brook lampreys are even more genetically divergent from other species within the genus *Eudontomyzon* (e.g., 19.4–20.3% divergent from the Carpathian lamprey). As with the Western Transcaucasian and Turkish brook lampreys, the Macedonia and Epirus brook lampreys are genetically similar to one another but exhibit morphological differences and have disjunct distributions, and it is unknown if they were derived independently from a recently extinct parasitic ancestor or if recent vicariant speciation followed divergence from a common ancestor.

4.6.3 Paired Species: Update on the Update

The concept of paired lamprey species has been discussed and reviewed by numerous authors over the years, including Hubbs (1925), Zanandrea (1959), Hardisty and Potter (1971a), Potter (1980), Salewski (2003), Hardisty (2006), and Docker (2009). In this section, we continue these discussions, in particular, providing an update on Docker (2009)'s "update on the paired species concept," and we attempt to clarify previous uncertainties or misconceptions resulting from earlier molecular phylogenetic studies. The lack of fixed differences in mtDNA sequence in many paired species, although rare among "good" vertebrate species (e.g., Johns and Avise 1998), is not in itself evidence for phenotypic plasticity (i.e., where the different feeding types are produced from a single genotype under the induction of an environmental cue). Conversely, demonstration of fixed genetic differences in some pairs does not indicate that these differences are species-level differences, nor can these findings be extrapolated to conclude that all paired species are distinct species. We review and update the "speciation continuum" discussed by Docker (2009), with a particular focus on the better-studied species pairs where recent population genetic and genomic studies both confirm that there are genome-level differences between the feeding types (and thus refute the hypothesis of phenotypic plasticity; Mateus et al. 2013b; Rougemont et al. 2017) and show significant gene flow between them where they co-occur (i.e., refuting suggestions of immediate reproductive isolation between feeding types; Rougemont et al. 2015, 2016, 2017). We do not try to answer conclusively the long-standing question "are paired species 'real' species?" (e.g., Salewski 2003; Docker 2009). Rather, we try to show the complexity of the issue and emphasize that there is not a universal "one size fits all" answer to this question (see Chap. 7). By appreciating that different species and populations represent different stages in the evolution of non-parasitism, we will better understand the process by which the parasitic feeding phase has been eliminated in different lamprey taxa. Taxonomic changes should only be made, if warranted, based on a more complete understanding of the process and its outcome in different pairs.

4.6.3.1 Not All Paired Species are Equivalent

Paired non-parasitic species are generally defined as those that are morphologically similar to a particular parasitic species in all aspects other than body size and that are assumed to have evolved from that parasitic species (Potter et al. 2015). However, it is becoming apparent that there are different degrees of morphological and genetic divergence between parasitic lampreys and their presumed non-parasitic derivatives (Fig. 4.1) with the distinction between paired and relict species sometimes unclear (see Sect. 4.6.2.1), and variation within each category becoming evident. Potter et al. (2015) considered 15 of the 23 recognized non-parasitic species as being paired with a congeneric parasitic species. An additional two species (Northern California brook lamprey *Entosphenus folletti* and Pit-Klamath brook lamprey *En. lethophagus*) also appear to be recent non-parasitic derivatives, but it is not clear whether Pacific lamprey or one of its freshwater parasitic derivatives (e.g., the Klamath lamprey) is the ancestor (Potter et al. 2015). Docker (2009) included 14 non-parasitic species as paired species. Three species considered paired species by Potter et al. (2015) were omitted from her list: Northern California and Pacific brook lampreys because they were, at the time, considered to be synonymous with Pit-Klamath and western brook lampreys, respectively, and Turkish brook lamprey, which Docker (2009) considered a relict species.

As outlined in Sect. 4.6.2.1, we have used here a combination of genetic divergence from the presumed parasitic ancestor and reduction in the number of trunk myomeres as proxies of time since divergence and degree of morphological divergence, respectively (Fig. 4.2 and references therein). To focus on the evolutionary processes rather than the taxonomy, we included the three Portuguese species described by Mateus et al. (2013a), the recently described *Lampetra soljani* (Tutman et al. 2017), and the undescribed *Lethenteron* sp. N and sp. S (Yamazaki et al. 2006). We found that these characters divided the extant non-parasitic "species" into three categories:

(1) Species (n = 8) showing little or no genetic divergence (≤0.4% K2P) and no reduction in number of trunk myomeres (i.e., 0–1.5 more trunk myomeres) compared to the presumed parasitic ancestors. This category includes non-parasitic species whose distributions are largely sympatric with that of their parasitic ancestor and who appear to have either diverged recently or still experience gene flow:

- the three *Ichthyomyzon* brook lamprey species (northern brook, southern brook, and mountain brook lampreys *I. fossor*, *I. gagei*, and *I. greeleyi*, respectively) which show no species-specific differences in mtDNA gene sequence compared to their respective parasitic ancestors (silver, chestnut, and Ohio lampreys; Docker et al. 2012: Ren et al. 2016) or even haplotype or allele frequency differences where they occur sympatrically (Docker et al. 2012), have virtually identical myomere counts (Hubbs and Trautman 1937), and appear to show considerably less reduction in dentition (in terms of number and sharpness of cusps or teeth) relative to more divergent non-parasitic species;
- the three non-parasitic derivatives of Arctic lamprey that occur within the range of the ancestor (Alaskan brook lamprey, Far Eastern brook lamprey *Lethenteron reissneri*, and Siberian brook lamprey *Le. kessleri;* Yamazaki et al. 2006; Renaud and Naseka 2015; Yamazaki and Goto 2016; Sutton 2017);
- European brook lamprey which, where it occurs sympatrically with the ancestral European river lamprey, usually shows evidence of contemporary gene flow (e.g., Rougemont et al. 2015, 2016, 2017; see Sect. 4.6.3.2);
- Australian brook lamprey.

(2) Species (n = 9) showing intermediate genetic divergence and/or trunk myomere reduction (i.e., 0.2–2.9% K2P and 1–8.5 fewer myomeres) compared to the presumed parasitic ancestors. This category includes non-parasitic species whose range no longer overlaps with that of their parasitic ancestor or where the parasitic ancestor appears to have a more restricted distribution within the range of the non-parasitic derivative:

- American brook lamprey, which is now allopatric with its Arctic lamprey ancestor and shows slightly more differentiation than the other descendants of this species (Li 2014);
- the three Portuguese brook lamprey species described by Mateus et al. (2013a), which likewise are no longer sympatric with the European river lamprey, and which show slight but species-specific differences in mtDNA sequence and slightly lower but overlapping trunk myomere counts (57–63 in the three brook lamprey species versus 58–67 in European river lamprey);
- Pit-Klamath and Northern California brook lampreys, which are genetically similar to, or indistinguishable from, the Pacific lamprey (assumed here, for

simplicity, to be the ancestor) but which have a reduced number of myomeres (i.e., 63–68 and 56–65, respectively) compared to Pacific lamprey (61–77; Vladykov and Kott 1976a; Docker 2009). However, if compared to the Klamath lamprey with 58–65 myomeres (Renaud 2011), these two species would be placed in category 1;

- Ukrainian brook lamprey and Drin brook lamprey *Eudontomyzon stankokaramani*, which show only a moderate reduction in number of myomeres (58–68 and 56–65, respectively) relative to the Carpathian lamprey (61–67; see Docker 2009), but >2% sequence divergence at the cytochrome b gene. These two species are more genetically distinct from their presumed ancestor than other paired species (Lang et al. 2009), although the widely distributed Ukrainian brook lamprey and Carpathian lamprey are still not reciprocally monophyletic (Levin et al. 2016);

- western brook lamprey, in which some populations (e.g., those in Alaska and British Columbia) are genetically indistinguishable from western river lamprey where their ranges overlap, but other populations (even excluding the four highly divergent populations from Oregon and California; Fig. 4.1) show increasingly greater divergence (up to 2.3%) from western river lamprey (Boguski et al. 2012). Number of trunk myomeres in this species also appears to differ among populations (Reid et al. 2011), although there are consistently fewer in western brook lamprey (57–67) relative to western river lamprey (63–71; Docker 2009; Reid et al. 2011).

(3) Species (n = 11) showing both higher genetic (≥2.9% K2P) and morphological (2.5–13 fewer myomeres) divergence when compared to the most closely related extant parasitic species (see Sect. 4.6.2.1).

We recognize that this categorization is imperfect because we are trying to divide a speciation continuum into discrete categories using only one gene and one morphological character. However, the first two categories include most or all of the recognized paired brook lamprey species, and the third category roughly corresponds with the relict species (see Sect. 4.6.2.1). The Mexican brook lamprey would presumably fall into category 1 or 2; it was excluded from analysis here because cytochrome b gene sequence is not available for the parasitic Mexican lamprey (Lang et al. 2009), and myomere counts are not available for each species individually (Renaud 2011). Thus, by this approach, 18 brook lamprey species would be considered as paired (or recently derived) species, including 15 of the 17 species recognized by Potter et al. (2015)—all but the Turkish and Pacific brook lampreys which were placed into category 3—and the three Portuguese brook lamprey species. It is not our intention here to redefine the terms "paired" and "relict" species per se, but merely to facilitate discussion regarding steps in the evolution of non-parasitism.

A continuum is evident within the categories as well. In addition to variation in the degree to which different brook lamprey species have diverged from their parasitic ancestor, there are likely also differences among populations within species. Some apparent intraspecific differences may be the result of unrecognized diversity within nominal species. Some brook lampreys currently considered a single widely

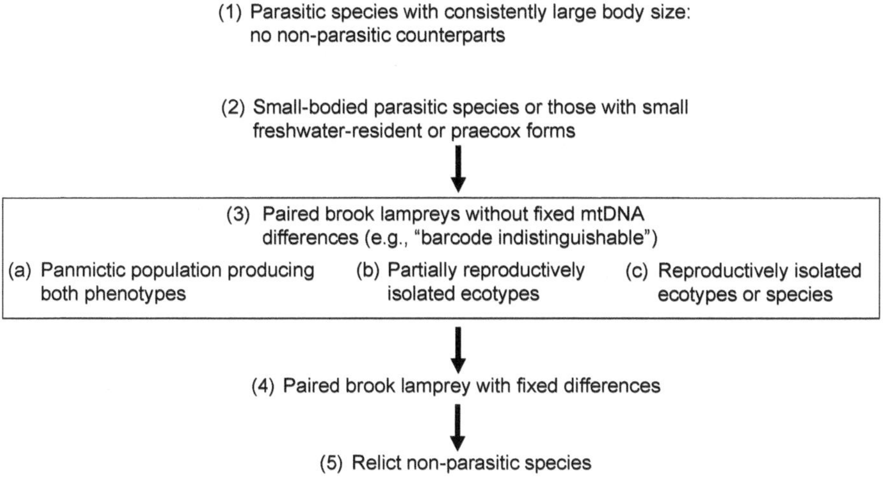

(1) Parasitic species with consistently large body size:
    no non-parasitic counterparts

(2) Small-bodied parasitic species or those with small
    freshwater-resident or praecox forms

$\downarrow$

┌───┐
│   (3) Paired brook lampreys without fixed mtDNA   │
│   differences (e.g., "barcode indistinguishable")   │
│ (a) Panmictic population producing   (b) Partially reproductively   (c) Reproductively isolated │
│   both phenotypes   isolated ecotypes   ecotypes or species │
└───┘

$\downarrow$

(4) Paired brook lamprey with fixed differences

$\downarrow$

(5) Relict non-parasitic species

Fig. 4.3 Representation of speciation continuum seen in lampreys, ranging from (1) parasitic species with no non-parasitic counterparts to (5) relict non-parasitic species that have long diverged from a parasitic counterpart. Updated from Docker (2009) to more accurately show that (3) paired species lacking fixed differences in mitochondrial DNA (mtDNA) sequence may also include different species and populations at different stages of speciation; resolution among these subcategories are possible only with higher-resolution markers

distributed species may consist of different populations that evolved from the parasitic ancestor at different times (i.e., are polyphyletic; see Docker 2009) or that diverged as the result of vicariance following the evolution of non-parasitism (see Sect. 4.6.2.3). Spatially disjunct non-parasitic derivatives of the Arctic lamprey are traditionally recognized as four distinct species (or three, with Far Eastern and Siberian brook lampreys perhaps being synonymous; see Renaud and Naseka 2015; Yamazaki and Goto 2016), but different populations of other widespread species are often "lumped" together. Mateus et al. (2013a) described three evolutionarily distinct units within the European brook lamprey as distinct species, and more such populations (whether or not they are considered distinct Linnean species) may also exist (e.g., Pereira et al. 2014; see Chap. 7). Similar "splitting" might also be warranted in the widely distributed western brook lamprey (Boguski et al. 2012) and Ukrainian brook lamprey (Levin et al. 2016). A genetically divergent brook lamprey from the Aegean Sea basin, for example, may represent a tentative new species (Levin et al. 2016), and, as with other widespread non-parasitic species, the *Eudontomyzon* species complex requires further taxonomic examination.

In other cases, however, it is not merely a matter of phylogenetically distinct populations having escaped taxonomic notice to date. Different populations may be at different stages of divergence as the result of different demographic histories and different levels of contemporary gene flow. Within category 1, for example, further subdivision appears to exist that is not evident with mtDNA sequence data alone (Fig. 4.3):

(a) Panmictic populations producing both phenotypes: There have been reports of parasitic individuals arising within otherwise non-parasitic populations, most notably, the "giant" American brook lamprey in the Great Lakes (Manion and Purvis 1971; Cochran 2008) and the Morrison Creek lamprey on Vancouver Island (Beamish 1985, 1987; Beamish et al. 2016), as well as parasitic populations which apparently produce non-parasitic individuals (e.g., Kucheryavyi et al. 2007; Yamazaki et al. 2011; see Docker 2009). However, we do not yet know whether these polymorphic populations result from phenotypic plasticity or whether feeding type represents a genetically based polymorphism within a population with no barriers to interbreeding (see Sect. 4.6.3.3).

(b) Partially reproductively isolated ecotypes, where there is evidence of contemporary gene flow between sympatric parasitic and non-parasitic lampreys (e.g., at microsatellite loci), but genomic regions that consistently differ between the forms suggest a genetic basis to feeding type (e.g., Rougemont et al. 2015, 2017; see Sect. 4.6.3.2). Distinctly different phenotypes are maintained, even with extensive gene flow. These cases may represent the early stages of (incipient) speciation if there is selection against intermediate phenotypes, although it is not a foregone conclusion that full reproductive isolation would eventually result, nor is this evidence for divergence in sympatry (see Sect. 4.6.3.4).

(c) Reproductively isolated ecotypes (e.g., in parapatric populations), where there is a genetic basis to feeding type (Mateus et al. 2013b; Rougemont et al. 2017) and significant differentiation at microsatellite loci (e.g., Rougemont et al. 2015; Mateus et al. 2016) that indicates a lack of gene flow and separate evolutionary trajectories, but insufficient time has elapsed for morphological or genetic differences to become fixed.

The use of higher-resolution markers, especially non-neutral genome-wide markers, and modern population genetic and genomic analyses are allowing better resolution of these sub-categories. The remainder of this section will focus on reviewing these studies and their implications to our understanding of the evolution of non-parasitism in lampreys.

4.6.3.2 Incomplete Reproductive Isolation and Contemporary Gene Flow in Some Pairs

It has generally been thought that size-assortative mating would result in immediate reproductive isolation in most paired species (Hardisty and Potter 1971a; Beamish and Neville 1992) and that temporal, spatial, or behavioral isolation may further reduce or prevent gene flow (see Docker 2009). However, there is increasing evidence that size-assortative mating is an insufficient pre-zygotic barrier to hybridization and that reproductive isolation between paired species is not complete in sympatry. Previous reports of mixed-species spawning aggregations (Huggins and Thompson 1970; Kucheryavyi et al. 2007; Cochran et al. 2008) have been supported by further observations of paired species spawning together in the wild (Lasne et al. 2010).

Size-assortative mating may provide less of a barrier where size differences between species are relatively small (see Docker 2009; Rougemont et al. 2016), and, even with larger-bodied parasitic species, evidence of sneak (or satellite) male mating tactics may permit non-parasitic males to fertilize the eggs of large parasitic females. Satellite male mating behavior was reported previously within European and American brook lampreys (Malmqvist 1983; Cochran et al. 2008), and it has recently been demonstrated in mixed-species aggregations of European river and brook lampreys (Hume et al. 2013b). Lack of hybrid inviability in the early developmental stages (e.g., Piavis et al. 1970; Beamish and Neville 1992; see Docker 2009) has likewise been confirmed in recent studies (Hume et al. 2013c; Rougemont et al. 2015), showing lack of immediate and obvious post-zygotic reproductive isolation in paired species.

The study by Rougemont et al. (2015) was particularly interesting, because it used genetic parentage analysis to evaluate fertilization success when European river lamprey females (n = 2) were provided with simultaneous access to European brook (n = 4) and European river (n = 2) lamprey males. Reproductive success of the European brook lamprey males with European river lamprey females was relatively low, but it was not negligible. Of the 73 offspring assigned without ambiguity, 81% were sired by European river lamprey, and 19% were sired by European brook lamprey. Granted, without the presence of brook lamprey females, the proportion of interspecific matings may have been over-estimated if brook lamprey males were "forced" to mate with heterospecific females, and it is possible that some gamete mixing occurred during strict size-assortative mating. Nevertheless, the potential for substantial contemporary gene flow in sympatry was demonstrated. It is also interesting to speculate that, even with reduced fertilization success relative to European river lamprey males, siring 19% of the offspring from more fecund European river lamprey females (~20,000–35,000 eggs) could equate to higher reproductive success for European brook lamprey males (~3,800–6,650 offspring)—barring any selection against hybrids—than siring 100% of the offspring from less fecund European brook lamprey females (~1,500–2,000 eggs; see Chap. 1). However, we do not know if brook lamprey males could fertilize even close to all the eggs from a river lamprey female (i.e., whether sperm would be limited, whether they would be driven off by river lamprey males, or whether they would fail to induce female river lamprey to release their eggs). The relative testis size (the gonadosomatic index) of brook lamprey males is higher than that of parasitic species (10 and 4%, respectively), but, given the differences in body size, absolute testis size is still much greater in parasitic species (1.3 and 0.5 g in European river and brook lampreys, respectively; see Chap. 1).

Furthermore, "barring any selection against hybrids" is an important caveat. Many studies have demonstrated that the survival of hybrids between lamprey paired species is equivalent to that of pure individuals for the first few weeks following fertilization (e.g., Enequist 1937; Piavis et al. 1970), but there is virtually nothing known regarding possible selection against hybrids later in development. Only one study to date is known where hybrids between paired species were reared for more than a few weeks after hatch. Beamish and Neville (1992) reared hybrids between western river and western brook lampreys for 2.5 years, but, unfortunately, the experiment was terminated prior to metamorphosis. We therefore do not know what would happen in hybrids at metamorphosis (when the developmental trajectories of the parental species diverge so dramatically) or at spawning (e.g., in terms of mating behavior and viability). Furthermore, hybrid incompatibilities are best revealed in subsequent (F2) generations or in backcrosses (i.e., when the first generation, F1, hybrid is mated with one of parental species; Bierne et al. 2002, 2006). Therefore, it is clearly premature to suggest that there is no post-zygotic reproductive isolation in paired lamprey species. Evaluating the fitness of experimentally generated hybrids over at least two generations would be very difficult to accomplish entirely in the laboratory (see Chap. 2). However, Rougemont et al. (2017), using Restriction site Associated DNA Sequencing (RAD-Seq) to identify a small set of loci that were highly differentiated between European river and brook lampreys (40 of 8,962 SNPs; see Sect. 4.6.3.3), were able to identify putative hybrids among wild-caught individuals. Among 338 individuals genotyped, these authors found evidence of 22 hybrids (6.5%), 20–21 of which were F1 hybrids. The virtual absence of later-generation hybrids suggests some form of hybrid breakdown (e.g., reduced survival or fertility of the hybrids), and this warrants further study.

Nevertheless, behavioral studies showing the potential for interbreeding in paired species is consistent with recent population genetic studies that show contemporary gene flow in sympatry (e.g., Docker et al. 2012; Bracken et al. 2015; Rougemont et al. 2015). However, it is very important in such studies to distinguish between true sympatry, where the two species come into contact, and situations where they are found in the same basin or river systems but with no opportunity for contemporary gene flow (i.e., where they are parapatric). Using microsatellite loci, Docker et al. (2012) demonstrated a lack of significant genetic differentiation (F_{ST} 0) between silver and northern brook lampreys where they were collected from the same rivers in the Lake Huron basin, but the two species were significantly differentiated (F_{ST} 0.067) in the Lake Michigan basin where northern brook lamprey were collected almost exclusively from the eastern shores of the basin and silver lamprey were collected from the western arm of the lake. Similar patterns have been seen in recent studies investigating the level of gene flow between European river and brook lampreys from multiple locations in the British Isles and northern France that varied in their level of connectivity. Bracken et al. (2015) found evidence of ongoing gene flow between European river and brook lampreys where they occurred sympatrically (in the Loch Lomond basin, F_{ST} 0.019), but five parapatric populations (where brook lampreys were isolated above barriers to migration) showed higher levels of genetic differentiation (mean F_{ST} 0.073). Rougemont et al. (2015) sampled five sympatric and five parapatric European river and brook lamprey population pairs and likewise found little or no genetic differentiation where they occurred sympatrically (i.e., no significant genetic differentiation in one population, F_{ST} 0.008, and significant but low levels of differentiation in four populations, mean F_{ST} 0.055) and higher levels of differentiation (mean F_{ST} 0.113) in parapatry. The significant differentiation observed in all but one sympatric population argues against phenotypic plasticity in a completely panmictic population, but evidence of contemporary gene flow in sympatry shows that reproductive isolation is incomplete. Interestingly, in the one sympatric population in France (the Oir River population) where there was no significant genetic differentiation between the two forms, European river lamprey were much smaller (mean TL 225 mm) than at other sites (mean TL 303 mm). Rougemont et al. (2016) suggested that the smaller size difference between European river and brook lampreys in this river may have facilitated interbreeding of the two species.

Beamish et al. (2016) also presented evidence for gene flow in sympatry between the two forms of the western brook lamprey in Morrison Creek on Vancouver Island, where the normally non-parasitic population also produces a potentially parasitic "silver" form (Beamish 1985, 1987; see Docker 2009). Beamish et al. (2016) found no significant genetic differentiation between the forms (i.e., no evidence of even partial reproductive isolation).

Microsatellite loci provide much higher resolution than mtDNA loci (Selkoe and Toonen 2006). Even non-parasitic populations that appear to have been derived very recently show evidence of genetic differentiation at microsatellite loci when no longer in sympatry. Yamazaki et al. (2011) present evidence that non-parasitic lamprey populations have evolved from the anadromous Arctic lamprey in two rivers in the Agano River system in Japan when dam construction ~90 years ago isolated them from the anadromous population. These newly founded populations show strong genetic differentiation (F_{ST} 0.433–0.635) when compared to the parapatric Arctic lamprey, but a non-parasitic population that is not isolated above dams (i.e., is sympatric with the Arctic lamprey) was not significantly differentiated from Arctic lamprey.

Few examples exist of closely related parasitic and non-parasitic lampreys that show substantial barriers to gene flow in sympatry. One such example is the European river and brook lamprey pair in the Sorraia River in the Tagus River basin in southern Portugal (Mateus et al. 2016). Like all of the examples discussed in this section, this pair does not show species-specific differences in mtDNA sequence (Mateus et al. 2011), but significant and high levels of genetic differentiation (F_{ST} 0.317) have been demonstrated with the use of microsatellite loci (Mateus et al. 2016). This pair appears to be truly sympatric (i.e., collected from a common spawning site; Mateus et al. 2013b) but appears not to experience ongoing gene flow. The Tagus River basin is near the southern limit of distribution for European river and brook lampreys, and the climate here has been stable over longer periods of time than in the British Isles and northern France (Bracken et al. 2015; Rougemont et al. 2015). In northern Europe, recolonization following glacial retreat may have brought these species into contact before reproductive isolating mechanisms were fully established, enabling gene flow in sympatry (see Sect. 4.6.3.4). In contrast, reproductive isolating mechanisms may have had time to evolve in the Portuguese population. Evolutionary theory predicts that there will be selection for pre-zygotic reproductive isolating mechanisms when hybridization is maladaptive (Ortiz-Barrientos et al. 2009). Pre-zygotic reproductive isolation appears complete between *Lethenteron* sp. N and sp. S that occur sympatrically in the Gakko River in Japan (Yamazaki and Goto 2000, 2016). Although both species are non-parasitic (i.e., they are not paired species), they are genetically very divergent, and hybridization would presumably be maladaptive. Yamazaki and Goto (2016) observed no temporal isolation or size differences that would prevent interbreeding between these two non-parasitic species, but no mixed-species nests were observed (although communal single-species spawning was observed). The extent to which hybridization between closely related paired species might be maladaptive is unknown (see above).

Despite growing evidence for contemporary gene flow when paired parasitic and non-parasitic species occur sympatrically, at least in the more northerly parts of their range, it is important to recognize that they still maintain highly distinct phenotypes. Recent studies are showing that introgression at neutral markers (e.g., microsatellite loci) does not preclude differentiation at a restricted number of loci related to feeding type (see Sect. 4.6.3.3).

4.6.3.3 Genetic Basis of Feeding Type

Multiple phenotypes within a single species are common in a wide range of organisms. In some organisms, these do indeed represent phenotypic plasticity (i.e., polyphenisms) where the different phenotypes are produced from a single genotype under the induction of an environmental cue (e.g., Greene 1999; Hoffman and Pfenning 1999; Shine 2004; Podjasek et al. 2005). However, in many other cases (e.g., populations of rainbow trout where individuals adopt either a freshwater-resident or anadromous life history type), they appear to be (at least partially) genetically based polymorphisms (Hale et al. 2013; Hecht et al. 2013).

Phenotypic plasticity with respect to feeding type has been suggested in lampreys. For example, Kucheryavyi et al. (2007) suggested that larval growth conditions determine whether individuals in an Arctic lamprey population become parasitic or non-parasitic at metamorphosis. These authors proposed that individuals that accumulate a sufficient quantity of energy resources during the larval stage are able to mature without post-metamorphic feeding and become non-parasitic. A lack of species-specific differences in mtDNA gene sequence, although rare in vertebrates, is not evidence of phenotypic plasticity (see Docker 2009). Even the existence of polymorphic populations that appear panmictic (i.e., freely interbreeding, with no apparent barriers to gene flow) using microsatellite loci does not necessarily indicate that feeding type lacks a genetic basis. Mitochondrial and microsatellite markers are generally considered neutral markers (i.e., indicators of historical and recent or contemporary gene flow, respectively; Avise 2000; Selkoe and Toonen 2006), and introgression at neutral markers does not preclude differentiation at "genomic islands" related to feeding type (see below).

The western brook lamprey population in Morrison Creek on Vancouver Island produces both a potentially parasitic "silver" form and the typical non-parasitic form and appears to be panmictic (Beamish et al. 2016; see Sect. 4.6.3.2). However, we are not yet able to distinguish between phenotypic plasticity and a genetically based polymorphism within a freely interbreeding population. Interestingly, the abundance of the parasitic form has greatly diminished since the 1980s (i.e., comprising ~65% of the total catch in 1981 and 1987, but only 8% in 2011–2012). Over this time, the average length of the silver form has stayed the same (125 mm TL), but mean TL of the non-parasitic form has increased from 116 to 131 mm. Although we could argue that better growth during the larval stage permitted more individuals to mature without feeding, whether this shift in the proportion of the two phenotypes might be due to changes in the environmental cues inducing phenotype or selection on genetically based phenotypes is not resolvable at this point. Potentially parasitic individuals may "spontaneously" appear in other western brook lamprey populations. Jolley et al. (2016) reported capturing a western river lamprey outmigrant above the John Day Dam, the third upriver mainstem dam on the Columbia River located 348 rkm from the ocean, and a few transformed western river lamprey are collected each year in the lower Yakima River >530 km from the ocean (Ralph Lampman, Yakama Nation, Fisheries Resources Management Program, Toppenish, WA, personal communication, 2018). No freshwater-resident parasitic river lamprey have been observed in these areas, and the return of anadromous western river lamprey adults would be highly unlikely. However, we cannot distinguish between phenotypic plasticity and expression of an otherwise recessive genetic trait. Likewise, the appearance of a self-sustaining, non-migratory, non-parasitic population when anadromous Arctic lamprey are isolated above dams (Yamazaki et al. 2011) is not evidence of phenotypic plasticity, because existing genetic variation within the population cannot be excluded (see Sect. 4.6.3.4).

A few recent studies have found putative functional loci that differ between paired parasitic and non-parasitic species, providing evidence for a genetic basis for life history type. Yamazaki and Nagai (2013) found a significant signature of directional selection in a non-parasitic lamprey population that has been recently derived from the anadromous Arctic lamprey (see Sect. 4.6.3.2); one microsatellite locus exhibited a much higher degree of differentiation (F_{ST} 0.701–0.914) between life history types than the other six loci tested. This locus was estimated to be ~5,800 nucleotides from the vasotocin precursor gene, which plays an important role in osmoregulation. Yamazaki and Nagai (2013) thus suggested that there has been recent and strong natural selection related to the transition from anadromy to freshwater residency (i.e., that this particular difference was related to migratory rather than feeding type), and that selection was detected at the microsatellite locus due to a "hitchhiking effect" of the selective forces around the gene region.

In a groundbreaking study, Mateus et al. (2013b) used RAD-Seq to survey for genome-wide differences in European river and brook lampreys from the Sorraia River in Portugal. RAD-Seq is a reduced-representation genome sequencing strategy (i.e., rather than whole genome sequencing) designed to interrogate ~0.1–10% of the genome. Mateus et al. (2013b) recovered >8,000 polymorphic RAD loci and almost 14,700 SNPs. Of these, they found 166 loci fixed for different alleles between European river and brook lampreys (i.e., 166 species-specific differences in the genomes of these two species). This was the first study to show species-specific differences between European river and brook lampreys, at a time when the observation that most lamprey species pairs were "barcode indistinguishable" was sometimes interpreted as meaning that there were no genetic differences between paired species (see Docker 2009; Artamonova et al. 2011). However, a subsequent study by these authors showed that European river and brook lampreys from this population could also be differentiated at neutral microsatellite loci (F_{ST} 0.317; Mateus et al. 2016), suggesting that not all of the 166 fixed loci were necessarily correlated with life history type. With barriers to gene flow, genetic differentiation due to drift or selection on other traits would also be expected. Nevertheless, Mateus et al. (2013b) were able to link 12 of the 166 loci to genes that had been annotated in the sea lamprey genome (Smith et al. 2013): the vasotocin gene, the same gene implicated in migratory type adaptation by Yamazaki and Nagai (2013); gonadotrophin-releasing hormone 2 (GnRH2) precursor; four genes related to immune function; three genes related to axial patterning; a pineal gland-specific opsin; a voltage-gated sodium channel gene; and a tyrosine phosphate gene. Evidence of a species-specific SNP in the GnRH2 precursor was interesting, because GnRH is found at the top of the hypothalamic-pituitary axis in all vertebrates and is a key regulator of gonadal development and differentiation (Sower 2015), the timing of which differs between parasitic and non-parasitic lampreys (see Docker 2009; Chap. 1). Differences in genes related to axial patterning could potentially be related to differences in the number of trunk myomeres in parasitic and non-parasitic lampreys or other aspects of development related to the ultimate differences in adult body size (see Irvine et al. 2002; Childs 2013). The pineal gland-specific opsin gene may be an important regulator of the photosensitive pineal gland which is involved in the photoperiodic control of sexual maturation in adult lampreys (Joss 1973; Yokoyama and Zhang 1997) and which might play a role in metamorphosis (Cole and Youson 1981; see Manzon et al. 2015). Most of the 166 species-specific loci still remain to be annotated, but this preliminary list of candidates serves as a very important first step in identifying genes involved in evolution of the non-parasitic European brook lamprey from the parasitic anadromous European river lamprey.

In a subsequent study, Rougemont et al. (2017) likewise used RAD-Seq data from European river and brook lampreys, but they performed population genomic analyses using nine replicated pairs experiencing different degrees of gene flow (F_{ST} 0.008–0.189; Rougemont et al. 2015). This approach allowed these authors to disentangle the effects of selection from those of genetic drift. In sympatric pairs showing high genetic connectivity, most of the genome would be expected to show strong introgression (i.e., with little or no differentiation at neutral loci), and only regions of the genome involved in reproductive isolation and local adaptation would be expected to show strong differentiation. Rougemont et al. (2017) identified 40 SNPs that were highly differentiated between European river and brook lampreys (i.e., a small number of highly differentiated "genomic islands") amid a background or "sea" of less differentiated loci. Furthermore, 28 outlier loci (i.e., those most highly differentiated between life history types) were shared in the four population pairs showing high genetic connectivity, and this amount of sharing was higher than expected by chance alone. Homology searches for these outlier loci identified some of the same candidate genes as those found by Mateus et al. (2013b), that is, the GnRH2 precursor gene, the pineal gland-specific opsin gene, and genes involved in immunity and axial patterning. However, that these genes were correlated with the brook lamprey phenotype in multiple pairs is not necessarily the result of parallel and independent evolution in each pair. Rather, the apparent genetic parallelism is likely the result of a common history of divergence initiated in allopatry followed by secondary contact in the different populations. Secondary gene flow would have eroded past divergence at variable rates across the genome, but those loci associated with life history type appear to have resisted introgression (see Sect. 4.6.3.4).

In the above studies, feeding type and migratory type were always confounded, because the evolution of non-parasitism also involved a switch from anadromy to freshwater residency. In contrast, Hume et al. (2018) recently used RAD-Seq data to infer the demographic history of three life history types in Loch Lomond: anadromous European river lamprey, freshwater-resident European river lamprey, and European brook lamprey. In this manner, it might be possible to disentangle the effects of the anadromous to freshwater transition from the parasitic to non-parasitic transition. Outlier genes associated with the migratory type transition included those related to immune function (*nckap-1*; Zhou et al. 2017) and growth (*cd109*; Hockla et al. 2010), and genes broadly associated with embryonic development (e.g., *reck, scn4aa, rev31*; Wittschieben et al. 2000; Yamamoto et al. 2012) were implicated in the transition from parasitism to non-parasitism. This does not mean, of course, that the SNPs identified in these outlier genes are the causal mechanisms for the transition from an anadromous or freshwater-resident parasitic lamprey to a non-parasitic lamprey, but these three RAD-Seq studies serve as important first steps in elucidating the genetic mechanism of life history evolution in lampreys.

Although the above studies have refuted the hypothesis of phenotypic plasticity, at least in the European river and brook lampreys populations examined, the results of a 10-year study designed to test the heritability of feeding type in silver and northern brook lampreys deserves mention here (Neave et al. 2019). These authors tested for feeding type plasticity using two approaches. The first approach used a common garden experiment to determine if raising offspring from each species under common laboratory conditions would produce the parental phenotype regardless of conditions (suggesting a genetic component to feeding type) or induce the alternative feeding type (indicating phenotypic plasticity). The second approach used a transplant experiment to determine whether placing larvae of known parentage into streams which appear conducive to the development of the alternative feeding type (as determined by comparison of abiotic and biotic characteristics in streams inhabited by the two species) would result in production of the alternative feeding type. In short, 100% larval mortality by 3 months post-hatch in each of 3 years necessitated termination of the common garden experiment. In the transplant experiments, >12,000 larvae were stocked into 10 stream reaches in the Lake Huron basin, and post-metamorphic individuals of the alternative feeding type were recaptured 4–5 years later in two streams. However, transplantation was only permitted in streams already containing *Ichthyomyzon* larvae, and genetic parentage analysis indicated that the recovered individuals were not offspring of the original known-phenotype parents. Thus, phenotypic plasticity was not demonstrated. This was the first known study to attempt a common garden experiment or transplant study through metamorphosis. Even with improvements to artificial propagation procedures (see Chap. 2), a direct repeat of this study is likely not warranted, given the recent studies indicating a genetic basis to feeding type. However, rearing studies combined with a genomic approach could be powerful. For example, with the identification of specific loci associated with life history type, we could more easily study the phenotype and fate of hybrids. Of the 22 hybrids detected by Rougemont et al. (2017), 64% displayed the river lamprey phenotype, but whether this is due to inheritance patterns or selection against hybrids with the brook lamprey phenotype is unknown. Whether parasitic lamprey parents could produce non-parasitic offspring (e.g., if both parents possessed recessive genes for "non-parasitism") is likewise unknown.

Furthermore, although the emphasis to date has largely been on the morphological and developmental "parting of the ways" observed at metamorphosis, parasitic and non-parasitic lampreys show different developmental trajectories related to ovarian development, and these differences precede metamorphosis by several years (see Docker 2009). In most species studied to date, ovarian differentiation occurs at ~1 year of age in non-parasitic species and at 2–3+ years of age in parasitic species; the larger size at the onset of oogenesis in parasitic species is presumably responsible for the greater number of oocytes elaborated in these ultimately larger-bodied species (see Chap. 1). Thus, as suggested by Hardisty (1964) and Beamish and Thomas (1983), fecundity differences among species are very likely genetically based and largely determined at or before sex differentiation. Rearing offspring from paired species and hybrids through ovarian differentiation would shed light on the earliest point at which the developmental trajectories of paired species diverge, the genes involved in this process, and possible genetic incompatibilities in hybrids (see Mavarez et al. 2009; Renaut and Bernatchez 2011).

In addition, in recently diverged pairs or those experiencing gene flow, there may be a gradual (rather than immediate) acquisition of traits associated with non-parasitism. Despite the general observation that non-parasitic species initiate oogenesis earlier than their parasitic counterpart, a recent study by Spice and Docker (2014) suggested that some species are polymorphic with respect to timing of oogenesis and the resulting number of oocytes produced. In northern brook and chestnut lampreys, ovarian differentiation occurred in age classes I and II in both species (with no significant differences in the timing between species), and northern brook and chestnut lampreys had similar minimum oocyte counts. Granted, maximum oocyte counts were higher in chestnut lamprey larvae, a similar pattern observed by Neave et al. (2007) when comparing presumptive northern brook and silver lampreys, but these results suggest that all of the changes associated with the evolution of non-parasitism may not occur simultaneously. Changes in the phasing of gonadogenesis might not always coincide with the elimination of the parasitic feeding phase. Parasitic lamprey populations may already be polymorphic for this trait (i.e., with evolution of non-parasitism drawing on existing genetic variation within the population), or, conversely, selection for earlier ovarian differentiation and lower potential fecundity may follow the elimination of the parasitic feeding phase. The sequence of changes associated with the evolution of non-parasitism—and a better understanding of the mechanisms of life history evolution in lampreys—can be better resolved now with our improved understanding of where different species and populations fit on the continuum.

4.6.3.4 Demographic History of Divergence and Origin of Genetic Parallelism

Given the morphological similarity and lack of mtDNA sequence differentiation generally observed between lamprey species pairs in sympatry, particularly those in recently deglaciated regions, it is often assumed that divergence happened very recently (e.g., Hubbs and Potter 1971; Salewski 2003; Docker 2009). Although molecular phylogenetic studies that fail to find species-specific genetic differences and lack of reciprocal monophyly in sympatric pairs often add the caveat that genetic similarity may also result from introgression following secondary contact after divergence in allopatry (see Taylor 1999; Espanhol et al. 2007), this caveat is sometimes forgotten. Thus, it is often thought that most paired brook lamprey species have evolved from their parasitic ancestor in post-glacial times (e.g., Beamish and Withler 1986; Docker 2009), with the two distinct phenotypes rapidly differentiating in sympatry as a result of disruptive selection (Salewski 2003).

Previous studies have lacked the resolution to distinguish between primary, but recent or ongoing, divergence in sympatry and secondary gene flow following initial divergence in allopatry. However, recent studies by Rougemont et al. (2016, 2017) have used population genetic and genomic analyses to reconstruct the demographic history of divergence between European river and brook lampreys, and their results make us question past hypotheses regarding modes of speciation in lampreys. Using 13 microsatellite loci and an approximate Bayesian computational (ABC) approach combined with a random forest model, Rougemont et al. (2016) tested different scenarios of divergence in six replicated populations of European river and brook lampreys from northern France. These six pairs were all highly connected by gene flow: five populations were either truly sympatric or the two species were found in close proximity and not separated by permanent barriers to migration, and one population was parapatric but showed gene flow comparable to that of the sympatric populations. These authors statistically compared five alternative models for the divergence between each pair of species: (1) a model of panmixia (PAN) in which European river and brook lampreys within a population constitute a single gene pool; (2) a strict isolation model (SI) in which European river and brook lampreys diverged T_{DIV} years ago with no subsequent gene flow; (3) an ancient migration model (AM) in which European river and brook lampreys diverged T_{DIV} years ago but continued to experience gene flow up until T_{ISOL} years ago; (4) an isolation with migration model (IM) in which European river and brook lampreys diverged T_{DIV} years ago but continue to experience contemporary gene flow; and (5) a model of secondary contact (SC) after past isolation, where European river and brook lampreys diverged in allopatry T_{DIV} years ago and then started experiencing gene flow following secondary contact T_{SC} years ago (Fig. 4.4). The strict isolation (SI) and ancient migration (AM) models were both rejected for all six populations. Rejection of the SI model refutes the hypothesis that reproductive isolation between parasitic and non-parasitic lampreys would be immediate and complete as soon as elimination of the adult feeding phase produced differences in body size at maturity. Rejection of the AM model is consistent with the results showing that these populations are experiencing ongoing (contemporary) gene

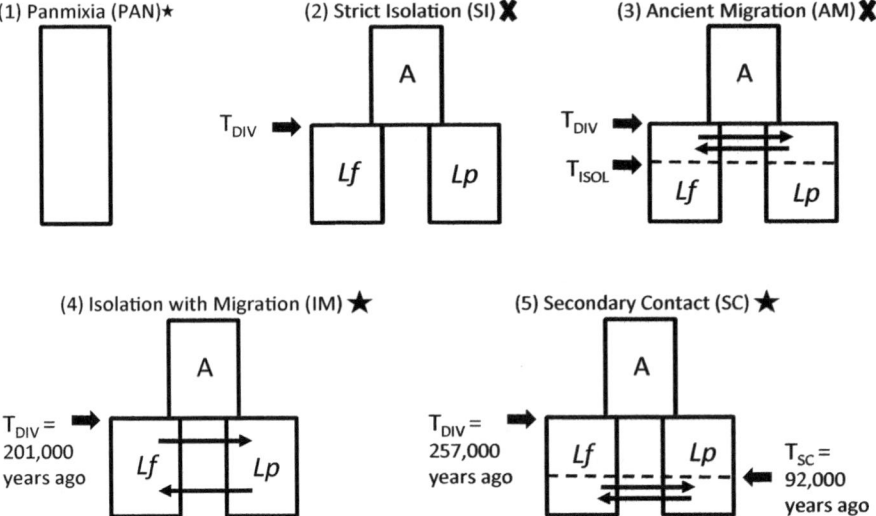

Fig. 4.4 Models of divergence between European river lamprey *Lampetra fluviatlis* (*Lf*) and European brook lamprey *La. planeri* (*Lp*) tested by Rougemont et al. (2016). T_{DIV} is number of generations since divergence; T_{ISOL} is generations since the two life history types stopped exchanging genes; T_{SC} is generations since the life history types entered into secondary contact following a period of isolation. PAN was the best-supported model for one of population (*asterisk*), and IM and SC were both well-supported models for five populations (*star*); SI and AM were rejected for all six populations (*X*). Estimates of T_{DIV} (expressed in years, assuming generation time of 5 years) and T_{SC} under the IM and SC models (median for the five populations) are given

flow; in fact, these populations were chosen because they were highly connected by gene flow. Of the remaining models, panmixia (PAN) was the best-supported model for one population; this was the Oir River population, where no significant genetic differentiation was previously found between the forms (Rougemont et al. 2015; see Sect. 4.6.3.2). Isolation with migration (IM) and secondary contact (SC) were the best-supported models in the remaining five populations, although it was not possible to distinguish between them. These two models could also not be ruled out in the Oir River population (at least not with neutral markers; see below).

The divergence time estimates generated by Rougemont et al. (2016) provided exciting insights into the tempo and mode of evolution in paired lamprey species. Assuming a generation time of 5 years for these species, these authors estimated that European river and brook lampreys in the five populations best characterized by the IM and SC models diverged on average 201,760 and 257,040 years ago, respectively (or 282,464 and 359,856 years ago, respectively, assuming a generation time of 7 years). This is completely inconsistent with recent and rapid divergence following the recent glacial retreats ~10,000–15,000 years ago. Even the bottom end of the 95% confidence interval from each of the individual populations (24,200 years ago) shows T_{DIV} higher than predicted if divergence occurred in post-glacial times. Also surprisingly, in the SC model, the median time at which gene flow resumed following secondary contact (T_{SC}) was 92,960 years ago. This suggests that not even secondary contact occurred in post-glacial times. The ancient secondary contact suggested by the T_{SC} estimate implies that the genetic signature of historical geographic isolation carried by neutral markers has been lost, which would explain the difficulty these authors experienced distinguishing between the SC and IM models (Rougemont et al. 2016). It also suggests that the period of isolation (i.e., between T_{DIV} ~200,000–250,000 years ago and T_{SC} ~90,000 years ago) was too short to allow genetic incompatibilities to accumulate and strong barriers to gene flow to develop.

However, with neutral markers, it can be difficult to distinguish between primary divergence (i.e., the SI, IM, or AM models) and secondary contact (SC), because both scenarios tend to converge to the same equilibrium (Bierne et al. 2013). Thus, Rougemont et al. (2017) subsequently used RAD-Seq data and a diffusion approximation approach to infer the demographic history of each of four sympatric and five parapatric European river–brook lamprey pairs. In all of the sympatric pairs, their analyses supported a model of secondary contact (SC) after initial divergence in allopatry (including the Oir River pair where their 2016 study suggested panmixia). In contrast, parapatric pairs have retained a signal of ancient migration (AM). The AM model was rejected in Rougemont et al. (2016) where only sympatric populations were included, but Rougemont et al. (2017) indicate that sympatric versus parapatric populations do not necessarily have radically different divergence histories; the signal of past secondary contact may have been lost or obscured in parapatric populations as result of recent drift.

In all nine pairs examined by Rougemont et al. (2017), models accounting for differential introgression among loci (i.e., incorporating heterogeneity in divergence along the genome) outperformed homogeneous migration models. As discussed above (see Sect. 4.6.3.3), there does not appear to be uniform gene flow across the genome; rather, regions of the genome involved in reproductive isolation and local adaptation appear to resist the homogenizing effect of introgression. Rougemont et al. (2017) found that 6–12% of loci in the most genetically connected pairs displayed a reduced effective migration rate between the life history types; during secondary contact, erosion of past genetic differentiation outside the direct vicinity of these "barrier loci" would result in low levels of genetic differentiation elsewhere in the genome. Only European river and brook lampreys have been studied to date using this approach, but the demographic histories of divergence should be tested in other paired species using this approach. Interestingly, Hubbs and Trautman (1937) suggested that the *Ichthyomyzon* brook lamprey species originated before the last glacial advance in North America, although Hubbs and Potter (1971) subsequently considered that a more recent origin was equally possible.

An extension of the "recent divergence in sympatry" hypothesis (although less explicit) is the assumption that widely distributed brook lamprey species evolved through this mechanism independently and repeatedly in disjunct locations (e.g., Beamish and Withler 1986; Docker 2009). Molecular phylogenetic analyses suggest that many widespread brook lamprey species are polyphyletic (i.e., derived from two or more ancestral sources), and their distribution in disjunct drainages separated by salt water argued against dispersal (at least via current connections) following a single brook lamprey origin. However, it is important to recognize that current connections between drainages do not reflect past connections. For example, stream capture can transfer some portion of an aquatic fauna into a new drainage, and rivers isolated from each other by marine or estuarine conditions can join farther out on the continental shelf if sea levels are lowered (Hughes et al. 2009). Furthermore, the current geographic distribution of contemporary species may not reflect the initial conditions of divergence (Bierne et al. 2011), and the results of Rougemont et al. (2016, 2017) remind us that repeated and independent evolution of non-parasitism in each location (i.e., in sympatry) or a single origin of each non-parasitic species in allopatry followed by subsequent brook lamprey dispersal are not the only two options. Their demographic models support divergence in allopatry, but with subsequent gene flow following secondary contact. Gene flow between the life history types in sympatry gives the erroneous appearance of divergence in sympatry and multiple independent and parallel origins of each brook lamprey species.

This point is worth emphasizing as we begin to explore the genetic basis of feeding type in lampreys, particularly with respect to understanding the extent to which the genetic changes are parallel among different species pairs and populations (e.g., whether they involve the same mutations in the same gene). There was considerable overlap in the list of genes that were highly differentiated between European river and brook lampreys in northern France and in southern Portugal (see Sect. 4.6.3.3). Is the apparent parallelism the result of parallel selection on standing genetic variation in the ancestral European river lamprey population, the result of a single speciation event in allopatry followed by dispersal, or the result of mutations occurring in these genes independently in disjunct locations? Rougemont et al. (2017) evaluated the extent of genetic parallelism among replicate European river and brook lamprey pairs in northern France, and they suggested that ancestral variation related to life history type arose in allopatry so that the "brook lamprey background" existed before the recent colonization of rivers. The brook lamprey phenotype may have arisen either through hybrid genotypes colonizing fresh water or through transport of alleles broken up by recombination and at low frequency in the river lamprey background into the freshwater populations. There would presumably be selection for the brook lamprey alleles in fresh water, and these rare alleles would be driven to high frequencies in multiple different rivers. This would be similar to the "transporter hypothesis" proposed by Schluter and Conte (2009) to explain the rapid and repeated evolution of multiple freshwater-resident threespine stickleback populations from the anadromous or marine form. Schluter and Conte (2009) proposed that ecological speciation has occurred multiple times in parallel when selection in freshwater environments repeatedly acts on standing genetic variation that is maintained in the marine population when freshwater-adapted alleles from elsewhere in the range are exported back into the marine population. In a similar manner, brook lamprey alleles are likely transported among disjunct rivers, mediated by the river lamprey which shows few barriers to gene flow among locations, at least within a region (Bracken et al. 2015; Rougemont et al. 2015; Mateus et al. 2016; see Sect. 4.7.5). Selection within each river on existing genetic variation would help account for the apparent rapidity of speciation (e.g., following northward range expansion and post-glacial colonization).

Further study is clearly required, but the results to date emphasize the need to avoid overly simple explanations or broad generalizations when trying to understand the evolution of non-parasitism in different lamprey species and throughout the range of each. It is unknown if all populations of a particular brook lamprey species share the same "brook lamprey background" derived from a single ancestral population, and the extent of genetic parallelism in different species pairs and different genera is likewise unknown. Even distantly related brook lampreys show obvious phenotypic parallelism, but whether their independent evolution has involved, for example, different mutations in the same genes or different genes in same developmental pathway (see Arendt and Reznick 2008), has yet to be explored.

4.7 Factors that Promote or Constrain Evolution of Non-parasitism

The conditions under which brook lampreys evolve have been reviewed previously by several authors (e.g., Salewski 2003; Hardisty 2006; Docker 2009). Two scenarios have generally been proposed: (1) that parasitic lampreys become non-parasitic in habitats where there is an insufficient prey base, perhaps as the result of barriers to migration that prevent access to the ocean or large lakes (e.g., as the result of glaciation and deglaciation events during the past 10,000–15,000 years); or (2) that parasitic lampreys become non-parasitic as the result of events that alter the relative benefits and costs of migration (e.g., Zanandrea 1959; Espanhol et al. 2007). The two options are not entirely mutually exclusive, although the first implies that non-parasitism evolved to "make the best of a bad situation" when post-metamorphic feeding was poor or not possible, and the second suggests that non-parasitism is a valid "choice" that, under certain conditions, confers greater (rather than merely adequate) fitness as a result of the trade-off between reduced fecundity and reduced mortality. Our intention in this section is not to repeat past reviews but, rather, to continue the discussion introduced in Sect. 4.4 regarding factors that promote and constrain freshwater residency in lampreys. Loss of anadromy almost invariably leads to a reduction in the duration of the parasitic feeding phase and size at maturity, and we contend that the complete elimination of the parasitic feeding phase is an extension of this process. The emphasis in this section will be less on the environmental conditions that might have promoted elimination of the parasitic feeding phase, and more on characteristics of parasitic lampreys that permitted or constrained evolution of brook lamprey derivatives. Some parasitic species are clearly much more prolific in terms of producing non-parasitic offshoots than others. Although there is not universal agreement on the "true" number of brook lamprey species, and there may never be an exact, objectively definable number (see Sect. 4.6.3; Chap. 7), it is apparent that the two anadromous parasitic *Lampetra* species are the "mothers" of many brook lampreys. At minimum, 35% of the 23 brook lamprey species recognized by Potter et al. (2015) are thought to be derived from the European and western river lampreys or European and western river lamprey-like ancestors (Fig. 4.1). At the other end of the spectrum, there is a conspicuous absence of brook lamprey derivatives from the large-bodied sea, pouched, and Caspian lampreys. The following discussion will largely try to address why there are no brook lamprey derivatives from this latter group, and we will speculate as to whether freshwater-resident sea lamprey might represent a "jumping-off point" for the evolution of non-parasitism.

4.7.1 Phylogenetic Constraints

Sea lamprey, pouched lamprey, and Caspian lamprey are each the sole species in their respective genera. Thus, the lack of brook lamprey derivatives from these species is potentially different from that of other parasitic species for which no non-parasitic derivatives have been described. No brook lamprey derivatives have been attributed to the Chilean lamprey, Korean lamprey, Vancouver lamprey, and perhaps the Klamath lamprey (depending on whether it or the Pacific lamprey is the ancestor to the two *Entosphenus* brook lamprey species; Sect. 4.6.3.1), but non-parasitic lampreys are found within each of these genera (see Fig. 4.1). Thus, "phylogenetic constraints" could potentially be used to explain the absence of non-parasitic lampreys in *Petromyzon*, *Geotria*, and *Caspiomyzon*—that is, that these lineages simply do not have the "wherewithal" to develop brook lampreys. However, *Petromyzon*, *Geotria*, and *Caspiomyzon* are well-distributed throughout the lamprey phylogenetic tree (based on both morphological and molecular characters) and are intermixed with the seven genera containing brook lamprey species (Potter et al. 2015). Brook lampreys are found in two of the three families of extant lampreys and, within the Petromyzontidae, in both or all three of the proposed subfamilies (Vladykov 1972; see Potter et al. 2015). In subfamily Petromyzontinae, all three parasitic species of *Ichthyomyzon* have given rise to brook lamprey derivatives, even if *Petromyzon* and *Caspiomyzon* have not. Phylogenetic constraint has been invoked in a variety of contexts, but as yet there is no consensus on its definition, and it was described by Alexander (1989) as "an argument of last resort." McKitrick (1993) defined phylogenetic constraint as "any result or component of the phylogenetic history of a lineage that prevents an anticipated course of evolution in that lineage." Of course, we cannot rule out such a constraint. However, we argue below that, given their consistently large body size, evolution of non-parasitism would not be anticipated in these species (see Sect. 4.7.4). Thus, although there may be genetic constraints within each of the large-bodied species (see Sect. 4.7.5), we argue against phylogenetic constraints.

4.7.2 Ecological Constraints

Although the discussion of the factors that promote or constrain the evolution of non-parasitic lampreys often focuses on the ecological conditions under which non-parasitism arises, the dearth of non-parasitic species in some taxa appears not to be caused by lack of the appropriate ecological conditions. We recognize that the conditions under which even recent non-parasitic derivatives arose are not identical to conditions under which they are currently found, particularly if, as suggested by recent demographic analyses in European river and brook lampreys in northern France, divergence occurred in allopatry at least 202,000–257,000 years ago (Rougemont et al. 2016; see Sect. 4.6.3.4). However, those species that have no or few non-parasitic derivatives frequently overlap in their distribution with species that have given rise to non-parasitic derivatives. On either side of the Atlantic, anadromous sea lamprey co-occur in drainages with American brook lamprey (e.g., Aman et al. 2017; Evans 2017) and European river and brook lampreys (e.g., Maitland 1980; Taverny et al. 2012), and the Great Lakes sea lamprey overlaps in distribution with silver and chestnut lampreys, as well as northern brook and American brook lampreys (Renaud et al. 2009). The pouched lamprey co-occurs in Australian waters with short-headed lamprey which has given rise to a non-parasitic derivative. The Caspian lamprey has a more restricted distribution than the widespread sea and pouched lampreys but, nonetheless, overlaps somewhat in its distribution with *Eudontomyzon* brook lampreys (Levin and Holčík 2006; Potter et al. 2015; Levin et al. 2016).

There are clear ecological constraints related to the loss of anadromy in parasitic lampreys (see Sect. 4.4.2). Access to a sufficient prey base is obviously critical for the evolution and persistence of freshwater parasitic lampreys, and what constitutes sufficient varies considerably among species. The relative scarcity of large lakes within the range of the pouched lamprey, for example, may have prevented this species from establishing any freshwater-resident parasitic populations, but such constraints would not apply to the evolution of non-parasitism. Conversely, one could argue that non-parasitic species need to have a more productive larval environment to compensate for the lack of further growth following metamorphosis (Kucheryavyi et al. 2007). There is certainly a general trend showing that non-parasitic species typically have a longer larval stage and greater size at metamorphosis relative to their parasitic counterparts (Potter 1980; Hardisty 2006; Docker 2009), but duration of the larval stage and size at metamorphosis varies considerably among and within species. Large size at metamorphosis does not seem to be an absolute requirement for non-parasitic lampreys, nor is small size at metamorphosis the rule among parasitic species. Metamorphosing American brook lamprey as small as 100–109 mm TL have been reported (Hoff 1988), and sea lamprey average 130–140 mm TL at metamorphosis (Potter 1980). In contrast, pouched lamprey metamorphoses at relatively small sizes (~90–100 mm; Neira 1984; Potter and Hilliard 1986). Presumably, there is a lower size limit below which non-parasitic species would be unable to undergo both non-trophic metamorphosis and non-trophic sexual maturation. However, larval lampreys and post-metamorphic brook lampreys appear to be extremely energy efficient (Sutton and Bowen 1994; Beamish and Medland 1988; see Dawson et al. 2015), and it may be that the minimum size requirements for metamorphosis and sexual maturation in brook lampreys is not that much different than the minimum size requirements for metamorphosis and downstream migration in parasitic species. Differences may be apparent between the sexes (e.g., with larger size at metamorphosis being more important for female brook lampreys than for female parasitic lampreys; Docker 2009) and among some species or environments (e.g., where downstream migration is more arduous or the delay prior to parasitic feeding is longer), but it appears that brook lampreys should be able to persist anywhere that there is suitable spawning and rearing habitat for any lampreys. Even at high latitudes, brook lamprey larvae are able to grow well; Alaskan brook lamprey larvae as large as 144 mm TL were reported in the Martin River, Northwest Territories (61.924 °N; Renaud et al. 2016), and they were even larger (up to 214–215 mm TL) in the Chatanika and Chena rivers, Alaska (65.281 °N; Sutton 2017).

This is not to say that ecological conditions will not influence the relative costs and benefits of parasitism versus non-parasitism, but rather that evolution of non-parasitism does not appear to require particular ecological conditions. Certainly non-parasitic lampreys will be favored above barriers to migration that prevent access to a sufficient prey base, but the co-occurrence of parasitic and non-parasitic lampreys downstream of barriers in countless streams and rivers indicate that this is not the only factor to consider. Rather, it appears that non-parasitism represents an evolutionarily stable strategy in situations where the relative costs of migration and feeding outweigh their benefits. The balance of this trade-off will depend on ecological conditions, but it appears to vary considerably among species and even among populations or individuals (see Sect. 4.7.4).

4.7.3 Osmoregulatory Ability and the Importance of Freshwater Intermediates

It has been suggested that freshwater parasitic lampreys may be important intermediaries in the transition from anadromous parasitic to freshwater non-parasitic lampreys (Hubbs and Potter 1971; Beamish 1985; Salewski 2003; Hardisty 2006). Beamish (1985) reasoned that the many changes that must occur in this transition are too major to occur in a single step. Instead, he proposed that the first step in this evolutionary pathway was the ability to osmoregulate in fresh water during the parasitic feeding phase, possibly acquired as the result of gradual changes in salinity. By extension, one could argue that those parasitic species that are poorly adapted for feeding in a freshwater environment during the parasitic phase would be less likely to give rise to non-parasitic derivatives. However, an inability to osmoregulate during the juvenile feeding phase is not likely relevant once the juvenile feeding phase has been eliminated. Even anadromous lampreys exhibit a breakdown of their saltwater osmoregulatory mechanisms during upstream migration and sexual maturation (e.g., Pickering and Morris 1970; Beamish et al. 1978; Ferreira-Martins et al. 2016). Hence, elimination of the parasitic feeding phase should not require an intermediate freshwater form as means of adapting to fresh water during sexual maturation. Moreover, even though there is a general correlation between those parasitic species with freshwater forms and those that have given rise to non-parasitic derivatives, it is not absolute (Fig. 4.1). Most notably, freshwater-resident populations of western river lamprey are rare at best. Only one viable population has been reported (in Lake Washington; see Sect. 4.3.3.2), and yet this species has given rise to western brook lamprey in innumerable Pacific drainages from Alaska to California, and at least two older brook lamprey derivatives in Oregon and California.

Recent demographic reconstructions, using genome-wide markers in the three lamprey life history types from Loch Lomond (anadromous and freshwater-resident European river lamprey and European brook lamprey; see Sect. 4.3.4.2), compared 12 hypothetical evolutionary scenarios for divergence of the three forms, and the two models that were best supported both suggested a common ancestry for the two freshwater forms (Hume et al. 2018). These models suggested either a hybrid speciation scenario, by which hybridization between the freshwater and anadromous parasitic forms gave rise to the non-parasitic form, or a scenario by which the anadromous ancestor gave rise to both freshwater parasitic and non-parasitic forms. However, it is important to recognize that a linear progression was not supported (i.e., anadromous parasitic → freshwater parasitic → freshwater non-parasitic), and there was clear evidence of subsequent gene flow between the non-parasitic and anadromous parasitic forms (see Sect. 4.6.3.2). Interestingly, the freshwater parasitic form was more genetically differentiated from the anadromous and non-parasitic forms than either of these latter forms were from each other (Bracken et al. 2015; Hume et al. 2018), despite the two parasitic (river lamprey) forms being considered a single species distinct from the non-parasitic European brook lamprey.

4.7.4   Life History Trade-Offs

Life history theory seeks to explain the major demographic traits in an organism's life cycle (e.g., growth rate, age and size at maturity, number and size of offspring, age- and size-specific mortality rates) and understand the trade-offs and fitness consequences associated with the different traits. Lampreys are excellent models to study life history evolution; they show a highly conserved body plan but a diversity of life history types, and, because they are semelparous, lifetime reproductive success is easily quantifiable. As articulated by Hubbs and Potter (1971), brook lampreys and the largest anadromous forms represent two extreme forms of adaptation, exhibiting the range of life history trade-offs seen in lampreys. It has long been suggested that the viability of non-parasitic populations depends on low mortality resulting from the elimination of parasitic feeding and migration balancing the reduction in fecundity caused by the resulting decrease in size at maturity (Hubbs and Potter 1971; Potter 1980; Hardisty 2006; Docker 2009). At the other end of the spectrum, the large anadromous species combine maximum body size, wide-ranging migration, extended adult life (and presumably relatively high total mortality) with exceptionally high egg numbers. In the middle are the majority of parasitic species, those that are intermediate in body size (or, at least, with segments of their populations that are smaller bodied) and with more restricted migrations. Significantly, it is from this latter group that the vast majority of brook lampreys have been derived (Hubbs and Potter 1971). Thus, as suggested previously, it appears that selective constraints related to life history trade-offs have limited evolution of non-parasitism in the largest anadromous species and promoted their evolution in small-bodied parasitic species.

The latter group of parasitic species includes freshwater parasitic lampreys and the smaller anadromous lampreys. The smaller anadromous lampreys, given their need for a less extensive prey base than the very large anadromous lampreys, have generally been successful at colonizing fresh water (see Sects. 4.4.1 and 4.4.2). This gives the appearance of a correlation between the ability to feed parasitically in fresh water and the ability to give rise to non-parasitic forms, but the key appears to be reduced duration of the feeding phase and body size at maturity and not osmoregulatory ability. The anadromous sea lamprey is the largest extant lamprey (up to 800–900 mm TL at maturity) and has produced highly successful freshwater-resident populations (Sect. 4.3.4.4), but it has yielded no brook lampreys. In contrast, western river lamprey is the smallest anadromous lamprey (~200 mm TL at maturity); few or no parasitic populations have become established in fresh water (see Sect. 4.3.3.2), but this species has given rise to an abundance of brook lamprey populations.

The other two species with a conspicuous absence of brook lamprey derivatives or relatives are likewise those that are consistently large bodied (Fig. 4.5). In Australia, for example, pouched lamprey measure 530–740 mm TL at maturity, and no small-bodied forms are known (Potter et al. 1983). Small-bodied (praecox) anadromous Caspian lamprey have been reported, but they are apparently rare. Likewise, although there are reports of landlocked Caspian lamprey following dam construction, there is no evidence that they have become established in fresh water (see Sect. 4.3.3.1). Anadromous sea and pouched lampreys also tend to show more wide-ranging offshore feeding relative to smaller-bodied species (Potter et al. 2015). This suggests little limitation in terms of prey availability, so they would not have been under selection to reduce the duration of their parasitic feeding phase. For these large, ocean-going lampreys, the decrease in mortality rates that would accompany the shift to non-parasitism could not possibly offset the very drastic reduction in fecundity (Hardisty 2006). As discussed in Sect. 4.4.4, fecundity increases approximately with the cubic power of length, and reducing size at maturity from, for example, 600–800 mm TL to 120 mm TL would result in a reduction of fecundity from 149,240–342,690 to 1,430 eggs (Fig. 4.5; see Chap. 1), a reduction of more than 99%. The Caspian lamprey differs in having a more restricted distribution and more confined food resource than these other two large-bodied lampreys. However, its possible mode of carrion feeding (see Chap. 3) might have helped reduce the pressure on local host populations, so there was likewise little pressure for this species to reduce the duration of its feeding phase. The shift from a large-bodied parasitic form directly to a non-parasitic form would represent a significant reduction in fecundity.

In contrast, although Pacific and Arctic lampreys are also known to reach large sizes as the result of extended feeding at sea, both species show considerable intraspecific variation in size at maturity and duration of the parasitic feeding phase, and Pacific and Arctic lampreys are each recognized as having given rise to two or

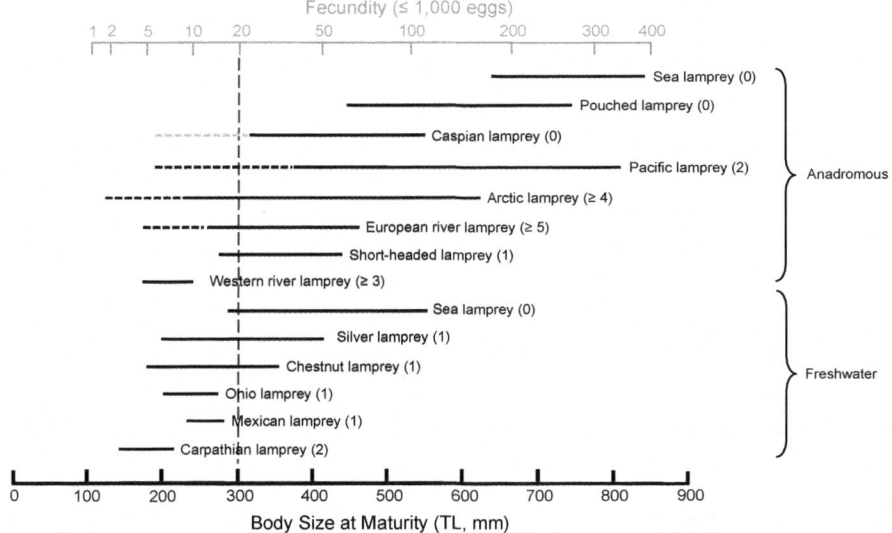

Fig. 4.5 Body size (total length, TL) at maturity for eight anadromous and six freshwater parasitic lamprey species and the number of extant non-parasitic lamprey species to which they are thought to have given rise (in *parentheses*, see Fig. 4.1). Size range at maturity is from Sect. 4.3, Docker (2009), and Chap. 1 (Table 1.11 and references therein); *black dotted lines* indicate size of well-characterized anadromous praecox or freshwater-resident individuals; *gray dotted line* indicates less well-known praecox populations. The relationship between mean fecundity and mean TL calculated across species and populations (see Chap. 1) was used to estimate total number of eggs produced by parasitic lampreys across this size range; mean fecundity for non-parasitic lampreys averages ~1,770 eggs. It appears that only parasitic species with at least some individuals smaller than ~300 mm TL at maturity (*red dotted line*) have given rise to non-parasitic species, resulting in a 10–12× reduction in fecundity

more non-parasitic species (Fig. 4.5). Hubbs and Potter (1971) described the Pacific lamprey as a "vast complex of very small to large forms." Praecox or dwarf anadromous populations are known in both Pacific and Arctic lampreys, and smaller-bodied freshwater-resident Arctic lamprey populations are also known (see Sects. 4.3.4.1 and 4.3.4.3). Compared to sea and pouched lampreys, these species feed more in coastal waters and at shallower depths, and they appear to show more limited dispersal at sea (Spice et al. 2012). Consequently, these species would be more vulnerable to declines in local fish stocks with a concomitant selection for the praecox form (see Sect. 4.4.2). These smaller-bodied individuals would then have been better poised to abandon parasitic feeding altogether than sea or pouched lampreys, with much more modest reductions in fecundity (Fig. 4.5).

Among the smaller anadromous species, the European river and western river lampreys have been particularly prolific, giving rise to at least five and three recognized brook lamprey species, respectively, and potentially many more (Fig. 4.1). European river lamprey, in addition to its typical anadromous form, also exists as even smaller-bodied praecox anadromous and freshwater forms (Sect. 4.3.4.2), and it could be argued that all western river lamprey correspond to an anadromous praecox form (Sect. 4.3.3.2). Freshwater parasitic species (with the exception of the landlocked sea lamprey) are likewise relatively small-bodied. The three parasitic *Ichthyomyzon* species, Carpathian lamprey, and Mexican lamprey generally feed parasitically for ≤1 year, and all have given rise to non-parasitic forms (Fig. 4.5). In these species, the reduction in fecundity resulting from elimination of the parasitic feeding phase will be limited. In the western river lamprey, for example, bypassing the brief parasitic feeding phase and the associated down- and upstream migrations results in a considerable reduction in exposure to predators (e.g., R. J. Beamish 1980; Roffe and Mate 1984; see Docker et al. 2015) with only a modest reduction in size at maturity and fecundity (Fig. 4.5). We thus suggest that western river lamprey rarely, if ever, establish viable freshwater parasitic populations, not because of osmoregulatory constraints, but rather as the result of selective constraints. Western river lamprey, even as anadromous parasites, are already so small at maturity that "skipping" right to a non-parasitic life history type represents a more profitable trade-off between mortality and fecundity than freshwater parasitism would.

Our comparisons indicate that all parasitic species that have given rise to non-parasitic derivatives have at least some individuals measuring ≤300 mm TL at maturity, and that those where large segments of the species are below this cut-off are more likely to have produced more brook lamprey derivatives. Not all parasitic species that are ≤300 mm TL have non-parasitic derivatives, but all that have non-parasitic derivatives are ≤300 mm TL. All parasitic lampreys that are ≤300 mm TL at maturity, and do not have known non-parasitic derivatives, belong to genera in which non-parasitic species are found. Chilean, Korean, Vancouver, and Klamath lampreys do not have known brook lamprey derivatives, but they belong to genera in which non-parasitic species are found, and our analysis suggests that they could give rise to non-parasitic derivatives. Hubbs and Potter (1971) indicated that *Mordacia* in Chile may have produced a brook lamprey in parallel to that observed in Australia, although there is no evidence that the Chilean lamprey has given rise to such a species (Potter and Strahan 1968).

Hardisty (2006) predicted that, in smaller parasitic species with lower levels of fecundity, the balance would be "more delicately poised" such that unstable environments that reduced feeding opportunities or increased mortality during the adult phase could tip the balance towards non-parasitism. The point at which the balance between reduced mortality and reduced fecundity tips in favor of non-parasitism is presumably subject to some variation, based on the relative costs and benefits of feeding and migration under different circumstances, but our comparison suggests that 300 mm TL (with a 10–12× reduction in fecundity) is the cut-off above which shifts to non-parasitism would not be beneficial. Mortality during the migratory and juvenile feeding stages in parasitic species has not been quantified, but we would predict that it is no more than 10–12× higher than that observed following metamorphosis in non-parasitic species.

Based on these comparisons, we predict that Great Lakes sea lamprey are not likely to give rise to a non-parasitic derivative, despite the occurrence of an "intermediate" freshwater parasitic form. Despite the consistently large size of anadromous sea lamprey, this species was apparently able to colonize fresh water because of the abundant prey resources and large size of the Great Lakes. Even in fresh water, size at maturity (mean TL ~395–500 mm) and fecundity (mean ~70,000 eggs; see Chap. 1) is still very high. Complete elimination of the parasitic feeding phase would represent too large (~40×) a reduction in fecundity.

4.7.5 Genetic Factors

Selective and genetic constraints on the evolution of non-parasitism are not mutually exclusive. Although it makes intuitive sense that the transition to non-parasitism would be disadvantageous to large-bodied, highly fecund species, we do not know if their presumably genetic propensity for consistently large size and a long, wide-ranging feeding phase would preclude evolution of non-parasitism. We know virtually nothing regarding the genetic factors governing differences in size at maturity among species or what factors regulate length of the feeding phase (see Moser et al. 2015). However, the above discussion suggests that the ranges in body size of fully grown anadromous sea and pouched lampreys (and perhaps Caspian lamprey, to a lesser extent) are genetically fixed, while species such as Pacific, Arctic, and European river lampreys show more intraspecific genetic variation in these traits. Recent genomic studies in Pacific lamprey have identified loci associated with intraspecific differences in size at maturity, and they appear to show regional heterogeneity associated with this trait (Hess et al. 2013, 2014). Population genetic studies suggest that sea lamprey on each side of the Atlantic constitute largely panmictic populations (Bryan et al. 2005), while evidence is accruing that Pacific, Arctic, and European river lampreys represent more heterogeneous gene pools. Lampreys do not home to their natal streams (see Moser et al. 2015), so there is no evidence of local adaptation to specific stream or river basins, but it appears that dispersal distance is related to body size and smaller-bodied anadromous lamprey typically show evidence of isolation by distance. This is the case in Pacific lamprey (Spice et al. 2012), and Rougemont et al. (2015) and Mateus et al. (2016) likewise found some genetic structuring among European river lamprey populations, at least among regions. In the Arctic lamprey from Japan and eastern Russia, significant genetic differences were found between three populations located in the northern part of the study area and the remaining nine more southerly populations; however, there was no significant isolation by distance, and there was evidence of gene flow among the populations (Yamazaki et al. 2014). Overall, it seems reasonable to infer that the regional differences in size at maturity that have been reported in Pacific, Arctic, and European river lampreys (Sect. 4.3.3) have some genetic basis and that evolution of non-parasitism involves further selection on individuals already predisposed to small size at maturity.

However, the additional genetic changes required for the complete elimination of the parasitic feeding phase are unknown. We do not know what factors prompt parasitic species to stop feeding and initiate sexual maturation (or vice versa, i.e., to initiate sexual maturation and thus stop feeding), and we certainly do not know what factors initiate sexual maturation without any parasitic feeding whatsoever. Whether the evolution of non-parasitism drew on existing (standing) genetic variation within these small-bodied parasitic lamprey populations or whether it depended on de novo mutations is unknown. With the new genomic technologies available, it is now possible to address some of these questions. Some candidate genes have been identified (see Sect. 4.6.3.3), but there is still considerable work required to understand the genetic and developmental changes associated with the evolution of non-parasitism.

Furthermore, whether other life history features associated with non-parasitism (e.g., earlier onset of gonadogenesis and lower potential fecundity) are also polymorphic in the ancestral parasitic population is unknown. The results from Spice and Docker (2014) suggest that some parasitic species are polymorphic with respect to the timing of oogenesis and the resulting number of oocytes produced (see Sect. 4.6.3.3; Chap. 1), and Hardisty (1964) suggested that non-parasitic forms may have evolved from an ancestral parasitic population whose fecundity was comparatively low. However, the more dramatic changes evident in non-parasitic species presumably evolved after establishment of fully non-parasitic populations, as the result of relaxation on selection for morphological (e.g., dentition, number of trunk myomeres; see Sect. 4.6.2.2) and developmental characters. The genetic basis of these traits is entirely unknown at present but will no doubt be elucidated in the future.

4.8 Conclusions

Early lampreys appear to have been small direct developers (i.e., with only gradual ontogenetic change occurring during development), with the characteristic larval "ammocoetes" stage and dramatic metamorphosis evolving later. The prolonged larval stage is now conserved among all extant lampreys, but there has been considerable diversification of life history types on two main axes related to post-metamorphic feeding and migration. The parasitic species delay sexual maturation until after the juvenile feeding phase, while the non-parasitic species initiate sexual maturation during metamorphosis and eliminate the parasitic feeding phase. All non-parasitic species are freshwater resident with the same post-metamorphic life span (~6–8 months) and similar body sizes (~110–150 mm TL) and fecundities (mean 1,770). In contrast, parasitic species are either anadromous or freshwater resident and vary widely in all three of these traits, presenting a continuum of trade-offs related to post-metamorphic growth opportunities that will increase an individual's reproductive output and the costs incurred while taking advantage of these opportunities. Duration of the parasitic feeding phase ranges from a single summer (3–4 months) to perhaps as much as 4 years, giving a post-metamorphic life span estimated to range from ~1.5–1.8 years to as much as ~4.5–5.8 years (i.e., accounting for the non-trophic periods of downstream and upstream migration). Body size at maturity ranges from ~125 mm in the freshwater Miller Lake lamprey to 800–900 mm in anadromous sea lamprey, and fecundity (which varies with the cubic power of TL) ranges from a few thousand to 150,000–300,000 eggs. The life history trade-offs associated with anadromy and freshwater residency have long been discussed in other fishes (e.g. Gross et al. 1988; Jonsson and Jonsson 1993, 2006; Fleming 1996), and those associated with the evolution of non-parasitism have been discussed in lampreys (e.g., Hubbs and Potter 1971; Potter 1980; Hardisty 2006). Nevertheless, there is much that we still do not know. Many of these knowledge gaps have been identified in the sections above. Here, we briefly highlight three of the overarching questions related to life history evolution in lampreys that remain to be answered; perhaps some of these questions will be answered by the next "update." Lampreys are becoming an important model for "evo-devo" research, helping to reconstruct some of the major events in the evolution of vertebrates (see Docker et al. 2015; McCauley et al. 2015; Chap. 6). With their conserved body form, but rich diversity of feeding and migratory types in replicate taxa, they are also an excellent model system for examining microevolutionary processes.

4.8.1 What are the Proximate Mechanisms Determining the Duration of the Parasitic Feeding Phase?

We do not yet know what factors prompt parasitic species to stop feeding and initiate sexual maturation or, conversely, to initiate sexual maturation and thus stop feeding. The relatively strict adherence to fixed maturation schedules in many lamprey species (e.g., anadromous sea lamprey and western river lamprey, most or all freshwater species) suggests that differences among species are "hardwired." However, intraspecific differences have been observed or inferred in other species (i.e., based on differences in size at maturation), and whether these differences have a genetic basis or whether cessation of the feeding phase is triggered by environmental or endogenous (e.g., body condition, growth rate) cues is unknown. In Pacific lamprey, the correlation between certain loci and body size (Hess et al. 2013, 2014) suggests at least some genetic component to size at maturity, but whether this is tied to differences in duration of the feeding phase or to other factors (e.g., metabolic efficiency, effectiveness in finding prey, habitat selection) is unknown. In sea lamprey, the transition to fresh water has involved a 1-year reduction in the duration of the parasitic phase (Bergstedt and Swink 1995), but whether this involved rapid evolution following colonization of large-bodied anadromous lamprey via canals, more gradual evolution following post-Pleistocene colonization, colonization via canals of smaller-bodied individuals already pre-adapted for feeding in a more confined environment, or phenotypic plasticity has yet to be resolved.

Factors controlling the onset of maturation have been studied in salmonids (e.g., to understand alternative male life history tactics and to reduce the incidence of precocious maturation in aquaculture settings; Paéz et al. 2011; Good and Davidson 2016). They are far from being fully understood, but it is clearly a complex situation. In Atlantic salmon, for example, the onset of maturation is governed by a variety of heritable, physiological, biochemical, and environmental cues and their interactions (Good and Davidson 2016). In one recent and exciting study, a single locus was found to have a highly significant role in age at maturation in Atlantic salmon (Ayllon et al. 2015). Four SNPs, including two mutations in the Vestigial Like Family Member 3 (*vgll3*) gene, explained 33–36% of the variation in age at maturation. Interestingly, a SNP in proximity to this gene in humans has been linked to age at puberty. Prince et al. (2017) found that premature migration in both Chinook salmon *Oncorhynchus tshawytscha* and steelhead was associated with mutations in the *GREB1L* locus, a paralog (i.e., arising during gene duplication) of the *GREB1* (Growth Regulating Estrogen Receptor Binding 1) gene. In a recent study in mice, *GREB1L* was found to be differentially regulated in neurons of the hypothalamic arcuate nucleus as a result of feeding versus fasting (Henry et al. 2015). Prince et al. (2017) suggested that the premature migration alleles arose from a single evolutionary event within each species and subsequently spread to other populations through straying and positive selection. Additional genomics studies are sure to further contribute to our understanding of the factors that control the onset of maturation in lampreys and other anadromous fishes.

4.8.2 What are the Proximate Mechanisms Underlying Non-parasitism?

When trying to unravel the genetic and developmental basis of non-parasitism in lampreys, it is harder to draw inspiration from genetic and genomic studies in other fishes. Since no other vertebrates are known to have a non-feeding adult stage, there would appear to be no obvious parallels to the evolution of non-parasitism. However, many other anadromous fishes (most conspicuously, semelparous Pacific salmon) cease feeding at sexual maturation. Because acceleration of sexual maturation at metamorphosis in brook lampreys results in the merging of metamorphosis with sexual maturation without an intervening feeding phase, non-parasitism may simply represent an extreme trade-off between size at maturity and mortality. Thus, the genetic mechanisms underlying this acceleration of sexual maturation may not be unique, and different life history trajectories (e.g., related to the phasing of gonadogenesis and differences in potential fecundity) may indeed be initiated well in advance of metamorphosis (Docker 2009; see Chap. 1). Pioneering genomic studies on different European river and brook lamprey populations have identified loci that differ between life history types (Mateus et al. 2013b; Rougemont et al. 2017; Hume et al. 2018), but interpretation of the results have been hampered by the difficulty in assigning a definitive function to these loci. Continuing improvements to the assembly and annotation of the sea lamprey genome (e.g., Smith et al. 2018) and developing genomic resources for other lamprey species will help these efforts. Furthermore, similar research in other species pairs will help determine the extent to which the genetic changes associated with non-parasitism are parallel among different independently derived species, and whether "brook lamprey alleles" exist (i.e., as standing genetic variation) within most parasitic lamprey populations. We predict that such alleles would be more common in those species or populations that most readily give rise to non-parasitic derivatives.

The "paired species problem" is definitely not unique to lampreys. A number of other postglacial fish species exhibit alternative migratory types or resource use polymorphisms that have arisen in parallel (within and among taxa) and fail to show reciprocal monophyly (see Taylor 1999; Docker 2009). Although taxonomic distinctions have been made between many of them in the past, most are now considered ecotypes of a single Linnaean species. This does not mean, of course, that ecotypes do not represent "evolutionarily significant units" (see Fraser and Bernatchez 2001) and certainly not that they are merely the product of phenotypic plasticity. There is now compelling evidence that European river and brook lampreys are partially reproductively isolated ecotypes (to variable degrees in different river systems) that nevertheless maintain distinct phenotypes in sympatry. However, we cannot assume that all paired species (or populations) will be the same, and it is important to avoid overly simplistic explanations. Major taxonomic revisions in lampreys at this point would be premature (see Chap. 7).

4.8.3 Can We Predict the Potential for Anadromous Lampreys to Become Invasive in Fresh Water?

In addition to its evolutionary significance, understanding the ease with which anadromous lampreys can become freshwater resident has important conservation and management implications. For example, large-bodied species of conservation concern (e.g., Pacific lamprey, anadromous sea lamprey, Caspian lamprey) are generally extirpated when dams prevent movement between upstream spawning habitats and the sea (see Maitland et al. 2015). This means that lamprey passage (e.g., Moser et al. 2015) or dam removal (e.g., Hogg et al. 2013) will be necessary when conservation is a priority. Alternatively, if a population is able to establish above the dams, selection will almost certainly lead to smaller-bodied, less fecund lampreys. Loss of anadromy is also detrimental where large-bodied anadromous forms are exploited for human harvest (e.g., Pacific lamprey by Native Americans) and through loss of marine-derived nutrients in freshwater systems (see Docker et al. 2015).

Conversely, successful establishment of anadromous lampreys in new inland water bodies could lead to serious management concerns, and it is important to be able to predict the potential for anadromous lampreys to become invasive in fresh water (i.e., which ones and where?). A non-anadromous Pacific lamprey-like form is already known to feed on an endangered sockeye salmon stock in Sakinaw Lake in British Columbia (COSEWIC 2016), and Farlinger and Beamish (1984) raised concerns regarding the possibility that Pacific lamprey could become freshwater resident in Babine Lake. If so, a major impact on salmon in this lake would likely result (see

Sect. 4.3.4.1). Establishment of both pink and coho salmon in the Great Lakes, despite previous views that these species require salt water for completion of their life cycle, makes it clear that we should "never say never" (see Sect. 4.4.3). Our discussions above and studies in other species suggest that standing genetic variation in the source population is important for rapid colonization of new environments and that the number of individuals that gain access to a new region will thus increase the chance of survival and establishment (e.g., Blackburn et al. 2015). Williamson (1996) coined the term "propagule pressure" (which incorporates estimates of the absolute number of individuals involved in any one invasion event and the number of such events) to predict species invasiveness. Although this suggests that a few large-bodied anadromous lampreys that undergo long migrations in manmade canals or are inadvertently transferred to fresh water would be unlikely to establish (i.e., if relying on freshwater alleles at relatively low frequency in the anadromous population), chances would increase if pre-adapted individuals show increased survival rates or preferential colonization (Briskie et al. 2017). Many anadromous lampreys appear capable of some parasitism in fresh water en route to sea (see Sect. 4.3.4.4), and some researchers have hypothesized that this feeding behavior can lead to adaptation to freshwater environments (Potter and Beamish 1977). Sea lamprey rapidly spread and soon reached pest proportions in the upper Great Lakes once access was permitted, and there is evidence of a secondary "landlocking" event in the upper Cheboygan River system (Johnson et al. 2016). Predicting the potential for further spread is important. Although sea lamprey depend on large productive lakes for establishment, there are several large inland lakes (e.g., Lake Nipigon in Ontario, 4,850 km^2) that could support sea lamprey if they were to gain access. Spread to Lakes Simcoe (745 km^2) and Winnebago (560 km^2), in Ontario and Wisconsin, respectively, is of particular concern. Two locks and a boat lift are the only obstacles on the Trent-Severn Canal between Lake Simcoe and Lake Huron, although they are maintained and operated to prevent movement of sea lamprey into Lake Simcoe. Lake Winnebago was connected to Lake Michigan via a chain of locks on the Fox River, but the lock system was closed in 1983 and a barrier was installed in 1988 (see Chap. 5). It is also important to ask if the likelihood of establishment in fresh water would increase with climate change or other ecosystem alterations (e.g., salmonid stocking or colonization by other invasive species). Expected climate-driven changes to the relative productivity of marine and freshwater systems, for example, are expected to alter the prevalence and distribution of anadromy in salmonids (Finstad and Hein 2012) and could similarly alter the relative benefits and costs of freshwater residency in lampreys.

Acknowledgments We gratefully acknowledge Dr. John B. Hume for providing insightful comments and discussion on an earlier draft of this chapter.

References

Abou-Seedo FS, Potter IC (1979) Estuarine phase in the spawning run of the river lamprey *Lampetra fluviatilis*. J Zool 188:5–25

Adams CE, Bissett N, Newton J, Maitland PS (2008) Alternative migration and host parasitism strategies and their long-term stability in river lampreys from the River Endrick, Scotland. J Fish Biol 72:2456–2466

Aguirre WE, Bell MA (2012) Twenty years of body shape evolution in a threespine stickleback population adapting to a lake environment. Biol J Linn Soc 105:817–831

Ahmadi M, Amiri BM, Abdoli A, Fakharzade SME, Hoseinifar SH (2011) Sex steroids, gonadal histology and biological indices of fall and spring Caspian lamprey (*Caspiomyzon wagneri*) spawning migrants in the Shirud River, Southern Caspian Sea. Environ Biol Fishes 92:229–235

Alexander RD (1989) Evolution of the human psyche. In: Mellars P, Stringer C (eds) The human revolution. University of Edinburgh Press, Edinburgh, pp 455–513

Aman JT, Docker MF, Grimes KW (2017) New England range extension of American Brook Lamprey (*Lethenteron appendix*), as confirmed by genetic analysis. Northeast Nat 24:536–543

Applegate VC (1950) Natural history of the sea lamprey (*Petromyzon marinus*) in Michigan. US Fish Wildl Serv Spec Sci Rep Fish 55:1–237

Arendt J, Reznick D (2008) Convergence and parallelism reconsidered: what have we learned about the genetics of adaptation? Trends Ecol Evol 23:26–32

Aron WI, Smith SH (1971) Ship canals and aquatic ecosystems. Science 174:13–20

Artamonova VS, Kucheryavyy AV, Pavlov DS (2011) Nucleotide sequences of the mitochondrial cytochrome oxidase subunit I (COI) gene of lamprey classified with *Lethenteron camtschaticum* and the *Lethenteron reissneri* complex show no species-level differences. Dokl Biol Sci 437:113–118

Avise JC (2000) Phylogeography: the history and formation of species. Harvard University Press, Cambridge, MA

Ayllon F, Kjaerner-Semb E, Furmanek T et al (2015) The *vgll3* locus controls age at maturity in wild and domesticated Atlantic salmon (*Salmo salar* L.) males. PLoS Genet 11:e1005628

Baer J, Hartmann F, Brinker A (2018) Abiotic triggers for sea and river lamprey spawning migration and juvenile outmigration in the River Rhine, Germany. Ecol Freshw Fish 27:988–998

Bahr K (1952) Beitrage zur Biologie des Flussneunauges *Petromyzon fluviatilis* L. (Lebensraum und Ernahrung). Zool Jb 81:408–436

Bailey RM, Smith GR (1981) Origin and geography of the fish fauna of the Laurentian Great Lakes basin. Can J Fish Aquat Sci 38:1539–1561

Baldwin CM, McLellan JG, Polacek MC, Underwood K (2003) Walleye predation on hatchery releases of kokanees and rainbow trout in Lake Roosevelt, Washington. N Am J Fish Manag 23:660–676

Bardack D, Zangerl R (1968) First fossil lamprey: a record from Pennsylvanian of Illinois. Science 162:1265–1267

Bardack D, Zangerl R (1971) Lampreys in the fossil record. In: Hardisty MW, Potter IC (eds) The biology of lampreys, vol 1. Academic Press, London, pp 67–84

Barker LA, Beamish FWH (2000) Gonadogenesis in landlocked and anadromous forms of the sea lamprey, *Petromyzon marinus*. Environ Biol Fish 59:229–234

Barker LA, Morrison BJ, Wicks BJ, Beamish FWH (1998) Potential fecundity of landlocked sea lamprey larvae, *Petromyzon marinus*, with typical and atypical gonads. Copeia 1998:1070–1075

Barrett RDH, Rogers SM, Schluter D (2008) Natural selection on a major armor gene in threespine stickleback. Science 322:255–257

Bartels H, Fazekas U, Youson JH, Potter IC (2011) Changes in the cellular composition of the gill epithelium during the life cycle of a nonparasitic lamprey: functional and evolutionary implications. Can J Zool 89:538–545

Bartels H, Docker MF, Fazekas U, Potter IC (2012) Functional and evolutionary implications of the cellular composition of the gill epithelium of feeding adults of a freshwater parasitic species of lamprey, *Ichthyomyzon unicuspis*. Can J Zool 90:1278–1283

Bartels H, Docker MF, Krappe M et al (2015) Variations in the presence of chloride cells in the gills of lampreys (Petromyzontiformes) and their evolutionary implications. J Fish Biol 86:1421–1428

Bartels H, Wrede C, Przybylski M, Potter IC, Docker MF (2017) Implications of absence of seawater-type mitochondria-rich cells and results of molecular analyses for derivation of the non-parasitic Ukrainian brook lamprey *Eudontomyzon mariae*. Environ Biol Fish 100:509–518

Beamish FWH (1980a) Biology of the North American anadromous sea lamprey, *Petromyzon marinus*. Can J Fish Aquat Sci 37:1924–1943

Beamish FWH (1980b) Osmoregulation in juvenile and adult lampreys. Can J Fish Aquat Sci 37:1739–1750

Beamish FWH, Medland TE (1988) Metamorphosis of the mountain brook lamprey *Ichthyomyzon greeleyi*. Environ Biol Fish 23:45–54

Beamish FWH, Potter IC (1975) The biology of the anadromous sea lamprey (*Petromyzon marinus*) in New Brunswick. J Zool 177:57–72

Beamish FWH, Thomas EJ (1983) Potential and actual fecundity of the "paired" lampreys *Ichthyomyzon gagei* and *I. castaneus*. Copeia 1983:367–374

Beamish FWH, Strachan PD, Thomas E (1978) Osmotic and ionic performance of the anadromous sea lamprey, *Petromyzon marinus*. Comp Biochem Physiol 60A:435–443

Beamish FWH, Potter IC, Thomas E (1979) Proximate composition of the adult anadromous sea lamprey, *Petromyzon marinus*, in relation to feeding, migration and reproduction. J Anim Ecol 48:1–19

Beamish RJ (1980) Adult biology of the river lamprey (*Lampetra ayresi*) and the Pacific lamprey (*Lampetra tridentata*) from the Pacific coast of Canada. Can J Fish Aquat Sci 37:1906–1923

Beamish RJ (1982) *Lampetra macrostoma*, a new species of freshwater parasitic lamprey from the west coast of Canada. Can J Fish Aquat Sci 39:736–747

Beamish RJ (1985) Freshwater parasitic lamprey on Vancouver Island and a theory of the evolution of the freshwater parasitic and nonparasitic life history types. In: Foreman RE, Gorbman A, Dodd JM, Olsson R (eds) Evolutionary biology of primitive fishes. Plenum Publ Corp, New York, pp 123–140

Beamish RJ (1987) Evidence that parasitic and nonparasitic life history types are produced by one population of lamprey. Can J Fish Aquat Sci 44:1779–1782

Beamish RJ (2001) Updated status of the Vancouver Island Lake Lamprey, *Lampetra macrostoma*, in Canada. Can Field-Nat 115:127–130

Beamish RJ (2010) The use of gill pore papillae in the taxonomy of lampreys. Copeia 2010:618–628

Beamish R (2016) The structure and taxonomy of the gill pore in lampreys of the genus *Entosphenus*. In: Orlov A, Beamish R (eds) Jawless fishes of the world, vol 1. Cambridge Scholars Publishing, Newcastle upon Tyne, pp 126–153

Beamish RJ, Levings CD (1991) Abundance and freshwater migrations of the anadromous parasitic lamprey, *Lampetra tridentata*, in a tributary of the Fraser River, British Columbia. Can J Fish Aquat Sci 48:1250–1263

Beamish RJ, Neville CM (1992) The importance of size as an isolating mechanism in lampreys. Copeia 1992:191–196

Beamish RJ, Neville CM (1995) Pacific salmon and Pacific herring mortalities in the Fraser River plume caused by river lamprey (*Lampetra ayresii*). Can J Fish Aquat Sci 52:644–650

Beamish RJ, Northcote TG (1989) Extinction of a population of anadromous parasitic lamprey, *Lampetra tridentata*, upstream of an impassable dam. Can J Fish Aquat Sci 46:420–425

Beamish RJ, Withler RE (1986) A polymorphic population of lampreys that may produce parasitic and a nonparasitic varieties. In: Uyeno T, Arai R, Taniuchi T, Matsuura K (eds) Indo-Pacific fish biology, Proceedings of the second international conference on Indo-Pacific fishes. Ichthyological Society of Japan, Tokyo, pp 31–49

Beamish RJ, Youson JH (1987) Life history and abundance of young adult *Lampetra ayresi* in the Fraser River and their possible impact on salmon and herring stocks in the Strait of Georgia. Can J Fish Aquat Sci 44:525–537

Beamish R, Withler R, Wade J, Beacham T (2016) A nonparasitic lamprey produces a parasitic life history type: the Morrison Creek lamprey enigma. In: Orlov A, Beamish R (eds) Jawless fishes of the world, vol 1. Cambridge Scholars Publishing, Newcastle upon Tyne, pp 191–230

Bell MA, Aguirre WE, Buck NJ (2004) Twelve years of contemporary armor evolution in a three-spine stickleback population. Evolution 58:814–824

Benesh DP, Chubb JC, Parker GA (2013) Complex life cycles: why refrain from growth before reproduction in the adult niche? Am Nat 181:39–51

Berezina NA, Strelnikova AP (2010) The role of the introduced amphipod *Gmelinoides fasciatus* and native amphipods as fish food in two large-scale north-western Russian inland water bodies: Lake Ladoga and Rybinsk Reservoir. J Appl Ichthyol 26:89–95

Berg LS (1931) A review of the lampreys of the northern hemisphere. Annuaire du Musée Zoologique de l'Académie des Sciences de l'URSS 32:87–116+8 pls

Berg LS (1948) Freshwater fishes of the USSR and adjacent countries. Guide to the Fauna of the USSR 1(27):1–504

Bergstedt RA, Swink WD (1995) Seasonal growth and duration of the parasitic life stage of the landlocked sea lamprey (*Petromyzon marinus*). Can J Fish Aquat Sci 52:1257–1264

Bierne N, David P, Boudry P, Bonhomme F (2002) Assortative fertilization and selection at larval stage in the mussels *Mytilus edulis* and *M. galloprovincialis*. Evolution 56:292–298

Bierne N, Bonhomme F, Boudry P, Szulkin M, David P (2006) Fitness landscapes support the dominance theory of post-zygotic isolation in the mussels *Mytilus edulis* and *M. galloprovincialis*. Proc R Soc B Biol Sci 273:1253–1260

Bierne N, Welch J, Loire E, Bonhomme F, David P (2011) The coupling hypothesis: why genome scans may fail to map local adaptation genes. Mol Ecol 20:2044–2072

Bierne N, Gagnaire PA, David P (2013) The geography of introgression in a patchy environment and the thorn in the side of ecological speciation. Curr Zool 59:72–86

Bigelow HB, Schroeder WC (1948) Cyclostomes. Memoir of Sears Foundation for Marine Research No. 1, Part 1:29–58

Bishop CD, Erezyilmaz DF, Flatt T et al (2006) What is metamorphosis? Integr Comp Biol 46:655–661

Blackburn TM, Lockwood JL, Cassey P (2015) The influence of numbers on invasion success. Mol Ecol 24:1942–1953

Bogaevskii VT (1949) On possibilities of commercial fishery of lamprey in Amur region. Rybn Khoz (Moscow) 7:22–24 [in Russian]

Boguski DA, Reid SB, Goodman DH, Docker MF (2012) Genetic diversity, endemism and phylogeny of lampreys within the genus *Lampetra sensu stricto* (Petromyzontiformes: Petromyzontidae) in western North America. J Fish Biol 81:1891–1914

Bond CE, Kan TT (1973) *Lampetra (Entosphenus) minima* n. sp., a dwarfed parasitic lamprey from Oregon. Copeia 1973:568–574

Bond CE, Kan TT (1986) Systematics and evolution of the lampreys of Oregon. In: Uyeno T, Arai R, Taniuchi T, Matsuura K (eds) Indo-Pacific fish biology, Proceedings of the second international conference on Indo-Pacific fishes. Ichthyological Society of Japan, Tokyo, p 919

Bracken FS, Hoelzel A, Hume JB, Lucas MC (2015) Contrasting population genetic structure among freshwater-resident and anadromous lampreys: the role of demographic history, differential dispersal and anthropogenic barriers to movement. Mol Ecol 24:1188–1204

Briski E, Chan FT, Darling JA et al (2018) Beyond propagule pressure: importance of selection during the transport stage of biological invasions. Front Ecol Environ 16:345–353

Bryan MB, Zalinksi D, Filcek KB et al (2005) Patterns of invasion and colonization of the sea lamprey (*Petromyzon marinus*) in North America as revealed by microsatellite genotypes. Mol Ecol 14:3757–3773

Chang MM, Zhang JY, Miao DD (2006) A lamprey from the Cretaceous Jehol biota of China. Nature 441:972–974

Chang MM, Wu F, Miao D, Zhang J (2014) Discovery of fossil lamprey larva from the Lower Cretaceous reveals its three-phased life cycle. Proc Natl Acad Sci USA 111:15486–15490

Chapman B, Skov C, Hulthen K et al (2012) Partial migration in fishes: definitions, methodologies and taxonomic distribution. J Fish Biol 81:479–499

Childs D (2013) Genomic and phylogenetic assessment of sea lamprey (*Petromyzon marinus*) Hox genes and analysis of Hox genes in association with myomeres across multiple lamprey genera. MSc thesis, University of Manitoba, Winnipeg, MB

Clarke WC, Beamish RJ (1988) Response of recently metamorphosed anadromous parasitic lamprey (*Lampetra tridentata*) to confinement in fresh water. Can J Fish Aquat Sci 45:42–47

Clemens BJ, Mesa MG, Magie RJ, Young DA, Schreck CB (2012) Pre-spawning migration of adult Pacific lamprey, *Entosphenus tridentatus*, in the Willamette River, Oregon, U.S.A. Environ Biol Fish 93:245–254

Clemens BJ, van de Wetering S, Sower SA, Schreck CB (2013) Maturation characteristics and life-history strategies of the Pacific lamprey, *Entosphenus tridentatus*. Can J Zool 91:775–788

Cochran PA (2008) Observations on giant American brook lampreys (*Lampetra appendix*). J Freshw Ecol 23:161–164

Cochran PA (2009) Predation on lampreys. In: Brown LR, Chase SD, Mesa MG, Beamish RJ, Moyle PB (eds) Biology, management, and conservation of lampreys in North America. Am Fish Soc Symp 72:139–151

Cochran PA, Lyons J (2016) The silver lamprey and the paddlefish. In: Orlov A, Beamish RJ (eds) Jawless fishes of the world, vol 2. Cambridge Scholars Publishing, Newcastle upon Tyne, pp 214–233

Cochran PA, Lyons J, Gehl MR (2003) Parasitic attachments by overwintering silver lampreys, *Ichthyomyzon unicuspis*, and chestnut lampreys, *Ichthyomyzon castaneus*. Environ Biol Fish 68:65–71

Cochran PA, Bloom DD, Wagner RJ (2008) Alternative reproductive behaviors in lampreys and their significance. J Freshw Ecol 23:437–444

Cole WC, Youson JH (1981) The effect of pinealectomy, continuous light, and continuous darkness on metamorphosis of anadromous sea lampreys, *Petromyzon marinus* L. J Exp Zool 218:397–404

Colosimo PF, Hosemann KE, Balabhadra S et al (2005) Widespread parallel evolution in sticklebacks by repeated fixation of ectodysplasin alleles. Science 307:1928–1933

COSEWIC (2008) COSEWIC assessment and update status report on the Vancouver Lamprey *Lampetra macrostoma* in Canada. Committee on the Status of Endangered Wildlife in Canada, Ottawa

COSEWIC (2016) COSEWIC assessment and status report on the Sockeye Salmon *Oncorhynchus nerka*, Sakinaw population, in Canada. Committee on the Status of Endangered Wildlife in Canada, Ottawa

Czesny S, Epifanio J, Michalak P (2012) Genetic divergence between freshwater and marine morphs of alewife (*Alosa pseudoharengus*): a 'next-generation' sequencing analysis. PLoS ONE 7:e31803

D'Aloia CC, Azodi CB, Sheldon SP, Trombulak SC, Ardren WR (2015) Genetic models reveal historical patterns of sea lamprey population fluctuations within Lake Champlain. PeerJ 3:e1369

Dalley EL, Andrews CW, Green JM (1983) Precocious male Atlantic salmon parr (*Salmo salar*) in insular Newfoundland. Can J Fish Aquat Sci 40:647–652

Daniels RA (2001) Untested assumptions: the role of canals in the dispersal of sea lamprey, alewife, and other fishes in the eastern United States. Environ Biol Fish 60:309–329

Davis RM (1967) Parasitism by newly-transformed anadromous sea lampreys on landlocked salmon and other fishes in a coastal Maine lake. Trans Am Fish Soc 96:11–16

Dawson HA, Quintella BR, Almeida PR, Treble AJ, Jolley JC (2015) The ecology of larval and metamorphosing lampreys. In: Docker MF (ed) Lampreys: biology, conservation and control, vol 1. Springer, Dordrecht, pp 75–137

Dempson JB, Porter TR (1993) Occurrence of sea lamprey, *Petromyzon marinus*, in a Newfoundland River, with additional records from the Northwest Atlantic. Can J Fish Aquat Sci 50:1265–1269

Dicken SN, Dicken EF (1985) The legacy of ancient Lake Modoc: a historical geography of the Klamath Lakes basin. Dicken and Dicken, Eugene, OR

Diogo R, Ziermann JM (2015) Development, metamorphosis, morphology, and diversity: the evolution of chordate muscles and the origin of vertebrates. Dev Dyn 244:1046–1057

Docker MF (2009) A review of the evolution of nonparasitism in lampreys and an update of the paired species concept. In: Brown LR, Chase SD, Mesa MG, Beamish RJ, Moyle PB (eds) Biology, management, and conservation of lampreys in North America. Am Fish Soc Symp 72:71–114

Docker MF, Beamish FWH (1991) Growth, fecundity, and egg size of least brook lamprey, *Lampetra aepyptera*. Environ Biol Fish 31:219–227

Docker MF, Youson JH, Beamish RJ, Devlin RH (1999) Phylogeny of the lamprey genus *Lampetra* inferred from mitochondrial cytochrome *b* and ND3 gene sequences. Can J Fish Aquat Sci 56:2340–2349

Docker MF, Mandrak NE, Heath DD (2012) Contemporary gene flow between "paired" silver (*Ichthyomyzon unicuspis*) and northern brook (*I. fossor*) lampreys: implications for conservation. Conserv Genet 13:823–835

Docker MF, Hume JB, Clemens BJ (2015) Introduction: a surfeit of lampreys. In: Docker MF (ed) Lampreys: biology, conservation and control, vol 1. Springer, Dordrecht, pp 1–34

Dodson JJ, Laroche J, Lecomte F (2009) Contrasting evolutionary pathways of anadromy in euteleostean fishes. In: Haro A, Smith KL, Rulifson RA et al (eds) Challenges for diadromous fishes in a dynamic global environment. Am Fish Soc Symp 69:63–77

Dymond JR, Hart JL, Pritchard AL (1929) The fishes of the Canadian waters of Lake Ontario. University of Toronto Studies, Publication of the Ontario Fisheries Research Laboratory 37, Toronto

Emery L (1981) Range extension of pink salmon (*Oncorhynchus gorbuscha*) into the lower Great Lakes. Fisheries 6:7–10

Enequist P (1937) Das Bachneunauge als Ökologische Modifikation des Flussneunauges—Über die Fluss—und Bachneunaugen Schwedens Vorläufige Mitteilung. Ark Zool 29:1–22

Engel MS (2015) Insect evolution. Curr Biol 25:868–872

Ermolin VP (2010) Composition of the ichthyofauna of the Saratov Reservoir. J Ichthyol 50:211–215

Eshenroder RL (2009) Comment: mitochondrial DNA analysis indicates sea lampreys are indigenous to Lake Ontario. Trans Am Fish Soc 138:1178–1189

Eshenroder RL (2014) The role of the Champlain Canal and Erie Canal as putative corridors for colonization of Lake Champlain and Lake Ontario by Sea Lampreys. Trans Am Fish Soc 143:634–649

Eshenroder R, Amatangelo KL (2002) Reassessment of the lake trout population collapse in Lake Michigan during the 1940s. Great Lakes Fish Comm Tech Rep 65:1–32

Espanhol R, Almeida PR, Alves MJ (2007) Evolutionary history of lamprey paired species *Lampetra fluviatilis* (L.) and *Lampetra planeri* (Bloch) as inferred from mitochondrial DNA variation. Mol Ecol 16:1909–1924

Evans TM (2017) Are lampreys homebodies? Studying ammocoetes with open population models. Ecol Freshw Fish 26:168–180

Evans TM, Janvier P, Docker MF (2018) The evolution of lamprey (Petromyzontida) life history and the origin of metamorphosis. Rev Fish Biol Fish 28:825–838

Farlinger SP, Beamish RJ (1984) Recent colonization of a major salmon-producing lake in British Columbia by Pacific lamprey (*Lampetra tridentata*). Can J Fish Aquat Sci 41:278–285

Ferreira-Martins D, Coimbra J, Antunes C, Wilson J (2016) Effects of salinity on upstream-migrating, spawning sea lamprey, *Petromyzon marinus*. Conserv Physiol 4:cov064

Finstad AG, Hein CL (2012) Migrate or stay: terrestrial primary productivity and climate drive anadromy in Arctic char. Global Change Biol 18:2487–2497

Fleming IA (1996) Reproductive strategies of Atlantic salmon: ecology and evolution. Rev Fish Biol Fish 6:379–416

Fontaine M (1930) Recherches sur le milieu interieur de la lamproie marine *(Petromyzon marinus)*. Ses variations en fonction de celles du milieu extérieur. C R Hebd Séances Acad Sci 191:680–682

Foote CJ, Brown GS (1998) Ecological relationship between freshwater sculpins (genus *Cottus*) and beach-spawning sockeye salmon *(Oncorhynchus nerka)* in Iliamna Lake, Alaska. Can J Fish Aquat Sci 55:1524–1533

Fraser DJ, Bernatchez L (2001) Adaptive evolutionary conservation: towards a unified concept for defining conservation units. Mol Ecol 10:2741–2752

Freyhof J, Kottelat M (2008a) *Eudontomyzon* sp. nov. 'migratory'. The IUCN Red List of Threatened Species 2008: e.T135505A4134478. http://dx.doi.org/10.2305/IUCN.UK.2008. RLTS.T135505A4134478.en. Accessed 02 August 2018

Freyhof J, Kottelat M (2008b) *Caspiomyzon wagneri*. The IUCN Red List of Threatened Species 2008: T135706A4187207. http://dx.doi.org/10.2305/IUCN.UK.2008.RLTS. T135706A4187207.en. Accessed 02 August 2018

Gage SH (1893) The lake and brook lampreys of New York, especially those of Cayuga and Seneca lakes. Wilder quarter-century book. Comstock, Ithaca, NY, pp 421–493

Gage SH (1928) A biological survey of the Oswego River system–life history and economics. Supplemental to the seventeenth annual report of the New York State Conservation Department, 1927. New York State Conservation Department, Albany, pp 158–191

Galtier N, Nabholz B, Glemin S, Hurst G (2009) Mitochondrial DNA as a marker of molecular diversity: a reappraisal. Mol Ecol 18:4541–4550

Gaygalas KS, Matskevichyus AP (1968) Fishing for the river lamprey [*Lampetra fluviatilis* (L.)] in the basin of the Nyamunas River; some features and potentialities. Prob Ichthyol 8:169–176

Gelmond O, von Hippel F, Christy M (2009) Rapid ecological speciation in three-spined stickleback *Gasterosteus aculeatus* from Middleton Island, Alaska: the roles of selection and geographic isolation. J Fish Biol 75:2037–2051

Gess RW, Coates MI, Rubidge BS (2006) A lamprey from the Devonian period of South Africa. Nature 443:981–984

Gharrett AJ, Thomason MA (1987) Genetic changes in pink salmon *(Oncorhynchus gorbuscha)* following their introduction into the Great Lakes. Can J Fish Aquat Sci 44:787–792

Gibling MR, Davies NS (2012) Palaeozoic landscapes shaped by plant evolution. Nat Geosci 5:99–105

Gill HS, Renaud CB, Chapleau F, Mayden RL, Potter IC (2003) Phylogeny of living parasitic lampreys (Petromyzontiformes) based on morphological data. Copeia 2003:687–703

Gompel N, Prud'homme B (2009) The causes of repeated genetic evolution. Dev Biol 332:36–47

Good C, Davidson J (2016) A review of factors influencing maturation of Atlantic salmon, *Salmo salar*, with focus on water recirculation aquaculture system environments. J World Aquac Soc 47:605–632

Goodman DH, Kinziger AP, Reid SB, Docker MF (2009) Morphological diagnosis of *Entosphenus* and *Lampetra* ammocoetes (Petromyzontidae) in Washington, Oregon, and California. In: Brown LR, Chase SD, Moyle PB, Beamish RJ, Mesa MG (eds) Biology, management, and conservation of lampreys in North America. Am Fish Soc Symp 72:223–232

Goodwin CE, Griffiths D, Dick TA, Elwood RW (2006) A freshwater-feeding *Lampetra fluviatilis* L. population in Lough Neagh, Northern Ireland. J Fish Biol 68:628–633

Graur D, Martin W (2004) Reading the entrails of chickens: molecular timescales of evolution and the illusion of precision. Trends Genet 20:80–86

Greeley JR (1930) Fishes of the Lake Champlain watershed. A biological survey of the Champlain watershed (1929) In: Supplemental to the nineteenth annual report of the New York State Conservation Department. New York State Conservation Department, Albany, pp 44–87

Greene E (1999) Phenotypic variation in larval development and evolution: polymorphism, polyphenism, and developmental reaction norms. In: Hall BK, Wake MH (eds) The origin and evolution of larval forms. Academic Press, San Diego, pp 379–410

Griffith RW (1987) Freshwater or marine origin of the vertebrates. Comp Biochem Physiol 87A:523–531

Griffith RW (1994) The life of the first vertebrates. BioScience 44:408–417

Gross MR, Coleman RM, McDowall RM (1988) Aquatic productivity and the evolution of diadromous fish migration. Science 239:1291–1293

Hale MC, Thrower FP, Berntson EA, Miller MR, Nichols KM (2013) Evaluating adaptive divergence between migratory and non-migratory ecotypes of a salmonid fish, *Oncorhynchus mykiss*. Genes Genomes Genet 3:1273–1285

Hall JD (1963) An ecological study of the chestnut lamprey, *Ichthyomyzon castaneus* Girard, in the Manistee River, Michigan. PhD thesis, University of Michigan, Ann Arbor, MI

Halliday RG (1991) Marine distribution of the sea lamprey (*Petromyzon marinus*) in the northwest Atlantic. Can J Fish Aquat Sci 48:832–842

Halstead LB (1985) The vertebrate invasion of fresh water. Phil Trans R Soc Lond 309B:243–258

Hamilton JB, Curtis GL, Snedaker SM, White DK (2005) Distribution of anadromous fishes in the Upper Klamath River watershed prior to hydropower dams—a synthesis of the historical evidence. Fisheries 30:10–20

Hardisty MW (1964) The fecundity of lampreys. Arch Hydrobiol 60:340–357

Hardisty MW (1969) A comparison of gonadal development in the ammocoetes of the landlocked and anadromous forms of the sea lamprey *Petromyzon marinus* L. J Fish Biol 2:153–166

Hardisty MW (1971) Gonadogenesis, sex differentiation and gametogenesis. In: Hardisty MW, Potter IC (eds) The biology of lampreys, vol 1. Academic Press, London, pp 295–360

Hardisty MW (1986a) *Lampetra fluviatilis* (Linnaeus, 1758). In: Holčík J (ed) The freshwater fishes of Europe, part I, vol 1, Petromyzontiformes. Aula, Wiesbaden, pp 249–278

Hardisty MW (1986b) *Petromyzon marinus* Linnaeus, 1758. In: Holčík J (ed) The freshwater fishes of Europe, vol 1, part 1, Petromyzontiformes. Aula, Wiesbaden, pp 92–116

Hardisty MW (2006) Lampreys. Life without jaws. Forrest Text, Tresaith, UK

Hardisty MW, Huggins RJ (1973) Lamprey growth and biological conditions in the Bristol Channel region. Nature 243:229–231

Hardisty MW, Potter IC (1971a) Paired species. In: Hardisty MW, Potter IC (eds) The biology of lampreys, vol 1. Academic Press, London, pp 249–277

Hardisty MW, Potter IC (1971b) The general biology of adult lampreys. In: Hardisty MW, Potter IC (eds) The biology of lampreys, vol 1. Academic Press, London, pp 127–205

Hardisty MW, Potter IC, Hilliard RW (1989) Physiological adaptations of the living agnathans. Trans R Soc Edinb Earth Sci 80:241–254

Hart MW, Grosberg RK (2009) Caterpillars did not evolve from onychophorans by hybridogenesis. Proc Natl Acad Sci USA 106:19906–19909

Haug JT, Haug C, Garwood RJ (2016) Evolution of insect wings and development—new details from Palaeozoic nymphs. Biol Rev 91:53–69

Hauser DD, Allen CS, Rich HB, Quinn TP (2008) Resident harbor seals (*Phoca vitulina*) in Iliamna Lake, Alaska: summer diet and partial consumption of adult sockeye salmon (*Oncorhynchus nerka*). Aquat Mamm 34:303–309

Heard WR (1966) Observations on lampreys in the Naknek River system of southwest Alaska. Copeia 1966:332–339

Heath DD, Heath JW, Iwama GK (1991) Maturation in chinook salmon, *Oncorhynchus tshawytscha* (Walbaum): early identification based on the development of a bimodal weight frequency distribution. J Fish Biol 39:565–575

Hecht BC, Campbell NR, Holecek DE, Naru SR (2013) Genome-wide association reveal genetic basis for the propensity to migrate in wild populations of rainbow and steelhead trout. Mol Ecol 22:3061–3076

Hendler G, Dojiri M (2009) The contrariwise life of a parasitic, pedomorphic copepod with a non-feeding adult: ontogenesis, ecology, and evolution. Invertebr Biol 128:65–82

Henry FE, Sugino K, Tozer A, Branco T, Sternson SM (2015) Cell type-specific transcriptomics of hypothalamic energy-sensing neuron responses to weight-loss. eLife 4:e09800

Hess JE, Campbell NR, Close DA, Docker MF, Narum SR (2013) Population genomics of Pacific lamprey: adaptive variation in a highly dispersive species. Mol Ecol 22:2898–2916

Hess JE, Caudill CC, Keefer ML et al (2014) Genes predict long distance migration and large body size in a migratory fish, Pacific lamprey. Evol Appl 7:1192–1208

Heyland A, Hodin J, Reitzel AM (2005) Hormone signaling in evolution and development: a non-model system approach. BioEssays 27:64–75

Hockla A, Radisky DC, Radisky ES (2010) Mesotrypsin promotes malignant growth of breast cancer cells through shedding of CD109. Breast Cancer Res Treat 124:27–38

Hoff JG (1988) Some aspects of the ecology of the American brook lamprey, *Lampetra appendix*, in the Mashpee River, Cape Cod, Massachusetts. Can Field-Nat 102:735–737

Hoffman EA, Pfennig DW (1999) Proximate causes of cannibalistic polyphenism in larval tiger salamanders. Ecology 80:1076–1080

Hogg R, Coghlan SM Jr, Zydlewski J (2013) Anadromous sea lampreys recolonize a Maine coastal river tributary after dam removal. Trans Am Fish Soc 142:1381–1394

Hohenlohe PA, Hand BK, Andrews KR, Luikart G (2019) Population genomics provides key insights in ecology and evolution. In: Rajora OP (ed) Population genomics: concepts, approaches and applications. Springer, Dordrecht, pp 483–510

Holbrook CM, Johnson NS, Steibel JP et al (2014) Estimating reach-specific fish movement probabilities in rivers with a Bayesian state-space model: application to sea lamprey passage and capture at dams. Can J Fish Aquat Sci 71:1713–1729

Holčík J (1986) *Lethenteron japonicum* (Martens, 1868). In: Holčík J (ed) The freshwater fishes of Europe, vol 1, part I, Petromyzontiformes. Aula, Wiesbaden, pp 198–219

Holecek DE, Scarnecchia DL, Miller SE (2012) Smoltification in an impounded, adfluvial Redband Trout population upstream from an impassable dam: does it persist? Trans Am Fish Soc 141:68–75

Holland ND, Chen JY (2001) Origin and early evolution of the vertebrates: new insights from advances in molecular biology, anatomy, and palaeontology. BioEssays 23:142–151

Hubbs CL (1925) The life cycle and growth of lampreys. Pap Mich Acad Sci Arts Lett 4:587–603

Hubbs C, Lagler K (1947) Fishes of the Great Lakes region. Cranbrook Inst Sci Bull 26:1–186

Hubbs CL, Potter IC (1971) Distribution, phylogeny and taxonomy. In: Hardisty MW, Potter IC (eds) The biology of lampreys, vol 1. Academic Press, London, pp 1–65

Hubbs CL, Trautman MB (1937) A revision of the lamprey genus *Ichthyomyzon*. Misc Publ Mus Zool Univ Mich 35:7–109

Huggins RJ, Thompson A (1970) Communal spawning of brook and river lampreys, *Lampetra planeri* Bloch and *Lampetra fluviatilis*. J Fish Biol 2:53–54

Hughes JM, Schmidt DJ, Finn DS (2009) Genes in streams: using DNA to understand the movement of freshwater fauna and their riverine habitat. BioScience 59:573–583

Hume JB (2013) The evolutionary ecology of lampreys (Petromyzontiformes). PhD thesis, University of Glasgow, Glasgow

Hume JB, Adams CE, Bean CW, Maitland PS (2013a) Evidence of a recent decline in river lamprey *Lampetra fluviatilis* parasitism of a nationally rare whitefish *Coregonus lavaretus*: is there a diamond in the ruffe *Gymnocephalus cernuus*? J Fish Biol 82:1708–1716

Hume JB, Adams CE, Mable B, Bean CW (2013b) Sneak male mating tactics between lampreys (Petromyzontiformes) exhibiting alternative life-history strategies. J Fish Biol 82:1093–1100

Hume JB, Adams CE, Mable B, Bean C (2013c) Post-zygotic hybrid viability in sympatric species pairs: a case study from European lampreys. Biol J Linn Soc 108:378–383

Hume JB, Recknagel H, Bean CW, Adams CE, Mable BK (2018) RADseq and mate choice assays reveal unidirectional gene flow among three lamprey ecotypes despite weak assortative mating: insights into the formation and stability of multiple ecotypes in sympatry. Mol Ecol 27:4572–4590

Inger R, McDonald RA, Rogowski D et al (2010) Do non-native invasive fish support elevated lamprey populations? J Appl Ecol 47:121–129

Irvine SQ, Carr JL, Bailey WJ et al (2002) Genomic analysis of Hox clusters in the sea lamprey *Petromyzon marinus*. J Exp Zool Mol Evol Dev 294:47–62

Iwata A, Hamada K (1986) A dwarf male of the Arctic lamprey *Lethenteron japonicum* from the Assabu River Hokkaido Japan. Bull Fac Fish Hokkaido Univ 37:17–22

Janvier P (1996) Early vertebrates. Oxford University Press, Oxford

Janvier P (2007) Living primitive fishes and fishes from deep time. In: McKenzie D, Farrell A, Brauner C (eds) Fish physiology: primitive fishes, vol 26. Academic Press, New York, pp 1–51

Janvier P (2008) Early jawless vertebrates and cyclostome origins. Zool Sci 25:1045–1056

Janvier P, Lund R (1983) *Hardistiella montanensis* n. gen. et sp. (Petromyzontida) from the lower carboniferous of Montana, with remarks on the affinities of the lampreys. J Vertebr Paleontol 2:407–413

Janvier P, Lund R, Grogan ED (2004) Further consideration of the earliest known lamprey, *Hardistiella montanensis* Janvier and Lund, 1983, from the carboniferous of bear gulch, Montana, USA. J Vertebr Paleontol 24:742–743

Janzen D (1984) Two ways to be a tropical big moth: Santa Rosa saturniids and sphingids. Oxf Surv Evol Biol 1:85–140

Johns GC, Avise JC (1998) A comparative summary of genetic distances in the vertebrates from the mitochondrial cytochrome *b* gene. Mol Biol Evol 15:1481–1490

Johnson CK, Voss SR (2013) Salamander paedomorphosis: linking thyroid hormone to life history and life cycle evolution. Curr Top Dev Biol 103:229–258

Johnson NS, Buchinger TJ, Li W (2015) Reproductive ecology of lampreys. In: Docker MF (ed) Lampreys: biology, conservation and control, vol 1. Springer, Dordrecht, pp 265–303

Johnson NS, Twohey MB, Miehls SM et al (2016) Evidence that sea lampreys (*Petromyzon marinus*) complete their life cycle within a tributary of the Laurentian Great Lakes by parasitizing fishes in inland lakes. J Great Lakes Res 42:90–98

Johnson W (1982) Body lengths, body weight and fecundity of sea lampreys (*Petromyzon marinus*) from Green Bay, Lake Michigan. Trans Wis Acad Sci Arts Lett 70:73–77

Jonsson B, Jonsson N (1993) Partial migration: niche shift versus sexual maturation in fishes. Rev Fish Biol Fish 3:348–365

Jonsson B, Jonsson N (2006) Life-history effects of migratory costs in anadromous brown trout. J Fish Biol 69:860–869

Jolley JC, Kovalchuk G, Docker MF (2016) River lamprey *Lampetra ayresii* outmigrant upstream of the John Day Dam in the Mid-Columbia River. Northwest Nat 97:48–52

Joss JMP (1973) Pineal-gonad relationships in lamprey *Lampetra fluviatilis*. Gen Comp Endocrinol 21:112–118

Kan TT (1975) Systematics, variation, distribution, and biology of lampreys of the genus *Lampetra* in Oregon. PhD thesis, Oregon State University, Corvallis, OR

Kan TT, Bond CE (1981) Notes on the biology of the Miller Lake lamprey *Lampetra (Entosphenus) minima*. Northwest Sci 55:70–74

Keefer ML, Taylor GA, Garletts DF et al (2013) High-head dams affect downstream fish passage timing and survival in the Middle Fork Willamette River. River Res Appl 29:483–492

Kelly FL, King JJ (2001) A review of the ecology and distribution of three lamprey species, *Lampetra fluviatilis* (L.), *Lampetra planeri* (Bloch) and *Petromyzon marinus* (L.): a context for conservation and biodiversity considerations in Ireland. Proc R Irish Acad 101B:165–185

Kim J, Mandrak NE (2016) Assessing the potential movement of invasive fishes through the Welland Canal. J Great Lakes Res 42:1102–1108

King JJ, O'Gorman (2018) Initial observations on feeding juvenile sea lamprey (*Petromyzon marinus* L.) in Irish lakes. Biol Environ Proc R Ir Acad 118B:113–120

Klemetsen A, Amundsen PA, Dempson JB et al (2003) Atlantic salmon *Salmo salar* L., brown trout *Salmo trutta* L. and Arctic charr *Salvelinus alpinus* (L.): a review of aspects of their life histories. Ecol Freshw Fish 12:1–59

Kostow K (2002) Oregon lampreys: natural history, status, and analysis of management issues. Oregon Department of Fish and Wildlife, Portland, OR

Kottelat M, Bogutskaya N, Freyhof J (2005) On the migratory Black Sea lamprey and the nomenclature of the ludoga, Peipsi and ripus whitefishes (Agnatha: Petromyzontidae; Teleostei: Coregonidae). Zoosyst Rossica 14:181–186

Kucheryavyi AV, Savvaitova KA, Pavlov DS et al (2007) Variations of life history strategy of the Arctic lamprey *Lethenteron camtschaticum* from the Utkholok River (Western Kamchatka). J Ichthyol 47:37–52

Kunnasranta M, Hyvarinen H, Sipila T, Medvedev N (2001) Breeding habitat and lair structure of the ringed seal (*Phoca hispida ladogensis*) in northern Lake Ladoga in Russia. Polar Biol 24:171–174

Kuraku S, Kuratani S (2006) Time scale for cyclostome evolution inferred with a phylogenetic diagnosis of hagfish and lamprey cDNA sequences. Zool Sci 23:1053–1064

Kuznetsov Y, Mosyagina M, Zelennikov O (2016) The formation of fecundity in ontogeny of lampreys. In: Orlov A, Beamish R (eds) Jawless fishes of the world, vol 1. Cambridge Scholars Publishing, Newcastle upon Tyne, pp 323–344

Lampman RT (2011) Passage, migration behavior, and autoecology of adult Pacific lamprey at Winchester Dam and within the North Umpqua River Basin, Oregon, USA. MS thesis, Oregon State University, Corvallis, OR

Lancaster J, Downes BJ (2013) Aquatic entomology. Oxford University Press, Oxford

Lang NJ, Roe KJ, Renaud CB et al (2009) Novel relationships among lampreys (Petromyzontiformes) revealed by a taxonomically comprehensive molecular dataset. In: Brown LR, Chase SD, Mesa MG, Beamish RJ, Moyle PB (eds) Biology, management, and conservation of lampreys in North America. Am Fish Soc Symp 72:41–55

Lanzing WJR (1959) Studies on the river lamprey, *Lampetra fluviatilis*, during its anadromous migration. Uitgeversmaatschappij, Utrecht, Utrecht

Laporte M, Dalziel AC, Martin N, Bernatchez L (2016) Adaptation and acclimation of traits associated with swimming capacity in Lake Whitefish (*Coregonus clupeaformis*) ecotypes. BMC Evol Biol 16:160

Lark JGI (1973) Early record of sea lamprey (*Petromyzon marinus*) from Lake Ontario. J Fish Res Board Can 30:131–133

Larson GL, Christie GC, Johnson DA et al (2003) The history of sea lamprey control in Lake Ontario and updated estimates of suppression targets. J Great Lakes Res 29(Suppl 1):637–654

Lasne E, Sabatie M-R, Evanno G (2010) Communal spawning of brook and river lampreys (*Lampetra planeri* and *L. fluviatilis*) is common in the Oir River (France). Ecol Freshw Fish 19:323–325

Laudet V (2011) The origins and evolution of vertebrate metamorphosis. Curr Biol 21:R726–R737

Lescak EA, Bassham SL, Catchen J et al (2015) Evolution of stickleback in 50 years on earthquake-uplifted islands. Proc Nat Acad Sci USA 112:E7204–E7212

Levin BA, Holčík J (2006) New data on the geographic distribution and ecology of the Ukrainian brook lamprey, *Eudontomyzon mariae* (Berg, 1931). Folia Zool 55:282–286

Levin B, Ermakov A, Ermakov O et al (2016) Ukrainian brook lamprey *Eudontomyzon mariae* (Berg): phylogenetic position, genetic diversity, distribution, and some data on biology. In: Orlov A, Beamish R (eds) Jawless fishes of the world, vol 1. Cambridge Scholars Publishing, Newcastle upon Tyne, pp 58–82

Li Y (2014) Phylogeny of the lamprey genus *Lethenteron* Creaser and Hubbs 1922 and closely related genera using the mitochondrial cytochrome b gene and nuclear gene introns. MSc thesis, University of Manitoba, Winnipeg, MB

Loman JCC (1912) Über die Naturgeschichte des Bachneunauges *Lampetra planeri*. Zool Jahrb (Suppl) 15:243–270

Lorion CM, Markle DF, Reid SB, Docker MF (2000) Redescription of the presumed-extinct Miller Lake lamprey, *Lampetra minima*. Copeia 2000:1019–1028

Lund R, Janvier P (1986) A second lamprey from the Lower Carboniferous (Namurian) of Bear Gulch, Montana (U.S.A.). Geobios 19:647–652

Lutz PL (1975) Adaptive and evolutionary aspects of the ionic content of fishes. Copeia 1975:369–373

Lyons J, Polaco OJ, Cochran PA (1996) Morphological variation among the Mexican lampreys (Petromyzontidae: *Lampetra*: subgenus *Tetrapleurodon*). Southwest Nat 41:365–374

MacKay HH, MacGillivray E (1949) Recent investigations on the sea lamprey, *Petromyzon marinus*, in Ontario. Trans Am Fish Soc 76:148–159

Maitland PS (1980) Review of the ecology of lampreys in northern Europe. Can J Fish Aquat Sci 37:1944–1952

Maitland PS, Morris KH, East K (1994) The ecology of lampreys (Petromyzonidae) in the Loch Lomond area. Hydrobiologia 290:105–120

Maitland PS, Renaud CB, Quintella BR, Close DA, Docker MF (2015) Conservation of native lampreys. In: Docker MF (ed) Lampreys: biology, conservation, and control, vol 1. Springer. Dordrecht, Netherlands, pp 375–427

Mallatt J (1984) Feeding ecology of the earliest vertebrates. Zool J Linn Soc 82:261–272

Mallatt J (1985) Reconstructing the life cycle and the feeding of ancestral vertebrates. In: Foreman RE, Gorbman A, Dodd JM, Olsson R (eds) Evolutionary biology of primitive fishes. Plenum, New York, pp 59–68

Malmqvist B (1983) Breeding behavior of brook lampreys *Lampetra planeri*: experiments on mate choice. Oikos 41:43–48

Mandrak NE, Crossman EJ (1992) Postglacial dispersal of freshwater fishes into Ontario. Can J Zool 70:2247–2259

Manion PJ (1972) Fecundity of the sea lamprey (*Petromyzon marinus*) in Lake Superior. Trans Am Fish Soc 101:718–720

Manion PJ, Purvis HA (1971) Giant American brook lampreys, *Lampetra lamottei*, in the upper Great Lakes. J Fish Res Board Can 28:616–620

Manzon RG, Youson JH, Holmes JA (2015) Lamprey metamorphosis. In: Docker MF (ed) Lampreys: biology, conservation and control, vol 1. Springer, Dordrecht, pp 139–214

Mariussen E, Fjeld E, Breivik K et al (2008) Elevated levels of polybrominated diphenyl ethers (PBDEs) in fish from Lake Mjøsa, Norway. Sci Total Environ 390:132–141

Marsden JE, Langdon RW (2012) The history and future of Lake Champlain's fishes and fisheries. J Great Lakes Res 38:19–34

Martin H, White MM (2008) Intraspecific phylogeography of the Least Brook Lamprey (*Lampetra aepyptera*). Copeia 2008:579–585

Mateus CS, Almeida PR, Quintella BR, Alves MJ (2011) MtDNA markers reveal the existence of allopatric evolutionary lineages in the threatened lampreys *Lampetra fluviatilis* (L.) and *Lampetra planeri* (Bloch) in the Iberian glacial refugium. Conserv Genet 12:1061–1074

Mateus CS, Alves MJ, Quintella BR, Almeida PR (2013a) Three new cryptic species of the lamprey genus *Lampetra* Bonnaterre, 1788 (Petromyzontiformes: Petromyzontidae) from the Iberian Peninsula. Contrib Zool 82:37–53

Mateus CS, Stange M, Berner D et al (2013b) Strong genome-wide divergence between sympatric European river and brook lampreys. Curr Biol 23:R649–R650

Mateus CS, Almeida PR, Mesquita N, Quintella BR, Alves MJ (2016) European lampreys: new Insights on postglacial colonization, gene flow and speciation. PLoS ONE 11:e0148107

Mathers JS, Beamish FWH (1974) Changes in serum osmotic and ionic concentration in landlocked *Petromyzon marinus*. Comp Biochem Physiol 49:677–688

Mavarez J, Audet C, Bernatchez L (2009) Major disruption of gene expression in hybrids between young sympatric anadromous and resident populations of brook charr (*Salvelinus fontinalis* Mitchill). J Evol Biol 22:1708–1720

May-McNally SL, Quinn TP, Woods PJ, Taylor EB (2015) Evidence for genetic distinction among sympatric ecotypes of Arctic char (*Salvelinus alpinus*) in south-western Alaskan lakes. Ecol Freshw Fish 24:562–574

McCauley DW, Docker MF, Whyard S, Li W (2015) Lampreys as diverse model organisms in the genomic era. BioScience 65:1046–1056

McCormick SD, Naiman RJ, Montgomery ET (1985) Physiological smolt characteristics of anadromous and non-anadromous brook trout (*Salvelinus fontinalis*) and Atlantic salmon (*Salmo salar*). Can J Fish Aquat Sci 42:529–538

McDowall RM (1988) Diadromy in fishes: migrations between freshwater and marine environments. Croom Helm, London

McDowall RM (1993) A recent marine ancestry for diadromous fishes? Sometimes yes, but mostly no. Environ Biol Fish 37:329–335

McKitrick MC (1993) Phylogenetic constraint in evolutionary theory: has it any explanatory power? Annu Rev Ecol Syst 24:307–330

McLaughlin RL, Smyth ER, Castro-Santos T et al (2013) Unintended consequences and trade-offs of fish passage. Fish Fish 14:580–604

McMahon DP, Hayward A (2016) Why grow up? A perspective on insect strategies to avoid metamorphosis. Ecol Entomol 41:505–515

McPhail JD, Lindsey CC (1970) Freshwater fishes of northwestern Canada and Alaska. Fish Res Board Can Bull 173, Ottawa

Meek SE (1889) The fishes of the Cayuga Lake basin. Ann N Y Acad Sci 4:297–316

Minelli A (2010) The origins of larval forms: what the data indicate, and what they don't. BioEssays 32:5–8

Morris K (1989) A multivariate morphometric and meristic description of a population of freshwater-feeding river lampreys Lampetra fluviatilis (L.), from Loch Lomond, Scotland. Zool J Linn Soc 96:357–371

Moser ML, Almeida PR, Kemp PS, Sorensen PW (2015) Lamprey spawning migration. In: Docker MF (ed) Lampreys: biology, conservation and control, vol 1. Springer, Dordrecht, pp 215–263

Moyle PB (2002) Inland fishes of California, revised and expanded. University of California Press, Berkeley, CA

Naseka AM, Tuniyev SB, Renaud CB (2009) Lethenteron ninae, a new nonparasitic lamprey species from the north-eastern Black Sea basin (Petromyzontiformes: Petromyzontidae). Zootaxa 2198:16–26

Nazari H, Abdoli A, Kiabi B, Renaud CB (2017) Biology and conservation status of the Caspian lamprey in Iran: a review. Bull Lampetra 8:6–32

Near TJ, Eytan RI, Dornburg A et al (2012) Resolution of ray-finned fish phylogeny and timing of diversification. Proc Natl Acad Sci USA 109:13698–13703

Neave FB, Mandrak NE, Docker MF, Noakes DL (2007) An attempt to differentiate sympatric Ichthyomyzon ammocoetes using meristic, morphological, pigmentation, and gonad analyses. Can J Zool 85:549–560

Neave FB, Steeves TB, Pratt TC et al (2019) Stream characteristics associated with feeding type in silver (Ichthyomyzon unicuspis) and northern brook (I. fossor) lampreys and tests for phenotypic plasticity. Environ Biol Fish (in press)

Neira FJ (1984) Biomorfologia de las lampreas parasitas chilenas Geotria australis Gray, 1851 y Mordacia lapicida (Gray, 1851) (Petromyzoniformes). Gayana Zool 48:3–40

Nelson JS, Crossman EJ, Espinosa-Pérez H et al (2004) Common and scientific names of fishes from the United States, Canada, and Mexico, 6th edition. American Fisheries Society Special Publication 29, Bethesda, MD

Nelson TC, Cresko WA (2018) Ancient genomic variation underlies repeated ecological adaptation in young stickleback populations. Evol Lett 2:9–21

Nilsen TO, Ebbesson LOE, Stefansson SO (2003) Smolting in anadromous and landlocked strains of Atlantic salmon (Salmo salar). Aquaculture 222:71–82

Nilsen TO, Ebbesson LO, Madsen SS et al (2007) Differential expression of gill Na+, K+ -ATPase α- and β-subunits, Na+, K+, 2Cl− cotransporter and CFTR anion channel in juvenile anadromous and landlocked Atlantic salmon Salmo salar. J Exp Biol 210:2885–2896

Northcutt RG, Gans C (1983) The genesis of neural crest and epidermal placodes: a reinterpretation of vertebrate origins. Q Rev Biol 58:1–28

Nursall JR, Buchwald D (1972) Life history and distribution of the Arctic lamprey (Lethenteron japonicum [Martens]) of Great Slave Lake, N.W.T. Fish Res Board Can Tech Rep 304:1–28

O'Connor LM (2001) Spawning success of introduced sea lampreys (Petromyzon marinus) in two streams tributary to Lake Ontario. MSc thesis, University of Guelph, Guelph, ON

Olsson IC, Greenberg LA, Bergman E, Wysujack K (2006) Environmentally induced migration: the importance of food. Ecol Lett 9:645–651

Ortiz-Barrientos D, Grealy A, Nosil P (2009) The genetics and ecology of reinforcement implications for the evolution of prezygotic isolation in sympatry and beyond. Ann N Y Acad Sci 1168:156–182

Orlov AM, Savinyh VF, Pelenev DV (2008) Features of the spatial distribution and size structure of the Pacific lamprey *Lampetra tridentata* in the North Pacific. Russ J Mar Biol 34:276–287

Orlov A, Baitalyuk A, Pelenev DV (2014) Distribution and size composition of the Arctic lamprey *Lethenteron camtschaticum* in the North Pacific. Oceanology 54:180–194

Ozerov MY, Veselov AJ, Lumme J, Primmer CR (2010) Genetic structure of freshwater Atlantic salmon (*Salmo salar* L.) populations from the lakes Onega and Ladoga of northwest Russia and implications for conservation. Conserv Genet 11:1711–1724

Paéz DJ, Bernatchez L, Dodson JJ (2011) Alternative life histories in the Atlantic salmon: genetic covariances within the sneaker sexual tactic in males. Proc Roy Soc B 278:2150–2158

Page RB, Boley MA, Smith JJ, Putta S, Voss SR (2010) Microarray analysis of a salamander hopeful monster reveals transcriptional signatures of paedomorphic brain development. BMC Evol Biol 10:199

Parker KA (2018) Evidence for the genetic basis and inheritance of ocean and river-maturing ecotypes of Pacific lamprey (*Entosphenus tridentatus*) in the Klamath River, California. MS thesis, Humboldt State University, Arcata, CA

Parkinson EA, Perrin CJ, Ramos-Espinoza D, Taylor EB (2016) Evidence for freshwater residualism in Coho Salmon, *Oncorhynchus kisutch*, from a watershed on the north coast of British Columbia. Can Field-Nat 130:336–343

Paris M, Laudet V (2008) The history of a developmental stage: metamorphosis in chordates. Genesis 46:657–672

Parry G (1960) The development of salinity tolerance in the salmon, *Salmo salar* (L.) and some related species. J Exp Biol 37:425–435

Patterson WP (1998) North American continental seasonality during the last millennium: high-resolution analysis of sagittal otoliths. Palaeogeogr Palaeoclimatol Palaeoecol 138:271–303

Pearce WA, Braem RA, Dustin SM, Tibbles JJ (1980) Sea lamprey (*Petromyzon marinus*) in the lower Great Lakes. Can J Fish Aquat Sci 37:1802–1810

Pereira A, Doadrio I, Robalo JI, Almada V (2014) Different stocks of brook lamprey in Spain and their origin from *Lampetra fluviatilis* at two distinct times and places. J Fish Biol 85:1793–1798

Piavis GW, Howell JH, Smith AJ (1970) Experimental hybridization among five species of lampreys from the Great Lakes. Copeia 1970:29–37

Pickering AD, Morris R (1970) Osmoregulation of *Lampetra fluviatilis* L. and *Petromyzon marinus* (Cyclostomata) in hyperosmotic solutions. J Exp Biol 53:231–243

Piersma T, Drent J (2003) Phenotypic flexibility and the evolution of organismal design. Trends Ecol Evol 18:228–233

Pletcher FT (1963) The life history and distribution of lampreys in the Salmon and certain other rivers in British Columbia, Canada. PhD thesis, University of British Columbia, Vancouver

Podjasek J, Bosnjak L, Brooker D, Mondor E (2005) Alarm pheromone induces a transgenerational wing polyphenism in the pea aphid, *Acyrthosiphon pisum*. Can J Zool 83:1138–1141

Polacek MC, Baldwin CM, Knuttgen K (2006) Status, distribution, diet, and growth of burbot in Lake Roosevelt, Washington. Northwest Sci 80:153–164

Potter IC (1980) The Petromyzoniformes with particular reference to paired species. Can J Fish Aquat Sci 37:1595–1615

Potter IC, Beamish FWH (1977) Freshwater biology of adult anadromous sea lampreys *Petromyzon marinus*. J Zool 181:113–130

Potter IC, Hilliard RW (1987) A proposal for the functional and phylogenetic significance of differences in the dentition of lampreys (Agnatha: Petromyzontiformes). J Zool 212:713–737

Potter IC, Strahan R (1968) The taxonomy of the lamprey *Geotria* and *Mordacia* and their distribution in Australia. Proc Linn Soc Lond 179:229–240

Potter IC, Lanzing WJR, Strahan R (1968) Morphometric and meristic studies on populations of Australian lampreys of the genus *Mordacia*. Zool J Linn Soc 47:533–546

Potter IC, Wright GM, Youson JH (1978) Metamorphosis in the anadromous sea lamprey, *Petromyzon marinus* L. Can J Zool 56:561–570

Potter IC, Hilliard RW, Bird DJ, Macey DJ (1983) Quantitative data on morphology and organ weights during the protracted spawning-run period of the Southern Hemisphere lamprey *Geotria australis*. J Zool 200:1–20

Potter IC, Gill HS, Renaud CB (2014) Petromyzontidae: lampreys. In: Burr BM, Warren ML Jr (eds) North American freshwater fishes: natural history, ecology, behavior, and conservation. Johns Hopkins University Press, Baltimore, MD, pp 105–139

Potter IC, Gill HS, Renaud CB, Haoucher D (2015) The taxonomy, phylogeny, and distribution of lampreys. In: Docker MF (ed) Lampreys: biology, conservation and control, vol 1. Springer, Dordrecht, pp 35–73

Prince DJ, O'Rourke SM, Thompson TQ et al (2017) The evolutionary basis of premature migration in Pacific salmon highlights the utility of genomics for informing conservation. Sci Adv 3:e1603198

Quinn TP, Sergeant CJ, Beaudreau AH, Beauchamp DA (2012) Spatial and temporal patterns of vertical distribution for three planktivorous fishes in Lake Washington. Ecol Freshw Fish 21:337–348

Quinn TP, McGinnity P, Reed TE (2016) The paradox of "premature migration" by adult anadromous salmonid fishes: patterns and hypotheses. Can J Fish Aquat Sci 73:1015–1030

Reid SB, Boguski DA, Goodman DH, Docker MF (2011) Validity of *Lampetra pacifica* (Petromyzontiformes: Petromyzontidae), a brook lamprey described from the lower Columbia River Basin. Zootaxa 3091:42–50

Ren J, Buchinger T, Pu J, Jia L, Li W (2016) Complete mitochondrial genomes of paired species northern brook lamprey (*Ichthyomyzon fossor*) and silver lamprey (*I. unicuspis*). Mitochondrial DNA 27:1862–1863

Renaud CB (2008) Petromyzontidae, *Entosphenus tridentatus*: southern distribution record, Isla Clarión, Revillagigedo Archipelago, Mexico. Check List 4:82–85

Renaud CB (2011) Lampreys of the world. An annotated and illustrated catalogue of the lamprey species known to date. FAO Species Cat Fish Purp 5:1–109

Renaud CB, Economidis PS (2010) *Eudontomyzon graecus*, a new nonparasitic lamprey species from Greece (Petromyzontiformes: Petromyzontidae). Zootaxa 2477:37–48

Renaud CB, Holčík J (1986) *Eudontomyzon danfordi* Regan, 1911. In: Holčík J (ed) The freshwater fishes of Europe, vol 1, part I, Petromyzontiformes. Aula, Weisbaden, pp 146–164

Renaud CB, Naseka AM (2015) Redescription of the Far Eastern brook lamprey *Lethenteron reissneri* (Dybowski, 1869) (Petromyzontidae). Zookeys 75–93

Renaud CB, Docker MF, Mandrak NE (2009) Taxonomy, distribution, and conservation of lampreys in Canada. In: Brown LR, Chase SD, Mesa MG, Beamish RJ, Moyle PB (eds) Biology, management, and conservation of lampreys in North America. Am Fish Soc Symp 72:293–309

Renaud C, Naseka A, Alfonso N (2016) Description of the larval stage of the Alaskan Brook Lamprey, *Lethenteron alaskense* Vladykov & Kott 1978. In: Orlov A, Beamish R (eds) Jawless fishes of the world, vol 1. Cambridge Scholars Publishing, Newcastle upon Tyne, pp 231–250

Renaut S, Bernatchez L (2011) Transcriptome-wide signature of hybrid breakdown associated with intrinsic reproductive isolation in lake whitefish species pairs (*Coregonus* spp. Salmonidae). Heredity 106:1003–1011

Richards JE, Beamish FWH (1981) Initiation of feeding and salinity tolerance in the Pacific lamprey *Lampetra tridentata*. Mar Biol 63:73–77

Roffe TJ, Mate BR (1984) Abundances and feeding habits of pinnipeds in the Rogue River, Oregon. J Wildl Manag 48:1262–1274

Romer AS (1955) Fish origins—fresh or salt water? Pap Mar Biol Oceanogr 3:261–280

Rooney SM, Wightman G, Ó'Conchúir R, King JJ (2015) Behaviour of sea lamprey (*Petromyzon marinus* L.) at man-made obstacles during upriver spawning migration: use of telemetry to assess weir modifications for improved passage. Biol Environ Proc R Ir Acad 115:125–136

Rougemont Q, Gaigher A, Lasne E et al (2015) Low reproductive isolation and highly variable levels of gene flow reveal limited progress towards speciation between European river and brook lampreys. J Evol Biol 28:2248–2263

Rougemont Q, Roux C, Neuenschwander S et al (2016) Reconstructing the demographic history of divergence between European river and brook lampreys using approximate Bayesian computations. PeerJ 4:e1910

Rougemont Q, Gagnaire PA, Perrier C et al (2017) Inferring the demographic history underlying parallel genomic divergence among pairs of parasitic and nonparasitic lamprey ecotypes. Mol Ecol 26:142–162

Ruchin A, Klevakin A, Semenov D, Artaev O (2012) Long-term dynamics and modern species composition of lampreys and fish of the Sura River basin. Izvestiya Samarskogo Nauchnogo Centra Rossiskaya Akademii Nauk 14:26–35 [in Russian]

Salewski V (2003) Satellite species in lampreys: a worldwide trend for ecological speciation in sympatry? J Fish Biol 63:267–279

Sandercock FK (1991) Life history of coho salmon (*Oncorhynchus kisutch*). In: Groot C, Margolis L (eds) Pacific salmon life histories. University of British Columbia Press, Vancouver, pp 395–445

Sandlund OT, Kjellberg G, Norheim G (1987) Mercury in fish and invertebrates in Lake Mjøsa Norway. Fauna 40:10–15

Schluter D, Conte GL (2009) Genetics and ecological speciation. Proc Natl Acad Sci USA 106:9955–9962

Schoch RR (2009) Evolution of life cycles in early amphibians. Annu Rev Earth Planet Sci 37:135–162

Schoch RR (2014) Life history evolution. In: Schoch RP (ed) Amphibian evolution: the life of early land vertebrates. Wiley, New York, pp 208–221

Schoch RR, Fröbisch NB (2006) Metamorphosis and neoteny: alternative pathways in an extinct amphibian clade. Evolution 60:1467–1475

Selkoe KA, Toonen RJ (2006) Microsatellites for ecologists: a practical guide to using and evaluating microsatellite markers. Ecol Lett 9:615–629

Seversmith HF (1953) Distribution, morphology, and life history of *Lampetra aepyptera*, a brook lamprey, in Maryland. Copeia 1953:225–232

Shine R (2004) Seasonal shifts in nest temperature can modify the phenotype of hatchling lizards, regardless of overall mean incubation temperature. Funct Ecol 18:43–49

Shortreed K, Morton K (2000) An assessment of the limnological status and productive capacity of Babine Lake, 25 years after the inception of the Babine Lake Development Project. Can Tech Rep Fish Aquat Sci 2316:1–52

Sidorov LK, Pichugin MY (2005) Lampreys of genus *Lethenteron* from Sopochnoe Lake (the Iturup Lake, Southern Kuril Islands). Vopr Ikhtiol 45:423–426 [in Russian]

Silva S, Servia MJ, Vieira-Lanero R, Barca S, Cobo F (2013a) Life cycle of the sea lamprey *Petromyzon marinus*: duration of and growth in the marine life stage. Aquat Biol 18:59–62

Silva S, Servia M, Vieira-Lanero R, Nachon D, Cobo F (2013b) Haematophagous feeding of newly metamorphosed European sea lampreys *Petromyzon marinus* on strictly freshwater species. J Fish Biol 82:1739–1745

Silva S, Servia M, Vieira-Lanero R, Cobo F (2013c) Downstream migration and hematophagous feeding of newly metamorphosed sea lampreys (*Petromyzon marinus* Linnaeus, 1758). Hydrobiologia 700:277–286

Silva S, Vieira-Lanero R, Sanchez-Hernandez J, Servia MJ, Cobo F (2014) Accidental introduction of anadromous sea lampreys (*Petromyzon marinus* Linnaeus, 1758) into a European reservoir. Limnetica 33:41–46

Sly BJ, Snoke MS, Raff RA (2003) Who came first—larvae or adults? Origins of bilaterian metazoan larvae. Int J Dev Biol 47:623–632

Smith CL (1985) The inland fishes of New York State. New York State Department of Environmental Conservation, Albany, NY

Smith HW (1953) From fish to philosopher. Little, Brown and Co, Boston

Smith JJ, Kuraku S, Holt C et al (2013) Sequencing of the sea lamprey (*Petromyzon marinus*) genome provides insights into vertebrate evolution. Nat Gen 45:415–421

Smith JJ, Timoshevskaya N, Ye C et al (2018) The sea lamprey germline genome provides insights into programmed genome rearrangement and vertebrate evolution. Nat Genet 50:270–277

Smith SJ, Marsden JE (2007) Predictive morphometric relationships for estimating fecundity of sea lampreys from Lake Champlain and other landlocked populations. Trans Am Fish Soc 136:979–987

Snäll N, Tammaru T, Wahlberg N et al (2007) Phylogenetic relationships of the tribe Operophterini (Lepidoptera, Geometridae): a case study of the evolution of female flightlessness. Biol J Linn Soc 92:241–252

Snoeks J, Laleye P, Contreras-MacBeath T (2009) *Lampetra spadicea*. The IUCN Red List of Threatened Species 2009: e.T169396A6617258. http://dx.doi.org/10.2305/IUCN.UK.2009-2. RLTS.T169396A6617258.en. Accessed 31 July 2018

Sower SA (2015) The reproductive hypothalamic-pituitary axis in lampreys. In: Docker MF (ed) Lampreys: biology, conservation and control, vol 1. Springer, Dordrecht, pp 305–373

Spice EK, Docker MF (2014) Reduced fecundity in non-parasitic lampreys may not be due to heterochronic shift in ovarian differentiation. J Zool 294:49–57

Spice EK, Goodman DH, Reid SB, Docker MF (2012) Neither philopatric nor panmictic: microsatellite and mtDNA evidence suggests lack of natal homing but limits to dispersal in Pacific lamprey. Mol Ecol 21:2916–2930

Spice EK, Whitesel TA, Silver GS, Docker MF (2019) Contemporary and historical river connectivity influence population structure in western brook lamprey in the Columbia River Basin. Conserv Genet 20:299–314

Staurnes M, Sigholt T, Lysfjord G, Gulseth OA (1992) Difference in the seawater tolerance of anadromous and landlocked populations of Arctic char (*Salvelinus alpinus*). Can J Fish Aquat Sci 49:443–447

Stein WE, Mannolini F, Hernick LV, Landing E, Berry CM (2007) Giant cladoxylopsid trees resolve the enigma of the Earth's earliest forest stumps at Gilboa. Nature 446:904–907

Strathmann RR, Eernisse DJ (1994) What molecular phylogenies tell us about the evolution of larval forms. Am Zool 34:502–512

Sutton T (2017) Distribution and ecology of lampreys *Lethenteron* spp. in interior Alaskan rivers. J Fish Biol 90:1196–1213

Sutton TM, Bowen SH (1994) Significance of organic detritus in the diet of larval lampreys in the Great Lakes basin. Can J Fish Aquat Sci 51:2380–2387

Suzuki DG, Grillner S (2018) The stepwise development of the lamprey visual system and its evolutionary implications. Biol Rev Camb Philos Soc 93:1461–1477

Taugbol T (1994) The brown trout of Mjøsa. Fauna 47:60–65

Taverny C, Lassalle G, Ortusi I et al (2012) From shallow to deep waters: habitats used by larval lampreys (genus *Petromyzon* and *Lampetra*) over a western European basin. Ecol Freshw Fish 21:87–99

Taylor EB (1999) Species pairs of north temperate freshwater fishes: evolution, taxonomy, and conservation. Rev Fish Biol Fish 9:299–324

Taylor EB, Harris LN, Spice EK, Docker MF (2012) Microsatellite DNA analysis of parapatric lamprey (*Entosphenus* spp.) populations: implications for evolution, taxonomy, and conservation of a Canadian endemic. Can J Zool 90:291–303

Thompson Z (1842) History of Vermont, natural, civil, and statistical. Clancy Goodrich, Burlington, VT

Thrower FP, Joyce JE (2004) Effects of 70 years of freshwater residency on survival, growth, early maturation, and smolting in a stock of anadromous rainbow trout from southeast Alaska. In:

Nickum MJ, Mazik PM, Nickum JG, Mackinlay DD (eds) Propagated fish in resource management. Am Fish Soc Symp 44:485–496

Truman JW, Riddiford LM (1999) The origins of insect metamorphosis. Nature 401:447–452

Tsimbalov IA, Kucheryavyi AV, Veselov AE, Pavlov ADS (2015) Description of the European river lamprey *Lampetra fluviatilis* (L., 1758) from the Lososinka River (Onega Lake Basin). Doklady Biol Sci 462:124–127

Tuniyev S, Naseka A, Renaud C (2016) Review of the Western Transcaucasian brook lamprey, *Lethenteron ninae* Naseka, Tuniyev & Renaud, 2009 (Petromyzontidae). In: Orlov A, Beamish R (eds) Jawless fishes of the world, vol 1. Cambridge Scholars Publishing, Newcastle upon Tyne, pp 154–190

Tutman P, Freyhof J, Dulcic J, Glamuzina B, Geiger M (2017) *Lampetra soljani*, a new brook lamprey from the southern Adriatic Sea basin (Petromyzontiformes: Petromyzontidae). Zootaxa 4273:531–548

Tuunainen P, Ikonen E, Auvinen H (1980) Lampreys and lamprey fisheries in Finland. Can J Fish Aquat Sci 37:1953–1959

Underhill JC (1986) The fish fauna of the Laurentian Great Lakes, the St. Lawrence lowlands, Newfoundland, and Labrador. In: Hocutt CH, Wiley EO (eds) The zoogeography of North American freshwater fishes. Wiley, New York, pp 105–136

United States Fish and Wildlife Service (2004) Endangered and threatened wildlife and plants; 90-day finding on a petition to list three species of lampreys as threatened or endangered. Fed Reg 69:77158–77167

Vamosi SM (2006) Contemporary evolution of armour and body size in a recently introduced population of threespine stickleback *Gasterosteus aculeatus*. Acta Zool Sinica 52:483–490

Veale AJ, Russell JC, King CM (2018) The genomic ancestry, landscape genetics and invasion history of introduced mice in New Zealand. R Soc Open Sci 5:170879

Vernygora O, Murray AM, Wilson MV (2016) A primitive clupeomorph from the Albian Loon River Formation (Northwest Territories, Canada). Can J Earth Sci 53:331–342

Vladykov VD (1951) Fecundity of Quebec lampreys. Can Fish Cult 10:1–14

Vladykov VD (1955) *Lampetra zanandreai*, a new species of lamprey from northern Italy. Copeia 1955:215–223

Vladykov VD (1972) Sous-division en trois sous-familles des lamproies de l'hémisphère-nord de la famille Petromyzonidae. ACFAS 39:148

Vladykov VD (1985) Does neoteny occur in Holarctic lampreys (Petromyzontidae)? Syllogeus 57:1–13

Vladykov VD, Follett WI (1958) Redescription of *Lampetra ayresii* (Günther) of Western North America, a species of lamprey (Petromyzontidae) distinct from *Lampetra fluviatilis* (Linnaeus) of Europe. J Fish Res Board Can 15:47–77

Vladykov VD, Kott E (1976a) A second nonparasitic species of *Entosphenus* Gill, 1862 (Petromyzonidae) from Klamath River System, California. Can J Zool 54:974–989

Vladykov VD, Kott E (1976b) A new nonparasitic species of lamprey of the genus *Entosphenus* Gill, 1862 (Petromyzonidae) from south central California. Bull South Calif Acad Sci 75:60–67

Vladykov VD, Kott E (1978) A new nonparasitic species of the holarctic lamprey genus *Lethenteron* Creaser and Hubbs, 1922, (Petromyzonidae) from northwestern North America with notes on other species of the same genus. Biol Pap Univ Alsk 19:1–74

Vladykov VD, Kott E (1979a) A new parasitic species of the Holarctic lamprey genus *Entosphenus* Gill, 1862 (Petromyzonidae) from Klamath River, in California and Oregon. Can J Zool 57:808–823

Vladykov VD, Kott E (1979b) Satellite species among the holarctic lampreys (Petromyzonidae). Can J Zool 57:860–867

Vladykov VD, Kott E (1984) A second record for California and additional morphological information on *Entosphenus hubbsi* Vladykov and Kott 1976 (Petromyzontidae). Calif Fish Game 70:121–127

Vladykov VD, Roy J-M (1948) Biologie de la lamproie d'eau douce (*Ichthyomyzon unicuspis*) après la métamorphose. Rev Can Biol 7:483–485

Vladykov VD, Kott E, Pharand-Coad S (1975) A new nonparasitic species of lamprey, genus *Lethenteron* (Petromyzonidae), from eastern tributaries of the Gulf of Mexico, U.S.A. Publ Zool Natn Mus Nat Sci Can 12:1–36

Waldman JR, Grunwald C, Roy NK, Wirgin II (2004) Mitochondrial DNA analysis indicates sea lampreys are indigenous to Lake Ontario. Trans Am Fish Soc 133:950–960

Waldman JR, Grunwald C, Wirgin I (2006) Evaluation of the native status of sea lampreys in Lake Champlain based on mitochondrial DNA sequencing analysis. Trans Am Fish Soc 135:1076–1085

Waldman J, Daniels R, Hickerson M, Wirgin I (2009) Mitochondrial DNA analysis indicates sea lampreys are indigenous to Lake Ontario: response to comment. Trans Am Fish Soc 138:1190–1197

Wall GR, LaFleur RG (1995) The paleofluvial record of glacial Lake Iroquois in the eastern Mohawk Valley, New York. In: Garver JL, Smith JA (eds) Field trips for the 67th annual meeting of the New York State Geological Association. Union College, Schenectady, NY, pp 173–203

Wallace RL, Ball KW (1978) Landlocked parasitic Pacific lamprey in Dworshak Reservoir, Idaho. Copeia 1978:545–546

Weitkamp LA, Hinton SA, Bentley PJ (2015) Seasonal abundance, size, and host selection of western river (*Lampetra ayresii*) and Pacific (*Entosphenus tridentatus*) lampreys in the Columbia River estuary. Fish Bull 113:213–226

Wigley RL (1959) Life history of the sea lamprey of Cayuga Lake, New York. U S Fish Wildl Serv Fish Bull 59:559–617

Wilkie MP, Turnbull S, Bird J et al (2004) Lamprey parasitism of sharks and teleosts: high capacity urea excretion in an extant vertebrate relic. Comp Biochem Physiol 138A:485–492

Williamson DI (2001) Larval transfer and the origins of larvae. Zool J Linn Soc 131:111–122

Williamson DI (2012) The origins of chordate larvae. Cell Dev Biol 1:1

Williamson DI (2009) Caterpillars evolved from onychophorans by hybridogenesis. Proc Natl Acad Sci USA 106:19901–19905

Williamson MH (1996) Biological invasions. Chapman and Hall, London

Witkowski A, Jęsior M (2000) Fecundity of river lamprey *Lampetra fluviatilis* (L.) in Drweca River (Vistula Basin, northern Poland). Arch Ryb Polskiego 8:225–232

Wittschieben J, Shivji MKK, Lalani E et al (2000) Disruption of the developmentally regulated *Rev3 l* gene causes embryonic lethality. Curr Biol 10:1217–1220

Wood CM, Bucking C (2011) The role of feeding in salt and water balance. In: Grosell M, Farrell AP, Brauner CJ (eds) Fish physiology: the multifunctional gut of fish, vol 30. Academic Press, New York, pp 165–212

Yamamoto M, Matsuzaki T, Takahashi R et al (2012) The transformation suppressor gene *Reck* is required for postaxial patterning in mouse forelimbs. Biol Open 1:458–466

Yamazaki Y, Goto A (1996) Genetic differentiation of *Lethenteron reissneri* populations, with reference to the existence of discrete taxonomic entities. Ichthyol Res 43:283–299

Yamazaki Y, Goto A (1998) Genetic structure and differentiation of four *Lethenteron* taxa from the Far East, deduced from allozyme analysis. Environ Biol Fish 52:149–161

Yamazaki Y, Goto A (2000) Breeding season and nesting assemblages in two forms of *Lethenteron reissneri*, with reference to reproductive isolating mechanisms. Ichthyol Res 47:271–276

Yamazaki Y, Goto A (2016) Molecular phylogeny and speciation of East Asian lampreys (genus *Lethenteron*) with reference to their life-history diversification. In: Orlov A, Beamish R (eds) Jawless fishes of the world, vol 1. Cambridge Scholars Publishing, Newcastle upon Tyne, pp 17–57

Yamazaki Y, Nagai T (2013) Directional selection against different life histories in the Arctic lamprey (*Lethenteron camtschaticum*): identification by microsatellite analysis. Can J Fish Aquat Sci 70:825–829

Yamazaki Y, Sugiyama H, Goto A (1998) Mature dwarf males and females of the Arctic lamprey, *Lethenteron japonicum*. Ichthyol Res 45:404–408

Yamazaki Y, Konno S, Goto A (2001) Interspecific differences in egg size and fecundity among Japanese lampreys. Fish Sci 67:375–377

Yamazaki Y, Yokoyama R, Nishida M, Goto A (2006) Taxonomy and molecular phylogeny of *Lethenteron* lampreys in eastern Eurasia. J Fish Biol 68:251–269

Yamazaki Y, Yokoyama R, Nagai T, Goto A (2011) Formation of a fluvial non-parasitic population of *Lethenteron camtschaticum* as the first step in petromyzontid speciation. J Fish Biol 79:2043–2059

Yamazaki Y, Yokoyama R, Nagai T, Goto A (2014) Population structure and gene flow among anadromous Arctic lamprey (*Lethenteron camtschaticum*) populations deduced from polymorphic microsatellite loci. Environ Biol Fish 97:43–52

Yokoyama S, Zhang H (1997) Cloning and characterization of the pineal gland-specific opsin gene of marine lamprey (*Petromyzon marinus*). Gene 202:89–93

Youson JH (2004) The impact of environmental and hormonal cues on the evolution of fish metamorphosis. In: Hall BK, Pearson RD, Muller GB (eds) Environment, development, and evolution: towards a synthesis. Massachusetts Institute of Technology Press, Cambridge, pp 239–278

Youson JH, Beamish RJ (1991) Comparison of the internal morphology of adults of a population of lampreys that contains a nonparasitic life-history type, *Lampetra richardsoni*, and a potentially parasitic form, *L. richardsoni* var. *marifuga*. Can J Zool 69:628–637

Youson JH, Freeman PA (1976) Morphology of gills of larval and parasitic adult sea lamprey, *Petromyzon marinus* L. J Morphol 149:73–103

Youson JH, Sower SA (2001) Theory on the evolutionary history of lamprey metamorphosis: role of reproductive and thyroid axes. Comp Biochem Physiol 129B:337–345

Zanandrea G (1957) Neoteny in a lamprey. Nature 179:925–926

Zanandrea G (1959) Speciation among lampreys. Nature 184:360

Zhou T, Liu S, Geng X et al (2017) GWAS analysis of QTL for enteric septicemia of catfish and their involved genes suggest evolutionary conservation of a molecular mechanism of disease resistance. Mol Genet Genomics 292:231–242

Chapter 5
Control of Invasive Sea Lamprey in the Great Lakes, Lake Champlain, and Finger Lakes of New York

J. Ellen Marsden and Michael J. Siefkes

Abstract Sea lamprey invaded the Laurentian Great Lakes above Lake Ontario in the 1900s, and were a factor in the collapse of several major fish stocks. Whether the species is native to Lake Ontario, Lake Champlain, and the Finger Lakes of New York is debated; nevertheless, it is considered to be a nuisance species in these waters. Early control of sea lamprey included use of barriers in streams to prevent upstream spawning, and use of lampricides to kill larvae; these methods are the mainstay of the current control program. Sterile males were used to reduce spawning success for several years in the St. Marys River, but this practice has been discontinued due to challenges with evaluating its success and the availability of improved lampricide options. Success of the control program is measured as reduction in number of spawning adults that ascend streams, and reduction in lake trout wounding. Both metrics have been substantially reduced and either meet or are close to targets for each of the Great Lakes; targets for the Finger Lakes and Lake Champlain are considerably higher than in the Great Lakes, and have not yet been met. Concerns about the effect of lampricides and barriers on non-target species and on ecosystem integrity have prompted a search for additional control methods. Research has focused on the use of pheromones and repellants, genetic strategies, and improving the specificity and efficacy of existing control methods. Prevention of the spread of sea lamprey into new bodies of water is also a priority. Over 60 years of sea lamprey control efforts have considerably advanced our understanding of sea lamprey behavior, physiology, genomics, and chemical communication.

Keywords Barriers · Cayuga Lake · Control · Genetics · Invasive species · Lampricide · Pest control · Seneca Lake

J. E. Marsden (✉)
University of Vermont, 81 Carrigan Drive, Burlington, VT 05405, USA
e-mail: ellen.marsden@uvm.edu

M. J. Siefkes
Great Lakes Fishery Commission, 2100 Commonwealth Blvd, Suite 100, Ann Arbor, MI 48105, USA
e-mail: msiefkes@glfc.org

© Springer Nature B.V. 2019
M. F. Docker (ed.), *Lampreys: Biology, Conservation and Control*,
Fish & Fisheries Series 38, https://doi.org/10.1007/978-94-024-1684-8_5

...from every economical standpoint it would appear to be advantageous to rid the world entirely of the lampreys.... Naturally, however, the student of biology must mourn the loss of a form so interesting and so instructive.

Gage (1893)

5.1 Introduction

Sea lamprey *Petromyzon marinus* control in the Laurentian Great Lakes was implemented over 60 years ago, and is one of the largest and most intensive efforts to control a vertebrate predator ever attempted. The history, status, and future of this control program have been reviewed elsewhere (e.g., Christie and Goddard 2003; Siefkes et al. 2013). In this chapter, in addition to outlining the progress of the control program since those publications, we focus on additional methods of control, possible compensatory changes in sea lamprey populations as the result of management, and ways in which research on control strategies has advanced our understanding of lamprey biology, particularly behavior and chemical communication. We begin with a brief review of sea lamprey biology, the history of the invasion of the Great Lakes, and the debate concerning their endemicity in Lake Ontario, Lake Champlain, and the Finger Lakes of New York. We then review the current control techniques and their application in each of the Great Lakes, and compare and contrast the more recent sea lamprey control programs in Lake Champlain and the Finger Lakes of New York with that in the Great Lakes. Emerging control techniques are then reviewed, including efforts to reduce further spread of sea lamprey into additional inland waters.

5.1.1 Sea Lamprey Biology and Life History

The native range of sea lamprey in North America extends along the Atlantic coast from the Gulf of St. Lawrence to northern Florida; in Europe, they range from Iceland and the Faroe Islands to northwestern Africa, and are found on the west coast of Greenland and in the Baltic, Mediterranean, and Adriatic seas and occasionally in the western Barents Sea (Scott and Crossman 1973; Potter et al. 2015; Novikov and Kharlamova 2018). Introduced populations have been present in the upper four Great Lakes since at least 1921; the endemicity of landlocked populations in Lake Ontario, the Finger Lakes of New York, and Lake Champlain is debated (see Chap. 4). The sea lamprey is a species of conservation concern throughout most of its European range (see Maitland et al. 2015), and is considered a nuisance species in its landlocked range in North America due to its parasitism on commercial and recreationally harvested fishes, and ecological impacts.

Sea lamprey hatch in streams, and spend the majority of their lives as blind, burrowing larvae (Fig. 5.1). They prefer silt-sand habitats with particle sizes <0.5 mm in diameter (Manion and McLain 1971; Potter 1980), and feed on organic detritus and algae, particularly diatoms (Manion 1967; Moore and Mallatt 1980; Sutton and Bowen 1994; see Dawson et al. 2015). Metamorphosis into the juvenile stage generally begins when larvae reach 120 mm, or 3.0 g, with a condition factor >1.50 (Holmes and Youson 1994; Henson et al. 2003; Dawson et al. 2015), but its onset is also influenced by factors such as latitude and stream temperature (Treble et al. 2008; Manzon et al. 2015). In most streams, metamorphosis occurs at 4–6 years, but can take as long as 12 years (Manion and Smith 1978) and as few as 2 years (Morkert et al. 1998). Metamorphosis commences in mid-summer, and is completed by October (Youson 1980). Recently metamorphosed juveniles begin outmigration to the lake or ocean in fall, but may remain in streams until spring (Applegate and Brynildson 1952). The adaptive function of this split strategy is unknown, as differences in survival and growth of fall and spring outmigrants have not been demonstrated (Swink and Johnson 2014; see Manzon et al. 2015). Landlocked sea lamprey generally do not feed in streams after metamorphosis, though a population of sea lamprey appear to complete their life cycle within the Cheboygan River (Michigan) system of Lake Huron (Johnson et al. 2016a; see Chap. 4). In contrast, newly metamophosed anadromous sea lamprey do feed prior to or during their downstream migration in fresh water (e.g., Silva et al. 2013) and at least a few individuals appear to reach 300–400 mm total length (TL) before entering the sea (see Chap. 4).

Juvenile sea lamprey are parasitic (Fig. 5.1); in freshwater systems their preferred prey are small-scaled salmonids, primarily lake trout *Salvelinus namaycush* and Atlantic salmon *Salmo salar*, which share their preferred temperature range (15–20 °C; Farmer 1980). However, they have been known to parasitize a wide range of species in fresh water, including burbot *Lota lota*, lake whitefish *Coregonus clupeaformis* and other coregonines, walleye *Sander vitreus*, lake sturgeon *Acipenser fulvescens*, freshwater drum *Aplodinotus grunniens*, ictalurid catfishes, catostomids, esocids, and (rarely) rainbow smelt *Osmerus mordax* (Surface 1898; Marsden et al. 2003; see Chap. 3). Choice of some of these species has been presumed to be a consequence of scarcity of preferred prey species, but may also be related to the availability of hosts in the vicinity of the natal stream (Young et al. 2003). However, lake trout in particular have been caught with several sea lamprey attached to them, so the role of competition for hosts is not understood. Sea lamprey remained attached to their hosts in tank experiments for 1–13 days, after which they detached voluntarily and sought another host; this behavior reduces the mortality of host fish by allowing recovery from a short-term feeding bout (Farmer 1980). During their lifetime, sea lamprey can kill an estimated 19 kg of fish, though this estimate may be biased high because it was derived from experiments in tanks (Swink 2003). Sea lamprey are more likely to attach to larger hosts, and survival of the host fish is related to their size (Farmer and Beamish 1973; see Chap. 3). Estimates from laboratory experiments with captive lake trout yielded an estimated probability of survival of 0.55 (Swink 1990, 2003). Madenjian et al. (2008) re-estimated this probability for wild fish to be 0.66, using estimates of annual survival, fishing mortality, and sea lamprey wounding rates on adult lake trout in Lake Champlain. Landlocked sea lamprey are considerably smaller than the anadromous form. Mean size at maturity in the Great Lakes, Lake Champlain, and Finger Lakes ranges from ~395 to 500 mm (e.g.,

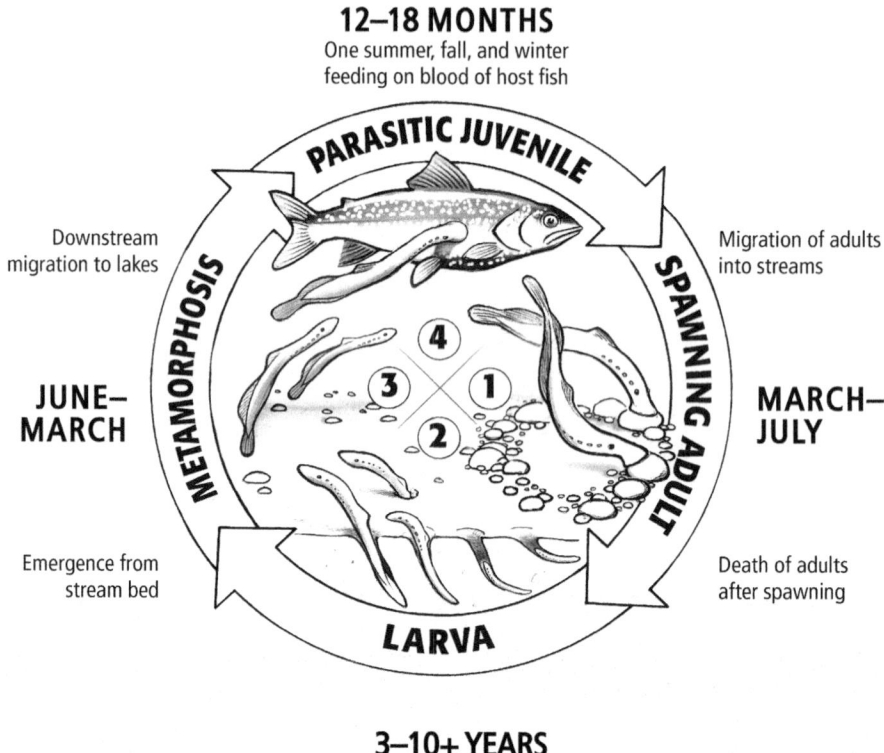

12–18 MONTHS
One summer, fall, and winter
feeding on blood of host fish

PARASITIC JUVENILE

Downstream
migration to lakes

Migration of adults
into streams

METAMORPHOSIS

SPAWNING ADULT

**JUNE–
MARCH**

**MARCH–
JULY**

Emergence from
stream bed

Death of adults
after spawning

LARVA

3–10+ YEARS

Fig. 5.1 The life cycle of sea lamprey in the Great Lakes (*Image* © GLFC)

Applegate 1950; Wigley 1959; Johnson 1982; Smith and Marsden 2007; see Chap. 4). The maximum size reported for the landlocked form in recent years is ~570 mm (Johnson 1982), although MacKay and MacGillivray (1949) reported that the majority of upstream migrants in the Little Thessalon River (Lake Huron) in 1946 were ~610 mm in length, with a maximum length of 762 mm. Anadromous individuals average 743 mm and can reach 800–900 mm at maturity (Vladykov 1949; Rooney et al. 2015; see Chap. 4).

In the spring after their first full year as a juvenile (18 months after metamorphosis), landlocked sea lamprey stop feeding and ascend streams for spawning as migrating sub-adults (Fig. 5.1). Streams are selected in part by detection of larval pheromones; the presence of larvae infers availability of suitable spawning and nursery habitat (see Moser et al. 2015; Sect. 5.5.1). Migrating adults will generally move upstream until blocked by natural or man-made barriers, seeking spawning substrates comprised of gravel and small rocks 0.9–5.2 cm in diameter with water velocities 0.5–1.5 m/s (Manion and Hanson 1980). Males begin construction of a nest comprised of a semi-circular ridge of rocks encircling a depression, with the open side facing upstream (see Johnson et al. 2015a). Sexual maturation occurs during migration (see Chap. 1), and fully mature males begin to release a sex pheromone that speeds maturation of conspecifics and attracts mature females (Chung-Davidson et al. 2013a, b). Once a mature female finds a male, or more than one male, they engage in sequences of nest building and spawning, lasting for 2–3 days (Manion and Hanson 1980). During spawning, the female attaches to a rock placed in the center of the nest, the male attaches himself to the back of her head, and their bodies become entwined. Spawning is accompanied by vigorous movements of both bodies, resulting in roiling of the substrate, water, sperm, and eggs; this serves to bury many of the eggs, but a large proportion, estimated as high as 85%, are dislodged from the nest and drift downstream (Manion and Hanson 1980). The majority of eggs that are not retained in the nest are consumed by crayfish and fishes, or are suffocated if they land on silt substrate (Applegate 1950; Smith and Marsden 2009). Spawning is most often between a pair, but multiple females and, in some cases, multiple males have been observed in a single nest and contribute to the resulting progeny (Applegate 1950; Hanson and Manion 1980; Kelso et al. 2001; see Johnson et al. 2015a). Spawning occurs in short bursts, 2–3 s long, every 4–5 min. Fecundity ranges from 45,000 to 100,000 eggs per female, and varies among lakes (Applegate 1950; Heinrich et al. 1980; Manion and Hanson 1980; Smith and Marsden 2007; Gambicki and Steinhart 2017; see Chap. 1). Larvae hatch in 7–13 days (Piavis 1971), and the prolarvae reside in the spawning gravel for 17–33 days, until they are 9–10 mm long (see Chap. 2). At this point, they have developed a functional mouth and begin to burrow into soft substrate (Piavis 1971). Survival to this burrowing stage is highly variable, with estimates ranging from 0.7 to 80% (Applegate 1950; Manion and Hanson 1980; Manion and McLain 1971; Jones et al. 2009; Dawson et al. 2015).

5.1.2   Sea Lamprey Invasion and Fisheries Collapse

Sea lamprey gained access to the upper Great Lakes with the opening of the Welland Canal in 1829 (Fig. 5.2). They were first noted in Lake Ontario in 1835 (Lark 1973), although Eshenroder (2014) argued that the 1835 record is suspect and that the first credible report of sea lamprey in Lake Ontario was in 1888 (see Chap. 4). Sea lamprey were not seen in Lake Erie until 1921 (Applegate 1950). Poor water quality, blockage by locks, and winter dewatering of the locks likely inhibited sea lamprey passage in the early canal system. However, the building of the second Welland Canal in 1914–1932 may have facilitated their migration into the upper lakes. Thereafter, the invasion progressed steadily upstream, with sea lamprey appearing in Lake St. Clair in 1934, Lake Michigan in 1936, Lake Huron in 1937, and Lake Superior in 1938 (Dymond 1922; Trautman 1949; Applegate 1950; Lawrie 1970; Smith 1971; Pearce et al. 1980; Smith and Tibbles 1980). Rapid geographic expansion was presumably facilitated by the absence of natal homing (Bergstedt and Seelye 1995; Howe et al. 2006; Waldman et al. 2008; Moser et al. 2015). Sea lamprey actively seek spawning tributaries that are already occupied by larvae, as determined by detecting the presence of larval pheromones (Moser et al. 2015; Johnson et al.

Table 5.1 Numbers of tributaries to the Laurentian Great Lakes, Finger Lakes (Cayuga and Seneca), and Lake Champlain, including the number of tributaries that have been infested with sea lamprey, have been treated at least once with lampricides, and have sea lamprey barriers and traps. Wounding targets for each lake are shown

Lake	Total tributaries	Tributaries with sea lamprey	Lampricide used at least once	Tributaries with barriers	Tributaries with traps	Wounding target (A1–A3 wounds per 100 lake trout)
Superior	1,566	165	113	18	22	5
Huron	1,761	127	83	17	15	5
Michigan	511	128	90	15	17	5
Erie	842	29	13	6	6	5
Ontario	659	66	38	16	11	2[a]
Cayuga	12	6	1	1	1	20
Seneca	2	2	2	2	0	150[b]
Champlain	35	26	14	3	12	25

[a]Target for Lake Ontario includes only A1 wounds
[b]Target for Seneca Lake includes A1–A4 wounds

2015a), and successful spawning is facilitated by evidence of prior spawning (Vrieze et al. 2010; Meckley et al. 2012, 2014). Although the attraction to pre-existing larval populations would seem to be contrary to rapid invasiveness, migrating sea lamprey are also attracted to river water (Vrieze and Sorensen 2001) and to odors released by native lampreys (Fine et al. 2004), both of which would facilitate rapid invasion of streams with suitable habitat. Sea lamprey have been found in 516 (9.7%) of the 5,339 tributaries in the Great Lakes basin, 26 (74%) of 35 tributaries surveyed in the Lake Champlain basin, both principal tributaries in Seneca Lake, and six of 12 tributaries (50%) in Cayuga Lake (Table 5.1).

Whether the sea lamprey is endemic in the lakes below the Welland Canal is unresolved. Lake Ontario is directly accessible to sea lamprey from the Atlantic Ocean via the St. Lawrence River, so there is no clear rationale for why they would have invaded in the 1800s and not earlier. Similarly, sea lamprey had direct access to the former Champlain Sea while it was an embayment of the Atlantic Ocean between 13,000 and 10,000 years ago (Cronin et al. 2008), and Lake Champlain (Fig. 5.3) remains accessible from the Atlantic Ocean via the St. Lawrence and Richelieu rivers (Marsden and Langdon 2012). From Lake Ontario, sea lamprey would have had access to Seneca and Cayuga lakes in New York via the Oswego and Seneca rivers (Fig. 5.4). Native lake trout in Seneca Lake have been shown to have a behavioral resistance to sea lamprey attacks, suggesting that this population may have evolved in the presence of sea lamprey (Schneider et al. 1983; Swink and Hanson 1986). Early genetic work identified similarities between sea lamprey in the Finger Lakes of New York and the Atlantic populations (Brussard et al. 1981). More recently, genetic studies using different portions of the genome suggested that sea lamprey were native to both Lake Ontario and Lake Champlain (Bryan et al. 2005; Waldman

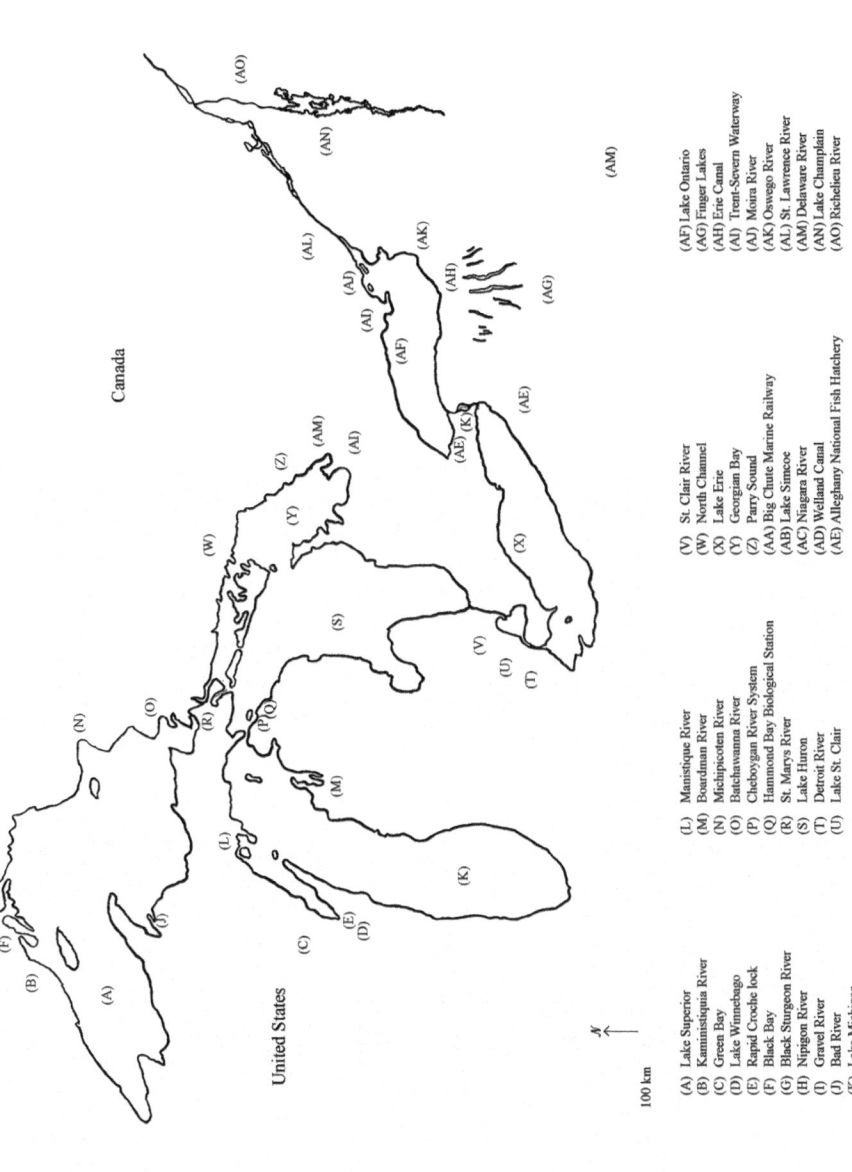

Fig. 5.2 Map of the Great Lakes, showing locations referred to in the text; detailed maps of the Finger Lakes and Lake Champlain are provided in Figs. 5.3 and 5.4, respectively (*Image* © GLFC)

(A) Lake Superior
(B) Kaministiquia River
(C) Green Bay
(D) Lake Winnebago
(E) Rapid Croche lock
(F) Black Bay
(G) Black Sturgeon River
(H) Nipigon River
(I) Gravel River
(J) Bad River
(K) Lake Michigan

(L) Manistique River
(M) Boardman River
(N) Michipicoten River
(O) Batchawana River
(P) Cheboygan River System
(Q) Hammond Bay Biological Station
(R) St. Marys River
(S) Lake Huron
(T) Detroit River
(U) Lake St. Clair

(V) St. Clair River
(W) North Channel
(X) Lake Erie
(Y) Georgian Bay
(Z) Parry Sound
(AA) Big Chute Marine Railway
(AB) Lake Simcoe
(AC) Niagara River
(AD) Welland Canal
(AE) Alleghany National Fish Hatchery

(AF) Lake Ontario
(AG) Finger Lakes
(AH) Erie Canal
(AI) Trent-Severn Waterway
(AJ) Moira River
(AK) Oswego River
(AL) St. Lawrence River
(AM) Delaware River
(AN) Lake Champlain
(AO) Richelieu River

et al. 2004, 2006). This conclusion was based on large genetic distance between Lake Ontario and Lake Champlain populations and sea lamprey from the Atlantic Ocean, and the presence in these populations of two alleles and two mitochondrial haplotypes that were not found in Atlantic Ocean sea lamprey. Subsequently, D'Aloia et al. (2015) used population genetic modeling to infer endemicity, although they note that the model results are also consistent with a recent invasion (see Chap. 4).

The primary argument against native status in Lake Ontario rests on the lack of historic observations of sea lamprey or wounding on salmonids before the mid- to late 1800s—1835 according to Lark (1973) or 1888 according to Eshenroder (2014)—and there were similarly no historic records in the Finger Lakes prior to the late 1800s. The first credible account of sea lamprey in Cayuga Lake was in 1875 and they were abundant by 1886 (Eshenroder 2014). Surface (1898) reported wounds on a number of species in Cayuga Lake (e.g., lake trout, lake sturgeon, brown bullhead *Ameiurus nebulosus*, lake whitefish, muskellunge *Esox masquinongy*, pike *E. lucius*, white sucker *Catostomus commersonii*, and yellow perch *Perca flavescens*), including fatal wounding of brown bullhead and a reported case of a lake sturgeon with 21 sea lamprey attached; this level of wounding seems unlikely to have gone unremarked if sea lamprey were historically present in these lakes. The timing of these observations support the hypothesis that sea lamprey are non-native to Lake Ontario and the Finger Lakes and invaded the watershed through a diversion linking Lake Ontario with the Susquehanna River (Atlantic drainage) where sea lamprey larvae were known to be present (Eshenroder 2014; see Chap. 4). Adult sea lamprey could have also entered the Lake Ontario watershed from the Hudson River via the Erie Canal, opened in 1825. Likewise, there are no reports of sea lamprey or sea lamprey wounding in Lake Champlain prior to opening of the Champlain Barge Canal in 1916, which offered access to Lake Champlain via the Hudson River in the Atlantic drainage (Eshenroder 2009, 2014). The first confirmed observation of sea lamprey in Lake Champlain was in 1929, and is consistent with the time required for sea lamprey to reach detectable levels following colonization of the upper Great Lakes.

While sea lamprey were undoubtedly a factor in the collapse of major fish stocks in the Great Lakes, the extent of their role is debated. Lake trout, lake whitefish, and burbot were already depleted by commercial fishing prior to the arrival of sea lamprey into the upper Great Lakes. Christie (1973) argued that the fishery, by targeting the largest individuals in a population, left populations vulnerable to sea lamprey attacks on smaller individuals that were more likely to die prior to reaching sexual maturity. In addition to the salmonids and burbot, high wounding rates were seen on walleye and catostomids (*Catostomus* spp. and *Moxostoma* spp.; Smith and Tibbles 1980). Coregonines, including cisco *Coregonus artedi*, bloater *C. hoyi*, kiyi *C. kiyi*, and shortnose cisco *C. reighardi*, became the target of sea lamprey after the preferred prey, lake trout, lake whitefish, and burbot, were depleted or extirpated. Native species,

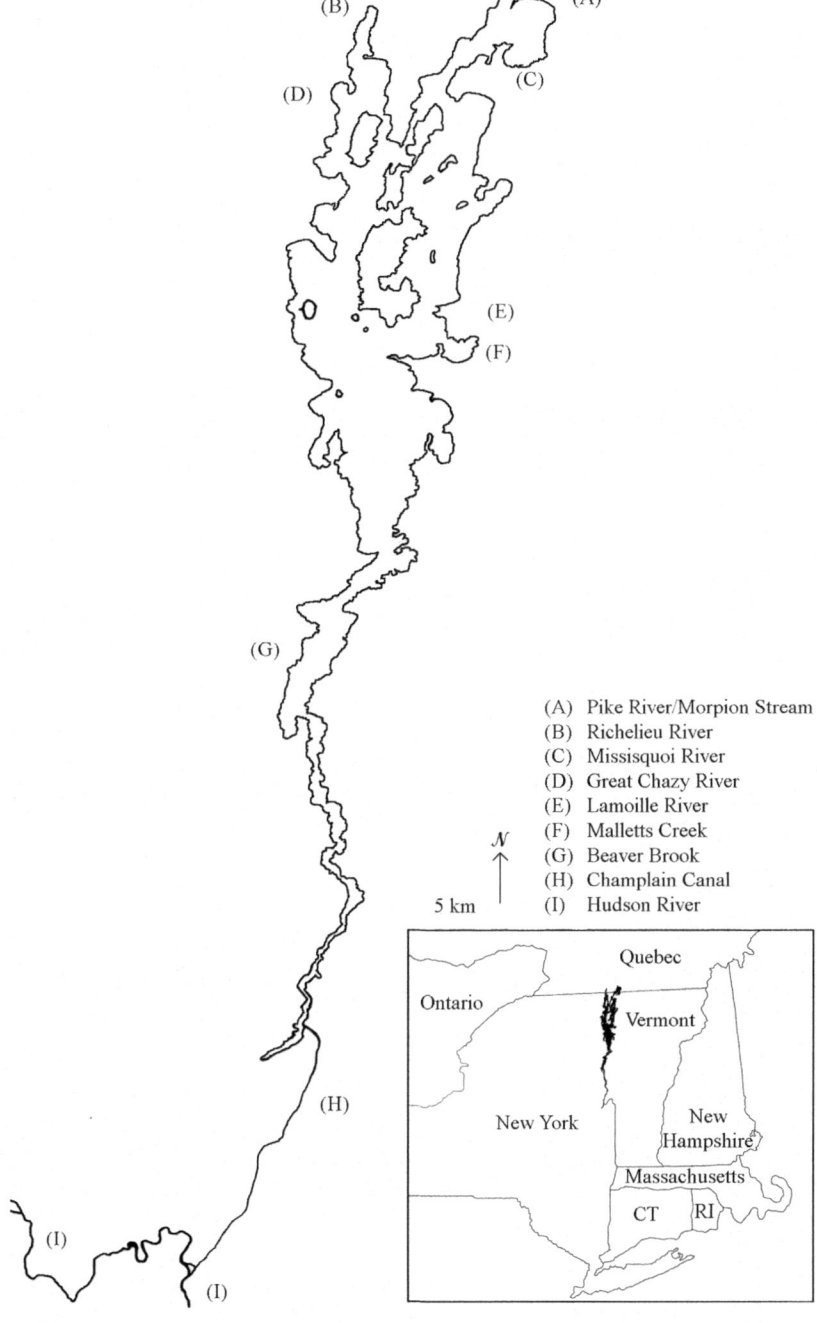

Fig. 5.3 Map of Lake Champlain (*Image* © GLFC)

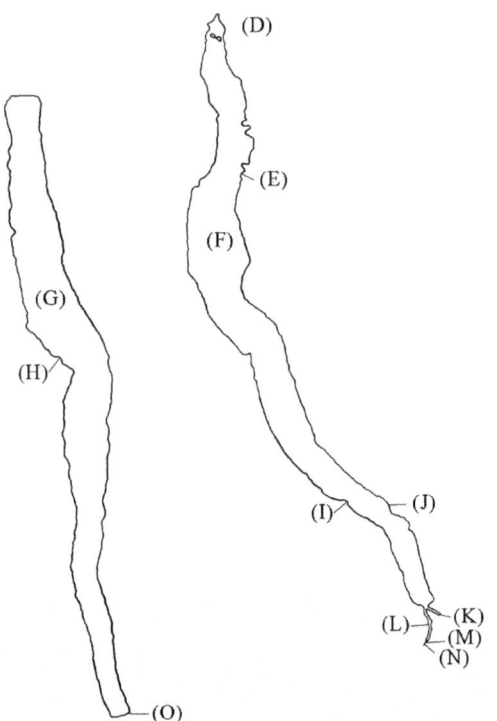

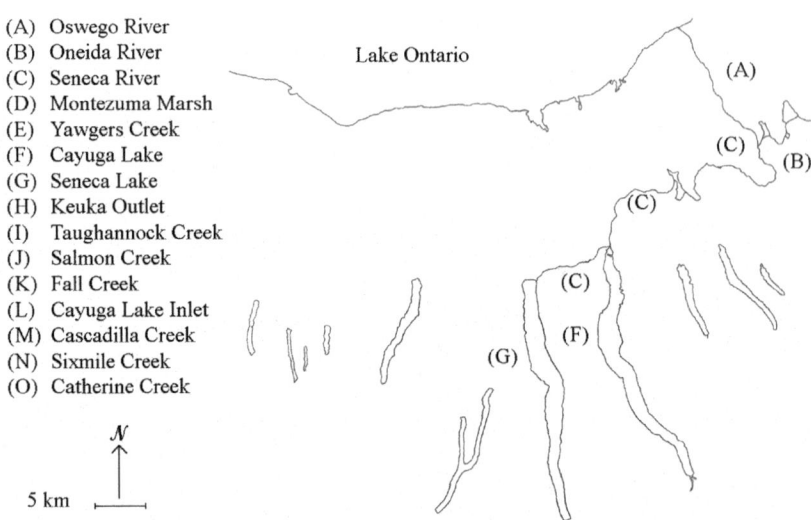

(A) Oswego River
(B) Oneida River
(C) Seneca River
(D) Montezuma Marsh
(E) Yawgers Creek
(F) Cayuga Lake
(G) Seneca Lake
(H) Keuka Outlet
(I) Taughannock Creek
(J) Salmon Creek
(K) Fall Creek
(L) Cayuga Lake Inlet
(M) Cascadilla Creek
(N) Sixmile Creek
(O) Catherine Creek

5 km

Fig. 5.4 Map of the Finger Lakes (*Image* © GLFC)

particularly the planktivores, were also challenged by the invasion of rainbow smelt in 1912 and alewife *Alosa pseudoharengus* in the 1930s. By 1960, lake trout were extirpated from Lakes Michigan, Ontario, and Erie, and lake whitefish populations were severely depressed (Eshenroder 1992; Cornelius et al. 1995; Elrod et al. 1995; Hansen et al. 1995; Muir et al. 2013). Commercial fishing for lake trout was closed in most areas of the lakes, but attempts to restore lake trout by stocking, prior to sea lamprey control, failed; mortality of stocked fish was extremely high once they reached a size at which they became vulnerable to sea lamprey (e.g., Elrod et al. 1995; Eshenroder et al. 1995).

5.2 Sea Lamprey Control in the Laurentian Great Lakes

5.2.1 Development of the Control Program

The sea lamprey invasion precipitated the formation of international entities to manage and regulate fisheries (Fetterolf 1980; Gaden et al. 2013), activities that were previously the responsibility of the Great Lakes states and province of Ontario (Dochoda and Koonce 1994). The first of these entities, the Great Lakes Sea Lamprey Committee, was established in 1946 and focused primarily on determining the life history and ecosystem effects of the sea lamprey in the Great Lakes (Smith et al. 1974). The Convention on Great Lakes Fisheries was signed by Canada and the United States on 10 September 1954, establishing the Great Lakes Fishery Commission (GLFC; GLFC 1955), which is still in existence today. The primary responsibilities of the GLFC are "to formulate and implement a comprehensive program for the purpose of eradicating or minimizing the sea lamprey populations" in the Great Lakes, and to coordinate research on economically and ecologically important fish stocks (GLFC 1955). The GLFC contracts Fisheries and Oceans Canada (formerly Department of Fisheries and Oceans, DFO) and the United States Fish and Wildlife Service (USFWS) to implement a coordinated sea lamprey control program throughout the Great Lakes (Christie and Goddard 2003), and is guided by an advisory committee and several task forces consisting of government staff, sea lamprey control agents, fisheries managers, researchers, and other relevant experts (Koonce et al. 1982; Spangler and Jacobson 1985; Christie and Goddard 2003).

Important in the development of effective sea lamprey control tactics were the description of the sea lamprey life cycle (Fig. 5.1) and the identification of tributaries used for spawning. The life stages in which sea lamprey occupy tributaries (larvae and adults), and thus are most concentrated and easily accessible to humans, were determined to be the best targets for sea lamprey control. Early sea lamprey control efforts began in the 1940s, targeting adults and their strong drive to migrate upstream to spawn. Mechanical and electrical weirs, low-head barriers, and traps were employed to reduce reproduction by disrupting spawning migrations and preventing access to spawning habitat (Hunn and Youngs 1980;

Lavis et al. 2003a; Siefkes et al. 2013). Sea lamprey control biologists, however, quickly learned that building and maintaining weirs and barriers was an expensive and sometimes dangerous endeavor that could negatively impact stream ecosystems (Hunn and Youngs 1980). Additionally, biologists observed behaviors that would potentially limit the effectiveness of barriers; adults would sometimes spawn below barriers or would leave a stream with a barrier and enter an adjacent stream that was barrier-free to spawn (Applegate and Smith 1951). Furthermore, biologists discovered that trapping could not remove enough of the spawning population to impact sea lamprey production, as a single female can produce up to 100,000 eggs (Manion and Hanson 1980). Recognizing the limitations of barriers and traps, biologists concluded that targeting multiple year classes of larvae with a lamprey-specific pesticide (lampricide) before they metamorphose into juveniles and enter the lakes to feed on fish would likely be the most effective means to reduce sea lamprey populations in the lakes (Siefkes et al. 2013).

5.2.2 Current Control Techniques

5.2.2.1 Lampricides

Control directed at the stream-resident life stages of sea lamprey could take two approaches: kill larvae while they are resident in tributaries or prevent metamorphosed juveniles from moving out of tributaries into the lakes. The effort required to capture juveniles during the protracted period of outmigration, especially during the winter months, rendered it the more difficult strategy. Use of weirs and inclined plane traps in the early phase of sea lamprey control removed several thousand juveniles (Applegate 1950; Applegate and Brynildson 1952), but was highly labor-intensive. Because larvae typically take 4 or more years to reach the juvenile stage, larval reduction generally needs only to be done once every 3 years in each stream to prevent outmigration of juveniles, substantially reducing the costs and ecological effects of lampricide treatment. Treatment intervals may be shorter if treatment efficiency is low, or longer in streams in which recruitment does not occur annually. Larvae are also found in delta areas of a few tributaries (Dawson et al. 2015; see Chap. 7), so a strategy to treat lentic as well as lotic habitats was needed.

Intensive efforts in the 1950s were focused on finding a chemical that would: kill sea lamprey larvae with minimal effects on non-target species, be effective at low concentrations over a short period of time, not persist in the environment, and be relatively inexpensive to produce and apply (Siefkes et al. 2013). After evaluating more than 6,000 compounds, 3-trifluoromethyl-4-nitrophenol (TFM) was discovered to meet the necessary criteria (Applegate et al. 1961) and experimental treatments using a liquid form of TFM were conducted on Lake Huron tributaries during 1957. Treatment of tributaries began in Lake Superior in 1958 (Smith and Tibbles 1980). TFM is applied as a liquid upstream of the highest larval populations, with additional application points to boost the in-stream concentration as needed. In 1963, a second lampricide, niclosamide (2′,5-dichloro-4′-nitrosalicylanilide, registered as Bayer-73), was used in small quantities to reduce the amount of TFM required for a given target mortality, thereby reducing costs of treatment (Howell et al. 1964). A different approach was needed for treatment of delta-resident populations, where the liquid TFM would disperse too rapidly to be effective. Beginning in 1966, niclosamide was applied in a granular form that would sink to the substrate, allowing spot applications and treatment of deltas (Applegate et al. 1961; Howell et al. 1964; Smith and Tibbles 1980). In recent years (2008–2017), lampricide treatments have been applied to about 6.5% of Great Lakes tributaries: 119 of 1,566 tributaries in Lake Superior, 92 of 511 tributaries to Lake Michigan, 84 of 1,761 tributaries to Lake Huron, 17 of 842 tributaries to Lake Erie, and 36 of 659 tributaries to Lake Ontario (Sullivan and Mullett 2018).

The early control program used TFM treatment concentrations of at least twice the Minimum Lethal Concentration (MLC, defined as the concentration of TFM required to produce 99.9% mortality in a 9-h exposure), with a goal of eradicating sea lamprey (Brege et al. 2003). However, the cost of lampricides, concerns about effects on non-target organisms (see Sects. 5.5 and 5.6), and changes in the focus of the control program motivated a reduction in TFM use to around 1.5 MLC.

Lampricide applications are scheduled based on assessment of larval sea lamprey populations in each stream (Fig. 5.5; Christie et al. 2003; Hansen and Jones 2008). The primary metric used for prioritizing streams for treatment is the number of larvae that have a high likelihood of metamorphosing in the following year. Larvae are concentrated in substrates composed of silty sand, so assessment of larval densities involves initial surveys of stream substrates, stratified larval sampling in each habitat type, and extrapolation of larval densities in sampled areas to the total amount of preferred habitats (Slade et al. 2003). Lampricide treatments are designed to maximize mortality of larvae while minimizing non-target mortality. Surviving larvae, or residuals, are usually found during post-treatment assessments (Brege et al. 2003); these larvae contribute to the juvenile population, and their production of migratory pheromone draws in adults the following year.

Currently, sea lamprey control funding levels preclude the treatment of all sea lamprey-producing tributaries in the Great Lakes in a single year. Typically, about 160 tributaries are treated regularly (about every 3 years) with TFM or a mixture of TFM and niclosamide (Fig. 5.5). Additionally, about 45 large or slow-moving tributaries and connecting channels (including associated inland lakes), and areas near the mouths of tributaries are treated regularly with granular niclosamide (Fig. 5.5).

Early work with lampricides focused on maximizing kill of sea lamprey while minimizing harm to non-target species; thus, describing efficacy of TFM and niclosamide alone and in combination, and in relation to temperature and pH, were priorities. Research has described the physiological mode of action of these chemicals. TFM impairs ATP production by uncoupling oxidative phosphorylation (Applegate et al. 1966; Niblett and Ballantyne 1976; Howell et al. 1980; Birceanu et al. 2009, 2011), but the mode of action of niclosamide is not well understood (Dawson 2003). Non-target effects vary among teleost species; centrarchids appear to be the least affected, whereas mortality has been noted in brown bullhead, brown trout *Salmo trutta*, log-

◄**Fig. 5.5** Lampricide treatment begins with assessing the density and upstream extent of larval sea lamprey populations in tributaries; electrofishing is the primary method used for these assessments (*first row*; Photos: © GLFC). The lampricide 3-trifluoromethyl-4-nitrophenol (TFM) is applied to most sea lamprey-infested areas (*second row*; Photos: © GLFC). 2′,5-dichloro-4′-nitrosalicylanilide (niclosamide) can be used as an additive to TFM and is formulated into granules and applied via a spray boat to treat estuaries of infested tributaries and the large connecting waterways of the Great Lakes (*third row*; Photos: © GLFC). Lampricide treatments are designed to maximize mortality of larvae while minimizing non-target mortality (*fourth row*; Photos: © Left—GLFC; Right—Chris Sierzputowski)

perch *Percina caprodes*, northern pike, rainbow trout *Oncorhynchus mykiss*, trout perch *Percopsis omiscomaycus*, and walleye (Boogaard et al. 2003). Mortality has also been noted in amphibians and aquatic invertebrates (Maki et al. 1975; Gilderhus and Johnson 1980; Waller et al. 2003; Weisser et al. 2003; Boogaard et al. 2015; Newton et al. 2017). Species of particular concern are lake sturgeon, stonecat *Noturus flavus*, mudpuppy *Necturus maculosus*, and unionid mussels, due to their susceptibility to lampricide and status in some areas as Threatened or Endangered. Despite concerns regarding some non-target mortality in individuals of these species, there is no evidence that intermittent lampricide applications have had long-term effects on these non-target species at the population level (Marsden et al. 2003; Siefkes et al. 2013). However, given that TFM is largely lamprey-specific, native lampreys in the Great Lakes (northern brook lamprey *Ichthyomyzon fossor*, silver lamprey *I. unicuspis*, chestnut lamprey *I. castaneus*, and American brook lamprey *Lethenteron appendix*) are particularly susceptible to non-target effects. In toxicity trials, northern and American brook lampreys were less susceptible to TFM than sea lamprey larvae (King and Gabel 1985), but the difference is generally insufficient to allow for selective control of sea lamprey where their distribution overlaps with that of the native species. American brook lamprey, which often inhabit upstream reaches not inhabited by sea lamprey, appear to have been less affected by sea lamprey control efforts, but vulnerability to lampricides is considered a threat to northern brook and silver lampreys in the Great Lakes basin (see Maitland et al. 2015). More work is needed to further understand the impacts of lampricides on non-target species and how lampricide treatments can be better targeted to minimize non-target impacts (McDonald and Kolar 2007).

5.2.2.2 Barriers

The sea lamprey is semelparous, meaning their entire lifetime fitness is invested in achieving just one terminal reproductive event. Sea lamprey spawning is also focused during a relatively small window of time during the spring of the year (Johnson et al. 2015a), so small "missteps" by an individual can exclude them from the spawning act. Therefore, disrupting reproductive behaviors such as migration can be a viable sea lamprey control tactic that significantly decreases reproduction and the subsequent recruitment of parasitic juveniles to the lakes. Preventing sea lamprey infestation of suitable tributary habitat by using barriers to block adult sea lamprey access is an obvious control method, a point that was perceived as early as 1893 (Gage 1893). Even though sea lamprey engage in strong rheotactic movements during their spawning migrations (Manion and Hanson 1980), they are relatively poor swimmers and jumpers (Beamish 1978; Youngs 1979; Reinhardt et al. 2009; Almeida and Quintella 2013), and do not use their suction cup mouths to climb (Reinhardt et al. 2009), unlike the Pacific lamprey *Entosphenus tridentatus* (Reinhardt et al. 2008; see Moser et al. 2015). These limitations allow for the effective use of barriers to stop sea lamprey migrations. Barriers do not necessarily prevent spawning, as sea lamprey may, after encountering a barrier, return downstream to spawn or enter another stream to spawn (Applegate and Smith 1951). However, use of traps integrated with barriers overcomes this problem by intercepting and then removing sea lamprey. Escapement past old or temporary barriers may occur (see Manistique River example in Sect. 5.2.3.3) and allow upstream spawning; nevertheless, these barriers reduce spawning and can increase the interval between treatments, thus reducing overall lampricide use. However, the primary use of barriers is to reduce the number of river miles that need to be treated with lampricide; downstream spawning or escapement to other streams does not affect this function.

The history of sea lamprey barriers is well documented (Hunn and Youngs 1980; Lavis et al. 2003a; Siefkes et al. 2013). Sea lamprey barriers may be permanent or seasonal, fixed-crest or adjustable-crest, low-head (also known as weirs) or barrier dams, mechanical or electrical, and constructed with or without traps (Fig. 5.6). Barriers built in strategic locations specifically for sea lamprey control (purpose-built) are present on 68 Great Lakes tributaries and six tributaries in the Finger Lakes and Lake Champlain (Table 5.1). Purpose-built sea lamprey barriers in the Great Lakes eliminate the need for lampricide treatment in an estimated 1,400 km of stream, and reduce access by sea lamprey to an estimated 15% of available type I (i.e., preferred) larval habitat (Lavis et al. 2003a; McLaughlin et al. 2007). In addition to purpose-built sea lamprey barriers, dams built for other purposes (e.g., power generation), but that also block sea lamprey and are serendipitously located in useful locations on critical sea lamprey-producing tributaries, are also important to sea lamprey control (Smith and Tibbles 1980). Termed de facto sea lamprey barriers (Siefkes et al. 2013), there are nearly 900 such barriers across the Great Lakes (Peter Hrodey, USFWS, Marquette, MI, personal communication, 2016; data.glfc.org). The length of stream protected and the reduction in available larval rearing habitat by sea lamprey barriers is far greater when considering the impacts of de facto sea lamprey barriers. Barriers also create upstream refugia for native lampreys from lampricide treatments, an important function considering the significant decline in native lamprey distribution where they overlap with sea lamprey (Schuldt and Goold 1980; see Sect. 5.2.2.1).

Although barriers are effective at blocking sea lamprey access to suitable habitat, they also have significant impacts on stream ecosystems and non-target species. Larger barriers have severe effects on lotic ecosystems including impoundment of water, habitat fragmentation, and significant shifts in fish assemblages (McLaughlin et al. 2003). Smaller sea lamprey barriers typically do not create large impoundments and therefore do not significantly alter habitat in terms of substrate and temperature

◀**Fig. 5.6** Sea lamprey barriers and traps: barriers built for other purposes that also block sea lamprey (i.e., de facto sea lamprey barriers) (*top row*; Photos: © Left—GLFC; Right—U.S. Army Corps of Engineers Buffalo District); low-head seasonal barrier with removeable stop logs (*middle row; left*; Photo: © NYSDEC); hybrid low-head/electrical barrier, with electrodes visible under the water above the barrier lip (*middle row; right*; Photo: © GLFC); low-head seasonal barrier with a vertical slot fishway (*lower row; left*; Photo: © GLFC); and closeup of adult sea lamprey in a barrier-integrated trap (*lower row, right*; Photo: © USFWS)

(Dodd et al. 2003). Nevertheless, smaller sea lamprey barriers do impact the fish assemblages of tributaries by influencing species richness upstream of barriers (Porto et al. 1999; Dodd et al. 2003; McLaughlin et al. 2006). Changes in fish assemblages caused by sea lamprey barriers are the result of impeding fish passage, but in general, larger fish are better able to traverse a barrier compared to smaller fish (Porto et al. 1999). Jumping fish can also more easily traverse barriers than non-jumping fish and jumping pools can be added to barrier designs to assist in their passage (Pratt et al. 2009; Siefkes et al. 2013). Because the ability to move upstream and downstream is an important life history component of many of the more than 90 native fish species and associated species, such as mussels, that reside in Great Lakes tributaries, the impacts of sea lamprey barriers on fish passage will continue to be issues to consider (McLaughlin et al. 2007). The use of seasonal barriers may mitigate some of these concerns, but questions remain about their effectiveness: early- and late-run sea lamprey have a higher probability of escaping upstream, potentially creating selection pressure for this trait (McLaughlin et al. 2007), and many non-target species migrate during the same time as sea lamprey (McLaughlin et al. 2007; Vélez-Espino et al. 2011).

The use of sea lamprey barriers is also constrained by their construction and maintenance costs. The construction of new purpose-built barriers and maintenance of aging de facto sea lamprey barriers is an expensive endeavor (Siefkes et al. 2013), with projects on larger tributaries costing several millions of dollars to complete. This represents a large portion of the annual sea lamprey control budget in the Great Lakes. De facto sea lamprey barriers present a unique set of issues in that nearly all are owned and operated outside of the sea lamprey control program and many are aging and in disrepair. Because repairing or replacing a failing dam is expensive, dam removal is often the only economical option for owners without financial assistance from the sea lamprey control program. The sea lamprey control program monitors the condition of purpose-built and de facto sea lamprey barriers across the Great Lakes basin and prepares funding strategies for the repair or replacement of barriers. This strategy weighs construction costs with the estimated costs of lampricide treatment over the expected lifespan of the barrier (i.e., barrier construction costs should be less than lifetime lampricide treatment costs). These funding issues, coupled with the ecological impacts highlighted in the previous paragraph, have limited the use of sea lamprey barriers to only the most critical sea lamprey-producing tributaries.

Funding constraints and ecological impacts of sea lamprey barriers have also led to research initiatives to overcome these limitations. New barrier designs are being explored, including strategies to design and operate adjustable and seasonal barriers to be more effective and selective to sea lamprey (McLaughlin et al. 2007); use of electricity (Johnson et al. 2014a) and velocity to block sea lamprey is also being tested (Andrew Muir, GLFC, Ann Arbor, MI, personal communication, 2016). The use and improved efficiency of "trap-and-sort" fishways to facilitate selective fish passage and mitigate the ecological effects of barriers has also occurred, but more research is needed to understand their impacts on non-target species, including the effects of delays at barriers and in traps prior to sorting, the associated energetic costs, and the potential for reduction in reproductive output (Pratt et al. 2009). Recently, developing selective fish passage technologies and tactics outside of trap-and-sort fishways and understanding sea lamprey and non-target fish behavior, movement, and swim performance has become a priority of the GLFC and its partners; the construction of a dedicated fish passage research facility is being planned for the Boardman River in Traverse City, Michigan (Andrew Muir, personal communication, 2017). In the event that tough decisions need to be made between sea lamprey control and the restoration of aquatic connectivity, decision tools that balance the needs of sea lamprey control and non-target species are being developed (McLaughlin et al. 2003) and structured decision-making is being used (Dale Burkett, GLFC, Ann Arbor, MI, personal communication, 2015). Overall, the ecological trade-offs between sea lamprey control and aquatic habitat connectivity need to be considered in a systematic way to ensure that the best decisions possible are being made regarding sea lamprey barrier construction or removal.

5.2.2.3   Traps

In addition to sea lamprey barriers, traps that capture adult sea lamprey also exploit the strong rheotactic, chemoattractive, and social behaviors associated with reproduction. Early sea lamprey trapping efforts used several methods (Applegate and Smith 1951; Smith and Elliot 1953; Wigley 1959; McLain et al. 1965). The most successful traps were operated in conjunction with sea lamprey barriers, which increase the probability of capture by forcing sea lamprey to congregate below the barrier and repeatedly interact with traps as they attempt passage around the dam. Early trapping operations were expensive to operate (Hunn and Youngs 1980), but the development of a portable trap (Schuldt and Heinrich 1982) and the strategic integration of permanent traps into the construction of new and existing barriers, both of which allowed for operation by smaller crews, made sea lamprey trapping more cost-effective (Fig. 5.6).

Traps on 37 tributaries (Table 5.1) removed ~50,000 adult sea lamprey from spawning populations in 2017. On average, ~40% of the spawning population in a trapped tributary is removed annually. This level of trapping does not likely affect the overall sea lamprey population due to their high fecundity (Manion and Hanson 1980; see Chap. 1) and density-independent recruitment variation that can lead to strong year classes from small adult populations (Jones et al. 2003; Dawson and Jones 2009). Increases in trapping efficiency to a level that offsets these factors are needed before trapping can become a viable sea lamprey control tactic in the Great

Lakes. For instance, Young (2005) predicted that removing 50–60% of the adult population before spawning would successfully control sea lamprey in Lake Huron. Low trapping efficiency is likely a result of low trap encounter rate (Bravener and McLaughlin 2013). Trap encounter rates could be increased by focusing trapping effort early in the spawning season when sea lamprey are more abundant and active (Dawson et al. 2017), by placing traps in areas known to be used more frequently by sea lamprey (Holbrook et al. 2015; Rous et al. 2017), and by manipulating behavior using pheromone attractants (Li et al. 2003; Johnson et al. 2013), repellents (Bals and Wagner 2012; Luhring et al. 2016), and electricity (Johnson et al. 2014a) to guide sea lamprey towards traps.

Adult sea lamprey trapping operations allow for the assessment of spawning populations and, since the late 1970s, have provided population data throughout the Great Lakes as a means to assess the success of the sea lamprey control program (Mullett et al. 2003). Currently, mark-recapture estimates using a modified Schaefer estimate (Schaefer 1951) are conducted on index tributaries to each lake. A lake-wide adult sea lamprey abundance index is calculated by summing the abundance estimates of individual index tributaries to each lake, and are used as the key measure of sea lamprey control program success.

Trapping outmigrating juvenile sea lamprey with nets or rotary screw traps is also conducted in the Great Lakes. Like targeting larval sea lamprey with lampricides, trapping outmigrating juveniles removes sea lamprey before they harm fish, but also sea lamprey that have the highest probability of harming fish (i.e., after they have survived through the larval stage and metamorphosis). Currently, trapping outmigrating juveniles is inefficient and therefore only conducted ad hoc, and on a limited basis when lampricide treatments have failed or been deferred and there is substantial risk of sea lamprey escaping to the lakes. Research to develop improved methods to capture outmigrating juveniles is needed (see Sect. 5.5.5).

Despite current limitations as a control technique, trapping will remain an important assessment tool for the sea lamprey control program, but will also increase in value as a control technique as trapping technologies advance and sea lamprey behavioral research solves the mysteries of migration and movement from the lake, into spawning tributaries, and through reproduction. Current research relevant to trapping is advancing on several fronts including understanding migratory behavior (Holbrook et al. 2015), identifying reproductive cues and pheromones and characterizing associated behaviors (Buchinger et al. 2015), and developing new trapping protocols and devices that target different behaviors and/or life stages (e.g., metamorphosing sea lamprey; Sotola et al. 2018) and that target different habitats that are currently difficult to trap (e.g., large rivers and rivers without barriers; McLaughlin et al. 2007). Taking advantage of sea lamprey behavior to increase trapping efficiency may improve assessment by providing more accurate and precise population estimates of adult sea lamprey using the same or less trapping effort. Additionally, a better understanding of sea lamprey behavior may improve the capacity of traps to affect sea lamprey control by increasing the number of sea lamprey removed from spawning populations or from the parasitic juvenile population.

5.2.2.4   Sterile-Male-Release Technique (SMRT)

The SMRT was first developed for insect pest control in the 1950s (Knipling 1968) and investigation of its potential for sea lamprey control began in the 1970s (Hanson and Manion 1978, 1980). The intent is to reduce the reproductive output of a population by releasing sterilized individuals of one sex (typically the "calling" sex in pheromone-producing individuals), and overwhelming the population with sterilized adults that compete successfully for matings. Importantly, the SMRT is species-specific and benign to non-target species. Additionally, the SMRT is most effective for species in which the female mates only once and when population densities are low, either naturally or when reduced with pesticides, such as with the sea lamprey in the Great Lakes.

An effective SMRT depends on successful sterilization without causing adverse effects on male competitiveness. The chemical P,P-bis(1-aziridinyl)-N-methylphosphinothioic amide (bisazir; Chang et al. 1970) was determined to effectively sterilize male sea lamprey (Hanson and Manion 1978; Hanson 1981) without affecting male competitiveness and spawning behavior (Hanson and Manion 1978, 1980). Siefkes et al. (2003) later determined that bisazir also did not affect sex pheromone production in male sea lamprey. Bisazir is an effective sterilant because of its mutagenic properties. As applied in sea lamprey control, bisazir damages the genetic material present in sperm (Hanson 1990); however, sperm concentration, motility, and ability to fertilize eggs are not affected by bisazir exposure; sperm from sterilized males can fertilize eggs, but nearly all fertilized eggs die before hatching (Ciereszko et al. 2002). Despite the effectiveness of bisazir as a sterilant, it is extremely hazardous to humans (Rudrama and Reddy 1985; Hanson 1990; Ciereszko et al. 2003; Sower 2003). Therefore, other potential sterilants have been explored such as Cobalt-60 and Cesium-137 radiation (Hanson 1990), several spermicidal compounds (Ciereszko et al. 2003), and gonadotropin releasing hormone (GnRH) agonists and antagonists (Sower 2003). Unfortunately, these potential sterilants either do not effectively sterilize sea lamprey or negatively affect sea lamprey health, competitiveness, and spawning behaviors. Because bisazir is the only effective sterilant currently identified, a specialized sterilization facility was constructed during 1991 at the U.S. Geological Survey Hammond Bay Biological Station in Millersburg, Michigan, to contain the hazards of bisazir. Additionally, a unique auto-injector was engineered to administer an accurate and precise dose of bisazir to male sea lamprey while minimizing staff exposure to bisazir (Twohey et al. 2003a).

The SMRT was first tested in sea lamprey in 1991–1996 in 33 Lake Superior tributaries and the St. Marys River, which connects Lake Superior and Lake Huron (Twohey et al. 2003a). Lake Superior was chosen because of its relatively low adult sea lamprey population and isolation from the other Great Lakes, and the St. Marys River was selected because of its status as a major uncontrolled source of sea lamprey in northern Lakes Huron and Michigan (Schleen et al. 2003). The huge size of this river challenged existing methods to control sea lamprey because treating the entire river with lampricides on a regular basis would be prohibitively expensive and the river is too wide to use barriers effectively. Additionally, the large sea lamprey trapping network on the St. Marys River could be used to further enhance the SMRT by combining sterilization and release of the captured males with removal of the captured females. In Lake Superior, tributaries suspected to be the primary sources of sea lamprey were selected for application of the SMRT. During this 6-year time period, an average of ~16,000 sterile males were released into 10–27 tributaries per year, resulting in an estimated average sterile to fertile male ratio of 1.5:1. However, this did not produce the expected lake-wide reduction in adult sea lamprey abundance and lake trout wounding rates. Researchers concluded that, although the logistics of the study (collection of males, sterilization, and release) were successful, the number of sterile males released was not adequate to affect lake-wide populations and wounding rates, and that the SMRT—given the limited number of males each year for sterilization—should be applied on a smaller scale.

Because of the limited success of the SMRT in Lake Superior, SMRT application was focused entirely on the St. Marys River after 1997 (Twohey et al. 2003a). Initial releases in the St. Marys River in 1991–1996 had primarily relied on males captured from the river; an average of 4,600 sterile males were released annually, producing an estimated average ratio of 0.6:1 sterile to fertile males. However, in 1997–2011, when all available sterilized males were released in this river alone, annual releases averaged 26,000 sterile males, increasing the expected average ratio of sterile to fertile males to 3.4:1. In conjunction with the SMRT, a large granular niclosamide treatment was conducted during 1999 to significantly reduce the larval sea lamprey population in the river. After the treatment, the SMRT was expected to further reduce sea lamprey production from the river without the need for further lampricide treatment. As a result of these efforts, populations of larvae, juveniles, and adults were reduced, as was the sea lamprey wounding rate on lake trout in northern Lake Huron (Bergstedt and Twohey 2007). The larval sea lamprey population, however, began to increase shortly thereafter and, by 2009, was not statistically different than the pre-SMRT larval population. A review of the SMRT found that sterile males were not observed on the spawning grounds or on nests at the expected sterile to normal ratio, the viability of eggs found in nests was not different than that expected in normal nests (Bergstedt et al. 2003a; Bravener and Twohey 2016), and the adult population in the river was much larger than originally reported (Holbrook et al. 2016). These results suggested that the number of sterile males released into the river was still inadequate to overcome the compensatory mechanisms of the sea lamprey and reduce recruitment. Additionally, a review of the St. Marys River sea lamprey control strategy decision analysis indicated that increased annual treatment of the river with granular niclosamide could drastically reduce the larval sea lamprey population in the river and the SMRT had little effect on sea lamprey production (Jones et al. 2012). Therefore, the SMRT was discontinued after the 2011 application, and sea lamprey control in the river shifted to more intensive granular niclosamide treatments (Bravener and Twohey 2016).

The SMRT still remains a potentially viable sea lamprey control technique, especially in circumstances where it could be used in an Integrated Pest Management (IPM) strategy (i.e., in combination with other techniques), or for an extremely low-density population. Currently, the SMRT is being tested on the Cheboygan River, a tributary to Lake Huron that contains a low-density sea lamprey population that appears to complete its life cycle above a sea lamprey barrier without apparent recruitment from Lake Huron (Johnson et al. 2016a; see Chap. 4). Nevertheless, more work needs to be done to overcome the challenges to its implementation, most notably the density-independent drivers of recruitment success (Jones et al. 2003). Additionally, the dynamics of the target population need to be well understood to predict the effects of the application of the SMRT and confidently set appropriate suppression targets (Siefkes et al. 2013). Bisazir also remains a human health hazard and constrains the technique by requiring sterilization to occur in a contained facility, increasing operational costs and the costs to transport males to and from the facility. Development of a mobile sterilization unit might alleviate some of these constraints, but the development of a safe sterilant would be a better solution. Inadvertently transferring fish diseases (e.g., bacterial kidney disease and viral hemorrhagic septicemia) through the redistribution of sterile males between basins is also a concern that needs to be addressed (Bergstedt and Twohey 2007).

5.2.3 Evaluation of Sea Lamprey Control

Performance of the sea lamprey control program in the Great Lakes is measured on each lake using three metrics: index estimates of adult sea lamprey abundance when they ascend tributaries during their upstream migration (Fig. 5.7); the wounding rate observed on lake trout (their preferred host; Fig. 5.8); and the relative abundance of lake trout (Fig. 5.9; Siefkes et al. 2013). Targets for adult indices and wounding rates are established for each lake, and performance relative to targets is evaluated using 3-year averages (ignoring confidence intervals) to accommodate annual variation. Targets have not been set for lake trout relative abundance in the context of sea lamprey control. Trends for each metric are evaluated based on linear regressions of the most recent 5 years of data, judged at the 5% significance level. Lake-wide adult sea lamprey abundance, compared to population targets specific to each lake, is the primary metric in which sea lamprey control efforts are measured.

Index estimates of adult sea lamprey abundance are calculated by summing individual index stream population estimates in a given basin (Fig. 5.7). Population estimates are generated through mark-recapture using a modified Schaeffer method (Mullett et al. 2003). Population targets are established by calculating the mean adult index value over a 5-year period when wounding rates on lake trout averaged five per 100 fish or lower, except in Lakes Huron and Michigan where this wounding rate was not achieved in any 5-year period. For Lake Huron, the target is set at 25% of pre-control abundance; for Lake Michigan, the target is set at approximately half the abundance that corresponds with the lowest 5-year period of wounding (Siefkes et al. 2013). The second metric, lake trout wounding, is evaluated in all lakes except Lake Ontario as annual estimates of fresh sea lamprey wounds per 100 lake trout >532 mm TL captured in standardized gill net surveys (Fig. 5.8), where fresh wounds are classified as A1–A3, with A1 representing the freshest wounds and A2 and A3

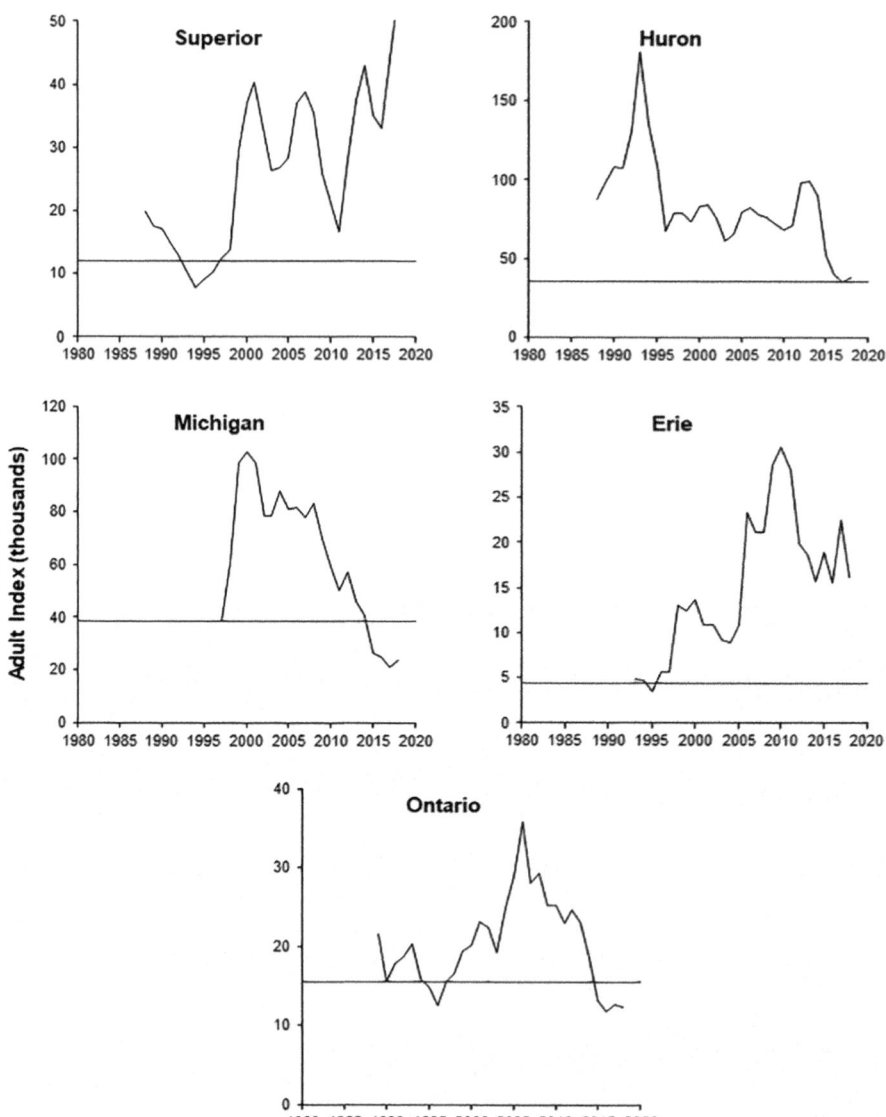

Fig. 5.7 Adult sea lamprey index estimates for each of the Great Lakes compared to the population targets (*black horizontal lines*). Data from the Great Lakes Fishery Commission

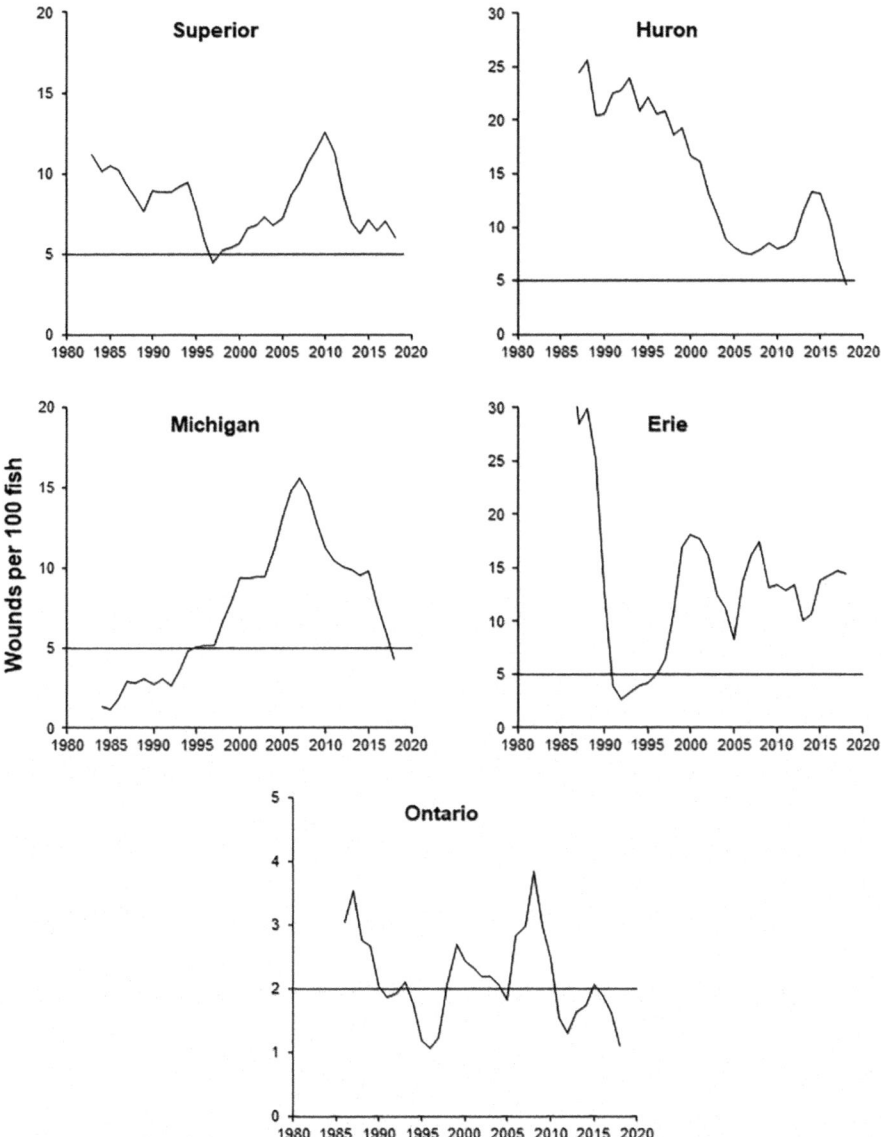

Fig. 5.8 Sea lamprey wounding rates per 100 lake trout >532 mm total length (TL) compared to the five-wound target (*black horizontal lines*); Lake Ontario: 100 lake trout >431 mm compared to a two-wound target. Data from the Great Lakes Fishery Commission

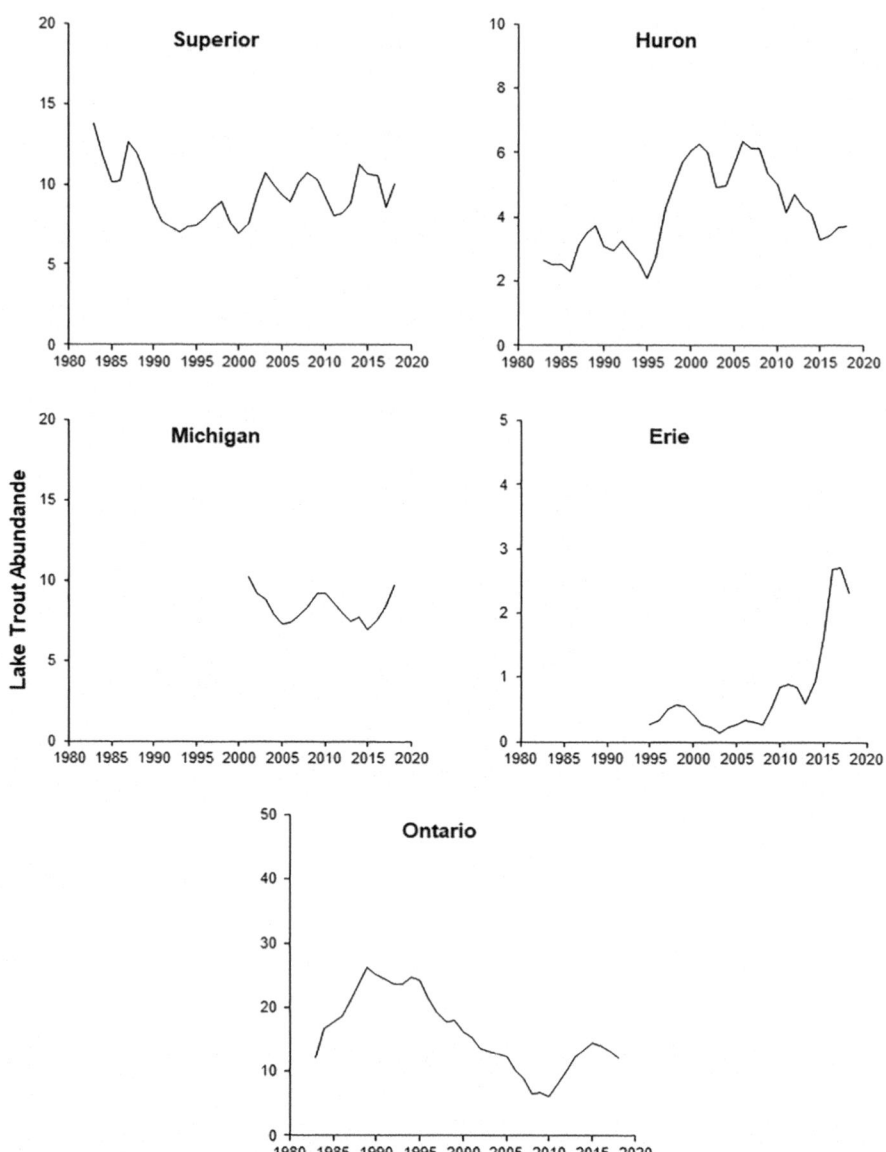

Fig. 5.9 Lake trout relative abundance estimates as catch per effort (CPE). Lakes Superior, Michigan, and Huron: CPE = fish/km/net night of lean lake trout >532 mm TL. Lake Erie: CPE = number of age 5 and older lean lake trout per lift. Lake Ontario: CPE = fish/km/net night of lean lake trout >431 mm. Data from the Great Lakes Fishery Commission

representing fresh wounds that have begun to heal (King 1980; Ebener et al. 2003). In Lake Ontario, A1 wounds on lake trout >431 mm are used; only A1 wounds were found to correlate with lake trout mortality in this lake (Schneider et al. 1996). Wounding rates measure the harm caused by sea lamprey that elude control activities each year and generally correlate positively with adult sea lamprey abundance, but this relationship can vary and needs further exploration (Siefkes et al. 2013). Furthermore, sea lamprey do not feed solely on lake trout, but are opportunistic (Farmer and Beamish 1973; Harvey et al. 2008), targeting warmer water and lower trophic level species when available or necessary (see Chap. 3). Therefore, sea lamprey predation on species other than lake trout should be assessed to provide a more complete measure of sea lamprey control success. Nevertheless, a target of five A1–A3 wounds per 100 lake trout was selected to protect adequate numbers of mature fish to ensure natural reproduction in all lakes except Ontario, where two A1 wounds per 100 fish is the target. Targets related to the third metric, lake-wide population estimates of lake trout (Fig. 5.9), have not been established, but lake trout abundance is considered when interpreting lake trout wounding rates, as an inverse relationship between lake trout abundance and wounding rate is expected if sea lamprey populations and lake trout stocking are held constant.

Program success is also measured on individual tributaries through post-treatment larval sea lamprey surveys. Ninety-five to 99% of larval sea lamprey in a stream are expected to be killed by a lampricide treatment (Heinrich et al. 2003), but treatment effectiveness can vary greatly due to many factors that vary both spatially and temporally. Understanding the factors that affect treatment success and knowing the population structure of larval sea lamprey in tributaries post-treatment is critical to understanding the effectiveness of lampricide treatment.

Sea lamprey control needs for each of the Great Lakes differ based on the production potential of their respective tributaries and lentic areas, which is dependent on such factors as the number and size of infested tributaries and the environmental factors that influence reproductive success, larval survival and growth, and recruitment of parasitic juveniles to feeding populations (Siefkes et al. 2013). The effectiveness of treatment options also varies and can affect sea lamprey control needs. Thus, each lake constitutes a unique sea lamprey control scenario, and these scenarios are important to consider when interpreting the effects of the sea lamprey control program in each lake.

The fish community responses to sea lamprey control have been positive; wounding rates have declined (Fig. 5.8), and abundance and growth of stocked and naturally self-sustaining lake trout has increased in most of the lakes (Fig. 5.9). However, direct attribution of changes in the lakes to the success of sea lamprey control is complicated by many simultaneously occurring factors that affect fish populations in the lake communities. Since sea lamprey control was initiated, the lakes have been invaded by dreissenid mussels and several exotic plankton species (Mills et al. 1993; Ricciardi 2006); populations of the small crustacean *Diporeia*, an important food source to many fish species, have collapsed (Bunnell et al. 2018); and abundant alewife populations led to suppression of several native fish species, although the subsequent collapse of this invasive species in Lakes Huron and Ontario have allowed population recovery of the native species (Bunnell et al. 2018). Nutrient and energy flows have been altered by dreissenids, particularly in the nearshore zone (Bunnell et al. 2018).

5.2.3.1   Lake Superior

Lake Superior was the focus of much of the early sea lamprey control efforts (e.g., electrical and mechanical weirs) in an attempt to salvage the remnant lake trout population that had already been extirpated from Lakes Michigan, Erie, Ontario, and much of Lake Huron (Smith 1971; Heinrich et al. 2003). Nevertheless, reductions in lake-wide adult sea lamprey abundance and the wounding rate on lake trout, and subsequent increases in lake trout survival to older age classes, were not realized until the advent of lampricide treatments in the late 1950s (Pycha and King 1975; Smith and Tibbles 1980). Currently, adult sea lamprey abundance in Lake Superior is greater than the target and stable, but remains at a level >90% lower than peak abundance (Heinrich et al. 2003; Fig. 5.7). The sea lamprey wounding rate on lake trout is greater than the target and stable (Fig. 5.8) and lake trout abundance is stable (Fig. 5.9).

Current fish community objectives for Lake Superior specify a reduction in sea lamprey abundance "to population levels that cause only insignificant mortality on adult lake trout," with insignificant mortality defined as <5% of total mortality (Horns et al. 2003). Despite >90% reduction in sea lamprey abundance, sea lamprey mortality on lake trout was estimated at 16% in 1985–1994 (Horns et al. 2003) and as high as 59% in 1995–1999 (Heinrich et al. 2003). Nevertheless, fishery managers have ceased stocking salmonids in Lake Superior due to a high occurrence of natural recruitment (Krueger and Ebener 2004); populations of lake trout and lake whitefish may be approaching abundance levels similar to those seen prior to the sea lamprey invasion (Schreiner and Schram 1997; Bronte et al. 2003; Wilberg et al. 2003). Lake trout stocks are sufficiently robust to support gillnetting and tribal harvest (Hansen et al. 1995). Although the successful rehabilitation of some Lake Superior fish stocks is due in part to the remnant populations that remained at the time that coordinated, lake-wide fishery management was implemented (Krueger and Ebener 2004; Muir et al. 2013), sea lamprey control also played a major role.

After a period of decline in Lake Superior in ~2005–2010, adult sea lamprey abundance has been increasing, indicating there are still many challenges to face. Tribal interests on the Bad River in Wisconsin, a system with significant sea lamprey production potential, restrict the use of lampricides to every few years, leaving little flexibility to retreat if an ineffective treatment occurs. Additionally, of all the Great Lakes, Lake Superior contains the majority of the areas (estuaries and embayments associated with the Kaministiquia, Nipigon, Gravel, Michipicoten, and Batchawana rivers) requiring treatment with granular niclosamide, which is less effective than conventional TFM treatments; only the St. Marys River has a higher abundance of sea lamprey targeted with granular niclosamide. Balancing sea lamprey control with concerns about aquatic habitat connectivity is a growing issue throughout the basin when considering dam construction and removal. For example, sea lamprey control and concerns over rehabilitation of walleye populations are competing management trade-offs for the Black Sturgeon River, which flows into Black Bay on the north shore of Lake Superior (McLaughlin et al. 2012). Taken together, these challenges pose serious threats to sea lamprey control and rehabilitation of fish populations in Lake Superior. Creative solutions need to be achieved to balance conflicting objectives and ensure the success of sea lamprey control for years to come.

5.2.3.2 Lake Huron

Like Lake Superior, Lake Huron also received a high share of the early sea lamprey control efforts, but success was again not achieved until after the implementation of lampricide control (Smith and Tibbles 1980; Morse et al. 2003). Currently, adult sea lamprey abundance in Lake Huron is greater than the target and stable, but remains at a level >90% lower than peak abundance (Fig. 5.7). The sea lamprey wounding rate on lake trout is greater than the target and stable (Fig. 5.8) and lake trout abundance is stable (Fig. 5.9).

Despite the reduction in sea lamprey abundance from its peak, and re-establishment of tribal fisheries and commercial fishing for lake trout in Canadian waters, sea lamprey-induced mortality of lake trout is still a major concern in Lake Huron. High localized wounding in areas of the North Channel and increased wounding in the Parry Sound area in Ontario has been reported (Dave Gonder and Adam Cottrill, Ontario Ministry of Natural Resources, Owen Sound, ON, personal communication, 2015). Protecting lake trout from sea lamprey mortality is a particular concern, as lake trout natural reproduction has been occurring at an unprecedented level during recent years (J. E. Johnson et al. 2015).

Success of sea lamprey control in Lake Huron is dependent on suppression in the St. Marys River, the large connecting channel between Lakes Superior and Huron. Sea lamprey production potential from the St. Marys River has grown since the 1980s, whether because of improvements to habitat and water quality that have enhanced larval production and survival (Smith and Tibbles 1980; Eshenroder 1987; Young et al. 1996) or due to a decline in juvenile mortality resulting from greater abundance and availability of host species (Eshenroder et al. 1995; Young et al. 1996). During the 1990s, the larval sea lamprey population in the river was estimated to be 5.2 million (Fodale et al. 2003). In response, an integrated approach to sea lamprey control that included granular niclosamide treatments, the release of sterile males, and adult trapping was implemented in 1997 (Schleen et al. 2003). Although larval sea lamprey populations were initially reduced to ~1.4 million following a large-scale (880 ha) treatment of the river with granular niclosamide (Bergstedt and Twohey 2007; Robinson et al. 2013), larvae rebounded to a level requiring annual granular niclosamide treatments in an attempt to further reduce populations (Bravener and Twohey 2016). The increasing larval sea lamprey population in the St. Marys River since the 1999 large-scale granular niclosamide treatment prompted a review of the integrated approach in 2009. Based on the results of the review, control efforts in the St. Marys River shifted entirely to granular niclosamide treatments, and sterile-male release was discontinued after the 2011 application. Larval sea lamprey abundance in the St. Marys River responded quickly to another large-scale granular niclosamide treatment (875 ha) in 2011 and declined to the lowest level on record (~350,000 larvae) in 2012. Annual granular niclosamide treatments of ~300 ha from 2012 to present have kept larval sea lamprey abundance at ~1 million larvae on average (Kevin Tallon, Fisheries and Oceans Canada, Sea Lamprey Control Centre, Sault Ste. Marie, ON, personal communication, 2017). Nevertheless, egg viability in nests observed on the St. Marys River rapids has increased since the discontinuation of sterile male releases (Bravener and Twohey 2016). The St. Marys River will continue to be monitored to evaluate the effectiveness of granular niclosamide treatment and the impacts of discontinuing the sterile male releases.

Control efforts in Lake Michigan may also influence sea lamprey abundance in Lake Huron as sea lamprey migrate freely between the two lakes (Moore et al. 1974; Bergstedt et al. 2003b), and sea lamprey production from the Manistique River, a large northern Lake Michigan tributary, likely contributes to the Lake Huron population (Bence et al. 2008). Construction of a new sea lamprey barrier on the Manistique River is expected to reduce sea lamprey abundance in Lake Huron. Throughout the rest of Lake Huron, lampricide control efforts increased starting in 2006 (Siefkes et al. 2013). In 2010, a focused lampricide treatment effort targeted sea lamprey-producing tributaries that contribute to the northern Lake Huron population, including tributaries to the North Channel and other northern Lake Huron tributaries, the St. Marys River, and Lake Michigan tributaries that likely contribute sea lamprey to northern Lake Huron. Lake-wide adult sea lamprey populations appear to have responded to these increases in sea lamprey control efforts as adult sea lamprey abundance in 2015–2017 was at historic lows (Fig. 5.7).

5.2.3.3   Lake Michigan

Sea lamprey control was critical to the recovery and maintenance of the Lake Michigan fishery after its collapse in the mid-1900s (Fetterolf 1980; Eshenroder 1987; Holey et al. 1995; Lavis et al. 2003b). Currently, adult sea lamprey abundance in Lake Michigan remains at a level >90% below peak abundance and is at target and stable (Fig. 5.7). The sea lamprey wounding rate on lake trout is greater than the target, but decreasing (Fig. 5.8) and lake trout abundance is stable (Fig. 5.9).

Despite the reduction in sea lamprey abundance from its peak and the establishment of lake trout populations that allow harvest by sport and tribal fisheries in addition to commercial fishing bycatch (Holey et al. 1995), sea lamprey-induced mortality is still a major concern in Lake Michigan (Bronte et al. 2008) and harvestable lake trout populations are only achieved through stocking (Bunnell 2012). Additionally, much is still unknown about sea lamprey-host interactions on species other than lake trout (a knowledge gap present on all lakes; see Chap. 3); information is still needed to fully understand the impacts of sea lamprey across the Great Lakes. Work is currently being conducted to better understand the effects of sea lamprey-induced mortality on the entire fish community.

Further improvements to sea lamprey control in Lake Michigan will primarily rely on sea lamprey control in the Manistique River, which was identified as a major source of sea lamprey in the early 2000s (Siefkes et al. 2013). Prior to this time, control efforts were confined to the lower kilometer of river due to an old mill dam that effectively functioned as a sea lamprey barrier. Deterioration of this dam led to the infestation of >500 km of river with larval sea lamprey numbering in the millions. Making matters worse is the remote and dendritic nature of the watershed, which increases treatment costs (over $800,000 USD per treatment) and impairs treatment efficacy, resulting in large residual populations that recruit parasitic juveniles to the lake. The Manistique River has been treated with lampricides seven times since 2003, which has likely driven the recent reductions in sea lamprey abundance and the wounding rate on lake trout. Nevertheless, further reductions will likely not be achieved until the aging mill dam is replaced with an effective sea lamprey barrier; construction of a new barrier has been planned and is currently awaiting final approval (Michael Siefkes, personal observation). The treatment of large tributaries in the northern portion of Lake Michigan and the St. Marys River, which also contributes sea lamprey to Lake Michigan (Siefkes et al. 2013), is also required to keep sea lamprey abundance and the damage they cause to fish in check. Focused lampricide treatments in these areas have been conducted since 2006, and likely contributed to reducing sea lamprey abundance to target levels and reducing the wounding rate on lake trout.

5.2.3.4   Lake Erie

Although sea lamprey were first observed in Lake Erie in 1921, they did not become abundant for nearly six decades, likely due to a lack of preferred prey, degraded water quality and habitat in spawning streams, and eutrophication of the lake (Pearce et al. 1980; Sullivan et al. 2003). The initiation of programs in the 1970s to improve water quality, restore lake trout populations, and create a sport fishery for Pacific salmonids (Sullivan and Fodale 2009) likely led to the marked increase in sea lamprey abundance during the 1970s and 1980s. Sea lamprey control began in Lake Erie in 1986 (Sullivan et al. 2003). Adult sea lamprey abundance in Lake Erie is currently greater than the target and stable (Fig. 5.7). The sea lamprey wounding rate on lake trout is stable, but greater than the target (Fig. 5.8) and lake trout abundance is stable and there is no commercial fishery (Cornelius et al. 1995; Fig. 5.9).

A low and stable sea lamprey population is a prerequisite to the rehabilitation of the native coldwater fish community in Lake Erie's eastern basin. Current sea lamprey populations prevent the achievement of this goal. Mortality from sea lamprey wounding on lake trout, lake whitefish, burbot, and steelhead *Oncorhynchus mykiss* may impact populations of these species (Markham and Knight 2017).

Lake Erie has the fewest suitable sea lamprey spawning tributaries of the five Great Lakes and therefore should be the easiest to achieve sea lamprey control targets (Sullivan et al. 2003). Nevertheless, sea lamprey control has been hard to achieve due to a number of issues including habitat and water quality improvements, greater host fish abundance, and changes in sea lamprey control protocols (Markham and

Knight 2017). Sea lamprey control efforts have intensified recently with back-to-back lampricide treatments of all known sea lamprey-producing tributaries in 2008–2010. This effort, however, has failed to reduce sea lamprey abundance from historically high levels or return the wounding rate on lake trout to the target. Extensive assessment efforts on nearly all Lake Erie tributaries following the back-to-back treatments failed to find any major sources of sea lamprey. Assessment efforts have most recently turned to the St. Clair River, Lake St. Clair, and the Detroit River (the large connecting waterway between Lakes Huron and Erie) and results suggest that this system could be the primary source of sea lamprey in Lake Erie (Mullett and Sullivan 2017). Like the St. Marys River, it was thought that the St. Clair and Detroit River system historically did not produce many sea lamprey (Pearce et al. 1980; Smith and Tibbles 1980; Sullivan et al. 2003), but recent improvements in habitat and water quality may have made conditions more favorable for sea lamprey production. Assessment efforts will continue on the St. Clair and Detroit River system to further define the extent of sea lamprey infestation and determine if a lampricide treatment strategy could be deployed effectively. The key to sea lamprey control success in Lake Erie could possibly depend on successful treatment of the St. Clair and Detroit River system.

5.2.3.5   Lake Ontario

Sea lamprey were first observed in Lake Ontario during the mid- to late 1800s; although their endemicity to the lake is unclear (Eshenroder 2009, 2014; see Sect. 5.1.2), they are injurious to the coldwater fish community. Sea lamprey control was first implemented in Lake Ontario during 1971 with the treatment of 23 Canadian tributaries infested with sea lamprey (Pearce et al. 1980). In the U.S., 20 tributaries infested with sea lamprey were treated the following year, but reductions in sea lamprey abundance were not observed until nearly a decade later (Pearce et al. 1980; Larson et al. 2003). Currently, adult sea lamprey abundance in Lake Ontario is at target and steady (Fig. 5.7). The sea lamprey wounding rate on lake trout is also at target and steady (Fig. 5.8); recent declines in lake trout abundance (Fig. 5.9) are not fully understood, but do not appear to be attributable to sea lamprey.

Sea lamprey control in Lake Ontario has increased the survival of lake trout and salmon (Pearce et al. 1980; Elrod et al. 1995), and the current level of sea lamprey control is conducive to further rehabilitation of the coldwater fish community. Additionally, fishery pressure on lake trout populations is likely low, as there is no commercial fishery for lake trout in New York; bycatch, however, is allowed in Ontario fisheries (Elrod et al. 1995). Nevertheless, lake trout populations had been on the decline from the mid-1990s to the late 2000s (O'Gorman 2017). This trend was likely due to decreased stocking efforts resulting from production limitations associated with the Allegheny National Fish Hatchery in New York, declines in survival of stocked lake trout, and increased sea lamprey-induced mortality (O'Gorman 2017). Stocking levels were restored in the mid-2000s, survival began to increase, and sea lamprey were brought under control, likely causing lake trout abundance to rise. Current levels of sea lamprey control should also enable restoration efforts underway for

Atlantic salmon, and the maintenance of other coldwater fish populations including burbot, lake whitefish, and Pacific salmonids (O'Gorman 2017).

There are no sources of sea lamprey that are of particular concern in Lake Ontario (Paul Sullivan, Fisheries and Oceans Canada, Sea Lamprey Control Centre, Sault Ste. Marie, ON, personal communication, 2016). Current sources of sea lamprey include residual populations that survive lampricide treatments and intermittent production of untreated large tributaries including the Niagara and Moira rivers. Lampricide control effort has remained steady on Lake Ontario since the mid-1980s and has maintained sea lamprey abundance and the wounding rate on lake trout near target levels for three decades. Current sea lamprey control efforts will likely be maintained and sea lamprey abundance and the wounding rate on lake trout are expected to remain at or near targets for the foreseeable future.

5.3 Sea Lamprey Control in Lake Champlain and the Finger Lakes of New York

The endemicity of sea lamprey in the Finger Lakes (i.e., Seneca and Cayuga Lakes) and Lake Champlain is debated (see Sect. 5.1.2; Chap. 4), and its role in the collapse of salmonines in Cayuga Lake and Lake Champlain is unclear. Sea lamprey populations in these lakes, prior to human influences on the fish community, were inflicting at least as much or more damage than the invasive populations in the Great Lakes prior to control, yet neither of the two native salmonine species collapsed until after European settlers began to influence the lakes (Marsden et al. 2010; Marsden and Langdon 2012). Atlantic salmon and lake trout were extirpated by 1838 and 1900, respectively, the former due to extensive construction of dams on spawning streams. Consequently, the negative effects of sea lamprey were not perceived until salmonine stocking began in 1973, and sea lamprey must have subsisted on non-salmonine prey for several decades until stocking began. The virtual extirpation of lake trout from Cayuga Lake in the 1930s is hypothesized to be due to siltation of spawning areas, caused by extensive logging of the watershed (Youngs and Oglesby 1972). Much of the Lake Champlain watershed was also deforested in the 1800s, resulting in lake-wide accumulation of silt that may have encompassed natural spawning areas; this may explain the early demise of lake trout populations (Marsden and Langdon 2012). Commercial fishing for salmonids was absent in the Finger Lakes, and was confined to shoreline seining in Lake Champlain. Unlike in the Great Lakes, invasive plantivores played a minor role in the extirpation of native fish species in Lake Champlain; rainbow smelt are native, and alewife were absent until 2003. However, both smelt and alewife are invasive in the Finger Lakes (Hubbs et al. 2004). Lake trout populations were re-established by stocking in both Cayuga Lake and Lake Champlain prior to sea lamprey control.

Control of sea lamprey in the Finger Lakes and Lake Champlain began later than in the Great Lakes (Bishop and Chiotti 1996). In both cases, development of these new control programs was stimulated by the success and benefited from the experience in the Great Lakes, but was challenged by new regulatory and permitting restrictions and increased sensitivity of the public toward use of pesticides in natural waters. Adult abundance targets are not used in the Finger Lakes or Lake Champlain, as abundance data have not been collected in any of the lakes except at Cayuga Inlet in Cayuga Lake (Bishop and Chiotti 1996). Due to substantially higher wounding rates than in the Great Lakes, and slower progress toward population reduction, wounding targets are higher.

5.3.1 Finger Lakes

Sea lamprey are present in only the two largest of the 11 Finger Lakes, Seneca and Cayuga, that both drain into Lake Ontario. The Seneca River at the north end of Seneca Lake flows eastward into the Montezuma Marsh at the north end of Cayuga Lake, then joins the Oneida and Oswego rivers and flows into Lake Ontario at Oswego, New York. The region of the Seneca River that connects the lake was expanded into a canal in 1921 and subsequently connected to the Erie Canal in 1928 (http://www.nycanals.com/Cayuga-Seneca_Canal). Seneca Lake has a native population of lake trout that is now heavily supplemented by stocking; lake trout in Cayuga Lake have been primarily supported by stocking of fish originating from Seneca Lake since the mid-1970s (Bishop 1992).

Prior to control, sea lamprey wounding on lake trout was much higher in Cayuga Lake than Seneca Lake, with 86% of lake trout 533–813 mm TL bearing wounds (at an average of 134 wounds per 100 fish) in Cayuga Lake and only 18.2% of lake trout with wounds (averaging 28 per 100 fish) in Seneca Lake (Wigley 1959). Sea lamprey control was planned for both lakes in 1982, and began that year in Seneca Lake with application of TFM in Catherine Creek and Keuka Outlet, the only two tributaries to the lake. A legal challenge, implemented by concerned citizens who objected to the use of chemicals in the environment, delayed treatment in Cayuga Lake until 1986 (Bishop and Chiotti 1996).

Sea lamprey spawning in Cayuga Lake is largely confined to a single tributary, the Cayuga Inlet at the south end of the lake; six other tributaries (Salmon, Yawgers, Fall, Sixmile, Taughannock, and Cascadilla creeks) produce less than 10% of the total population (Wigley 1959; Bishop and Chiotti 1996). Spawning migrations in Cayuga Inlet and access to larval habitat are limited by a low-head dam 2.5 km upstream from the lake. Installation of a sea lamprey barrier and trap in the fishway in 1969 suppressed population growth, so that sea lamprey now enter Cayuga Inlet only during periods of high water when they are able to pass over the low-head dam (Bishop and Chiotti 1996). Response to the lampricide treatment was immediate and substantial; the number of adult sea lamprey captured in the Cayuga Inlet dropped from $2{,}712 \pm 739$ (mean $\pm$ SD) annually to 75 in 1987 (Fig. 5.10; Bishop and Chiotti 1996). Escapements over the dam occurred in 1993 and 1994, requiring a second lampricide treatment of the inlet in 1996. Subsequent escapements in 2007 and

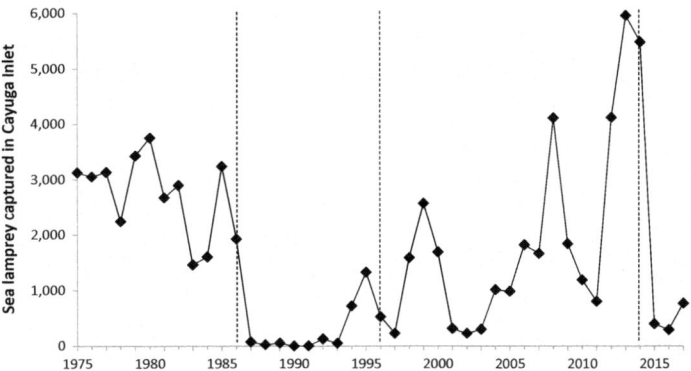

Fig. 5.10 Number of sea lamprey captured at the Cayuga Inlet fishway trap, Cayuga Lake, 1975–2017. *Vertical dashed lines* indicate TFM treatments. Data from Bishop and Chiotti (1996) and Emily Zollweg-Horan (NYSDEC)

2011 did not result in a sufficient population increase to warrant further treatments; however, record high numbers of spawning adults, double the highest pre-control abundance, were collected in Cayuga Inlet in 2013, likely attracted by odor cues from the newly established larval populations (see Sect. 5.5.1). Spawning occurred above the dam again in 2014, and the inlet was treated with TFM in August 2014, again resulting in a dramatic reduction in the number of adults ascending the inlet in 2015 (Fig. 5.10). In Seneca Lake, sea lamprey were found only in Catherine Creek prior to 1970 but, after abatement of pollution from domestic and industrial sources, sea lamprey were discovered in Keuka Outlet (Hammers et al. 2010). Lake trout abundance increased in both lakes within 3 years of the initial treatment, and growth of age 6, 7, and 8 year old lake trout increased in Seneca Lake after 1999 (Bishop 1992; Hammers and Kosowski 2011).

The target of the control program in Cayuga Lake is 22 A1–A3 wounds per 100 fall lake trout 650–699 mm TL and summer lake trout 600–649 mm, and 27 A1–A3 wounds per spring rainbow trout (Hammers et al. 2010). Wounding of fall lake trout fell from an average of 80 ± 18 wounds per hundred fish in 1981–1986 to 15 ± 5 wounds per hundred fish in 1987–1991 (Bishop and Chiotti 1996). Wounding assessment of lake trout resumed in 2013, and dropped from 83 wounds per 100 lake trout in 2013 to a post-treatment average of 15 ± 7.5. Rainbow trout wounding was much higher, with an average of 112 ± 41 wounds per 100 fish in 1980–1986, prior to the first treatment. After the 1986 treatment, rainbow trout wounding dropped below the target to an average of 11 ± 11 wounds per 100 fish (Fig. 5.11). In Seneca Lake, the targets are considerably higher and the metric includes A4 (i.e., healed) wounds, which are not used in other lakes: the targets are 150 A1–A4 wounds per 100 lake trout 600–699 mm TL, 100 wounds per 100 rainbow trout 500–599 mm, and 100 wounds per 100 brown trout 400–499 mm (Hammers et al. 2010). A1–A4 wounds on lake trout dropped substantially in the first 6 years following treatment in 1982, but have risen steadily since then and remain above the target, whereas A1–A3 wounding

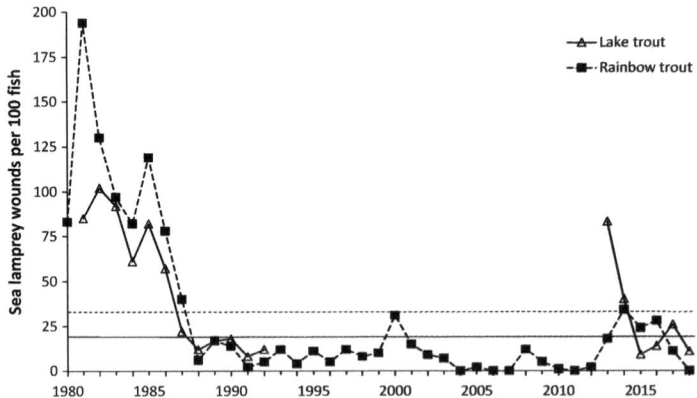

Fig. 5.11 Sea lamprey wounds on lake trout (*triangle and solid line*) and rainbow trout (*square and dashed line*) in Cayuga Lake, 1979–2017. *Horizontal lines* indicate wounding targets for the two fish species. Data from Bishop and Chiotti (1996) and Emily Zollweg-Horan (NYSDEC)

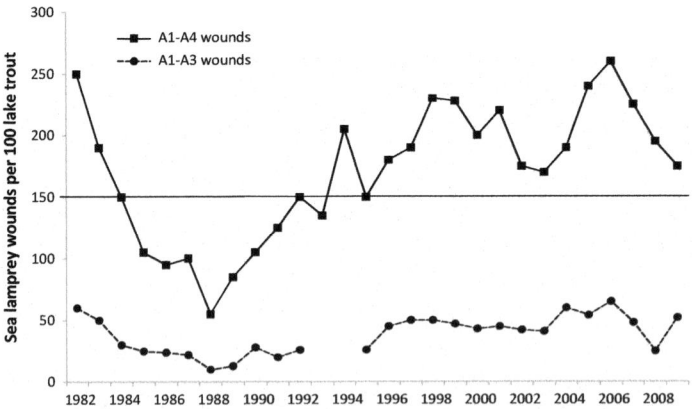

Fig. 5.12 Total A1–A4 and A1–A3 sea lamprey wounds on lake trout 600–699 mm TL collected during a Memorial Day fishing Derby, Seneca Lake, 1982–2009. *Horizontal line* indicates the target of 150 type A1–A4 wounds per 100 lake trout 600–699 mm. Data from Hammers and Kosowski (2011)

levels have remained close to 50 wounds per 100 lake trout after an initial decline to <20 (Fig. 5.12). The substantial contribution of A4 wounds to total wounding may be due to their persistence on fish; A1–A3 wounds heal progressively to A4 within 1–2 years, but A4 wounds may be visible and therefore scored for an extended period (Ebener et al. 2003, 2006).

Drawing from information gained in the Great Lakes, Engstrom-Heg (1990) postulated that the reduction in adult sea lamprey in Cayuga Inlet after the first TFM application may have been due to the lack of a detectable odor signal from larvae. In an effort to draw sea lamprey into the Cayuga Inlet and away from other streams,

200 lamprey larvae were collected from the Delaware River and held in a cage in the inlet in 1992; the number of adult sea lamprey increased that year to 129, from 7 the previous year (Bishop and Chiotti 1996). Thus, control efforts in the Finger Lakes anticipated both the use of weirs and the use of pheromones to control sea lamprey substantially earlier than in the Great Lakes.

5.3.2   Lake Champlain

The status of sea lamprey in Lake Champlain is controversial, as discussed earlier (Eshenroder 2014; see Chap. 4). If sea lamprey are native to the lake, an understanding of why sea lamprey populations and wounding rates were so high in the 1980s through to the early 2000s compared to the Great Lakes is needed. The absence of any historic record of sea lamprey wounds prior to the 1900s suggests that wounding was not as prevalent as it was after salmonid stocking began in the 1970s (Marsden and Langdon 2012). Changes in the landscape, dominated by extensive deforestation in the 1800s, added quantities of silt and organic matter to streams and improved their suitability as sea lamprey nursery habitat. Changes in the fish fauna of the basin, particularly the severe declines of lake sturgeon and American eel *Anguilla rostrata*, may have removed important predators of larval sea lamprey (see Sect. 5.5.7). Stocking of lake trout and Atlantic salmon has provided an adequate food supply for the parasitic juveniles. If sea lamprey are native, then the native salmon and trout populations must have been able to co-exist with their predator; similar to Seneca Lake, native salmonids could have avoided sea lamprey attacks by inhabiting deeper areas at temperatures which are not preferred by sea lamprey (Bergstedt et al. 2007).

Regardless of their origin, fisheries managers have designated sea lamprey as a nuisance species in Lake Champlain, and sea lamprey are considered to be an impediment to restoration of native salmonids (lake trout and Atlantic salmon). Atlantic salmon disappeared from the lake by 1838, and lake trout by 1900 (Marsden and Langdon 2012). A sustained stocking program began in 1972 for Atlantic salmon and rainbow trout, in 1973 for lake trout, and in 1977 for brown trout. Annual stocking rates of lake trout since 1973 have ranged from 39,000 to 272,000 yearling equivalents (Marsden et al. 2010). High wounding rates, with up to 100% of lake trout >635 mm TL in the main lake exhibiting fresh wounds (types A1–A3), stimulated development of an experimental control program to evaluate the potential to reduce wounding and increase salmonid survival and growth. The experimental program, involving barriers and the use of TFM and niclosamide, was conducted from 1990 to 1998 (Marsden et al. 2003). Two treatments were implemented on 13 rivers and four deltas, 4 years apart, although four of the rivers were not retreated due to low recolonization of sea lamprey (Marsden et al. 2003). The majority of treatments reduced catch per unit effort (CPUE) of larvae by >91%. Assessment of the experimental control program indicated that lake trout catch increased significantly in index gillnets, and total catch in the sport fishery also increased. Survival of lake trout older than 2 years (i.e., above the age where they are fully recruited to the sampling gear) increased significantly for most age classes, and the number of age classes present increased from eight to 12 (Marsden et al. 2003). Wounding of all size classes of lake trout also decreased significantly. Similar increases in catch and number of age classes were seen in Atlantic salmon. As a result of this evaluation, a long-term control program was initiated in 2002. Since implementation of the long-term control program, five additional streams were found to contain sea lamprey populations and were added to the control program (Brad Young, USFWS, Essex Junction, VT, personal communication, 2018). These streams either had low densities of larvae, which may have been missed in earlier surveys, or were colonized since the inception of the experimental program. As of 2014, 19 stream systems and five deltas had been treated at least once with lampricides.

Permanent barriers were present on four streams in the Champlain basin when sea lamprey control began, and only one of these dams, on the Great Chazy River, New York, incorporates a trap for removal of adults that might otherwise spawn elsewhere (Brad Young, personal communication, 2018). A seasonal barrier with an integrated sea lamprey trap was installed in Beaver Brook, New York, in 2009, and successfully reduced sea lamprey migrations such that subsequent treatments were cancelled. The steep elevation of the New York shoreline and the presence of a fall line (i.e., a geomorphologic break resulting in a rapid change in elevation, often producing rapids or waterfalls) 10–20 km from the lake on most major rivers in Vermont limit the extent of river miles colonized by larvae. On the Vermont side, two hydroelectric dams, on the Lamoille and Missisquoi rivers, are present below the fall line and restrict sea lamprey movement upstream.

Wounding targets were set after the experimental control program at 25 A1–A3 wounds per 100 lake trout 533–633 mm TL, and 15 A1–A3 wounds per 100 Atlantic salmon 432–533 mm TL (Fisheries Technical Committee 2017). Wounding rates rebounded after termination of the experimental sea lamprey control program, despite the short (2-year) gap before long-term control was initiated. Lake trout wounding peaked at 99 wounds per 100 fish in the 533–633 mm size class in 2006; wounding dropped to an average of 42 wounds per 100 lake trout in 2007–2017, but the target of 25 wounds has not yet been reached in any year (Fisheries Technical Committee 2017). Wounds on Atlantic salmon dropped in 2008 and have remained between 15 and 32 wounds per 100 Atlantic salmon since 2009 (Fig. 5.13). Sea lamprey wounds have also been found on a wide range of non-salmonine species, including lake whitefish, walleye, northern pike, smallmouth bass *Micropterus dolomieu*, rainbow smelt, lake sturgeon and freshwater drum (Howe et al. 2006).

Sea lamprey control in Lake Champlain was initiated after the establishment of the United States Environmental Protection Agency (in 1970), significant expansion of the Clean Water Act (in 1972), and enactment of the Endangered Species Act (in 1973). Consequently, National Pollutant Discharge Elimination System (NPDES) permits are required for use of lampricides. Applications are restricted to 1.5 MLC for 9 h and, in one case, treatment was restricted to 0.8 MLC due to the presence of endangered mussels (Marsden et al. 2003). Lampricide use is prohibited in Malletts Creek, Vermont, which contains northern brook lamprey which is state-listed as Endangered. Treatments have been postponed, cancelled, or held below 1 MLC in

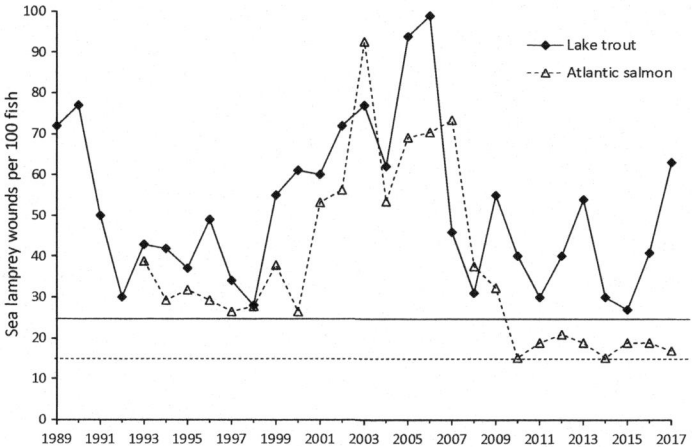

Fig. 5.13 Type A1–A3 sea lamprey wounds per 100 lake trout 533–633 mm TL (1989–2017) and Atlantic salmon 432–533 mm TL (1993–2017) in Lake Champlain. *Solid horizontal line* is the target wounding rate for lake trout (25 wounds per 100 fish). *Dashed horizontal line* is the target wounding rate for Atlantic salmon (15 wounds per 100 fish). Data from Brad Young, USFWS

other tributaries due to the presence of threatened or rare species including mudpuppies, stonecat, and several mussel species. The Pike River/Morpion Stream system in Quebec is not treated because provincial statute prohibits use of pesticides in Quebec waters (Brad Young, personal communication, 2018). Consequently, a sea lamprey barrier was constructed on Morpion Stream in 2013; adult sea lamprey captures declined from 248 in 2015 to 40 in 2018. Development and use of alternatives to chemical control is a high priority in the basin (Marsden et al. 2010).

To date, the only alternative that has been used successfully is seasonal trapping in conjunction with barriers. Portable assessment traps, with wings set across the entire stream width, are placed annually in eight to 11 small streams (<15 m wide, <1 m deep) in the Lake Champlain basin prior to the spawning season (i.e., early to mid-April) and removed in early June after catches decline to zero (Brad Young, personal communication, 2018). Traps are set below spawning substrate, and are used to extend the interval between lampricide treatments. Eel pots have been used in conjunction with portable assessment traps in eight streams, and contribute 2–40% of the total adult catch (B. J. Allaire, USFWS, Essex Junction, VT, personal communication, 2017). In four streams, larval catches declined to zero after 1–5 years of trapping; capture of adults declined to zero in only one of these streams (Wayne Bouffard, USWFS, Essex Junction, VT, personal communication, 2010). Evaluation of trap capture effectiveness, using a second upstream trap in two streams, indicated that the downstream trap captured 39–98% (mean 72%) of the adults moving upstream (Wayne Bouffard, personal communication, 2010). Non-target species are released from the traps when they are checked at least three times per week. Assessment of non-target mortality in traps in three streams in 2001–2004 showed that only 8 of

23 fish species and two of four amphibian species captured suffered mortality, with overall mortality over that period averaging 6% for fishes (5–26% for individual species, total n = 4,355) and 18% for amphibians (17–50% for individual species, n = 11). In contrast, an early single lampricide treatment in one of these streams killed 6,329 fishes and 2,669 amphibians (Marsden et al. 2003). Thus, use of seasonal traps has reduced lampricide use and non-target mortality, while blocking non-target fish movements for only a few days per individual during only a portion of the year. Effectiveness of trapping has been improved in recent years by installation of permanent structures with seasonal traps in two streams; these structures are less subject to failure during high water events, and have floating, self-cleaning debris racks (Brad Young, personal communication, 2018).

5.4 Biological Response of Sea Lamprey to Control

Use of lampricides raises several concerns, including development of resistance, increased larval recruitment in lentic areas, and compensatory shifts in life history parameters. The potentially strong selective pressure imposed by 95–99% mortality in each generation may lead to resistance to TFM or niclosamide (Dunlop et al. 2018). Although 57 years of toxicity data showed that the lethal concentration of TFM has not increased over time, Dunlop et al. (2018) stressed that this is only indirect evidence that Great Lakes sea lamprey have not become more resistant to TFM. They suggested that direct tests for lampricide resistance be conducted and that continued vigilance is necessary. The evolution of lampricide resistance would result in the need to increase the concentration or frequency of applications and would likely result in increased sea lamprey abundance. Behavioral resistance, including selection for individuals that rear in untreated lentic areas, is also a concern (Dunlop et al. 2018). Although the majority of larval lampreys in deepwater environments are thought to be a consequence of natural downstream drift during the protracted larval stage (Hansen and Hayne 1962; Lee and Weise 1989; Fodale et al. 2003; Dawson et al. 2015; see Chap. 7), it is not known to what extent these habitats might be contributing to the juvenile population (Johnson et al. 2016b) or if there has been selection for earlier or more extensive downstream migration to these harder-to-treat areas (see Sect. 5.2.2.1).

Alternatively, sea lamprey could evolve compensatory mechanisms to offset population reductions that result from control activities (Jones et al. 2003). Because metamorphosis is dependent on reaching a size threshold (length and weight; Youson et al. 1993), response to reduced density might be observed both as increased larval growth and metamorphosis at a younger age. Accelerated growth and time to metamorphosis has been observed in some populations following lampricide application (e.g., Purvis 1979; Weise and Pajos 1998; Morkert et al. 1998; see Dawson et al. 2015), but no relationship between density and metamorphosis was observed in laboratory experiments (Youson 2003; see Manzon et al. 2015). Comparison of treated and untreated streams in the Lake Champlain drainage showed that, despite higher densities and slower growth of age-1 larvae in the treated stream, transformation of the first recolonizing year class occurred earlier than in the untreated streams (Zerrenner and Marsden 2005). In cage experiments, however, density did not have a significant effect on growth or survival of larvae (Zerrenner 2004), in contrast to experiments conducted by Morman (1987) and Rodríguez-Muñoz et al. (2003) in which growth and survival were affected by density (see Dawson et al. 2015). In adults, mean length and weight increased in each of the three upper Great Lakes after populations declined due to control (Heinrich et al. 1980; Houston and Kelso 1991)—increasing the capacity for harm to host populations (e.g., Madenjian et al. 2008) and leading to increases in female fecundity (Gambicki and Steinhart 2017; see Chap. 1). Populations may also respond to declines in numbers by increasing the sex ratio to favor females; if larval production is correlated with the number of females (i.e., the supply of eggs) and the supply of males does not limit recruitment, then a shift in sex ratio to a preponderance of females could compensate for overall declines in adult abundance (Jones et al. 2003). Indeed, the sex ratio of larval and adult sea lamprey populations shifted from a predominance of males to a predominance of females in the three upper Great Lakes after the control program drastically reduced densities in the 1960s (Purvis 1979; Heinrich et al. 1980; Torblaa and Westman 1980; see Chap. 1). Subsequent increases in prey abundance, due to stocking, and sea lamprey size have been linked with the increasing proportion of males in Lakes Huron and Superior (Houston and Kelso 1991). Sex in sea lamprey appears to be related to larval growth and condition, and thus indirectly related to environmental productivity (Johnson et al. 2017), so a mechanism may exist for sex ratio to respond to a change in the environment of the larvae, including changes in density. Mean length and proportion of females increased in Cayuga Lake after treatment and the increase in number of salmonid hosts (Bishop and Chiotti 1996). However, in the Great Lakes, adult sex ratios returned to parity or a slight excess of males by the mid-1990s, despite continued low densities, and there is no recent evidence of a relationship between larval density and sex ratio (Wicks et al. 1998; see Chap. 1). Furthermore, review of historical data and stock-recruitment analysis by Jones et al. (2003) indicated that the evidence for compensatory effects is equivocal, and suggested that such effects might be in part masked by high variability in recruitment. From the standpoint of control programs, studies of compensatory effects have primarily been valuable in defining areas of uncertainty in population modeling, rather than predicting clear responses to population control.

5.5 Emerging Control Techniques

The successful suppression of sea lamprey populations throughout their invasive range still relies heavily on use of pesticides. Concerns about non-target effects and public perception have been a significant aspect of sea lamprey control since inception of the program (see Sect. 5.2.2.1), and extensive research has been focused on measuring and reducing susceptibility of non-target species (e.g., Gilderhus et al.

1975; Maki et al. 1975; Dahl and McDonald 1980; Boogaard et al. 2003; O'Connor et al. 2017). Methods to reduce non-target mortality include: maintaining the lampricide concentration at the MLC for the minimum exposure period (9 h); monitoring pH during treatments to maintain, but not exceed, lethal activity of the lampricides; treatment at lower water temperatures that reduces non-target mortality without impacting lethality to sea lamprey; and scheduling treatment application later in the year on some tributaries to avoid affecting juvenile lake sturgeon while they are still stream residents (e.g., Johnson et al. 1999; Boogaard et al. 2003). Native lamprey species are particularly susceptible to TFM treatments and population declines have been noted in Great Lakes tributaries (Schuldt and Goold 1980; see Maitland et al. 2015). In New York and Vermont lakes where sea lamprey control began after establishment of the Clean Water Act and NPDES discharge permit requirements, treatment of streams that contain state-listed endangered lamprey species is not permitted (Marsden et al. 2003; see Sect. 5.3.2). Although niclosamide granules have higher non-target toxicity than TFM, particularly for molluscs and soft-bodied invertebrates, formulations that combine niclosamide with TFM have a lower LC50 for many non-target species (Boogaard et al. 2003). Nevertheless, the cost of lampricides and their periodic re-registration with EPA, and the potential for cumulative environmental effects, has prompted the search for increased efficiency in the use of lampricides and alternatives to lampricide control (McDonald and Kolar 2007).

In the 1970s, the GLFC adopted an IPM approach to sea lamprey control. The concept of IPM was developed for use in combating agricultural insect pests in the 1950s, and its application to sea lamprey—a novel approach to control a vertebrate pest—was outlined at the first Sea Lamprey International Symposium in 1979 (Sawyer 1980). IPM promotes the use of multiple tactics to control pests to tolerable levels of damage, rather than attempt eradication. The GLFC adopted a policy for integrated management of sea lamprey (IMSL) in the 1980s (Christie and Goddard 2003). The IMSL strategy focused on increasing the efficiency and effectiveness of current methods, particularly use of barriers and traps, and developing other alternative approaches to lampricides. The initial goals of the GLFC were to "accomplish at least 50% of sea lamprey suppression with alternative technologies while reducing TFM use by 20% through use of at least one new alternative-control method, and increased use of current methods such as sterile-male-release, trapping, and barrier deployment" by 2010 (GLFC 2001). In 2011, the GLFC's IMSL strategy was renamed Integrated Sea Lamprey Control (ISLC). The goals of the control program were changed to recognize the continued importance of lampricides and barriers, and did not include reductions in use of lampricide (GLFC 2011). The focus of ISLC continues to be suppression of sea lamprey populations to target levels and increasing the effectiveness and efficiency of sea lamprey control; suppression is achieved through integration of targeted adult and larval sea lamprey assessment, continued use of lampricides and barriers, and the development of additional control tactics including those based on semiochemicals, that is, migratory and sex pheromones and alarm cues (GLFC 2011). Additional approaches to enhance ISLC are discussed below.

5.5.1 Pheromones and Other Chemical Cues

The sea lamprey has an acutely sensitive olfactory system and relies extensively on chemical information, including pheromones and other chemical cues (Buchinger et al. 2015; Green et al. 2017). Pheromones are chemicals released by an individual that induce a behavioral or developmental response in other members of the same species (Karlson and Luscher 1959). Pheromones convey specific messages in specific contexts (e.g., sex pheromones), are shaped by natural selection acting on both the sender and receiver, and are typically used in direct communication. Chemical cues, on the other hand, typically represent public information and have not been molded by natural selection to carry information (Hume and Wagner 2018). Rather, the receiver has evolved an ability to perceive and respond to useful cues that are incidentally available in the environment (e.g., evolving recognition of chemicals inadvertently released when a fish is attacked or when a fish dies, termed alarm cues; Brown et al. 2011).

The use of sex-attractant pheromones has been a major tool in insect pest management since the 1970s (Baker 2008 and references therein). Pheromones have several features that make them singularly advantageous for use in pest control; they are non-toxic, biodegradable, highly species-specific, potent attractants that are detectable in minute quantities and capable of inducing responses at long distances. Unlike pesticides or pathogens, whose continued use may induce rapid selection for resistance, the response to pheromones is biologically useful and loss of efficacy is unlikely. In insect pest management, pheromones are primarily used to monitor the presence of a pest population to better target pesticide application, to remove insects by attracting them en masse into traps, or to disrupt mating activity by overwhelming the olfactory system or using a false scent to distract responders (typically males) away from senders (typically females). These applications have been significantly improved by advances in spectrometry that enabled separation of constituent chemicals in scents emitted by insects, by the ability to synthesize active compounds in pheromones, and the discovery of mixtures that induce the most consistent behavioral response (not always the natural ratio).

The importance of odors in the biology of sea lamprey is recognizable by the size and sensitivity of the olfactory system, which undergoes a radical increase in complexity during metamorphosis (VanDenBossche et al. 1995). Use of scent to detect prey is common for many species, particularly for parasites for which suitable prey are restricted to a few species (e.g., Dillman et al. 2012). Sea lamprey are known to attack a wide range of host species (see Chap. 3), but they do show preferences (e.g., for lake trout) and are able to orient to host odors (Kleerekoper and Mogensen 1963). Several other life history characteristics of the sea lamprey dictate the advantages of using semiochemicals. Sea lamprey are semelparous and access to spawning areas may require an arduous upstream migration (see Johnson et al. 2015a; Moser et al. 2015). Therefore, correct choice of an appropriate spawning stream and the ability to find mates is vital, as a mistake will cost an individual most or all of its reproductive potential. Two functional groups of semiochemicals have been identified in sea lamprey: migratory pheromones to find suitable spawning streams (see Sorensen et al. 2005; Moser et al. 2015) and sex pheromones to find mates (Li et al. 2002; see Johnson et al. 2015a, b).

Semelparous stream-spawning salmonids choose a spawning stream based on their own personal history, that is, they home to the stream where they were successfully hatched and reared (Dittman and Quinn 1996). This strategy would not be as advantageous for sea lamprey, because they are vulnerable to being substantially displaced from their natal stream by the hosts on which they attach (Waldman et al. 2008). Consequently, an optimal cue to identify a spawning stream with suitable habitat would be the scent of "success," that is, the presence of live larvae and the odor they emit. This cue would be particularly useful in sea lamprey given their long residence in streams as larvae; in any given year or season, a good spawning stream may be inhabited by at least three to four year classes of larvae that would be emitting a cue, increasing its strength and representing multiple years of suitable environmental conditions for reproduction and larval rearing. The use of a migratory pheromone produced by larvae and used by adults to find suitable spawning streams was first suggested by Moore and Schleen (1980) who noted a positive relationship between the number of larvae in a stream and the number of migrating adults that enter the stream. The hypothesis was further supported through research that showed that the bile acids petromyzonol sulfate (PZS or PS), petromyzonamine disulfate (PADS), petromyzosterol disulfate (PSDS), and 3-keto petromyzonol sulfate (3KPZS) emitted by stream-resident larvae stimulated electro-olfactogram (EOG) reactivity (Li and Sorensen 1997; Siefkes and Li 2004; Sorensen et al. 2005, Fine and Sorensen 2008; Yeh et al. 2012; Buchinger et al. 2015) and appeared to induce stream finding behaviors in the lake (Meckley et al. 2014) and influence in-stream behavior of pre-spawning adult sea lamprey (Teeter 1980; Bjerselius et al. 2000; Sorensen et al. 2005; Johnson et al. 2013). However, field experiments show that the whole larval extract is still more attractive than the tested combinations of these compounds, suggesting that there are still more active compounds that have not yet been identified (Meckley et al. 2012, 2014). Recent research has identified several novel putative migratory pheromone compounds that appear to influence in-stream behaviors (Li et al. 2013a, 2014; Brant et al. 2016).

After entering the spawning tributary, sea lamprey would further benefit from a cue that efficiently brought sexually mature females and males together to spawn before their energy reserves have been depleted. The first indications that sea lamprey use sex pheromones dates from observations by fishermen who used ripe, or spermiated, males as "bait" to attract females into traps (Fontaine 1938). Female sea lamprey may also release a pheromone that attracts males (Teeter 1980), but the putative female sex pheromone has not been extensively explored. Once mature adults have entered a stream and males have begun nest construction, a sex pheromone emitted by spermiated males attracts ovulated females to a nest for spawning (Johnson et al. 2015b). Two-choice maze experiments confirm that female sea lamprey are attracted to washings from spermiated males (Li et al. 2002). Several bile acids produced by spermiated male sea lamprey, including 3-keto petromyzonol sulfate (3KPZS), 3-keto allocholic acid (3KACA), diketo petromyzonene sulfate (DKPES), and 3-keto

1-ene petromyzonol sulphate (3K1CS), stimulated EOG reactivity in adult female sea lamprey (Li et al. 2002, 2013b; Siefkes and Li 2004). Evidence suggests that 3KPZS and 3KACA stimulate sexual maturation, that is, synchronizing maturation in females and males (Chung-Davidson et al. 2013a, b). 3KPZS was shown to elicit upstream movement in ovulated females (Li et al. 2002; Siefkes et al. 2005; Johnson et al. 2009), ultimately assisting them in finding a mate. 3KPZS was also shown to induce pair maintenance behaviors such as nest construction, retention, and cleaning (Johnson et al. 2012). Recent research has shown that DKPES and 3K1CS also function in mate finding (Li et al. 2013b; Johnson et al. 2014b) and that DKPES enhances behavioral responses to 3KPZS (Li et al. 2013b). Like the whole larval extract, whole spermiated male extract is still more attractive than the tested combinations of these compounds, suggesting that there are still more active compounds to be identified (Johnson et al. 2015b).

In contrast to control strategies that use lethal chemicals, the intent of odor-mediated control tactics is not to reduce established larval populations, but to reduce recruitment by removing reproductive adults prior to spawning (Twohey et al. 2003b; Buchinger et al. 2015). The primary strategy that is currently being investigated involves using captive males or male pheromones to attract females into traps (Johnson et al. 2013). Attractive odors could also be used to deflect adult sea lamprey away from a stream that is challenging to treat with lampricides into a stream that is easier to treat, or easier to trap, by manipulating habitat selection behavior. This tactic requires first decreasing the larval densities and thus the strength of the larval cue in the river water, then introducing a strong signal in the target tributary by adding whole larval extract of synthesized components of the cue. Additional strategies that use odors have been proposed, including enhancement of attraction to sterile males, and disruption of spawning behaviors by either flooding a stream with sex pheromone to overwhelm the females' ability to locate males, or by finding antagonists to pheromones that would reduce sensitivity to specific odors (Li et al. 2003; Klassen et al. 2005; Buchinger et al. 2015).

Recently, two probable components of the male sex pheromone were tested as trap attractants for their ability to enhance trapping of migratory adult sea lamprey. Use of 3KACA did not increase trapping success, and use of 3KPZS resulted in a marginal (~10%) increase (Luehring et al. 2011; Johnson et al. 2013). Research to better understand the factors influencing the success of pheromone-baited traps and what trap designs are best suited for use with pheromone bait are needed and underway (Johnson et al. 2015b, c). Similar to lampricides, the use of pheromones and cues requires ongoing research to identify additional chemical components, produce synthetic analogs, and elucidate behavioral responses, and work with federal and state regulatory agencies to register compounds for use in control applications (Li et al. 2007; Siefkes 2017). Progress has been made with registration of the sex pheromone component 3KPZS with the EPA and the Canadian Pest Management Regulatory Agency for use in management applications (Siefkes 2017). Attention must also be paid to non-target effects; while pheromones are usually highly species-specific, some of the compounds released by sea lamprey are also released by both parasitic and non-parasitic native lamprey species throughout North America, as well as European and New Zealand species (Fine et al. 2004; Stewart and Baker 2012; Buchinger et al. 2015). Adults of all species tested using EOG recordings responded to bile acid components common to sea lamprey. Thus, use of these compounds to induce behavioral responses in sea lamprey may also disrupt migratory and possibly also mating behaviors in native species, of which some are listed as Threatened or Endangered in certain jurisdictions (see Maitland et al. 2015). Production of components of the migratory pheromone by native species has the potential to confound control efforts, as complete removal of sea lamprey larvae may leave a stream still attractive to adults if pheromone-producing native species are present. These same native odors may have facilitated the rapid dispersal of sea lamprey into suitable streams during the initial invasion.

5.5.2   Repellants

A second class of chemosensory cues, repellants, has recently been proposed as a potential control tool (Imre et al. 2010, 2014; Wagner et al. 2011). Possible repellant compounds include alarm cues produced from injured conspecifics, necromones (odors of decaying conspecifics that indirectly signal a threat), or noxious chemicals including non-biological compounds. Interest in repellants has been stimulated by the curious observation that the introduction of human saliva into a tank of sea lamprey elicits a rapid, vigorous response in which all of the sea lamprey attempt to leap out of the water (Michael B. Twohey, USFWS, Marquette, MI, personal communication 2010). A similar reaction has been reported for stream-resident sea lamprey when decaying sea lamprey were placed in the water (William D. Swink, U.S. Geological Survey, Hammond Bay Biological Station, personal communication, 2007). In laboratory experiments, migratory sea lamprey strongly avoided extracts of freshly killed and decaying conspecifics (Wagner et al. 2011; Bals and Wagner 2012) and, in field experiments, migratory sea lamprey also adjusted their swimming route and speed to avoid the extracts (Luhring et al. 2016). Avoidance of extracts from sea lamprey and white sucker (simulating an injured conspecific and a potential predator) and human saliva was demonstrated in semi-natural enclosures, suggesting that sea lamprey respond to various cues that indicate the presence of predators (Imre et al. 2014). In contrast, Hume and Wagner (2018) observed no response to chemical cues from injured white sucker; however, they determined that sea lamprey respond to alarm cues from five other North American lamprey species, with weaker responses from more phylogenetically distant species (i.e., American brook and Pacific lampreys) than the more closely related *Ichthyomyzon* species. Repellants could be used to induce complete avoidance of a spawning stream if they are species-specific, inexpensive to produce, and do not induce habituation. Alternatively, repellants could be used to "push" sea lamprey toward traps. In field experiments, sea lamprey were trapped more rapidly when they avoided an area of stream treated with repellant; however, the overall trapping rate did not increase in the presence of a repellant (Hume et al. 2015). A more likely use of repellants for control is in a "push-pull" scenario, that is, in combination with attractants to enhance trapping while inducing avoidance of areas distal to the traps or to enhance lampricide control by pulling sea lamprey to areas that are easy to treat while pushing them from areas that are difficult to treat (Buchinger et al. 2015; Hume et al. 2015).

5.5.3 Genes and Genomes

Recently, potential use of genetic and genomic technology to develop alternative control tactics for sea lamprey, as well as other invasive species, has been explored (e.g., Thresher et al. 2014; McCauley et al. 2015). The mitochondrial genome of sea lamprey was sequenced by Lee and Kocher (1995), and the nuclear genome was subsequently sequenced by Smith et al. (2013). Assembly of the nuclear genome sequence has proved to be challenging due to the large number of duplicated genes, which may in part date from two genome duplication events that occurred early in the vertebrate lineage (Putnam et al. 2009; Smith et al. 2013; see Docker et al. 2015). Sea lamprey also undergo substantial DNA reorganization, in which large portions of the genome are discarded during an extended period beginning soon after fertilization (Smith et al. 2009; Bryant et al. 2016). Currently, researchers are looking for genes or genetic functions that may lead to use in the control program. Early work with the genome has revealed unique characteristics of sea lamprey genetics; for example, identification of the genes for chemosensory reception in sea lamprey has shown that the repertoire of these genes is relatively small compared with teleost fishes (Libants et al. 2009). Heath et al. (2014) demonstrated that short strands of double-stranded RNA (interfering RNA), either injected into sea lamprey embryos or fed to larvae, effectively destroyed matching RNA transcripts and increased larval mortality. This work may lead to development of highly species-specific control based on unique gene sequences within sea lamprey. An improved understanding of the genetic basis of sea lamprey developmental biology, immunology, and physiology may substantially assist progress in finding opportunities for control strategies (McCauley et al. 2015; see Chap. 6).

Two particular areas of interest are the use of "daughterless" technology and "Trojan" genes to manipulate sex ratios in Great Lakes sea lamprey. Trojan genes are a general category of genetic modifications in which an inserted or modified gene produces a fitness advantage in one portion of the phenotype but a disadvantage in another, resulting in population decline to extinction (Muir and Howard 1999). In the daughterless gene example, a female fish may be engineered with two Y chromosomes, so that her progeny are all male; half would be sterile YY males (Thresher et al. 2009). Models predict that continued introduction of a small percentage of Trojan-Y females will result in eventual population extinction (Gutierrez and Teem 2006; Thresher et al. 2009). Because the genetic basis of sex determination in lampreys is as yet unknown (see Chap. 1), considerably more work would be required to develop such a system in lampreys, but its potential to be a highly effective and species-specific alternative to lampricides holds promise (Thresher et al.

2019). Induced triploidy is a similar technology that has been used for stocking grass carp *Ctenopharyngodon idella* to control aquatic macrophytes (Zajicek et al. 2011). Triploids can be produced when heat shock or pressure shock applied at a prescribed interval after fertilization prevents extrusion of the second polar body; because the second meiotic division in fish eggs occurs after fertilization, the resulting zygote carries two copies of the female genome and one copy of the male genome. In most species the developed adult is capable of mating, but the gametes are not viable due to unequal division of chromosomes among the eggs that are produced by meiotic division. An advantage of this method is that large numbers of eggs can be exposed to heat or pressure shock simultaneously, allowing rapid production of sterile individuals for stocking. However, recent discovery of diploid grass carp in Lake Erie indicates that this tactic can fail (Chapman et al. 2013).

Use of environmental DNA (eDNA) is successful for detecting larval sea lamprey presence in streams, and could be a significant tool for discovering new populations or recolonized streams (Gingera et al. 2016). Development of dependable quantitative eDNA could replace substantial time and effort spent sampling tributaries throughout the Great Lakes to assess whether populations are sufficiently large to warrant lampricide treatment (Schloesser et al. 2018; see Chap. 7).

5.5.4 Tagging and Tracking

Control of larval sea lamprey populations is predicated on the assumption that streams with large larval populations produce correspondingly large numbers of parasitic juveniles (Jones 2007). However, although consistent evidence for compensatory mechanisms has not been found (Jones et al. 2003; see Sect. 5.4), compensatory increases in growth or survival could lead to higher parasitic production from streams with lower larval densities (Zerrenner and Marsden 2005, 2006). Consequently, research has been focused on determining the natal origins of parasitic sea lamprey, with the goal of targeting control on streams that produce the most parasitic juveniles. Outmigrating juvenile sea lamprey have been tagged in Lake Huron and Lake Champlain using coded-wire tags and recaptured as parasitic juveniles in the lakes; these studies have led to a better understanding of sea lamprey movements, but returns to date have been too low to be informative about survival rates from individual streams (e.g., Bergstedt et al. 2003b; Howe et al. 2006).

Use of the microelemental chemistry of sea lamprey statoliths (a calcium carbonate structure suspended in the the inner ear, analogous to the otolith of teleost fishes, that incorporates the inorganic chemical signature of natal waters) initially showed promise as a method to identify the natal stream of parasitic juveniles (see Dawson et al. 2015). Statoliths incorporate certain chemicals from natal stream water during the protracted larval stage, and, like otoliths, are presumed to be inert. Thus, the core region of a parasite's statoliths where material was deposited during the larval period should reflect the chemistry of the natal stream. Research in Lake Huron and Lake Champlain showed that larval sea lamprey from individual tributaries, or clusters of adjacent tributaries, have different microchemical signatures (Hand et al. 2008; Howe et al. 2013). However, parasitic juveniles of known origin were not successfully identified to the correct natal stream. Recent research indicates that the statolith is not inert, but is reworked during metamorphosis (Lochet et al. 2013); thus, tracking of natal origins may require identifying the statolith signature from post-metamorphic rather than larval sea lamprey (Lochet et al. 2014).

Recent advances in tagging technologies have improved the ability to follow fine-scale movements of large larvae and post-metamorphic sea lamprey in streams (e.g., Quintella et al. 2005; Wagner et al. 2006, 2009; Johnson et al. 2009; see Chap. 7). Likewise, advances in acoustic telemetry and radio tagging have been used to describe migratory movements of spawning-stage sea lamprey in rivers (Holbrook et al. 2015), and work on tracking sea lamprey in open lake waters is in progress (Andrew Muir, GLFC, Ann Arbor, MI, personal communication, 2017).

5.5.5 *Improvement of Existing Control Methods*

Sea lamprey control is under constant internal and external pressure to improve effectiveness, minimize costs, and address non-target effects. Lampricide use has received particular attention, motivating development of methods to reduce toxicity to non-target species and reduce the amount used (see Sects. 5.2.2.1 and 5.5). Additional lampricide application strategies to lower non-target mortality have been discussed, including treatments with lower concentrations but longer application duration (Dale Burkett, GLFC, Ann Arbor, MI, personal communication, 2017). Research continues to focus on finding new lampricides with better attributes than TFM or niclosamide. Use of GPS (global positional system) to spot-treat deltas with granular niclosamide, using fine-resolution mapping of larval distribution and densities, has improved treatment efficiency.

Use of barriers for sea lamprey control has received increasing scrutiny as fisheries managers and the public weigh trade-offs between benefits of sea lamprey control and the negative consequences of dams as impediments for native fish migrations and other ecosystem functions (McLaughlin et al. 2007; Jensen and Jones 2018). For example, the Black Sturgeon River, a tributary to Lake Superior in Ontario, is particularly controversial (McLaughlin et al. 2012); a dam located 16 km from the river mouth restricts movements of walleye and provincially-listed lake sturgeon. However, removal of the dam would result in substantial increase in costs for chemical control, and expose upstream populations of northern brook lamprey, which is listed as a species of Special Concern under Canada's Species at Risk Act (Maitland et al. 2015), to TFM. A preferred alternative is design of a selective barrier that blocks sea lamprey but allows for passage or transport of desirable fishes. The GLFC is currently investing in developing selective fish passage through a better understanding of sea lamprey and non-target fish behavior, movement, and swim performance, and the use of current and emerging technologies (e.g., image recognition and robotic automation; McCann et al. 2018; see Chap. 7). The construction of a dedicated fish passage research facility is being planned for the Boardman River in Traverse City, Michigan (see Sect. 5.2.2.2).

Methods to improve barrier and trap effectiveness have concentrated on developing a better understanding of sea lamprey behavior and movement (e.g., McLean et al. 2015; Rous et al. 2017). Current methods use "trap-and-sort," in which non-target organisms are released manually from traps on a near-daily basis, requiring substantial effort. Ideally, technology could be developed for "sort-and-trap," whereby non-target organisms either are not caught or can be mechanically released while sea lamprey are selectively trapped and retained. Recent research on anguilliform locomotion of sea lamprey led to the concept of using eel ladders—sloped or wetted surfaces with a staggered array of protrusions as "push-off" points—that could be incorporated into a barrier design to lure sea lamprey into a trap (Pete Hrodey, USFWS, Marquette, MI, personal communication, 2018). Ascent rate is maximized by incorporation of high flow and a 45° incline (McDonald and Desrochers 2012). Additional investigations have used eel pots, which target bottom-dwelling species that seek dark, confined spaces, in conjunction with standard portable assessment traps (Nicholas Johnson, U.S. Geological Survey, Hammond Bay Biological Station, personal communication, 2017). Research is being conducted in the use of pulsed direct current electric fences to block access to spawning streams and direct adult sea lamprey toward traps (Johnson et al. 2014a).

New technologies have contributed to significantly increasing the effectiveness and reducing the costs of controlling delta populations of sea lamprey. Remote substrate classification using RoxAnn^{TM} can substantially improve quantitative estimates and mapping of larval substrate (Fodale et al. 2003; see Chap. 7). Advances in remote sensing, electronic navigation, and geo-spatial modeling have greatly aided in mapping of larval distributions (Aaron Jubar, USFWS, Ludington, MI, personal communication, 2018). This combination of technologies has led to development of a highly efficient, automated lampricide application boat which is guided by a GPS map to release granular niclosamide only where larvae are present (Michael Siefkes, personal observation). Reductions in lampricide use can be realized using this technology, resulting in lower costs and impacts on non-target species.

The potential to use capture of outmigrating juvenile sea lamprey as a tool for reducing parasitic populations has received renewed attention (e.g., Sotola et al. 2018). This portion of the life history has been considered somewhat intractable because outmigration occurs throughout the winter months, with peaks of movement in fall and spring (Applegate and Brynildson 1952); high water flows and accumulation of leaf litter, debris, and ice present hazards to in-stream trapping operations. However, if the factors that motivate movement can be better defined, then trapping could be focused during the diurnal period, season, flow regimes, and stream locations when and where outmigration is concentrated (see Dawson et al. 2015). The goal of capturing outmigrating juveniles is not specifically to affect population growth by reducing recruitment, but instead to directly reduce the number of parasitic juveniles entering a lake to feed on host fishes. Therefore, success is not necessarily contingent on capturing a very high proportion of the population.

5.5.6 Spread Prevention

The sea lamprey invasion of the Great Lakes (at least above Niagara Falls) was recent, rapid, and spectacular in the rate and extent of its impacts on native fish species (see Sect. 5.1.2). Despite their use of the scent of stream-resident larvae to select a spawning stream (see Sect. 5.5.1), the colonization potential of sea lamprey does not appear to be limited to streams with larvae. Although establishment in new water bodies will depend on access to a sufficient prey base (see Chap. 4), the recent study by Johnson et al. (2016a) showing that sea lamprey may complete their life cycle in the Cheboygan River system, upstream of a dam intended to prevent spawning-phase sea lamprey access from Lake Huron, indicates that further spread is possible. Given the cost of annual suppression of established populations, preventing their spread to additional lakes is a high priority. Fortunately, their vectors for spread are fairly limited: they are not used as a food fish in the United States or Canada, they are not a desirable target for anglers, and their use as bait is no longer legal (Maitland et al. 2015), so unauthorized introduction of sea lamprey is unlikely. Unlike many aquatic invasive invertebrate species (e.g., Kelly et al. 2013; Jansen et al. 2017), the larval stage in lampreys is not likely to be picked up in ballast water or by recreational boats or bait buckets, and the mobile stages (parasitic and migratory stages) are relatively short-lived. The major vector for spread to new lakes may be canals, either by direct movement of sea lamprey or as hitchhikers attached to boats or host fish. The potential for the spread of sea lamprey through canals has been debated, as historically canals were often highly polluted, stagnant, and drained for portions of the year (Daniels 2001). However, modern canals have water quality that is typically adequate for survival and transit of sea lamprey (e.g., Marsden and Ladago 2017).

Several inland lakes in the Great Lakes basin that do not contain sea lamprey are as large or larger than some of those that do, and could presumably support sea lamprey populations should they gain access. For example, Lakes Simcoe and Winnebago (area ~745 and 560 km^2, respectively) are larger than Cayuga and Seneca lakes (each ~170 km^2) and the two lakes in the upper Cheboygan River system (each ~70 km^2; see Chap. 4). Lake Simcoe lies between Lake Ontario and Georgian Bay in Lake Huron and is connected to both lakes via the Trent-Severn Waterway. Natural invasion into Lake Simcoe from Lake Ontario is improbable given the distance and number of natural barriers between the lakes. However, two locks and a boat lift are the only obstacles on the Trent-Severn canal between Georgian Bay and Lake Simcoe. The boat lift, at Big Chute, is a marine railway constructed in 1917; boats drive into a carriage and are supported on slings while the carriage is transported up- or downhill on rails to the next portion of the canal (Beahen 1980). An advantage of this system, however, is that the boat hull can be inspected for attached organisms, including sea lamprey. Plans to replace the old marine railway with a single lock at Big Chute were cancelled in the mid-1960s when sea lamprey were found at the lower end of the railway. The Big Chute Marine Railway was later enlarged and improved, and has been maintained to prevent the spread of sea lamprey into Lake Simcoe. Similar to Lake Simcoe, Lake Winnebago is connected to Green Bay,

Lake Michigan, via a chain of 17 locks on the Fox River system. The lock system was closed in 1983 and a barrier was installed at the Rapide Croche Lock in 1988 to prevent the spread of sea lamprey into Lake Winnebago; no sea lamprey were found during recent monitoring efforts for aquatic invasive species above and below the Rapide Croche Lock (De Stasio 2013). Although plans are underway to open portions of the waterway and refurbish most of the locks, the lock at Rapide Croche will remain closed; installation of a boat lift to transfer boats over the lock is planned.

5.5.7 Natural Population Regulation

Factors that naturally limit landlocked sea lamprey populations are not well understood. Sea lamprey require a prey population, ideally salmonids, and streams that contain areas of spawning habitat upstream of good larval habitat (see Dawson et al. 2015). The parasitic juvenile stage is not known to be highly vulnerable to predators, although predation in some regions (e.g., by abundant coolwater fish in Lake St. Clair and the western basin of Lake Erie) may keep their numbers in check (Sullivan et al. 2003), and insufficient availability of suitable hosts might limit growth and survival of this stage (see Chaps. 3 and 4). Downstream-migrating juveniles and upstream migrants and spawning adults are much more accessible to predators, and adult sea lamprey have been observed with marks that appear to be stab wounds from herons or other large birds (see Docker et al. 2015; Maitland et al. 2015). The burrowing habit of larvae must have a role in avoiding predation; during delta treatments, dying sea lamprey that emerge to the surface are rapidly taken by gulls and terns, indicating their attraction as a food (Stephen J. Smith, USFWS, Essex Junction, VT, personal communication, 2018). The two main fish species able to detect and access buried prey in the Great Lakes area are lake sturgeon and American eel; Siberian and white sturgeons (*Acipenser baeri* and *A. transmontanus*, respectively) have been observed feeding on other lamprey species (Cochran 2009) and American eel were observed to capture and consume brook lamprey larvae in laboratory experiments (Perlmutter 1951). Lake sturgeon and American eel have both suffered catastrophic population declines throughout the region, however, and are now listed as Threatened, Endangered, or extirpated in most regions (e.g., Jelks et al. 2008; COSEWIC 2012, 2017). Their declines may have hampered sea lamprey control efforts, although lake sturgeon and American eel recovery in some areas (e.g., Welsh et al. 2019) could eventually aid in sea lamprey population regulation. Conversely, increases in the quantity and quality of larval stream habitat as a consequence of land use practices and increased erosion may have historically improved larval survival (Marsden and Langdon 2012), and more recent pollution abatement efforts similarly may have favored sea lamprey, particularly in some of the interconnecting waterways in the Lake Erie basin (Sullivan et al. 2003). Larval incubation temperature appears to be an important element that may restrict use of some potential spawning streams (Rodríguez-Muñoz et al. 2001), and it will be important to integrate climate change scenarios into future sea lamprey control strategies. Potential for application of density-dependent suppression in larval growth in sea lamprey by larvae of native lamprey species (Murdoch et al. 1991) has not been studied.

If, as several authors have hypothesized, sea lamprey are native to Lake Ontario, the Finger Lakes, or Lake Champlain (see Sect. 5.1.2), they and their prey species had strategies that enabled coexistence. For example, lake trout from Seneca Lake are significantly less likely than lake trout from the Great Lakes to be attacked by sea lamprey or suffer mortality as a consequence (Schneider et al. 1996), likely due to low spatial overlap (Bergstedt et al. 2007). Seneca Lake lake trout stocked in the Great Lakes have higher survival and reproductive success than other strains, including strains derived from Great Lakes populations (Marsden et al. 1989; Page et al. 2003). Significantly, wounding rates in Lake Champlain and the Finger Lakes, even on the stable native population of lake trout in Seneca Lake, were substantially higher than those observed in the Great Lakes prior to sea lamprey control, when lake trout populations were depressed by fishing. In Lake Champlain, lake trout reproduction was high even during periods of wounding in excess of 50 wounds per 100 lake trout, and substantial lake trout recruitment began in 2012 when wounding was 40 wounds per 100 lake trout (Marsden et al. 2017). Thus, the extent to which sea lamprey are an impediment to lake trout recruitment in Lake Champlain is unclear; lake trout populations may be able to coexist with sea lamprey despite wounding rates that are higher than the established targets. However, sea lamprey control is still required to support a viable fishery of interest to anglers, and to support restoration in the Great Lakes.

5.6 Conclusions

After sea lamprey predation contributed to the devastating collapse of fish stocks and fisheries in the Great Lakes, Lake Champlain, and the Finger Lakes, sea lamprey control has been successful in reducing sea lamprey populations by up to 90%, enabling recovery of these valuable resources. Nevertheless, annual and diligent sea lamprey control efforts are needed to maintain sea lamprey populations at these reduced levels. Improvements in lampricide treatment efficacy, barrier effectiveness, and capture efficiency of traps, and the development of innovative approaches in the use of lampricides and additional controls, are required to further reduce sea lamprey populations to targets established to further enable rehabilitation of fish stocks and fisheries. At the same time, the control program faces issues of increasing costs of lampricides, impacts of lampricides and barriers to non-target species (especially species of conservation concern), lampricide re-registration, and local permitting restrictions. Importantly, attitudes of the public, regulators, and resource managers towards sea lamprey control and its importance in maintaining and rehabilitating fisheries and supporting healthy ecosystems are subject to change. Loss of the social license would significantly impact the sea lamprey control program if lampricide treatments are not permitted or important sea lamprey barriers are removed or new construction is not permitted. Efforts to balance effective control while minimizing non-target and ecological impacts have stimulated a broad range of research to "know thy enemy," and have substantially advanced our knowledge of sea lamprey biology, behavior, chemical communication, and genomics (e.g., Christie and Goddard 2003; Krueger and Marsden 2007; Binder and McDonald 2007, 2008; Binder et al. 2010; McCauley et al. 2015; Siefkes 2017). Understanding the toxic effect of lampricides on non-target species, describing optimal parameters for application of existing lampricides, and development of new lampricides are still foci of research (Waller et al. 2003; McDonald and Kolar 2007). Progress is underway to develop sea lamprey-specific barriers and traps that use specific movement patterns of sea lamprey and minimize non-target impacts (McLaughlin et al. 2007; McLean et al. 2015). Research related to sea lamprey control has also had synergistic benefits ranging from decision analysis (Haeseker et al. 2007) and understanding movements of teleosts in relation to barriers (e.g., Pratt et al. 2009), to evolutionary genomics and vertebrate evolution (e.g., Libants et al. 2009; McCauley et al. 2015; Smith et al. 2018). Just as research has been used since the onset of sea lamprey control to adapt the program to be more diverse, effective, and efficient, agencies involved in control have increasingly recognized and addressed the need for public outreach to maintain an understanding of the goals and outcomes of the control program.

Acknowledgements We thank Emily Zollweg-Horan and Brad Hammers (New York State Department of Environmental Conservation, NYSDEC), for their assistance in accessing unpublished data from the Finger Lakes; B. J. Allaire and Brad Young (USWFS) for supplying Lake Champlain sea lamprey assessment data; Brian Chipman (Vermont Department of Fish and Wildlife) for data on lake trout wounding in Lake Champlain; and Jill Wingfield (Great Lakes Fishery Commission, GLFC) for creating maps and photo figures.

References

Almeida PR, Quintella BR (2013) Sea lamprey migration: a millenial journey. In: Ueda H, Tsukamoto K (eds) Physiology and ecology of fish migration. CRC Press, Boca Raton, FL, pp 105–131

Applegate VC (1950) Natural history of the sea lamprey (*Petromyzon marinus*) in Michigan. US Fish Wildl Serv Spec Sci Rep Fish 55:1–237

Applegate VC, Brynildson CL (1952) Downstream movement of recently transformed sea lamprey, *Petromyzon marinus,* in the Carp Lake River, Michigan. Trans Am Fish Soc 81:275–290

Applegate VC, Smith BR (1951) Movement and dispersion of a blocked spawning run of sea lamprey in the Great Lakes. Trans N Am Wildl Conf 16:243–251

Applegate VC, Howell JH, Moffett JW et al (1961) Use of 3-trifluoromethyl-4-nitrophenol as a selective sea lamprey larvicide. Great Lakes Fish Comm Tech Rep 1:1–35

Applegate VC, Johnson BGH, Smith MA (1966) The relation between molecular structure and biological activity among mono-nitrophenols containing halogens. Great Lakes Fish Comm Tech Rep 11:1–29

Baker TC (2008) Use of pheromones in IPM. In: Radcliffe EB, Hutchinson WD, Cancelado RE (eds) Integrated pest management. Cambridge University Press, Cambridge, pp 273–285

Bals JD, Wagner CM (2012) Behavioral responses of sea lamprey (*Petromyzon marinus*) to a putative alarm cue derived from conspecific and heterospecific sources. Behaviour 149:901–923

Beahen W (1980) Development of the Severn River and Big Chute lock station (history and archaeology). National Historic Parks and Sites Branch, Parks Canada, Environment Canada, Ottawa, ON

Beamish FWH (1978) Swimming capacity. In: Hoar WS, Randall DJ (eds) Fish physiology, vol 7. Academic Press, New York, pp 101–187

Bence JR, Johnson JE, He J et al (2008) Offshore predators and their fish community. In: Bence JR, Mohr LC (eds) The state of Lake Huron in 2004. Great Lakes Fish Comm Spec Publ 09-01:11–36

Bergstedt RA, Seelye JG (1995) Evidence for lack of homing by sea lamprey. Trans Am Fish Soc 124:235–239

Bergstedt RA, Twohey MB (2007) Research to support sterile-male release and genetic alteration techniques for sea lamprey control. J Great Lakes Res 33(Spec Issue 2):48–69

Bergstedt RA, McDonald RB, Twohey MB et al (2003a) Reduction in sea lamprey hatching success due to release of sterilized males. J Great Lakes Res 29(Suppl 1):435–444

Bergstedt RA, McDonald RB, Mullett KM et al (2003b) Mark-recapture population estimates of parasitic sea lamprey (*Petromyzon marinus*) in Lake Huron. J Great Lakes Res 29(Suppl 1):283–296

Bergstedt RA, Argyle RL, Krueger CC et al (2007) Seasonal and diel bathythermal habitat use and habitat overlap of sea lamprey (*Petromyzon marinus*) and lake trout (*Salvelinus namaycush*) as determined with implanted archival tags. Great Lakes Fishery Commission Completion Report, Ann Arbor, MI

Binder TR, McDonald DG (2007) Is there a role for vision in the behaviour of sea lampreys (*Petromyzon marinus*) during their upstream spawning migration? Can J Fish Aquat Sci 64:1403–1412

Binder TR, McDonald DG (2008) The role of dermal photoreceptors during the sea lamprey (*Petromyzon marinus*) spawning migration. J Comp Physiol A 194:921–928

Binder TR, McLaughlin RL, McDonald DG (2010) Relative importance of water temperature, water level, and lunar cycle to migratory activity in spawning-phase sea lampreys in Lake Ontario. Trans Am Fish Soc 139:700–712

Birceanu O, McClelland GB, Wang YS, Wilkie MP (2009) Failure of ATP supply to match ATP demand: the mechanism of toxicity of the lampricide 3-trifluoromethyl-4-nitrophenol (TFM), used to control sea lamprey (*Petromyzon marinus*) populations in the Great Lakes. Aquat Toxicol 94:265–274

Birceanu O, McClelland GB, Wang YS, Brown JCL, Wilkie MP (2011) The lampricide 3-trifluoromethyl-4-nitrophenol (TFM) uncouples mitochondrial oxidative phosphorylation in both sea lamprey (*Petromyzon marinus*) and TFM-tolerant rainbow trout (*Oncorhynchus mykiss*). Comp Biochem Physiol C 153:342–349

Bishop DL (1992) Interaction between adult lake trout abundance and recruitment of stocked salmonines in Cayuga and Seneca Lakes. MS thesis. SUNY, Syracuse, NY

Bishop DL, Chiotti DL (1996) Evaluation of the experimental sea lamprey control program in Cayuga Lake, New York. Final Report, New York State Department of Environmental Conservation, Cortland, NY

Bjerselius RW, Li W, Teeter JH et al (2000) Direct behavioural evidence that unique bile acids released by larval sea lamprey (*Petromyzon marinus*) function as a migratory pheromone. Can J Fish Aquat Sci 57:557–569

Boogaard MA, Bills TD, Johnson DA (2003) Acute toxicity of TFM and a TFM/niclosamide mixture to selected species of fish, including lake sturgeon (*Acipenser fulvescens*) and mudpuppies (*Necturus maculosus*), in laboratory and field exposures. J Great Lakes Res 29(Suppl 1):529–541

Boogaard MA, Newton TJ, Hubert TD, Kaye CA, Barnhart MC (2015) Evaluation of short term 12 hour toxicity of 3-trifluoromethyl-4-nitrophenol (TFM) to multiple life stages of *Venustaconcha ellipsiformis* and *Epioblasma triquetra* and its host fish (*Percina caprodes*). Environ Toxicol Chem 34:1634–1641

Brant CO, Huertas M, Li K, Li W (2016) Mixtures of two bile alcohol sulfates function as a proximity pheromone in sea lamprey. PLoS ONE 11:e0149508

Bravener GA, McLaughlin RL (2013) A behavioural framework for trapping success and its application to invasive sea lamprey. Can J Fish Aquat Sci 70:1438–1446

Bravener GA, Twohey M (2016) Evaluation of a sterile-male release technique: a case study of invasive sea lamprey control in a tributary of the Laurentian Great Lakes. N Am J Fish Manag 36:1125–1138

Brege DC, Davis DM, Genovese JH et al (2003) Factors responsible for the reduction in quantity of the lampricide, TFM, applied annually in streams tributary to the Great Lakes from 1979 to 1999. J Great Lakes Res 29(Suppl 1):500–509

Bronte CR, Ebener MP, Schreiner DR et al (2003) Fish community change in Lake Superior, 1970–2000. Can J Fish Aquat Sci 60:1552–1574

Bronte CR, Krueger CC, Holey ME et al (2008) A guide for the rehabilitation of lake trout in Lake Michigan. Great Lakes Fish Comm Misc Publ 2008–01:1–40

Brown GW, Ferrari MCO, Chivers DP (2011) Learning about danger: chemical alarm cues and threat-sensitive assessment of predation risk by fishes. In: Brown C, Laland K, Krause J (eds) Fish cognition and behavior, 2nd edn. Blackwell Publishing, Oxford, pp 59–80

Brussard PF, Hall MC, Wright J (1981) Structure and affinities of freshwater sea lamprey (*Petromyzon marinus*) populations. Can J Fish Aquat Sci 38:1708–1714

Bryan MB, Zalinski D, Filcek KB et al (2005) Patterns of invasion and colonization of sea lamprey (*Petromyzon marinus*) in North America as revealed by microsatellite genotypes. Mol Ecol 14:3757–3773

Bryant SA, Herdy JR, Amemiya CT, Smith JJ (2016) Characterization of somatically-eliminated genes during development of the sea lamprey (*Petromyzon marinus*). Mol Biol Evol 33:2337–2344

Buchinger TJ, Siefkes MJ, Zielinski BS, Brant CO, Li W (2015) Chemical cues and pheromones in the sea lamprey (*Petromyzon marinus*). Front Zool 12:32

Bunnell DB (ed) (2012) The state of Lake Michigan in 2011. Great Lakes Fish Comm Spec Publ 12-01:1–70

Bunnell DB, Carrick HJ, Madenjian CP et al (2018) Are changes in lower trophic levels limiting prey-fish biomass and production in Lake Michigan? Great Lakes Fish Comm Misc Publ 2018–1:1–41

Chang SC, Woods CW, Borkovec AB (1970) Sterilizing activity of bis(1-aziridinyl)phosphine oxides and sulfides in male houseflies. J Econ Entomol 63:1744–1746

Chapman DC, Davis JJ, Jenkins JA et al (2013) First evidence of grass carp recruitment in the Great Lakes Basin. J Great Lakes Res 39:547–554

Christie GC, Goddard CI (2003) Sea Lamprey International Symposium (SLIS II): advances in the integrated management of sea lamprey in the Great Lakes. J Great Lakes Res 29(Suppl 1):1–14

Christie GC, Adams JV, Steeves TB et al (2003) Selecting Great Lakes streams for lampricide treatment based on larval sea lamprey surveys. J Great Lakes Res 29(Suppl 1):152–160

Christie WJ (1973) A review of the changes in the fish species composition of Lake Ontario. Great Lakes Fish Comm Tech Rep 23:1–65

Chung-Davidson Y-W, Wang H, Siefkes MJ et al (2013a) Pheromonal bile acid 3-ketopetromyzonol sulfate primes the neuroendocrine system in sea lamprey. BMC Neurosci 14:11

Chung-Davidson Y-W, Wang H, Bryan MB, Johnson NS, Li W (2013b) An anti-steroidogenic inhibitory primer pheromone in male sea lamprey (*Petromyzon marinus*). Gen Comp Endocrinol 189:24–31

Ciereszko A, Dabrowski K, Toth GP, Christ SA, Glogowski J (2002) Factors affecting motility characteristics and fertilizing ability of sea lamprey spermatozoa. Trans Am Fish Soc 131:193–202

Ciereszko A, Babiak I, Dabrowski K (2003) Efficacy of animal anti-fertility compounds against sea lamprey (*Petromyzon marinuns*) spermatozoa. Theriogenology 61:1039–1050

Cochran PA (2009) Predation on lampreys. In: Brown LR, Chase SD, Mesa MG, Beamish RJ, Moyle PB (eds) Biology, management, and conservation of lampreys in North America. Am Fish Soc Sym 72:139–151

Cornelius FC, Muth KM, Kenyon R (1995) Lake trout rehabilitation in Lake Erie: a case history. J Great Lakes Res 21(Suppl 1):65–82

COSEWIC (2012) COSEWIC assessment and status report on the American eel *Anguilla rostrata* in Canada. Committee on the Status of Endangered Wildlife in Canada, Ottawa, ON

COSEWIC (2017) COSEWIC assessment and status report on the Lake Sturgeon *Acipenser fulvescens*, Western Hudson Bay populations, Saskatchewan-Nelson River populations, Southern Hudson Bay-James Bay populations and Great Lakes-Upper St. Lawrence populations in Canada. Committee on the Status of Endangered Wildlife in Canada, Ottawa, ON

Cronin TM, Manley PL, Brachfeld S et al (2008) Impacts of post-glacial lake drainage events and revised chronology of the Champlain Sea episode 13–9 ka. Palaeogeogr Palaeoclimatol Palaeoecol 262:46–60

D'Aloia CC, Azodi CB, Sheldon SP, Trombulak SC, Ardren WR (2015) Genetic models reveal historical patterns of sea lamprey population fluctuations within Lake Champlain. PeerJ 3:e1369

Dahl FH, McDonald RB (1980) Effects of control of the sea lamprey (*Petromyzon marinus*) on migratory and resident fish populations. Can J Fish Aquat Sci 37:1886–1894

Daniels RA (2001) Untested assumptions: the role of canals in the dispersal of sea lamprey, alewife, and other fishes in the eastern United States. Environ Biol Fishes 60:309–329

Dawson HA, Jones MJ (2009) Factors affecting recruitment dynamics of Great Lakes sea lamprey (*Petromyzon marinus*) populations. J Great Lakes Res 35:353–360

Dawson HA, Quintella BR, Almeida PR, Treble AJ, Jolley JC (2015) The ecology of larval and metamorphosing lampreys. In: Docker MF (ed) Lampreys: biology, conservation and control, vol 1. Springer, Dordrecht, pp 75–137

Dawson HA, Bravener G, Beaulaurier J et al (2017) Contribution of manipulable and non-manipulable environmental factors to trapping efficiency of invasive sea lamprey. J Great Lakes Res 43:172–181

Dawson VK (2003) Environmental fate and effects of the lampricide niclosamide: a review. J Great Lakes Res 29(Suppl 1):475–492

De Stasio B (2013) Aquatic invasive species monitoring project. Year 2013 report to the Fox River Navigational System Authority. Department of Biology, Lawrence University, Appleton, WI

Dillman AR, Guillermin ML, Lee JH et al (2012) Olfaction shapes host-parasite interactions in parasitic nematodes. Proc Natl Acad Sci U S A 109:E2324–E2333

Dittman AH, Quinn TP (1996) Homing in Pacific salmon: mechanisms and ecological basis. J Exp Biol 199:83–91

Dochoda MR, Koonce JF (1994) A perspective on progress and challenges under a joint strategic plan for management of Great Lakes fisheries. Univ Toledo Law Rev 25:425–442

Docker MF, Hume JB, Clemens BJ (2015) Introduction: a surfeit of lampreys. In: Docker MF (ed) Lampreys: biology, conservation and control, vol 1. Springer, Dordrecht, pp 1–34

Dodd HR, Hayes DB, Baylis JR et al (2003) Low-head sea lamprey barrier effects on stream habitat and fish communities in the Great Lakes basin. J Great Lakes Res 29(Suppl 1):386–402

Dunlop ES, McLaughlin RL, Adams JV et al (2018) Rapid evolution meets invasive species control: the potential for pesticide resistance in sea lamprey. Can J Fish Aquat Sci 75:152–168

Dymond JR (1922) A provisional list of the fishes of Lake Erie. Univ Toronto Stud Biol Ser 20:57–73

Ebener MP, Bence JR, Bergstedt RA, Mullett KM (2003) Classifying sea lamprey marks on Great Lakes lake trout: observer agreement, evidence on healing times between classes, and recommendations for reporting of marking statistics. J Great Lakes Res 29(Suppl 1):283–296

Ebener MP, King EL Jr, Edsall TA (2006) Application of a dichotomous key to the classification of sea lamprey marks on Great Lakes fish. Great Lakes Fish Comm Misc Publ 2006-2:1–21

Elrod JH, O'Gorman R, Schneider CP et al (1995) Lake trout rehabilitation in Lake Ontario. J Great Lakes Res 21(Suppl 1):83–107

Engstrom-Heg R (1990) Finger Lakes lamprey control evaluation, Cayuga Lake update. New York State Department of Environmental Conservation, Albany, NY

Eshenroder RL (1987) Socioeconomic aspects of lake trout rehabilitation in the Great Lakes. Trans Am Fish Soc 116:309–313

Eshenroder RL (1992) Decline of lake trout in Lake Huron. Trans Am Fish Soc 121:548–554

Eshenroder RL (2009) Comment: Mitochondrial DNA analysis indicates sea lamprey are indigenous to Lake Ontario. Trans Am Fish Soc 138:1178–1189

Eshenroder RL (2014) The role of the Champlain Canal and Erie Canal as putative corridors for colonization of Lake Champlain and Lake Ontario by sea lampreys. Trans Am Fish Soc 143:634–649

Eshenroder RL, Payne NR, Johnson JE, Bowen C II, Ebener MP (1995) Lake trout rehabilitation in Lake Huron. J Great Lakes Res 21(Suppl 1):108–127

Farmer GJ (1980) Biology and physiology of feeding in adult lamprey. Can J Fish Aquat Sci 37:1751–1761

Farmer GJ, Beamish FWH (1973) Sea lamprey (*Petromyzon marinus*) predation on freshwater teleosts. J Fish Res Board Can 30:601–605

Fetterolf CM Jr (1980) Why a Great Lakes Fishery Commission and why a Sea Lamprey International Symposium. Can J Fish Aquat Sci 37:1588–1593

Fine JM, Sorensen PW (2008) Isolation and biological activity of the multi-component sea lamprey migratory pheromone. J Chem Ecol 34:1259–1267

Fine JM, Vrieze LA, Sorensen PW (2004) Evidence that petromyzontid lampreys employ a common migratory pheromone that is partially comprised of bile acids. J Chem Ecol 30:2091–2110

Fisheries Technical Committee (2017) 2016 Annual Report. Lake Champlain Fish and Wildlife Management Cooperative, Essex Junction, VT

Fodale MF, Bronte CR, Bergstedt RA (2003) Classification of lentic habitat for sea lamprey (*Petromyzon marinus*) larvae using a remote seabed classification device. J Great Lakes Res 29(Suppl 1):190–203

Fontaine M (1938) La lamproie marine. Sa peche et son importance economique. Bull Soc Ocean Fr 17:1681–1687

Gaden M, Goddard C, Read J (2013) Multi-jurisdictional management of the shared Great Lakes fishery: transcending conflict and diffuse political authority. In: Taylor WW, Lynch AJ, Leonard NJ (eds) Great Lakes fisheries policy and management: a binational perspective, 2nd edn. Michigan State University Press, East Lansing, MI, pp 305–338

Gage SH (1893) The lake and brook lampreys of New York, especially those of Cayuga and Seneca Lakes. Wilder quarter-century book. Comstock, Ithaca, NY, pp 421–493

Gambicki S, Steinhart GB (2017) Changes in sea lamprey size and fecundity through time in the Great Lakes. J Great Lakes Res 43:209–214

GLFC (1955) Convention on Great Lakes Fisheries. Great Lakes Fishery Commission, Ann Arbor, MI

GLFC (2001) Strategic vision of the Great Lakes Fishery Commission for the first decade of the new millennium. Great Lakes Fishery Commission, Ann Arbor, MI

GLFC (2011) Strategic vision of the Great Lakes Fishery Commission 2011–2020, Ann Arbor, MI

Gilderhus PA, Sills JB, Allen JL (1975) Residues of 3-trifluoromethyl-4-nitrophenol (TFM) in a stream ecosystem after treatment for control of sea lamprey. US Fish Wildl Serv Invest Fish Control 66:1–5

Gilderhus PA, Johnson BGH (1980) Effects of sea lamprey (*Petromyzon marinus*) control in the Great Lakes on aquatic plants, invertebrates, and amphibians. Can J Fish Aquat Sci 37:1895–1905

Gingera TD, Steeves TB, Boguski DA et al (2016) Detection and identification of lampreys in Great Lakes streams using environmental DNA. J Great Lakes Res 42:649–659

Green WW, Boyes K, McFadden et al (2017) Odorant organization in the olfactory bulb of the sea lamprey. J Exp Biol 220:1350–1359

Gutierrez JB, Teem JL (2006) A model describing the effect of sex-reversed YY fish in an established wild population: The use of a Trojan Y chromosome to cause extinction of an introduced exotic species. J Theor Biol 241:333–341

Haeseker SL, Jones ML, Peterman RM et al (2007) Explicit consideration of uncertainty in Great Lakes fisheries management: decision analysis of sea lamprey (*Petromyzon marinus*) control in the St Marys River. Can J Fish Aquat Sci 64:1456–1468

Hammers B, Kosowski DH (2011) Summary of salmonine monitoring in Seneca Lake, 1999–2009. Final report. New York State Department of Environmental Conservation, Avon, NY

Hammers BE, Pearsall W, Robins S et al (2010) Control of sea lamprey populations using lampricides in Seneca and Cayuga lakes and their tributaries. NEPA Categorical Exclusion Support Document, New York State Department of Environmental Conservation, Avon, NY

Hand CH, Ludsin SA, Fryer BJ, Marsden JE (2008) Development of statolith microchemistry as a technique for discriminating among sea lamprey (*Petromyzon marinus*) spawning tributaries in the Great Lakes. Can J Fish Aquat Sci 65:1–12

Hansen GJA, Jones ML (2008) A rapid assessment approach to prioritizing streams for control of Great Lakes sea lampreys (*Petromyzon marinus*): a case study in adaptive management. Can J Fish Aquat Sci 65:2471–2484

Hansen MJ, Hayne DW (1962) Sea lamprey larvae in Ogontz Bay and Ogontz River, Michigan. J Wild Manag 26:237–247

Hansen MJ, Peck JW, Schorfaar RG et al (1995) Lake trout (*Salvelinus namaycush*) populations in Lake Superior and their restoration in 1959–1993. J Great Lakes Res 21(Suppl 1):152–175

Hanson LH (1981) Sterilization of sea lamprey (*Petromyzon marinus*) by immersion in an aqueous solution of Bisazir. Can J Fish Aquat Sci 38:1285–1289

Hanson LH (1990) Sterilizing effects of Cobalt-60 and Cesium-137 radiation on male sea lamprey. N Am J Fish Manag 10:352–361

Hanson LH, Manion PJ (1978) Chemosterilization of sea lamprey (*Petromyzon marinus*). Great Lakes Fish Comm Tech Rep 29:1–15

Hanson LH, Manion PJ (1980) Sterility method of pest control and its potential role in an integrated sea lamprey (*Petromyzon marinus*) control program. Can J Fish Aquat Sci 37:2108–2117

Harvey CJ, Ebener MP, White CK (2008) Spatial and ontogenetic variability of sea lamprey diets in Lake Superior. J Great Lakes Res 34:434–449

Heath G, Childs D, Docker MF, McCauley DW, Whyard S (2014) RNA interference technology to control pest sea lampreys—a proof-of-concept. PLoS ONE 9:e88387

Heinrich JW, Weise JG, Smith BR (1980) Changes in biological characteristics of the sea lamprey (*Petromyzon marinus*) as related to lamprey abundance, prey abundance, and sea lamprey control. Can J Fish Aquat Sci 37:1861–1871

Heinrich JW, Mullett KM, Hansen MJ et al (2003) Sea lamprey abundance and management in Lake Superior 1957–1999. J Great Lakes Res 29(Suppl 1):566–583

Henson MP, Bergstedt RA, Adams JV (2003) Comparison of spring measures of length, weight, and condition factor for predicting metamorphosis in two populations of sea lamprey (*Petromyzon marinus*) larvae. J Great Lakes Res 29(Suppl 1):204–213

Holbrook CM, Bergstedt RA, Adams NS, Hatton TW, McLaughlin RL (2015) Fine-scale pathways used by adult Sea Lampreys during riverine spawning migrations. Trans Am Fish Soc 144:549–562

Holbrook CM, Bergstedt RA, Barber J et al (2016) Evaluating harvest-based control of invasive fish with telemetry: performance of sea lamprey traps in the Great Lakes. Ecol Appl 26:1595–1609

Holey ME, Rybicki RW, Eck GW et al (1995) Progress toward lake trout restoration in Lake Michigan. J Great Lakes Res 21(Suppl 1):128–151

Holmes JA, Youson JH (1994) Fall condition factor and temperature influence the incidence of metamorphosis in sea lamprey, *Petromyzon marinus*. Can J Zool 72:1134–1140

Horns WH, Bronte CR, Busiahn TR et al (2003) Fish community objectives for Lake Superior. Great Lakes Fish Comm Spec Publ 03–01:1–78

Houston KA, Kelso JRM (1991) Relation of sea lamprey size and sex ratio to salmonid availability in three Great Lakes. J Great Lakes Res 17:270–280

Howe EA, Marsden JE, Bouffard W (2006) Movement of sea lamprey in the Lake Champlain basin. J Great Lakes Res 32:776–787

Howe EA, Lochet A, Hand CP et al (2013) Tributary contributions to the parasitic and spawning adult population of sea lamprey (*Petromyzon marinus*) in Lake Champlain using elemental signatures. J Great Lakes Res 39:239–246

Howell JH, King EL Jr, Smith AJ, Hanson LH (1964) Synergism of 5,2'-dichloro-4'-nitrosalicylanilide and 3-trifluoromethyl-4-nitrophenol in a selective lamprey larvicide. Great Lakes Fish Comm Tech Rep 8:1–21

Howell JH, Lech JJ, Allen JL (1980) Development of sea lamprey (*Petromyzon marinus*) larvicides. Can J Fish Aquat Sci 37:2103–2107

Hubbs C, Lagler CF, Smith GR (2004) Fishes of the Great Lakes region. University of Michigan Press, Ann Arbor, MI

Hume JB, Wagner M (2018) A death in the family: sea lamprey (*Petromyzon marinus*) avoidance of confamilial alarm cues diminishes with phylogenetic distance. Ecol Evol 8:3751–3762

Hume B, Meckley TD, Johnson NS et al (2015) Application of a putative alarm cue hastens the arrival of invasive sea lamprey (*Petromyzon marinus*) at a trapping location. Can J Fish Aquat Sci 72:1799–1806

Hunn JB, Youngs WD (1980) Role of physical barriers in the control of sea lamprey (*Petromyzon marinus*). Can J Fish Aquat Sci 37:2118–2122

Imre I, Brown GE, Bergstedt RA, McDonald R (2010) Use of chemosensory cues as repellents for sea lamprey: potential directions for population management. J Great Lakes Res 36:790–793

Imre I, Di Rocco RT, Belanger CF, Brown GE, Johnson NS (2014) The behavioural response of adult *Petromyzon marinus* to damage-released alarm and predator cues. J Fish Biol 84:1490–1502

Jansen W, Gill G, Hann BJ (2017) Rapid geographic expansion of spiny water flea (*Bythotrephes longimanus*) in Manitoba, Canada, 2009–2015. Aquat Invasions 12:287–297

Jelks HL, Walsh SJ, Burkhead NM et al (2008) Conservation status of imperiled North American freshwater and diadromous fishes. Fisheries 33:372–407

Jensen JA, Jones ML (2018) Forecasting the response of Great Lakes sea lamprey (*Petromyzon marinus*) to barrier removals. Can J Fish Aquat Sci 75:1415–1426

Johnson DA, Weisser JW, Bills TD (1999) Sensitivity of lake sturgeon (*Acipenser fulvescens*) to the lampricide 3-trifluoromethyl-4-nitrophenol (TFM) in field and laboratory exposures. Great Lakes Fish Comm Tech Rep 62:1–23

Johnson JE, He J, Fielder DG (2015) Rehabilitation stocking of walleyes and lake trout: restoration of reproducing stocks in Michigan waters of Lake Huron. N Am J Aquacult 77:396–408

Johnson NS, Yun SS, Thompson HT, Brant CO, Li W (2009) A synthesized pheromone induces upstream movement in female sea lamprey and summons them into traps. Proc Natl Acad Sci U S A 106:1021–1026

Johnson NS, Yun SS, Buchinger TJ, Li W (2012) Multiple functions of a multi-component mating pheromone in sea lamprey *Petromyzon marinus*. J Fish Biol 80:538–554

Johnson NS, Siefkes MJ, Wagner CM et al (2013) A synthesized mating pheromone component increases adult sea lamprey (*Petromyzon marinus*) trap capture in management scenarios. Can J Fish Aquat Sci 70:1101–1108

Johnson NS, Thompson HT, Holbrook C, Tix JA (2014a) Blocking and guiding adult sea lamprey with pulsed direct current from vertical electrodes. Fish Biol 150:38–48

Johnson NS, Yun S-S, Li W (2014b) Investigations of novel unsaturated bile salts of male sea lamprey as potential chemical cues. J Chem Ecol 40:1152–1160

Johnson NS, Buchinger TJ, Li W (2015a) Reproductive ecology of lampreys. In: Docker MF (ed) Lampreys: biology, conservation and control, vol 1. Springer, Dordrecht, pp 265–304

Johnson NS, Tix JA, Hlina BL et al (2015b) A sea lamprey (*Petromyzon marinus*) sex pheromone mixture increases trap catch relative to a single synthesized component in specific environments. J Chem Ecol 41:311–321

Johnson NS, Siefkes MJ, Wagner CM et al (2015c) Factors influencing capture of invasive sea lamprey in traps baited with a synthesized sex pheromone component. J Chem Ecol 41:913–923

Johnson NS, Twohey MB, Miehls SM et al (2016a) Evidence that sea lampreys (*Petromyzon marinus*) complete their life cycle within a tributary of the Laurentian Great Lakes by parasitizing fishes in inland lakes. J Great Lakes Res 42:90–98

Johnson NS, Brenden TO, Swink WD, Lipps MA (2016b) Survival and metamorphosis of larval sea lamprey (*Petromyzon marinus*) residing in Lakes Michigan and Huron near river mouths. J Great Lakes Res 42:1461–1469

Johnson NS, Swink WD, Brenden TO (2017) Field study suggests that sex determination in sea lamprey is directly influenced by larval growth rate. Proc R Soc B 284:20170262

Johnson W (1982) Body lengths, body weight and fecundity of sea lampreys (*Petromyzon marinus*) from Green Bay, Lake Michigan. Trans Wis Acad Sci Arts Lett 70:73–77

Jones ML (2007) Toward improved assessment of sea lamprey population dynamics in support of cost-effective sea lamprey management. J Great Lakes Res(Spec Issue 2):35–47

Jones ML, Bergstedt RA, Twohey MB et al (2003) Compensatory mechanisms in Great Lakes sea lamprey populations: implications for alternative control strategies. J Great Lakes Res 29(Suppl 1):113–129

Jones ML, Irwin BJ, Hansen GJA et al (2009) An operating model for the integrated pest management of Great Lakes sea lampreys. Open Fish Sci J 2:59–73

Jones ML, Brenden TO, Irwin BJ (2012) Evaluating integrated pest management in the St Marys River. Great Lakes Fishery Commission Completion Report, Ann Arbor, MI

Karlson P, Luscher M (1959) 'Pheromones': a new term for a class of biologically active substances. Nature 183:55–56

Kelly NE, Wantola K, Weisz E, Yan ND (2013) Recreational boats as a vector of secondary spread for aquatic invasive species and native crustacean zooplankton. Biol Invasions 15:509–519

Kelso JRM, Gardner WM, McDonald RB (2001) Interactions among fertile male, female, and sterile male sea lampreys during spawning in the Carp River, Lake Superior. N Am J Fish Manag 21:904–910

King EL Jr (1980) Classification of sea lamprey (*Petromyzon marinus*) attack marks on Great Lakes lake trout (*Salvelinus namaycush*). Can J Fish Aquat Sci 31:1989–2006

King EL Jr, Gabel JA (1985) Comparative toxicity of the lampricide 3-trifluoromethyl-4-nitrophenol to ammocoetes of three species of lampreys. Great Lakes Fish Comm Tech Rep 47:1–5

Kleerekoper H, Mogensen J (1963) Role of olfaction in the orientation of *Petromyzon marinus*. I. Response to a single amine in prey's body odor. Physiol Zool 36:347–360

Klassen W, Adams JV, Twohey MB (2005) Modeling the suppression of sea lamprey populations by use of the male sex pheromone. J Great Lakes Res 31:166–173

Knipling EF (1968) The potential role of sterility for pest control. In: LaBrecque GC, Smith CN (eds) Principles of insect chemosterilization. Appleton-Century-Crofts, NY, pp 7–40

Koonce JF, Greig LA, Henderson BA et al (eds) (1982) A review of the adaptive management workshop addressing salmonid/lamprey management in the Great Lakes. Great Lakes Fish Comm Spec Publ 82-2:1–58

Krueger CC, Ebener M (2004) Rehabilitation of lake trout in the Great Lakes: past lessons and future challenges. In: Gunn JM, Steedman RJ, Ryder RA (eds) Boreal shield watersheds: lake trout systems in a changing environment. CRC Press, Boca Raton, FL, pp 37–56

Krueger CC, Marsden JE (2007) Sea lamprey research: balancing basic and applied research, and using themes to define priorities. J Great Lakes Res 33(Spec Issue 2):1–6

Lark JGI (1973) An early record of the sea lamprey (*Petromyzon marinus*) from Lake Ontario. J Fish Res Board Can 30:131–133

Larson GL, Christie GC, Johnson DA et al (2003) The history of sea lamprey control in Lake Ontario and updated estimates of suppression targets. J Great Lakes Res 29(Suppl 1):637–654

Lavis DS, Hallett A, Koon EM, McAuley TC (2003a) History of and advances in barriers as an alternative method to suppress sea lamprey in the Great Lakes. J Great Lakes Res 29(Suppl 1):362–372

Lavis DS, Henson MP, Johnson DA, Koon EM, Ollila DJ (2003b) A case history of sea lamprey control in Lake Michigan: 1979 to 1999. J Great Lakes Res 29(Suppl 1):584–598

Lawrie AH (1970) The sea lamprey in the Great Lakes. Trans Am Fish Soc 99:766–775

Lee DS, Weise JG (1989) Habitat selection of lentic larval lampreys: preliminary analysis based on research with a manned submersible. J Great Lakes Res 15:156–163

Lee WJ, Kocher TD (1995) Complete sequence of a sea lamprey (*Petromyzon marinus*) mitochondrial genome: early establishment of the vertebrate genome organization. Genetics 139:873–887

Libants S, Carr KN, Wu H et al (2009) The sea lamprey *Petromyzon marinus* genome reveals the early origin of several chemosensory receptor families in the vertebrate lineage. BMC Evol Biol 9:180

Li K, Brant CO, Huertas M, Hur SK, Li W (2013a) Petromyzonin, a hexahydrophenanthrene sulfate isolated from the larval sea lamprey (*Petromyzon marinus* L.). Org Lett 15:5924–5927

Li K, Brant CO, Siefkes MJ, Kruckman HG, Li W (2013b) Characterization of a novel bile alcohol sulfate released by sexually mature male sea lamprey (*Petromyzon marinus*). PLoS ONE 8:e68157

Li K, Huertas M, Brant CO et al (2014) (+) and (−) petromyroxols: antipodal tetrahydrofurandiols from larval sea lamprey (*Petromyzon marinus* L.) that elicit enantioselective olfactory responses. Org Lett 17:286–289

Li W, Sorensen PW (1997) Highly independent olfactory receptor sites for naturally occurring bile acids in the sea lamprey, *Petromyzon marinus*. J Comp Physiol A 180:429–438

Li W, Scott AP, Siefkes MJ et al (2002) Bile acid secreted by male sea lamprey that acts as a sex pheromone. Science 296:138–141

Li W, Seifkes MJ, Scott AP, Teeter JH (2003) Sex pheromone communication in sea lamprey: implications for integrated management. J Great Lakes Res 29(Suppl 1):85–94

Li W, Twohey M, Jones M, Wagner M (2007) Research to guide use of pheromones to control sea lamprey. J Great Lakes Res 33(Spec Issue 2):70–86

Lochet A, Marsden JE, Fryer BJ, Ludsin SA (2013) Instability of statolith elemental signatures revealed in newly-metamorphosed sea lamprey (*Petromyzon marinus*). Can J Fish Aquat Sci 70:565–573

Lochet A, Fryer BJ, Ludsin SA, Howe EA, Marsden JE (2014) Identifying natal origins of spawning adult sea lamprey (*Petromyzon marinus*): reevaluation of the statolith microchemistry approach. J Great Lakes Res 40:763–770

Luehring MA, Wagner CM, Li W (2011) The efficacy of two synthesized sea lamprey sex pheromone components as a trap lure when placed in direct competition with natural male odors. Biol Invasions 13:1589–1597

Luhring TM, Meckley TD, Johnson NS et al (2016) A semelparous fish continues upstream migration when exposed to alarm cue, but adjusts movement speed and timing. Anim Behav 121:41–51

MacKay HH, MacGillivray E (1949) Recent investigations on the sea lamprey, *Petromyzon marinus*, in Ontario. Trans Am Fish Soc 76:148–159

Maki AW, Geissel L, Johnson HE (1975) Comparative toxicity of larval lampricide (TFM: 3-trifluoromethyl-4-nitrophenol) to selected benthic macroinvertebrates. J Fish Res Board Can 32:1455–1459

Madenjian CP, Chipman BD, Marsden JE (2008) Estimate of lethality of sea lamprey attacks in Lake Champlain: implications for fisheries management. Can J Fish Aquat Sci 65:535–542

Maitland PS, Renaud CB, Quintella BR, Close DA, Docker MF (2015) Conservation of native lampreys. In: Docker MF (ed) Lampreys: biology, conservation and control, vol 1. Springer, Dordrecht, pp 375–428

Manion PJ (1967) Diatoms as food of larval sea lamprey in a small tributary of northern Lake Michigan. Trans Am Fish Soc 96:224–226

Manion PJ, Hanson LH (1980) Spawning behavior and fecundity of lampreys from the upper three lakes. Can J Fish Aquat Sci 37:1635–1640

Manion PJ, McLain AL (1971) Biology of larval sea lamprey *Petromyzon marinus* of the 1960 year class, isolated in the Big Garlic River, Michigan 1960–65. Great Lakes Fish Comm Tech Rep 16:1–35

Manion PJ, Smith BR (1978) Biology of larval and metamorphosing sea lampreys, *Petromyzon marinus*, of the 1960 year class in the Big Garlic River, Michigan, Part II, 1966–72. Great Lakes Fish Comm Tech Rep 30:1–35

Manzon RG, Youson JH, Holmes JA (2015) Lamprey metamorphosis. In: Docker MF (ed) Lampreys: biology, conservation and control, vol 1. Springer, Dordrecht, pp 139–214

Markham JL, Knight RL (eds) (2017) The state of Lake Erie in 2009. Great Lakes Fish Comm Spec Publ 2017-01:1–140

Marsden JE, Ladago BJ (2017) The Champlain Canal as a non-indigenous species corridor. J Great Lakes Res 43:1173–1180

Marsden JE, Langdon RW (2012) History and future of Lake Champlain's fishes and fisheries. J Great Lakes Res 38(Suppl 1):19–34

Marsden JE, Krueger CC, May B (1989) Identification of parental origins of naturally produced lake trout fry in Lake Ontario: application of mixed stock analysis to a second generation. N Am J Fish Manag 9:257–268

Marsden JE, Chipman BD, Nashett LJ et al (2003) Sea lamprey control in Lake Champlain. J Great Lakes Res 29(Suppl 1):655–676

Marsden JE, Chipman BD, Pientka B, Schoch WF, Young BA (2010) Strategic plan for Lake Champlain fisheries. Great Lakes Fish Comm Misc Publ 2010–03:1–54

Marsden JE, Kozel CL, Chipman BD (2017) Lake trout recruitment in Lake Champlain. J Great Lakes Res 44:166–173

McCann EL, Johnson NS, Hrodey PJ, Pangle KL (2018) Characterization of sea lamprey stream entry using dual-frequency identification sonar. Trans Am Fish Soc 147:514–524

McCauley DW, Docker MF, Whyard S, Li W (2015) Lampreys as diverse model organisms in the genomics era. BioScience 65:1046–1056

McDonald DG, Desrochers D (2012) Using an eel ladder-trap to trap sea lampreys: proof of concept. Great Lakes Fishery Commission Completion Report, Ann Arbor, MI

McDonald DG, Kolar CS (2007) Research to guide the use of lampricides for controlling sea lamprey. J Great Lakes Res 33(Spec Issue 2):20–34

McLain AL, Smith BR, Moore HH (1965) Experimental control of sea lamprey with electricity on the south shore of Lake Superior, 1953–60. Great Lakes Fish Comm Tech Rep 10:1–48

McLaughlin RL, Marsden JE, Hayes DB (2003) Achieving the benefits of sea lamprey control while minimizing the effects on non-target species: conceptual synthesis and proposed policy. J Great Lakes Res 29(Suppl 1):755–765

McLaughlin RL, Porto L, Noakes DLG et al (2006) Effects of low-head barriers on stream fishes: taxonomic affiliations and morphological correlates of sensitive species. Can J Fish Aquat Sci 63:766–779

McLaughlin RL, Hallett A, Pratt TC, O'Connor LM, McDonald DG (2007) Research to guide use of barriers, traps, and fishways to control sea lamprey. J Great Lakes Res 33(Spec Issue 2):7–19

McLaughlin RL, Smyth ERB, Castro-Santos T et al (2012) Unintended consequences and trade-offs of fish passage. Fish Fish 14:580–604

McLean AR, Barber J, Bravener G, Rous AM, McLaughlin RL (2015) Understanding low success trapping invasive sea lampreys: an entry-level analysis. Can J Fish Aquat Sci 72:1876–1885

Meckley TD, Wagner DM, Luehring MA (2012) Field evaluation of larval odor and mixtures of synthesized pheromone components for attracting migrating sea lamprey in rivers. J Chem Ecol 38:1062–1069

Meckley TD, Wagner DM, Gurarie E (2014) Coastal movements of migrating sea lamprey (*Petromyzon marinus*) in response to a partial pheromone added to river water: implications for management of invasive populations. Can J Fish Aquat Sci 71:533–544

Mills EL, Leach JH, Carlton JT, Secor CL (1993) Exotic species in the Great Lakes: a history of biotic crises and anthropogenic introductions. J Great Lakes Res 19:1–54

Morkert SB, Swink WD, Seelye JG (1998) Evidence for early metamorphosis of sea lamprey in the Chippewa River, Michigan. N Am J Fish Manag 18:966–971

Morman RH (1987) Relationship of density to growth and metamorphosis of caged larval sea lampreys, *Petromyzon marinus* Linnaeus, in Michigan streams. J Fish Biol 30:173–181

Moore JW, Mallatt JM (1980) Feeding of larval lamprey. Can J Fish Aquat Sci 37:1658–1664

Moore HH, Schleen LP (1980) Changes in spawning runs of sea lamprey (*Petromyzon marinus*) in selected streams of Lake Superior after chemical control. Can J Fish Aquat Sci 37:1851–1860

Moore HH, Dahl FH, Lamsa AK (1974) Movement and recapture of parasitic sea lamprey (*Petromy-zon marinus*) tagged in the St Marys River and lakes Huron and Michigan, 1963–67. Great Lakes Fish Comm Tech Rep 27:1–20

Morse TJ, Ebener MP, Koon EM et al (2003) A case history of sea lamprey control in Lake Huron: 1979 to 1999. J Great Lakes Res 29(Suppl 1):599–614

Moser ML, Almeida PR, Kemp PS (2015) Lamprey spawning migration. In: Docker MF (ed) Lampreys: biology, conservation and control, vol 1. Springer, Dordrecht, pp 215–263

Muir AM, Krueger CC, Hansen MJ (2013) Re-establishing lake trout in the Laurentian Great Lakes: past, present, and future. In: Taylor WW, Lynch AJ, Leonard NJ (eds) Great Lakes fisheries policy and management: a binational perspective, 2nd edn. Michigan State University Press, East Lansing, MI, pp 533–588

Muir WM, Howard RD (1999) Possible ecological risks of transgenic organism release when transgenes affect mating success: sexual selection and the Trojan gene hypothesis. Proc Natl Acad Sci U S A 96:13853–13856

Mullett K, Sullivan P (2017) Sea lamprey control in the Great Lakes 2016. In: Minutes of the 2017 annual meeting. Great Lakes Fishery Commission, Ann Arbor, MI

Mullett KM, Heinrich JW, Adams JV et al (2003) Estimating lake-wide abundance of spawning phase sea lamprey (*Petromyzon marinus*) in the Great Lakes: extrapolating from sampled streams using regression models. J Great Lakes Res 29(Suppl 1):240–252

Murdoch SP, Beamish FH, Docker MF (1991) Laboratory study of growth and interspecific competition in larval lampreys. Trans Am Fish Soc 120:653–656

Newton TJ, Boogaard MA, Gray BR, Hubert TD, Schloesser NA (2017) Lethal and sub-lethal responses of native freshwater mussels exposed to granular Bayluscide®, a sea lamprey larvicide. J Great Lakes Res 43:370–378

Niblett PD, Ballantyne JS (1976) Uncoupling of oxidative phosphorylation in rat liver mitochondria by the lamprey larvicide TFM (3-trifluoromethyl-4-nitrophenol). Pest Biochem Physiol 6:363–366

Novikov MA, Kharlamova MN (2018) New data on the distribution of lamprey *Petromyzon marinus* and *Lethenteron camtschaticum* (Petromyzontidae) in Barents and White seas. J Ichthyol 58:296–302

O'Connor LM, Pratt TC, Steeves TB et al (2017) *In situ* assessment of lampricide toxicity to age-0 lake sturgeon. J Great Lakes Res 43:189–198

O'Gorman R (ed) (2017) The state of Lake Ontario in 2014. Great Lakes Fish Comm Spec Publ 2017-02:1–140

Page KS, Scribner KT, Bennett KR, Garzel LM, Burnham-Curtis MK (2003) Genetic assessment of strain-specific sources of lake trout recruitment in the Great Lakes. Trans Am Fish Soc 132:877–894

Pearce WA, Braem RA, Dustin SM, Tibbles JJ (1980) Sea lamprey (*Petromyzon marinus*) in the lower Great Lakes. Can J Fish Aquat Sci 37:1802–1810

Perlmutter A (1951) An aquarium experiment on the American eel as a predator on larval lampreys. Copeia 1951:173–174

Piavis GW (1971) Embryology. In: Hardisty MW, Potter IC (eds) The biology of lampreys, vol 1. Academic Press, London, pp 361–400

Porto LM, McLaughlin RL, Noakes DLG (1999) Low-head barrier dams restrict the movements of fishes in two Lake Ontario streams. N Am J Fish Manag 19:1028–1036

Potter IC (1980) Ecology of larval and metamorphosing lampreys. Can J Fish Aqat Sci 37:1641–1657

Potter IC, Gill HS, Renaud CB, Haoucher D (2015) The taxonomy, phylogeny, and distribution of lampreys. In: Docker MF (ed) Lampreys: biology, conservation and control, vol 1. Springer, Dordrecht, pp 35–73

Pratt TC, O'Connor LM, Hallett AG et al (2009) Balancing aquatic habitat fragmentation and control of invasive species: enhancing selective fish passage at sea lamprey control barriers. Trans Am Fish Soc 138:652–665

Purvis HA (1979) Variations in growth, age at transformation, and sex ratio of sea lamprey reestablished in chemically treated tributaries of the Upper Great Lakes. Great Lakes Fish Comm Tech Rep 35:1–49

Putnam NH, Butts T, Ferrier DEK et al (2009) The amphioxus genome and the evolution of the chordate karyotype. Nature 453:1064–1071

Pycha RL, King GR (1975) Changes in the lake trout population of southern Lake Superior in relation to the fishery, the sea lamprey, and stocking, 1950–70. Great Lakes Fish Comm Tech Rep 28:1–34

Quintella BR, Andrade NO, Espanhol R, Almeida PR (2005) The use of PIT telemetry to study movements of ammocoetes and metamorphosing sea lampreys in river beds. J Fish Biol 66:97–106

Reinhardt UG, Eidietis L, Friedl SE, Moser ML (2008) Pacific lamprey climbing behavior. Can J Zool 86:1264–1272

Reinhardt UG, Binder T, McDonald DG (2009) Ability of adult sea lamprey to climb inclined surfaces. In: Brown LR, Chase SD, Mesa MG, Beamish RJ, Moyle PB (eds) Biology, management, and conservation of lampreys in North America. Am Fish Soc Sym 72:125–138

Ricciardi A (2006) Patterns of invasion in the Laurentian Great Lakes in relation to changes in vector activity. Divers Distrib 12:425–433

Robinson JM, Wilberg MJ, Adams JV, Jones ML (2013) A spatial age-structured model for describing sea lamprey (Petromyzon marinus) population dynamics. Can J Fish Aquat Sci 70:1709–1722

Rodríguez-Muñoz R, Nicieza AG, Braña F (2001) Effects of temperature on developmental performance, survival and growth of sea lamprey embryos. J Fish Biol 58:475–486

Rodríguez-Muñoz R, Nicieza AG, Braña F (2003) Density-dependent growth of Sea Lamprey larvae: evidence for chemical interference. Funct Ecol 17:403–408

Rooney SM, Wightman G, Ó'Conchúir R, King JJ (2015) Behaviour of sea lamprey (Petromyzon marinus L.) at man-made obstacles during upriver spawning migration: use of telemetry to assess weir modifications for improved passage. Biol Environ Proc R Ir Acad 115:125–136

Rous AM, McLean AR, Barber J et al (2017) Spatial mismatch between sea lamprey behavior and trap location explains low success at trapping for control. Can J Fish Aquat Sci 74:2085–2097

Rudrama K, Reddy PP (1985) Bisazir induced cytogenetic damage and sperm head abnormalities in mice. IRCS Med Sci 13:1245–1246

Sawyer AJ (1980) Prospects for integrated pest management of the sea lamprey (Petromyzon marinus). Can J Fish Aqat Sci 37:2081–2092

Schaefer MB (1951) Estimation of size of animal populations by marking experiments. US Dept Inter Fish Wildl Serv Fish Bull 69:187–203

Schleen LP, Christie GC, Heinrich RA et al (2003) Development and implementation of an integrated program for control of sea lamprey in the St Marys River. J Great Lakes Res(Suppl 1):677–693

Schloesser NA, Merkes CM, Rees CB et al (2018) Correlating sea lamprey density with environmental DNA detections in the lab. Manag Biol Invasions 9:483–495

Schneider CP, Kolenosky DP, Goldwaite DB (1983) A joint plan for the rehabilitation of lake trout in Lake Ontario. In: Minutes of the 1983 meeting of the Lake Ontario Committee. Great Lakes Fishery Commission, Ann Arbor, MI

Schneider CP, Owens RW, Bergstedt RA, O'Gorman R (1996) Predation by sea lamprey (Petromyzon marinus) on lake trout (Salvelinus namaycush) in southern Lake Ontario, 1982–1992. Can J Fish Aquat Sci 53:1921–1932

Schreiner DR, Schram ST (1997) Lake trout rehabilitation in Lake Superior. Fisheries 22:12–14

Schuldt RJ, Goold R (1980) Changes in the distribution of native lampreys in Lake Superior tributaries in response to sea lamprey (Petromyzon marinus) control, 1953–77. Can J Fish Aquat Sci 37:1872–1885

Schuldt RJ, Heinrich JW (1982) Portable trap for collecting adult sea lamprey. Prog Fish Cult 44:220–221

Scott WB, Crossman EJ (1973) Freshwater fishes of Canada. Bull Fish Res Board Can 184, Ottawa, ON

Siefkes MJ (2017) Use of physiological knowledge to control the invasive sea lamprey (*Petromyzon marinus*) in the Laurentian Great Lakes. Conserv Physiol 5:cox031

Siefkes MJ, Li W (2004) Electrophysiological evidence for detectioin and discrimination of pheromonal bile acids by the olfactory epithelium of female sea lampreys (*Petromyzon marinus*). J Comp Physiol A 190:193–199

Siefkes MJ, Bergstedt RA, Twohey MB, Li W (2003) Chemosterilization of male sea lamprey (*Petromyzon marinus*) does not affect sex pheromone release. Can J Fish Aquat Sci 60:23–31

Siefkes MJ, Winterstein S, Li W (2005) Evidence that 3-keto petromyzonol sulfate specifically attracts ovulating female sea lamprey (*Petromyzon marinus*). Anim Behav 70:1037–1045

Siefkes MJ, Steeves TB, Sullivan WP, Twohey MB, Li W (2013) Sea lamprey control: past, present, and future. In: Taylor WW, Lynch AJ, Leonard NJ (eds) Great Lakes fisheries policy and management: a binational perspective, 2nd edn. Michigan State University Press, East Lansing, MI, pp 651–704

Silva S, Servia MJ, Vieira-Lanero R, Nachon D, Cobo F (2013) Haematophagous feeding of newly metamorphosed European sea lampreys *Petromyzon marinus* on strictly freshwater species. J Fish Biol 82:1739–1745

Slade JW, Adams JV, Christie GC et al (2003) Techniques and methods for estimating abundance of larval and metamorphosed sea lampreys in Great Lakes tributaries, 1995 to 2001. J Great Lakes Res 29(Suppl 1):137–151

Smith BR, Elliot OR (1953) Movement of parasitic-phase sea lamprey in lakes Huron and Michigan. Trans Am Fish Soc 82:123–128

Smith BR (1971) Sea lamprey in the Great Lakes of North America. In: Hardisty MW, Potter IC (eds) The biology of lampreys, vol 1. Academic Press, London, pp 207–247

Smith BR, Tibbles JJ (1980) Sea lamprey (*Petromyzon marinus*) in lakes Huron, Michigan, and Superior: history of invasion and control, 1936–1978. Can J Fish Aquat Sci 37:1780–1801

Smith BR, Tibbles JJ, Johnson BGH (1974) Control of the sea lamprey (*Petromyzon marinus*) in Lake Superior, 1953–1970. Great Lakes Fish Comm Tech Rep 26:1–60

Smith JJ, Antonacci F, Eichler EE et al (2009) Programmed loss of millions of base pairs from a vertebrate genome. Proc Natl Acad Sci U S A 106:112–127

Smith JJ, Kurako S, Holt C et al (2013) Sequencing of the sea lamprey (*Petromyzon marinus*) genome provides insights into vertebrate evolution. Nat Genet 45:415–421

Smith JJ, Timoshevskaya N, Ye C et al (2018) The sea lamprey germline genome provides insights into programmed genome rearrangement and vertebrate evolution. Nat Genet 50:270–277

Smith SJ, Marsden JE (2007) Predictive morphometric relationships for estimating fecundity of sea lamprey from Lake Champlain and other landlocked populations. Trans Am Fish Soc 136:979–987

Smith SJ, Marsden JE (2009) Factors affecting sea lamprey egg survival. N Am J Fish Manag 29:859–868

Sorensen PW, Fine JM, Dvornikovs V et al (2005) Mixture of new sulfated steroids functions as a migratory pheromone in the sea lamprey. Nat Chem Biol 2005:324–328

Sotola VA, Miehls SM, Simard LG, Marsden JE (2018) Lateral and vertical distribution of downstream migrating juvenile Sea Lamprey. J Great Lakes Res 44:491–496

Sower SA (2003) The endocrinology of reproduction in lampreys and applications for male lamprey sterilization. J Great Lakes Res 29(Suppl 1):50–65

Spangler GR, Jacobson LD (eds) (1985) A workshop concerning the application of integrated pest management (IPM) to sea lamprey control in the Great Lakes. Great Lakes Fish Comm Spec Publ 85-2:1–98

Stewart M, Baker CF (2012) A sensitive analytical method for quantifying petromyzonol sulfate in water as a potential tool for population monitoring of the southern pouched lamprey, *Geotria australis*, in New Zealand streams. J Chem Ecol 38:135–144

Sullivan WP, Fodale MF (2009) Past, present, and future of integrated sea lamprey management in Lake Erie. In: Tyson JT, Stein RA, Dettmers JM (eds) The state of Lake Erie 2004. Great Lakes Fish Comm Spec Publ 09-02:65–70

Sullivan P, Mullett K (2018) Sea lamprey control in the Great Lakes 2017. In: Minutes of the 2018 annual meeting. Great Lakes Fishery Commission, Ann Arbor, MI

Sullivan WP, Christie GC, Cornelius C et al (2003) The sea lamprey in Lake Erie: a case history. J Great Lakes Res 29(Suppl 1):615–636

Surface HA (1898) The lampreys of central New York. Bull US Fish Comm 17:209–215

Sutton TM, Bowen SH (1994) Significance of organic detritus in the diet of larval lampreys in the Great Lakes basin. Can J Fish Aqat Sci 51:2380–2387

Swink WD (1990) Effect of lake trout size on survival after a single sea lamprey attack. Trans Am Fish Soc 119:966–1002

Swink WD (2003) Host selection and lethality of attacks by sea lamprey (*Petromyzon marinus*) in laboratory studies. J Great Lakes Res 29(Suppl 1):307–319

Swink WD, Hanson LH (1986) Survival from sea lamprey (*Petromyzon marinus*) predation by two strains of lake trout (*Salvelinus namaycush*). Can J Fish Aqat Sci 43:2528–2531

Swink WD, Johnson NS (2014) Growth and survival of sea lampreys from metamorphosis to spawning in Lake Huron. Trans Am Fish Soc 143:380–386

Teeter J (1980) Pheromone communication in sea lamprey (*Petromyzon marinus*): implications for population management. Can J Fish Aquat Sci 37:2123–2132

Thresher RE, Grewe P, Patil JG et al (2009) Development of repressible sterility to prevent the establishment of feral populations of exotic and genetically modified animals. Aquaculture 290:104–109

Thresher RE, Hayes K, Bax NJ et al (2014) Genetic control of invasive fish: technological options and its role in integrated pest management. Biol Invasions 16:1201–1216

Thresher RE, Jones M, Drake DAR (2019) Evaluating active genetic options for the control of Sea Lampreys (*Petromyzon marinus*) in the Laurentian Great Lakes. Can J Fish Aquat Sci (in press)

Torblaa RL, Westman RW (1980) Ecological impacts of lampricide treatments on sea lamprey (*Petromyzon marinus*) ammocoetes and metamorphosed individuals. Can J Fish Aquat Sci 37:1835–1850

Trautman MB (1949) The invasion, present status, and life history of the sea lamprey in the waters of the Great Lakes, especially the Ohio waters of Lake Erie. Contrib Franz Theodore Stone Lab Ohio State Univ 1949:1–9

Treble AJ, Jones ML, Steeves TB (2008) Development and evaluation of a new predictive model for metamorphosis of Great Lakes larval sea lamprey (*Petromyzon marinus*) populations. J Great Lakes Res 34:404–417

Twohey MB, Heinrich JW, Seelye JG et al (2003a) The sterile-male-release technique in Great Lakes sea lamprey management. J Great Lakes Res 29(Suppl 1):410–423

Twohey MB, Sorensen PW, Li W (2003b) Possible applications of pheromones in an integrated sea lamprey management program. J Great Lakes Res 29(Suppl 1):794–800

VanDenBossche J, Seelye JG, Zielinski BS (1995) The morphology of the olfactory epithelium in larval, juvenile and upstream migrant stages of the sea lamprey, *Petromyzon marinus*. Brain Behav Evol 45:19–24

Vélez-Espino LA, McLaughlin RL, Jones ML, Pratt TC (2011) Demographic analysis of trade-offs with deliberate fragmentation of streams: control of invasive species versus protection of native species. Biol Conserv 144:1068–1080

Vladykov VD (1949) Quebec lampreys (Petromyzonidae). 1. List of species and their economical importance. Contrib Dép Pêcheries Québec 26:7–67

Vrieze LA, Sorensen PW (2001) Laboratory assessment of the role of a larval pheromone and natural stream odor in spawning stream localization by migratory sea lamprey (*Petromyzon marinus*). Can J Fish Aquat Sci 58:2374–2385

Vrieze LA, Bjerselius R, Sorensen PW (2010) Importance of the olfactory sense to migratory sea lampreys *Petromyzon marinus* seeking riverine spawning habitat. J Fish Biol 76:949–964

Wagner CM, Jones ML, Twohey MB, Sorensen PW (2006) A field test verifies that pheromones can be useful for sea lamprey (*Petromyzon marinus*) control in the Great Lakes. Can J Fish Aquat Sci 63:475–479

Wagner CM, Twohey MB, Fine JM (2009) Conspecific cueing in the sea lamprey: do reproductive migrations consistently follow the most intense larval odour? Anim Behav 78:593–599

Wagner CM, Stroud EM, Meckley TD (2011) A deathly odor suggests a new sustainable tool for controlling a costly invasive species. Can J Fish Aquat Sci 68:1157–1160

Waldman JR, Grunwald C, Roy NK et al (2004) Mitochondrial DNA analysis indicates sea lampreys are indigenous to Lake Ontario. Trans Am Fish Soc 133:950–960

Waldman JR, Grunwald C, Wirgin I (2006) Evaluation of the native status of sea lamprey *Petromyzon marinus* in Lake Champlain based on mitochondrial DNA sequencing analysis. Trans Am Fish Soc 135:1076–1085

Waldman J, Grunwald C, Wirgin I (2008) Sea lamprey *Petromyzon marinus*: an exception to the rule of homing in anadromous fishes. Biol Lett 4:659–662

Waller DL, Bills TD, Boorgaard MA, Johnson DA, Doolittle TCJ (2003) Effects of lampricide exposure on the survival, growth, and behavior of the unionid mussels *Elliptio complanata* and *Pyganadon cataracta*. J Great Lakes Res 29(Suppl 1):542–551

Weise JG, Pajos TA (1998) Intraspecific competition between larval sea lamprey year-classes as Salem Creek was recolonized, 1990–1994, after a lampricide application. N Am J Fish Manag 18:561–568

Weisser JW, Adams JV, Schuldt RJ et al (2003) Effects of repeated TFM applications on riffle macroinvertebrate communities in four Great Lakes tributaries. J Great Lakes Res 29(Suppl 1):552–565

Welsh AB, Schumacher L, Quinlan HR (2019) A reintroduced lake sturgeon population comes of age: a genetic evaluation of stocking success in the St. Louise River. J Appl Ichthyol 35:149–159

Wicks BJ, Morrison BJ, Barker LA, Beamish FWH (1998) Unusual sex ratios in larval sea lamprey, *Petromyzon marinus*, from Great Lakes tributaries. Great Lakes Fisheries Commission Completion Report, Ann Arbor, MI

Wigley RL (1959) Life history of the sea lamprey of Cayuga Lake, New York. U S Fish Wildl Serv Fish Bull 59:561–617

Wilberg M, Hansen MJ, Bronte CR (2003) Historic and modern abundance of wild lean lake trout in Michigan waters of Lake Superior: implications for restoration goals. N Am J Fish Manag 23:100–108

Yeh C-Y, Chung-Davidson Y-W, Wang H, Li K, Li W (2012) Intestinal synthesis and secretion of bile salts as an adaptation to developmental biliary atresia in the sea lamprey. Proc Natl Acad Sci U S A 109:11419–11424

Young RJ (2005) Integrating heterogenous survey data to characterize the success of the Lake Huron sea lamprey (*Petromyzon marinus*) control program. PhD thesis, Michigan State University, East Lansing, MI

Young RJ, Christie GC, McDonald RB et al (1996) Effects of habitat change in the St Marys River and northern Lake Huron on sea lamprey (*Petromyzon marinus*) populations. Can J Fish Aquat Sci 53:99–104

Young RJ, Jones ML, Bence JR et al (2003) Estimating parasitic sea lamprey abundance in Lake Huron from heterogenous data sources. Great Lakes Res 29(Suppl 1):214–225

Youngs WD (1979) Evaluation of barrier dams to adult sea lamprey migration. Great Lakes Fishery Commission Completion Report, Ann Arbor, MI

Youngs WD, Oglesby RT (1972) Cayuga Lake: effects of exploitation and introductions of the salmonid community. J Fish Res Board Can 29:787–794

Youson JH (1980) Morphology and physiology of lamprey metamorphosis. Can J Fish Aquat Sci 37:1687–1710

Youson JH (2003) The biology of metamorphosis in sea lampreys: endocrine, environmental, and physiological cues and events, and their potential application to lamprey control. J Great Lakes Res 29(Suppl 1):26–49

Youson JH, Holmes JA, Guchardi J et al (1993) The importance of condition factor and the influence of water temperature and photoperiod in metamorphosis of sea lampreys, *Petromyzon marinus*. Can J Fish Aquat Sci 50:2448–2456

Zajicek P, Goodwin AE, Weier T (2011) Triploid grass carp: triploid induction, sterility, reversion, and certification. N Am J Fish Manag 31:614–618

Zerrenner A (2004) Effect of density and age on larval sea lamprey growth and survival in three Lake Champlain streams. J Freshw Ecol 19:515–519

Zerrenner A, Marsden JE (2005) Influence of larval sea lamprey density on transformer life history characteristics in Lewis Creek, Vermont. Trans Am Fish Soc 134:687–696

Zerrenner A, Marsden JE (2006) Comparison of larval sea lamprey life history characteristics in a lampricide-treated tributary and untreated tributary system of Lake Champlain. Trans Am Fish Soc 135:1301–1311

Chapter 6
The Lamprey as a Model Vertebrate in Evolutionary Developmental Biology

Joshua R. York, Eric Myung-Jae Lee and David W. McCauley

Abstract The development of lampreys has fascinated evolutionary developmental (evo-devo) biologists for a long time. Lampreys, as one of the two surviving members of an ancient group of jawless vertebrates, have long been recognized as key for understanding vertebrate evolution due to their basal position in vertebrate phylogeny. While classical descriptions of lamprey development have uncovered many similarities in development among the few lamprey species that have been studied, these studies, together with modern techniques, have provided key insights for understanding how developmental changes have been important for vertebrate evolution. In recent years, the sea lamprey *Petromyzon marinus* has moved to the forefront of studies on lamprey development due to its invasion into the Great Lakes, and the critical need to understand its biology for management purposes. The sea lamprey genome has also been published and these two developments, taken together, facilitate the use of lampreys in evo-devo investigations. Here we provide a current overview of contributions of lamprey developmental studies for understanding vertebrate evolution, a summary of modern molecular and genetic tools and methods that have been applied in lamprey evo-devo research. Finally, we provide information to facilitate setting up the lamprey as a model organism in a modern research laboratory setting.

Keywords Development · Embryos · Evo-devo · Evolution · Gene knockdown · Jaws · Lamprey culture · Molecular genetics · Neural crest · Paired appendages · Placode · Skeleton · Vertebrates

J. R. York · E. M.-J. Lee · D. W. McCauley (✉)
Department of Biology, University of Oklahoma, 730 van Vleet Oval,
Norman, OK 73019, USA
e-mail: dwmccauley@ou.edu

J. R. York
e-mail: joshuayork@ou.edu

E. M.-J. Lee
e-mail: ericmlee@ou.edu

© Springer Nature B.V. 2019   481
M. F. Docker (ed.), *Lampreys: Biology, Conservation and Control*,
Fish & Fisheries Series 38, https://doi.org/10.1007/978-94-024-1684-8_6

6.1   Introduction

"The development of the Lamprey has occupied the attention of many embryologists during the last fifty years". This introductory sentence to Arthur Shipley's description of development in the European river lamprey *Lampetra fluviatilis* was published in 1887 and highlights a longstanding interest in lamprey development among biologists that continues more than a century later (Shipley 1887). Lampreys have become an important model organism to evolutionary developmental biologists, primarily due to their phylogenetic position at the base of the vertebrates. Because of this basal position among extant vertebrates, studies of lamprey development can inform our understanding of vertebrate evolution, and may provide insight into the origin of numerous vertebrate character traits.

Approximately 40 lamprey species exist worldwide (Renaud 2011; Potter et al. 2015; see Chap. 7), but modern developmental studies have been limited largely to four species, based in part on availability of embryonic material to researchers. Two species, sea lamprey *Petromyzon marinus* and Arctic lamprey *Lethenteron camtschaticum* (formerly recognized as *Lethenteron japonicum* or *Lampetra japonica* and sometimes still called the Japanese lamprey; Renaud 2011; Potter et al. 2015), account for many recent developmental studies. Numerous early descriptions of lamprey development were made using the European river lamprey or the European brook lamprey *Lampetra planeri* (Schultze 1856; Shipley 1887; Damas 1944; Newth 1956; Akoev and Muraveiko 1984; Kuratani et al. 1997). Such descriptions were often dependent on obtaining embryos by collecting gravid adults during the spawning phase in the spring of the year, and then rearing artificially fertilized eggs through embryogenesis in the laboratory, or removing embryos from nest sites following natural spawning activity. Starting in the 1950s, efforts to control the invasive sea lamprey in the Great Lakes bordering the United States and Canada (see Chap. 5) have led to increased use of the sea lamprey as a research organism. Advances in resources to study the biology of the sea lamprey have in turn resulted in the expansion of tools to study the development of this species.

6.1.1   Historical Context for Sea Lamprey Developmental Studies

Niagara Falls is a natural barrier to the upstream inland migration of the parasitic sea lamprey beyond Lake Ontario. However, with construction of the Welland Canal beginning in 1824 (Aitken 1954), sea lamprey were able to gain access above Niagara Falls. Following improvements to the Welland Canal in 1919, sea lamprey were observed in Lake Erie by 1921, and by the late 1930s, sea lamprey had been observed in all of the Great Lakes, causing severe damage to the Great Lakes fish stocks (Applegate 1950). The Great Lakes Fishery Commission was established in 1955 with the suppression of sea lamprey populations in the Great Lakes as one of its primary responsibilities (Crowe 1975; Fetterolf 1980; see Chap. 5). In seeking efforts to control sea lamprey population numbers, it was recognized that sea lamprey embryology was one aspect of its life history that was not well understood. A description of in vitro fertilization for the sea lamprey was first published in 1955 (Lennon 1955) and, in work conducted at the Hammond Bay Laboratory (now the Hammond Bay Biological Station) on Lake Huron beginning in 1954, Piavis (1961) described methods for culturing sea lamprey embryos, and developed a staging table of normal sea lamprey development. Because development is temperature-dependent, staging of embryos is based on morphological criteria rather than time post-fertilization. Staging tables have also been constructed for the Far Eastern brook lamprey *Lethenteron reissneri* (Tahara 1988) and the Pacific lamprey *Entosphenus tridentatus* (Yamazaki et al. 2003; see Chap. 2). Although developmental stages seem to differ slightly in timing, the general patterns of embryonic development appear to be very similar between these species. Developmental stages have also been described for European river lamprey (Damas 1944) and comparisons have been drawn among these staging tables (Kuratani et al. 1997; McCauley and Bronner-Fraser 2002).

With the published description of its early embryology, efforts to control sea lamprey populations began to focus on the selective vulnerability of lamprey larvae to specific compounds (Applegate et al. 1961; Applegate and King 1962). Such efforts required access to large numbers of embryos and larvae that could be provided by facilities at Hammond Bay. While studies on the development of other lamprey species were often limited by the availability of source material, landlocked spawning sea lamprey were easily obtained in abundance during the spawning runs in the spring each year. In contrast, decreases in the abundance of many other lamprey species have resulted in conservation efforts, including protected status (Renaud 2011; Maitland et al. 2015), making them difficult to obtain for research purposes.

Because of the abundance of spawning habitat for sea lamprey in the tributaries near Hammond Bay on Lake Huron, large numbers of migratory sea lamprey were trapped, to be either disposed of or used in research. Sea lamprey also began to be shipped throughout the United States for various research purposes. However, and unfortunately, it was recognized that embryos did not develop from gravid adults that had been packaged and shipped in the same manner, so developmental studies remained limited to researchers able to work at Hammond Bay Biological Station or those with access themselves to spawning sea lamprey. Recent efforts have resulted in the successful development of methods to obtain gametes and to culture sea lamprey embryos in the laboratory (Nikitina et al. 2009a, b), thus reducing the need to conduct studies near the source of spawning sea lamprey. These methods are included at the conclusion of this chapter (see Sect. 6.4); efforts to artificially propagate other lamprey species (e.g., Pacific and European river lampreys for restoration purposes) are described in Chap. 2.

Going forward, the sea lamprey is likely to see increased use as an evo-devo model due in large part to the sea lamprey genome project funded by the National Human Genome Research Institute, of the National Institutes of Health, and also due to availability of the Arctic lamprey genome. The sea lamprey genome, published in February 2013 (Smith et al. 2013) and recently updated (Smith et al. 2018), has become an invaluable resource to evo-devo investigators as genetic methods are adapted for use in lampreys. The Arctic lamprey genome has provided insights into vertebrate genome evolution (Mehta et al. 2013; Manousaki et al. 2016) and will be an important reference for comparison with the better annotated sea lamprey genome.

6.1.2   Lampreys and Vertebrate Phylogeny

Since the time of Darwin and Haeckel, lampreys and hagfishes have been considered to be primitive members of the vertebrates (Haeckel 1866; Gee 2007). However, the precise positioning of lampreys and hagfishes in vertebrate phylogeny has long been a source for debate (for a historical review, please see Janvier 2008). Early on, lampreys were considered to be the sister taxon to hagfishes, the other extant group of jawless fishes. Lampreys and hagfishes, based on a series of shared characters, together formed the Cyclostomata (Duméril 1806). Throughout the 19th and into the 20th century, arguments persisted over the monophyly of cyclostomes and their relationship to other fossil jawless fishes (Agnatha) (Cope 1889). With the increased interest in systematics and its use of parsimony analysis beginning in the mid-20th century (Hennig 1950), the question of cyclostome monophyly returned. In the mid-20th century, comparative anatomists, based primarily on comparisons of morphological traits, suggested that a lack of numerous vertebrate characters in hagfishes placed them basal to lampreys and gnathostome vertebrates; Løvtrup, for example, based on parsimony and outgroup comparison of anatomical and physiological characters, concluded that cyclostomes are paraphyletic (Løvtrup 1977). This view, which suggests that lampreys are sister to the jawed vertebrates, is still accepted by those who consider only lampreys and gnathostomes as vertebrates, with hagfishes considered as non-vertebrate craniates (e.g., Nelson et al. 2016; see Docker et al. 2015).

Following the advent of molecular phylogenetics, multiple studies showed renewed support for cyclostome monophyly (Stock and Whitt 1992; Kuraku et al. 1999; Delarbre et al. 2002; Furlong and Holland 2002). Cyclostome monophyly is now also supported by the discovery of novel families of micro-RNAs that are restricted to hagfishes and lampreys, as well as shared mechanisms of variable lymphocyte receptor (VLR) development and programmed genome rearrangement (Smith et al. 2009; Heimberg et al. 2010; Boehm et al. 2018). New evidence has also begun to suggest that some characters lacking in hagfishes are likely to have been lost, rather than to be missing ancestrally (Ota et al. 2011, 2013; Gabbott et al. 2016; Kuratani et al. 2016), further shifting support toward cyclostome monophyly. Thus, whereas cyclostome paraphyly suggested that lampreys were the basal-most group of extant vertebrates, cyclostome monophyly now indicates that lampreys and hagfishes share this position (Docker et al. 2015). The use of hagfishes in developmental studies is increasing, following a long "drought" caused by the difficulty in obtaining hagfish embryos (Holland 2007; Ota et al. 2007), but the relative ease with which lamprey embryos can be obtained and the genomic resources now available still make lampreys the ideal cyclostome model organisms for studying the origin and development of vertebrate-specific traits.

6.2 Lampreys and Comparative Studies of Vertebrate Evolution

Lampreys are popularly referred to as "living fossils" (Eisner 2003; McCauley et al. 2015). However, they cannot be considered as a direct proxy for the ancestral vertebrate; since cyclostomes and gnathostomes diverged from a common ancestor ~500–600 million years ago (Janvier 1996; Hedges et al. 2015), lampreys are no more closely related to this common ancestor than are the gnathostomes. Nevertheless, extant lampreys are remarkably similar in appearance to fossil lampreys (Bardack and Zangerl 1968; Gess et al. 2006; Chang et al. 2014). In the Foreword to Hardisty and Potter's *The Biology of Lampreys*, Volume 1, Young pointed out that the interest in lampreys among zoologists stems from the observation that "Lampreys and hagfishes retain more features of the presumed ancestral craniate than do any other members of the group" (Young 1971). However, as basal vertebrates, lampreys contain characters that are defining for vertebrates, including an axial skeleton, tripartite brain complexity, placode-derived sensory ganglia, and neural crest cells and their derivatives (Green and Bronner 2014; McCauley et al. 2015; Sugahara et al. 2016). As the sister taxon to gnathostome vertebrates, lampreys and hagfishes can be used in comparative studies with model gnathostomes to differentiate the origins of developmental mechanisms for characters that are shared among all vertebrates from those that may be derived in gnathostomes. Each of these characters represents an avenue of investigation for understanding vertebrate development and evolution. Among these, development of the neural crest has gained perhaps the most interest among evolutionary developmental biologists, owing to the hypothesized critical importance of the neural crest for vertebrate origins (Gans and Northcutt 1983; Trainor 2013; Green et al. 2015).

Early in the 21st century, the advent of new molecular, cellular, and genetic tools, and their application to lamprey development, has led to increasing interest among evolutionary developmental biologists investigating the evolutionary origin of many vertebrate traits. The balance of this chapter provides an updated overview of the contribution of lampreys to current understanding of the evolution of vertebrate development. Topics are grouped into vertebrate characters informed by lamprey developmental studies, and techniques available for exploitation. Finally, some methods are presented to encourage use of lampreys as an evo-devo model in a modern laboratory environment.

6.2.1 Neural Crest Cells

The neural crest is a transient population of multipotent cells that migrate through tissues of the early embryo and contribute or give rise to numerous derivatives critical to vertebrate development. These include such defining features as the peripheral nervous system with contributions to cranial ganglia, craniofacial cartilage, and most notably the jaws (Hall 1999; Le Douarin and Kalchiem 1999; Trainor 2013).

Since its discovery by Wilhelm His in 1868 (Hörstadius 1950; His 1868), the neural crest has been of interest to embryologists and evolutionary developmental biologists due to its intimate link to the vertebrate transition from sedentary to predatory lifestyles (Gans and Northcutt 1983). Over the past decade, increasing knowledge of the neural crest induction process at the molecular and genetic level (discussed below) suggests that the origin of neural crest cells predates vertebrates (Donoghue et al. 2008). Critical support for this idea comes from the discovery and investigation of rudimentary neural crest-like cells (NCLC) in urochordates (tunicates). In 2004, Jeffery and colleagues discovered NCLC in the ascidian *Ecteinascidia turbinata* that originate near the neural tube, undergo extensive migration, express the HNK-1 antigen, and differentiate into pigment cells (Jeffery et al. 2004). Subsequent studies showed that this cell line originates from mesoderm flanking the neural tube, but nonetheless, expresses a host of key neural crest markers (*Twist, AP2, FoxD,* and *Myc*) reminiscent of vertebrate neural crest cells (Jeffery 2006; Jeffery et al. 2008). Similarly, Abitua et al. (2012) identified a cephalic melanocyte lineage in *Ciona intestinalis.* This cell line originates at the neural plate border, expresses neural crest specification genes (*Id, Snail, Ets,* and *FoxD*), and can be reprogrammed into migrating "ectomesenchyme" by targeted missexpression of *Twist* driven by a *Mitf* enhancer. More recently, it was shown that cells expressing the neural crest transcription factors *Snail, Msx and Pax3/7* in the caudal neural plate border in *C. intestinalis* could migrate and differentiate into bipolar tail neurons, a cell population that is strikingly similar to neural crest-derived spinal ganglia in vertebrates (Stolfi et al. 2015).

While these studies reinforce the idea that a rudimentary neural crest gene regulatory network (NC-GRN) existed prior to the emergence of the neural crest, true neural crest cells still remain a vertebrate innovation (Hall and Gillis 2013; Medeiros 2013; Green et al. 2015). Evidence to support this theory comes from lamprey developmental studies that employ molecular techniques. As a basal vertebrate, lampreys possess a well-developed bona fide neural crest population, and although they lack major neural crest derivatives, such as the jaws and sympathetic chain ganglia of gnathostomes, lamprey neural crest development follows that of other vertebrates (Johnels 1956; Horigome et al. 1999; Tomsa and Langeland 1999; McCauley and Bronner-Fraser 2003). Limitations of earlier studies of lamprey neural crest development, using purely descriptive or experimental embryology (Newth 1950, 1951; Langille and Hall 1988), have been overcome by using molecular techniques. Investigations using lipophilic DiI-labeling experiments show that lamprey neural crest cells take migratory routes similar to those seen in gnathostomes, with the exception of the migratory pattern of neural crest originating from the hindbrain, and timing differences of migration into the presumptive pharyngeal region (Horigome et al. 1999; McCauley and Bronner-Fraser 2003; Green et al. 2017).

Gene expression studies laid the initial groundwork for comparisons to be made between lamprey and gnathostome neural crest regulation (Tomsa and Langeland 1999; Myojin et al. 2001; Neidert et al. 2001; Meulemans and Bronner-Fraser 2002; Meulemans et al. 2003). Subsequent studies used synthetic antisense morpholino oligonucleotides (morpholinos) to assess the effects of knocking-down lamprey neural crest specifier genes in the formation of cartilage of the pharyngeal arches (McCauley and Bronner-Fraser 2006). Morpholinos have also been used in conjunction with messenger RNA (mRNA) rescue experiments to carefully dissect the lamprey NC-GRN (Sauka-Spengler et al. 2007; Nikitina et al. 2008). With the advent of genome editing techniques (see Sect. 6.3.4), it is now possible to induce precise mutations in targeted genomic regions in lamprey embryos to dissect the molecular-genetic control of neural crest development (Square et al. 2015; Zu et al. 2016; York et al. 2017). These studies revealed that the underlying NC-GRN is conserved between lampreys and higher vertebrates, albeit with differences in the spatiotemporal expression of neural crest specifiers such as *Twist* and *Ets1* (Sauka-Spengler et al. 2007; Nikitina et al. 2008; Sauka-Spengler and Bronner-Fraser 2008a; Nikitina and Bronner-Fraser 2009).

Lampreys, and in particular sea lamprey, are well suited for these studies due to their relatively slow rate of development; fertilization to hatching occurs over 11 days (Piavis 1971), and neural crest migration can be observed by the 6th day of development (McCauley and Bronner-Fraser 2003). This slower rate of development allows for investigators to more precisely observe the timing of gene expression and the effects of gene knockdown on putative gene targets, which may have otherwise been missed in a more rapidly developing model system. The construction of the lamprey neural crest gene regulatory network (NC-GRN) has also opened doors for comparisons to be made to invertebrate chordates (Yu et al. 2008). With release of the lamprey genome (Smith et al. 2013, 2018), lampreys will continue to be a valuable model for studying the evolution and diversification of neural crest cells (Green and Bronner 2013).

6.2.2   Placodes

Like neural crest cells, the emergence of cranial placodes is central to the evolution of vertebrate sensory systems (Gans and Northcutt 1983; Baker and Bronner-Fraser 1997). Despite the fact that the term "placodes" was coined by von Kupffer more than a century ago (van Wijhe 1883; Beard 1885; von Kupffer 1891), much of our understanding of placode development comes from recent studies. Cranial placodes are transient ectodermal thickenings of columnar epithelial cells with defined boundaries that form in stereotypic regions of the vertebrate embryonic head. Together with contributions from neural crest cells, placodes make contributions to numerous cranial paired sensory organs of the vertebrate embryo, including the nose, ears, eyes, and sensory ganglia, as well as the lateral line system (Le Douarin 1986; Le Douarin et al. 1992; Vogel and Davies 1993; Webb and Noden 1993; Northcutt 1996; Graham and Begbie 2000). Much like the various derivatives of neural crest cells, individual placode lineages that give rise to different derivatives are thought to have evolved at different times (Baker and Bronner-Fraser 1997; Graham and Begbie 2000; Shimeld and Holland 2000; Graham and Shimeld 2013).

Placode development has been studied in several vertebrate species including zebrafish *Danio rerio*, African clawed frog *Xenopus laevis*, chick *Gallus gallus domesticus*, and mouse *Mus musculus* (Baker and Bronner-Fraser 2001). Vertebrate placodal differentiation is thought to originate from a common pan-placodal primordium located at the border of the neural plate and future epidermis (Schlosser and Northcutt 2000; Baker and Bronner-Fraser 2001; Noramly and Grainger 2002; Schlosser 2002; Toro and Varga 2007). Initial differentiation requires the expression of general placode markers *Six1/2*, *Six4/5* (*sine oculis*) and *Eya* (eyes absent) families of transcription factors, while later expression of the *Pitx*, *Sox*, *Dlx*, *Fox*, and *Pax* families of transcription factors is required for lineage-specific differentiation (Schlosser 2005, 2006, 2010; Schlosser and Ahrens 2004; Ladher et al. 2010; Sato et al. 2012). Placode development has also been studied in urochordates (Wada et al. 1998; Meinertzhagen and Okamura 2001; Manni et al. 2004; Mazet et al. 2005; Gasparini et al. 2013; Abitua et al. 2015), cephalochordates (Manzanares et al. 2000; Holland and Holland 2001; Kozmik et al. 2007; Meulemans and Bronner-Fraser 2007; Schlosser 2017), and other invertebrates (Hill et al. 2010; Posnien et al. 2011; Schlosser 2015). A global comparison of these studies suggests that during the course of chordate evolution, a pre-existing gene regulatory network for sensory epidermal cell formation was co-opted for placode formation (Schlosser 2006; Bertrand and Escriva 2011; Abitua et al. 2015; Schlosser 2016). From recent work in tunicates, however, it is also possible that a "proto-placodal ectoderm" already existed in the last common ancestor of tunicates and vertebrates (i.e., Olfactores), and was subsequently elaborated upon during the evolution of early vertebrates (Abitua et al. 2015).

Lampreys possess sensory organs and cranial ganglia that are derived from placodes as in higher vertebrates, and may provide key insights into the origin of such placode-derived features as ears and the lateral line system, as well as developmental mechanisms important for origins of diplorhiny. Here we highlight the current understanding of these features.

6.2.2.1 Lateral Line and Otic Placodes

Similar to gnathostomes, the lamprey lateral line contains both mechanosensory neuromasts and electroreceptive epidermal "end bud" organs, suggesting that the vertebrate acquisition of the lateral line predates the gnathostome-agnathan divergence (Akoev and Muraveiko 1984; Gelman et al. 2007; Baker et al. 2013; Modrell et al. 2014). The lateral line and ears together form the acoustico-lateralis system that originates from a common placode; a system that possesses mechanoreceptive hair cells (Schlosser 2002; Gelman et al. 2007; Baker et al. 2013; Piotrowski and Baker 2014). While the otic placode is believed to be common to all chordates (Shimeld and Holland 2000), its origin remains a mystery. In order to address questions regarding vertebrate ear evolution, it is also important to understand the development of its critical components, namely hair cells and sensory neurons (Fritzsch and Beisel 2001), all of which are derived from the otic placode (Barald and Kelley 2004; Fritzsch et al. 2006). Recent studies using light and electron microscopy have shown that tunicates possess secondary sensory cells located on the coronal organ that resemble vertebrate hair cells, suggesting that hair cells originated in the chordate common ancestor (Burighel et al. 2003, 2008; Manni et al. 2004, 2006; Caicci et al. 2007, 2010, 2013; Rigon et al. 2013, 2018).

The gnathostome inner ear is a complex sensory organ that is responsible for hearing, balance, and spatial orientation in three-dimensional space. It is comprised of the cochlea of the auditory system along with the semicircular canals and otolith organs (utricle, saccule, lagena) of the vestibular system (Rinkwitz et al. 2001). Angular acceleration causes the displacement of endolymph contained throughout the three semicircular canals. This displacement is detected by mechanoreceptive hair cells of the crista ampullaris located at the base of each canal. Therefore, each semicircular canal detects a major axis of movement. Development of the vertebrate inner ear begins during gastrulation as surface ectoderm that thickens to form the otic placodes at either side of the neural tube (Rinkwitz et al. 2001). A signaling cascade, involving fibroblast growth factors (FGFs), bone morphogenetic proteins (BMPs), sonic hedgehog (Shh), and Wnts, has been described for otic placode induction and inner ear morphogenesis (Chatterjee et al. 2010; Ladher et al. 2010; Groves and Fekete 2012; Chen and Streit 2013; Kiernan 2013). Lamprey otic vesicle development follows that of gnathostomes; however, lamprey ears possess only two semicircular canals, as a third canal never seems to appear during development (Scott 1887; Shipley 1887; Richardson et al. 2010). The process of patterning and morphogenesis of the three semicircular canals from the dorsal otic placode is not fully understood (Martin and Swanson 1993; Bok et al. 2007). Recent studies have shown that *Otx1* may account for all major differences between gnathostome and lamprey otic vesicles, suggesting that lamprey ears may represent a primitive version of gnathostome inner ears (Fritzsch et al. 2001; Hammond and Whitfield 2006). Further studies have highlighted the importance of *bmp2b* and Wnt/β-catenin signaling specifically during morphogenesis of semicircular canals in zebrafish and mice (Hammond et al. 2009; Rakowiecki and Epstein 2013), but this has yet to be examined in lampreys.

6.2.2.2 Nasohypophyseal Placode

Unlike gnathostomes, lampreys possess a single nostril (monorhiny) that develops from a median domain of the rostral ectoderm called the nasohypophyseal placode (Kleerekoper and Erkel 1960). The solid nasohypophyseal plate precludes the rostromedial growth of premandibular ectomesenchyme, which forms major components of the gnathostome jaw. It is therefore hypothesized that the heterotopic separation of the nasal and hypophyseal placodes may have been a prerequisite to the emergence of the jaw (Kuratani et al. 2001, 2013; Uchida et al. 2003; Kuratani 2005, 2012; Gai et al. 2011; Oisi et al. 2013; Dupret et al. 2014).

Despite the evolutionary significance of placode-derived features, little is known as to whether the developmental and molecular mechanisms of early placode development are conserved between lampreys and gnathostomes. In gene expression studies, it was shown that placodes present in the developing lamprey embryo express *Dlx* and *Pax* transcription factors, likely reflecting an ancient role of *Dlx* and *Pax* genes in fate specification of placodes that extends to the base of vertebrates (Neidert et al. 2001; McCauley and Bronner-Fraser 2002). The authors of a recent fate map and gene expression analysis of cranial ganglia development in lamprey support this notion by positing a combinatorial "Pax code" that governs formation and patterning of placode-derived elements of cranial sensory ganglia (Modrell et al. 2014). Nonetheless, our current understanding of pan-vertebrate mechanisms of placode development remains poor, and comparative analyses focusing on the evolution of vertebrate placode development is therefore ripe for investigation using the lamprey as a model.

6.2.3 Paired Appendages

Another key vertebrate innovation is the emergence of paired lateral appendages. Lateral appendages are important for locomotive stability and sophisticated maneuvering (Breder 1926; Drucker and Lauder 2002). After over 150 years of research (Owen 1849), the vertebrate limb has garnered a long standing interest from evolutionary and developmental biologists (Coates 1994; Coates and Cohn 1999; Ruvinsky and Gibson-Brown 2000). Modern molecular techniques can now be used to address questions regarding the evolutionary origin of vertebrate paired appendages (Niswander 1997; Tickle 2003; Tanaka and Onimaru 2012; Adachi et al. 2016; Gehrke and Shubin 2016). All gnathostomes possess paired appendages; the paired sets of pectoral and pelvic fins in bony and cartilaginous fishes are homologous to the forelimbs and hindlimbs of tetrapods, respectively (Carroll 1988; Shubin et al. 1997). Snakes, caecilians, and eels have undergone secondary loss of paired appendages, while some other aquatic species (e.g., whales, dolphins, pufferfishes) have lost pelvic fins that in some cases exist as vestigial structures (Cohn and Tickle 1999; Bejder and Hall 2002; Tanaka et al. 2005; Don et al. 2013; Dial et al. 2015).

In contrast, lampreys are ancestrally limbless, having diverged from the rest of the vertebrate lineage prior to the emergence of paired appendages over 360 million years ago, and represent the plesiomorphic condition limited to median fins (Donoghue et al. 2000; Gess et al. 2006). A study by Freitas and colleagues shows the shared expression of two genes implicated in limb development (*Hox* and *Tbx*) in both median and paired fins of the developing catshark (Scyliorhinidae) (Freitas et al. 2006). These genes are also expressed in lamprey median fins, suggesting that the developmental mechanism responsible for the paired appendages of gnathostomes may have its origins in the median fin of the ancestral vertebrate (Freitas et al. 2006). More recent analysis by Freitas et al. (2012) shows that *hoxd13a* activity promotes distal proliferation of zebrafish fins, suggesting that the modulation of *5'Hoxd* gene expression through novel enhancer elements may have facilitated the evolution of fins. Similarly, work by Adachi et al. (2016) revealed that a highly conserved gnathostome enhancer regulating expression of the limb specifier, *Tbx5*, is not activated in lamprey embryos, providing strong evidence that cis-regulatory turnover was seminal in establishing a limb outgrowth program in jawed vertebrates. Further, analysis of the recently sequenced lamprey genome revealed a lack of the long range cis-acting enhancer *Shh* appendage-specific regulatory element (ShARE), which is required for limb-specific expression of *Shh*. The authors suggest that this regulatory element required for patterning the anteroposterior axis of limbs evolved independently in the gnathostome lineage (Smith et al. 2013). Finally, recent work by Letelier et al. (2018) revealed that both median and paired fin development in gnathostomes requires the activity of *Shh*, which is activated by a shared enhancer (the so-called ZPA regulatory sequence, or ZRS). This suggests that paired fins may have emerged in part by the co-option of this enhancer from a plesiomorphic function in median fin development. Interestingly, however, the ZRS-mediated development of fins appears to be a gnathostome innovation, as neither the ZRS enhancer, nor *Hh* gene expression, is active in lamprey median fins (Letelier et al. 2018).

Comparative studies of lampreys have also elucidated our understanding of the tissue context in which paired fins first appeared. In gnathostomes, the generation of fin/limb buds from the somatic mesoderm (somatopleure) includes multiple developmental steps. First, the lateral plate mesoderm divides into cardiac mesoderm (CM) and posterior lateral plate mesoderm (LPM). Hox genes have been shown to play a crucial role in defining the anterior-posterior axis of the LPM, where they show nested expression in co-linear fashion (Ruvinsky and Gibson-Brown 2000). Second, the LPM thickens before further splitting into the somatopleure and splanchnopleure. Genes involved in the generation of fin/limb bud-forming fields are expressed in the somatopleure, which gives rise to fin/limb buds that develop into paired appendages (Logan 2003). Recent investigations reveal that although nested Hox gene expression is present in the LPM of lamprey embryos, histological evidence shows that the LPM does not split into the somatopleure and the splanchnopleure (Onimaru et al. 2011). This is supported by lipophilic DiI-labeling showing that the somatopleure is eliminated during the course of lamprey embryonic development (Tulenko et al. 2013). These results suggest that innovations of the nested Hox gene expression patterns in the LPM and the formation of the somatopleure facilitated the emergence of fin/limb buds after the agnathan-gnathostome transition. Future advancements of molecular techniques will allow for the dissection of gene regulatory interactions in lampreys to further our understanding of vertebrate paired appendage evolution.

6.2.4 Skeleton

Vertebrate cartilage and mineralized bone are used for structural support, protection, and predation. Cartilaginous structures and cell types similar to that in vertebrates have been found in a wide range of invertebrates (Cole and Hall 2004; Cole 2011). Although the degree of homology between invertebrate and vertebrate cartilage remains unclear, the gene regulatory network underlying cartilage development is evolutionarily ancient, tracing back to the last common ancestor of all bilaterian animals (Tarazona et al. 2016). The evolution of the vertebrate skeleton has long been a subject of interest to biologists (Hertwig 1874; Kingsley 1894; De Beer 1924, 1937; Gadow 1933; Reif 1982; Smith and Hall 1990). In regards to the origin of the vertebrae, a complex patterning of vertebrate somites creates separate compartments that form the dermatome, myotome, and sclerotome. It is the ventromedial somites (sclerotome) that give rise to the vertebrate axial skeleton. The mechanism of vertebrate sclerotome induction involves an interplay between hedgehog signals from the notochord and antagonistic Bmp signaling from more lateral mesoderm to subdivide the somite (Shimeld 1999; Shimeld and Holland 2000; Christ et al. 2004). Gnathostome vertebrae differentiate from the sclerotome, and consist of two axial elements that form both dorsally and ventrally along the notochord (Goodrich 1930; Janvier 1996).

While the spinal cord of lampreys is not ensheathed within a vertebral column, they do possess neural crest-derived and sclerotome-derived axial cartilage nodules dorsally along the notochord, which are thought to be homologous to gnathostome vertebral elements (Tretjakoff 1927; Zhang 2009; Shimeld and Donoghue 2012). Recently, sclerotome-derived axial cartilage nodules have been found in ventral aspects of the notochord of the inshore hagfish *Eptatretus burgeri* (Ota et al. 2011, 2013; Kuratani et al. 2016). The evolutionary sequence that led to these cartilage nodules in lampreys and hagfishes remains a mystery. Questions regarding the evolution of skeletal tissues and their mineralization have also been addressed using lampreys and hagfishes. Studies have shown that the cartilage of lampreys and hagfishes share similar gene expression profiles (*SoxD*, *SoxE*, and *Runx*) with that of gnathostomes, while additional studies in cephalochordates (amphioxus) suggest that a primitive cellular cartilage program—and even bona fide cellular cartilage—predates vertebrate origins (McCauley and Bronner-Fraser 2006; Zhang and Cohn 2006; Zhang et al. 2006; Hecht et al. 2008; McCauley 2008; Ohtani et al. 2008; Wada 2010; Cattell et al. 2011; Jandzik et al. 2015).

Lampreys are known to have structurally distinct cartilage types not found in gnathostomes; elastin-like proteins known as lamprins serve as the major extracellular matrix component in contrast to that of gnathostome cellular cartilage composed mainly of fibrillar collagen (Wright et al. 1983, 2001; Wright and Youson 1983; Robson et al. 1993; Ohtani et al. 2008; Yao et al. 2008; Lakiza et al. 2011; Jandzik et al. 2014). Lamprey craniofacial cartilage is composed of elements that support and protect the brain, and also a viscerocranial skeleton made up of cartilage elements that form a fused pharyngeal basket to support the seven gill arches and associated lamellibranchs (Martin et al. 2009; Jandzik et al. 2014). The lamprey trabecular cartilage forms as paired cartilage rod-like elements that form laterally alongside the adenohypophysis to support the brain (Johnels 1948; Langille and Hall 1988; Kuratani et al. 2004; Martin et al. 2009).

Mucocartilage is another lamprey-specific type of cartilage that supports most of the anterior head structures of the ammocoete larva (Martin et al. 2009; Yao et al. 2011). Whereas the elastin-like cartilage of the branchial basket supports the pharynx and gill openings, mucocartilage supports the lamprey upper and lower lips, the ventral pharynx, and the first and second arches. This histologically distinct cartilage shares major similarities with gnathostome cellular cartilage in that it expresses *FGFR*, *RunxA*, *Barx*, and *Alx* genes, and is patterned along the dorso-ventral axis by endothelin signaling (Wright and Youson 1982; Cattell et al. 2011; Jandzik et al. 2014; Square et al. 2016a, b; Yao et al. 2011). Given their possession of diverse and unique cartilage types, dissecting the genetic basis underlying the diverse cartilage types in lampreys may elucidate our understanding of vertebrate skeletal evolution.

6.2.5 Articulated Jaws

The acquisition of articulated jaws during vertebrate evolution is thought to have led to their explosive adaptive radiation (Gans and Northcutt 1983; Janvier 1996; Mallatt 1996). Advantages conferred by jaws include the improvement of the branchial respiration system via the musculature of the upper and lower skeletal elements, and the ability to occupy entirely new niches via predation. The vertebrate head is comprised of the neurocranium (dorsal), viscerocranium (ventral), and mandibular arch. With the exception of the neurocranium, all of these structures are derived exclusively from the neural crest (Noden 1988; Le Douarin and Kalchiem 1999). The development of the jaw requires the dorsal-ventral subdivision of the embryonic rostral-most pharyngeal arch, the mandibular arch (Kuratani and Ota 2008; Mallatt 2008). The mandibular arch formed the palatoquadrate of the upper jaw and Meckel's cartilage of the lower jaw in ancient placoderm fish (Sienknecht 2013; Miyashita 2016; Zhu et al. 2016).

The classic theory by Carl Gegenbaur postulated that evolution of the jaw and hyoid arch was facilitated by the transformation of a rostral gill arch (Gegenbaur et al. 1878; Gillis et al. 2013). Mallatt (1996) theorized that the original mandibular arch first functioned in ventilation before moving rostrally towards the old mouth to form a "new mouth." Janvier (1996) hypothesized that the mandibular arch arose through modification of the velar skeleton (found in cephalochordates and lampreys), because the velar skeleton in extinct and modern lampreys is comprised of articulated upper and lower elements. While fossil intermediaries to support these theories are lacking, Gegenbaur's original theory is supported by molecular evidence to suggest the importance of the Distal-less homologs, *Dlx* genes, in the dorso-ventral (DV) patterning of the first pharyngeal arch, and ultimately in the evolutionary acquisition of jaws (Simeone et al. 1994; Qiu et al. 1997; Depew et al. 2002; Panganiban and Rubenstein 2002; Gillis et al. 2013). The advent of the segmented branchial bars and jaws is assumed to have occurred in the vertebrate lineage after the agnathan-gnathostome divergence, and that gradual changes in the interaction between migrating neural crest cells and surrounding pharyngeal tissues could account for the evolution of the mandibular arch (Shigetani et al. 2002). Alternatively, a new hypothesis has been proposed in which a differentiated mandibular arch first appeared in stem vertebrates, prior to the divergence of agnathans and gnathostomes (Miyashita 2016). Under this scenario, the gnathostome jaw is thought to be derived from an already differentiated mandibular arch by confinement and structural organization of mandibular mesenchyme (Miyashita 2016). The mandibular confinement hypothesis is therefore fundamentally different from classical hypotheses for vertebrate jaw evolution, which suggested that a differentiated mandibular arch (including articulated jaws) was derived from an ancestrally homonomous (i.e., having similar structure) series of pharyngeal arches. Regardless of exactly which scenario is correct, the differences and similarities in pharyngeal development between lampreys and gnathostomes make lampreys an attractive model for studying vertebrate jaw evolution (Kuratani and Ota 2008; McCauley et al. 2015).

As discussed above, lampreys possess an upper lip, lower lip, first arch, and second arch that consist of mucocartilage, and a fused branchial basket composed of seven pharyngeal arches that consist of cellular cartilage. Studies have reported a conserved nested pattern of *Dlx* expression in the pharyngeal arch of gnathostomes, suggesting that a "*Dlx* code" was co-opted for the dorsoventral patterning of the jaw during vertebrate evolution (Minoux and Rijli 2010; Talbot et al. 2010; Zuniga et al. 2011; Medeiros and Crump 2012; Takechi et al. 2013). While initial studies using lampreys showed expression of *Dlx* throughout the proximodistal axis of the pharyngeal arches (Neidert et al. 2001; Kuraku et al. 2010), a subsequent study showed a nested expression of *Dlx* genes, together with dynamic expression of *Msx*, *Hand*, and *Gsc* genes, along the dorsoventral axis of the lamprey pharyngeal arch (Cerny et al. 2010). This suggests that the pharyngeal arch dorsoventral polarity already existed in the vertebrate common ancestor (Medeiros and Crump 2012; Square et al. 2016b). Furthermore, recent studies have reported the nested expression of *Dlx* genes in the pharyngeal arch of elasmobranchs and paddlefish *Polyodon spathula* (i.e., a basal actinoptergyian fish), suggesting a minimal degree of neofunctionalization of *Dlx* genes over gnathostome evolution and further supporting the theory of a pharyngeal arch-derived jaw by the cooption of an ancient "*Dlx* code" (Compagnucci et al. 2013; Debiais-Thibaud et al. 2013; Takechi et al. 2013; Gillis et al. 2013; Frisdal and Trainor 2014; Square et al. 2016b).

While these studies suggest that the core components of the dorsoventral patterning program already existed in a jawless vertebrate ancestor, several key differences have also been noted. Key regulators of joint formation (*Bapx* and *Gdf5/6/7*) were found to be missing in the rostral-most pharyngeal arch of lampreys, whereas *Barx1*, which is a known repressor of joint formation, was expressed in the intermediate first arch of lampreys (Cerny et al. 2010; Kuraku et al. 2010). Similarly, an analysis of endothelin signaling in sea lamprey suggests that endothelin-mediated neural crest patterning may have functioned ancestrally to broadly pattern the dorsoventral identity of posterior pharyngeal arches, and acquired a unique function in jaw joint placement only in gnathostomes (Square et al. 2016a; Square 2017). These observations suggest that a pre-existing pharyngeal dorsoventral patterning program was co-opted to work in conjunction with novel *Bapx, Gdf5/6/7, Barx1* and *Endothelin* expression domains to give rise to articulated jaws (Medeiros and Crump 2012; Nichols et al. 2013). Given the current level of understanding, further investigations are required in order to establish a precise evolutionary relationship between lamprey and gnathostome *Dlx/Msx/Hand/Endothelin* dorsoventral patterning programs, and to determine the functional roles of *Bapx, Gdf5/6/7,* and *Barx1* during lamprey skeletal development.

6.2.6   Myelination of Vertebrate Nerves

The axons of gnathostome vertebrate nerve cells are capable of high velocity saltatory conduction due to the insulation provided by myelinated membranous sheaths that surround them. Myelination may have enhanced predatory abilities and escape response times in early vertebrates (Gans and Northcutt 1983; Ritchie 1984; Zalc and Colman 2000; Salzer and Zalc 2016). Consistent with this notion, recent work has demonstrated that chondrichthyans have true myelinated axons, and histological analysis of fossil impressions suggests that myelin may have originated in placoderms and other stem gnathostomes (de Bellard 2016; Zalc 2016). Interestingly, however, myelinated axons are absent in lampreys (Bullock et al. 1984).

In the peripheral nervous system (PNS), axons are ensheathed by myelinating Schwann cells that originate from neural crest cells (Geren 1954; Dupin et al. 1990; Le Douarin et al. 1991; Salzer and Zalc 2016). Schwann cell development involves three phases; migrating neural crest cells give rise to precursor Schwann cells; these give rise to immature Schwann cells; and finally, Schwann cells mature into myelinating and non-myelinating Schwann cells (Jessen and Mirsky 2005). Myelination requires the continuous contact and interaction between axons and Schwann cells, whereby axonal cues such as neuregulin-1 (Nrg1) are detected by the ErbB family of tyrosine kinase receptors located on Schwann cells (Meyer et al. 1997; Jessen and Mirsky 2005). Nrg1 type-III binding to ErbB2/3 receptors activates signal transduction cascades that are essential for myelination of axons (Lemke and Chao 1988; Leimeroth et al. 2002; Taveggia et al. 2005; Nave and Salzer 2006; Brinkmann et al. 2008; Birchmeier 2009; Newbern and Birchmeier 2010). One study has also highlighted the function of a G-protein coupled receptor, Gpr126, that plays a role during development in elevating cAMP levels in Schwann cells after axonal contact to trigger myelination (Monk et al. 2009).

Tetrapod peripheral myelin is characterized by the presence of highly compact regions held together by cell-cell adhesion transmembrane proteins identified as myelin protein zero (P_0). P_0 is encoded by the *myelin protein zero (mpz)* gene. The extracellular domain of P_0 adheres to other P_0 molecules across the extracellular matrix at cell-cell interfaces (Lemke et al. 1988). Myelin is generally considered to be a vertebrate innovation, although myelin-like sheaths that appear to be structurally and functionally similar have arisen independently in crustaceans and annelids through convergent evolution (Waehneldt 1990; Roots 2008). The initial steps in the evolution of myelin may have incorporated a homophilic P_0 analog to achieve an early version of an electrical seal between glial and axonal membranes (Hartline and Colman 2007). While P_0 is not essential for peripheral myelination due to its functional redundancy with *Pmp2* (peripheral myelin protein 2), it is thought to have been a key molecule for the emergence of myelin within the gnathostome lineage (Nawaz et al. 2013).

There are no extant species or fossil records that exhibit the primitive condition of myelination (Hartline and Colman 2007). However, while lampreys and hagfishes do not possess myelin, they do possess axon-neighboring glial cells that maintain close cellular contact (Bullock et al. 1984) and show P_0 immunoreactivity in the central nervous system (Waehneldt et al. 1987). Furthermore, analysis of the lamprey genome revealed the presence of a number of genes associated with myelin formation, including *Pmp22* (peripheral myelin protein 22), *Mpz* (myelin protein zero, P_0), *Mbp* (myelin basic protein), *Plp* (myelin proteolipid protein), *Mal* (myelin and lymphocyte protein), and *Myt1l* (myelin transcription factor 1-like) (Smith et al. 2013). Originally, the authors of this study suggested two evolutionary scenarios: (1) the ancestral vertebrate already possessed the molecular components of myelination and these were adapted by glial cells to form myelin in the gnathostome lineage, or alternatively, (2) the ancestral vertebrate possessed oligodendrocyte-like glial cells that were secondarily lost in the lamprey lineage (Smith et al. 2013). However, a followup study on some of these putative myelin-specific gene orthologs casts doubt on their identity. For example, the so-called *Mbp* gene in the sea lamprey genome was unlikely to be a homolog to gnathostome *Mbp*, and instead was more likely to be a *gene-of-the-oligodendrocyte-lineage* (GOLLI) family member (Werner 2013). This reinforces the notion that myelin, along with a myelin gene regulatory program, likely first appeared in gnathostomes (Werner 2013).

6.2.7 Adaptive Immunity

Lampreys have provided recent insights into evolution of the vertebrate adaptive immune system (Pancer et al. 2004, 2005; Alder et al. 2008; Guo et al. 2009; Kasamatsu et al. 2010; Litman et al. 2010; Boehm et al. 2012). Adaptive immunity in vertebrates is characterized by the presence of two types of lymphocytes, B-cells derived from bone marrow and equivalent tissues, and T-cells that develop in the thymus. B-cells produce billions of unique immunoglobulin proteins (antibodies) that recognize and bind foreign antigens. T-cells interact with cells that express a foreign antigen at their surface to elicit an immune response, dependent on expression of T-cell receptors (TCRs). The diversity of antibodies and TCRs are both dependent on activity of recombination-activating gene (RAG) proteins (Nagaoka et al. 2000; Cannon et al. 2004).

Over the past decade, independent emergence of adaptive immunity has been demonstrated in agnathans, where variable lymphocyte receptors (VLRs) are encoded at three discrete loci in the lamprey genome, known as VLRA, VLRB and VLRC (Das et al. 2013; Sutoh and Kasahara 2016; Boehm et al. 2018). Interestingly, these three VLR paralogs have also been identified in hagfish, further strengthening cyclostome monophyly (Pancer et al. 2005; Li et al. 2013; Holland et al. 2014). VLR-based adaptive immunity is similar to the TCR receptors of gnathostome vertebrates in that VLR assembly involves genetic rearrangement dependent on a cytosine deaminase (CDA) instead of RAG (Rogozin et al. 2007; Sutoh and Kasahara 2016; Boehm et al. 2018). Though these two systems arose independently in agnathans and gnathostomes, their functions depend on the activity of lymphocytes in both groups, suggesting that the evolution and development of adaptive immunity was likely dependent on cell regulatory networks present in the vertebrate common ancestor (Rast and Buckley 2013; Boehm et al. 2018). Thus, it seems that the presence of distinct T-cell and B-cell lineages for immune function may in fact be an ancestral feature for vertebrates. On the other hand, ability to undergo somatic diversification of numerous antigen receptor genes appears to have evolved independently in agnathans and gnathostomes (Boehm et al. 2018). Going forward, comparative investigations of the sea lamprey genome (Smith et al. 2013, 2018) with gnathostomes may provide additional insight into the evolution of the vertebrate adaptive immune system.

6.2.8 Programmed Genome Rearrangement

One of the assumptions in genome biology is that the large-scale structural preservation of an organism's genome is essential for proper genomic function, which in turn maintains normal development, physiology and behavior. And yet, there are a handful of metazoan species that contradict this expectation. Groups as diverse as copepods (phylum Arthropoda), roundworms (phylum Nematoda), ciliates (phylum Ciliophora), hagfishes, zebra finch *Taeniopygia guttata*, and the bandicoot *Isoodon macrourus* (a marsupial mammal), all appear to undergo large-scale changes in the structural organization of their genomes, a phenomenon termed programmed genome rearrangement, PGR (Sémon et al. 2012; Smith 2018; and references therein). In 2009, Smith and colleagues described PGR for the first time in the sea lamprey (Smith et al. 2009), and have recently found that PGR occurs in other lamprey species (Timoshevskiy et al. 2017). Remarkably, lampreys jettison ~20% of their germline genome from somatic cell lineages, with many of these eliminated fragments being hundreds of thousands of base pairs in length, and potentially include entire chromosomes (Smith et al. 2012, 2018). At the cellular level, some of these fragments appear to be marked for elimination by specific epigenetic tags (Timoshevskiy et al. 2016). Once marked, these fragments do not migrate in synchrony with the rest of the somatic genome and are subsequently packaged into small organelles and ejected from the dividing cells. Follow-up studies employing gene ontology analy-

sis have found that the majority of the eliminated genomic sequences during PGR
are related to germline development and pluripotency (Bryant et al. 2016). Recent
completion of the sea lamprey germline genome has further revealed that many of
these eliminated genomic regions are typically shut down during development in
jawed vertebrates by the polycomb repressive complex (PRC) of proteins, rather
than by means of PGR (Smith et al. 2018). This raises the interesting possibility
that the common developmental goal of targeted gene silencing during development
may have evolved by very different molecular and cellular mechanisms in jawed and
jawless vertebrates: PGR-mediated silencing in lampreys (Smith et al. 2018), and
PRC-mediated silencing in jawed vertebrates. Functional analysis of genome rearrangement during lamprey embryogenesis will help shed light on exactly how PGR
may influence developmental-genetic programs in lampreys and how this compares
with similar mechanisms of gene silencing in gnathostomes.

6.3   Techniques to Study Lamprey Development

Over the past few decades, numerous experimental techniques developed for use in
model vertebrates have been adapted for use in lampreys. These techniques have
contributed to our understanding of the basic developmental biology of lampreys,
and coupled with classical observations of lamprey development (Schultze 1856;
Scott 1887; Koltzoff 1902; Damas 1944), demonstrate that development across the
few species that have been observed is strikingly similar, often with species-specific
differences being related to differential timing of developmental events (Damas 1944;
Tahara 1988; Kuratani 1997; McCauley and Bronner-Fraser 2002). Here we highlight
some of the experimental techniques that have been adapted for use with lampreys
and have been critical for gaining insight into the evolution and development of
vertebrates.

6.3.1   Extirpation, Ablation and Transplantation

A classical technique, first used in the 19th century, has been to remove a cell or
cells of interest to determine how an embryo develops in their absence, or how the
cells develop in isolation (Chabry 1887). This information can be used to determine
the necessity of specific cells during development. Extirpation of embryonic tissue
and removal, coupled with transplantation of cells to a foreign environment, has also
been used to determine the contribution and requirement for specific populations of
cells during development. Transplantation experiments, especially, have the power
to reveal if tissues are competent to differentiate under heterotopic conditions and
are useful in determining if mechanisms that regulate development are conserved
across the agnathan-gnathostome divergence. This information can be used to infer
timing of the evolutionary origin of particular developmental mechanisms.

Ablation and transplantation of lamprey tissues were first described in a series of experiments conducted in the 1950s in which David Newth removed specific neural crest populations from embryos of the European brook lamprey (Newth 1950, 1951, 1956). Ablation resulted in a reduction in the size of cranial nerves, suggesting a neural crest contribution to cranial ganglia. Extirpation was less informative on crest contribution to the head skeleton; Newth (1951) found no reduction in development of the head skeleton, suggesting the possibility of a mesodermal origin for the lamprey viscerocranial skeleton. However, subsequent experiments in which lamprey neural crest was transplanted into a urodele host resulted in the formation of cartilage nodules originating from the lamprey cells, while extirpation of cranial neural crest resulted in absence of the branchial basket (Newth 1956). This result was subsequently confirmed by Langille and Hall (1988), supporting the neural crest origin of the lamprey pharyngeal skeleton. Similarly, ablation of the dorsal neural tube in lamprey by McCauley and Bronner-Fraser (2003) confirmed a neural crest origin for cranial melanocytes, and a more recent study involving neural tube ablation in lamprey demonstrated a trunk neural crest origin of enteric neurons (Green et al. 2017). These extirpation and heterospecific transplantation experiments suggested that the neural crest origin of the vertebrate viscerocranial skeleton predates the divergence of agnathans and gnathostome vertebrates.

6.3.2 Pharmacological and Implant Techniques

Implantation of protein-soaked beads into the lamprey embryo has been used to determine if signaling cascades present in vertebrate development are likely to be conserved in lampreys (Shigetani et al. 2002). When beads soaked in either bone morphogenetic protein 4 (BMP4) or fibroblast growth factor 8 (FGF8) were implanted into the oral region of developing Arctic lamprey, both proteins were able to upregulate endogenous lamprey putative target genes. This suggests that epithelial-mesenchymal interactions important for oral development in gnathostomes are likely to also occur in lampreys, in spite of morphological differences in these groups. However, a heterotopic shift in epithelial-mesenchymal interactions among lampreys and gnathostomes is likely to account for observed differences in the expression domains of conserved genes (Shigetani et al. 2002).

The application of pharmacological agents has also been used to decipher developmental events. The application of Retinoic Acid (RA) has been shown to affect Hox gene expression, resulting in anteroposterior transformation of rhombomere identity in the vertebrate hindbrain (Morrisskay et al. 1991; Conlon and Rossant 1992; Wood et al. 1994; Hill et al. 1995; Lopez et al. 1995; Alexandre et al. 1996). Murakami et al. (2004) used RA treatment on Arctic lamprey to suggest that positioning of branchial motoneurons is coordinated with Hox gene expression in common with other vertebrates. Similarly, pharmacological inhibition of endothelin signaling in lamprey embryos by Bosentan treatment revealed that the first pharyngeal arch in lampreys is patterned by Endothelin-mediated signaling, a mechanism similar to that which patterns the gnathostome jaw (Yao et al. 2011). This suggests that vertebrate jaw evolution was not driven by novel deployment of Endothelin signaling, but rather by changes in downstream programs (Yao et al. 2011). In a study on the evolution of FGF-mediated signaling in vertebrate pharyngeal arch development, incubation of lamprey embryos in the FGF inhibitor SU5402 resulted in impaired pharyngeal pouch outpocketing, reduction of cartilage gene expression markers (*SoxE1, Endothelin Receptor2*), and loss of alcian-blue staining cellular cartilage. This work suggests that FGF-mediated signaling during neural crest and pharyngeal arch development dates back to the last common vertebrate ancestor (Jandzik et al. 2014). In another study on evolution of patterning in the vertebrate forebrain, roles of hedgehog (Hh) and FGF signaling were tested using pharmacological inhibitors specific for Hh (cyclopamine) and FGF (SU5402) activity (Sugahara et al. 2011). Among vertebrates, the Hh paralog, Sonic Hedgehog (Shh), is expressed in prechordal mesoderm and is involved in dorsoventral patterning of the overlying telencephalon (Gunhaga et al. 2000; Fuccillo et al. 2004; Danesin et al. 2009). When Hh signaling was blocked in the lamprey using the inhibitor cyclopamine, the ventral telencephalon was reduced in size while the dorsal region was enlarged. There was also ventral expansion in expression of the dorsal specifier genes *Pax6* and *Gli*. These results suggest that dorsoventral patterning mechanisms of the telencephalon that involve hedgehog signaling may be conserved among vertebrates.

6.3.3   Lineage Tracing

Fate map studies are useful for understanding the developmental origins of tissues. Fate maps can be created by using lineage tracers to mark cells of interest, and following their developmental progression. Early studies used dyes placed on the surface of embryos to record cell movements (Vogt 1925). Modern lineage tracers are often fluorescent dextrans that can be injected into a cell and are large and/or charged and therefore unable to pass through the cell membrane. Other fluorescent dyes are lipophilic and incorporate into the cell membrane. DiI is a lipophilic fluorescent dye that has been used to follow cells for long term cell tracing both in vivo and in vitro (Honig and Hume 1986; Markus et al. 1997).

Several studies have used fluorescent dyes to examine the contributions of cranial neural crest to development of the lamprey head (Horigome et al. 1999; McCauley and Bronner-Fraser 2003; Martin et al. 2009; Häming et al. 2011; Modrell et al. 2014; Green et al. 2017). McCauley and Bronner-Fraser (2003) used DiI to demonstrate that while lamprey cranial neural crest cells migrate along three pathways, as in other vertebrates, the migration of neural crest into the presumptive branchial arches to form the pharyngeal skeleton, or branchial basket, occurs prior to formation of pharyngeal pouches such that presumptive skeletogenic neural crest cells are able to migrate along the rostrocaudal axis. Martin et al. (2009) used DiI-labeling to demonstrate the contribution of cranial neural crest to cartilage bars of the lamprey branchial basket, confirming previous extirpation and transplantation studies (Newth 1956; Langille and Hall 1988).

Among gnathostomes, pouch formation occurs prior to neural crest migration such that cells migrating into a specific pouch are prevented from migrating along the rostrocaudal axis, and are thus limited in their contribution to the pharyngeal skeleton, dependent on their location along the rostrocaudal axis. In the lamprey, neural crest cells migrating into the pharyngeal region are not initially restricted in their rostrocaudal movements, and may not be restricted in their potential to contribute to the branchial skeleton. Instead, these cells may be able to contribute to any skeletal rod that arises within the pharyngeal pouch that forms surrounding their current location following neural crest migration (McCauley and Bronner-Fraser 2003). Results from these DiI-labeling experiments suggested that the morphology of the lamprey branchial basket may not depend on the identity of specific cells contributing to a particular skeletal rod. Instead, this may suggest a key difference from gnathostomes where the identity of cells forming the pharyngeal skeletal elements may be crucial since these elements give rise to morphologically distinct skeletal structures along the rostrocaudal axis.

Häming and colleagues also used DiI labeling to examine the contribution of trunk neural crest in the developing sea lamprey (Häming et al. 2011). They showed that trunk neural crest cells form dorsal root ganglia, but there does not appear to be ventral migration of these cells to form sympathetic ganglia. Thus, previous reports that lampreys do not contain sympathetic chain ganglia (Nicol 1952) are supported by modern lineage tracing methods. Lineage tracing has the potential to inform on the developmental origin of numerous characters in the lamprey that may be key features for understanding vertebrate evolution.

More recent fate mapping studies in the Baker and Bronner laboratories have extended our understanding of neural crest development in lampreys (Modrell et al. 2014; Green et al. 2017). Modrell et al. (2014) used DiI to trace cranial ganglia development in lamprey and found that lamprey cranial ganglia appear to be patterned by a combinatorial "Pax code" and receive contributions from both neural crest and placode cells, similar to gnathostomes. Until recently, relatively little was known about the role of trunk neural crest cells during lamprey development. By combining DiI lineage tracing experiments with ablation experiments, Green et al. (2017) demonstrated that lampreys appear to lack a vagal neural crest population, which in gnathostomes contributes neural crest cells to the heart and gut. Moreover, it was found that caudal trunk neural crest cells in lampreys can form enteric neurons, just like gnathostomes, yet these cells may populate the agnathan gut in a way that is quite different from gnathostomes (Green et al. 2017). This points to a gnathostome origin for vagal neural crest cells and a new scenario for the stepwise acquisition of specialized neural crest sub-populations along the vertebrate anteroposterior axis (Green et al. 2017).

6.3.4 Molecular Tools

Prior to sequencing of the sea lamprey genome (Smith et al. 2013), identification of lamprey gene sequences required either screening genomic or complementary DNA (cDNA) libraries using heterologous probes, or by polymerase chain reaction (PCR) amplification of lamprey gene fragments using degenerate oligonucleotides. With the availability of the sea lamprey genome, identification and isolation of gene sequences has been simplified. Availability of lamprey gene sequences will facilitate the development of molecular and genetic tools to investigate the evolution of developmental gene networks in lampreys.

6.3.4.1 Detecting mRNA

Wholemount In Situ Hybridization

In situ hybridization is a tool that is used to determine the spatiotemporal pattern of genes expressed during development, and has been especially useful for evo-devo studies. In situ hybridization methods have been adapted for use in at least three lamprey species (sea lamprey, Arctic lamprey, and European river lamprey), and have allowed investigators to infer important insights into the evolution of developmental mechanisms in early vertebrates (Swain et al. 1994; Tomsa and Langeland 1999; Ogasawara et al. 2000; Myojin et al. 2001; Murakami et al. 2001; Boorman and Shimeld 2002; Derobert et al. 2002; McCauley and Bronner-Fraser 2002, 2006; Zhang et al. 2006; Sauka-Spengler et al. 2007; Nikitina et al. 2009c; Rahimi et al. 2009; Lakiza et al. 2011; Sugahara et al. 2015). In one important study, Guèrin et al. (2009) determined the developmental expression patterns of 43 genes in the developing forebrain and compared these to model organisms to show that while conserved expression patterns likely reflect features shared among all vertebrates, expression pattern differences pointed to possible changes in signaling mechanisms that were likely important in the evolution of the forebrain.

Additionally, there is evidence that sequences isolated from sea lamprey may in some cases cross-hybridize to sequences from other lampreys. For example, a riboprobe constructed from a sea lamprey *Pax3/7* gene sequence was found to hybridize to mRNA in Arctic lamprey embryos, while a muscle actin probe from Arctic lamprey cross-hybridized to transcripts in sea lamprey (Kusakabe et al. 2004; McCauley and Bronner-Fraser 2006). Guèrin et al. (2009) also found that heterologous riboprobes generated from European river lamprey and sea lamprey were able to cross-hybridize to produce identical and specific signals. These observations of cross-hybridization are consistent with conserved gene sequences for these targets and relatively recent divergence times between sea lamprey and the other two species (~16 million years; Kumar et al. 2017).

Quantitative Real Time PCR

Quantitative real time qPCR has also been used to demonstrate relative changes in gene expression levels following gene perturbation. Lakiza et al. (2011) showed that morpholino-induced knockdown of SoxE genes in sea lamprey resulted in reduced expression of the cartilage effector protein, Col2a1, confirming that SoxE regulation of Type II collagen is conserved among agnathan and gnathostome vertebrates (Zhang et al. 2006). While the cartilage found in lampreys is dependent on the presence of elastin-like lamprey-specific lamprin proteins (Wright and Youson 1983; Wright et al. 1988, 2001; Robson et al. 1993, 2000; McBurney et al. 1996), these results suggest that SoxE-dependent chondrogenic mechanisms likely arose prior to the divergence of agnathans and gnathostome vertebrates.

6.3.4.2   Genetic Tools

Due to their long lifespan and semelparous mode of reproduction (Cole 1954), lampreys are not amenable to classical "Mendelian" forward genetic studies to determine the roles of genes with developmental importance. Since lampreys die soon after spawning, germline transmission of gene constructs is not practical for establishing stable germline transgenic animals. In addition, there is an interval of at least 5 years between embryogenesis and reproduction (Potter et al. 2015). These biological constraints have limited the use of genetics as a tool to understand their biology. However, molecular tools that have been developed in other model organisms have been adapted for use in lampreys and have been particularly useful for evo-devo studies. In the following section, we highlight how several modern molecular tools have been adapted for use in lampreys to provide insight into the evolution of vertebrate development.

Transgenesis

Transgenesis is the technique of introducing exogenous DNA into an organism, either to determine the spatiotemporal expression of a gene through the use of an enhancer-reporter construct, or to introduce a gene sequence that will produce a phenotypic effect. Transgenesis may be transient, in which expression of the transgene is limited to somatic cells, such that phenotypic effects are manifest only within organisms undergoing the transgenesis procedure. Alternatively, germline transgenesis involves incorporation of the exogenous sequence into the germ line of an organism such that the transgene is heritable. Germline transgenics can be maintained as stable lines for genetic analyses.

With the advent of reverse genetic techniques for developmental studies, lampreys have become more tractable as an evo-devo model. Kuratani's group was the first to show that transient transgenic lampreys could be made to express a green fluorescent protein (GFP) reporter gene under the control of a gene-specific (actin) promoter (Kusakabe et al. 2003). More recently, a gene reporter assay was optimized for use in lampreys, which is especially useful for analyzing the potential activity of genomic cis-regulatory elements (Parker et al. 2014a). This technique was used with great success to show that a Hox-mediated gene regulatory network specifying rhombomere identity in the hindbrain is conserved between lampreys and gnathostomes (Parker et al. 2014b). The forced expression of genes under tissue-specific promoters, as well as tests for conservation and/or divergence of cis-regulatory elements, has the potential to broaden understanding of evolutionary changes in the developmental roles of genes and gene regulatory networks (Sauka-Spengler et al. 2007; Sauka-Spengler and Bronner-Fraser 2008a).

Gene Knockdown and Knockout

Gene knockdown is an experimental technique that has been adapted for use in lampreys and is useful for understanding the developmental roles of specific genes in an evolutionary context (McCauley and Bronner-Fraser 2006; Lakiza et al. 2011). Gene knockdown techniques are examples of "reverse genetics" in which genes of interest are perturbed in function by preventing accumulation of specific protein products in order to determine the phenotypes that arise from specific gene sequences. Two such knockdown techniques that have been used in lampreys include microinjection of morpholinos and RNA interference (RNAi).

Morpholinos are synthetic oligonucleotides that disrupt translation initiation of the mRNA message, or alternatively can be engineered to disrupt proper splicing of a pre-messsenger RNA sequence. McCauley and Bronner-Fraser (2006) first showed that morpholinos could be used in lampreys to perturb a specific gene of interest (*SoxE1*) that is required for development of the cartilaginous branchial basket. Lakiza et al. (2011) used morpholinos to show that closely related gene duplicates (*SoxE1, SoxE2* and *SoxE3*) maintain specific roles during development of the craniofacial skeleton. Sauka-Spengler and colleagues used morpholino knockdown of key gene sequences to demonstrate that a neural crest gene regulatory network is present in lampreys, indicating that the NC-GRN predated the divergence of agnathan and gnathostome vertebrates (Sauka-Spengler et al. 2007; Sauka-Spengler and Bronner-Fraser 2008a, b). By taking advantage of the slow development of lamprey embryos, morpholinos have also been used to tease apart some of the earliest stages of neural crest development in agnathans (Nikitina et al. 2008).

RNA interference (RNAi) is an endogenous intracellular mechanism to regulate gene expression via the targeted degradation of specific mRNA transcripts (Mello and Conte 2004). RNAi has gained widespread use as a tool for understanding gene function and has recently been shown to perturb lamprey development (Heath et al. 2014). Investigated as a possible species-specific tool for use in sea lamprey control measures in the Great Lakes, small interfering RNA (siRNA) sequences injected into lamprey embryos were found to reduce target transcript levels by more than 50% (Heath et al. 2014). In the same study, it was also found that delivery of siRNA to lamprey larvae via feeding caused increased mortality. Further advances in RNAi technology as a tool for understanding lamprey development may provide important new insight into the evolution of vertebrate developmental mechanisms.

In contrast to gene knockdown via morpholinos or RNAi, recent advances in genetic technology have allowed researchers for the first time to precisely induce transient or trans-generational changes to an organism's genome. Although several approaches were initially developed along these lines (e.g., TALEN, zinc-finger nucleases), the CRISPR/Cas9 system has emerged as the most effective, cost-efficient and adaptable means by which to alter genomic DNA (Kunin et al. 2007; Urnov et al. 2010; Joung and Sander 2013; Ran et al. 2013; Doudna and Charpentier 2014; Sander and Joung 2014). Based on a relatively simple experimental design, it is now possible to use CRISPR/Cas9 technology to probe gene function in both standard animal models and non-traditional models, including lampreys. Given its relative ease of implementation, cost-effectiveness, and the ability to efficiently induce direct genomic modifications, the CRISPR/Cas9 system has gradually replaced morpholino and RNAi technology as the first choice for investigations of gene function and is quickly becoming an indispensable technique in the experimental toolkit of developmental biologists.

The Medeiros laboratory was the first to successfully demonstrate the application of CRISPR/Cas9 in lamprey embryos (Square et al. 2015). They first targeted the genomic coding sequence of *Tyrosinase* (*Tyr*), an enzyme that catalyzes the production of melanin in pigment cells (Square et al. 2015). By disrupting *Tyr*, they obtained a relatively high number of albino larvae and linked these phenotypes to mutations at the targeted *Tyr* locus (Square et al. 2015). It was then shown that targeted genomic disruptions via CRISPR could be easily extended to other loci, including *FGF8/17/18*, the mutation of which resulted in reduced expression of *SoxE1* and *Mef2* in pharyngeal chondrocytes and muscle, respectively (Square et al. 2015). Similarly, Weiming Li and colleagues showed the versatility of the CRISPR/Cas9 system in lamprey embryos by efficiently inducing indels in the *Golden* (*gol*) *Kctd10*, *Wee1*, *SoxE2*, and *Wnt7b* loci (Zu et al. 2016). The McCauley laboratory recently used CRISPR in lamprey embryos to test the functional role of the *Snail* gene during neural crest development and found that *Snail* was essential for activation of genes governing early neural crest migration and the formation of neural crest derivatives including cranial sensory neurons and head skeleton (York et al. 2017). These early studies demonstrate the power of the CRISPR/Cas9 system for investigating gene function in lamprey embryos.

6.4 Laboratory Culture of Sea Lamprey Embryos

One of the impediments to the widespread use of lampreys as a model for evo-devo research has been the relative difficulty in obtaining embryos for research purposes. Many lamprey species appear to be in decline, making the collection of spawning adults either impractical, or impermissible (Renaud 2011; Maitland et al. 2015). While sea lamprey are abundant in certain locations, their spawning season along the Atlantic coast and throughout the Great Lakes occurs over a brief period in the spring. Historically, raising lamprey embryos from newly fertilized eggs to the pro-ammocoete burrowing larval stage (Piavis 1961, 1971) has been tedious, requiring daily care in order to maintain healthy embryos. Culture methods used by Piavis and colleagues were sufficient to raise several hundred embryos in a monolayer on the bottom of a single dish for use in small-scale experiments (Piavis 1961; Smith et al. 1968; Piavis and Howell 1969; Piavis et al. 1970). However, as the importance of lampreys has increased as a model for understanding the evolution of vertebrate developmental mechanisms (Shimeld and Donoghue 2012; McCauley et al. 2015), methods are needed for cultivating larger numbers of embryos from the tens of thousands of eggs that can be obtained from an individual female sea lamprey (Hardisty 1971; see Chap. 1). In the following section, we describe techniques to obtain spawning adults, and to fertilize and rear large numbers of sea lamprey embryos under ordinary laboratory conditions. Additional methods to culture embryos of other lamprey species are provided in this volume by Moser and colleagues (see Chap. 2).

6.4.1 Procurement of Spawning Adults

Several methods have been used to obtain sea lamprey gametes from spawning adult animals in the United States and Canada. During the spawning season, it is possible to collect adult sea lamprey directly from nests, return to the laboratory, and obtain eggs and sperm for artificial in vitro fertilization. Embryos are then reared under conditions as described previously (Piavis 1961, 1971). However, a limitation of this method is the inaccessibility to nest sites for researchers whose laboratories are not located near spawning habitats (e.g., the Great Lakes and Lake Champlain).

Sea lamprey complete gamete maturation in response to elevated water temperatures in the spring (Johnson et al. 2015). Prior to maturation and spawning, invasive sea lamprey undergo upstream migration into streams and tributaries of the Great Lakes to reach spawning habitats. Through cooperation with Fisheries and Oceans Canada and the U.S. Fish and Wildlife Service, migratory pre-spawning animals are trapped as a part of management practices that are overseen by the Great Lakes Fishery Commission (GLFC). Following their collection, animals are transported to the Hammond Bay Biological Station (Millersburg, MI). Animals that are not required for GLFC-funded research are then made available to researchers upon request, pending their availability. Once the proper import permits have been obtained, where required, these animals can be shipped and held under conditions that promote gamete maturation (Nikitina et al. 2009b). Upon receipt, pre-spawning adults are housed at ~12 °C and allowed to acclimate to their new environment. The water temperature is then gradually raised to 20 °C. As a result, sea lamprey will complete gamete maturation, with males undergoing spermiation and females becoming ovulatory (see Chap. 1). If multiple tanks are available to house adults, a stock population can be held at 12 °C and then allowed to mature as needed. An advantage to this strategy is that the timing of maturation can be controlled. Nikitina et al. (2009b) also describe a commercially available system for housing adult lamprey.

A different strategy to obtain spawning adults was developed based on field observations that temperature control was the key to shipping spawning lamprey. The routine method for shipment of sea lamprey from Hammond Bay Biological Station has been to pack the animals in ice with a minimal amount of water. While this method has proved successful for shipping non-spawning (parasitic and migratory) sea lamprey, eggs exposed to near-freezing conditions can be fertilized and will raise a fertilization membrane (suggesting the initial events of fertilization occur) but do not cleave (David W. McCauley, personal observation). It was noted that nighttime stream temperatures are documented to reach 8 °C during the spawning season (Erik Larson, U.S. Geological Survey, Hammond Bay Biological Station, Millersburg, MI, personal communication, 2008), suggesting spawning sea lamprey can tolerate this temperature without detrimental effects on gametes (see also Chap. 2). We have found that spawning sea lamprey can be shipped successfully with minor adjustment to shipping conditions. The shipping water temperature is adjusted to a minimum of 8 °C, and then heavily oxygenated before sealing the shipping bag. We have also found that heavily insulated coolers rated to maintain ice for up to 5 days at 90 °F (e.g., Igloo Maxcold® or Coleman Extreme®) are ideal shipping containers and are able to support successful shipping of up to six spawning adults in a single container. Anecdotal evidence suggests female sea lamprey are more sensitive to shipping conditions and as a result are less hardy and show higher transit mortality.

Interestingly, we have found that in many instances, eggs removed from a recently deceased female can still be fertilized successfully, likely dependent on the time interval between death and egg removal. Eggs from recently deceased females (and also healthy live females) can be expressed dry directly into a petri dish (by first drying the exterior of the female around the cloaca so that the eggs are not activated by exposure to water). The petri dish is then sealed with parafilm and stored at 11 °C. Under these conditions, we have found that eggs may remain fertile for up to 4 days, but with decreasing viability (see also Ciereszko et al. 2000).

6.4.2 In Vitro Fertilization

Methods for in vitro fertilization of sea lamprey eggs are provided elsewhere (Ciereszko et al. 2000; Nikitina et al. 2009a; see also Chap. 2). However, differences in handling conditions related to successful cultivation of embryos (Nikitina et al. 2009a), from those described here, suggest that individual experience may differ for successful cultivation of lamprey embryos. Nevertheless, the successful fertilization and cultivation of lamprey embryos is straightforward and can be accomplished where basic aquatic resources are available.

6.4.3   High-Density Embryo Culture

Increased viability of embryos, and an increase in the density of embryos per dish, is possible over methods described in the past (Piavis 1961; Piavis and Howell 1969; Piavis et al. 1970; Nikitina et al. 2009a; see also Chap. 2). Previous methods emphasized the need to keep embryos undisturbed and at low density (a monolayer on bottom of a dish). In contrast, we have found that following gastrulation (Piavis stage 9, P9; Tahara stage 12, T12) (Piavis 1961; Tahara 1988), embryos held within an enclosed recirculating dish, in which the water flow creates continual turbulence, are able to be raised at a density of ~5,000 embryos inside a single Pyrex® custard dish (236 mL) with >90% viability (Fig. 6.1). Hatching jars can be made by modifying the plastic lid of the Pyrex® custard dishes that may be purchased locally. A small hole drilled into the center of the lid allows for insertion of a tube for incurrent water, while larger holes are drilled and covered with fine nylon mesh to allow release of excurrent water and prevent escape of embryos. This creates turbulence that results in constant rotation of eggs within the dish. Continuous rotation of embryos in these hatching jars is similar to the action of McDonald-type hatching jars used in aquaculture labs and allows for constant oxygenation of embryos. Hatching dishes can be maintained in troughs that allow for constant water flow over embryos, and recirculation through a filtration system. After hatching (~10 days post-fertilization at 18–19 °C), prolarvae (Piavis stages 14–17; Piavis 1961) are transferred into larger dishes with low-flow running water which permits constant water flow over the larvae without the turbulence. In addition to the higher numbers of embryos that can be reared under turbulent conditions, they also are less susceptible to fungus that can form on the eggs in the absence of flow.

6.4.4   Flow-Through Recirculating System for Embryo Culture

A small footprint flow-through recirculating system can be constructed using parts purchased from local suppliers, with the only specialized components being Plexiglas® or acrylic sheeting for assembly of water troughs to hold the rearing dishes, and an aquarium chiller to maintain the system at a constant 18–19 °C. We have found that landscape drip irrigation components may be easily adapted for use in constructing a recirculating system. A diagram of one possible recirculating system configuration is shown in Fig. 6.2. All components can be housed on steel shelving (A) (48″ W × 18″ D × 72″ H). Filtration equipment is contained on the bottom shelf. A 20-gal aquarium or other suitable container may be used as a sump (B), from which water is drawn into a pump (C) and forced through particle (D) and carbon (E) filters. Filter housings can be adapted for use from whole-house filters available from plumbing suppliers. Filter media may be obtained from aquatic suppliers (e.g., Pentair-Aquatic Ecosystems®). A UV sterilizer (F) is placed in front of a chiller (G)

Fig. 6.1 Lamprey hatching dish modified from a 470-mL custard dish (Pyrex® 7200). Views are shown from above (**a**) and from the side (**b**). Two 25-mm holes are drilled through the lid and Nitex mesh is glued over both openings on the inside of the lid. A 1.5-cm^2 piece of acrylic is glued onto the center of the lid. A 4-mm hole drilled through the center of the acrylic and lid is used to support a length of 1-mL serological pipette attached to ¼″ tubing through which water enters the dish. Water exits through the two holes while embryos are prevented from escape by the Nitex mesh. Flow rate is adjusted so that embryos circulate continuously within the dish

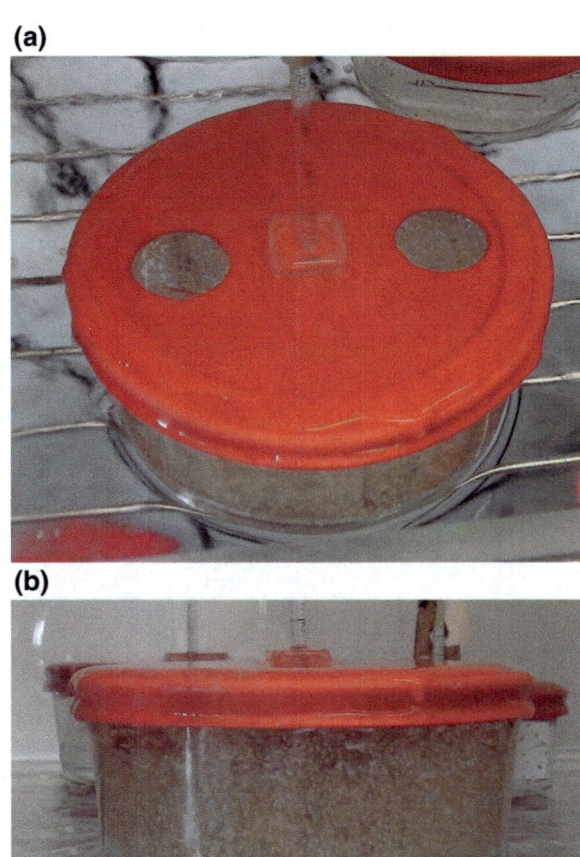

to maintain temperature at a constant 18–19 °C. The riser and distribution manifolds (*H*) above each shelf are made from ½″ thin-wall landscape irrigation tubing joined using ½″ poly lock T-fittings (*I*) and elbows (*J*) and capped at each end with a ½″ poly tubing end fitting (*L*). Each distribution manifold contains ¼″ barb × barb couplers (*M*) fitted into the ½″ tubing using a hole punch. Individual couplers are connected to ¼″ drip irrigation tubing (*O*) to distribute water to individual hatching dishes. The flow rate can be regulated to each hatching dish (*P*) using ¼″ on/off valves (*N*) inserted in each out-flow tube. Hatching dishes on each shelf sit within an acrylic trough to receive the outflow from each dish. A 1″ PVC pipe is inserted into a hole drilled into the bottom of each trough which allows water to drain into the trough immediately below, with water returned to the sump through the 1″ PVC drain in the bottom trough. Alternatively, we have also fabricated a recirculating system from a repurposed zebrafish rack (Pentair-Aquatic Ecosystems®) as shown in Fig. 6.3, for

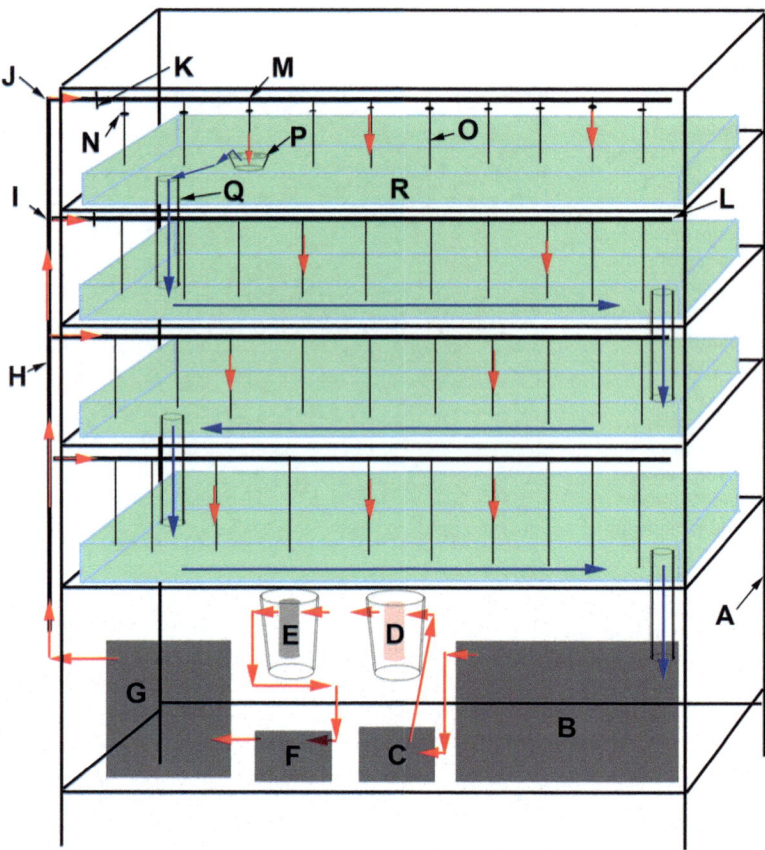

A	Stainless steel rack with wire shelving	J	1/2"Poly locking collar elbow fitting
B	Sump/holding tank (20 gal aquarium)	K	1/2" poly on/off valve
C	Pump (Little Giant 4-MDQX-SC)	L	1/2" poly tubing end fitting
D	Housing for particle filter (Whirlpool WHKF-DWHV)	M	1/4" barb x barb coupler
E	Housing for carbon filter (Whirlpool WHKF-WHWC)	N	1/4" on/off valve
F	UV sterilizer	O	1/4" vinyl outflow tubing
G	Chiller (Oceanic systems 1/6 HP)	P	Pyrex dish (hatching dish)
H	1/2" thin wall tubing (Mister Landscaper)	Q	1" PVC drain pipe
I	1/2"Poly locking collar Tee fitting	R	Water tight Acrylic shelf tray

Fig. 6.2 Recirculating water system for raising lamprey embryos. Pyrex® hatching dish is described in Fig. 6.1; arrows indicate the direction of water flow. *Red arrows* represent water flow through the filtration equipment into the distribution manifolds above each shelf. *Blue arrows* indicate the pathway of effluent water returning to the sump following passage through hatching dishes. Acrylic effluent catch troughs are colored *green*

Fig. 6.3 Zebrafish rack (Pentair-Aquatic Ecosystems®) repurposed to raise lamprey embryos. Stock fish tanks were removed from the rack and replaced with custom-fabricated sealed acrylic troughs. Individual lamprey dishes are fed with water distributed from each manifold through the original tubing. Water is drained from each trough through 3″ bulkhead fittings and then passed into a 4″ return that empties into the sump (white PVC pipe shown at right). Constant water temperature (18 °C) is maintained using an Aqua Logic Cyclone® 1/4 HP chiller with a drop-in titanium coil placed in the sump

which acrylic shelves, return drains to the sump tank, and a chiller, were the only additional modifications required.

The number of hatching dishes that can be held on each shelf is dependent on the number of outlets added to each distribution manifold. Using the system shown in Fig. 6.2, we have reared between 12 and 20 dishes per shelf simultaneously. The system shown in Fig. 6.3 is able to accommodate 84 hatching dishes simultaneously, with each dish capable of holding genetically distinct embryos, and each containing up to 5,000 embryos. Thus, this system has the capacity for rearing ~400,000 sea lamprey embryos simultaneously through the hatching stage, and should be able to accommodate the research needs of a small laboratory.

6.5 Conclusions

Lampreys have already made important contributions to our understanding of vertebrate evolution, and the origins of numerous characters. Going forward, removing the barriers that have prevented easy access to embryonic and larval lampreys, coupled with facilitating the development and adaptation of modern genetic techniques for use in lampreys, will be critical to advance the widespread adoption of this animal as an evo-devo model. We are optimistic that removing these barriers will catalyze rapid growth in the lamprey evo-devo community and we suggest that the lamprey is poised to make a powerful impact in decoding the evolution of vertebrate developmental mechanisms in the coming decades.

Acknowledgments DWM has received funding from the National Science Foundation, National Institutes of Health, and the University of Oklahoma.

References

Abitua PB, Wagner E, Navarrete IA, Levine M (2012) Identification of a rudimentary neural crest in a non-vertebrate chordate. Nature 492:104–107

Abitua PB, Gainous TB, Kaczmarcyk AN et al (2015) The pre-vertebrate origins of neurogenic placodes. Nature 524:462–465

Adachi N, Robinson M, Goolsbee A, Shubin NH (2016) Regulatory evolution of Tbx5 and the origin of paired appendages. Proc Natl Acad Sci USA 113:10115–101120

Aitken GJ (1954) The Welland Canal Company. A study in Canadian enterprise. Studies in Entrepreneurial History. Harvard University Press, Cambridge, MA

Akoev GN, Muraveiko VM (1984) Physiological properties of lateral line receptors of the lamprey. Neurosci Lett 49:171–173

Alder MN, Herrin BR, Sadlonova A et al (2008) Antibody responses of variable lymphocyte receptors in the lamprey. Nat Immunol 9:319–327

Alexandre D, Clarke JDW, Oxtoby E et al (1996) Ectopic expression of Hoxa-1 in the zebrafish alters the fate of the mandibular arch neural crest and phenocopies a retinoic acid-induced phenotype. Development 122:735–746

Applegate VC (1950) Natural history of the sea lamprey, *Petromyzon marinus*, in Michigan. US Fish Wildl Serv Spec Sci Rep Fish 55:1–237

Applegate VC, King EL (1962) Comparative toxicity of 3-trifluoromethyl-4-nitrophenol (TFM) to larval lamprey and eleven species of fishes. Trans Am Fish Soc 91:342–345

Applegate VC, Howell JH, Moffett JW, Johnson GH, Smith MA (1961) Use of 3-trifluoromethyl-4-nitrophenol as a selective sea lamprey larvicide. Great Lakes Fish Comm Tech Rep 1:1–35

Baker CV, Bronner-Fraser M (1997) The origins of the neural crest. Part II: an evolutionary perspective. Mech Dev 69:13–29

Baker CV, Bronner-Fraser M (2001) Vertebrate cranial placodes I. Embryonic induction. Dev Biol 232:1–61

Baker CV, Modrell MS, Gillis JA (2013) The evolution and development of vertebrate lateral line electroreceptors. J Exp Biol 216:2515–2522

Barald KF, Kelley MW (2004) From placode to polarization: new tunes in inner ear development. Development 131:4119–4130

Bardack D, Zangerl R (1968) First fossil lamprey: a record from the Pennsylvanian of Illinois. Science 162:1265–1267

Beard J (1885) The system of branchial sense organs and their associated ganglia in Ichthyopsida. A contribution to the ancestral history of vertebrates. Q J Microsc Sci 26:95–156

Bejder L, Hall BK (2002) Limbs in whales and limblessness in other vertebrates: mechanisms of evolutionary and developmental transformation and loss. Evol Dev 4:445–458

Bertrand S, Escriva H (2011) Evolutionary crossroads in developmental biology: amphioxus. Development 138:4819–4830

Birchmeier C (2009) ErbB receptors and the development of the nervous system. Exp Cell Res 315:611–618

Boehm T, McCurley N, Sutoh Y et al (2012) VLR-based adaptive immunity. Annu Rev Immunol 30:203–220

Boehm T, Hirano M, Holland SJ et al (2018) Evolution of alternative adaptive immune systems in vertebrates. Annu Rev Immunol 36:19–42

Bok J, Chang W, Wu DK (2007) Patterning and morphogenesis of the vertebrate inner ear. Int J Dev Biol 51:521–533

Boorman CJ, Shimeld SM (2002) Cloning and expression of a Pitx homeobox gene from the lamprey, a jawless vertebrate. Dev Genes Evol 212:349–353

Breder CM (1926) The locomotion of fishes. Zoologica 4:159–297

Brinkmann BG, Agarwal A, Sereda MW et al (2008) Neuregulin-1/ErbB signaling serves distinct functions in myelination of the peripheral and central nervous system. Neuron 59:581–595

Bryant SA, Herdy JR, Amemiya CT, Smith JJ (2016) Characterization of somatically-eliminated genes during development of the sea lamprey (*Petromyzon marinus*). Mol Biol Evol 33:2337–2344

Bullock TH, Moore JK, Fields RD (1984) Evolution of myelin sheaths: both lamprey and hagfish lack myelin. Neurosci Lett 48:145–148

Burighel P, Lane NJ, Fabio G et al (2003) Novel, secondary sensory cell organ in ascidians: in search of the ancestor of the vertebrate lateral line. J Comp Neurol 461:236–249

Burighel P, Caicci F, Zaniolo G et al (2008) Does hair cell differentiation predate the vertebrate appearance? Brain Res Bull 75:331–334

Caicci F, Burighel P, Manni L (2007) Hair cells in an ascidian (Tunicata) and their evolution in chordates. Hear Res 231:63–72

Caicci F, Degasperi V, Gasparini F et al (2010) Variability of hair cells in the coronal organ of ascidians (Chordata, Tunicata). Can J Zool 88:567–578

Caicci F, Gasparini F, Rigon F et al (2013) The oral sensory structures of Thaliacea (Tunicata) and consideration of the evolution of hair cells in Chordata. J Comp Neurol 521:2756–2771

Cannon JP, Haire RN, Rast JP, Litman GW (2004) The phylogenetic origins of the antigen-binding receptors and somatic diversification mechanisms. Immunol Rev 200:12–22

Carroll RL (1988) Vertebrate paleontology and evolution. WH Freeman & Company, New York

Cattell M, Lai S, Cerny R, Medeiros DM (2011) A new mechanistic scenario for the origin and evolution of vertebrate cartilage. PLoS ONE 6:e22474

Cerny R, Cattell M, Sauka-Spengler T et al (2010) Evidence for the prepattern/cooption model of vertebrate jaw evolution. Proc Natl Acad Sci USA 107:17262–17267

Chabry L (1887) Embryologie normale et teratologique des Ascidies simples. J Anat Physiol 23:167–319

Chang M-m WuF, Miao D, Zhang J (2014) Discovery of fossil lamprey larva from the Lower Cretaceous reveals its three-phased life cycle. Proc Natl Acad Sci USA 111:15486–15490

Chatterjee S, Kraus P, Lufkin T (2010) A symphony of inner ear developmental control genes. BMC Genet 11:68

Chen J, Streit A (2013) Induction of the inner ear: stepwise specification of otic fate from multipotent progenitors. Hear Res 297:3–12

Christ B, Huang R, Scaal M (2004) Formation and differentiation of the avian sclerotome. Anat Embryol 208:333–350

Ciereszko A, Glogowski J, Dabrowski K (2000) Fertilization in landlocked sea lamprey: storage of gametes, optimal sperm: egg ratio, and methods of assessing fertilization success. J Fish Biol 56:495–505

Coates MI (1994) The origin of vertebrate limbs. Dev Suppl 1994:169–180

Coates MI, Cohn MJ (1999) Vertebrate axial and appendicular patterning: the early development of paired appendages. Am Zool 39:676–685

Cohn MJ, Tickle C (1999) Developmental basis of limblessness and axial patterning in snakes. Nature 399:474–479

Cole AG (2011) A review of diversity in the evolution and development of cartilage: the search for the origin of the chondrocyte. Eur Cell Mater 21:122–129

Cole AG, Hall BK (2004) Cartilage development: insights from studies on invertebrate taxa. Dev Biol 271:601–601

Cole LC (1954) The population consequences of life history phenomena. Q Rev Biol 29:103–137

Compagnucci C, Debiais-Thibaud M, Coolen M et al (2013) Pattern and polarity in the development and evolution of the gnathostome jaw: both conservation and heterotopy in the branchial arches of the shark, Scyliorhinus canicula. Dev Biol 377:428–448

Conlon RA, Rossant J (1992) Exogenous retinoic acid rapidly induces anterior ectopic expression of murine Hox-2 genes in vivo. Development 116:357–368

Cope ED (1889) Synopsis of the families of Vertebrata. Am Nat 23:1–29

Crowe WR (1975) Great Lakes Fishery Commission history, program, and progress. Great Lakes Fishery Commission, Ann Arbor, MI

Damas H (1944) Recherches sur le developpment de Lampetra fluviatilis. Archs Biol Paris 55:1–284

Danesin C, Peres JN, Johansson M et al (2009) Integration of telencephalic Wnt and hedgehog signaling center activities by Foxg1. Dev Cell 16:576–587

Das S, Hirano M, Aghaallaei N, Bajoghli B, Boehm T, Cooper MD (2013) Organization of lamprey variable lymphocyte receptor C locus and repertoire development. Proc Natl Acad Sci USA 110:6043–6048

De Beer GR (1924) Growth. E Arnold & Co, London

De Beer GR (1937) The development of the vertebrate skull. Oxford University Press, London

de Bellard ME (2016) Myelin in cartilaginous fish. Brain Res 1641:34–42

Debiais-Thibaud M, Metcalfe CJ, Germon I et al (2013) Heterogeneous conservation of Dlx paralog co-expression in jawed vertebrates. PLoS ONE 8:e68182

Delarbre C, Gallut C, Barriel V, Janvier P, Gachelin G (2002) Complete mitochondrial DNA of the hagfish, Eptatretus burgeri: the comparative analysis of mitochondrial DNA sequences strongly supports the cyclostome monophyly. Mol Phylogenet Evol 22:184–192

Depew MJ, Lufkin T, Rubenstein JLR (2002) Specification of jaw subdivisions by Dix genes. Science 298:381–385

Derobert Y, Baratte B, Lepage M, Mazan S (2002) Pax6 expression patterns in Lampetra fluviatilis and Scyliorhinus canicula embryos suggest highly conserved roles in the early regionalization of the vertebrate brain. Brain Res Bull 57:277–280

Docker MF, Hume JB, Clemens BJ (2015) Introduction: a surfeit of lampreys. In: Docker MF (ed) Lampreys: biology, conservation and control, vol 1. Springer, Dordrecht, pp 1–34

Dial KP, Shubin N, Brainerd EL (2015) Great transformations in vertebrate evolution. University of Chicago Press, Chicago, IL

Don EK, Currie PD, Cole NJ (2013) The evolutionary history of the development of the pelvic fin/hindlimb. J Anat 222:114–133

Donoghue PCJ, Forey PL, Aldridge RJ (2000) Conodont affinity and chordate phylogeny. Biol Rev 75:191–251

Donoghue PC, Graham A, Kelsh RN (2008) The origin and evolution of the neural crest. BioEssays 30:530–541

Doudna JA, Charpentier E (2014) The new frontier of genome engineering with CRISPR-Cas9. Science 346:1258096

Drucker EG, Lauder GV (2002) Wake dynamics and locomotor function in fishes: interpreting evolutionary patterns in pectoral fin design. Integr Comp Biol 42:997–1008

Duméril AMC (1806) Zoologie analytique, ou méthode de classification des animaux; rendue plus facile à l'aide de tableaux synoptiques. Didot, Paris

Dupin E, Baroffio A, Dulac C, Cameron-Curry P, Le Douarin NM (1990) Schwann-cell differentiation in clonal cultures of the neural crest, as evidenced by the anti-Schwann cell myelin protein monoclonal antibody. Proc Natl Acad Sci USA 87:1119–1123

Dupret V, Sanchez S, Goujet D, Tafforeau P, Ahlberg PE (2014) A primitive placoderm sheds light on the origin of the jawed vertebrate face. Nature 507:500–503

Eisner T (2003) Living fossils: on lampreys, baronia, and the search for medicinals. BioScience 53:265–269

Fetterolf CM (1980) Why a Great Lakes Fishery Commission and why a Sea Lamprey International Symposium. Can J Fish Aquat Sci 37:1588–1593

Freitas R, Zhang GJ, Cohn MJ (2006) Evidence that mechanisms of fin development evolved in the midline of early vertebrates. Nature 442:1033–1037

Freitas R, Gomez-Marin C, Wilson JM, Casares F, Gomez-Skarmeta JL (2012) Hoxd13 contribution to the evolution of vertebrate appendages. Dev Cell 23:1219–1229

Frisdal A, Trainor PA (2014) Development and evolution of the pharyngeal apparatus. Wiley Interdiscip Rev Dev Biol 3:403–418

Fritzsch B, Beisel KW (2001) Evolution and development of the vertebrate ear. Brain Res Bull 55:711–721

Fritzsch B, Signore M, Simeone A (2001) Otx1 null mutant mice show partial segregation of sensory epithelia comparable to lamprey ears. Dev Genes Evol 211:388–396

Fritzsch B, Pauley S, Beisel KW (2006) Cells, molecules and morphogenesis: the making of the vertebrate ear. Brain Res 1091:151–171

Fuccillo M, Rallu M, McMahon AP, Fishell G (2004) Temporal requirement for hedgehog signaling in ventral telencephalic patterning. Development 131:5031–5040

Furlong RF, Holland PWH (2002) Bayesian phylogenetic analysis supports monophyly of ambulacraria and of cyclostomes. Zool Sci 19:593–599

Gabbott SE, Donoghue PC, Sansom RS et al (2016) Pigmented anatomy in Carboniferous cyclostomes and the evolution of the vertebrate eye. Proc R Soc B 283:20161151

Gadow H (1933) The evolution of the vertebral column. The University Press, Cambridge

Gai ZK, Donoghue PCJ, Zhu M, Janvier P, Stampanoni M (2011) Fossil jawless fish from China foreshadows early jawed vertebrate anatomy. Nature 476:324–327

Gans C, Northcutt RG (1983) Neural crest and the origin of vertebrates: a new head. Science 220:268–273

Gasparini F, Degasperi V, Shimeld SM, Burighel P, Manni L (2013) Evolutionary conservation of the placodal transcriptional network during sexual and asexual development in chordates. Dev Dyn 242:752–766

Gee H (2007) Before the backbone: views on the origin of the vertebrates. Springer, Dordrecht

Gegenbaur C, Bell FJ, Lankester ER (1878) Elements of comparative anatomy. Macmillan and Company, London

Gehrke AR, Shubin NH (2016) Cis-regulatory programs in the development and evolution of vertebrate paired appendages. Semin Cell Dev Biol 57:31–39

Gelman S, Ayali A, Tytell ED, Cohen AH (2007) Larval lampreys possess a functional lateral line system. J Comp Physiol A Neuroethol Sens Neural Behav Physiol 193:271–277

Geren BB (1954) The formation from the Schwann cell surface of myelin in the peripheral nerves of chick embryos. Exp Cell Res 7:558–562

Gess RW, Coates MI, Rubidge BS (2006) A lamprey from the Devonian period of South Africa. Nature 443:981–984

Gillis JA, Modrell MS, Baker CV (2013) Developmental evidence for serial homology of the vertebrate jaw and gill arch skeleton. Nat Commun 4:1436

Goodrich ES (1930) Studies on the structure and development of the vertebrates. Macmillan, London

Graham A, Begbie J (2000) Neurogenic placodes: a common front. Trends Neurosci 23:313–316

Graham A, Shimeld SM (2013) The origin and evolution of the ectodermal placodes. J Anat 222:32–40

Green SA, Bronner ME (2013) Gene duplications and the early evolution of neural crest development. Semin Cell Dev Biol 24:95–100

Green SA, Bronner ME (2014) The lamprey: a jawless vertebrate model system for examining origin of the neural crest and other vertebrate traits. Differentiation 87:44–51

Green SA, Simoes-Costa M, Bronner ME (2015) Evolution of vertebrates as viewed from the crest. Nature 520:474–482

Green SA, Uy BR, Bronner ME (2017) Ancient evolutionary origin of vertebrate enteric neurons from trunk-derived neural crest. Nature 544:88–91

Groves AK, Fekete DM (2012) Shaping sound in space: the regulation of inner ear patterning. Development 139:245–257

Guèrin A, d'Aubenton-Carafa Y, Marrakchi E et al (2009) Neurodevelopment genes in lampreys reveal trends for forebrain evolution in craniates. PLoS ONE 4:e5374

Gunhaga L, Jessell TM, Edlund T (2000) Sonic hedgehog signaling at gastrula stages specifies ventral telencephalic cells in the chick embryo. Development 127:3283–3293

Guo P, Hirano M, Herrin BR et al (2009) Dual nature of the adaptive immune system in lampreys. Nature 459:796–801

Haeckel E (1866) Generelle Morphologie der Organismen. Georg Reimer, Berlin

Hall BK (1999) The neural crest in development and evolution. Springer, New York

Hall BK, Gillis JA (2013) Incremental evolution of the neural crest, neural crest cells and neural crest-derived skeletal tissues. J Anat 222:19–31

Häming D, Simoes-Costa M, Uy B et al (2011) Expression of sympathetic nervous system genes in lamprey suggests their recruitment for specification of a new vertebrate feature. PLoS ONE 6:e26543

Hammond KL, Whitfield TT (2006) The developing lamprey ear closely resembles the zebrafish otic vesicle: otx1 expression can account for all major patterning differences. Development 133:1347–1357

Hammond KL, Loynes HE, Mowbray C et al (2009) A late role for bmp2b in the morphogenesis of semicircular canal ducts in the zebrafish inner ear. PLoS ONE 4:e4368

Hardisty MW (1971) Gonadogenesis, sex differentiation, and gametogenesis. In: Hardisty MW, Potter IC (eds) The biology of lampreys, vol 1. Academic Press, London, pp 295–359

Hartline DK, Colman DR (2007) Rapid conduction and the evolution of giant axons and myelinated fibers. Curr Biol 17:R29–35

Heath G, Childs D, Docker MF, McCauley DW, Whyard S (2014) RNA interference technology to control pest sea lampreys–a proof-of-concept. PLoS ONE 9:e88387

Hecht J, Stricker S, Wiecha U et al (2008) Evolution of a core gene network for skeletogenesis in chordates. PLoS Genet 4:e1000025

Hedges SB, Marin J, Suleski M, Paymer M, Kumar S (2015) Tree of life reveals clock-like speciation and diversification. Mol Biol Evol 32:835–845

Heimberg AM, Cowper-Sal-lari R, Sémon M, Donoghue PC, Peterson KJ (2010) microRNAs reveal the interrelationships of hagfish, lampreys, and gnathostomes and the nature of the ancestral vertebrate. Proc Natl Acad Sci USA 107:19379–19383

Hennig W (1950) Grundzüge einer Theorie der Phylogenetischen Systematik. Deutscher Zentralverlag, Berlin

Hertwig O (1874) Über Bau und Entwicklung der Placoidschuppen und der Zähne der Selachier. Jen Z Naturwiss 8:331–404

Hill J, Clarke JDW, Vargesson N, Jowett T, Holder N (1995) Exogenous retinoic acid causes specific alterations in the development of the midbrain and hindbrain of the zebrafish embryo including positional respecification of the Mauthner neuron. Mech Dev 50:3–16

Hill A, Boll W, Ries C et al (2010) Origin of Pax and Six gene families in sponges: single PaxB and Six1/2 orthologs in Chalinula loosanoffi. Dev Biol 343:106–123

His W (1868) Die erste Entwicklung des Hühnchens im Ei: Untersuchungen über die erste Anlage des Wirbelthierleibes. Vodel, Liepzig, Germany

Holland ND (2007) Hagfish embryos again—the end of a long drought. BioEssays 29:833–836

Holland LZ, Holland ND (2001) Evolution of neural crest and placodes: amphioxus as a model for the ancestral vertebrate? J Anat 199:85–98

Holland SJ, Gao M, Hirano M et al (2014) Selection of the lamprey VLRC antigen receptor repertoire. Proc Natl Acad Sci USA 111:14834–14839

Honig MG, Hume RI (1986) Fluorescent carbocyanine dyes allow living neurons of identified origin to be studied in long-term cultures. J Cell Biol 103:171–187

Horigome N, Myojin M, Ueki T et al (1999) Development of cephalic neural crest cells in embryos of *Lampetra japonica*, with special reference to the evolution of the jaw. Dev Biol 207:287–308

Hörstadius OS (1950) The neural crest. Oxford University Press, London

Jandzik D, Hawkins MB, Cattell MV et al (2014) Roles for FGF in lamprey pharyngeal pouch formation and skeletogenesis highlight ancestral functions in the vertebrate head. Development 141:629–638

Jandzik D, Garnett AT, Square TA et al (2015) Evolution of the new vertebrate head by co-option of an ancient chordate skeletal tissue. Nature 518:534–537

Janvier P (1996) Early vertebrates. Oxford University Press, Oxford

Janvier P (2008) Early jawless vertebrates and cyclostome origins. Zool Sci 25:1045–1056

Jeffery WR (2006) Ascidian neural crest-like cells: phylogenetic distribution, relationship to larval complexity, and pigment cell fate. J Exp Zool B Mol Dev Evol 306:470–480

Jeffery WR, Strickler AG, Yamamoto Y (2004) Migratory neural crest-like cells form body pigmentation in a urochordate embryo. Nature 431:696–699

Jeffery WR, Chiba T, Krajka FR et al (2008) Trunk lateral cells are neural crest-like cells in the ascidian *Ciona intestinalis*: insights into the ancestry and evolution of the neural crest. Dev Biol 324:152–160

Jessen KR, Mirsky R (2005) The origin and development of glial cells in peripheral nerves. Nat Rev Neurosci 6:671–682

Johnels AG (1948) On the development and morphology of the skeleton of the head of *Petromyzon*. Acta Zool 29:139–279

Johnels AG (1956) On the peripheral autonomic nervous system of the trunk region of *Lampetra planeri*. Acta Zool 37:251–286

Johnson NS, Buchinger TJ, Li W (2015) Reproductive ecology of lampreys. In: Docker MF (ed) Lampreys; biology, conservation and control, vol 1. Springer, Dordrecht, pp 265–303

Joung JK, Sander JD (2013) TALENs: a widely applicable technology for targeted genome editing. Nat Rev Mol Cell Biol 14:49–55

Kasamatsu J, Sutoh Y, Fugo K et al (2010) Identification of a third variable lymphocyte receptor in the lamprey. Proc Natl Acad Sci USA 107:14304–14308

Kiernan AE (2013) Notch signaling during cell fate determination in the inner ear. Semin Cell Dev Biol 24:470–479

Kingsley JS (1894) The origin of the vertebrate skeleton. Am Nat 28:635–640

Kleerekoper H, Erkel GAV (1960) The olfactory apparatus of *Petromyzon marinus* L. Can J Zool 38:209–223

Koltzoff NK (1902) Entwicklungsgeschichte des Kopfes von *Petromyzon planeri*. Bull Soc Nat Moscou 15:259–589

Kozmik Z, Holland ND, Kreslova J et al (2007) *Pax-Six-Eya-Dach* network during amphioxus development: conservation *in vitro* but context specificity *in vivo*. Dev Biol 306:143–159

Kumar S, Stecher G, Suleski M, Hedges SB (2017) TimeTree: a resource for timelines, timetrees, and divergence times. Mol Biol Ecol 34:1812–1819

Kunin V, Sorek R, Hugenholtz P (2007) CRISPR–a widespread system that provides acquired resistance against phages in bacteria and archaea. Nat Rev Microbiol 6:181–186

Kuraku S, Hoshiyama D, Katoh K, Suga H, Miyata T (1999) Monophyly of lampreys and hagfishes supported by nuclear DNA-coded genes. J Mol Evol 49:729–735

Kuraku S, Takio Y, Sugahara F, Takechi M, Kuratani S (2010) Evolution of oropharyngeal patterning mechanisms involving *Dlx* and *endothelins* in vertebrates. Dev Biol 341:315–323

Kuratani S (1997) Spatial distribution of postotic crest cells defines the head/trunk interface of the vertebrate body: embryological interpretation of peripheral nerve morphology and evolution of the vertebrate head. Anat Embryol 195:1–13

Kuratani S (2005) Developmental studies of the lamprey and hierarchical evolutionary steps towards the acquisition of the jaw. J Anat 207:489–499

Kuratani S (2012) Evolution of the vertebrate jaw from developmental perspectives. Evol Dev 14:76–92

Kuratani S, Ota KG (2008) Primitive versus derived traits in the developmental program of the vertebrate head: views from cyclostome developmental studies. J Exp Zool B Mol Dev Evol 310:294–314

Kuratani S, Ueki T, Aizawa S, Hirano S (1997) Peripheral development of cranial nerves in a cyclostome, *Lampetra japonica*: morphological distribution of nerve branches and the vertebrate body plan. J Comp Neurol 384:483–500

Kuratani S, Nobusada Y, Horigome N, Shigetani Y (2001) Embryology of the lamprey and evolution of the vertebrate jaw: insights from molecular and developmental perspectives. Philos Trans R Soc Lond B Biol Sci 356:1615–1632

Kuratani S, Murakami Y, Nobusada Y, Kusakabe R, Hirano S (2004) Developmental fate of the mandibular mesoderm in the lamprey, *Lethenteron japonicum*: comparative morphology and development of the gnathostome jaw with special reference to the nature of the trabecula cranii. J Exp Zool B Mol Dev Evol 302:458–468

Kuratani S, Adachi N, Wada N, Oisi Y, Sugahara F (2013) Developmental and evolutionary significance of the mandibular arch and prechordal/premandibular cranium in vertebrates: revising the heterotopy scenario of gnathostome jaw evolution. J Anat 222:41–55

Kuratani S, Oisi Y, Ota KG (2016) Evolution of the vertebrate cranium: viewed from hagfish developmental studies. Zool Sci 33:229–238

Kusakabe R, Tochinai S, Kuratani S (2003) Expression of foreign genes in lamprey embryos: an approach to study evolutionary changes in gene regulation. J Exp Zool B Mol Dev Evol 296:87–97

Kusakabe R, Takechi M, Tochinai S, Kuratani S (2004) Lamprey contractile protein genes mark different populations of skeletal muscles during development. J Exp Zool B Mol Dev Evol 302:121–133

Ladher RK, O'Neill P, Begbie J (2010) From shared lineage to distinct functions: the development of the inner ear and epibranchial placodes. Development 137:1777–1785

Lakiza O, Miller S, Bunce A, Lee EM, McCauley DW (2011) SoxE gene duplication and development of the lamprey branchial skeleton: insights into development and evolution of the neural crest. Dev Biol 359:149–161

Langille RM, Hall BK (1988) Role of the neural crest in development of the trabeculae and branchial arches in embryonic sea lamprey, *Petromyzon marinus* (L). Development 102:301–310

Le Douarin NM, Kalchiem C (1999) The neural crest. Cambridge University Press, Cambridge, UK

Le Douarin NM, Fontaine-Pérus J, Couly G (1986) Cephalic ectodermal placodes and neurogenesis. Trends Neurosci 9:175–180

Le Douarin N, Dulac C, Dupin E, Cameron-Curry P (1991) Glial cell lineages in the neural crest. Glia 4:175–184

Le Douarin NM, Dupin E, Baroffio A, Dulac C (1992) New insights into the development of neural crest derivatives. Int Rev Cytol 138:269–314

Leimeroth R, Lobsiger C, Lussi A et al (2002) Membrane-bound neuregulin1 type III actively promotes Schwann cell differentiation of multipotent Progenitor cells. Dev Biol 246:245–258

Lemke G, Chao M (1988) Axons regulate Schwann cell expression of the major myelin and NGF receptor genes. Development 102:499–504

Lemke G, Lamar E, Patterson J (1988) Isolation and analysis of the gene encoding peripheral myelin protein zero. Neuron 1:73–83

Lennon RE (1955) Artificial propagation of the sea lamprey, *Petromyzon marinus*. Copeia 1955:235–236

Letelier J, de la Calle-Mustienes E, Pieretti J et al (2018) A conserved *Shh* cis-regulatory module highlights a common developmental origin of unpaired and paired fins. Nat Genet 50:504–509

Li J, Sabyasachi D, Herrin BR, Masayuki H, Cooper M (2013) Definition of a third VLR gene in hagfish. Proc Natl Acad Sci USA 110:15013–15018

Litman GW, Rast JP, Fugmann SD (2010) The origins of vertebrate adaptive immunity. Nat Rev Immunol 10:543–553

Logan M (2003) Finger or toe: the molecular basis of limb identity. Development 130:6401–6410

Lopez SL, Dono R, Zeller R, Carrasco AE (1995) Differential effects of retinoic acid and a retinoid antagonist on the spatial distribution of the homeoprotein Hoxb-7 in vertebrate embryos. Dev Dyn 204:457–471

Løvtrup S (1977) The phylogeny of the vertebrata. Wiley, New York

Maitland PS, Renaud CB, Quintella BR, Close DA, Docker MF (2015) Conservation of native lampreys. In: Docker MF (ed) Lampreys: biology, conservation and control, vol 1. Springer, Dordrecht, pp 375–428

Mallatt J (1996) Ventilation and the origin of jawed vertebrates: a new mouth. Zool J Linn Soc 117:329–404

Mallatt J (2008) The origin of the vertebrate jaw: neoclassical ideas versus newer, development-based ideas. Zool Sci 25:990–998

Manousaki T, Qiu H, Noro M et al (2016) Molecular evolution in the lamprey genomes and its relevance to the timing of whole genome duplications. In: Orlov A, Beamish R (eds) Jawless fishes of the world, vol 1. Cambridge Scholars Publishing, Newcastle upon Tyne, pp 2–16

Manni L, Caicci F, Gasparini F, Zaniolo G, Burighel P (2004) Hair cells in ascidians and the evolution of lateral line placodes. Evol Dev 6:379–381

Manni L, Mackie GO, Caicci F, Zaniolo G, Burighel P (2006) Coronal organ of ascidians and the evolutionary significance of secondary sensory cells in chordates. J Comp Neurol 495:363–373

Manzanares M, Wada H, Itasaki N et al (2000) Conservation and elaboration of Hox gene regulation during evolution of the vertebrate head. Nature 408:854–857

Markus PM, Koenig S, Krause P, Becker H (1997) Selective intraportal transplantation of DiI-marked isolated rat hepatocytes. Cell Transplant 6:455–462

Martin P, Swanson GJ (1993) Descriptive and experimental analysis of the epithelial remodellings that control semicircular canal formation in the developing mouse inner ear. Dev Biol 159:549–558

Martin WM, Bumm LA, McCauley DW (2009) Development of the viscerocranial skeleton during embryogenesis of the sea lamprey, *Petromyzon marinus*. Dev Dyn 238:3126–3138

Mazet F, Hutt JA, Milloz J et al (2005) Molecular evidence from *Ciona intestinalis* for the evolutionary origin of vertebrate sensory placodes. Dev Biol 282:494–508

McBurney KM, Keeley FW, Kibenge FS, Wright GM (1996) Spatial and temporal distribution of lamprin mRNA during chondrogenesis of trabecular cartilage in the sea lamprey. Anat Embryol 193:419–426

McCauley DW (2008) SoxE, Type II collagen, and evolution of the chondrogenic neural crest. Zool Sci 25:982–989

McCauley DW, Bronner-Fraser M (2002) Conservation of Pax gene expression in ectodermal placodes of the lamprey. Gene 287:129–139

McCauley DW, Bronner-Fraser M (2003) Neural crest contributions to the lamprey head. Development 130:2317–2327

McCauley DW, Bronner-Fraser M (2006) Importance of SoxE in neural crest development and the evolution of the pharynx. Nature 441:750–752

McCauley DW, Docker MF, Whyard S, Li W (2015) Lampreys as diverse model organisms in the genomics era. BioScience 65:1046–1056

Medeiros DM (2013) The evolution of the neural crest: new perspectives from lamprey and invertebrate neural crest-like cells. Wiley Interdiscip Rev Dev Biol 2:1–15

Medeiros DM, Crump JG (2012) New perspectives on pharyngeal dorsoventral patterning in development and evolution of the vertebrate jaw. Dev Biol 371:121–135

Mehta TK, Ravi V, Yamasaki S et al (2013) Evidence for at least six Hox clusters in the Japanese lamprey (*Lethenteron japonicum*). Proc Natl Acad Sci USA 110:16044–16049

Meinertzhagen LA, Okamura Y (2001) The larval ascidian nervous system: the chordate brain from its small beginnings. Trends Neurosci 24:401–410

Mello CC, Conte D (2004) Revealing the world of RNA interference. Nature 431:338–342

Meulemans D, Bronner-Fraser M (2002) Amphioxus and lamprey AP-2 genes: implications for neural crest evolution and migration patterns. Development 129:4953–4962

Meulemans D, Bronner-Fraser M (2007) The amphioxus SoxB family: implications for the evolution of vertebrate placodes. Int J Biol Sci 3:356–364

Meulemans D, McCauley D, Bronner-Fraser M (2003) Id expression in amphioxus and lamprey highlights the role of gene cooption during neural crest evolution. Dev Biol 264:430–442

Meyer D, Yamaai T, Garratt A et al (1997) Isoform-specific expression and function of neuregulin. Development 124:3575–3586

Minoux M, Rijli FM (2010) Molecular mechanisms of cranial neural crest cell migration and patterning in craniofacial development. Development 137:2605–2621

Miyashita T (2016) Fishing for jaws in early vertebrate evolution: a new hypothesis of mandibular confinement. Biol Rev 91:611–657

Modrell MS, Hockman D, Uy B et al (2014) A fate-map for cranial sensory ganglia in the sea lamprey. Dev Biol 385:405–416

Monk KR, Naylor SG, Glenn TD et al (2009) A G protein-coupled receptor is essential for Schwann cells to initiate myelination. Science 325:1402–1405

Morrisskay GM, Murphy P, Hill RE, Davidson DR (1991) Effects of retinoic acid excess on expression of Hox-2.9 and Krox-20 and on morphological segmentation in the hindbrain of mouse embryos. EMBO J 10:2985–2995

Murakami Y, Ogasawara M, Sugahara F et al (2001) Identification and expression of the lamprey *Pax6* gene: evolutionary origin of the segmented brain of vertebrates. Development 128:3521–3531

Murakami Y, Pasqualetti M, Takio Y et al (2004) Segmental development of reticulospinal and branchiomotor neurons in lamprey: insights into the evolution of the vertebrate hindbrain. Development 131:983–995

Myojin M, Ueki T, Sugahara F et al (2001) Isolation of *Dlx* and *Emx* gene cognates in an agnathan species, *Lampetra japonica*, and their expression patterns during embryonic and larval development: conserved and diversified regulatory patterns of homeobox genes in vertebrate head evolution. J Exp Zool 291:68–84

Nagaoka H, Yu W, Nussenzweig MC (2000) Regulation of RAG expression in developing lymphocytes. Curr Opin Immunol 12:187–190

Nave KA, Salzer JL (2006) Axonal regulation of myelination by neuregulin 1. Curr Opin Neurobiol 16:492–500

Nawaz S, Schweitzer J, Jahn O, Werner HB (2013) Molecular evolution of myelin basic protein, an abundant structural myelin component. Glia 61:1364–1377

Neidert AH, Virupannavar V, Hooker GW, Langeland JA (2001) Lamprey *Dlx* genes and early vertebrate evolution. Proc Natl Acad Sci USA 98:1665–1670

Nelson JS, Grande TC, Wilson MVH (2016) Fishes of the world, 5th edn. Wiley, Hoboken, NJ

Newbern J, Birchmeier C (2010) Nrg1/ErbB signaling networks in Schwann cell development and myelination. Semin Cell Dev Biol 21:922–928

Newth DR (1950) Fate of the neural crest in lampreys. Nature 165:284–284

Newth DR (1951) Experiments on the neural crest of the lamprey embryo. J Exp Biol 28:247–260

Newth DR (1956) On the neural crest of the lamprey embryo. J Embryol Exp Morph 4:358–375

Nichols JT, Pan LY, Moens CB, Kimmel CB (2013) *barx1* represses joints and promotes cartilage in the craniofacial skeleton. Development 140:2765–2775

Nicol JAC (1952) Autonomic nervous systems in lower chordates. Biol Rev 27:1–50

Nikitina NV, Bronner-Fraser M (2009) Gene regulatory networks that control the specification of neural-crest cells in the lamprey. Biochim Biophys Acta 1789:274–278

Nikitina N, Sauka-Spengler T, Bronner-Fraser M (2008) Dissecting early regulatory relationships in the lamprey neural crest gene network. Proc Natl Acad Sci USA 105:20083–20088

Nikitina N, Bronner-Fraser M, Sauka-Spengler T (2009a) Culturing lamprey embryos. Cold Spring Harb Protoc 2009:pdb.prot5122

Nikitina N, Bronner-Fraser M, Sauka-Spengler T (2009b) The sea lamprey Petromyzon marinus: a model for evolutionary and developmental biology. Cold Spring Harb Protoc 2009:pdb.emo113

Nikitina N, Bronner-Fraser M, Sauka-Spengler T (2009c) Whole-mount in situ hybridization on lamprey embryos. Cold Spring Harb Protoc 2009:pdb prot5125

Niswander L (1997) Limb mutants: what can they tell us about normal limb development? Curr Opin Genet Dev 7:530–536

Noden DM (1988) Interactions and fates of avian craniofacial mesenchyme. Development 103:121–140

Noramly S, Grainger RM (2002) Determination of the embryonic inner ear. J Neurobiol 53:100–128

Northcutt RG (1996) The agnathan ark: the origin of craniate brains. Brain Behav Evol 48:237–247

Ogasawara M, Shigetani Y, Hirano S, Satoh N, Kuratani S (2000) *Pax1/Pax9*-related genes in an agnathan vertebrate, *Lampetra japonica*: expression pattern of *LjPax9* implies sequential evolutionary events toward the gnathostome body plan. Dev Biol 223:399–410

Ohtani K, Yao T, Kobayashi M et al (2008) Expression of Sox and fibrillar collagen genes in lamprey larval chondrogenesis with implications for the evolution of vertebrate cartilage. J Exp Zool B Mol Dev Evol 310:596–607

Oisi Y, Ota KG, Kuraku S, Fujimoto S, Kuratani S (2013) Craniofacial development of hagfishes and the evolution of vertebrates. Nature 493:175–180

Onimaru K, Shoguchi E, Kuratani S, Tanaka M (2011) Development and evolution of the lateral plate mesoderm: comparative analysis of amphioxus and lamprey with implications for the acquisition of paired fins. Dev Biol 359:124–136

Ota KG, Kuraku S, Kuratani S (2007) Hagfish embryology with reference to the evolution of the neural crest. Nature 446:672–675

Ota KG, Fujimoto S, Oisi Y, Kuratani S (2011) Identification of vertebra-like elements and their possible differentiation from sclerotomes in the hagfish. Nat Commun 2:373

Ota KG, Fujimoto S, Oisi Y, Kuratani S (2013) Late development of hagfish vertebral elements. J Exp Zool B Mol Dev Evol 320:129–139

Owen R (1849) On the nature of limbs. John Van Voorst, London

Pancer Z, Amemiya CT, Ehrhardt GR et al (2004) Somatic diversification of variable lymphocyte receptors in the agnathan sea lamprey. Nature 430:174–180

Pancer Z, Saha NR, Kasamatsu J et al (2005) Variable lymphocyte receptors in hagfish. Proc Natl Acad Sci USA 102:9224–9229

Panganiban G, Rubenstein JLR (2002) Developmental functions of the *Distal-less*/Dlx homeobox genes. Development 129:4371–4386

Parker HJ, Sauka-Spengler T, Bronner M, Elgar G (2014a) A reporter assay in lamprey embryos reveals both functional conservation and elaboration of vertebrate enhancers. PLoS ONE 9:e85492

Parker HJ, Bronner ME, Krumlauf R (2014b) A Hox regulatory network of hindbrain segmentation is conserved to the base of vertebrates. Nature 514:490–493

Piavis GW (1961) Embryological stages in the sea lamprey and effect of temperature on development. US Fish Wildl Serv Fish Bull 61:111–143

Piavis GW (1971) Embryology. In: Hardisty MW, Potter IC (eds) The biology of lampreys, vol 1. Academic Press, London, pp 361–400

Piavis GW, Howell JH (1969) Rearing of sea lamprey *Petromyzon marinus* embryos in distilled water. Copeia 1969:204–205

Piavis GW, Howell JH, Smith AJ (1970) Experimental hybridization among five species of lampreys from the Great Lakes. Copeia 1970:29–37

Piotrowski T, Baker CV (2014) The development of lateral line placodes: taking a broader view. Dev Biol 389:68–81

Posnien N, Koniszewski N, Bucher G (2011) Insect *Tc-six4* marks a unit with similarity to vertebrate placodes. Dev Biol 350:208–216

Potter IC, Gill HS, Renaud CB, Haoucher D (2015) The taxonomy, phylogeny, and distribution of lampreys. In: Docker MF (ed) Lampreys: biology, conservation and control, vol 1. Springer, Dordrecht, pp 35–73

Qiu MS et al (1997) Role of the Dlx homeobox genes in proximodistal patterning of the branchial arches: mutations of Dlx-1, Dlx-2, and Dlx-1 and -2 alter morphogenesis of proximal skeletal and soft tissue structures derived from the first and second arches. Dev Biol 185:165–184

Rahimi RA, Allmond JJ, Wagner H, McCauley DW, Langeland JA (2009) Lamprey *snail* highlights conserved and novel patterning roles in vertebrate embryos. Dev Genes Evol 219:31–36

Rakowiecki S, Epstein DJ (2013) Divergent roles for Wnt/β-catenin signaling in epithelial maintenance and breakdown during semicircular canal formation. Development 140:1730–1739

Ran FA, Hsu PD, Lin C-Y et al (2013) Double nicking by RNA-guided CRISPR Cas9 for enhanced genome editing specificity. Cell 154:1380–1389

Rast JP, Buckley KM (2013) Lamprey immunity is far from primitive. Proc Natl Acad Sci USA 110:5746–5747

Reif W (1982) Evolution of dermal skeleton and dentition in vertebrates. The odontode regulation theory. Evol Biol 15:287–368

Renaud CB (2011) Lampreys of the world. An annotated and illustrated catalogue of lamprey species known to date. FAO Species Cat Fish Purp 5, FAO, Rome

Richardson MK, Admiraal J, Wright GM (2010) Developmental anatomy of lampreys. Biol Rev 85:1–33

Rigon F, Stach T, Caicci F et al (2013) Evolutionary diversification of secondary mechanoreceptor cells in tunicata. BMC Evol Biol 13:112

Rigon F, Gasparini F, Shimeld SM, Candiani S, Manni L (2018) Developmental signature, synaptic connectivity and neurotransmission are conserved between vertebrate hair cells and tunicate coronal cells. J Comp Neurol 526:957–971

Rinkwitz S, Bober E, Baker R (2001) Development of the vertebrate inner ear. Ann N Y Acad Sci 942:1–14

Ritchie JM (1984) Physiological basis of conduction in myelinated nerve fibers. In: Morell P (ed) Myelin, 2nd edn. Plenum Press, New York, pp 117–145

Robson P, Wright GM, Sitarz E et al (1993) Characterization of lamprin, an unusual matrix protein from lamprey cartilage. Implications for evolution, structure, and assembly of elastin and other fibrillar proteins. J Biol Chem 268:1440–1447

Robson P, Wright GM, Youson JH, Keeley FW (2000) The structure and organization of lamprin genes: multiple-copy genes with alternative splicing and convergent evolution with insect structural proteins. Mol Biol Evol 17:1739–1752

Rogozin IB, Iyer LM, Liang L et al (2007) Evolution and diversification of lamprey antigen receptors: evidence for involvement of an AID-APOBEC family cytosine deaminase. Nat Immunol 8:647–656

Roots BI (2008) The phylogeny of invertebrates and the evolution of myelin. Neuron Glia Biol 4:101–109

Ruvinsky I, Gibson-Brown JJ (2000) Genetic and developmental bases of serial homology in vertebrate limb evolution. Development 127:5233–5244

Salzer J, Zalc B (2016) Myelination. Curr Biol 26:R971–R975

Sander JD, Joung JK (2014) CRISPR-Cas systems for editing, regulating and targeting genomes. Nat Biotechnol 32:347–355

Sato S, Ikeda K, Shioi G et al (2012) Regulation of *Six1* expression by evolutionarily conserved enhancers in tetrapods. Dev Biol 368:95–108

Sauka-Spengler T, Bronner-Fraser M (2008a) A gene regulatory network orchestrates neural crest formation. Nat Rev Mol Cell Bio 9:557–568

Sauka-Spengler T, Bronner-Fraser M (2008b) Insights from a sea lamprey into the evolution of neural crest gene regulatory network. Biol Bull 214:303–314

Sauka-Spengler T, Meulemans D, Jones M, Bronner-Fraser M (2007) Ancient evolutionary origin of the neural crest gene regulatory network. Dev Cell 13:405–420

Schlosser G (2002) Development and evolution of lateral line placodes in amphibians I. Dev Zool 105:119–146

Schlosser G (2005) Evolutionary origins of vertebrate placodes: insights from developmental studies and from comparisons with other deuterostomes. J Exp Zool B Mol Dev Evol 304:347–399

Schlosser G (2006) Induction and specification of cranial placodes. Dev Biol 294:303–351

Schlosser G (2010) Making senses development of vertebrate cranial placodes. Int Rev Cell Mol Biol 283:129–234

Schlosser G (2015) Vertebrate cranial placodes as evolutionary innovations—the ancestor's tale. Curr Top Dev Biol 111:235–300

Schlosser G (2016) Evolution of neural crest and cranial placodes. In: Striedter G (ed) Evolution of nervous systems, vol 1, 2nd edn. Elsevier, New York, pp 25–35

Schlosser G (2017) From so simple a beginning–what amphioxus can teach us about placode evolution. Int J Dev Biol 61:633–648

Schlosser G, Ahrens K (2004) Molecular anatomy of placode development in *Xenopus laevis*. Dev Biol 271:439–466

Schlosser G, Northcutt RG (2000) Development of neurogenic placodes in *Xenopus laevis*. J Comp Neurol 418:121–146

Schultze MS (1856) Die Entwickelungs-geschichte von *Petromyzon planeri*. Loosjes, Haarlem

Scott WB (1887) Notes on the development of *Petromyzon*. J Morphol 1:253–310

Sémon M, Schubert M, Laudet V (2012) Programmed genome rearrangements: in lampreys, all cells are not equal. Curr Biol 22:1524–1529

Shigetani Y, Sugahara F, Kawakami Y et al (2002) Heterotopic shift of epithelial-mesenchymal interactions in vertebrate jaw evolution. Science 296:1316–1319

Shimeld SM (1999) The evolution of dorsoventral pattern formation in the chordate neural tube. Am Zool 39:641–649

Shimeld SM, Donoghue PCJ (2012) Evolutionary crossroads in developmental biology: cyclostomes (lamprey and hagfish). Development 139:2091–2099

Shimeld SM, Holland PWH (2000) Vertebrate innovations. Proc Natl Acad Sci USA 97:4449–4452

Shipley AE (1887) On some points in the development of *Petromyzon fluviatilis*. Q J Microsc Sci 27:325–370

Shubin N, Tabin C, Carroll S (1997) Fossils, genes and the evolution of animal limbs. Nature 388:639–648

Sienknecht UJ (2013) Developmental origin and fate of middle ear structures. Hear Res 301:19–26

Simeone A, Acampora D, Pannese M et al (1994) Cloning and characterization of two members of the vertebrate Dlx gene family. Proc Natl Acad Sci USA 91:2250–2254

Smith AJ, Howell JH, Piavis GW (1968) Comparative embryology of five species of lampreys of the Upper Great Lakes. Copeia 1968:461–469

Smith JJ (2018) Programmed DNA elimination: keeping germline genes in their place. Curr Biol 28:R601–R603

Smith JJ, Antonacci F, Eichler EE, Amemiya CT (2009) Programmed loss of millions of base pairs from a vertebrate genome. Proc Natl Acad Sci USA 106:11212–11217

Smith JJ, Baker C, Eichler EE, Amemiya CT (2012) Genetic consequences of programmed genome rearrangement. Curr Biol 22:1524–1529

Smith JJ, Kuraku S, Holt C et al (2013) Sequencing of the sea lamprey (*Petromyzon marinus*) genome provides insights into vertebrate evolution. Nat Genet 45:415–421

Smith JJ, Timoshevskaya N, Ye C et al (2018) The sea lamprey germline genome provides insights into programmed genome rearrangement and vertebrate evolution. Nat Genet 50:270–277

Smith MM, Hall BK (1990) Development and evolutionary origins of vertebrate skeletogenic and odontogenic tissues. Biol Rev 65:277–373

Square TA (2017) Neural crest cell development and evolution. PhD thesis, University of Colorado at Boulder, Boulder, CO

Square T, Romášek M, Jandzik D et al (2015) CRISPR/Cas9-mediated mutagenesis in the sea lamprey *Petromyzon marinus*: a powerful tool for understanding ancestral gene functions in vertebrates. Development 142:4180–4187

Square T, Jandzik D, Cattell M, Hansen A, Medeiros DM (2016a) Embryonic expression of *endothelins* and their receptors in lamprey and frog reveals stem vertebrate origins of complex Endothelin signaling. Sci Rep 6:34282

Square T, Jandzik D, Romášek M, Cerny R, Medeiros DM (2016b) The origin and diversification of the developmental mechanisms that pattern the vertebrate head skeleton. Dev Biol 427:219–229

Stock DW, Whitt GS (1992) Evidence from 18S ribosomal-RNA sequences that lampreys and hagfishes form a natural group. Science 257:787–789

Stolfi A, Ryan K, Meinertzhagen IA, Christiaen L (2015) Migratory neuronal progenitors arise from the neural plate borders in tunicates. Nature 527:371–374

Sugahara F, Aota S, Kuraku S et al (2011) Involvement of Hedgehog and FGF signalling in the lamprey telencephalon: evolution of regionalization and dorsoventral patterning of the vertebrate forebrain. Development 138:1217–1226

Sugahara F, Murakami Y, Kuratani S (2015) Gene expression analysis of lamprey embryos. In: Hauptmann G (ed) In situ hybridization methods, Neuromethods, vol 99. Humana Press, New York, pp 263–278

Sugahara F, Pascual-Anaya J, Oisi Y et al (2016) Evidence from cyclostomes for complex regionalization of the ancestral vertebrate brain. Nature 531:97–100

Sutoh Y, Kasahara M (2016) Lymphocyte populations in jawless vertebrates: insights into the origin and evolution of adaptive immunity. In: Malagoli D (ed) The evolution of the immune system: conservation and diversification. Elsevier, New York, pp 51–67

Swain GP, Jacobs AJ, Frei E, Selzer ME (1994) A method for in-situ hybridization in wholemounted lamprey brain: neurofilament expression in larvae and adults. Exp Neurol 126:256–269

Tahara Y (1988) Normal stages of development in the lamprey, *Lampetra reissneri* (Dybowski). Zool Sci 5:109–118

Takechi M, Adachi N, Hirai T, Kuratani S, Kuraku S (2013) The Dlx genes as clues to vertebrate genomics and craniofacial evolution. Semin Cell Dev Biol 24:110–118

Talbot JC, Johnson SL, Kimmel CB (2010) *hand2* and *Dlx* genes specify dorsal, intermediate and ventral domains within zebrafish pharyngeal arches. Development 137:2506–2516

Tanaka M, Hale LA, Amores A et al (2005) Developmental genetic basis for the evolution of pelvic fin loss in the pufferfish *Takifugu rubripes*. Dev Biol 281:227–239

Tanaka M, Onimaru K (2012) Acquisition of the paired fins: a view from the sequential evolution of the lateral plate mesoderm. Evol Dev 14:412–420

Tarazona OA, Slota LA, Lopez DH, Zhang G, Cohn MJ (2016) The genetic program for cartilage development has deep homology within Bilateria. Nature 533:86–89

Taveggia C, Zanazzi G, Petrylak A et al (2005) Neuregulin-1 type III determines the ensheathment fate of axons. Neuron 47:681–694

Tickle C (2003) Patterning systems—from one end of the limb to the other. Dev Cell 4:449–458

Tomsa JM, Langeland JA (1999) *Otx* expression during lamprey embryogenesis provides insights into the evolution of the vertebrate head and jaw. Dev Biol 207:26–37

Timoshevskiy VA, Herdy JR, Keinath MC, Smith JJ (2016) Cellular and molecular features of developmentally programmed genome rearrangement in a vertebrate (sea lamprey: *Petromyzon marinus*). PLoS Genet 12:e1006103

Timoshevskiy VA, Lampman RT, Hess JE, Porter LL, Smith JJ (2017) Deep ancestry of programmed genome rearrangement in lampreys. Dev Biol 429:31–34

Toro S, Varga ZM (2007) Equivalent progenitor cells in the zebrafish anterior preplacodal field give rise to adenohypophysis, lens, and olfactory placodes. Semin Cell Dev Biol 18:534–542

Trainor PA (2013) Neural crest cells: evolution, development and disease. Academic Press, Cambridge, MA

Tretjakoff D (1927) Die Chordascheiden der Urodelen. Z Zellforsch 5:174–207

Tulenko FJ, McCauley DW, MacKenzie EL et al (2013) Body wall development in lamprey and a new perspective on the origin of vertebrate paired fins. Proc Natl Acad Sci USA 110:11899–11904

Uchida K, Murakami Y, Kuraku S, Hirano S, Kuratani S (2003) Development of the adenohypophysis in the lamprey: evolution of epigenetic patterning programs in organogenesis. J Exp Zool B Mol Dev Evol 300:32–47

Urnov FD, Rebar EJ, Holmes MC, Zhang HS, Gregory PD (2010) Genome editing with engineered zinc finger nucleases. Nat Rev Genet 11:636

van Wijhe JW (1883) Uber die Mesodermsegmente und die Entwicklung der Nerven des Selachierkopfes. Verh K Akad Wetenschappen 22:1–50

Vogel KS, Davies AM (1993) Heterotopic transplantation of presumptive placodal ectoderm changes the fate of sensory neuron precursors. Development 119:263–276

Vogt W (1925) Analysis of the shape of amphibian germs using localised colour staining. Foreword on the methods and results. I. Part. Method and effectiveness of localised vital colour staining using agar as a colour carrier. Roux Arch Dev Biol 106:542–610

von Kupffer C (1891) The development of the cranial nerves of vertebrates. J Comp Neurol 1:246–264

Wada H (2010) Origin and genetic evolution of the vertebrate skeleton. Zool Sci 27:119–123

Wada H, Saiga H, Satoh N, Holland PW (1998) Tripartite organization of the ancestral chordate brain and the antiquity of placodes: insights from ascidian *Pax-2/5/8*, *Hox* and *Otx* genes. Development 125:1113–1122

Waehneldt TV (1990) Phylogeny of myelin proteins. Ann N Y Acad Sci 605:15–28

Waehneldt TV, Matthieu JM, Stoklas S (1987) Immunological evidence for the presence of myelin-related integral proteins in the CNS of hagfish and lamprey. Neurochem Res 12:869–873

Webb JF, Noden DM (1993) Ectodermal placodes: contributions to the development of the vertebrate head. Am Zool 33:434–447

Werner HB (2013) Do we have to reconsider the evolutionary emergence of myelin? Front Cell Neurosci 7:217

Wood H, Pall G, Morrisskay G (1994) Exposure to retinoic acid before or after the onset of somitogenesis reveals separate effects on rhombomeric segmentation and 3' HoxB gene-expression domains. Development 120:2279–2285

Wright GM, Youson JH (1982) Ultrastructure of mucocartilage in the larval anadromous sea lamprey, *Petromyzon marinus* L. Am J Anat 165:39–51

Wright GM, Youson JH (1983) Ultrastructure of cartilage from young adult sea lamprey, *Petromyzon marinus* L: a new type of vertebrate cartilage. Am J Anat 167:59–70

Wright GM, Armstrong LA, Jacques AM, Youson JH (1988) Trabecular, nasal, branchial, and pericardial cartilages in the sea lamprey, *Petromyzon marinus*: fine structure and immunohistochemical detection of elastin. Am J Anat 182:1–15

Wright GM, Keeley FW, Robson P (2001) The unusual cartilaginous tissues of jawless craniates, cephalochordates and invertebrates. Cell Tissue Res 304:165–174

Yamazaki Y, Fukutomi N, Takeda K, Iwata A (2003) Embryonic development of the Pacific lamprey, *Entosphenus tridentatus*. Zool Sci 20:1095–1098

Yao T, Ohtani K, Kuratani S, Wada H (2011) Development of lamprey mucocartilage and its dorsal-ventral patterning by endothelin signaling, with insight into vertebrate jaw evolution. J Exp Zool B Mol Dev Evol 316:339–346

Yao T, Ohtani K, Wada H (2008) Whole-mount observation of pharyngeal and trabecular cartilage development in lampreys. Zool Sci 25:976–981

York JR, Yuan T, Zehnder K, McCauley DW (2017) Lamprey neural crest migration is Snail-dependent and occurs without a differential shift in cadherin expression. Dev Biol 428:176–187

Young JZ (1971) Foreword. In: Hardisty MW, Potter IC (eds) The biology of lampreys, vol 1. Academic Press, London, pp vii– viii

Yu JK, Meulemans D, McKeown SJ, Bronner-Fraser M (2008) Insights from the amphioxus genome on the origin of vertebrate neural crest. Genome Res 18:1127–1132

Zalc B (2016) The acquisition of myelin: an evolutionary perspective. Brain Res 1641:4–10

Zalc B, Colman DR (2000) Origins of vertebrate success. Science 288:271–272

Zhang G (2009) An evo-devo view on the origin of the backbone: evolutionary development of the vertebrae. Int Comp Biol 49:178–186

Zhang G, Cohn MJ (2006) Hagfish and lancelet fibrillar collagens reveal that type II collagen-based cartilage evolved in stem vertebrates. Proc Natl Acad Sci USA 103:16829–16833

Zhang G, Miyamoto MM, Cohn MJ (2006) Lamprey type II collagen and *Sox9* reveal an ancient origin of the vertebrate collagenous skeleton. Proc Natl Acad Sci USA 103:3180–3185

Zhu M, Ahlberg PE, Pan Z et al (2016) A Silurian maxillate placoderm illuminates jaw evolution. Science 354:334–336

Zu Y, Zhang X, Ren J et al (2016) Biallelic editing of a lamprey genome using the CRISPR/Cas9 system. Sci Rep 6:23496

Zuniga E, Rippen M, Alexander C, Schilling TF, Crump JG (2011) Gremlin 2 regulates distinct roles of BMP and Endothelin 1 signaling in dorsoventral patterning of the facial skeleton. Development 138:5147–5156

Chapter 7
There and Back Again: Lampreys in the 21st Century and Beyond

Margaret F. Docker and John B. Hume

Abstract The 21st century is proving to be an exciting time to study lamprey biology. Lampreys have long provided important insights into key developments in vertebrate evolution; research in support of sea lamprey control in the Laurentian Great Lakes has made significant contributions to our understanding of lamprey biology; and there is now (near) global interest in the conservation of threatened lamprey species. Furthermore, we are beginning to see a convergence of these formerly discrete research areas, as well as greater interactions and knowledge exchange between researchers and managers from different geographic regions. In this conclusion to Volumes 1 and 2 of *Lampreys: Biology, Conservation and Control*, we provide an overview of some exciting advances in our knowledge of lamprey biology and potential challenges facing lampreys and lamprey biologists in the near future. Recent advances and remaining knowledge gaps in many aspects of fundamental lamprey biology are covered in other chapters in these two volumes; here, we focus on the intersection of biology, conservation, and control. For example, molecular analysis has resolved many of the previous uncertainties regarding lamprey phylogenetic relationships, but continued uncertainties (e.g., the relationship between "paired" parasitic and non-parasitic lampreys) and lack of an explicit phylogenetic framework contribute to ongoing confusion among biologists regarding correct lamprey nomenclature. Although lamprey taxonomy will no doubt continue to be revised as we refine our hypotheses regarding the evolutionary relationships among lampreys, it is important that we: (1) use consistent and accepted species names to enable accurate communication between researchers and managers from different regions; and (2) recognize that conservation legislation acknowledges biological diversity below the species level (i.e., evolutionarily significant units, ESUs) so that genetically or otherwise distinct lamprey populations are eligible for protection without prematurely or inconsistently

M. F. Docker (✉)
Department of Biological Sciences, University of Manitoba,
50 Sifton Road, Winnipeg, MB R3T 2N2, Canada
e-mail: Margaret.Docker@umanitoba.ca

J. B. Hume
Department of Fisheries and Wildlife, Michigan State University,
480 Wilson Road, East Lansing, MI 48824, USA
e-mail: jhume@msu.edu

describing each as a distinct species. Novel methodologies that are contributing to our understanding of lamprey biology and that have exciting applications to lamprey conservation and control include: (1) improvements to deepwater larval sampling methods to help evaluate the extent to which lentic and deep riverine habitats are used by different lamprey species; (2) improved tools for monitoring the spawning migration; (3) environmental DNA (eDNA) and pheromone detection assays that have the potential to provide cost-effective supplements to traditional lamprey survey methods; and (4) genetic and genomic tools that are being used in a variety of ways to help refine our understanding of lamprey biology (e.g., mating systems, larval dispersal and growth rates) and to aid conservation and control efforts (e.g., elucidating genetic stock structure, monitoring the success of translocation efforts). Not surprisingly, advances and challenges related to lamprey control and conservation are often "two sides of the same coin." This is particularly true with respect to passage of upstream migrants at anthropogenic barriers, and knowledge of lamprey behavior at barriers is being used to both block sea lamprey migration in Great Lakes tributaries and enhance passage efficiency for other lampreys elsewhere. Achieving successful lamprey conservation and control will also require positive public and legislative attitudes towards species in need of conservation and continued public support and acceptance of sea lamprey control efforts. Pursuit of genetic control options in particular will need to address ethical and societal concerns.

Keywords Barriers · Conservation · Dam removal · Deepwater sampling · DIDSON · Environmental DNA · Evolutionarily significant units · ESUs · Fishways · Genetic species ID · Genetic stock structure · Genetic pedigree analysis · Habitat connectivity · Invasive species · Japanese lamprey · Korean lamprey · Paired species · Pheromones · Sea lamprey control · Selective fish passage · Social license · Species delimitation · Taxonomy and nomenclature · Telemetry

7.1 Introduction

The 21st century is proving to be an exciting time to study lamprey biology. Lampreys have long provided important insights into key developments in vertebrate evolution (see Docker et al. 2015; Sower 2015; Chap. 6); research in support of sea lamprey *Petromyzon marinus* control in the Laurentian Great Lakes has been significantly contributing to our understanding of all aspects of the lamprey life cycle since the 1950s (see Dawson et al. 2015; Johnson et al. 2015a; Manzon et al. 2015; Moser et al. 2015; Chaps. 1 and 5); and there is now global (or near global; see Sect. 7.4.3) interest in the conservation and management of native lampreys (see Maitland et al. 2015). Importantly, we are also starting to see an exciting convergence of these research areas and increased interaction between researchers and managers. For example, knowledge and resources stemming from lamprey genomic and "evo-devo" studies are helping to develop innovative and integrative strategies to control sea lamprey

(McCauley et al. 2015; Thresher et al. 2019a; see Chaps. 5 and 6) and to manage species of conservation concern (e.g., Hess 2016). Genetic methods are also aiding in the resolution of phylogenetic relationships among lamprey taxa (Potter et al. 2015) and identification of evolutionarily significant units and management units for conservation and management (e.g., Spice et al. 2012; Mateus et al. 2013a). Artificial propagation of lampreys, first developed to produce specimens for the study of embryonic development in the laboratory, is now being used to provide larvae for a range of research and conservation needs and is enabling important insights into lamprey behavior, genetics, and early life history (see Chap. 2). Furthermore, there appears to be an increasing recognition that lamprey conservation and control are "two sides of the same coin," and new technologies developed for one are readily being applied to the other with exciting outcomes. Such synergies are particularly evident with respect to research in support of efforts to control landlocked sea lamprey in the Great Lakes and research to help conserve the anadromous form of this same species in its native range (Hansen et al. 2016). Knowledge of sea lamprey behavior at barriers, for example, is being used to impede sea lamprey migration in Great Lakes tributaries and enhance sea lamprey passage efficiency in Atlantic tributaries in Europe and the eastern United States.

In this conclusion to Volumes 1 and 2 of *Lampreys: Biology, Conservation and Control*, we provide an overview of some important and exciting developments in our knowledge of lamprey biology, particularly in regards to their applications to conservation and control, as well as identifying potential challenges facing lampreys and lamprey biologists in the near future. Recent advances and key knowledge gaps related to basic lamprey biology (e.g., lamprey development, sex determination, life history evolution) and a more detailed discussion of emerging sea lamprey control techniques are covered elsewhere (e.g., Chaps. 1, 4–6). In this chapter, we focus on the intersection of biology, conservation, and control. It is exciting to contemplate what further advances will be achieved by the end of the 21st century with continued cooperation and interaction among disciplines.

7.2 What's in a Name? Lamprey Taxonomy

Lamprey taxonomy has been in a state of flux for several decades. For example, the genus *Lampetra* has at times included *Lethenteron* and *Entosphenus* (Hubbs and Potter 1971) or *Lethenteron, Entosphenus, Eudontomyzon*, and *Tetrapleurodon* (Bailey 1980) as subgenera. Whether least brook lamprey *Lampetra aepyptera* should be included within *Lampetra* (sometimes in subgenus *Okkelbergia*; Creaser and Hubbs 1922) or placed in a genus of its own (Hubbs and Potter 1971) has also been debated (see Docker et al. 1999; Lang et al. 2009; Potter et al. 2015). Generic placement of other "relict" brook lamprey species (e.g., Po brook lamprey *Lampetra zanandreai*, Kern brook lamprey *Lampetra hubbsi*, and Western Transcaucasian brook lamprey *Lethenteron ninae*) has also changed over time or is still being debated (see Chap. 4). Some of the discrepancies have been resolved with molecular phylogenetic analysis (Potter et al. 2015); others (e.g., the "paired species problem") have intensified as the result of genetic analyses (Docker 2009).

The total number of species recognized by different authors can also vary, ranging—even in recent years—from 40 (Renaud 2011) to 41 (Potter et al. 2015) to 44 (Maitland et al. 2015). Conflicting "final tallies" are often based on different interpretations of new studies (Mateus et al. 2013a), different species concepts (Docker et al. 2015), and different levels of caution regarding adoption of changes prior to comprehensive investigation and comparison to type material (Renaud 2011; Potter et al. 2015). Even higher species counts may result from further "splitting" of existing taxa (e.g., a new brook lamprey distinct from the Po brook lamprey has been described in the southern Adriatic Sea basin; Tutman et al. 2017), and more as-yet-undescribed species may exist (e.g., Yamazaki et al. 2003, 2006; Boguski et al. 2012; Levin et al. 2016). At the same time, there have been calls for "lumping" of some lamprey paired or satellite species into a single species (e.g., Artamonova et al. 2011; see Docker 2009; Chap. 4).

The goal of this section is to briefly discuss how these debates, which may sometimes appear esoteric to all but the most ardent taxonomists, are relevant to lamprey managers and researchers. At issue is nomenclature and classification at both the lower ranks (what do I call this particular lamprey?) and at higher levels of classification (does the taxonomy accurately reflect the evolutionary relationships among lampreys?). Although taxonomic revisions can be confusing, they are necessary when new information allows us to refine our hypotheses regarding the evolutionary relationships among lampreys. A species' name should: (1) enable the accurate transfer of knowledge across time and space; (2) be founded on objective evidence of species boundaries yet remain open to challenge in light of newly acquired data; and (3) accurately reflect the species' evolutionary history, so that researchers can generate informed conclusions about its biology (e.g., based on knowledge of its closest relatives) and make accurate inferences regarding the way in which characters change over time (e.g., the direction and frequency of morphological or life history changes). Consistency of species names is also critical in management and conservation decision-making and legislation.

Nevertheless, it is reassuring that most conservation legislation acknowledges the importance of biological diversity below the species level (e.g., distinct population segments and designatable units). Thus, mechanisms exist to extend protection to distinct populations or forms without "oversplitting" or taxonomic inflation (Chaitra et al. 2004; Isaac et al. 2004). Following Potter et al. (2015), we recommend that there be no major revision of lamprey classification until more comprehensive studies can be completed and recommend against species-by-species ("piecemeal") revisions. Making changes that are short-lived has the effect of confusing rather than improving the situation (Page et al. 2013).

Taxonomic debates are not unique to lampreys. The American Fisheries Society's (AFS) *Common and Scientific Names of Fishes from the United States, Canada, and Mexico* requires regular updating (Page et al. 2013), and attempts are being made to develop a more explicitly phylogenetic classification of bony fishes (Bentacur-R et al. 2017). Similar species lists and classification efforts have been developed for other regions and taxa (e.g., Kottelat and Freyhof 2007; Crother et al. 2017). An explicitly phylogenetic classification of lampreys may be challenging given their dearth of taxon-distinctive morphological characters. On a positive note, however, the task may be manageable in lampreys given the relatively few species compared to other vertebrate classes.

7.2.1 Taxonomy and Communication

One fundamental role that species names play is to enable accurate communication between researchers and managers from different geographic regions. Consistent taxonomy is important to guarantee that we are all talking about the same organism and to ensure that existing and new information can be linked to the correct organism (Pante et al. 2014). Bad taxonomic practices can result in two parties referring to the same species with a different name or to different species with the same name. Accurate species lists are also necessary for thoughtful national and international planning (e.g., protecting areas with high species richness).

Although lamprey taxonomic work is still underway and not all authorities recognize the same species, this does not mean that all proposed classifications are equally valid. First and foremost, because classification changes over time, biologists need to be aware that older references likely used outdated classification schemes. For example, although *The Freshwater Fishes of Canada* by Scott and Crossman (1973) remains a treasure trove of information regarding the biology and distribution of many North American lamprey species, the scientific name for American brook lamprey *Lethenteron appendix* is no longer *Lampetra lamottei*, and some freshwater populations of Pacific lamprey *Entosphenus tridentatus* have since been described as distinct species (e.g., Vladykov and Kott 1979; Beamish 1982; see Chap. 4). When in doubt, biologists should refer to up-to-date and accepted classifications (e.g., Page et al. 2013; Potter et al. 2015; Froese and Pauly 2018), and, when there are discrepancies even among these authorities (e.g., regarding the number of recognized lamprey species), the source should be cited.

Among lampreys, one of the most notable examples of taxonomic confusion in recent years is the Korean lamprey (also referred to by some authors as the Northeast Chinese lamprey; Zu et al. 2016). The valid name of this species is *Eudontomyzon morii* (Berg 1931), but it has been variously referred to as *Lethenteron morii* (Lang et al. 2009; Pu et al. 2016; Zu et al. 2016), *Lampetra morii* (Li et al. 2016; Yan et al. 2016), and *Lampetra morri* (Peng et al. 2016). Phylogenetic analysis of mitochondrial DNA (mtDNA) suggests that the Korean lamprey is more closely related to the Arctic lamprey *Le. camtschaticum* and its satellite species than it is to other *Eudontomyzon* species (Lang et al. 2009; see Chap. 4), but this has not been fully resolved nor formally recognized (Potter et al. 2015). The Lang et al. (2009) study relied on a single metamorphosing individual with under-developed dentition, and Li (2014) and White (2014) used this same specimen. It is reassuring that the mtDNA sequence of this one specimen was sufficiently distinct from any known species to suggest that it was not merely a misidentified Arctic lamprey or *Lethenteron* brook lamprey, but independent confirmation on fully metamorphosed specimens is needed. The means by which specimens of Korean lamprey were identified in more recent studies were not reported. Based on their geographical distribution (Li et al. 2016) or near-identity to Arctic lamprey and its satellite species with respect to mtDNA sequence (Peng et al. 2016; Pu et al. 2016; Yan et al. 2016), these studies may have used specimens of Arctic lamprey (Peng et al. 2016; Pu et al. 2016; Yan et al. 2016) or Far Eastern brook lamprey (Li et al. 2016). It is exciting to see papers being published on this enigmatic lamprey species, but it is important that the new discoveries be unambiguously linked to the correct organism. Accurate communication with non-scientists, including policy makers, is also important. Peng et al. (2016) indicate that *Lampetra morri* [sic] is "one of the most important freshwater aquaculture species in China." If the authors are actually referring to the more widespread Arctic lamprey, policy makers might be misled into thinking that the Korean lamprey is more abundant than it is in reality.

Nomenclature of the Arctic lamprey itself is also confused. It is still frequently referred to as *Lampetra japonica* or *Lethenteron japonicum*, and by the common name Japanese lamprey or river lamprey (e.g., Mehta et al. 2013; Kawai et al. 2015). Kottelat (1997) showed *Le. japonicum* (von Martens 1868) to be a junior synonym of *Le. camtschaticum* (Tilesius von Tilenau 1811). Thus, according to the principle of priority in the International Commission on Zoological Nomenclature (ICZN), *camtschaticum* is the valid species epithet and reference to these other names should be discontinued to avoid confusion (see Renaud et al. 2009). Although there are no specific prohibitions against using the name Japanese lamprey, because there are no conventions regulating common names, Arctic lamprey is more appropriate given the broad circumpolar distribution of this species (Potter et al. 2015). Arctic lamprey is the common name accepted by the AFS, the International Union for the Conservation of Nature (IUCN), and the Food and Agriculture Organization (FAO) of the United Nations (Renaud 2011).

7.2.2 A New Taxonomic Framework

Like most other taxonomic groups, the existing classification of lampreys does not follow an explicitly phylogenetic framework, that is, one in which the classification intentionally reflects the evolutionary relationships among taxa. Like traditional fish classifications in general, lamprey classification mixes taxa with explicit phylogenetic support with traditional but subjective groupings based on "deep-rooted anatomical concepts" (Bentacur-R et al. 2017). Traditional lamprey classification schemes, based on morphological characters alone, have resulted in some non-monophyletic taxa. Monophyletic taxa (i.e., those that consist of all the descendants of a common ancestor) are considered the only "natural" taxa. In contrast, evolutionary relationships are obscured by paraphyletic taxa (i.e., those whose members are descended from a common ancestor but where other descendants of this ancestor are subjectively excluded from the taxon) and polyphyletic taxa (i.e., where members are derived from more than one common ancestor). For example, the genus *Lampetra* is non-monophyletic despite the rather stunning morphological similarity of European river lamprey *Lampetra fluviatilis* and western river lamprey *La. ayresii* (see Potter et al. 2015). Recognizing their phylogenetic distinctiveness will allow researchers to better understand the way in which dentition and other characters change over time. For example, it may be that these species developed the same dentition pattern independently as an adaptation to similar hosts or that both retained the ancestral pattern of dentition due to similar constraints related to feeding (Li 2014). An explicitly phylogenetic approach based on a comprehensive molecular phylogeny was recently applied to the bony fishes with success (Bentacur-R et al. 2017). An explicitly phylogenetic classification for lampreys will likewise rely heavily on molecular phylogenies, given the paucity of morphological and meristic characters (see Docker et al. 1999; Lang et al. 2009; Potter et al. 2015). However, one drawback of this approach is that molecular phylogenies to date in lampreys rely almost exclusively on mtDNA sequence data, which lack the necessary resolution to resolve some relationships (e.g., among *Lethenteron*, *Eudontomyzon*, and European *Lampetra*). A robust molecular phylogeny of lampreys will require the use of a wider range of genes and particularly of nuclear genes (Potter et al. 2015).

Likewise, resolution of the paired species problem, one of the major conundrums regarding lamprey taxonomy (see Docker 2009; Chap. 4), cannot be resolved with mtDNA data alone. Analysis of mtDNA sequence data is revealing that many paired species are not reciprocally monophyletic (i.e., with all non-parasitic members of a pair sharing an ancestor more recently with all other non-parasitic members of that pair than to any parasitic individuals and vice versa), but DNA barcode (i.e., cytochrome c oxidase sub-unit I, COI) sequencing is insufficient to resolve the issue. For example, individuals that are barcode indistinguishable should not be summarily lumped into a single species (e.g., Artamonova et al. 2011). Resolving this conundrum requires more than just lumping paired parasitic and non-parasitic species into one monophyletic taxon or splitting polyphyletic non-parasitic species into multiple monophyletic units. At the heart of the problem is an understanding of what constitutes a species, and we would do well in this regard to recognize that the "species problem" is not unique to lampreys (e.g., Taylor 1999; Vogler and Monaghan 2007; Hendry et al. 2009; Klemetsen 2010). Most biologists agree that species are evolutionarily independent units that are isolated by a lack of (significant) gene flow, but there is a lack of consensus on how, in practice, these evolutionarily independent units are recognized (Mayden 1997; de Queiroz 2007; see Docker et al. 2015). de Queiroz (2007) suggested that the root cause of the problem is confusing the issue of "species delimitation" with "species concepts." Thus, if we focus on species delimitation (i.e., the process by which populations are identified as being diagnosably distinct and reproductively isolated) rather than the subjective arguments regarding what constitutes a species, then the problem becomes focused on the methodology of inferring species boundaries.

An integrated framework that combines morphology, molecular phylogenetic and population genetic approaches, and behavioral ecology—while recognizing that all sources of data are not equal all of the time—is likely to be of greatest utility (e.g., Schlick-Steiner et al. 2010; Carstens et al. 2013; Harrison and Larson 2014; Pante et al. 2014, 2015). Rather than adopting a strictly morphological or strictly molecular phylogenetic approach, integrating both with population genetic and behavioral information should be considered. For example, ecological and genetic studies are showing that sympatric European river lamprey and European brook lamprey *Lampetra planeri* maintain distinct phenotypes even with contemporary gene flow (Rougemont et al. 2015, 2017), while some other paired species appear to exhibit temporal, spatial, or behavioral differences that prevent gene flow (see Docker 2009; Chap. 4). Such information will help us better recognize evolutionary independence. Therefore, more explicit convergence of previously disparate research fields will allow us to establish criteria for species boundaries based on objective evidence from a range of data sources, and it will provide a framework that explicitly considers both evolutionary history and the dynamic nature of species boundaries.

7.2.3   Conservation of Diversity Below the Species Level

Despite the critical importance of correct species identification and nomenclature, most conservation legislation also acknowledges the importance of biological diversity below the species level. This is reassuring given current taxonomic uncertainties (that may be resolved with time) and the fluidity of species boundaries within young species (that may not be easily resolvable). Recognition of biological diversity below the species level can ensure that genetically or morphologically distinct populations are afforded protection even if they fall under the umbrella of a widespread and abundant species (i.e., are identified by the same Latin binomial). Ryder (1986) coined the term "evolutionarily significant unit" (ESU) to recognize critical diversity below the species level in mammals, and Waples (1991, 1995) applied it to Pacific salmon *Oncorhynchus* spp. According to Waples (1991), an ESU is a population or group of populations that: (1) is substantially (but not necessarily completely) reproductively isolated from other conspecific population units; and (2) represents an important component of the evolutionary legacy of the species. ESUs now have important legal ramifications under the U.S. Endangered Species Act (ESA), the Canadian Species at Risk Act (SARA), and similar legislation in other countries (Fraser and Bernatchez 2001). Under the ESA, a distinct population segment (DPS) is a vertebrate population or group of populations that is discrete from other populations of the species and significant in relation to the entire species (United States Fish and Wildlife Service 1996). Similar designatable units (DUs) are recognized under SARA when a single status designation is thought not to reflect the extent of evolutionarily significant diversity within a species (COSEWIC 2015).

For example, western brook lamprey *Lampetra richardsoni* in Morrison Creek, British Columbia, is recognized as a distinct DU given its apparent ability to produce both non-parasitic and parasitic forms from a common gene pool (Beamish 1987; Docker 2009; see Chap. 4). Despite the fact that the species as a whole occurs in innumerable tributaries from Alaska to California, the Morrison Creek DU is listed as Endangered under SARA (Renaud et al. 2009). *Lethenteron* sp. N and sp. S are listed as Vulnerable in Japan, although they have yet to be formally described as species (Maitland et al. 2015). Other genetically distinct and geographically isolated brook lamprey populations likewise appear to represent non-interchangeable ESUs, regardless of whether they are recognized as distinct Linnean species (e.g., Martin and White 2008; Pereira et al. 2011, 2014; Mateus et al. 2013a; Tutman et al. 2017). In contrast, where brook lamprey populations occur sympatrically with their parasitic counterpart, the migratory parasitic species may facilitate gene flow among populations and between species, in which case all populations of both species in that region might be best managed as a single ESU. This would not preclude us from considering these same Linnean species as distinct ESUs in cases where they are highly differentiated and there is no evidence of contemporary gene flow between them. An integrated framework will allow us to recognize and protect those lamprey populations that are important to the evolutionary legacy of the species without prematurely or inconsistently describing each as a distinct species. Finer level subdivisions (e.g., stocks or management units) are also important to conservation and management (see Sect. 7.3.5), but they are often not afforded the legal protections enjoyed by listed entities.

7.3 New Tools for Lamprey Conservation and Control

As with many other aspects of science, a furthering of our understanding of lamprey biology goes hand in hand with the development of new technologies. Novel methodologies can reveal facets of lamprey biology not previously observed, and these observations can drive further improvements to methodology.

7.3.1 Deepwater Sampling

It is generally thought that larval lampreys most commonly inhabit shallow stream environments, typically in water depths <1 m (Dawson et al. 2015). Nevertheless, it has been observed for decades that larvae can also reside in deeper lake waters near river mouths (Hansen and Hayne 1962; Wagner and Stauffer 1962; Lee and Weise 1989; Bergstedt and Genovese 1994), and larval and metamorphosing lampreys are being increasingly detected in deep water in large river systems (e.g., Beamish and Youson 1987; Jolley et al. 2012; Taverny et al. 2012; Harris and Jolley 2017). Although the majority of larval lampreys in deepwater environments at river mouths are thought to be a consequence of natural downstream drift (Fodale et al. 2003; see Dawson et al. 2015), it is not clear to what extent these habitats contribute to recruitment. In species of conservation concern, dredging of large rivers for navigation could result in significant larval mortality and loss of larval habitat (Maitland et al. 2015). Conversely, lentic areas could provide "safe havens" for sea lamprey larvae in the Great Lakes basin, particularly if use of lampricides has resulted in earlier or more extensive migration of larvae out of more easily treated tributaries (Dunlop et al. 2018). Larval sea lamprey populations in large rivers also pose significant challenges to control in the Great Lakes basin. Fodale et al. (2003) estimated that the total population of sea lamprey larvae in the St. Marys River, the interconnecting waterway between Lake Superior and Lake Huron, was 5.2 million. The St. Marys River is more than 20× larger in volume than the largest tributary ever treated with the lampricide 3-trifluromethyl-4-nitrophenol (TFM), necessitating treatment with bottom-release lampricide formulations (Schleen et al. 2003) and prompting exploration of alternative control methods in this river (e.g., sterile-male-release technique; see Chap. 5).

Early attempts at sampling larvae in deep water largely relied on deepwater electrofishing to stimulate emergence of larvae from the substrate coupled with a trawl or suction pump to collect them (McLain and Dahl 1968; Bergstedt and Genovese 1994). These methods (or dredging alone; Beamish and Youson 1987) were costly in relation to backpack sampling and likely caused habitat destruction. Furthermore, lack of a standardized methodology and knowledge of detection efficiencies limited the application of these sampling methods by management agencies and researchers interested in estimating abundance. Subsequent improvements have included incorporation of a remote seabed classification device that uses acoustic sonar to identify substrate suitable to sea lamprey larvae (Fodale et al. 2003) and incorporation of an optical camera to detect larvae without the need to bring them to the surface (Mueller et al. 2012). In the western United States, initiatives to more accurately assess the distribution and abundance of Pacific lamprey larvae (Wang and Schaller 2015; Reid and Goodman 2015, 2016a; Clemens et al. 2017a) have seen a push to evaluate the detection efficiency of different deepwater sampling methods under different conditions and for different larval size classes. For example, Jolley et al. (2012) demonstrated that a deepwater electrofisher with a suction pump was able to detect larval lamprey measuring 20–144 mm total length (TL) at depths up to 16 m. Reach- and quadrat-specific detection probabilities were 0.07 and 0.23, detection probability did not differ with depth, and the sampling effort required for 80% confidence that larval lamprey were indeed absent when undetected was 20 quadrats, each 30 m^2. Using Pacific lamprey larvae seeded into hatchery chambers, Harris and Jolley (2017) experimentally estimated capture probability of their deepwater electrofishing method to be 0.70, which is comparable to the shocking efficiency (50–80%) reported by Bergstedt and Genovese (1994). Using their video-based deepwater electroshocking platform, Arntzen and Mueller (2017) reported that observation rates at three locations within the Columbia River basin (in water depths 0.8–4.5 m) ranged from 21 to 61% and averaged 56%. Video data analyzed in the laboratory allowed these authors to estimate sediment type and larval length. Observed larvae were estimated to range from <50 to 150 mm TL; 70% of larvae were ≤75 mm TL, although Arntzen and Mueller (2017) suggested that fewer large larvae may have been detected due to their tendency to burrow deeper in the substrate (see Dawson et al. 2015). This method appears to be useful for determining presence/absence of larval lampreys at moderate depths without requiring sediment dredging, but its usefulness will be limited under conditions of restricted visibility (e.g., high turbidity or at depths greater than ~6.5 m without artificial lighting). Further research evaluating sampling efficiency under different conditions (cf. Steeves et al. 2003 for backpack electrofishing) could encourage management agencies in other regions to consider these methods where practical.

Despite recent marked improvements in our ability to detect the presence of larval lampreys and even estimate abundance (Fodale et al. 2003; Harris and Jolley 2017), our general understanding of larval growth, survivorship, and persistence in deepwater habitats still lags behind that of stream populations. Johnson et al. (2016) suggested that larval annual survival rate was marginally higher in deepwater lentic areas (63%) compared to shallow tributary streams (40–57%; Jones et al. 2009; Irwin et al. 2012; Johnson et al. 2014), and larval survival in the St. Marys River was even higher (66–91%; Jones et al. 2015). However, larval growth rates in Great Lakes deepwater habitats is estimated to be 2–4× lower than in shallow streams (albeit from only six tagged individuals), resulting in a longer developmental period prior to metamorphosis and possibly higher overall larval mortality as a consequence (Johnson et al. 2016). Clearly, much is still unknown about the productivity of deepwater habitats and they warrant further study.

7.3.2   Tools to Monitor the Spawning Migration

The terminal reproductive migration is a comparatively well-studied part of the life cycle for several lamprey species, yet we still lack a deep understanding of their swimming capacity, motivation, and behavior during this period (see Moser et al. 2015). Understanding these aspects of the upstream migration is important to effectively mitigate the negative impact of anthropogenic barriers on species of conservation concern (Maitland et al. 2015; Kemp 2016) and to impede the spawning migration of sea lamprey in the Great Lakes (see Chap. 5). For example, until very recently, nothing was known of how lampreys that feed at sea or in large lakes locate coastal regions before entering rivers to spawn. Meckley et al. (2017) found that acoustically tagged sea lamprey in the Great Lakes located the coast by moving towards shallower water, presumably by sampling hydrostatic pressure during periodic dives to the lake bottom and then following decreases in depth. While at the surface, sea lamprey may detect and respond to olfactory cues emanating from river water, enabling them to orient towards the river mouth and initiate upstream movement to the spawning grounds (Meckley et al. 2014).

The timing of river entry is variable within and among species (Moser et al. 2015). Sea lamprey generally enter rivers a few months before spawning, but pouched lamprey *Geotria australis* and most Pacific lamprey generally enter fresh water ≥1 year prior to spawning. The reason for such "premature migration" in these latter species is not clear (Quinn et al. 2016; see Chap. 4). Early river entry reduces growth opportunities, but it may be favored when the freshwater environment provides moderate temperatures and flows, relative safety from predators, and reduced energetic expenditures compared to the sea or lake environment (Quinn et al. 2016). Climate change might affect the relative benefits and risks of this migration strategy in lampreys, for example, by accelerating the body shrinkage experienced by upstream migrants following freshwater entry (Clemens et al. 2009). Within species, timing of river entry often varies with environmental conditions such as water temperature and discharge (Moser et al. 2015), and knowledge of these triggers has important applications to conservation and control. For example, earlier sea lamprey migration in some Great Lakes tributaries may improve young-of-the-year growth and survival (McCann et al. 2018a).

Furthermore, the long-held belief that maturing sea lamprey in the Great Lakes "staged" off the mouths of spawning streams until a temperature threshold (10 °C) was reached (Applegate 1950) has, in part, guided the timing of setting barrier-associated traps to capture migrants to index their abundance. However, dual-frequency identification sonar (DIDSON) has recently revealed that sea lamprey enter streams at temperatures as low as 4 °C up to 6 weeks prior to their first appearance in traps (McCann et al. 2018b). This finding has major implications for sea lamprey control because seasonal barriers are sometimes used to block access to spawning habitat (McLaughlin et al. 2007). If seasonal barriers are not erected early enough, then a portion of adults may escape upstream and reproduce. DIDSON can produce high-resolution near-video-quality images in low-visibility water. Fishes as small as a few centimeters in length can be imaged, and they can be detected at distances exceeding 40 m. This method is increasingly being used in fisheries monitoring applications and to observe fish behavior (Martignac et al. 2015). However, although DIDSON works well for detecting sea lamprey at relatively low densities (McCann et al. 2018b), this technology appears susceptible to observer bias. Keefer et al. (2017) found that, when observing Pacific lamprey migrations using DIDSON, among-viewer variability in lamprey enumeration was high, and individual viewers scored only 32–63% of the total observations. Keefer et al. (2017) suggested that these differences were due to a combination of viewer familiarity, image duration and orientation, and lamprey density. Therefore, although DIDSON is a remarkable tool with positive applications to lamprey management, ensuring good quality control and proper set-up will be important.

Advances in telemetry methods, as well as more sophisticated statistical approaches and reduction of costs associated with these methods (DeCelles and Zemeckis 2014), are sure to result in more fascinating insights into lamprey spawning migrations in the near future. Early work by Almeida et al. (2002) using radio-tagged anadromous sea lamprey revealed in-stream barriers to be a serious impediment during the spawning migration of this valued species in Portugal, but it also identified areas that were important for completion of its life cycle. Rooney et al. (2015) similarly used radio telemetry in anadromous sea lamprey in Ireland to monitor passage success at artificial weirs prior to and following weir modifications. Pacific lamprey behavior at dams (e.g., Keefer et al. 2010, 2013; Mesa et al. 2010; see Moser et al. 2015) and in unobstructed rivers (Clemens et al. 2017b) has likewise been evaluated using radio telemetry. Using a combination of radio and acoustic telemetry, Holbrook et al. (2015) found that sea lamprey migrate close to the river substrate and often at channel edges, suggesting a preference for low water velocities. This is similar to findings from Pacific lamprey observed in artificial raceway environments (Reid and Goodman 2016b). These findings have implications for the positioning of traps to control sea lamprey (where acoustic telemetry has also confirmed spatial mismatch between trap positioning and lamprey movement tendencies; Rous et al. 2017) and for the location and design of fish passage devices for native lampreys (Reid and Goodman 2016b; Castro-Santos et al. 2017; Goodman and Reid 2017). Holbrook et al. (2016), concerned that sea lamprey recruitment from the St. Clair-Detroit River system was a significant new source of parasites in Lake Erie, used acoustic telemetry to help narrow the search for spawning locations in this large river system.

Acoustic telemetry has also shown that the anadromous European river lamprey, against expectation, ignores potential energetic savings during initial entry into rivers by not taking advantage of selective tidal stream transport (STST; Silva et al. 2017a). Other migratory fishes (especially those that, like lampreys, are poor swimmers) often take advantage of water currents to migrate. For example, fishes using STST to move upstream will move into strong inland currents when the water level is rising and take refuge out of the current to avoid the ebb (seaward) tide. Silva et al. (2017a) suggested that European river lamprey do not use STST due to fitness costs (e.g., risk of predation) separate from energetic expenditure, but the use of STST may vary under different environmental conditions.

Counts of lampreys undertaking spawning migrations are a useful metric to evaluate the effectiveness of control measures (see Chap. 5) and of conservation initiatives (e.g., whether population trends are positive or negative following intervention), and they are also important for estimating absolute abundance. Lake-wide sea lamprey abundance is estimated in each of the Great Lakes based on the number of upstream migrants captured in index streams (Mullett et al. 2003), and Pacific lamprey are enumerated from video recorded at "count stations" at dams in the Columbia River basin. However, the Pacific lamprey count stations were designed to enumerate salmonids during the day rather than anguilliform lampreys migrating at night, and manually counting lampreys from video is labor intensive and fraught with uncertainty due to observational error and complex movement patterns (Clabough et al. 2012). An advanced automated video counting system (FishTick), which builds on widely adopted systems of the 1990s that save images only when motion is detected, has been tested for Pacific lamprey (Fryer 2008). This design is intended to reduce the amount of data that must be reviewed by human observers, but Fryer (2008) reported that lamprey attached to the viewing window and those swaying passively in the current were not ignored by the computer algorithm (i.e., were counted as moving lamprey), and the abundance of lamprey was too high to reduce the human effort required to review the data. More recently, by using a new algorithm that removes empty video frames (in conjunction with submerged cameras), Negrea et al. (2014) showed that it was possible to reduce video observation time by >80% while detecting 99% of Pacific lamprey. Thus, rapid advancements are being made in this methodology, and its application could spread to other lamprey species.

7.3.3 Environmental DNA

Rapid advances in genetic methodologies are similarly providing significant opportunities to aid lamprey conservation and control efforts (see also Sect. 7.3.5). This includes the field of "environmental DNA" or eDNA, a term which was unfamiliar to most fish biologists 5–10 years ago but which is now becoming commonplace, as is evidenced by several special eDNA issues in journals related to ecology and conservation (e.g., Tablerlet et al. 2012; Goldberg et al. 2015) and by the newly launched Wiley journal *Environmental DNA*. Organisms shed DNA into their surrounding environment (e.g., via mucus, feces, gametes, or sloughed off cells), and detection of this species-specific genetic material in easily collected water samples offers a promising and economical alternative or supplement to traditional survey approaches (Rees et al. 2014; Goldberg et al. 2016; Port et al. 2016). Studies in other aquatic species have shown that eDNA surveys can be more sensitive than traditional electrofishing or other collection methods for determining species presence, particularly at low population densities (e.g., Biggs et al. 2015; Sigsgaard et al. 2015; McKelvey et al. 2016; Wilcox et al. 2016). This enables detection of rare endangered species or those in the early stages of invasion with relatively little field effort and without the need for specialized equipment and experienced operators. Such non-invasive sampling also reduces impacts on species or habitats sensitive to physical disturbance. However, the relative effectiveness of eDNA and traditional surveys are often not rigorously compared (Roussel et al. 2015), and some studies have concluded that traditional techniques are more effective than eDNA at detecting rare aquatic species, particularly in flowing waters (e.g., Ulibarri et al. 2017).

Assays for eDNA have been developed for anadromous sea lamprey to monitor passage across in-stream barriers (Gustavson et al. 2015) and for Pacific lamprey (Carim et al. 2016) and *Lampetra* spp. (Ostberg et al. 2018) to assess their distribution and evaluate conservation status. eDNA assays have also been developed to detect invasive sea lamprey in the Great Lakes and distinguish it from the native lampreys in this basin (American brook lamprey, chestnut lamprey *Ichthyomyzon castaneus*, and northern brook or silver lampreys *I. fossor* and *I. unicuspis*) (Gingera et al. 2016; Schloesser et al. 2018). Monitoring with eDNA can reduce the level of effort traditionally required for assessing lamprey presence (e.g., its presence in inaccessible upstream reaches can be inferred by eDNA detection at downstream locations). In addition to detecting presence, Gustavson et al. (2015) and Gingera et al. (2016) noted that sea lamprey eDNA concentration or detection frequency increased during the spawning season, and Ostberg et al. (2018) reported *Entosphenus* spp. and *Lampetra* spp. eDNA was similarly higher in the spring than in the fall. This is likely because the free-swimming, large-bodied adults, their gametes, and later their carcasses likely release more DNA into the water than the small, burrowed larvae (Gingera et al. 2016). Such eDNA "spikes" coincident with spawning could be used in monitoring programs (for both conservation and control) to identify streams with reproducing lampreys. Although spawning lampreys are visible in many stream systems (i.e., when they spawn in shallow riffle areas of clear streams), they are far more difficult to detect in larger and more turbid river systems or lentic areas (Johnson et al. 2015a).

Nevertheless, despite its great promise and utility to date, study design and interpretation of eDNA results should be carefully considered (Goldberg et al. 2016). The strengths of eDNA (i.e., its high sensitivity and ability to detect species presence without capturing or even seeing the organisms) can also cause drawbacks. Thus, stringent quality assurance and quality control are very important if the results are to be useful. For example, false positives (i.e., when eDNA tests are positive but the species is not there) can be caused by methodological errors, such as contamination (where DNA detected in the assay came from a source outside of the system, such as when equipment or reagents came in contact with the organism or its DNA) or non-specific amplification (where the assay erroneously amplified DNA from one or more non-target species). False positives can also result when DNA of the species of interest is present in the water sampled and correctly identified, but the species itself is not present in the system (e.g., when DNA is transported into the system from another source such as water flow, boat movement, or a predator's feces; Mahon et al. 2013; Merkes et al. 2014). Nevertheless, false positives caused by methodological errors can be eliminated or reduced with adherence to strict "clean" procedures (e.g., ensuring that filtration equipment and collected samples do not come into contact with the organism or its DNA), and non-specification amplification can be prevented by careful development and testing of the assay to ensure that DNA from only the target species is amplified (Wilcox et al. 2013). False positives or false detections caused by transport of the species' DNA into the system from other sources are harder to detect. Such situations require follow-up sampling to assess reproducibility (Rees et al. 2014). This will be particularly critical when the cost of initiating management actions (e.g., applying lampricide to a large stretch of river) is high.

False negatives (i.e., failure of the assay to detect the DNA of the species when it is there) can result when the sensitivity of the assay is not sufficient for detection of low-quantity, low-quality DNA. Dilution of eDNA in the water sample (e.g., due to high flow rates), or the presence of inhibitors (e.g., humic acids) that interfere with the PCR (see Gingera et al. 2016), can also result in false negatives. Nevertheless, improved amplification methods (e.g., probe-based quantitative PCR or digital drop PCR), increased sampling effort (e.g., increasing the volume of water collected per sample and the total number of water samples), and use of methods that reduce or detect the effect of inhibitors on the PCR reaction can help to reduce the incidence of false negatives (e.g., Nathan et al. 2014; McKee et al. 2015). Distinguishing false from true negatives will be particularly critical if the cost of inaction (e.g., failing to treat an infested stretch of river or identifying critical habitat for an imperiled species) is high.

Methods are being developed to use eDNA to estimate relative abundance (e.g., Lodge et al. 2012; Lacoursière-Roussel et al. 2016), but the effects of different environmental variables (e.g., flow rate, temperature) on eDNA detectability and signal strength have yet to be fully elucidated (Deiner and Altermatt 2014; Deiner et al. 2015; Jane et al. 2015). Furthermore, because eDNA is an indirect method of detection, inferring the presence of the species "sight unseen" (Jerde et al. 2011), it does not provide information on the size, stage, or condition of the organisms present (e.g., whether sea lamprey detected might be young-of-the year larvae, transformers that will soon migrate out of the stream and become parasitic, or spawning adults).

Therefore, although eDNA methods have the potential to revolutionize monitoring of aquatic species (particularly cryptic organisms like larval lampreys), at present, lamprey eDNA assays would be best used as a supplemental larval assessment tool rather than as a replacement for traditional survey methods. For example, Gingera et al. (2016) suggested that eDNA detection could be used as a "red flag" warning system for sea lamprey detection in the Great Lakes. Preliminary eDNA surveys could economically test for sea lamprey presence across large scales with relatively little field effort (e.g., to test for presence above barriers or in tributaries not currently known to harbor larval lamprey populations). Positive detections would warrant further investigation with traditional electrofishing, although lack of eDNA detection would not ensure that sea lamprey larvae were absent (e.g., where the eDNA signal is diluted in large rivers; Gingera et al. 2016). Nevertheless, with increased sensitivity and further refinements to eDNA methodologies, it will be exciting to see what is possible within the next 10 years.

7.3.4 Pheromone Detection

Pheromones and other semiochemicals are organic compounds used to convey information (e.g., mate availability or presence of predators) between individuals. They often occur as species-specific mixtures in water, and this makes them useful for monitoring purposes (Sorensen and Johnson 2016). Similar to eDNA, these compounds can be extracted from water samples using a relatively simple and inexpensive process, and they can then be measured with high sensitivity. Like eDNA, detection of pheromones could allow inferences to be made regarding the distribution and perhaps abundance of lampreys at a particular time of year, although not always to species (Fine et al. 2004; Buchinger et al. 2017a). However, unlike eDNA, measuring the concentration of particular pheromones in a water sample could provide insights into the reproductive condition of the organism. For example, although the male sex pheromone 3-keto-petromyzonol sulfate (3kPZS) is also released by larvae at low rates, spikes in 3kPZS could be used to identify spawning times (Johnson et al. 2015a; Brant et al. 2016; Sorensen and Johnson 2016). Furthermore, although eDNA false positives might occur if, for example, a predator sheds feces containing its prey's DNA into the system (see Sect. 7.3.3), it is thought that pheromones cannot be vectored as easily by another animal given that they are produced during relatively short periods of time (Stacey and Sorensen 2009). Methods that have been developed to detect minute environmental concentrations of lamprey pheromones (e.g., Xi et al. 2011; Stewart and Baker 2012) reduce the chance of false negatives, particularly when samples are concentrated before analysis (Fine et al. 2006). Therefore, detection of eDNA and pheromones together—from a single water sample—could provide complementary alternatives or supplements to traditional sampling methods.

Our ecologically relevant knowledge of the pheromonal mixture released by larval lampreys is largely restricted to three compounds: petromyzonamine disulfate (PADS), petromyzosterol disulfate (PSDS), and petromyzonol sulfate (PZS) (Sorensen et al. 2005; Buchinger et al. 2015; see Moser et al. 2015). The ability to detect these compounds in low concentrations is now aiding conservation efforts for pouched lamprey in New Zealand streams. Stewart et al. (2011) reported on a methodology to rapidly extract and identify these larval pheromones from stream water, and Stewart and Baker (2012) subsequently developed a method whereby passive accumulation of pheromone compounds can be achieved in situ via sorbent cartridges. This new method collects and concentrates pheromones in stream water and provides quantitative data on release rates. Unlike Northern Hemisphere lampreys (Buchinger et al. 2015), pouched lamprey principally release PZS, and PADS is released at far lower concentrations and PSDS may not be released at all (Baker et al. 2009). Thus, not only does this type of work allow for remote detection of lamprey populations, it also has the potential to reveal deeper insights into the evolution of chemical communication in lampreys (Buchinger et al. 2017a, b; Hume and Wagner 2018). Finally, pheromone detection techniques have been further refined for more rapid detection of the sea lamprey sex pheromone 3kPZS, potentially enabling estimation of adult abundance based on the concentration of the compound in stream water (Wang et al. 2013). The rate at which new pheromone compounds are being characterized is increasing rapidly (e.g., Li et al. 2017; Scott et al. 2018), as is the potential to make inferences about lamprey distribution, physiology, and behavior from their analysis. More research is still required to fully understand the limitations of these techniques (e.g., the degree to which different Northern Hemisphere species could be differentiated and the effect of stream discharge, abundance, and degradation rates on detectability), as well as what biological factors (e.g., body size) are responsible for variation in pheromone release rates (Buchinger et al. 2017c). In general, the remote sensing of lamprey populations is an area with considerable potential to improve our understanding of lamprey biology and management.

7.3.5 Other Genetic and Genomic Tools

A wealth of genetic resources are being developed for lampreys, including sequencing of the complete mitochondrial genome for a number of species (Lee and Kocher 1995; Kawai et al. 2015; Ren et al. 2015, 2016; Pu et al. 2016), development of nuclear microsatellite loci for population genetic applications (Bryan et al. 2005; Takeshima et al. 2005; McFarlane and Docker 2009; Luzier et al. 2010; Spice et al. 2011; Gaigher et al. 2013; Schedina et al. 2014), and sequencing of the complete somatic and germline genome of the sea lamprey (Smith et al. 2013, 2018) and somatic genome of the Arctic lamprey (see Mehta et al. 2013; Manousaki et al. 2016; see Chap. 6). Next generation sequencing technologies such as Restriction site Associated DNA Sequencing (RAD-Seq), which sequences small segments (totaling ~0.1–10%) of the genome, are allowing researchers to identify large numbers of single nucleotide polymorphisms (SNPs) for various applications related to lamprey evolution, ecology, and conservation (e.g., Mateus et al. 2013b; Hess et al. 2013; Hess 2016; Rougemont et al. 2017; Hume et al. 2018a). At the same time, new molecular and genetic tools (e.g., transgenesis and gene knockdown) have been adapted for use in lampreys (e.g., Kusakabe et al. 2003; Square et al. 2015; Zu et al. 2016; see McCauley et al. 2015). These new tools have been particularly useful for evo-devo studies (see Chap. 6), but they may also have application to sea lamprey control (e.g., Heath et al. 2014; Thresher et al. 2019a; see Chaps. 1, 5 and 6). These methodologies are bringing together various fields of study related to the biology, conservation, and control of lampreys, and they will no doubt do so to an even greater degree in the next few decades. Managers of natural resources, such as those protecting species or controlling them, have never had such a diverse array of tools to bring to bear on the multitude of challenges they currently face (Shafer et al. 2015; Allendorf 2017; Corlett 2017).

We briefly review here a few of the applications of these genetic and genomic tools to lamprey conservation and control (see also Sect. 7.3.3). Development and application of other innovative approaches to sea lamprey control—for example, the sterile-male-release technique (Bergstedt and Twohey 2007; Bravener and Twohey 2016) and the use of pheromones and alarm cues to disrupt sea lamprey behavior or increase trapping efficiency (e.g., Hume et al. 2015; Sorensen and Johnson 2016)—are reviewed in Chap. 5.

7.3.5.1   Genetic Species Identification

Research and conservation efforts in lampreys are often hampered by the fact that lampreys can be difficult to morphologically identify to species (or sometimes even to genus) during their long larval stage. This is particularly true for smaller larvae (Neave et al. 2007; Docker et al. 2016), resulting in significant information deficits in regions where multiple species co-occur but where, by necessity, data for the larval stages are lumped together (e.g., Schuldt and Goold 1980). Morphological methods for species identification (ID) have been improved by rearing wild larvae through metamorphosis to confirm identify (Richards et al. 1982) or by artificially propagating larvae of known parentage (Meeuwig et al. 2006; see Chap. 2). Genetic species ID has enabled additional fine-tuning of morphological ID methods or have replaced such methods in situations when morphological ID is shown to be unreliable.

As a result of several taxonomically comprehensive datasets now being available, many lamprey species can have their identity confirmed through comparison of their mitochondrial cytochrome b (e.g., Lang et al. 2009; Boguski et al. 2012) or COI (e.g., Hubert et al. 2008; April et al. 2011) DNA sequence to that of specimens of known identity. However, there are two important caveats. First, few paired and satellite species possess diagnostic differences in their mtDNA sequences (e.g., April et al. 2011; Artamonova et al. 2011; Docker et al. 2012; see Chap. 4). Members of a pair may be distinguishable in some populations (e.g., non-sympatric populations of western river and brook lampreys; Boguski et al. 2012), but range-wide species-specific mtDNA markers do not exist for most pairs. Likewise, closely related parasitic *Entosphenus* species (e.g., Pacific lamprey and Vancouver lamprey *E. macrostomus*) lack diagnostic differences in their mtDNA sequences (Lang et al. 2009; Taylor et al. 2012). Genetic ID methods have been developed that can distinguish *Entosphenus* spp. from *Lampetra* spp., which in itself is very useful (see below), but they are not capable of distinguishing among closely related *Entosphenus* or *Lampetra* species (Docker et al. 2016). The second caveat is that the reliability of this method is only as good as the available reference sequences. Misidentification or miscommunication of the species in the database will lead to misleading results (see Sect. 7.2.1), and users of the database need to be aware that intraspecific (e.g., geographic) or otherwise unappreciated genetic diversity may confound results. For example, Lang et al. (2009), although including representatives of almost all known lamprey species, includes few representatives of each species, and their data set does not begin to capture the genetic diversity of *Lampetra* on the west coast of North America (Boguski et al. 2012). Where funds are available, best practice should include archiving small, non-lethal tissue samples of lampreys being surveyed. This would generate a bank of material useful in large-scale comparative studies and would aid in the identification of specimens with ambiguous species designations.

Genetic species ID has been used to good effect in recent years, even in very small larvae. Direct cytochrome b sequencing helped confirm the first known occurrence of American brook lamprey in Maine (Aman et al. 2017) and the identity of a single western river lamprey above three dams in the Columbia River (Jolley et al. 2016). In addition to direct sequencing, cost-effective restriction fragment length polymorphism (RFLP) assays have been developed that rapidly screen the amplified mtDNA fragment for species- or genus-specific differences at a restriction enzyme's recognition ("cut") site. An RFLP assay that unambiguously distinguished chestnut lamprey from northern brook and silver lampreys allowed researchers to determine at what size (≥ 105 mm TL) the presence or absence of pigmented lateral line organs definitively distinguished chestnut lamprey from the latter two species (Neave et al. 2007). Similarly, an RFLP assay unambiguously distinguished the genera *Entosphenus* and *Lampetra*, allowing Goodman et al. (2009) to test the ability of caudal fin pigmentation to identify larvae ≥ 60 mm TL to genus. Urdaci et al. (2014) developed genetic methods to distinguish sea lamprey from European river and brook lampreys, because morphological differences are often unclear in larvae <60 mm TL.

Other recent assays have been developed that are even more amenable for high-throughput applications. Docker et al. (2016) found that size differences in a microsatellite locus could rapidly and cost-effectively distinguish *Entosphenus* spp. from all known North American west coast *Lampetra* spp. Using this assay, Pacific lamprey larvae were detected upstream of the former site of the Condit Dam in the White Salmon River, Washington, 3 years after dam removal (Jolley et al. 2018). Larvae (n = 13) measured only 26–67 mm TL, which would have made it difficult or impossible to morphologically distinguish them from western brook lamprey that were already present in some areas above the dam. Genetic species ID was thus able to unambiguously confirm that Pacific lamprey had started to naturally recolonize this river following removal of the impassable dam (see Sect. 7.4.1). Using RAD-Seq,

Hess et al. (2015) identified two SNPs capable of separating *Entosphenus* spp. from *Lampetra*. These markers showed the presence of Pacific lamprey larvae (<46 mm TL) in the Hood River less than 2 years after removal of the Powerdale Dam. These markers have been included in a panel of 96 SNPs designed to also permit parentage analysis and characterization of neutral and adaptive variation in Pacific lamprey (Hess et al. 2015), offering an efficient method for genetically screening lampreys to address a range of conservation applications (see Sects. 7.3.5.2 and 7.3.5.3).

7.3.5.2   Genetic Stock Structure

Population genetic analyses have long been a cornerstone of management science, and modern analyses using genetic and genome-scale data now allow for better detection of population structure, connectivity, and adaptive variation than ever before (Palsbøll et al. 2007; Funk et al. 2012). Population genetic analyses are being used in lampreys for identifying management units within species (e.g., Spice et al. 2012; Hess et al. 2013), or to test for reproductive isolation among closely related species that are indistinguishable using mtDNA sequence data (e.g., Docker et al. 2012; Taylor et al. 2012: Rougemont et al. 2015). Because migratory lampreys do not home to their natal streams (Bergstedt and Seelye 1995; Waldman et al. 2008; see Moser et al. 2015), they generally show limited population structure relative to salmonids and other philopatric species (e.g., Bryan et al. 2005; Goodman et al. 2008; Yamazaki et al. 2014; Bracken et al. 2015). Nevertheless, application of increasingly higher-resolution markers is revealing some population heterogeneity in most wide-ranging species, and the occurrence of adaptive variation within species that disperse widely has significant management implications (Hess et al. 2013; Hess 2016).

Even using presumed neutral markers (i.e., microsatellite loci and SNPs that showed no evidence of being under selection), Spice et al. (2012) and Hess et al. (2013) found evidence of isolation by distance in anadromous Pacific lamprey (i.e., lamprey from geographically distant locations were more genetically differentiated from each other than were those in close proximity), especially among locations with smaller-bodied forms. Bracken et al. (2015) likewise detected significant isolation by distance in European river lamprey. Spice et al. (2012) suggested that limits to dispersal distance at sea, especially in the smaller-bodied forms, prevented complete panmixia (i.e., random mating among all individuals) and range-wide genetic homogeneity. In the relatively small-bodied European river lamprey, Mateus et al. (2016) similarly inferred gene flow among locations but not complete panmixia. Recent studies using adaptive SNP loci (i.e., those that appear to be under selection) suggest even greater genetic heterogeneity in Pacific lamprey (Hess et al. 2013), leading Hess and colleagues to conclude that local selection helps maintain high frequencies of particular genetic variants in specific locations despite extensive gene flow. This and subsequent studies have shown evidence for selection related to geography, body size, and run timing in Pacific lamprey (Hess et al. 2013, 2014; Parker 2018; see Chap. 4). These results have important management implications. The absence of natal homing and strong local adaptation means that Pacific lamprey should be able to naturally recolonize areas from which the species has been locally extirpated (see Sect. 7.4.1), and translocation efforts (i.e., human-mediated reintroductions) are less likely to disrupt stock structure (Spice et al. 2012; Hess 2016). Nevertheless, given the weak but significant isolation by distance observed over a wide latitudinal range, source populations should be from sites reasonably close to the recipient site. Furthermore, given the evidence of adaptive variation, it would be good to match source and recipient populations in terms of run timing or body size, and to ensure that genetic variability is maintained within the population (Hess et al. 2013, 2014; Parker 2018). However, it is also important to recognize that the lack of philopatry exhibited by migratory lampreys implies that the offspring of adults translocated to a particular site will not necessarily return to that site to spawn. Thus, conservation is a range-wide issue for Pacific lamprey (Spice et al. 2012; Wang and Schaller 2015) and other anadromous lamprey species.

In sea lamprey, Bryan et al. (2005) used microsatellite loci to identify significant genetic differentiation between individuals from the Laurentian Great Lakes and the anadromous population on the Atlantic coast, as well as between freshwater populations in the upper and lower Great Lakes. Finer-scale genetic structure (e.g., among Atlantic locations or among the upper Great Lakes) was not evident using these loci, leading Spice et al. (2012) to suggest that anadromous sea lamprey (given their large size and blood-feeding strategy) may remain attached to individual hosts longer than predominantly flesh-feeding Pacific and European river lampreys (see Chap. 3) and may thus be able to disperse greater distances during the parasitic feeding phase. However, dispersal distances in the anadromous sea lamprey are not limitless; fixed differences in mtDNA sequence between European and North American sea lamprey indicate that there is a complete lack of gene flow across the Atlantic Ocean (Rodríguez-Muñoz et al. 2004). With respect to detecting gene flow or local adaptation within the European and North American populations, higher-resolution markers and loci under selection likely will be required. Such analyses will be important to determine the extent to which migrants from one basin contribute to populations in other basins (e.g., from Lake Huron to Lake Erie via movement through the St. Clair-Detroit River system; Holbrook et al. 2016) and to test for gene flow with the Atlantic population. A complete lack of gene flow between the freshwater and anadromous populations will be required if genetic control methods are to be employed in the Great Lakes (Bergstedt and Twohey 2007; see Sect. 7.4.3). Modern population genomic analyses may also help resolve questions about the origins of sea lamprey in Lake Ontario and Lake Champlain, that is, whether they invaded via canals in historical times (as outlined by Eshenroder 2014) or whether they colonized post-glacially (e.g., Waldman et al. 2004, 2006, 2009; Bryan et al. 2005; see Chap. 4), and they can be used to test for evidence of selection related to lampricide resistance (Dunlop et al. 2018).

7.3.5.3   Other Applications

Application of genetic and genomic tools is improving our understanding of lamprey biology in myriad other ways. For example, genetic parentage analysis has allowed researchers to better assess lamprey mating systems, estimate effective spawner abundance, and monitor the success of transplantation efforts.

In sea lamprey, Gilmore (2004) used microsatellite loci to identify the parents of embryos produced by a known population of spawners, and she showed that female sea lamprey often visited multiple nests and that males and females both mated with several different partners. Parentage analysis using the set of 96 SNPs developed by Hess et al. (2015) similarly confirmed that Pacific lamprey were polygynandrous (i.e., where multiple males mate with multiple females; Johnson et al. 2015a). Hess et al. (2015) showed that 41% of Pacific lamprey parents in Newsome Creek, Idaho, spawned with 2–4 mates, and Whitlock et al. (2017) estimated that 29% of parents were involved in multiple matings in the Luckiamute River, Oregon. Whitlock et al. (2017) were able to make a number of other inferences about the reproductive ecology of this population of Pacific lamprey using genetic pedigree reconstruction. They found that: individual Pacific lamprey produced offspring in 1–6 redds; the maximum distance among redds used by a single individual was 815 m; 44% of the redds contained embryos from more than one pair of spawners; and western brook lamprey also spawned in redds created by Pacific lamprey. This study also supported previous suggestions that Pacific lamprey excavate more redds than they use (Moser and Close 2003), and it estimated effective spawner abundance per redd to be 0.48. This value is similar to that obtained by Farlinger and Beamish (1984) using visual surveys in the Skeena River, British Columbia. In contrast, Brumo et al. (2009) estimated spawner abundance per redd in the Coquille River, Oregon, to be only 0.18–0.48 in 2004 and 0.04–0.17 in 2005. Pacific lamprey spawning behavior may vary seasonally (Brumo et al. 2009) or in response to different environmental conditions or differences in the density or demography (e.g., sex ratio) of spawners (see Johnson et al. 2015a). Therefore, genetic pedigree analysis appears to be a very useful complement to traditional redd surveys, because effective spawner abundance may not always be predictably correlated with redd abundance.

Genetic techniques are also improving our knowledge of other aspects of lamprey biology. For example, using microsatellite loci, Derosier et al. (2007) reported that siblings were found up to 0.9 km from each other within 3 months of emergence, and they showed even greater dispersal by the end of 1 year. The success of Pacific lamprey translocation efforts is being monitored using a parentage based tagging (PBT) approach, which involves using a multifunctional 96 SNP panel (see Sect. 7.3.5.1) to non-lethally genotype all translocated adults and the larvae collected from the translocation sites in subsequent years (Hess et al. 2015). Viable offspring were produced by 54% of the adults released in 2007, and many of these offspring were recovered as downstream-migrating juveniles 5 years later. Incredibly, these outmigrants ranged in size from 74 to 145 mm TL. Without parentage analysis to place them in single year class, we would likely have assumed that they represent a large range of ages (Hess et al. 2015; Hess 2016). These results are reminiscent of the study by Manion and Smith (1978), where monitoring a single sea lamprey year class after it was isolated above a barrier dam revealed a range of ages at metamorphosis (5–12+ years) and showed that female sea lamprey metamorphosed at an older age than males (see Chap. 1).

Population genetic methods are also useful for estimating effective population size (N_e), a measure of the rate at which populations experience genetic drift. Estimates of N_e are important for management because they provide an assessment of the adaptive potential of the population. When N_e is low, the random process of drift becomes more powerful than selection; in contrast, a population has a high capacity to respond via selection when N_e is high (Kimura et al. 1963; Lynch and Gabriel 1990; Willi et al. 2013). Genetic and genomic tools are being used to investigate the genetic basis for feeding type, body size, and run timing (Mateus et al. 2013b; Hess et al. 2013, 2014; Rougemont et al. 2017; Hume et al. 2018a; Parker 2018) and to reconstruct the demographic history of European river and brook lampreys (Rougemont et al. 2016; see Chap. 4).

Genetic options for the control of invasive sea lamprey are also being discussed, including use of "daughterless" technologies and "Trojan" genes to manipulate sex ratio (Bergstedt and Twohey 2007; McCauley et al. 2015; Thresher et al. 2019a; see Chaps. 1 and 5). Targeted research using "omics" technologies (e.g., genomics, transcriptomics, proteomics) could also help identify molecular targets for highly selective "next generation" lampricides that could provide increased efficiency and reduced environmental impact relative to those currents used (see Chap. 5). Such "omics" tools are now being used to identify new targets for pharmaceuticals (e.g., Yan et al. 2015) and insecticides (e.g., Van Leeuwen et al. 2012), and concerns over the evolution of resistance to existing lampricides could also require development of alternative control tactics (Dunlop et al. 2018).

7.4 Recent Progress and Upcoming Challenges in Conservation and Control

This section will provide a selective overview of some of the exciting "wins" achieved recently related to lamprey conservation and control, and highlight some anticipated challenges. We focus largely on recent progress related to conservation, because an up-to-date review of sea lamprey control is provided in Chap. 5. However, as indicated above, issues related to conservation and control are often "two sides of the same coin" (e.g., with dam removal providing a "good news, bad news" scenario for passage of native fishes versus containment of invasive sea lamprey). We focus largely on efforts targeting passage of upstream migrants, but recent progress has also been made in terms of mitigating the impact of instream engineering and construction activities on larval lampreys (e.g., King et al. 2015) and water diversion structures on outmigrating juveniles (e.g., Goodman et al. 2017). Other issues "on the horizon" include preventing spread of sea lamprey into new bodies of water (either via canals or as new regions or habitats become available as a result of climate change or pollution abatement; see Chaps. 4 and 5) and concerns regarding range contractions in native species that might result from climate change or habitat loss (see Maitland et al. 2015; Reid and Goodman 2016a).

7.4.1 Restoration of Habitat Connectivity via Dam Removal

Most of the world's rivers have now been dammed to one extent or another, resulting in extensive habitat fragmentation for migratory fishes (World Commission on Dams 2000; Grill et al. 2015). For lampreys, impassable barriers sever connectivity between marine or lake habitats used during the parasitic feeding phase and their riverine spawning and larval rearing habitats. Such barriers are a major threat on a global scale (see Maitland et al. 2015).

The good news is that migratory lampreys are able to rapidly recolonize river reaches when habitat connectivity is restored (Farlinger and Beamish 1984; Lin et al. 2008; Hogg et al. 2013, 2015; see Maitland et al. 2015). This ability to rapidly colonize areas may be largely due to their lack of philopatry (see Sect. 7.3.5.2). In recent years, there has been a major push in some regions to restore longitudinal connectivity of riverine habitats. In the Scorff River in France, a 50-km stretch of river was recolonized by sea lamprey within a few years of dam removal, and spawning sites became both more abundant and more widely distributed upstream of the dam site (Lasne et al. 2014). Similarly, in the Connecticut River basin, Magilligan et al. (2016) found that anadromous sea lamprey had moved into the upstream reaches of Amethyst Brook, Massachusetts, in the year following removal of a large dam. However, dam removal alone may not be an immediate "fix" to the problem. Habitat availability may also be a limiting factor in these newly accessible streams following many decades of altered flow regimes. In the Mill River, Massachusetts, stream channel reconstruction accompanied dam removal, and anadromous sea lamprey (not previously recorded in this river) found and spawned in this river in the 2 years following dam removal (Livermore et al. 2017). Pacific lamprey is also benefiting from barrier removal initiatives in the northwestern United States, where larvae are being detected in suitable habitats that were formerly inaccessible due to the presence of large dams (Hess 2016; Moser and Paradis 2017; Jolley et al. 2018; see Sect. 7.3.5.1).

The effects of anthropogenic barriers on brook lampreys—or even a basic understanding of what constitutes barriers for brook lampreys—are less clear. Brook lampreys are so named because they remain resident within their natal streams, undergoing upstream migrations of only a few kilometers (see Moser et al. 2015). With their naturally limited dispersal, one might assume that anthropogenic barriers do not cause significant habitat fragmentation. Nevertheless, Bracken et al. (2015) found that genetic distance between European brook lamprey populations increased with the number of anthropogenic barriers separating them. These authors suggested that such barriers exacerbated isolation among disjunct brook lamprey populations by inhibiting gene flow mediated by its paired parasitic and migratory counterpart, the European river lamprey (see Chap. 4). Bracken et al. (2015) found that barriers also amplified the asymmetry of gene flow from upstream to downstream sites by allowing some passive downstream drift but preventing active upstream migration. Among western brook lamprey populations in the Columbia River basin, Spice et al. (2019) likewise observed that gene flow occurred primarily in a downstream direction, resulting in a decrease in genetic diversity in upstream sites. This suggests that western brook lamprey populations in upstream areas may be particularly vulnerable to local extinction. However, Spice et al. (2019) observed that most gene flow appeared to take place in tributaries rather than through the mainstem Columbia River (perhaps not surprising given the size of this river), suggesting that a better understanding of potential barriers encountered by this species in tributary streams is needed.

Dam removal, although restoring habitat connectivity for other fishes in the Great Lakes basin, would be "bad news" for sea lamprey control. Deterioration of the dam on the Manistique River, Michigan, in the early 2000s allowed sea lamprey access to >200 river km of previously unavailable spawning and larval rearing habitat, and this escapement is thought to have contributed to population increases in Lake Michigan (Klar and Young 2005). Jensen and Jones (2018) modeled the effects of barrier removal on sea lamprey abundance in Lake Michigan and concluded that larval production in newly available stream reaches would result in rapid and disproportionate increases in abundance if there were no increases in lampricide applications to offset the effect of the barrier removals. Lampricide applications to these upstream reaches would increase the cost of sea lamprey control and expose native lampreys in these reaches to lampricide treatment (see Maitland et al. 2015; Chap. 5).

7.4.2 Lamprey Passage at Barriers

Because dam removal is not always possible, efforts are increasing to improve passage of native lamprey species at these structures. Conversely, in the Great Lakes basin, research is ongoing to restore habitat connectivity for native fishes while impeding upstream migration of sea lamprey.

Most traditional fishways were designed to enhance passage of teleost fishes, but recent research is helping design "lamprey friendly" fishways for species of conservation concern (see Moser et al. 2015). In particular, there has been an increased emphasis on better incorporating knowledge of each species' biology into the design and on more rigorously evaluating the effectiveness of the remediation efforts (e.g., Rooney et al. 2015; Kemp 2016). This is exemplified by ubiquitous "pool-and-weir" style fishways, which can provide passage for some lamprey species but which appear ineffective at passing others. For example, at mainstem dams in the Columbia River, passage rates through pool-and-weir sections can exceed 90% for Pacific lamprey (Keefer et al. 2013), but Castro-Santos et al. (2017) reported passage rates as low as 29–55% and delays of up to 2 weeks for anadromous sea lamprey attempting to pass structures of similar design. Castro-Santos et al. (2017) speculated that the transition between natural rivers and highly engineered fishway environments is difficult for sea lamprey, as this is where the majority of failed passage attempts were detected. Similar observations have been made for Pacific lamprey that successfully enter fishways in the Columbia River basin but which subsequently exhibit high levels of fallback (Kirk et al. 2015).

Use of studded tiles is a common methodology for improving passage of anguilliform fishes over obstacles (e.g., European eel *Anguilla anguilla*; Vowles et al. 2015, 2017). However, thus far, they have proven far less effective for passing lampreys under the same environmental conditions. Kerr et al. (2015) showed that 77% of European eel successfully surmounted a Crump weir that included vertically oriented tiles covered with bristles. In contrast, only 37% of European river lamprey successfully passed under the same conditions. Similarly, but in the face of a different barrier type, Vowles et al. (2017) reported passage rates of 67–93% for European eel when using tiles studded with plastic pegs compared to only 20–22% for European river lamprey. These findings are consistent with field observations, where Tummers et al. (2016) reported European river lamprey passage efficiencies of just 17% when an existing fishway design within a natural stream was retrofitted with wall-mounted studded tiles. In a subsequent study, passage efficiency past a Crump weir on which studded tiles were fixed horizontally was 26%, although this was still better than the 9% passage observed via the tileless control route (Tummers et al. 2018). Anadromous sea lamprey showed even lower passage rates when using retrofitted studded tiles, with 8% efficiency reported from 36 radio-tagged individuals in the Mulkear River, Ireland (Rooney et al. 2015). This suggests that, although novel fishway designs such as climbing ramps are often better than traditional designs (Foulds and Lucas 2013; Tummers et al. 2016), there is still much we have to learn to design more appropriate and effective fishways for migratory lampreys.

A first step towards doing so should be the explicit consideration of lamprey behavior and not just their physical or physiological capabilities (e.g., Goodman and Reid 2017). Climbing structures specially designed for Pacific lamprey have been shown to increase rates of passage at large dams compared with pool-and-weir designs (Moser et al. 2015; Gallion et al. 2016; Frick et al. 2017; Goodman and Reid 2017). For example, Goodman and Reid (2017) found that passage efficiency was only 44% at an existing pool-and-weir fishway, but that two relatively simple in situ modifications (involving either a corrugated galvanized culvert or a simple Acrylonitrile Butadiene Styrene pipe) permitted 100% passage efficiency through the alternative routes. Moreover, passage rates were ~10 and 20× faster with the culvert and tube modifications, respectively, compared to the pool-and-weir structure. Further refinements to such vertical wetted surfaces, such as identifying the lower and upper bounds of flow rates, is generating consistently high passage rates for Pacific lamprey (e.g., 94% total passage, including 76% of first attempts) even in highly engineered experimental environments (Frick et al. 2017). Likewise, anecdotal evidence indicates that pouched lamprey are able to use ramp style designs by taking advantage of their strong climbing capability (Cindy F. Baker, National Institute of Water and Atmospheric Research, Hamilton, NZ, personal communication, 2018), which could improve their apparently low usage of nature-like fishways (Beatty et al. 2007).

Recent studies have revealed complex behaviors by migrating lampreys at fishways. Kirk et al. (2017) showed that Pacific lamprey exhibit both avoidance of, and attraction to, turbulence within a vertical slot fishway. Pacific lamprey were attracted to turbulence when flow rates were low, but they actively avoided turbulent areas when flow rates increased beyond a particular threshold. The former behavior is likely related to orientation within a stream and would aid lamprey in maintaining upstream directionality, while the latter behavior could help individuals reduce energetic costs in highly turbulent and fast flowing reaches (Kirk et al. 2016). Similar behaviors were shown by acoustically tagged European river lamprey as they approached a navigation lock (Silva et al. 2017b). Furthermore, it is now apparent that Pacific lamprey express much greater intraspecific variation in passage behavior compared to other anadromous fishes such as salmonids. Kirk and Caudill (2016) revealed, using network analysis, that some Pacific lamprey explore multiple potential routes through a barrier, while others select the first route detected, and either group may ultimately reject fishways entirely.

Ultimately, for species of conservation concern, we need a better understanding of the fitness consequences of passage through such highly engineered environments. For example, even if lampreys pass barriers, delays could disrupt the highly synchronized processes of migration, spawning, and embryonic development, which are all strongly dependent on stream temperatures (see Johnson et al. 2015a). The energetic costs of passage also remain poorly studied. Furthermore, it is not clear how many individuals need to achieve reproductive success in order to maintain the population. Following the installation of a vertical slot fishway in the Mondego River in Portugal, the abundance of larval sea lamprey increased 29-fold within a 2–5-year period, despite passage rates of just 31% (Pereira et al. 2017). Perhaps given the generally polygamous nature of lampreys (see Johnson et al. 2015a; Sect. 7.3.5.3) and the high fecundity of anadromous species (see Chap. 1), relatively few individuals passing barriers each year are sufficient to sustain the population as long as there is sufficient high-quality habitat upstream. This is consistent with suggestions that relatively low rates of escapement past sea lamprey control barriers can result in high larval abundances upstream (see Sect. 7.4.1), but this clearly requires further study.

The "other side of the coin" regarding lamprey-specific passage requirements is the ability to exploit differences in lamprey swimming ability and behavior at barriers for sea lamprey control purposes. Lamprey-specific barriers have the potential to minimize negative effects to other native fishes while still preventing sea lamprey from gaining access to spawning habitat (Porto et al. 1999; Dodd et al. 2003; McLaughlin et al. 2006; Pratt et al. 2009). Advances related to selective fish passage are reviewed in Chap. 5 (and references therein), and other innovative approaches may help achieve this goal. Using a combination of studded tile ramps and a putative alarm cue, Hume et al. (2018b) demonstrated that it is feasible to provision at least one Great Lakes native fish species (white sucker *Catostomus commersonii*) with a potential route through a barrier, while simultaneously excluding sea lamprey from the same fishway design. Unique features of lamprey climbing abilities and behaviors are also being exploited to increase trapping efficiency, and the use of pheromones to enhance trapping success is also being investigated (Johnson et al. 2015b, c; McLean et al. 2015; see Chap. 5).

Given the phylogenetic distinctiveness of lampreys (e.g., resulting in reduced "cross reactivity" of alarm cues or pheromones between lampreys and non-target bony fishes; Hume and Wagner 2018) and differences in swimming ability related to their body form and behaviors (see Moser et al. 2015; Sherburne and Reinhardt 2016; Sanchez et al. 2017), lamprey-specific barriers should be feasible. However, the biggest challenge may be achieving sea lamprey specificity. Two other migratory lampreys (silver and chestnut lampreys) are native to the Great Lakes basin and Lake Champlain, and the silver lamprey in particular seems to have been negatively impacted by sea lamprey control (Schuldt and Goold 1980; Maitland et al. 2015; see Chap. 5). Given the conserved nature of the larval stage (see Dawson et al. 2015), it may not be possible to achieve complete sea lamprey-specific control with the use of lampricides alone, but solutions that exploit species-specific differences in response to chemical cues or climbing abilities may achieve a higher measure of selectivity. Exploitation of differences in climbing ability may be particularly fruitful given that the native silver and chestnut lampreys are considerably smaller (mean TL at maturity 224 and 216 mm, respectively; Hubbs and Trautman 1937) than freshwater-resident sea lamprey (mean TL ~ 400–500 mm; see Chap. 4).

7.4.3 Public Perceptions and Social License

Achieving successful lamprey conservation and control goes beyond conducting more research and implementing informed management actions. Ongoing advances and response to future challenges must include positive public and legislative attitudes towards species in need of conservation, and "social license" (i.e., broad social support and ongoing acceptance within the local community and other stakeholders) for sea lamprey control efforts. In North America and New Zealand, conservation efforts have been led or propelled largely by Indigenous groups who have long valued Pacific lamprey and pouched lamprey, respectively (Close et al. 2002; Petersen Lewis 2009; Columbia River Inter-Tribal Fish Commission 2011; Stewart and Baker 2012; see Docker et al. 2015; Maitland et al. 2015). In the Great Lakes basin, conservation of native lamprey species generally has been overshadowed by the pressing need to control sea lamprey, but the recent increase in outreach and legislative efforts targeted at conserving native lampreys (particularly outside of the Great Lakes basin) suggests that lamprey conservation in North America has not been irrevocably damaged by the negative public image of invasive sea lamprey (Mesa and Copeland 2009; Moyle et al. 2009; Renaud et al. 2009; see Maitland et al. 2015). A management plan has recently been approved for the anadromous sea lamprey in the Connecticut River basin (Connecticut River Atlantic Salmon Commission 2018). It is believed to be the first management plan for sea lamprey in North America that is focused on restoration and recovery of this species rather than on control of pest populations, and the anadromous sea lamprey has been designated as a "Species of Greatest Conservation Need" by all four basin states (Connecticut River Atlantic Salmon Commission 2018). In contrast, there has long been an appreciation for lampreys in Europe and Japan, particularly for those species historically used for human consumption (see Docker et al. 2015), and this has generated considerable support for research, restoration, and artificial propagation efforts (see Maitland et al. 2015; Chap. 2). In fact, it is perhaps ironic that the recent invasion of pink salmon *Oncorhynchus gorbuscha* into Scottish rivers is raising concerns that they will negatively impact the valued native sea lamprey (Armstrong et al. 2018).

Given the ecosystem and economic damage caused by the sea lamprey in the Great Lakes and Lake Champlain drainages, good public support exists for the sea lamprey control program in the Great Lakes. Nevertheless, current control methods are not without concerns related, for example, to costs associated with ongoing control or the effects of barriers and lampricides on non-target fishes and other organisms (see Sect. 7.4.2; Chap. 5). Concerns appear to be particularly acute in the Lake Champlain basin, where sea lamprey control was initiated after the establishment of the U.S. Environmental Protection Agency and enactment of the Endangered Species Act, and after significant expansion of the Clean Water Act. National Pollutant Discharge Elimination System permits are required for use of lampricides, and development and use of alternatives to chemical control is a high priority in the basin (see Marsden et al. 2010; Chap. 5). Alternatives such as pheromones are generally viewed as being of less concern (because they are non-toxic, biodegradable, highly species-specific, and capable of inducing responses even in minute quantities), but work with federal and state regulatory agencies will still be required to register such compounds for use in control applications (Li et al. 2007; Siefkes 2017). Furthermore, although Eshenroder (2014) argued convincingly that sea lamprey are invasive throughout the Great Lakes and Lake Champlain (see Chaps. 4 and 5), public perception of sea lamprey control in the Lake Ontario and Lake Champlain drainages might change if sea lamprey are confirmed (e.g., through population genomic analyses) to be native to these regions. Waldman et al. (2004) suggested that sea lamprey control policies aimed at intense suppression might need re-evaluation if sea lamprey are shown to be native to Lake Ontario. However, there are precedents for—but also disagreements regarding—control measures aimed at other significant but native pest species (e.g., Rey et al. 2012; Dale et al. 2014).

Exploration of genetic options for the control of sea lamprey will need to address many ethical and societal concerns regarding the genetic manipulation of organisms released into the environment (Thresher et al. 2019a, b). With advances in molecular and genetic technologies, such options are becoming increasingly feasible (see Chap. 6), but the extent to which there is social license to develop and use such technologies will need to be evaluated. Similar discussions related to public acceptance of the use of gene drive and other technologies to suppress insect and other pest species are occurring worldwide (e.g., Esvelt et al. 2014; Lucht 2015; Webber et al. 2015; Dearden et al. 2018). With respect specifically to sea lamprey control, a survey of stakeholder attitudes (e.g., professional fishery managers and recreational anglers) showed strong support for initiation of research and development related to genetic control options (Thresher et al. 2019b). Further exploration of the feasibility and management applications of such technologies will need to go hand in hand with public consultation and will need to involve careful evaluation of the actual and perceived risks and benefits.

7.5 Conclusions

Lampreys have survived for hundreds of millions of years, persisting through at least four mass extinction events when many other lineages perished (Docker et al. 2015). As generalists, lampreys may have had "a leg up" on more specialized lineages. For example, locating spawning tributaries by attraction to larval pheromones, which provides lampreys with a good indicator of contemporary rather than historical larval rearing habitat quality, may make them better able to respond to changes in environmental conditions than species that exhibit strong site fidelity. Similarly, the lack of strong host specificity observed in parasitic species (see Chap. 3) means that lampreys are less likely to be impacted by changes in the distribution and abundance of individual host species. The apparent ease with which migratory and feeding type can evolve in most lamprey taxa (e.g., allowing populations to abandon their anadromous migration or even the entire parasitic feeding phase when the growth opportunities that this stage permits are no longer balanced by its costs; see Chap. 4) may also allow lampreys to readily adapt to changing environmental and climatic conditions. This apparent resiliency is "bad news" for sea lamprey control where compensatory responses to control measures (e.g., faster growth, earlier metamorphosis, shifts in sex ratio; see Chaps. 1 and 5) may confound attempts to suppress population levels in the Great Lakes. Nevertheless, research and management innovations and a diversity of weapons in the sea lamprey control arsenal are being used and developed to combat this invasive species. With respect to lamprey conservation, their flexibility and long evolutionary history should not allow us to be complacent. Although lampreys as a lineage have survived for long periods of evolutionary time, individual species have not. Species with restricted distributions (e.g., some brook lamprey species), as well as anadromous lampreys that are facing numerous anthropogenic threats despite their wider geographic ranges, are at risk of extirpation or extinction. However, with the recent innovations targeted at conserving native lampreys in many parts of the world, lampreys should continue to persist as important components of our marine and freshwater environments if they are afforded the proper management.

From their formerly exalted status as "the food of kings" in the Middle Ages to the beginnings of recovery following their decline from our collective conscience and our rivers during colonial expansions and the industrial revolution, lampreys have experienced quite a journey. It is also clear that lampreys hold a special place in the minds of researchers from a broad array of scientific disciplines and that they will continue to do so for decades to come. Although they are few in number and many of them are quite little, lampreys are a very fine group of fishes.

Acknowledgements We gratefully acknowledge Drs. Tyler J. Buchinger and Mary L. Moser for providing insightful comments on an earlier draft of this chapter.

References

Allendorf FW (2017) Genetics and the conservation of natural populations: allozymes to genomes. Mol Ecol 26:420–430

Almeida PR, Quintella BR, Dias NM (2002) Movement of radio-tagged anadromous sea lamprey during the spawning migration in the River Mondego (Portugal). Hydrobiology 483:1–8

Aman JT, Docker MF, Grimes KW (2017) New England range extension of American Brook Lamprey (*Lethenteron appendix*), as confirmed by genetic analysis. Northeast Nat 24:536–543

Applegate VC (1950) Natural history of the sea lamprey, *Petromyzon marinus,* in Michigan. US Fish Wildl Serv Spec Sci Rep Fish 55:1–237

April J, Mayden RL, Hanner RH, Bernatchez L (2011) Genetic calibration of species diversity among North America's freshwater fishes. Proc Natl Acad Sci USA 108:10602–10607

Armstrong JD, Bean CW, Wells A (2018) The Scottish invasion of pink salmon in 2017. J Fish Biol 93:8–11

Arntzen EV, Mueller RP (2017) Video-based electroshocking platform to identify lamprey ammocoete habitats: field validation and new discoveries in the Columbia River basin. N Am J Fish Manag 37:676–681

Artamonova VS, Kucheryavyy AV, Pavlov DS (2011) Nucleotide sequences of the mitochondrial cytochrome oxidase subunit I (COI) gene of lamprey classified with *Lethenteron camtschaticum* and the *Lethenteron reissneri* complex show no species-level differences. Doklad Biol Sci 437:113–118

Baker CF, Stewart M, Fine JM, Sorensen PW (2009) Partial evolutionary divergence of a migratory pheromone between northern and southern hemisphere lampreys. In: Haro A, Smith KL, Rulifson RA et al (eds) Challenges for diadromous fishes in a dynamic global environment. Am Fish Soc Symp 69:845–846

Bailey RM (1980) Comments on the classification and nomenclature of lampreys—an alternative view. Can J Fish Aquat Sci 37:1626–1629

Beamish RJ (1982) *Lampetra macrostoma*, a new species of freshwater parasitic lamprey from the west coast of Canada. Can J Fish Aquat Sci 39:736–747

Beamish RJ (1987) Evidence that parasitic and nonparasitic life history types are produced by one population of lamprey. Can J Fish Aquat Sci 44:1779–1782

Beamish RJ, Youson JH (1987) Life history and abundance of young adult *Lampetra ayresi* in the Fraser River and their possible impact on salmon and herring stocks in the Strait of Georgia. Can J Fish Aquat Sci 44:525–537

Beatty SJ, Morgan DL, Torre A (2007) Restoring ecological connectivity in the Margaret River: western Australia's first rock-ramp fishways. Ecol Manag Restor 8:224–228

Bentacur-R R, Wiley EO, Arratia G et al (2017) Phylogenetic classification of bony fishes. BMC Evol Biol 17:162

Berg LS (1931) A review of the lampreys of the northern hemisphere. Ann Mus Zool Acad Sci URSS 32:87–116

Bergstedt RA, Genovese JH (1994) New technique for sampling sea lamprey larvae in deepwater habitats. N Am J Fish Manag 14:449–452

Bergstedt RA, Seelye JG (1995) Evidence for lack of homing by sea lamprey. Trans Am Fish Soc 124:235–239

Bergstedt RA, Twohey MB (2007) Research to support sterile-male-release and genetic alteration techniques for sea lamprey control. J Great Lakes Res 33(Spec Issue 2):48–69

Biggs J, Ewald N, Valentini A et al (2015) Using eDNA to develop a national citizen science-based monitoring programme for the great crested newt (*Triturus cristatus*). Biol Conserv 183:19–28

Boguski DA, Reid SB, Goodman DH, Docker MF (2012) Genetic diversity, endemism and phylogeny of lampreys within the genus *Lampetra sensu stricto* (Petromyzontiformes: Petromyzontidae) in western North America. J Fish Biol 81:1891–1914

Bracken FS, Hoelzel A, Hume JB, Lucas MC (2015) Contrasting population genetic structure among freshwater-resident and anadromous lampreys: the role of demographic history, differential dispersal and anthropogenic barriers to movement. Mol Ecol 24:1188–1204

Brant CO, Johnson NS, Li K, Buchinger TJ, Li W (2016) Female sea lamprey shift orientation toward a conspecific chemical cue to escape a sensory trap. Behav Ecol 27:810–819

Bravener GA, Twohey M (2016) Evaluation of a sterile-male release technique: a case study of invasive sea lamprey control in a tributary of the Laurentian Great Lakes. N Am J Fish Manag 36:1125–1138

Brumo AF, Grandmontagne L, Namitz SN, Markle DF (2009) Approaches for monitoring Pacific lamprey spawning populations in a coastal Oregon stream. In: Brown LR, Chase SD, Mesa MG, Beamish RJ, Moyle PB (eds) Biology, management, and conservation of lampreys in North America. Am Fish Soc Symp 72:203–222

Bryan MB, Zalinski D, Filcek KB et al (2005) Patterns of invasion and colonization of the sea lamprey (*Petromyzon marinus*) in North America as revealed by microsatellite genotypes. Mol Ecol 14:3757–3773

Buchinger TJ, Siefkes MJ, Zielinski BS, Brant CO, Li W (2015) Chemical cues and pheromones in the sea lamprey (*Petromyzon marinus*). Front Zool 12:32

Buchinger TJ, Li K, Huertas M et al (2017a) Evidence for partial overlap of male olfactory cues in lampreys. J Exp Biol 220:497–506

Buchinger TJ, Bussy U, Li K et al (2017b) Phylogenetic distribution of a male pheromone that may exploit a nonsexual preference in lampreys. J Evol Biol 30:2244–2254

Buchinger TJ, Bussy U, Buchinger EG et al (2017c) Increased pheromone signaling by small male sea lamprey has distinct effects on female mate search and courtship. Behav Ecol Sociobiol 71:155

Carim KJ, Dysthe JC, Young MK, McKelvey KS, Schwartz MK (2016) A noninvasive tool to assess the distribution of Pacific lamprey (*Entosphenus tridentatus*) in the Columbia River basin. PLoS ONE 12:e0169334

Carstens BC, Pelletier TA, Reid NM, Satler JD (2013) How to fail at species delimitation. Mol Ecol 22:4369–4383

Castro-Santos T, Shi X, Haro A (2017) Migratory behavior of adult sea lamprey and cumulative passage performance through four fishways. Can J Fish Aquat Sci 74:790–800

Chaitra MS, Vasudevan K, Shanker K (2004) The biodiversity bandwagon: the splitters have it. Curr Sci 86:897–899

Clabough TS, Keefer ML, Caudill CC, Johnson CL, Peery CA (2012) Use of night video to enumerate adult Pacific lamprey passage at hydroelectric dams: challenges and opportunities to improve escapement estimates. N Am J Fish Manag 32:687–695

Clemens BJ, van de Wetering S, Kaufman J, Holt RA, Schreck CB (2009) Do summer temperatures trigger spring maturation in Pacific lamprey, *Entosphenus tridentatus*. Ecol Freshw Fish 18:418–426

Clemens BJ, Beamish RJ, Coates KC et al (2017a) Conservation challenges and research needs for Pacific Lamprey in the Columbia River basin. Fisheries 42:268–280

Clemens BJ, Wyss L, McCoun R et al (2017b) Temporal genetic population structure and interannual variation in migration behavior of Pacific Lamprey *Entosphenus tridentatus*. Hydrobiologia 794:223–240

Close DA, Fitzpatrick MS, Li HW (2002) The ecological and cultural importance of a species at risk of extinction, Pacific lamprey. Fisheries 27:19–25

Columbia River Inter-Tribal Fish Commission (2011) Tribal Pacific lamprey restoration plan for the Columbia River basin, Columbia River Inter-Tribal Fish Commission, Portland, OR. http://www.critfc.org/wp-content/uploads/2012/12/lamprey_plan.pdf. Accessed 12 Dec 2013

Connecticut River Atlantic Salmon Commission (2018) Connecticut River anadromous sea lamprey management plan. Connecticut River Atlantic Salmon Commission, Sunderland, MA. https://www.fws.gov/r5crc/pdf/CRASC-sea-lamprey-plan-final-2018-11-26-18.pdf. Accessed 02 Jan 2019

Corlett TR (2017) A bigger toolbox: biotechnology in biodiversity conservation. Trends Biotechnol 35:55–65

COSEWIC (2015) Guidelines for recognizing designatable units. Committee on the Status of Endangered Wildlife in Canada. https://www.canada.ca/en/environment-climate-change/services/committee-status-endangered-wildlife/guidelines-recognizing-designatable-units.html. Accessed 26 Nov 2018

Creaser CW, Hubbs CL (1922) A revision of the Holarctic lampreys. Occas Pap Mus Zool Univ Mich 120:1–14

Crother BI, Bonnett RM, Boundy J et al (2017) Scientific and standard English names of amphibians and reptiles of North America north of Mexico, with comments regarding confidence in our understanding, 8th edn. Herpetol Circ 43:1–102

Dale PE, Knight JM, Griffin L et al (2014) Multi-agency perspectives on managing mangrove wetlands and the mosquitoes they produce. J Am Mosq Control 30:106–115

Dawson HA, Quintella BR, Almeida PR, Treble AJ, Jolley JC (2015) The ecology of larval and metamorphosing lampreys. In: Docker MF (ed) Lampreys: biology, conservation and control, vol 1. Springer, Dordrecht, pp 75–137

Dearden PK, Gemmell NJ, Mercier OR et al (2018) The potential for the use of gene drives for pest control in New Zealand: a perspective. J R Soc N Z 48:225–244

DeCelles G, Zemeckis D (2014) Acoustic and radio telemetry. In: Cadrin SX, Kerr LA, Mariani S (eds) Stock identification methods: applications in fishery science, 2nd edn. Academic Press, New York, pp 397–428

Deiner K, Altermatt F (2014) Transport distance of invertebrate environmental DNA in a natural river. PLoS ONE 9:e88786

Deiner K, Walser JC, Mächler E, Altermatt F (2015) Choice of capture and extraction methods affect detection of freshwater biodiversity from environmental DNA. Biol Conserv 183(Spec Issue):53–63

de Queiroz K (2007) Species concepts and species delimitation. Syst Biol 6:879–886

Derosier AL, Jones ML, Scribner KT (2007) Dispersal of sea lamprey larvae during early life: relevance for recruitment dynamics. Environ Biol Fish 78:271–284

Docker MF (2009) A review of the evolution of nonparasitism in lampreys and an update of the paired species concept. In: Brown LR, Chase SD, Mesa MG, Beamish RJ, Moyle PB (eds) Biology, management, and conservation of lampreys in North America. Am Fish Soc Symp 72:71–114

Docker MF, Youson JH, Beamish RJ, Devlin RH (1999) Phylogeny of the lamprey genus *Lampetra* inferred from mitochondrial cytochrome *b* and ND3 gene sequences. Can J Fish Aquat Sci 56:2340–2349

Docker MF, Mandrak ME, Heath DD (2012) Contemporary gene flow between "paired" silver (*Ichthyomyzon unicuspis*) and northern brook (*I. fossor*) lampreys: implications for conservation. Conserv Genet 13:823–835

Docker MF, Hume JB, Clemens BJ (2015) Introduction: a surfeit of lampreys. In: Docker MF (ed) Lampreys: biology, conservation and control, vol 1. Springer, Dordrecht, pp 1–34

Docker MF, Silver GS, Jolley JC, Spice EK (2016) Simple genetic assay distinguishes lamprey genera *Entosphenus* and *Lampetra*: comparison with existing genetic and morphological identification methods. N Am J Fish Manag 36:780–787

Dodd HR, Hayes DB, Baylis JR et al (2003) Low-head sea lamprey barrier effects on stream habitat and fish communities in the Great Lakes basin. J Great Lakes Res 29(Suppl 1):386–402

Dunlop ES, McLaughlin R, Adams JV et al (2018) Rapid evolution meets invasive species control: the potential for pesticide resistance in sea lamprey control. Can J Fish Aquat Sci 75:152–168

Eshenroder RL (2014) The role of the Champlain Canal and Erie Canal as putative corridors for colonization of Lake Champlain and Lake Ontario by Sea Lampreys. Trans Am Fish Soc 143:634–649

Esvelt KM, Smidler AL, Catteruccia F, Church GM (2014) Concerning RNA-guided gene drives for the alteration of wild populations. eLife 3:e03401

Farlinger SP, Beamish RJ (1984) Recent colonization of a major salmon-producing lake in British Columbia by Pacific lamprey (*Lampetra tridentata*). Can J Fish Aquat Sci 41:278–285

Fine JM, Vrieze LA, Sorensen PW (2004) Evidence that petromyzontid lampreys employ a common migratory pheromone that is partially comprised of bile acids. J Chem Ecol 30:2091–2110

Fine JM, Sisler SP, Vrieze LA, Swink WD, Sorensen PW (2006) A practical method for obtaining useful quantities of pheromones from sea lamprey and other fishes for identification and control. J Great Lakes Res 32:832–838

Fodale MF, Bronte CR, Bergstedt RA, Cuddy DW, Adams JV (2003) Classification of lentic habitat for sea lamprey (*Petromyzon marinus*) larvae using a remote seabed classification device. J Great Lakes Res 29(Suppl 1):190–203

Foulds WL, Lucas MC (2013) Extreme inefficiency of two conventional, technical fishways used by European river lamprey (*Lampetra fluviatilis*). Ecol Eng 58:423–433

Fraser DJ, Bernatchez L (2001) Adaptive evolutionary conservation: towards a unified concept for defining conservation units. Mol Ecol 10:2741–2752

Frick KE, Corbett SC, Moser ML (2017) Climbing success of adult Pacific lamprey on a vertical wetted wall. Fish Manag Ecol 24:230–239

Froese R, Pauly D (eds) (2018) FishBase. http://www.fishbase.org, version 06/2018. Accessed 23 Nov 2018

Fryer JK (2008) Feasibility of using a computerized video fish counting system to estimate night time passage of salmon and lamprey at the Washington Shore Bonneville Dam fish counting station. Columbia River Inter-Tribal Fish Commission Completion Report, Portland, OR

Funk WC, McKay JK, Hohenlohe PA, Allendorf FW (2012) Harnessing genomics for delineating conservation units. Trends Ecol Evol 27:489–496

Gaigher A, Launey S, Lasne E, Besnard A-L, Evanno G (2013) Characterization of thirteen microsatellite markers in river and brook lampreys (*Lampetra fluviatilis* and *L. planeri*). Conserv Genet Res 5:141–143

Gallion DG, van Hevelingen TH, van der Leeuw BK (2016) Use of lamprey passage structures at Bonneville and John Day dams. U.S. Army Corps of Engineers 2016 Annual Report, Portland District, Fisheries Field Unit, Cascade Locks, OR

Gilmore SA (2004) Genetic assessment of adult mating success and accuracy of statolith aging in the sea lamprey *Petromyzon marinus*. MS thesis, Michigan State University, Lansing, MI

Gingera TD, Steeves TB, Boguski DA et al (2016) Detection and identification of lampreys in Great Lakes streams using environmental DNA. J Great Lakes Res 42:649–659

Goldberg C, Strickler K, Pilliod D (eds) (2015) Special issue: environmental DNA: a powerful new tool for biological conservation. Biol Conserv 183:1–102

Goldberg CS, Turner CR, Deiner K et al (2016) Critical considerations for the application of environmental DNA methods to detect aquatic species. Methods Ecol Evol 7:1299–1307

Goodman DH, Reid SB (2017) Climbing above the competition: innovative approaches and recommendations for improving Pacific lamprey passage at fishways. Ecol Eng 107:224–232

Goodman DH, Reid SB, Docker MF, Haas GR, Kinziger AP (2008) Mitochondrial DNA evidence for high levels of gene flow among populations of a widely distributed anadromous lamprey *Entosphenus tridentatus* (Petromyzontidae). J Fish Biol 72:400–417

Goodman DH, Kinziger AP, Reid SB, Docker MF (2009) Morphological diagnosis of *Entosphenus* and *Lampetra* ammocoetes (Petromyzontidae) in Washington, Oregon, and California. In: Brown LR, Chase SD, Mesa MG, Beamish RJ, Moyle PB (eds) Biology, management, and conservation of lampreys in North America. Am Fish Soc Symp 72:223–232

Goodman DH, Reid SB, Reyes RC, Wu BJ, Bridges BB (2017) Screen efficiency and implications for losses of lamprey macrophthalmia at California's largest water diversions. N Am J Fish Manag 37:30–40

Grill G, Lehner B, Lumsdon AE et al (2015) An index-based framework for assessing patterns and trends in river fragmentation and flow regulation by global dams at multiple scales. Environ Res Lett 10:015001

Gustavson MS, Collins PC, Finarelli JA et al (2015) An eDNA assay for Irish *Petromyzon marinus* and *Salmo trutta* and field validation in running water. J Fish Biol 87:1254–1262

Hansen MJ, Hayne DW (1962) Sea lamprey larvae in Ogontz Bay and Ogontz River, Michigan. J Wild Manag 26:237–247

Hansen MJ, Madenjian CP, Slade JW et al (2016) Population ecology of the sea lamprey (*Petromyzon marinus*) as an invasive species in the Laurentian Great Lakes and an imperiled species in Europe. Rev Fish Biol Fish 26:509–535

Harris JE, Jolley JC (2017) Estimation of occupancy, density, and abundance of larval lampreys in tributary river mouths upstream of dams on the Columbia River, Washington and Oregon. Can J Fish Aquat Sci 74:843–852

Harrison RG, Larson EL (2014) Hybridization, introgression, and the nature of species boundaries. J Hered 105:795–809

Heath G, Childs D, Docker MF, McCauley DW, Whyard S (2014) RNA interference technology to control pest sea lampreys—a proof-of-concept. PLoS ONE 9:e88387

Hendry AP, Bolnick DI, Berner D, Peichel CL (2009) Along the speciation continuum in sticklebacks. J Fish Biol 75:2000–2036

Hess J (2016) Insights gained through recent technological advancements for conservation genetics of Pacific lamprey (*Entosphenus tridentatus*). In: Orlov A, Beamish R (eds) Jawless fishes of the world, vol 2. Cambridge Scholars Publishing, Newcastle upon Tyne, pp 149–159

Hess JE, Campbell NR, Close DA, Docker MF, Narum SR (2013) Population genomics of Pacific lamprey: adaptive variation in a highly dispersive species. Mol Ecol 22:2898–2916

Hess JE, Caudill CC, Keefer ML et al (2014) Genes predict long distance migration and large body size in a migratory fish, Pacific lamprey. Evol Appl 7:1192–1208

Hess JE, Campbell NR, Docker MF et al (2015) Use of genotyping by sequencing data to develop a high-throughput and multifunctional SNP panel for conservation applications in Pacific lamprey. Mol Ecol 15:187–202

Hogg R, Coghlan SM Jr, Zydlewski J (2013) Anadromous sea lampreys recolonize a Maine coastal river tributary after dam removal. Trans Am Fish Soc 142:1381–1394

Hogg RS, Coghlan SM Jr, Zydlewski J, Gardner C (2015) Fish community response to a small-stream dam removal in a Maine coastal river tributary. Trans Am Fish Soc 144:467–479

Holbrook CM, Bergstedt R, Adams NS, Hatton TW, McLaughlin RL (2015) Fine-scale pathways used by adult sea lampreys during riverine spawning migrations. Trans Am Fish Soc 144:549–562

Holbrook CM, Jubar AK, Barber JM, Tallon K, Hondorp DW (2016) Telemetry narrows the search for sea lamprey spawning locations in the St. Clair-Detroit River system. J Great Lakes Res 42:1084–1091

Hubbs CL, Potter IC (1971) Distribution, phylogeny and taxonomy. In: Hardisty MW, Potter IC (eds) The biology of lampreys, vol 1. Academic Press, London, pp 1–65

Hubbs CL, Trautman MB (1937) A revision of the lamprey genus *Ichthyomyzon*. Misc Publ Mus Zool Univ Mich 35:7–109

Hubert N, Hanner R, Holm E et al (2008) Identifying Canadian freshwater fishes through DNA barcodes. PLoS ONE 3:e2490

Hume JB, Wagner CM (2018) A death in the family: sea lamprey (*Petromyzon marinus*) avoidance of confamilial alarm cues diminishes with phylogenetic distance. Ecol Evol 8:3751–3762

Hume B, Meckley TD, Johnson NS et al (2015) Application of a putative alarm cue hastens the arrival of invasive sea lamprey (*Petromyzon marinus*) at a trapping location. Can J Fish Aquat Sci 72:1799–1806

Hume JB, Recknagel H, Bean CW, Adams CE, Mable BK (2018a) RADseq and mate choice assays reveal unidirectional gene flow among three lamprey ecotypes despite weak assortative mating: insights into the formation and stability of multiple ecotypes in sympatry. Mol Ecol 27:4572–4590

Hume JB, Lucas MC, Reinhardt U et al (2018b) Selective removal of sea lamprey via behavioral guidance in a model fishway: a proof of concept test. Great Lakes Fishery Commission Completion Report, Ann Arbor, MI

Irwin BJ, Liu W, Bence JR, Jones ML (2012) Defining economic injury levels for sea lamprey control in the Great Lakes basin. N Am J Fish Manag 32:760–771

Isaac NJB, Mallet J, Mace GM (2004) Taxonomic inflation: influence on macroecology and conservation. Trends Ecol Evol 19:464–469

Jane SF, Wilcox TM, McKelvey KS et al (2015) Distance, flow and PCR inhibition: eDNA dynamics in two headwater streams. Mol Ecol Resour 15:216–227

Jensen JA, Jones ML (2018) Forecasting the response of Great Lakes sea lamprey (*Petromyzon marinus*) to barrier removals. Can J Fish Aquat Sci 75:1415–1426

Jerde CL, Mahon AR, Chadderton WL, Lodge DM (2011) "Sight-unseen" detection of rare aquatic species using environmental DNA. Conserv Lett 4:150–157

Johnson NS, Swink WD, Brenden TO et al (2014) Survival and metamorphosis of low-density populations of larval sea lampreys (*Petromyzon marinus*) in streams following lampricide treatment. J Great Lakes Res 40:155–163

Johnson NS, Buchinger TJ, Li W (2015a) Reproductive ecology of lampreys. In: Docker MF (ed) Lampreys: biology, conservation and control, vol 1. Springer, Dordrecht, pp 265–303

Johnson NS, Tix JA, Hlina BL et al (2015b) A sea lamprey (*Petromyzon marinus*) sex pheromone mixture increases trap catch relative to a single synthesized component in specific environments. J Chem Ecol 41:311–321

Johnson NS, Siefkes MJ, Wagner CM et al (2015c) Factors influencing capture of invasive sea lamprey in traps baited with a synthesized sex pheromone component. J Chem Ecol 41:913–923

Johnson NS, Brenden TO, Swink WD, Lipps MA (2016) Survival and metamorphosis of larval sea lamprey (*Petromyzon marinus*) residing in Lakes Michigan and Huron near river mouths. J Great Lakes Res 42:1461–1469

Jolley JC, Silver GS, Whitesel TA (2012) Occupancy and detection of larval Pacific lampreys and *Lampetra* spp. in a large river: the Lower Willamette River. Trans Am Fish Soc 141:305–312

Jolley JC, Kovalchuk G, Docker MF (2016) River lamprey (*Lampetra ayresii*) outmigrant upstream of the John Day dam in the mid-Columbia River. Northwest Nat 97:48–52

Jolley JC, Silver GS, Harris JE, Whitesel TA (2018) Pacific lamprey recolonization of a Pacific northwest river following dam removal. River Res Appl 34:44–51

Jones ML, Irwin BJ, Hansen GJA et al (2009) An operating model for the integrated pest management of Great Lakes sea lampreys. Open Fish Sci J 2:59–73

Jones ML, Brenden TO, Irwin BJ (2015) Re-examination of sea lamprey control policies for the St. Marys River: completion of an adaptive management cycle. Can J Fish Aquat Sci 72:1538–1551

Kawai YL, Yura K, Shindo M et al (2015) Complete genome sequence of the mitochondrial DNA of the river lamprey, *Lethenteron japonicum*. Mitochondrial DNA 26:863–864

Keefer ML, Daigle WR, Peery CA et al (2010) Testing adult Pacific lamprey performance at structural challenges in fishways. N Am J Fish Manag 30:376–385

Keefer ML, Caudill CC, Clabough TS et al (2013) Fishway passage bottleneck identification and prioritization: a case study of Pacific lamprey at Bonneville Dam. Can J Fish Aquat Sci 70:1551–1565

Keefer ML, Caudill CC, Johnson EL et al (2017) Inter-observer bias in fish classification and enumeration using dual-frequency identification sonar (DIDSON): a Pacific lamprey case study. Northwest Sci 91:41–53

Kemp PS (2016) Meta-analyses, metrics and motivation: mixed messages in the fish passage debate. River Res Appl 32:2116–2124

Kerr JR, Karageorgopoulos P, Kemp PS (2015) Efficacy of side-mounted vertically oriented bristle pass for improving upstream passage of European eel (*Anguilla anguilla*) and river lamprey (*Lampetra fluviatilis*) at an experimental Crump weir. Ecol Eng 85:121–131

Kimura M, Maruyama T, Crow JF (1963) Mutation load in small populations. Genetics 48:1303–1312

King JJ, Wightman GD, Hanna G, Gilligan N (2015) River engineering works and lamprey ammocoetes: impacts, recovery, mitigation. Water Environ J 29:482–488

Kirk MA, Caudill CC (2016) Network analyses reveal intra- and interspecific differences in behavior when passing a complex migration obstacle. J Appl Ecol 54:836–845

Kirk MA, Caudill CC, Johnson EL, Keefer ML, Clabough TS (2015) Characterization of adult Pacific lamprey swimming behavior in relation to environmental conditions within large-dam fishways. Trans Am Fish Soc 144:998–1012

Kirk MA, Caudill CC, Tonina D, Syms JC (2016) Effects of water velocity, turbulence and obstacle length on the swimming capabilities of adult Pacific lamprey. Fish Manag Ecol 23:356–365

Kirk MA, Caudill CC, Syms JC, Tonina D (2017) Context-dependent responses to turbulence for an anguilliform swimming fish, Pacific lamprey, during passage of an experimental vertical-slot weir. Ecol Eng 106:296–307

Klar GT, Young RJ (2005) Integrated management of sea lampreys in the Great Lakes 2004. Annual Report to the Great Lakes Fishery Commission, Ann Arbor, MI

Klemetsen A (2010) The charr problem revisited: exceptional phenotypic plasticity promotes ecological speciation in postglacial lakes. Freshw Rev 3:49–74

Kottelat M (1997) European freshwater fishes. An heuristic checklist of the freshwater fishes of Europe (exclusive of former USSR), with an introduction for non-systematists and comments on nomenclature and conservation. Biologia 52(Suppl 5):1–271

Kottelat M, Freyhof J (2007) Handbook of European freshwater fishes. Kottelat, Cornol, Switzerland

Kusakabe R, Tochinai S, Kuratani S (2003) Expression of foreign genes in lamprey embryos: an approach to study evolutionary changes in gene regulation. J Exp Zool B Mol Dev Evol 296:87–97

Lacoursière-Roussel A, Côté G, Leclerc V, Bernatchez L (2016) Quantifying relative fish abundance with eDNA: a promising tool for fisheries management. J Appl Ecol 53:1148–1157

Lang NJ, Roe KJ, Renaud CB et al (2009) Novel relationships among lampreys (Petromyzontiformes) revealed by a taxonomically comprehensive molecular data set. In: Brown LR, Chase SD, Mesa MG, Beamish RJ, Moyle PB (eds) Biology, management, and conservation of lampreys in North America. Am Fish Soc Symp 72:41–55

Lasne E, Sabatié M-R, Jeannot N, Cucherousset J (2014) The effects of dam removal on river colonization by sea lamprey *Petromyzon marinus*. Riv Res Appl 31:904–911

Lee DS, Weise JG (1989) Habitat selection of lentic larval lampreys: preliminary analysis based on research with a manned submersible. J Great Lakes Res 15:156–163

Lee WJ, Kocher TD (1995) Complete sequence of a sea lamprey (*Petromyzon marinus*) mitochondrial genome: early establishment of the vertebrate genome organization. Genetics 139:873–887

Levin B, Ermakov A, Ermakov O et al (2016) Ukrainian brook lamprey *Eudontomyzon mariae* (Berg): phylogenetic position, genetic diversity, distribution, and some data on biology. In: Orlov A, Beamish R (eds) Jawless fishes of the world, vol 1. Cambridge Scholars Publishing, Newcastle upon Tyne, pp 58–82

Li K, Scott AM, Riedy JJ et al (2017) Three novel bile alcohols of mature male sea lamprey (*Petromyzon marinus*) act as chemical cues for conspecifics. J Chem Ecol 43:543–549

Li W, Twohey M, Jones M, Wagner M (2007) Research to guide use of pheromones to control sea lamprey. J Great Lakes Res 33(Spec Issue 2):70–86

Li Y (2014) Phylogeny of the lamprey genus *Lethenteron* Creaser and Hubbs 1922 and closely related genera using the mitochondrial cytochrome *b* gene and nuclear gene introns. MSc thesis, University of Manitoba, Winnipeg, MB

Li Y, Xie W, Li Q (2016) Characterisation of the bacterial community structures in the intestine of *Lampetra morii*. Antonie Leeuwenhoek 109:979–986

Lin B, Zhang Z, Wang Y et al (2008) Amplified fragment length polymorphism assessment of genetic diversity in Pacific lampreys. N Am J Fish Manag 28:1182–1193

Livermore J, Trainor M, Bednarski MS (2017) Successful spawning of anadromous *Petromyzon marinus* L. (sea lamprey) in a restored stream channel following dam removal. Northeast Nat 24:380–390

Lodge DM, Turner CR, Jerde CL et al (2012) Conservation in a cup of water: estimating biodiversity and population abundance from environmental DNA. Mol Ecol 21:2555–2558

Lucht JM (2015) Public acceptance of plant biotechnology and GM crops. Viruses 7:4254–4281

Luzier CW, Docker MF, Whitesel TA (2010) Characterization of 10 microsatellite loci for western brook lamprey *Lampetra richardsoni*. Conserv Genet Resour 2:71–74

Lynch M, Gabriel W (1990) Mutation load and the survival of small populations. Evolution 44:1725–1737

Magilligan FJ, Nislow KH, Kynard BE, Hackman AM (2016) Immediate changes in stream channel geomorphology, aquatic habitat, and fish assemblages following dam removal in a small upland catchment. Geomorphology 252:158–170

Mahon AR, Jerde CL, Galaska M et al (2013) Validation of eDNA surveillance sensitivity for detection of Asian caps in controlled and field experiments. PLoS ONE e58316

Maitland PS, Renaud CB, Quintella BR, Close DA, Docker MF (2015) Conservation of native lampreys. In: Docker MF (ed) Lampreys: biology, conservation and control, vol 1. Springer, Dordrecht, pp 375–428

Manion PJ, Smith BR (1978) Biology of larval and metamorphosing sea lampreys, *Petromyzon marinus*, of the 1960 year class in the Big Garlic River, Michigan, Part II, 1966–72. Great Lakes Fish Comm Tech Rep 30:1–35

Manousaki T, Qiu H, Noro M et al (2016) Molecular evolution in the lamprey genomes and its relevance to the timing of whole genome duplications. In: Orlov A, Beamish R (eds) Jawless fishes of the world, vol 1. Cambridge Scholars Publishing, Newcastle upon Tyne, pp 2–16

Manzon RG, Youson JH, Holmes JA (2015) Lamprey metamorphosis. In: Docker MF (ed) Lampreys: biology, conservation and control, vol 1. Springer, Dordrecht, pp 139–214

Marsden JE, Chipman BD, Pientka B, Schoch WF, Young BA (2010) Strategic plan for Lake Champlain fisheries. Great Lakes Fish Comm Misc Publ 2010–03:1–54

von Martens E (1868) About some East Asian freshwater animals. Arch Naturgesch 34:1–64 [in German]

Martignac F, Daroux A, Baglinieri J-L, Ombredane D, Guillarde J (2015) The use of acoustic cameras in shallow waters: new hydroacoustic tools for monitoring migratory fish population: a review of DIDSON technology. Fish Fish 16:486–510

Martin H, White MM (2008) Intraspecific phylogeography of the Least Brook Lamprey (*Lampetra aepyptera*). Copeia 2008:579–585

Mateus CS, Alves MJ, Quintella BR, Almeida PR (2013a) Three new cryptic species of the lamprey genus *Lampetra* Bonnaterre, 1788 (Petromyzontiformes: Petromyzontidae) from the Iberian Peninsula. Contrib Zool 82:37–53

Mateus CS, Stange M, Berner D et al (2013b) Strong genome-wide divergence between sympatric European river and brook lampreys. Curr Biol 23:R649–R650

Mateus CS, Almeida PR, Mesquita N, Quintella BR, Alves MJ (2016) European lampreys: new insights on postglacial colonization, gene flow and speciation. PLoS ONE 11:e0148107

Mayden RL (1997) A hierarchy of species concepts: the denouement in the saga of the species problem. In: Claridge MF, Dawah HA, Wilson MR (eds) Species: the units of diversity. Chapman and Hall, London, pp 381–423

McCann EL, Johnson NS, Pangle KL (2018a) Corresponding long-term shifts in stream temperature and invasive fish migration. Can J Fish Aquat Sci 75:772–778

McCann EL, Johnson NS, Hrodey PJ, Pangle KL (2018b) Characterization of sea lamprey stream entry using dual-frequency identification sonar. Trans Am Fish Soc 147:514–524

McCauley DW, Docker MF, Whyard S, Li W (2015) Lampreys as diverse model organisms in the genomics era. BioScience 65:1046–1056

McFarlane CT, Docker MF (2009) Characterization of 14 microsatellite loci in the paired lamprey species *Ichthyomyzon unicuspis* and *I. fossor* and cross amplification in four other *Ichthyomyzon* species. Conserv Genet Resour 1:377–380

McKee AM, Spear SF, Pierson TW (2015) The effect of dilution and the use of a post-extraction nucleic acid purification column on the accuracy, precision, and inhibition of environmental DNA samples. Biol Conserv 183(Spec Issue):70–76

McKelvey KS, Young MK, Knotek EL et al (2016) Sampling large geographic areas for rare species using environmental DNA (eDNA): a study of bull trout *Salvelinus confluentus* occupancy in western Montana. J Fish Biol 88:1215–1222

McLain AL, Dahl FH (1968) An electric beam trawl for the capture of larval lampreys. Trans Am Fish Soc 97:289–293

McLaughlin RL, Porto L, Noakes DLG et al (2006) Effects of low-head barriers on stream fishes: taxonomic affiliations and morphological correlates of sensitive species. Can J Fish Aquat Sci 63:766–779

McLaughlin RL, Hallett A, Pratt TC, O'Conner LM, McDonald DG (2007) Research to guide use of barriers, traps, and fishways to control sea lamprey. J Great Lakes Res 33(Spec Iss 2):7–19

McLean AR, Barber J, Bravener G, Rous AM, McLaughlin RL (2015) Understanding low success trapping invasive sea lampreys: an entry-level analysis. Can J Fish Aquat Sci 72:1876–1885

Meckley TD, Wagner CM, Gurarie E (2014) Coastal movements of migrating sea lamprey (*Petromyzon marinus*) in response to a partial pheromone added to river water: implications for management of invasive populations. Can J Fish Aquat Sci 71:533–544

Meckley TD, Gurarie E, Miller JR, Wagner CM (2017) How fishes find the shore: evidence for orientation to bathymetry from the non-homing sea lamprey. Can J Fish Aquat Sci 74:2045–2058

Meeuwig MH, Bayer JM, Reiche RA (2006) Morphometric discrimination of early life stage *Lampetra tridentata* and *L. richardsoni* (Petromyzonidae) from the Columbia River Basin. J Morph 267:623–633

Mehta TK, Ravi V, Yamasaki S et al (2013) Evidence for at least six Hox clusters in the Japanese lamprey (*Lethenteron japonicum*). Proc Natl Acad Sci USA 110:16044–16049

Merkes CM, McCalla SG, Jensen NR, Gaikowski MP, Amberg JJ (2014) Persistence of DNA in carcasses, slime and avian feces may effect interpretation of environmental DNA data. PLoS ONE 9:e113346

Mesa MG, Copeland ES (2009) Critical uncertainties and research needs for the restoration and conservation of native lampreys in North America. In: Brown LR, Chase SD, Mesa MG, Beamish RJ, Moyle PB (eds) Biology, management, and conservation of lampreys in North America. Am Fish Soc Symp 72:311–321

Mesa MG, Magie RJ, Copeland ES (2010) Passage and behavior of radio-tagged adult Pacific lampreys (*Entosphenus tridentatus*) at the Willamette Falls project, Oregon. Northwest Sci 84:233–242

Moser ML, Close DA (2003) Assessing Pacific lamprey status in the Columbia River Basin. Northwest Sci 77:116–125

Moser ML, Paradis RL (2017) Pacific lamprey restoration in the Elwha River drainage following dam removals. Am Curr 42:3–8

Moser ML, Almeida PR, Kemp PS, Sorensen PW (2015) Lamprey spawning migration. In: Docker MF (ed) Lampreys: biology, conservation and control, vol 1. Springer, Dordrecht, pp 215–263

Moyle, PB, Brown LR, Chase SD, Quiñones RM (2009) Status and conservation of lampreys in California. In: Brown LR, Chase SD, Mesa MG, Beamish RJ, Moyle PB (eds) Biology, management, and conservation of lampreys in North America. Am Fish Soc Symp 72:279–292

Mueller R, Arntzen E, Nabelek M et al (2012) Laboratory testing of modified electroshocking system designed for deepwater juvenile lamprey sampling. Trans Am Fish Soc 141:841–845

Mullett KM, Heinrich JW, Adams JV et al (2003) Estimating lake-wide abundance of spawning-phase sea lampreys (*Petromyzon marinus*) in the Great Lakes: extrapolating from sampled streams using regression models. J Great Lakes Res 29(Suppl 1):240–252

Nathan LR, Simmons MD, Wegleitner B, Jerde CL, Mahon A (2014) Quantifying environmental DNA signals for aquatic invasive species across multiple detection platforms. Environ Sci Technol 48:12800–12806

Neave FB, Mandrak NE, Docker MF, Noakes DL (2007) An attempt to differentiate sympatric *Ichthyomyzon* ammocoetes using meristic, morphological, pigmentation, and gonad analyses. Can J Zool 85:549–560

Negrea C, Thompson DE, Juhnke SD, Fryer DS, Loge FL (2014) Automatic detection and tracking of adult Pacific lampreys in underwater video collected at Snake and Columbia River fishways. N Am J Fish Manag 34:111–118

Ostberg CO, Chase DM, Hayes MC, Duda JJ (2018) Distribution and seasonal differences in Pacific lamprey and *Lampetra* spp eDNA across 18 Puget Sound watersheds. PeerJ e4496

Page LM, Espinosa-Pérez H, Findley LT et al (2013) Common and scientific names of fishes from the United States, Canada, and Mexico, 7th edn. Am Fish Soc Spec Publ 34, Bethesda, MD

Palsbøll PJ, Bérubé M, Allendorf FW (2007) Identification of management units using population genetic data. Trends Ecol Evol 22:11–16

Pante E, Schoelinck C, Puillandre N (2014) From integrative taxonomy to species description: one step beyond. Syst Biol 64:152–160

Pante E, Puillandre N, Vircel A et al (2015) Species are hypotheses: avoid connectivity assessments based on pillars of sand. Mol Ecol 24:525–544

Parker KA (2018) Evidence for the genetic basis and inheritance of ocean and river-maturing ecotypes of Pacific lamprey (*Entosphenus tridentatus*) in the Klamath River, California. MS thesis, Humboldt State University, Arcata, CA

Peng L, Lu J, Sun X-W (2016) Mitochondrial DNA sequence of *Lampetra morri*. Mitochondrial DNA 27:1391–1392

Pereira A, Almada V, Doadrio I (2011) Genetic relationships of brook lampreys of the genus *Lampetra* in a Pyrenean stream in Spain. Ichthyol Res 58:278–282

Pereira A, Doadrio I, Robalo J, I, Almada V (2014) Different stocks of brook lamprey in Spain and their origin from *Lampetra fluviatilis* at two distinct times and places. J Fish Biol 85:1793–1798

Pereira E, Quintella BR, Mateus CS et al (2017) Performance of a vertical-slot fish pass for the sea lamprey *Petromyzon marinus* L. and habitat recolonization. Riv Res Appl 33:16–26

Petersen Lewis RS (2009) Yurok and Karuk traditional ecological knowledge: Insights into Pacific lamprey populations of the Lower Klamath Basin. In: Brown LR, Chase SD, Mesa MG, Beamish RJ, Moyle PB (eds) Biology, management, and conservation of lampreys in North America. Am Fish Soc Symp 72:1–39

Port JA, O'Donnell JL, Romero-Maraccini OC et al (2016) Assessing vertebrate biodiversity in kelp forest ecosystem using environmental DNA. Mol Ecol 25:527–541

Porto LM, McLaughlin RL, Noakes DLG (1999) Low-head barrier dams restrict movements of fishes in two Lake Ontario streams. N Am J Fish Manag 138:652–665

Potter IC, Gill HS, Renaud CB, Haoucher D (2015) The taxonomy, phylogeny, and distribution of lampreys. In: Docker MF (ed) Lampreys: biology, conservation and control, vol 1. Springer, Dordrecht, pp 35–73

Pratt TC, O'Conner LM, Hallett AG et al (2009) Balancing aquatic habitat fragmentation and control of invasive species: enhancing selective fish passage at sea lamprey control barriers. Trans Am Fish Soc 138:652–665

Pu J, Ren J, Zhang Z et al (2016) Complete mitochondrial genomes of Korean lamprey (*Lethenteron morii*) and American brook lamprey (*L. appendix*). Mitochondrial DNA 27:1860–1861

Quinn TP, McGinnity P, Reed TE (2016) The paradox of "premature migration" by adult anadromous salmonid fishes: patterns and hypotheses. Can J Fish Aquat Sci 73:1015–1030

Rees HC, Bishop K, Middleditch DJ et al (2014) The application of eDNA for monitoring of the great crested newt in the UK. Ecol Evol 4:4023–4032

Reid SB, Goodman DH (2015) Detectability of Pacific lamprey occupancy in western drainages: implications for distribution surveys. Trans Am Fish Soc 144:315–322

Reid SB, Goodman DH (2016a) Pacific lamprey in coastal drainages of California: occupancy patterns and contraction of the southern range. Trans Am Fish Soc 145:703–711

Reid SB, Goodman DH (2016b) Free-swimming speeds and behavior in adult Pacific lamprey, *Entosphenus tridentatus*. Environ Biol Fish 99:969–974

Ren J, Pu J, Buchinger T et al (2015) The mitogenomes of the pouched lamprey (*Geotria australis*) and least brook lamprey (*Lampetra aepyptera*) with phylogenetic considerations. Mitochondrial DNA 27:3560–3562

Ren J, Buchinger T, Pu J, Jia L, Li W (2016) Complete mitochondrial genomes of paired species northern brook lamprey (*Ichthyomyzon fossor*) and silver lamprey (*I. unicuspis*). Mitochondrial DNA 27:1862–1863

Renaud CB (2011) Lampreys of the world: an annotated and illustrated catalogue of lamprey species known to date. FAO Species Cat Fish Purp 5, FAO, Rome

Renaud CB, Docker MF, Mandrak NE (2009) Taxonomy, distribution, and conservation of lampreys in Canada. In: Brown LR, Chase SD, Mesa MG, Beamish RJ, Moyle PB (eds) Biology, management, and conservation of lampreys in North America. Am Fish Soc Symp 72:293–309

Rey JR, Walton WE, Wolfe RJ et al (2012) North American wetlands and mosquito control. Int J Environ Res Public Health 9:4537–4605

Richards JE, Beamish RJ, Beamish FWH (1982) Descriptions and keys for ammocoetes of lampreys from British Columbia, Canada. Can J Fish Aquat Sci 39:1484–1495

Rodríguez-Muñoz R, Waldman JR, Grunwald C, Roy NK, Wirgin I (2004) Absence of shared mitochondrial DNA haplotypes between sea lamprey from North American and Spanish rivers. J Fish Biol 64:783–787

Rooney SM, Wightman G, Ó'Conchúir R, King JJ (2015) Behaviour of sea lamprey (*Petromyzon marinus* L.) at man-made obstacles during upriver spawning migration: use of telemetry to assess weir modifications for improved passage. Biol Environ Proc R Ir Acad 115:125–136

Rougemont Q, Gaigher A, Lasne E et al (2015) Low reproductive isolation and highly variable levels of gene flow reveal limited progress towards speciation between European river and brook lampreys. J Evol Biol 28:2248–2263

Rougemont Q, Roux C, Neuenschwander S et al (2016) Reconstructing the demographic history of divergence between European river and brook lampreys using approximate Bayesian computations. PeerJ 4:e1910

Rougemont Q, Gagnaire PA, Perrier C et al (2017) Inferring the demographic history underlying parallel genomic divergence among pairs of parasitic and nonparasitic lamprey ecotypes. Mol Ecol 26:142–162

Rous AM, McLean AR, Barber J et al (2017) Spatial mismatch between sea lamprey behaviour and trap location explains low success at trapping for control. Can J Fish Aquat Sci 74:2085–2097

Roussel J-M, Paillison J-M, Tréguier A, Petit E (2015) The downside of eDNA as survey tool in water bodies. J Appl Ecol 52:823–826

Ryder OA (1986) Species conservation and systematics: the dilemma of subspecies. Trends Ecol Evol 1:9–10

Sanchez NK, Corniuk N, Reinhardt U (2017) Effects of submergence depths on swimming capacity of sea lamprey. McNair Sch Res J 10:11

Schedina IM, Pfautsch S, Hartmann S et al (2014) Isolation and characterization of eight microsatellite loci in the brook lamprey *Lampetra planeri* (Petromyzontiformes) using 454 sequence data. J Fish Biol 85:960–964

Schleen LP, Christie GC, Heinrich RA et al (2003) Development and implementation of an integrated program for control of sea lamprey in the St Marys River. J Great Lakes Res(Suppl 1):677–693

Schlick-Steiner BC, Steiner FM, Seifert B et al (2010) Integrative taxonomy: a multisource approach to exploring biodiversity. Annu Rev Entomol 55:421–438

Schloesser NA, Merkes CM, Rees CB et al (2018) Correlating sea lamprey density with environmental DNA detections in the lab. Manag Biol Invasion 9:483–495

Schuldt RJ, Goold R (1980) Changes in the distribution of native lampreys in Lake Superior tributaries in response to sea lamprey (*Petromyzon marinus*) control, 1953–77. Can J Fish Aquat Sci 37:1872–1885

Scott AM, Li K, Li W (2018) The identification of sea lamprey pheromones using bioassay-guided fractionation. J Vis Exp 137:e58059

Scott WB, Crossman EJ (1973) Freshwater fishes of Canada. Bull Fish Res Board Can 184, Ottawa

Shafer ABA, Wolf JBW, Alves PC et al (2015) Genomics and the challenging translation into conservation practice. Trends Ecol Evol 30:78–87

Sherburne S, Reinhardt UG (2016) First test of a species-selective adult sea lamprey migration barrier. J Great Lakes Res 42:893–898

Siefkes MJ (2017) Use of physiological knowledge to control the invasive sea lamprey (*Petromyzon marinus*) in the Laurentian Great Lakes. Conserv Physiol 5:cox031

Sigsgaard AA, Carl H, Moller P, Thomsen PF (2015) Monitoring the near-extinct European weather loach in Denmark based on environmental DNA from water samples. Biol Conserv 183:46–52

Silva S, Macaya-Solis C, Lucas MC (2017a) Energetically efficient behaviour may be common in biology, but it is not universal: a test of selective tidal stream transport in a poor swimmer. Mar Ecol Prog Ser 584:161–174

Silva S, Lowry M, Macaya-Solis C, Byatt B, Lucas MC (2017b) Can navigation locks be used to help migratory fishes with poor swimming performance pass tidal barrages? A test with lampreys. Ecol Eng 102:291–302

Smith JJ, Kuraku S, Holt C et al (2013) Sequencing of the sea lamprey (*Petromyzon marinus*) genome provides insights into vertebrate evolution. Nat Genet 45:415–421

Smith JJ, Timoshevskaya N, Ye C et al (2018) The sea lamprey germline genome provides insights into programmed genome rearrangement and vertebrate evolution. Nat Genet 50:270–277

Sorensen PW, Johnson NJ (2016) Theory and application of semiochemicals in nuisance fish control. J Chem Ecol 42:698–715

Sorensen PW, Fine JM, Dvornikovs V et al (2005) Mixture of new sulfated steroids functions as a migratory pheromone in the sea lamprey. Nat Chem Biol 1:324–328

Sower SA (2015) The reproductive hypothalamic-pituitary axis in lampreys. In: Docker MF (ed) Lampreys: biology, conservation and control, vol 1. Springer, Dordrecht, pp 305–373

Spice EK, Whitesel TA, McFarlane CT, Docker MF (2011) Characterization of 12 microsatellite loci for the Pacific lamprey, *Entosphenus tridentatus* (Petromyzontidae), and cross-amplification in five other lamprey species. Genet Mol Res 10:3246–3250

Spice EK, Goodman DH, Reid SB, Docker MF (2012) Neither philopatric nor panmictic: microsatellite and mtDNA evidence suggests lack of natal homing but limits to dispersal in Pacific lamprey. Mol Ecol 21:2916–2930

Spice EK, Whitesel TA, Silver GS, Docker MF (2019) Contemporary and historical river connectivity influence population structure in western brook lamprey in the Columbia River Basin. Conserv Genet 20:299–314

Square T, Romášek M, Jandzik D et al (2015) CRISPR/Cas9-mediated mutagenesis in the sea lamprey *Petromyzon marinus*: a powerful tool for understanding ancestral gene functions in vertebrates. Development 142:4180–4187

Stacey N, Sorensen P (2009) Hormonal pheromones in fish. In: Pfaff DW, Arnold AP, Etgen AM, Fahrback SE, Rubin RT (eds) Hormones, brain and behavior, vol 2, 2nd edn. Elsevier, San Diego, pp 639–682

Steeves TB, Slade JW, Fodale MF, Cuddy DW, Jones ML (2003) Effectiveness of using backpack electrofishing gear for collecting sea lamprey (*Petromyzon marinus*) larvae in Great Lakes tributaries. J Great Lakes Res 29(Suppl 1):161–173

Stewart M, Baker CF (2012) A sensitive analytical method for quantifying petromyzonol sulfate ins water as a potential tool for population monitoring of the southern pouched lamprey, *Geotria australis*, in New Zealand streams. J Chem Ecol 38:135–144

Stewart M, Baker CF, Cooney T (2011) A rapid, sensitive, and selective method for quantitation of lamprey migratory pheromones in river water. J Chem Ecol 37:1203–1207

Tablerlet P, Coissac R, Hajibabaei M, Rieseberg L (eds) (2012) Molecular ecology special issue on environmental DNA. Mol Ecol 2:1789–2050

Takeshima H, Yokoyama R, Nishida M, Yamazaki Y (2005) Isolation of microsatellite loci in the threatened brook lamprey *Lethenteron* sp. N. Mol Ecol Notes 5:812–814

Taverny C, Lassalle G, Ortusi I et al (2012) From shallow to deep waters: habitats used by larval lampreys (genus *Petromyzon* and *Lampetra*) over a western European basin. Ecol Freshw Fish 21:87–99

Taylor EB (1999) Species pairs of north temperate freshwater fishes: evolution, taxonomy, and conservation. Rev Fish Biol Fish 9:299–324

Taylor EB, Harris LN, Spice EK, Docker MF (2012) Microsatellite DNA analysis of parapatric lamprey (*Entosphenus* spp.) populations: implications for evolution, taxonomy, and conservation of a Canadian endemic. Can J Zool 90:291–303

Thresher RE, Jones M, Drake DAR (2019a) Evaluating active genetic options for the control of Sea Lampreys (*Petromyzon marinus*) in the Laurentian Great Lakes. Can J Fish Aquat Sci (in press)

Thresher RE, Jones M, Drake DAR (2019b) Stakeholder attitudes towards the use of recombinant technology to manage the impact of an invasive species: Sea Lamprey in the North American Great Lakes. Biol Invasion 21:575–586

Tilesius von Tilenau WG (1811) Piscium camtschaticorum descriptions et icones. Mém Acad Imp Sci St-Pétersbg 3:225–285

Tummers JS, Winter E, Silva S et al (2016) Evaluating the effectiveness of a Larinier super active baffle fish pass for European river lamprey *Lampetra fluviatilis* before and after modification with wall-mounted studded tiles. Ecol Eng 91:183–194

Tummers JS, Kerr JR, O'Brien P, Kemp P, Lucas MC (2018) Enhancing the upstream passage of river lamprey at a microhydropower installation using horizontally-mounted studded tiles. Ecol Eng 125:87–97

Tutman P, Freyhof J, Dulcic J, Glamuzina B, Geiger M (2017) *Lampetra soljani*, a new brook lamprey from the southern Adriatic Sea basin (Petromyzontiformes: Petromyzontidae). Zootaxa 4273:531–548

Ulibarri RM, Bonar SA, Rees C et al (2017) Comparing efficiency of American Fisheries Society standard snorkeling techniques to environmental DNA sampling techniques. N Am J Fish Manag 37:644–651

United States Fish and Wildlife Service (1996) Policy regarding the recognition of distinct vertebrate population segments under the Endangered Species Act. Federal Regist 61:4722–4725

Urdaci MC, Taverny C, Élie A-M, Élie P (2014) A genetic method to differentiate *Petromyzon marinus* ammocoetes from those of the paired species *Lampetra fluviatilis* and *L. planeri*. Cybium 38:3–7

Van Leeuwen T, Demaeght P, Osborne EJ et al (2012) Population bulk segregant mapping uncovers resistance mutations and the mode of action of a chitin synthesis inhibitor in arthropods. Proc Natl Acad Sci USA 109:4407–4412

Vladykov VD, Kott E (1979) A new parasitic species of the Holarctic lamprey genus *Entosphenus* Gill, 1862 (Petromyzonidae) from Klamath River, in California and Oregon. Can J Zool 57:808–823

Vogler AP, Monaghan MT (2007) Recent advances in DNA taxonomy. J Zool Syst Evol Res 45:1–10

Vowles AS, Don AM, Karageorgopoulos P, Worthington TA, Kemp PS (2015) Efficacy of a dual density studded fish pass designed to mitigate for impeded upstream passage of juvenile European eels (*Anguilla anguilla*) at a model Crump weir. Fish Manag Ecol 22:307–316

Vowles AS, Don AM, Karageorgopoulos P, Kemp PS (2017) Passage of European eel and river lamprey at a model weir provisioned with studded tiles. J Ecohydraul 2:88–98

Wagner WC, Stauffer TM (1962) Sea lamprey larvae in lentic environments. Trans Am Fish Soc 91:384–387

Waldman JR, Grunwald C, Roy NK, Wirgin II (2004) Mitochondrial DNA analysis indicates sea lampreys are indigenous to Lake Ontario. Trans Am Fish Soc 133:950–960

Waldman JR, Grunwald C, Wirgin I (2006) Evaluation of the native status of sea lampreys in Lake Champlain based on mitochondrial DNA sequencing analysis. Trans Am Fish Soc 135:1076–1085

Waldman J, Grunwald C, Wirgin I (2008) Sea lamprey *Petromyzon marinus*: an exception to the rule of homing in anadromous fishes. Biol Lett 4:659–662

Waldman J, Daniels R, Hickerson M, Wirgin I (2009) Mitochondrial DNA analysis indicates sea lampreys are indigenous to Lake Ontario: response to comment. Trans Am Fish Soc 138:1190–1197

Wang C, Schaller H (2015) Conserving Pacific Lamprey through collaborative efforts. Fisheries 40:72–79

Wang H, Johnson N, Bernardy J, Hubert T, Li W (2013) Monitoring sea lamprey pheromones and their degradation using rapid stream-side extraction coupled with UPLC-MS/MS. J Sep Sci 36:1612–1620

Waples RS (1991) Pacific salmon, *Oncorhynchus* spp., and the definition of "species" under the Endangered Species Act. Mar Fish Rev 53:11–22

Waples RS (1995) Evolutionarily significant units and the conservation of biological diversity under the Endangered Species Act. In: Nielsen JL Powers GA (eds) Evolution and the aquatic ecosystem: defining unique units in population conservation. Am Fish Soc Symp 17:8–27

Webber BL, Raghu S, Edwards OR (2015) Is CRISPR-based gene drive a biocontrol silver bullet or global conservation threat. Proc Natl Acad Sci USA 112:10565–10567

White MM (2014) Intraspecific phylogeography of the American Brook Lamprey, *Lethenteron appendix* (DeKay, 1842). Copeia 2014:513–518

Whitlock SL, Schultz LD, Schreck CB, Hess JE (2017) Using genetic pedigree reconstruction to estimate effective spawner abundance from redd surveys: an example involving Pacific lamprey (*Entosphenus tridentatus*). Can J Fish Aquat Sci 74:1646–1653

Wilcox TM, McKelvey KS, Young MK et al (2013) Robust detection of rare species using environmental DNA: the importance of primer specificity. PLoS ONE 8:e59520

Wilcox TM, McKelvey KS, Young MK et al (2016) Understanding environmental DNA detection probabilities: a case study using a stream-dwelling char *Salvelinus fontinalis*. Biol Conserv 194:209–216

Willi Y, Griffin P, Van Buskirk J (2013) Drift load in populations of small size and low density. Heredity 110:296–302

World Commission on Dams (2000) Dams and development: a new framework for decision-making. Earthscan Publications Ltd, London

Xi X, Johnson NS, Brant CO et al (2011) Quantification of a male sea lamprey pheromone in tributaries of Laurentian Great Lakes by liquid chromatography tandem mass spectrometry. Environ Health Tech 45:6437–6443

Yamazaki Y, Goto A, Nishida M (2003) Mitochondrial DNA sequence divergence between two cryptic species of *Lethenteron*, with reference to an improved identification technique. J Fish Biol 62:591–609

Yamazaki Y, Yokoyama R, Nishida M, Goto A (2006) Taxonomy and molecular phylogeny of *Lethenteron* lampreys in eastern Eurasia. J Fish Biol 68:251–269

Yamazaki Y, Yokoyama R, Nagai T, Goto A (2014) Population structure and gene flow among anadromous Arctic lamprey (*Lethenteron camtschaticum*) populations deduced from polymorphic microsatellite loci. Environ Biol Fish 97:43–52

Yan S-K, Liu R-H, Jin H-Z et al (2015) "Omics" in pharmaceutical research: overview, applications, challenges, and future perspectives. Chin J Nat Med 13:3–21

Yan X, Meng W, Wu F et al (2016) The nuclear DNA content and genetic diversity of *Lampetra morii*. PLoS ONE 11:e0157494

Zu Y, Zhang X, Ren J et al (2016) Biallelic editing of a lamprey genome using the CRISPR/Cas9 system. Sci Rep 6:23496

Species Index

Subject Index

A
Ancestral life history type, 292
Atresia, 48, 53, 58, 71–74, 77–82, 84, 90, 104, 116, 123–127, 142, 356

B
Barriers, 1, 3, 17, 20, 25, 27, 28, 33, 199, 222, 318, 320, 324, 325, 329, 341, 349, 364, 366–368, 370, 373, 375–377, 380, 414, 416, 421, 422, 426, 428–431, 447, 448, 452, 459, 461, 463, 464, 512, 529, 537, 538, 540, 542, 550, 551, 553, 555
Broodstock, 196, 199–203, 205, 208, 210

C
Cayuga Lake, 29, 32, 104, 110, 111, 135, 143–145, 341, 342, 345, 416, 418, 443–446, 451
Conservation, 3, 4, 120, 146, 188, 189, 195, 196, 200, 208–210, 224, 226, 229, 389, 412, 463, 464, 483, 504, 528–530, 532, 534, 535, 537–540, 543, 544, 546, 547, 549, 551, 553–556
Contemporary gene flow, 198, 361, 363–366, 368, 373, 534, 535
Control, 1–3, 5, 24, 32, 33, 36, 38, 45, 82, 85–89, 146, 161, 188, 189, 191, 196, 197, 209, 221, 222, 224, 226, 228, 229, 265, 317, 334, 358, 370, 388, 412, 422, 423, 425, 426, 430, 431, 433, 437, 439–445, 447–452, 455–459, 463, 464, 482, 483, 487, 503, 507, 528, 529, 535, 536, 538–541, 544, 547, 549, 552, 554–556

Culture, 188, 189, 208, 209, 216, 217, 219, 221, 225–227, 229, 483, 506, 508

D
Dam removal, 329, 389, 428, 545, 549–551
Deepwater sampling, 535, 536
Development, 1–3, 5, 6, 12, 40, 41, 45, 48, 51, 53–55, 58–60, 62, 69, 71–75, 77, 78, 80–82, 85, 88–95, 106, 108, 112, 118–120, 141, 146, 156, 160–162, 188–191, 193–195, 197, 198, 200, 202, 207, 211, 213, 215, 216, 218, 221, 224–226, 228, 247, 293, 294, 296, 297, 299, 301, 303, 336, 337, 347, 365, 370–372, 386, 421, 429, 433, 443, 447, 449, 450, 452, 457–460, 463, 464, 482–495, 497–505, 512, 529, 535, 541, 543, 544, 549, 553, 555
Dual Frequency Identification Sonar (DIDSON), 538

E
Ecological constraints, 334, 378, 379
Ecotypes, 305, 316, 337, 364, 389
Egg size, 65, 112, 120–122, 136
Embryos, 43, 85, 93, 140, 188, 195, 196, 211–217, 227, 228, 457, 482–484, 487, 491, 499, 500, 502, 504–511, 548
Environmental DNA, 458, 540
Environmental sex determination (ESD), 1, 2, 39, 41, 43, 45, 46, 49
Evo-devo, 193, 387, 483–485, 502, 503, 505, 512, 528, 544
Evolution, 3, 4, 40, 49, 113, 119, 120, 146, 149, 158, 189, 191, 253, 292, 296, 298,

© Springer Nature B.V. 2019
M. F. Docker (ed.), *Lampreys: Biology, Conservation and Control*,
Fish & Fisheries Series 38, https://doi.org/10.1007/978-94-024-1684-8

Printed by Printforce, the Netherlands